ADVANCES IN FERROELECTRICS

Edited by **Aimé Peláiz Barranco**

Advances in Ferroelectrics
http://dx.doi.org/10.5772/45744
Edited by Aimé Peláiz Barranco

Contributors

Takeo Sasaki, Jeffrey F. Webb, Hongyang Zhao, Shu-Tao Ai, Ashim Kumar Bain, Prem Chand, Ashok Kumar, Ram S. Katiyar, M Correa, Nora Ortega, Shalini Kumari, Yuriy Garbovskiy, Anatoliy Glushchenko, Olena Zribi, Akira Onodera, Desheng Fu, Hiroki Taniguchi, Mitsuru Itoh, Shigeo Mori, Amaury Suárez-Gómez, José-Manuel Saniger, Francisco Calderón-Piñar, Junhao Chu, Zhigao Hu, Yawei Li, Kai Jiang, Ziqiang Zhu, Aimé Peláiz-Barranco, Osmany García-Zaldívar, Yuslin González-Abreu, Dong Guo, Kai Cai, Hiroki Moriwake, Toshirou Yagi, Toshiro Kohmoto, Qiang Wan, Changqing Chen, Yapeng Shen, Koukou Suu, Nicu Doinel Scarisoreanu, Ruxandra Birjega, Andreea Andrei, Maria Dinescu, Carmen Galassi, Floriana Craciun, Kai-Huang Chen, Chien-Min Cheng, Sean Wu, Jen-Hwan Tsai, Kaoru Miura, Funakubo Hiroshi, Martun Hovhannisyan, Matias Nunez

Published by InTech

Janeza Trdine 9, 51000 Rijeka, Croatia

Publishing Process Manager Ana Pantar
Technical Editor InTech DTP team
Cover InTech Design team

Additional hard copies can be obtained from orders@intechopen.com

Advances in Ferroelectrics, Edited by Aimé Peláiz Barranco
p. cm.
ISBN 978-953-51-0885-6

For EU product safety concerns:
IN TECH d.o.o., Prolaz Marije Krucifikse Kozulić 3, 51000 Rijeka, Croatia,
info@intechopen.com or visit our website at intechopen.com.

Meet the editor

Dr. Aimé Peláiz Barranco is a professor in the Physics Faculty, Havana University, Cuba. Her activities range from teaching, to advising undergraduate and graduate students and doing high quality research. She is one of the leaders in the development of ferroelectrics and antiferroelectric materials research in Cuba. She has authored several scientific papers in refereed journals and presented an important number of research works at national and international conferences. Her results have been recognized in Cuba and worldwide through several awards, such as TWOWS Award for Young Women Scientists in Physics/Mathematics (2010) and TWAS-ROLAC Award for Young Scientists in Physics (2011). She is the international coordinator of the Latin American Network of Ferroelectrics Materials.

Contents

Preface

Since the discovery of the phenomenon of ferroelectricity, the continuous development of materials and technology has provided an important number of applications. The ferroelectricity has been the heart and soul of several industries to the development of piezoelectric transducers, pyroelectric sensors, medical diagnostic transducers, radio and communication filters, ultrasonic motors, electro-optical devices, ferroelectric memories, etc.

The most striking properties of the ferroelectric materials are strong coupling effects, strong hysteresis in the field polarization response, and extremely high dielectric permittivity. For the development of new technology by using ferroelectric materials, much research has been carried out from the perspective of theory and of experiment. These have provided revolutionary breakthroughs in the understanding of single crystals, ceramics, thin films and composites. The advances in the studies of these materials at nanoscale have also showed the potential applications of ferroelectric nanoparticles.

The purpose of this book is to offer an up-to-date view of ferroelectricity, covering a wide range of theoretical and experimental topics. Several researches on ceramics, thin films, ferroelectric liquids and ferroelectricity at nanoscale are included. The book is the result of an important number of contributions of researchers from the international scientific community, which work in different topics of ferroelectricity.

I would like to thank all the authors for their interesting contributions to this book and the InTech team for the outstanding support.

We hope that this book will be of real value to students and researchers moving into the ferroelectric field.

Dr. Aimé Peláiz Barranco
Applied Physics Department
Physics Faculty, Havana University
Cuba

Chapter 1

Electronic Structures of Tetragonal ABX3: Role of the B-X Coulomb Repulsions for Ferroelectricity and Piezoelectricity

Kaoru Miura and Hiroshi Funakubo

Additional information is available at the end of the chapter

http://dx.doi.org/10.5772/52187

1. Introduction

Since Cohen proposed an origin for ferroelectricity in perovskites (ABX_3) [1], investigations of ferroelectric materials using first-principles calculations have been extensively studied [2-20]. Currently, using the pseudopotential (PP) methods, most of the crystal structures in ferroelectric ABX_3 can be precisely predicted. However, even in $BaTiO_3$, which is a well-known ferroelectric perovskite oxide with tetragonal structure at room temperature, the optimized structure by the PP methods is strongly dependent on the choice of the Ti PPs as illustrated in Fig. 1; preparation for Ti 3s and 3p semicore states in addition to Ti 3d, 4s, and 4p valence states is essential to the appearance of the tetragonal structure. This is an important problem for ferroelectricity, but it has been generally recognized for a long time that this problem is within an empirical framework of the calculational techniques [21].

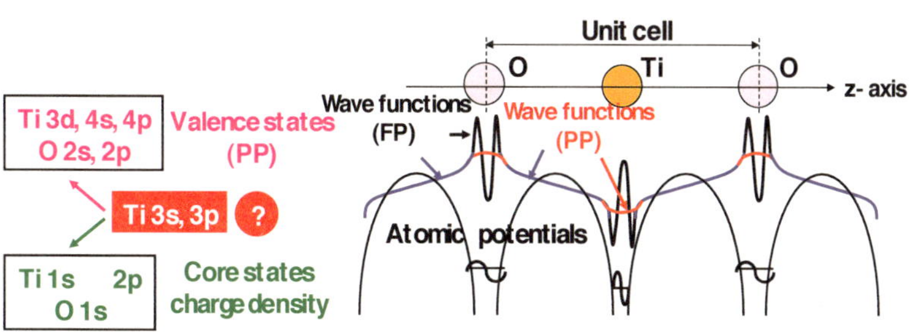

Figure 1. Illustration of the choice of Ti 3s and 3p states in pseudopotentials.

It is known that ferroelectric state appears when the long-range forces due to the dipole-dipole interaction overcome the short-range forces due to the Coulomb repulsions. Investigations about the relationship between the Ti-O Coulomb repulsions and the appearance of ferroelectricity in $ATiO_3$ (A = Ba, Pb) were reported both theoretically and experimentally. Theoretically, Cohen first proposed the hybridization between Ti 3d state and O 2p state (Ti 3d-O 2p) as an origin for ferroelectricity in $BaTiO_3$ and $PbTiO_3$ [1]. On the other hand, we investigated [20] the influence of the Ti-O_z Coulomb repulsions on Ti ion displacement in tetragonal $BaTiO_3$ and $PbTiO_3$, where O_z denotes the O atom to the z-axis (Ti is displaced to the z-axis). Whereas the hybridization between Ti 3d state and O_z $2p_z$ state stabilize Ti ion displacement, the strong Coulomb repulsions between Ti 3s and $3p_z$ states and O $2p_z$ states do not favourably cause Ti ion displacement. Experimentally, on the other hand, Kuroiwa *et al.* [22] showed that the appearance of ferroelectric state is closely related to the total charge density of Ti-O bonding in $BaTiO_3$. As discussed above, investigation about a role of Ti 3s and 3p states is important in the appearance of the ferroelectric state in tetragonal $BaTiO_3$, in addition to the Ti 3d-O 2p hybridization as an origin of ferroelectricity [1].

It seems that the strong B-X Coulomb repulsions affect the most stable structure of ABX_3. It has been well known that the most stable structure of ABX_3 is closely related to the tolerance factor t,

$$t = \frac{r_A + r_X}{\sqrt{2}\,(r_B + r_X)}, \tag{1}$$

where r_A, r_B, and r_X denote the ionic radii of A, B, and X ions, respectively [23]. In general ferroelectric ABX_3, the most stable structure is tetragonal for $t \gtrsim 1$, cubic for $t \approx 1$, and rhombohedral or orthorhombic for $t \lesssim 1$. In fact, $BaTiO_3$ with t = 1.062 shows tetragonal structure in room temperature. However, recently, $BiZn_{0.5}Ti_{0.5}O_3$ (BZT) with t = 0.935 was experimentally reported [24] to show a tetragonal $PbTiO_3$-type structure with high c/a ratio (1.211). This result is in contrast to that of $BiZn_{0.5}Mg_{0.5}O_3$ (BMT) with t = 0.939, *i.e.*, the most stable structure was reported to be the orthorhombic or rhombohedral structure [25, 26]. Several theoretical papers of BZT have been reported [4-6], but the role of the Zn-O Coulomb repulsions in the appearance of the tetragonal structure has not been discussed sufficiently.

Piezoelectric properties in ABX_3 are also closely related to the crystal structure. Investigations of the relationship between piezoelectric properties and the crystal structure of ABX_3 by first-principles calculations have been extensively studied [2-19]. Moreover, phenomenological investigations of the piezoelectric properties have been also performed [27, 28]. However, it seems that the piezoelectric properties in the atomic level have not been sufficiently investigated. Therefore, further theoretical investigation of the relationship between piezoelectric properties and the crystal structure of ABX_3, especially the B-X Coulomb repulsions, should be needed.

Recently, we investigated the roles of the Ti-O Coulomb repulsions in the appearance of a ferroelectric state in tetragonal $BaTiO_3$ by the analysis of a first-principles PP method [11-15]. We investigated the structural properties of tetragonal and rhombohedral $BaTiO_3$ with two kinds of Ti PPs, and propose the role of Ti 3s and 3p states for ferroelectricity. We also investigated the role of the Zn-O Coulomb repulsions in the appearance of a ferroelectric state in tetragonal BZT [10, 13]. Moreover, we also investigated the structural, ferroelectric, and piezoelectric properties of tetragonal ABX_3 and discussed the piezoelectric mechanisms based on the *B-X* Coulomb repulsions [12, 14, 15, 18, 19].

In this chapter, based on our recent papers and patents [10-19], we discuss a general role of *B-X* Coulomb repulsions for the appearance of the ferroelectric state in ABX_3. Then, we also discuss the relationship between the *B-X* Coulomb repulsions and the piezoelectric properties of tetragonal ABX_3.

2. Methodology

The calculations for ABX_3 were performed using the ABINIT code [29], which is one of the norm-conserving PP (NCPP) methods. Electron-electron interaction was treated in the local-density approximation (LDA) [30]. Pseudopotentials were generated using the OPIUM code [31]:

i. In $BaTiO_3$, 5s, 5p and 6s electrons for Ba PP, and 2s and 2p electrons for O PP were treated as semicore or valence electrons, respectively. Moreover, in order to investigate the role of Ti 3s and 3p states, two kinds of Ti PPs were prepared: the Ti PP with 3s, 3p, 3d and 4s electrons treated as semicore or valence electrons (Ti3spd4s PP), and that with only 3d and 4s electrons treated as valence electrons (Ti3d4s PP). In both PPs, the differences between the calculated result and experimental one are within 1.5 % of the lattice constant and within 10 % of the bulk modulus in the optimized calculation of bulk Ti. The cutoff energy for plane-wave basis functions was set to be 50 Hartree (Hr). The number of atoms in the unit cell was set to be five, and a 6×6×6 Monkhorst-Pack *k*-point mesh was set in the Brillouin zone of the unit cell. Positions of all the atoms were optimized within the framework of the tetragonal (*P4mm*) or rhombohedral (*R3m*) structure.

ii. In BZT and BMT, 5d, 6s, and 6p electrons for Bi PP, and 2s and 2p electrons for O PP were treated as semicore or valence electrons, respectively. Moreover, in order to investigate the roles of Zn and Ti 3s and 3p states, and Mg 2s and 2p states, two types of PPs were prepared: the PPs with only Zn and Ti 3d and 4s states, and Mg 3s states, considered as valence electrons (Case I), Zn and Ti 3s, 3p, 3d, and 4s states, and Mg 2s, 2p, and 3s states considered as semicore or valence electrons (Case II). The cutoff energy for plane-wave basis functions was set to be 70 Hr for Case I and 110 Hr for Case II. A 4×4×4 Monkhorst-Pack *k*-point mesh was set in the Brillouin zone of the unit cell. The calculated results can be discussed within 0.02 eV per formula unit (f.u.) using the above conditions. The present calculations were performed for the monoclinic, rhombohedral, and A-, C-, and G-type tetragonal structures. The number of atoms in the unit cell was set to be 10 for the rhombohedral and monoclinic structures, and 20 for the A-, C-, and G-type tetragonal structures. Positions of all the atoms were optimized within the framework of the rhombohedral (*R3*), monoclinic (*Pm*), and tetragonal (*P4mm*) structures.

iii. Relationship between the *B-X* Coulomb repulsions and the piezoelectric properties in tetragonal ABX_3 is investigated. The pseudopotentials were generated using the opium code [31] with semicore and valence electrons (e.g., Ti3spd4s PP), and the virtual crystal approximation [32] were applied to several ABX_3.

Spontaneous polarizations and piezoelectric constants were also evaluated, due to the Born effective charges [33]. The spontaneous polarization of tetragonal structures along the [001] axis, P_3, is defined as

$$P_3 = \sum_k \frac{ec}{\Omega} Z_{33}^*(k) u_3(k) , \tag{2}$$

where e, c, and Ω denote the charge unit, lattice parameter of the unit cell along the [001] axis, and the volume of the unit cell, respectively. $u_3(k)$ denotes the displacement along the [001] axis of the kth atom, and $Z_{33}^*(k)$ denotes the Born effective charges [33] which contributes to the P_3 from the $u_3(k)$.

The piezoelectric e_{33} constant is defined as

$$e_{3j} = \left(\frac{\partial P_3}{\partial \eta_3}\right)_u + \sum_k \frac{ec}{\Omega} Z_{33}^*(k) \frac{\partial u_3(k)}{\partial \eta_j} \qquad (j = 3,1), \tag{3}$$

where e and Ω denote the charge unit and the volume of the unit cell. P_3 and c denote the spontaneous polarization of tetragonal structures and the lattice parameter of the unit cell along the [001] axis, respectively. $u_3(k)$ denotes the displacement along the [001] axis of the kth atom, and $Z_{33}^*(k)$ denotes the Born effective charges which contributes to the P_3 from the $u_3(k)$. η_3 denotes the strain of lattice along the [001] axis, which is defined as $\eta_3 \equiv (c - c_0)/c_0$; c_0 denotes the c lattice parameter with fully optimized structure. On the other hand, η_1 denotes the strain of lattice along the [100] axis, which is defined as $\eta_1 \equiv (a - a_0)/a_0$; a_0 denotes the a lattice parameter with fully optimized structure. The first term of the right hand in Eq. (3) denotes the clamped term evaluated at vanishing internal strain, and the second term denotes the relaxed term that is due to the relative displacements.

The relationship between the piezoelectric d_{33} constant and the e_{33} one is

$$d_{33} \equiv \sum_{j=1}^{6} s_{3j}^E \times^T (e_{3j}), \tag{4}$$

where s_{3j^E} denotes the elastic compliance, and ``T " denotes the transposition of matrix elements. The suffix j denotes the direction-indexes of the axis, *i.e.*, 1 along the [100] axis, 2 along the [010] axis, 3 along the [001] axis, and 4 to 6 along the shear directions, respectively.

3. Results and discussion

3.1. Ferroelectricity

3.1.1. Role of Ti 3s and 3p states in ferroelectric $BaTiO_3$

Figures 2(a) and 2(b)show the optimized results for the ratio *c*/*a* of the lattice parameters and the value of the Ti ion displacement (δ_{Ti}) as a function of the *a* lattice parameter in tetragonal $BaTiO_3$, respectively. Results with arrows are the fully optimized results, and the others results are those with the *c* lattice parameters and all the inner coordination optimized for fixed *a*. Note that the fully optimized structure of $BaTiO_3$ is tetragonal with the Ti3spd4s PP, whereas it is cubic with the Ti3d4s PP. This result suggests that the explicit treatment of Ti 3s and 3p semicore states is essential to the appearance of ferroelectric states in $BaTiO_3$.

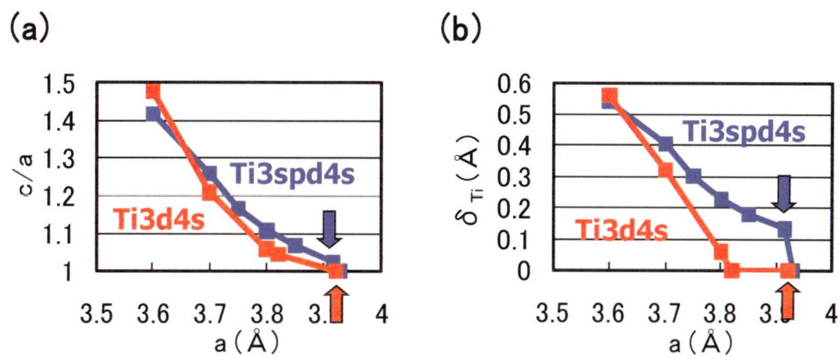

Figure 2. Optimized calculated results in tetragonal $BaTiO_3$. Results with arrows are the fully optimized results [11].

The calculated results shown in Fig. 2 suggest that the explicit treatment of Ti 3s and 3p semicore states is essential to the appearance of ferroelectric states in $BaTiO_3$. In the following, we investigate the role of Ti 3s and 3p states for ferroelectricity from two viewpoints.

One viewpoint concerns hybridizations between Ti 3s and 3p states and other states. Figure 3(a) and 3(b) shows the total density of states (DOS) of tetragonal $BaTiO_3$ with two Ti PPs. Both results are in good agreement with previous calculated results [7] by the full-potential linear augmented plane wave (FLAPW) method. In the DOS with the Ti3spd4s PP, the energy ``levels", not bands, of Ti 3s and 3p states, are located at -2.0 Hr and -1.2 Hr, respectively. This result suggests that the Ti 3s and 3p orbitals do not make any hybridization but only give Coulomb repulsions with the O orbitals as well as the Ba orbitals. In the DOS with the Ti3d4s PP, on the other hand, the energy levels of Ti 3s and 3p states are not shown because Ti 3s and 3p states were treated as the core charges. This result means that the Ti 3s and 3p orbitals cannot even give Coulomb repulsions with the O orbitals as well as the Ba orbitals.

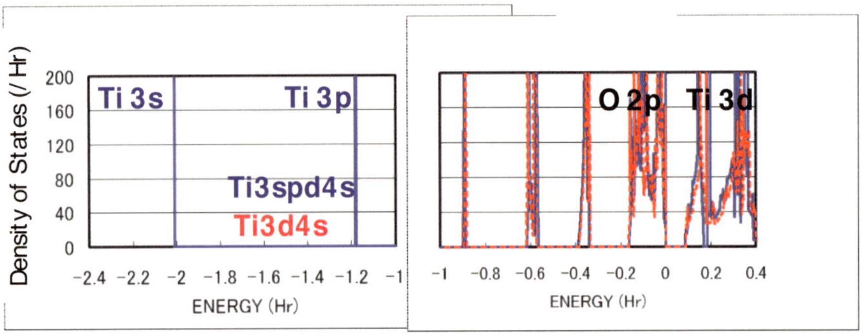

Figure 3. Total density of states (DOS) of fully optimized tetragonal $BaTiO_3$ with the Ti3spd4s PP (solid line) and cubic $BaTiO_3$ with the Ti3d4s PP (red dashed line) [11].

Another viewpoint is about the Coulomb repulsions between Ti 3s and $3p_{x(y)}$ states and $O_{x(y)}$ 2s and $2p_{x(y)}$ states in tetragonal $BaTiO_3$. Figure 4(a) and 4(b) show two-dimensional electron-density contour map on the *xz*-plane. These are the optimized calculated results with *a* fixed to be 3.8 Å, and the electron density in Fig. 4(a) is quantitatively in good agreement with the experimental result [22]. The electron density between Ti and O_x ions in Fig. 3(a) is larger than that in Fig. 4(b), which suggests that Ti ion displacement is closely related to the Coulomb repulsions between Ti 3s and $3p_{x(y)}$ states and $O_{x(y)}$ 2s and $2p_{x(y)}$ states; the Ti-O Coulomb repulsion is an important role in the appearance of the ferroelectric state in $BaTiO_3$.

The present discussion of the Coulomb repulsions is consistent with the previous reports. A recent soft mode investigation [8] of $BaTiO_3$ shows that Ba ions contribute little to the appearance of Ti ion displacement along the [001] axis. This result suggests that Ti ion displacement is closely related to the structural distortion of TiO_6 octahedra. In the present calculations, on the other hand, the only difference between $BaTiO_3$ with the Ti3spd4s PP and with the Ti3d4s PP is the difference in the expression for the Ti 3s and 3p states, *i.e.*, the explicit treatment and including core charges. However, our previous calculation [20] shows that the strong Coulomb repulsions between Ti 3s and $3p_z$ states and O_z 2s and $2p_z$ states do not favor Ti ion displacement along the [001] axis. This result suggests that the Coulomb repulsions between Ti 3s and $3p_{x(y)}$ states and $O_{x(y)}$ 2s and $2p_{x(y)}$ states would contribute to Ti ion displacement along the [001] axis, and the suggestion is consistent with a recent calculation [9] for $PbTiO_3$ indicating that the tetragonal and ferroelectric structure appears more favorable as the *a* lattice constant decreases.

http://dx.doi.org/10.5772/52187

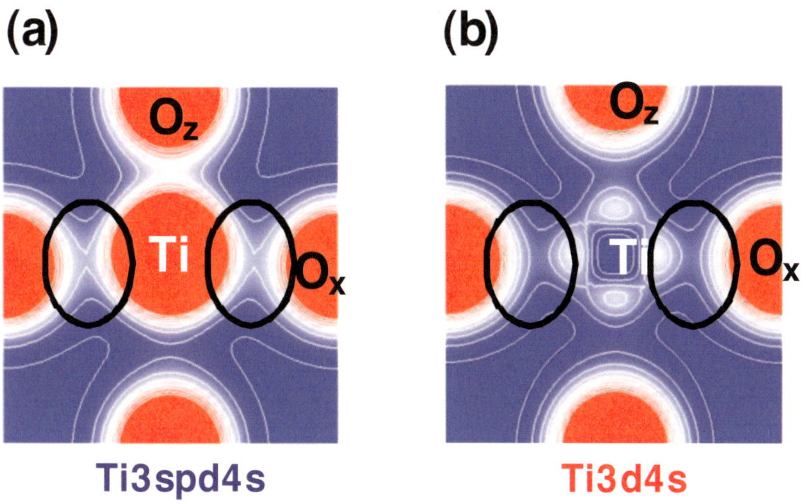

Figure 4. Two-dimensional electron-density contour map on the *xz*-plane for tetragonal $BaTiO_3$: (a) with the Ti3spd4s PP, and (b) with the Ti3d4s PP. The optimized calculated results with *a* fixed to be 3.8 Å are shown in both figures. The electron density increases as color changes from blue to red via white. Contour curves are drawn from 0.4 to 2.0 $e/Å^3$ with increments of 0.2 $e/Å^3$ [11].

Considering the above investigations, we propose the mechanism of Ti ion displacement as follows: Ti ion displacement along the *z*-axis appears when the Coulomb repulsions between Ti 3s and 3 $p_{x\,(y)}$ states and $O_{x\,(y)}$ 2s and 2 $p_{x\,(y)}$ states, in addition to the dipole-dipole interaction, overcome the Coulomb repulsions between Ti 3s and $3p_z$ states and O_z 2s and $2p_z$ states. An illustration of the Coulomb repulsions is shown in Fig. 5(a). In fully optimized $BaTiO_3$ with the Ti3spd4s PP, the Ti ion can be displaced due to the above mechanism. In fully optimized $BaTiO_3$ with the Ti3d4sPP, on the other hand, the Ti ion cannot be displaced due to the weaker Coulomb repulsions between Ti and $O_{x\,(y)}$ ions. However, since the Coulomb repulsion between Ti and O_z ions in $BaTiO_3$ with the Ti3d4s PP is also weaker than that in $BaTiO_3$ with the Ti3spd4s PP, the Coulomb repulsions between Ti and $O_{x\,(y)}$ ions in addition to the log-range force become comparable to the Coulomb repulsions between Ti and O_z ions both in Ti PPs, as the *a* lattice parameter becomes smaller. The above discussion suggests that the hybridization between Ti 3d and O_z 2s and $2p_z$ stabilizes Ti ion displacement, but contribute little to a driving force for the appearance of Ti ion displacement.

It seems that the above proposed mechanism for tetragonal BaTiO3 can be applied to the mechanism of Ti ion displacement in rhombohedral BaTiO3, as illustrated in Fig. 5(b). The strong isotropic Coulomb repulsions between Ti 3s and 3px (y, z) states and Ox (y, z) 2s and 2px (y, z) states yield Ti ion displacement along the [111] axis. On the other hand, when the isotropic Coulomb repulsions are weaker or stronger, the Ti ion cannot be displaced and therefore it is favoured for the crystal structure to be cubic.

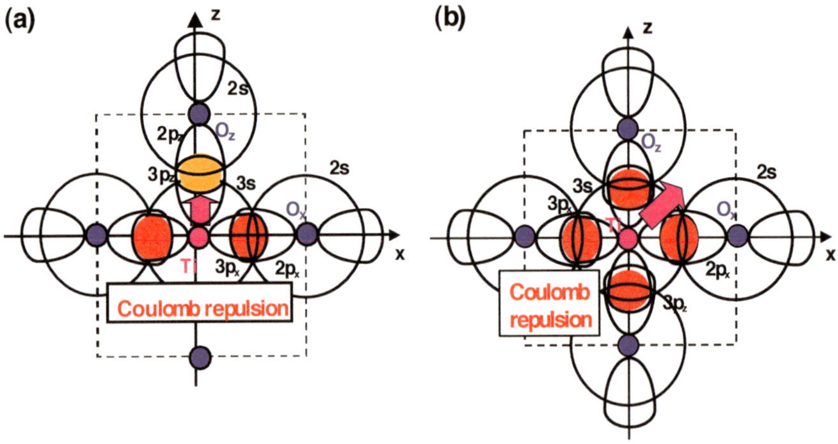

Figure 5. Illustrations of the proposed mechanisms for the Coulomb repulsions between Ti 3s and 3p states and O 2s and 2p states in $BaTiO_3$: (a) anisotropic Coulomb repulsions between Ti 3s and $3p_{x\,(y)}$ states and $O_{x\,(y)}$ 2s and $2p_{x\,(y)}$ states, and between Ti 3s and $3p_z$ states and O_z 2s and $2p_z$ states, in the tetragonal structure. (b) isotropic Coulomb repulsions between Ti 3s and $3p_{x\,(y,\,z)}$ states and $O_{x\,(y,\,z)}$ 2s and $2p_{x\,(y,\,z)}$ states, in the rhombohedral structure [11].

Let us investigate the structural properties of rhombohedral $BaTiO_3$. Figures 6(a) and 6(b) show the optimized results of the 90-α degree and δ_{Ti} as a function of fixed volumes of the unit cells in rhombohedral $BaTiO_3$, respectively, where α denotes the angle between two lattice vectors. In these figures, α denotes the angle between two crystal axes of rhombohedral $BaTiO_3$, and δ_{Ti} denotes the value of the Ti ion displacement along the [111] axis. Results with arrows are the fully optimized results; V_{opt} denote the volume of the fully optimized unit cell with the Ti3spd4s PP. The other results are those with all the inner coordination optimized with fixed volumes of the unit cells. The proposal mechanisms about the Coulomb repulsions seem to be consistent with the calculated results shown in Fig.6: For $V/V_{opt} \lesssim 0.9$ or $\gtrsim 1.3$, the isotropic Coulomb repulsions are weaker or stronger, and the Ti ion cannot be displaced along the [111] axis and therefore the crystal structure is cubic for both Ti PPs. For $0.9 \lesssim V/V_{opt} \lesssim 1.3$, on the other hand, the isotropic Coulomb repulsions are strong enough to yield Ti ion displacement for both Ti PPs. However, since the magnitude of the isotropic Coulomb repulsion is different in the two Ti PPs, the properties of the 90-α degree and δ_{Ti} are different quantitatively.

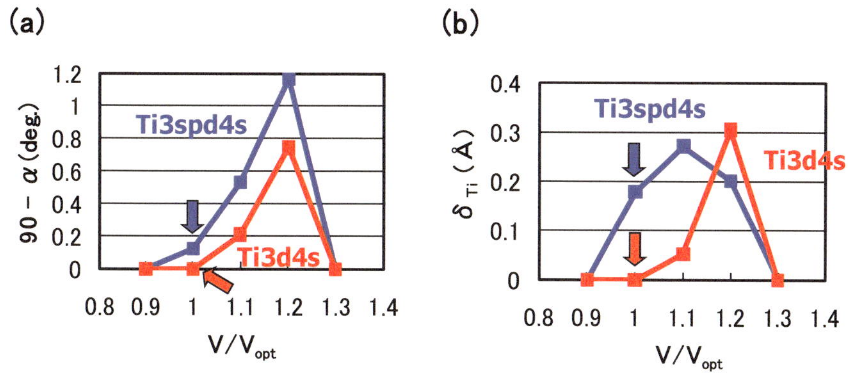

Figure 6. Optimized calculated results as a function of the fixed volumes of the unit cells in rhombohedral $BaTiO_3$: (a) 90-α degree and (b) δ_{Ti} to the [111] axis. Blue lines correspond to the results with the Ti3spd4s PP, and red lines correspond to those with the Ti3d4s PP. V_{opt} denote the volume of the fully optimized unit cell with the Ti3spd4s PP. Results with arrows are the fully optimized results, and the other results are those with all the inner coordination optimized for fixed volumes of the unit cells [11].

3.1.2. Role of Zn 3s, 3p and 3d states in ferroelectric $BiZn_{0.5}Ti_{0.5}O_3$

As discussed in Sec. 3.1.1, the Coulomb repulsions between Ti 3s and $3p_{x\,(y)}$ states and $O_{x\,(y)}$ 2s and $2p_{x\,(y)}$ states have an important role in the appearance of the ferroelectric state in tetragonal $BaTiO_3$. In this subsection, we discuss the role of Zn 3d (d^{10}) states in addition to 3s and 3p states for ferroelectricity in tetragonal BZT.

Table 1 shows a summary of the optimized results of BZT in Cases I and II. ΔE_{total} denotes the difference in total energy per f.u. between the rhombohedral and other structures. Although the lattice constant in each structure except the rhombohedral one seems to be quantitatively similar in both cases, properties of ΔE_{total} are different. In Case I, the rhombohedral structure is the most stable, which is in disagreement with the experimental result [24]. In Case II, on the other hand, the monoclinic structure, which is the ``pseudo-C-type-tetragonal'' structure, is the most stable. Unfortunately, this result seems to be in disagreement with the experimental result [24], but is in good agreement with the recent calculated result [6]. Note that the magnitude of ΔE_{total} in Case II is markedly smaller than that in Case I. In contrast to BZT, the rhombohedral structure is the most stable structure in both cases in BMT, which is consistent with the experimental result [26].

Figures 7(a) and 7(b) show two-dimensional electron density contour maps of the C-type tetragonal BZT in Cases I and II, respectively. The Coulomb repulsion of Zn-O_x in Case II is larger than that in Case I, and the Coulomb repulsion favorably causes Zn ion displacement to O_z in Case II. This result is consistent with Sec. 3.1.1. In contrast to the properties of Zn-O bonding, the inner coordination of the Ti ion is similar in both cases, although the electron densities are markedly different. This result suggests that the Coulomb repulsion magnitude of Ti-O_z is the same as that of Ti-O_x in small Ti-O bonding (≈ 1.8 Å), in both Cases I and II. Figures 7(c) and 7(d) show two-dimensional electron density contour maps of the C-type tetragonal BMT in Cases I and II, respectively. Although the electron densities in both cases are markedly different, the inner coordination of the Mg ion are similar. This result suggests that the Coulomb repulsion between Mg and O is not strong sufficiently for inducing Mg ion displacement even in Case II.

Structure	*a* (Å)	*c* (Å)	*c/a*	α (deg.)	ΔE_{total} (eV/f.u.)
A-type Tetra.	3.748	4.579	1.222	90	0.316
C-type Tetra.	3.681	4.784	1.299	90	0.240
G-type Tetra.	3.725	4.574	1.228	90	0.158
Monoclinic	3.735	4.741	1.269	β = 91.5	0.193
Rhombohedral	5.560		1	59.93	0
Experiment [24]	3.822	4.628	1.211	90	---

(a)

Structure	*a* (Å)	*c* (Å)	*c/a*	α (deg.)	ΔE_{total} (eV/f.u.)
A-type Tetra.	3.711	4.662	1.256	90	0.135
C-type Tetra.	3.670	4.789	1.305	90	0.091
G-type Tetra.	3.684	4.698	1.275	90	0.047
Monoclinic	3.726	4.740	1.272	β = 91.1	-0.021
Rhombohedral	5.590		1	59.90	0
Experiment [24]	3.822	4.628	1.211	90	---

(b)

Table 1. Summary of the optimized results of BZT in (a) Case I and (b) Case II. *a* and *c* denote the lattice parameters, and α and β denote angles between two lattice axes. ΔE_{total} denotes the difference in total energy per f.u. between the rhombohedral and other structures [10].

Finally in this subsection, we discuss the difference in the electronic structures between the C-type tetragonal and the monoclinic BZT. Figures 8(a) and 8(b) show the electron density contour maps of the C-type tetragonal BZT and that of the monoclinic BZT in Case II, respectively. This result suggests that the strong Coulomb repulsion between Zn and O_z causes the small Zn ion displacement in the [110] direction in the monoclinic BZT, which makes the Coulomb repulsion of Zn-O_z weaker than that in the C-type tetragonal BZT. As a result, this small Zn ion displacement makes the monoclinic BZT more stable than the C-type tetragonal structure.

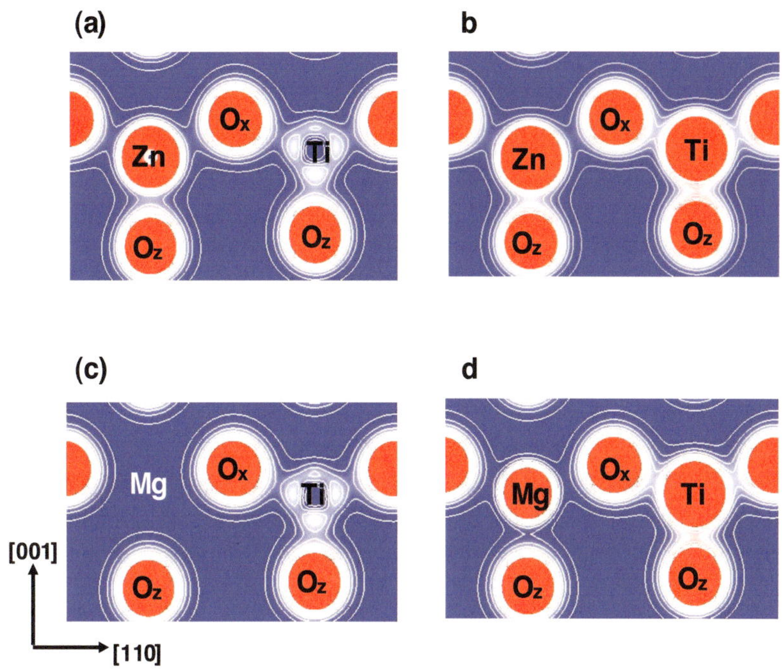

Figure 7. Two-dimensional electron density contour maps of monoclinic (a) BZT in Case I, (b) BZT in Case II, (c) BMT in Case I, and (d) BMT in Case II. The electron density increases as color changes from blue to red via white. Contour curves are drawn from 0.2 to 2.0 $e/Å^3$ with increments of 0.2 $e/Å^3$ [10].

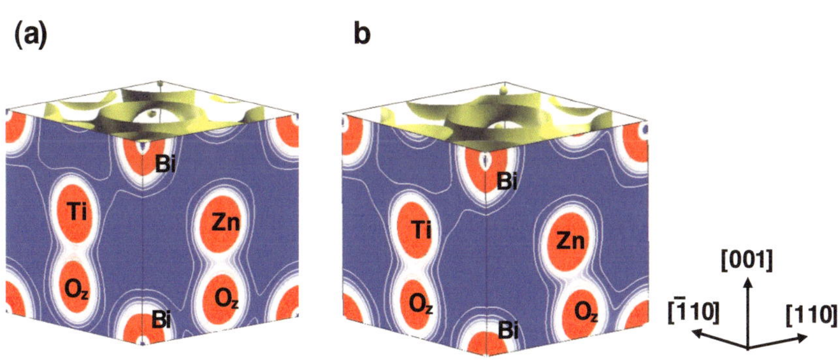

Figure 8. Two-dimensional electron density contour maps of BZT in Case II (a) C-type tetragonal and (b) monoclinic. The electron density increases as color changes from blue to red via white. Contour curves are drawn from 0.2 to 2.0 $e/Å^3$ with increments of 0.2 $e/Å^3$ [10].

3.2. Piezoelectricity

3.2.1. Role of the Ti-O Coulomb repulsions in tetragonal piezoelectric $SrTiO_3$ and $BaTiO_3$

As discussed in Sec. 3.1, the Coulomb repulsions between Ti 3s and $3p_{x\,(y)}$ states and $O_{x\,(y)}$ 2s and $2p_{x\,(y)}$ states have an important role in the appearance of the ferroelectric state in tetragonal $BaTiO_3$. In this subsection, we discuss the role of the Ti-O Coulomb repulsions for piezoelectric $SrTiO_3$ and $BaTiO_3$.

Figures 9(a) shows the optimized results for $c - c_{cub}$ as a function of the a lattice parameters in tetragonal $SrTiO_3$ and $BaTiO_3$, where c_{cub} denotes the c lattice parameter in cubic $SrTiO_3$ and $BaTiO_3$, respectively. These results are the fully optimized results and the results with the c lattice parameters and all the inner coordination optimized for fixed a. The fully optimized parameters of $SrTiO_3$ (a = 3.84 Å: cubic) and $BaTiO_3$ (a = 3.91 Å and c = 4.00 Å: tetragonal) are within 2.0 % in agreement with the experimental results in room temperature. Figure 9(b) shows the evaluated results for P_3 as a function of the a lattice parameters in tetragonal $SrTiO_3$ and $BaTiO_3$, where P_3, which is evaluated by Eq. (2), denotes the spontaneous polarization along the [001] axis. Note that the tetragonal and ferroelectric structures appear even in $SrTiO_3$ when the fixed a lattice parameter is compressed to be smaller than the fully-optimized a lattice parameter. As shown in Figs. 9(a) and 9(b), the tetragonal and ferroelectric structure appears more favorable as the fixed a lattice parameter decreases, which is consistent with previous calculated results [9, 11]. The results would be due to the suggestion discussed in the previous section that the large Coulomb repulsion of Ti-O bonding along the [100] axis (and the [010] axis) is a driving force of the displacement of Ti ions along the [001] axis, *i.e.*, the large Coulomb repulsion along the [100] axis (and the [010] axis) is essential for the appearance of the tetragonal structure.

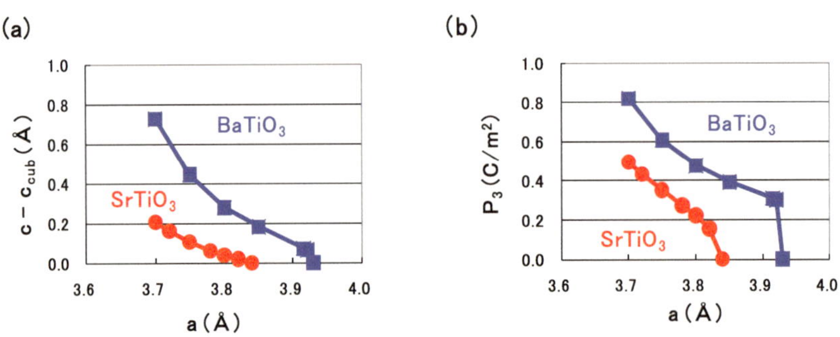

Figure 9. Optimized calculated results as a function of a lattice parameters in compressive tetragonal $SrTiO_3$ and $BaTiO_3$: (a) $c - c_{cub}$ and (b) P_3, *i.e.*, spontaneous polarization along the [001] axis [12].

In the following, we use $c - c_{cub}$ as a functional parameter, because $c - c_{cub}$ is closely related to η_3. Figures 10(a) and 10(b) shows the piezoelectric properties of e_{33} and e_{31} as a function of $c - c_{cub}$ in tetragonal $SrTiO_3$ and $BaTiO_3$. The value $c - c_{cub}$ is optimized value as shown in Fig. 9(a) and e_{33} and e_{31} are evaluated values in their optimized structures. Note that e_{33} become

http://dx.doi.org/10.5772/52187

larger at $c - c_{cub} \approx 0$, especially in $SrTiO_3$. These properties seem to be similar to the properties arond the Curie temperatures in piezoelectric ABO_3; Damjanovic emphasized the importance of the polarization extension as a mechanism of larger piezoelectric constants in a recent paper [28]. Contrary to e_{33}, on the other hand, the changes in e_{31} are much smaller than the changes in e_{33}, but note that e_{31} shows negative in $SrTiO_3$ while positive in $BaTiO_3$.

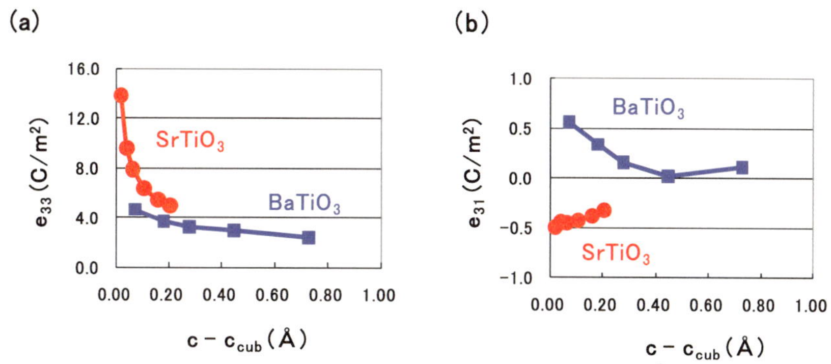

Figure 10. Evaluated piezoelectric constants as a function of $c - c_{cub}$ in optimized tetragonal $SrTiO_3$ and $BaTiO_3$: (a) e_{33} and (b) e_{31} [12].

As expressed in Eq. (3), e_{3j} is the sum of the contributions from the clamped term and the relaxed term. However, it has been generally known that the contribution to e_{3j} from the clamped term is much smaller than that from the relaxed term; in fact, the absolute values of the e_{33} clamped terms are less than 1 C/m^2 in both $SrTiO_3$ and $BaTiO_3$. We therefore investigate the contributions to the relaxed term of e_{33} and e_{31} in detail. As expressed in Eq. (3), the relaxed terms of e_{3j} are proportional to the sum of the products between the $Z_{33}^*(k)$ and $\partial u_3(k)/\partial \eta_j$ (j = 3 or 1) values. Let us show the evaluated results of $Z_{33}^*(k)$, $\partial u_3(k)/\partial \eta_3$, and $\partial u_3(k)/\partial \eta_1$ in the following. Figures 11(a) and 11(b) show the $Z_{33}^*(k)$ values in $SrTiO_3$ and $BaTiO_3$, respectively. Properties of the $Z_{33}^*(k)$ values are quantitatively similar in both $SrTiO_3$ and $BaTiO_3$. Therefore, the difference in the properties of e_{33} and e_{31} between $SrTiO_3$ and $BaTiO_3$ must be due to the difference in the properties of $\partial u_3(k)/\partial \eta_j$. Figures 12(a) and 12(b) show the $\partial u_3(k)/\partial \eta_3$ values in $SrTiO_3$ and $BaTiO_3$, respectively. In these figures, O_x and O_z denote oxygen atoms along the [100] and [001] axes, respectively. Clearly, the absolute values of $\partial u_3(k)/\partial \eta_3$ are different in between $SrTiO_3$ and $BaTiO_3$. On the other hand, Figs. 13(a) and 13(b) show the $\partial u_3(k)/\partial \eta_1$ values in $SrTiO_3$ and $BaTiO_3$, respectively. The absolute values of $\partial u_3(k)/\partial \eta_1$, especially for Ti, O_x, and O_z are different in between $SrTiO_3$ and $BaTiO_3$. As a result, the quantitative differences in e_{33} and e_{31} between $SrTiO_3$ and $BaTiO_3$ are due to the differences in the contribution of the $\partial u_3(k)/\partial \eta_3$ and $\partial u_3(k)/\partial \eta_1$ values, respectively.

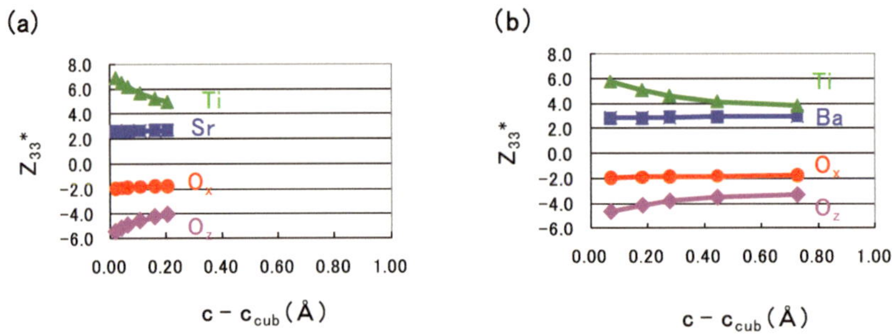

Figure 11. Evaluated Born effective charges Z_{33}^{*} (k) as a function of $c - c_{cub}$: (a) $SrTiO_3$ and (b) $BaTiO_3$. O_x and O_z denote oxygen atoms along the [100] axis and the [001] axis, respectively [12].

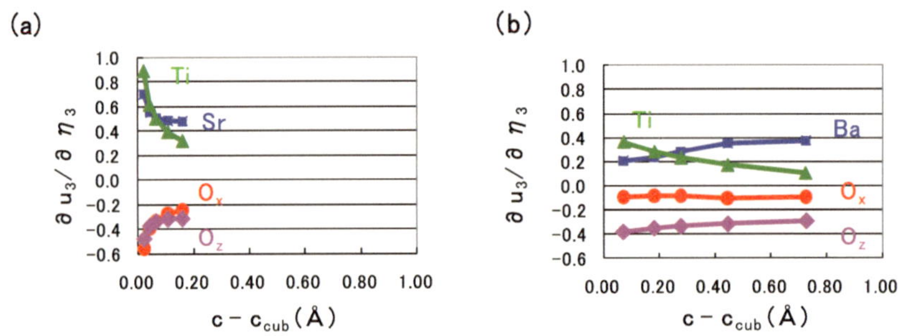

Figure 12. Evaluated values of $\partial u_3(k)/\partial \eta_3$ as a function of $c - c_{cub}$: (a) $SrTiO_3$ and (b) $BaTiO_3$ [12].

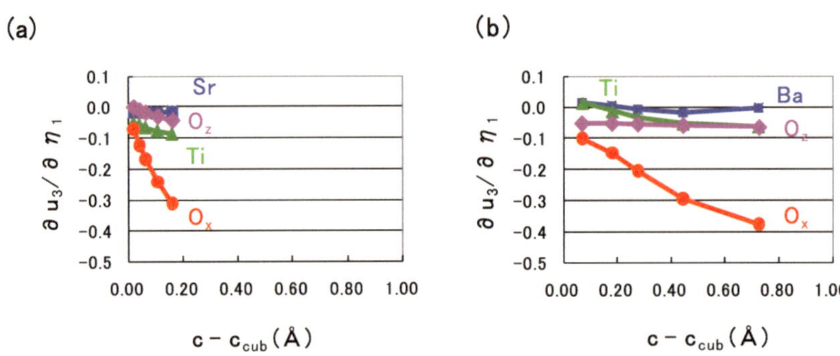

Figure 13. Evaluated values of $\partial u_3(k)/\partial \eta_1$ as a function of $c - c_{cub}$: (a) $SrTiO_3$ and (b) $BaTiO_3$ [12].

Let us discuss the reasons of the quantitative differences in e_{33} between $SrTiO_3$ and $BaTiO_3$. Figure 14(a) shows the difference between the Ti-O_z distance ($R_{Ti\text{-}Oz}$) and the sum of the r_{Ti}

and r_{Oz} ($r_{Ti} + r_{Oz}$) along the [001] axis as a function of $c - c_{cub}$. Note that $R_{Ti\text{-}Oz}$ is smaller than $r_{Ti} + r_{Oz}$ in both $SrTiO_3$ and $BaTiO_3$. However, the difference in absolute value between $R_{Ti\text{-}Oz}$ and $r_{Ti} + r_{Oz}$ in $SrTiO_3$ is smaller than the difference in $BaTiO_3$ for $0 \lesssim c - c_{cub} \lesssim 0.20$. This result suggests that the Ti-O_z Coulomb repulsion along the [001] axis in $SrTiO_3$ is smaller than that in $BaTiO_3$ and that therefore the Ti ion of $SrTiO_3$ can be displaced more easily along the [001] axis than that of $BaTiO_3$. This would be a reason why the absolute values of $\partial u_3(k)/\partial \eta_3$ of Ti and O_z ions in $SrTiO_3$ are larger than that in $BaTiO_3$. Figure 14(b) shows the difference between the A-O_x distance ($R_{A\text{-}Ox}$) and the sum of r_A and r_{Ox} ($r_A + r_{Ox}$) on the (100) plane as a function of $c - c_{cub}$, where the values of the ionic radii are defined as Shannon's ones [23]. Note that $R_{A\text{-}Ox}$ is smaller than $r_A + r_{Ox}$ in both $SrTiO_3$ and $BaTiO_3$. However, the difference in absolute value between $R_{A\text{-}Ox}$ and $r_A + r_{Ox}$ in $SrTiO_3$ is much smaller than the difference in $BaTiO_3$ for $0 \lesssim c - c_{cub} \lesssim 0.20$. This result suggests that the Sr-O_x Coulomb repulsion on the (100) plane in $SrTiO_3$ is much smaller than the Ba-O_x Coulomb repulsion in $BaTiO_3$ and that therefore Sr and O_x ions of $SrTiO_3$ can be displaced more easily along the [001] axis than Ba and O_x ions of $BaTiO_3$. This would be a reason why the absolute values of $\partial u_3(k)/\partial \eta_3$ of Sr and O_x ions in $SrTiO_3$ are larger than those of Ba and O_x ions in $BaTiO_3$.

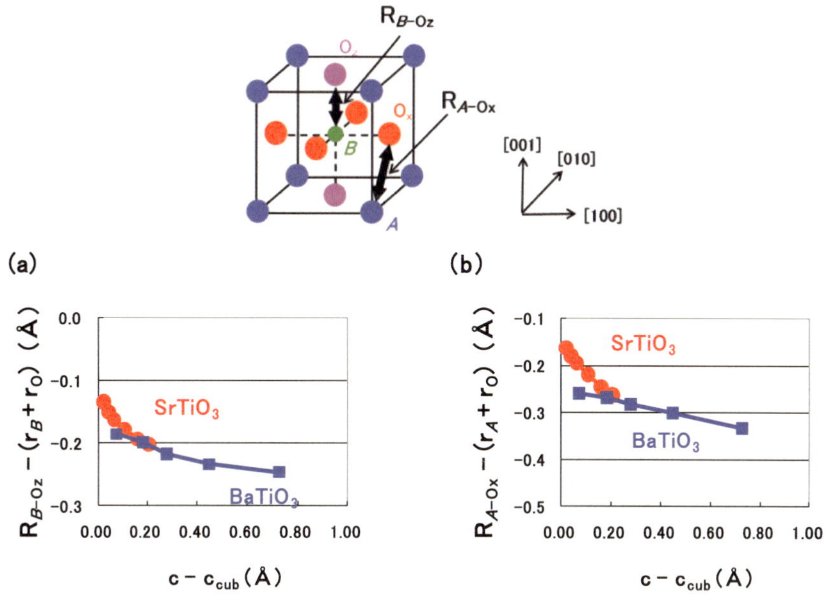

Figure 14. Evaluated values as a function of $c - c_{cub}$ in optimized tetragonal $SrTiO_3$ and $BaTiO_3$: (a) difference between the Ti-O_z distance ($R_{Ti\text{-}Oz}$) and $r_{Ti} + r_{Oz}$. (b) difference between the A-O_x distance ($R_{A\text{-}Ox}$) and $r_A + r_{Ox}$. $R_{A\text{-}Ox}$ and $R_{Ti\text{-}Oz}$ in $ATiO_3$ are also illustrated. Note that all the ionic radii are much larger and that A and Ti ions are displaced along the [001] axis in actual $ATiO_3$ [12].

Finally, in this subsection, we discuss the relationship between $\partial u_3(k)/\partial \eta_3$ and $c - c_{cub}$ in detail. Figure 15(a) shows the differences in the total energy (ΔE_{total}) as a function of u_3(Ti). In this figure, the properties of $SrTiO_3$ with η_3 = 0.011, $SrTiO_3$ with η_3 = 0.053 and fully optimized $BaTiO_3$ as a reference, are shown. Calculations of E_{total} were performed with the fixed crystal structures of previously optimized structures except Ti ions. Clearly, the magnitude of u_3(Ti) at the minimum points of the ΔE_{total} and the depth of the potential are closely related to the spontaneous polarization P_3 and the Curie temperature (T_C), respectively. On the other hand, e_{33} seems to be closely related to the deviation at the minimum points of the ΔE_{total}. Figure 15(b) shows illustrations of ΔE_{total} curves with deviations at the minimum points of the ΔE_{total} values, corresponding to the ΔE_{total} curves of $SrTiO_3$ in Fig. 15(a). Clearly, as η_3 becomes smaller, the deviated value at the minimum point of the ΔE_{total} values becomes smaller, *i.e.*, the Ti ion can be displaced more favourably. On the other hand, as shown in Fig. 12(a), the absolute value of $\partial u_3(\text{Ti})/\partial \eta_3$ becomes larger as η_3 becomes smaller.

Therefore, the Ti ion can be displaced more favourably as the deviated value at the minimum point of the ΔE_{total} values becomes smaller. The relationship between e_{33} and $\partial u_3(\text{Ti})/\partial \eta_3$ is discussed in Sec. 3.2.3.

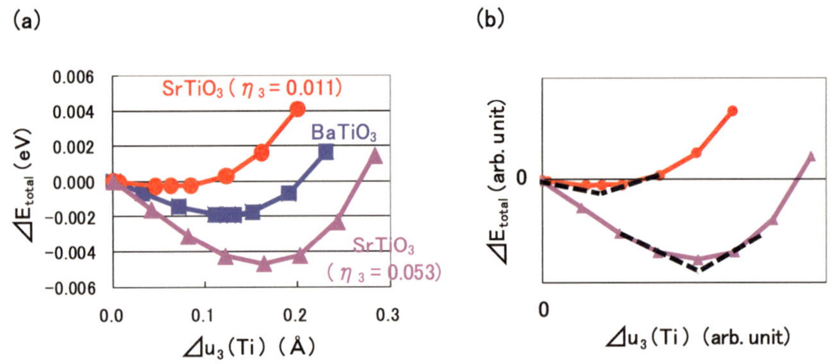

Figure 15. a) ΔE_{total} as a function of u_3(Ti) in tetragonal $SrTiO_3$ and $BaTiO_3$. (b) Illustration of the ΔE_{total} curves in tetragonal $SrTiO_3$ with η_3 = 0.011 and $SrTiO_3$ with η_3 = 0.053 with deviations at the minimum point of ΔE_{total} [14].

3.2.2. Proposal of new piezoelectric materials

The previous discussion in Sec. 3.2.1 suggests that the piezoelectric properties of e_{33} are closely related to the *B-X* Coulomb repulsions in tegtragonal ABX_3. In the viewpoint of the change of the *B-X* Coulomb repulsions, we recently proposed new piezoelectric materials [16, 17], *i.e.*, $BaTi_{1-x}Ni_xO_3$ and $Ba(Ti_{1-3z}Nb_{3z})(O_{1-z}N_z)_3$.

It has been known that $BaNiO_3$ shows the 2H hexagonal structure as the most stable structure in room temperature. Moreover, the ionic radius of Ni^{4+} (d^6) with the low-spin state in 2H $BaNiO_3$ is 0.48 Å, which is much smaller than that of Ti^{4+} (d^0), 0.605 Å, in $BaTiO_3$. Therefore, due to the drastic change in the $(Ti_{1-x}Ni_x)$-O Coulomb repulsions in tetragonal $BaTi_{1-x}Ni_xO_3$, the e_{33} piezoelectric values are expected to be larger than that in tetragonal $BaTiO_3$, especially around the morphotropic phase boundary (MPB). Figure 16(a) shows the total-en-

http://dx.doi.org/10.5772/52187

ergy difference ΔE_{total} between 2H and tetragonal structures of $BaTi_{1-x}Ni_xO_3$ as a function of x. The most stable structure changes at $x \approx 0.26$. Figure 16(b) shows $c - c_{\text{cub}}$ as a function of x. The $c - c_{\text{cub}}$ value shows 0 around $x = 0.26$, which suggests the appearance of the MPB, *i.e.*, the e_{33} piezoelectric value shows a maximum at $x \approx 0.26$.

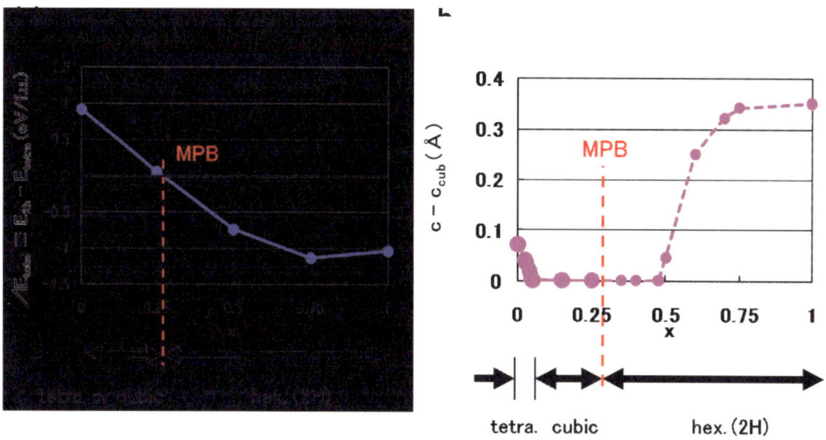

Figure 16. a) ΔE_{total} (total-energy difference between 2H and tetragonal structures), and (b) $c - c_{\text{cub}}$ of the tetragonal structure, as a function of x in $BaTi_{1-x}Ni_xO_3$ [16]. For or $0.26 \leq x \leqq 1$, the tetragonal structure is not the most stable one.

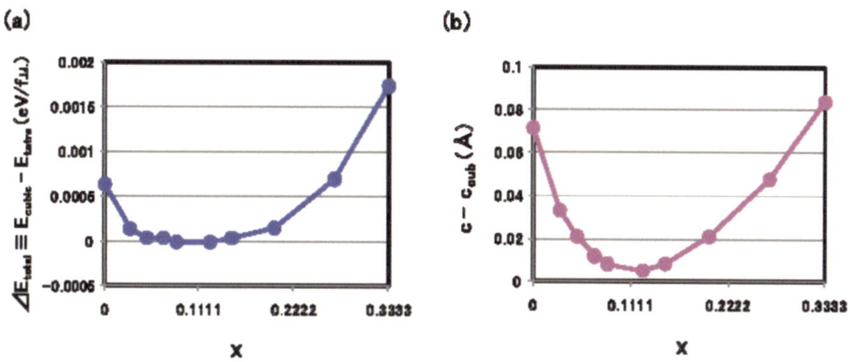

Figure 17. (a) ΔE_{total} (total-energy difference between cubic and tetragonal structures), and (b) $c - c_{\text{cub}}$, as a function of x in $Ba(Ti_{1-3z}Nb_{3z})(O_{1-z}N_z)_3$ [17].

Another proposal is tetragonal $Ba(Ti_{1-3z}Nb_{3z})(O_{1-z}N_z)_3$, which consists of $BaTiO_3$ and $BaNbO_2N$ [17]. Due to the change of $(Ti_{1-3z}Nb_{3z})$-$(O_{1-z}N_z)$ Coulomb repulsions, the e_{33} piezoelectric values are expected to be larger than that in tetragonal $BaTiO_3$. Recent experimental paper reported that the most stable structure of $BaNbO_2N$ is cubic in room temperature [34]. Contrary to the experimental result, however, our calculations suggest that the tetragonal structure will be more stable than the cubic one, as shown in Fig. 17(a). Figure 17(b) shows $c - c_{\text{cub}}$

as a function of x. The $c - c_{cub}$ value shows almost 0 at $x \approx 0.12$. Although the MPB does not appear in tetragonal $Ba(Ti_{1-3z}Nb_{3z})(O_{1-z}N_z)_3$, the e_{33} piezoelectric values are expected to show a maximum at $x \approx 0.12$.

3.2.3. Piezoelectric properties of in tetragonal ABX_3

In the following, we discuss the role of the B-X Coulomb repulsions in piezoelectric ABX_3.

Figures 18(a) and 18(b) show the piezoelectric properties of e_{33} as a function of the value $c - c_{cub}$ in tetragonal ABX_3, where c_{cub} denotes the c lattice parameter in cubic ABX_3; $c - c_{cub}$ is a closely related parameter to η_3. For ABX_3, $SrTiO_3$, $BaTiO_3$ and $PbTiO_3$ with the c lattice parameter and all the inner coordination optimized for fixed a, and $BaTi_{1-x}Ni_xO_3$ ($0 \leqq x \leqq 0.05$), $Ba(Ti_{1-3z}Nb_{3z})(O_{1-z}N_z)_3$ ($0 \leqq z \leqq 0.125$), $Ba_{1-y}Sr_yTiO_3$ ($0 \leqq y \leqq 0.5$), $BaTi_{1-x}Zr_xO_3$ ($0 \leqq x \leqq 0.06$), and $BiM'O_3$ (M' = Al, Sc) with fully optimized, were prepared [15]. Note that e_{33} becomes larger as $c - c_{cub}$ becomes smaller and that the trend of e_{33} is almost independent of the kind of A ions. Moreover, note also that e_{33} of $BaTi_{1-x}Ni_xO_3$ and that of $Ba(Ti_{1-3z}Nb_{3z})(O_{1-z}N_z)_3$ show much larger values than the other ABX_3.

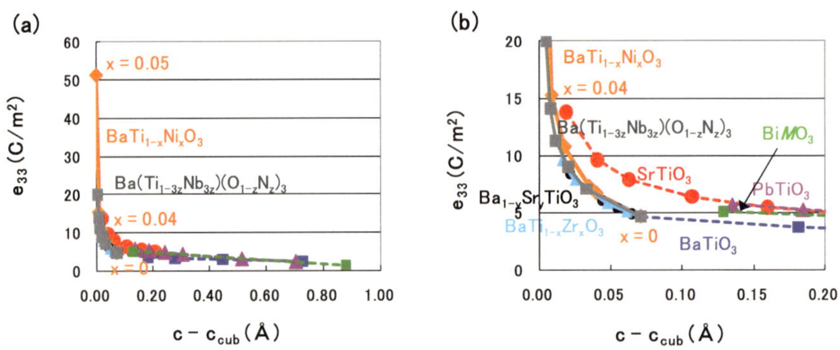

Figure 18. e_{33} as a function of $c - c_{cub}$ for different scales [15].

Let us discuss the relationship between $\partial u_3(k)/\partial \eta_3$ and $c - c_{cub}$ in $BaTi_{1-x}Ni_xO_3$ and $BaTiO_3$ in the following. Figures 19(a) and 19(b) show the $\partial u_3(k)/\partial \eta_3$ values. In these figures, O_x and O_z denote oxygen atoms along the [100] and [001] axes, respectively. Clearly, the absolute values of $\partial u_3(k)/\partial \eta_3$ in $BaTi_{1-x}Ni_xO_3$ are much larger than those in $BaTiO_3$. Moreover, in comparison with Fig.18, properties of e_{33} are closely related to those of $\partial u_3(k)/\partial \eta_3$. Figure 20(a) shows the difference between $R_{B\text{-}Oz}$ and $r_B + r_{Oz}$ along the [001] axis, and Fig. 20(b) shows the difference between $R_{A\text{-}Ox}$ and $r_A + r_{Ox}$ on the (100) plane for several ABO_3, as a function of $c - c_{cub}$. Clearly, the difference between $R_{B\text{-}Oz}$ and $r_B + r_{Oz}$ is closely related to e_{33} shown in Fig. 18, rather than the difference between $R_{A\text{-}Ox}$ and $r_A + r_{Ox}$. Moreover, note that the difference in absolute value between $R_{B\text{-}Oz}$ and $r_B + r_{Oz}$ in $BaTi_{1-x}Ni_xO_3$ is much smaller than that in $BaTiO_3$. This result suggests that the $(Ti_{1-x}Ni_x)$-O_z Coulomb repulsion along the [001] axis in $BaTi_{1-x}Ni_xO_3$ is much smaller than the Ti-O_z Coulomb repulsion in $BaTiO_3$ and that therefore $Ti_{1-x}Ni_x$ ion of $BaTi_{1-x}Ni_xO_3$ can be displaced more easily along the [001] axis than Ti ion of

$BaTiO_3$. This must be a reason why the absolute value of $\partial u_3(k)/\partial \eta_3$ of $Ti_{1-x}Ni_x$ and O_z ions in $BaTi_{1-x}Ni_xO_3$ is larger than those in $BaTiO_3$.

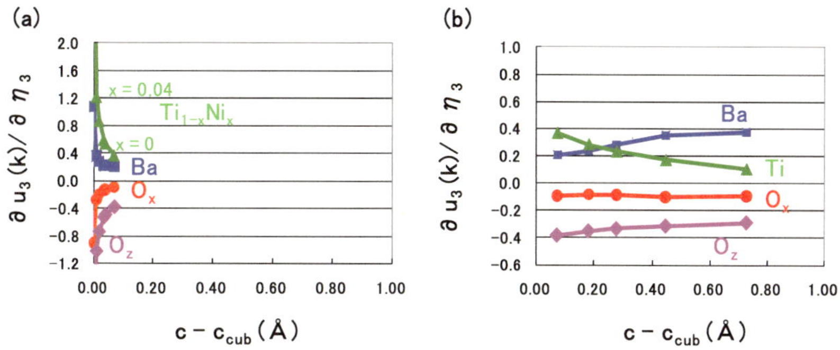

Figure 19. $\partial u_3(k)/\partial \eta_3$ as a function of $c - c_{cub}$: (a) $BaTi_{1-x}Ni_xO_3$ and (b) $BaTiO_3$ [15].

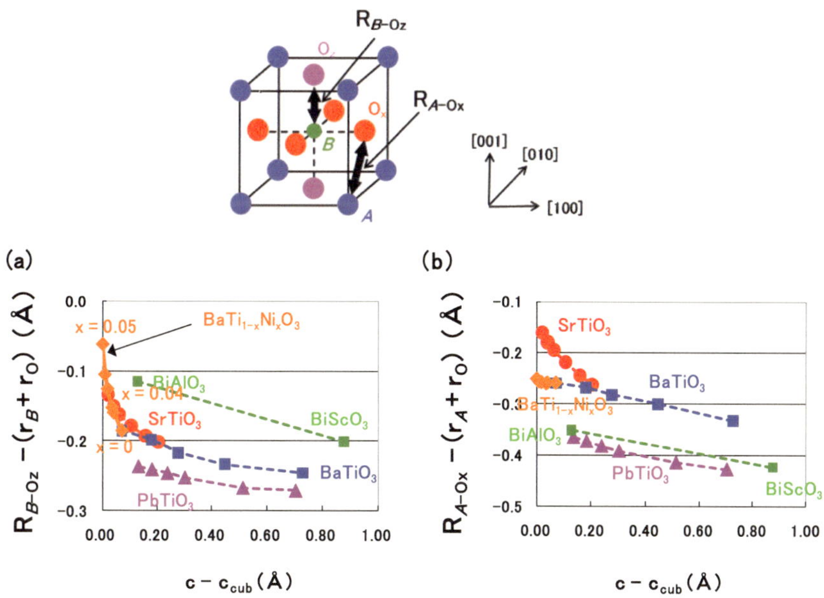

Figure 20. Evaluated values of in optimized tetragonal $BaTi_{1-x}Ni_xO_3$, $BaTiO_3$, and several ABO_3, as a function of $c - c_{cub}$: (a) $R_{B\text{-}Oz}$ - $(r_B + r_{Oz})$, and (b) $R_{A\text{-}Ox} - (r_A + r_{Ox})$ [15].

Figure 21(a) shows ΔE_{total} as a function of the displacement of the $Ti_{1-x}Ni_x$ ions with fixed crystal structures of fully-optimized $BaTi_{1-x}Ni_xO_3$. Calculations of E_{total} were performed with the fixed crystal structures of previously optimized structures except $Ti_{1-x}Ni_x$ ions. The deviated value at the minimum point of ΔE_{total}, *i.e.*, $\partial(\Delta E_{total})/\partial u_3(Ti_{1-x}Ni_x)$, becomes smaller as x becomes larger. Moreover, both e_{33} and $\partial u_3(\mathrm{Ti}_{1\text{-}x}\mathrm{Ni}_x)/\partial\eta_3$ become larger as x becomes larger, as shown in Figs. 18 and 19. This result is consistent with the result of $SrTiO_3$ shown in Fig. 15(a).

Let us discuss the above reasons in the following. $\partial(\Delta E_{\mathrm{total}})/\partial u_3(\mathrm{Ti}_{1\text{-}x}\mathrm{Ni}_x)$ can be written as

$$\left(\frac{\partial \Delta E_{\mathrm{total}}}{\partial u_3(\mathrm{Ti}_{1-x}\mathrm{Ni}_x)}\right)=\left(\frac{\partial \Delta E_{\mathrm{total}}}{\partial \eta_3}\right)\times\left(\frac{\partial u_3(\mathrm{Ti}_{1-x}\mathrm{Ni}_x)}{\partial \eta_3}\right)^{-1}. \tag{5}$$

As shown in Fig. 21(b), $\partial(\Delta E_{\mathrm{total}})/\ \partial\eta_3$ is almost constant, and therefore, $\partial(\Delta E_{\mathrm{total}})/\partial u_3(\mathrm{Ti}_{1\text{-}x}\mathrm{Ni}_x)$ is almost proportional to $(\partial u_3(\mathrm{Ti}_{1\text{-}x}\mathrm{Ni}_x)/\partial\eta_3)^{-1}$, *i.e.*,

$$\left(\frac{\partial \Delta E_{\mathrm{total}}}{\partial u_3(\mathrm{Ti}_{1-x}\mathrm{Ni}_x)}\right)\propto\left(\frac{\partial u_3\left(\mathrm{Ti}_{1-x}\mathrm{Ni}_x\right)}{\partial \eta_3}\right)^{-1}. \tag{6}$$

On the other hand, according to Eq. (3), e_{33} becomes larger as $\partial u_3(\mathrm{Ti}_{1\text{-}x}\mathrm{Ni}_x)/\partial\eta_3$ becomes larger. This is a reason why e_{33} becomes larger as $\partial(\Delta E_{\mathrm{total}})/\partial u_3(\mathrm{Ti}_{1\text{-}x}\mathrm{Ni}_x)$ becomes smaller. This result is consistent with the result of $SrTiO_3$ discussed in Sec. 3.2.1.

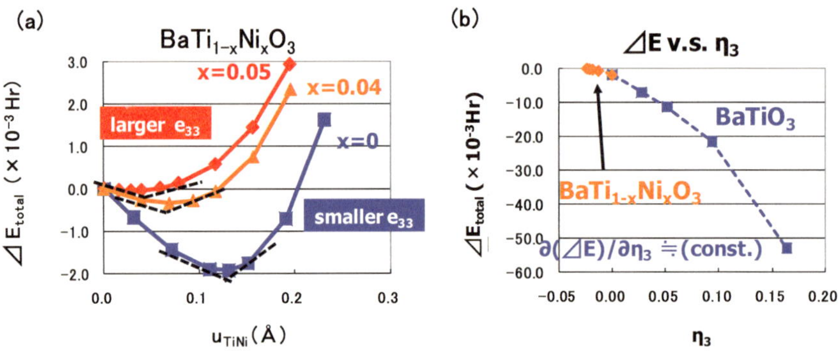

Figure 21. (a) $\Delta E_{\mathrm{total}}$ as a function of $u_3(\mathrm{Ti}_{1\text{-}x}\mathrm{Ni}_x)$ in $\mathrm{BaTi}_{1\text{-}x}\mathrm{Ni}_x\mathrm{O}_3$. Results with $x = 0.05$, 0.04, and 0 are shown. Dashed lines denote guidelines of $\partial(\Delta E_{\mathrm{total}})/\partial u_3(\mathrm{Ti}_{1\text{-}x}\mathrm{Ni}_x)$ for each x. (b) $\Delta E_{\mathrm{total}}$ as a function of η_3 for $\mathrm{BaTi}_{1\text{-}x}\mathrm{Ni}_x\mathrm{O}_3$ and $BaTiO_3$ [15].

Finally, we comment on the difference in the properties between e_{33} and d_{33} in tetragonal ABX_3. Figures 22(a) and 22(b) show the piezoelectric properties of d_{33} as a function of $c - c_{\mathrm{cub}}$. Note that the trend of d_{33} is closely dependent on the kind of A ions. This result is in contrast with the trend of e_{33} as shown in Fig. 18. As expressed in Eq. (4), d_{33} is closely related to the elastic compliance s_{3j}^{E} as well as e_{3j}. In fact, the absolute value of s_{3j}^{E} in $\mathrm{Bi}BX_3$ or $\mathrm{Pb}BX_3$ is generally larger than that in ABX_3 with alkaline-earth A ions. This result must be due to the larger Coulomb repulsion of Bi-X or Pb-X derived from 6s electrons in Bi (Pb) ion.

http://dx.doi.org/10.5772/52187

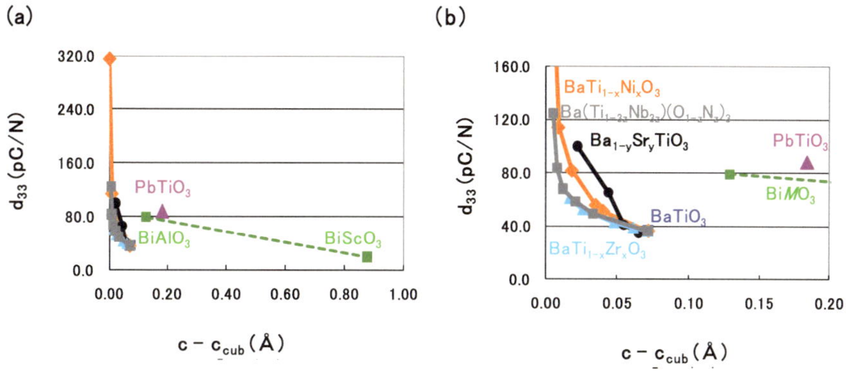

Figure 22. d_{33} as a function of $c - c_{cub}$ for different scales [15].

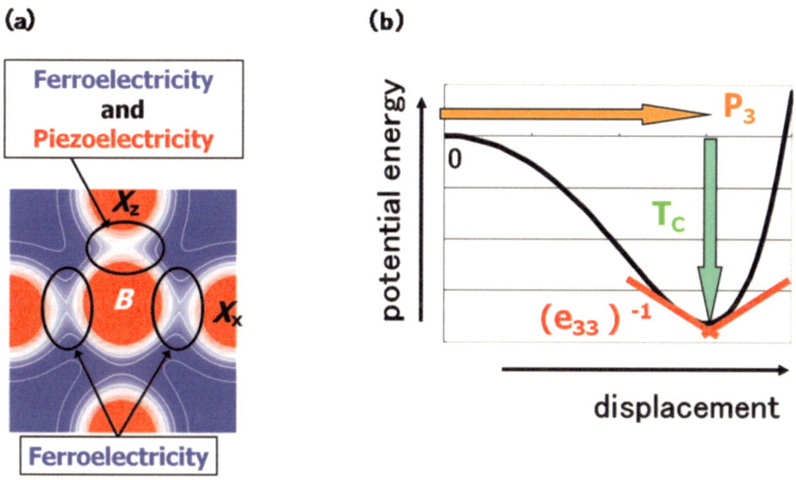

Figure 23. (a) Illustration of the relationship between the B-X Coulomb repulsions and the ferroelectric and piezoelectric states in tetragonal ABX_3. (b) Illustration of the relationship between e_{33} and the deviation. P_3 and T_C denote the spontaneous polarization and the Curie temperature, respectively.

4. Conclusion

We have discussed a general role of the *B-X* Coulomb repulsions for ferroelectric and piezoelectric properties of tetragonal ABX_3, based on our recent papers and patents. We have found that both ferroelectric state and piezoelectric state are closely related to the $B\text{-}X_z$ Coulomb repulsions as well as the $B\text{-}X_x$ ones, as illustrated in Fig. 23(a). Moreover, as illustrated in Fig. 23(b), we have also found that e_{33} is closely related to the deviation at the minimum point of the ΔE_{total}.

Acknowledgements

We thank Professor M. Azuma in Tokyo Institute of Technology for useful discussion. We also thank T. Furuta, M. Kubota, H. Yabuta, and T. Watanabe for useful discussion and computational support. The present work was partly supported by the Elements Science and Technology Project from the Ministry of Education, Culture, Sports, Science and Technology.

Author details

Kaoru Miura[1] and Hiroshi Funakubo[2]

1 Canon Inc., Tokyo, Japan

2 Tokyo Institute of Technology, Yokohama, Japan

References

[1] Cohen R E. Origin of ferroelectricity in perovskite oxides. Nature 1992; 451 (6382) : 136-8.

[2] Wu Z, Cohen R E. Pressure-induced anomalous phase transitions and colossal enhancement of piezoelectricity in $PbTiO_3$. Phys. Rev. Lett. 2005; 95 (3): 037601, and related references therein.

[3] Deguez O, Rabe K M, Vanderbilt D. First-principles study of epitaxial strain in perovskites. Phys. Rev. B 2005; 72 (14): 144101, and related references therein.

[4] Qi T, Grinberg I, Rappe A M. First-principles investigation of the high tetragonal ferroelectric material $BiZn_{1/2}Ti_{1/2}O_3$. Phys. Rev. B 2010; 79 (13): 134113, and related references therein.

[5] Wang H, Huang H, Lu W, Chan H L W, Wang B, Woo C H. Theoretical prediction on the structural, electronic, and polarization properties of tetragonal Bi_2ZnTiO_6. J. Appl. Phys. 2009; 105 (5): 053713.

[6] Dai J Q, Fang Z. Structural, electronic, and polarization properties of Bi_2ZnTiO_6 supercell from first-principles. J. Appl. Phys. 2012; 111 (11): 114101.

[7] Khenata R, Sahnoun M, Baltache H, Rerat M, Rashek A H, Illes N, Bouhafs B. First-principle calculations of structural, electronic and optical properties of $BaTiO_3$ and $BaZrO_3$ under hydrostatic pressure. Solid State Commun. 2005; 136 (2): 120-125.

[8] Oguchi T, Ishii F, Uratani Y. New method for calculating physical properties from first principles-piezoelectric and multiferroics. Butsuri 2009; 64 (4): 270-6 [in Japanese].

[9] Uratani Y, Shishidou T, Oguchi T. First-principles calculations of colossal piezoelectric response in thin film $PbTiO_3$. Jpn. Soc. Appl. Phys. 2008: conference proceedings, 27-30 March 2008, Funabashi, Japan [in Japanese].

[10] Miura K, Kubota M, Azuma M, and Funakubo H. Electronic and structural properties of $BiZn_{0.5}Ti_{0.5}O_3$. Jpn. J. Appl. Phys. 2009; 48 (9): 09KF05.

[11] Miura K, Furuta T, Funakubo H. Electronic and structural properties of $BaTiO_3$: A proposal about the role of Ti 3s and 3p states for ferroelectricity. Solid State Commun. 2010; 150 (3-4): 205-8.

[12] Furuta T, Miura K. First-principles study of ferroelectric and piezoelectric properties of tetragonal $SrTiO_3$ and $BaTiO_3$ with in-plane compressive structures. Solid State Commun. 2010; 150 (47-48): 2350-3.

[13] Miura K, Azuma M, Funakubo H. Electronic and structural properties of ABO_3: Role of the B-O Coulomb repulsions for ferroelectricity. Materials 2011; 4 (1): 260-73. http://www.mdpi.com/1996-1944/4/1/260 (accessed 20 July 2012).

[14] Miura K. First-principles study of ABO_3: Role of the B-O Coulomb repulsions for ferroelectricity and piezoelectricity. Lallart M. (ed.) Ferroelectrics - Characterization and Modeling. Rijeka: InTech; 2012. p395-410. Available from http://www.intechopen.com/books/ferroelectrics-characterization-and-modeling/first-principles-study-of-abo3-role-of-the-b-o-coulomb-repulsions-for-ferroelectricity-and-piezoelec (accessed 20 July 2012).

[15] Miura K, Funakubo H. First-principles analysis of tetragonal ABO_3: Role of the B-O Coulomb repulsions for ferroelectricity and piezoelectricity. Proceedings of 15th US-Japan Seminar on Dielectric and Piezoelectric Ceramics, 6-9 November 2011, Kagoshima, Japan.

[16] Miura K, Ifuku T, Kubota M, Hayashi J. Piezoelectric Material. Japan Patent 2011-001257 [in Japanese].

[17] Miura K, Kubota M, Hayashi J, Watanabe T. Piezoelectric Material. submitted to Japan Patent [in Japanese].

[18] Miura K, Furuta T. First-principles study of structural trend of $BiMO_3$ and $BaMO_3$: Relationship between tetragonal and rhombohedral structure and the tolerance factors. Jpn. J. Appl. Phys. 2010; 49 (3): 031501 (2010).

[19] Miura K, Kubota M, Azuma M, and Funakubo H. Electronic, structural, and piezoelectric properties of $BiFe_{1-x}Co_xO_3$. Jpn. J. Appl. Phys. 2010; 49 (9): 09ME07.

[20] Miura K, Tanaka M. Electronic structures of $PbTiO_3$: I. Covalent interaction between Ti and O ions. Jpn. J. Appl. Phys. 1998; 37 (12A): 6451-9.

[21] LDA_TM_psp1_data: ABINIT. http://www.abinit.org/downloads/psp-links/lda_tm_psp1_data/ (accessed 20 July 2012).

[22] Kuroiwa Y, Aoyagi S, Sawada A, Harada J, Nishibori E, Tanaka M, and Sakata M. Evidence for Pb-O covalency in tetragonal $PbTiO_3$. Phys. Rev. Lett. 2001; 87 (21): 217601.

[23] Shannon R D. Revised effective ionic radii and systematic studies of interatomic distances in halides and chalcogenides. Acta Crystallogr. Sect. A. 1976; 32 (5): 751-67.

[24] Suchomel M R, Fogg A M, Allix M, Niu H, Claridge J B, Rosseinsky M J. Bi_2ZnTiO_6: A lead-free closed-shell polar perovskite with a calculated ionic polarization of 150 $\mu C\ cm^{-2}$. Chem. Mater. 2006; 18 (21), 4987-9.

[25] Khalyavin D D, Salak A N, Vyshatko N P, Lopes A B, Olekhnovich N M, Pushkarev A V, Maroz I I, Radyush T V. Crystal structure of metastable perovskite $Bi(Mg_{1/2}Ti_{1/2})O_3$: Bi-based structural analogue of antiferroelectric $PbZrO_3$. Chem. Mater. 2006; 18 (21), 5104-10.

[26] Randall C A, Eitel R, Jones B, Shrout T R, Woodward D I, Reaney I M, Investigation of a high T_C piezoelectric system: $(1-x)Bi(Mg_{1/2}Ti_{1/2})O_3-(x)PbTiO_3$. J. Appl. Phys. 2004; 95 (7): 3633-9.

[27] Rossetti Jr G A, Khachaturyan A G. Concepts of morphotropism in ferroelectric solid solutions. Proceedings of 13th US-Japan Seminar on Dielectric and Piezoelectric Ceramics, 4-7 November 2007, Awaji, Japan, and related references therein.

[28] Damjanovic D. A morphotropic phase boundary system based on polarization rotation and polarization extension. Appl. Phys. Lett. 2010; 97 (6): 062906, and related references therein.

[29] Gonze X, Beuken J-M, Caracas R, Detraux F, Fuchs M, Rignanese G-M, Sindic L, Verstraete M, Zerah G, Jollet F, Torrent M, Roy A, Mikami M, Ghosez P, Raty J-Y, Allan D C. First-principles computation of material properties: the ABINIT software project. Comput. Mater. Sci. 2002; 25 (3), 478-92.

[30] Hohenberg P, Kohn W. Inhomogeneous electron gas. Phys. Rev. 1964; 136 (3B): B864-71.

[31] Opium - pseudopotential generation project. http://opium.sourceforge.net/index.html (accessed 20 July 2012).

[32] Ramer N J, Rappe A M. Application of a new virtual crystal approach for the study of disordered perovskites. J. Phys. Chem. Solids. 2000; 61 (2): 315-20.

[33] Resta R. Macroscopic polarization in crystalline dielectrics: the geometric phase approach. Rev. Mod. Phys. 1994; 66 (3): 899-915.

[34] Kim Y-I, Lee E. Constant-wavelength neutron diffraction study of cubic perovskites $BaTaO_2N$ and $BaNbO_2N$. J. Ceram. Soc. Jpn. 2011; 119 (5): 371-4.

Chapter 2

Advances in Thermodynamics of Ferroelectric Phase Transitions

Shu-Tao Ai

Additional information is available at the end of the chapter

http://dx.doi.org/10.5772/52089

1. Introduction

The thermodynamics of ferroelectric phase transitions is an important constituent part of the phenomenological theories of them, as well as the interface dynamics of them. In particular, if we confine the thermodynamics to the equilibrium range, we can say that the Landau-Devonshire theory is a milestone in the process of the development of ferroelectric phase transition theories. This can be found in many classical books such as [1,2]. Many studies centering on it, especially the size-effects and surface-effects of ferroelectric phase transitions, have been carried out. For the reason of simplicity, we just cite a few [3,4]. But we think these are a kind of technical but not fundamental progress.

Why we think so is based on that the Landau-Devonshire theory is confined to the equilibrium range in essene so it can't deal with the outstanding irreversible phenomonon of first-order ferroelectric phase transitions strictly, which is the „thermal hysteresis". The Landau-Devonshire theory attributes the phenomenon to a series of metastable states existing around the Curie temperature T_C. In principle, the metastable states are not the equilibrium ones and can not be processed by using the equilibrium thermodynamics. Therefore, we believe the Landau-Devonshire theory is problematic though it is successful in mathematics. The real processes of phase transition were distorted. In this contribution, the Landau-Devonshire theory will be reviewed critically, then the latest phenomenological theory of ferroelectric phase transitions will be established on the basis of non-equilibrium or irreversible thermodynamics.

This contribution are organized as the follows. In Section 2, we will show the unpleasant consequence caused by the metastable states hypothesis, and the evidence for the non-existence of metastable states, i.e. the logical conflict. Then in Section 3 and 4, we will give the non-equilibrium (or irreversible) thermodynamic description of ferroelectric phase transitions, which eliminates the unpleasant consequence caused by the metastable states hypothesis. In Section 5, we will give the non-equilibrium thermodynamic explanation of the irreversibility of ferroelectric phase transitions, i.e. the thermal hysteresis and the domain occurrences in ferroelectrics. At last, in Section 6 we will make some concluding remarks and look forward to some possible developments.

2. Limitations of Landau-Devonshire theory and demonstration of new approach

The most outstanding merit of Landau-Devonshire theory is that the Curie temperature and the spontaneous polarization at Curie temperature can be determined simply. However, in the Landau-Devonshire theory, the path of a first-order ferroelectric phase transition is believed to consist of a series of metastable states existing around the Curie temperature. This is too difficult to believe because of the difficulties encounted (just see the follows) T_C.

2.1. Unpleasant consequence caused by metastable states hypothesis

Basing on the Landau-Devonshire theory, we make the following inference. Because of the thermal hysteresis, a first-order ferroelectric phase transition must occur at another temperature, which is different from the Curie temperature [5]. The state corresponding to the mentioned temperature (i.e. actural phase transition temperature) is a metastable one. Since the unified temperature and spontanous polarization can be said about the metastable state, we neglect the heterogeneity of system actually. In other words, every part of the system, i.e. either the surface or the inner part, is of equal value physically. When the phase transition occurs at the certain temperature, every part of the system absorbs or releases the latent heat simultaneously by a kind of action at a distance. (The concept arose in the electromagnetism first. Here it maybe a kind of heat transfer.) Otherwise, the heat transfer in system, with a finite rate, must destroy the homogeneity of system and lead to a non-equilibrium thermodynamic approach. The unpleasant consequence, i.e. the action at a distance should be eliminated and the lifeforce should be bestowed on the non-equilibrium thermodynamic approach.

In fact, a first-order phase transition process is always accompanied with the fundamental characteristics, called the co-existence of phases and the moving interface (i.e. phase boundary). The fact reveals that the phase transition at various sites can not occur at the same time. Yet, the phase transition is induced by the external actions (i.e. absorption or release of latent heat). It conflicts sharply with the action at a distance.

2.2. Evidence for non-existence of metastable states: logical conflict

In the Landau-Devonshire theory, if we neglect the influence of stress, the elastic Gibbs energy G_1 can be expressed with a binary function of variables, namely the temperature T and the electric displacement D (As G_1 is independent of the orientation of D, here we are interested in the magnitude of D only)

$$G_1 = g_1(T,D) \tag{1}$$

The long-standing, close correlation between analytical dynamics and thermodynamics implies that Equation (1) can be taken as a scleronomic constraint equation

$$f_1(G_1,T,D) = G_1 - g_1(T,D) = 0 \tag{2}$$

where G_1, T, D are the generalized displacements. The possible displacements $\mathrm{d}G_1$, $\mathrm{d}T$ and $\mathrm{d}D$ satisfy the following equation

$$\mathrm{d}G_1 - \frac{\partial g_1}{\partial T}\mathrm{d}T - \frac{\partial g_1}{\partial D}\mathrm{d}D = 0 \tag{3}$$

In the Landau-Devonshire theory, the scleronomic constraint equation, i.e. Equation (1) is expressed in the form of the power series of D (For simplicity, only the powers whose orders are not more than six are considered)

$$G_1 = G_{10} + \frac{1}{2}\alpha D^2 + \frac{1}{4}\beta D^4 + \frac{1}{6}\gamma D^6 \tag{4}$$

where α, β, γ are the functions of T, and G_{10} is the elastic Gibbs energy of paraelectric phase. The relation between G_1 and D at various temperature, which belongs to first-order phase transition ferroelectrics is represented graphically in Figure 1. The electric displacements which correspond to the bilateral minima of G_1 are identified as $\pm D^*$, and the electric displacement which corresponds to the middle minimum of G_1 equals zero. The possible electric displacements should be the above ones which correspond to the minima of G_1.

Equivalently, imposed on the generalized displacements G_1, T, D is a constraint, which is

$$\frac{\partial G_1}{\partial D} = 0 \tag{5}$$

So, the possible displacement $\mathrm{d}D$ should be the follows $\mathrm{d}D^*$, $-\mathrm{d}D^*$, $0-(\pm D^*)= \mp D^*$, $\pm D^*-0= \pm D^*$, 0. After all, if our discussion are limited in the equilibrium thermodynamics strictly, there must be the third constraint, i.e. the equilibrium D and T should satisfy

$$h(D,T) = 0 \tag{6}$$

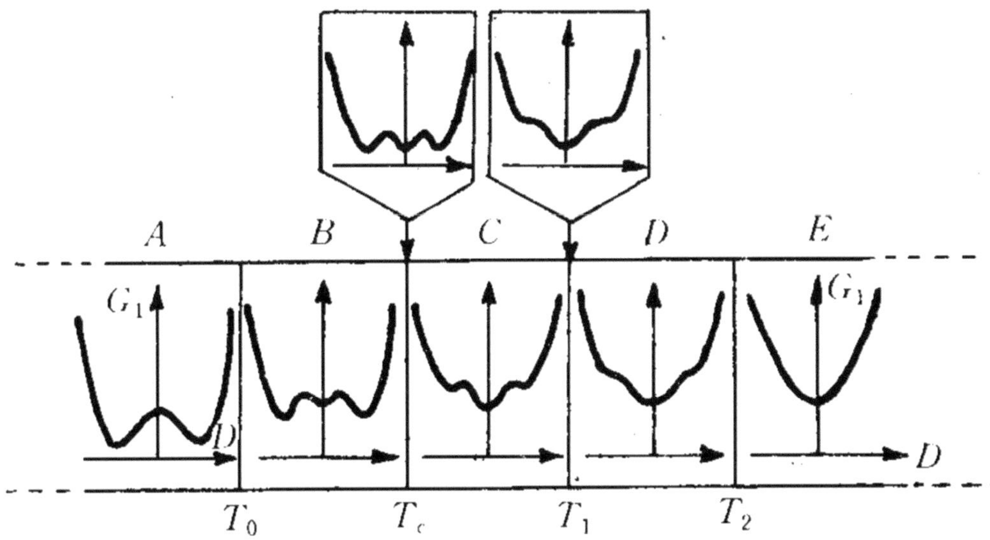

Figure 1. The relation between elastic Gibbs energy G_1 and zero field electric displacement D belongs to the ferroelectrics at various temperature, which undergoes a first-order phase transition [5]. T_0 :the lowest temperature at which the paraelectric phase may exist, i.e. the Curie-Weiss temperature; T_C :the phase transition temperature, i.e. the Curie temperature; T_1 : the highest temperature at which the ferroelectric phase may exist; T_2 : the highest temperature at which the ferroelectric phase may be induced by the external electric field.

where h is a binary function of the variables D and T. It can be determined by the principle of minimum energy

$$G_1 = \min \tag{7}$$

for certain T. Then, the metastable states are excluded. Thus, the thermal hysteresis does not come into being. The corollary conflicts with the fact sharply. This reveals that the first-order ferroelectric phase transition processes must not be reversible at all so as not to be dealt with by using the equilibrium thermodynamics.

How can this difficulty be overcome? An expedient measure adopted by Devonshire is that the metastable states are considered. However, do they really exist?

Because the metastable states are not the equilibrium ones, the relevant thermodynamic variables or functions should be dependent on the time t. In addition, the metastable states are close to equilibrium, so the heterogeneity of system can be neglected. Here, the elastic Gibbs energy G'_1 should be

$$G'_1 = g_2(T, D, t) \tag{8}$$

For the same reason as was mentioned above, Equation (8) can be regarded as a rheonomic constraint on the generalized displacements G'_1, T, D

$$f_2\left(G_1',T,D,t\right)=G_1'-g_2\left(T,D,t\right)=0 \tag{9}$$

In this case, the possible displacements $\mathrm{d}G'_1$, $\mathrm{d}T$ and $\mathrm{d}D$ satisfy the following equation

$$\mathrm{d}G_1'-\frac{\partial g_2}{\partial T}\mathrm{d}T-\frac{\partial g_2}{\partial D}\mathrm{d}D-\frac{\partial g_2}{\partial t}\mathrm{d}t=0 \tag{10}$$

Comparing Equation (3) with Equation (10), we may find that the possible displacements here are not the same as those in the former case which characterize the metastable states for they satisfy the different constraint equations, respectively. (In the latter case, the possible displacements are time-dependent, whereas in the former case they are not.) Yet, the integral of possible displacement $\mathrm{d}D$ is the possible electric displacement in every case. The possible electric displacements which characterize one certain metastable state vary with the cases. A self-contradiction arises. So the metastable states can not come into being.

What are the real states among a phase transition process? In fact, both the evolution with time and the spatial heterogeneity need to be considered when the system is out of equilibrium [6-9]. Just as what will be shown in Section 2.3, the real states should be the stationary ones, which do not vary with the time but may be not metastable.

2.3. Real path: Existence of stationary states

The real path of a first-order ferroelectric phase transition is believed by us to consist of a series of stationary states. At first, this was conjectured according to the experimental results, then was demonstrated reliable with the aid of non-equilibrium variational principles.

Because in the experiments the ferroelectric phase transitions are often achieved by the quasi-static heating or cooling, we conjetured that they are stationary states processes [8]. The results on the motion of interface in ferroelectrics and antiferroelectrics support our opinion [10-12]. From Figure 2, we may find that the motion of interface is jerky especially when the average velocity v_a is small. A sequence of segments of time corresponding to the states of rest may be found. The experimental results about other materials such as $PbTiO_3$ are alike [11]. This reveals that in these segments of time (i.e. characteristic time of phase transition) the stationary distributions of temperature, heat flux, stress, etc. may be established. Otherwise, if the motion of interface is continuous and smooth, with the unceasing moving of interface (where the temperature is T_C) to the inner part, the local temperature of outer part must change to keep the temperature gradient ∇T of this region unchanged for it is determined by $\pm l\rho v'=J_q^{diff}=-\kappa\nabla T$, where l is the latent heat (per unit mass), ρ is the mass density, v' is the velocity of interface (where the phase transition is occurring), J_q^{diff} is the diffusion of heat, i.e. heat conduction, κ is the thermal conductivity (and maybe a tensor.) Then, the states are not stationary.

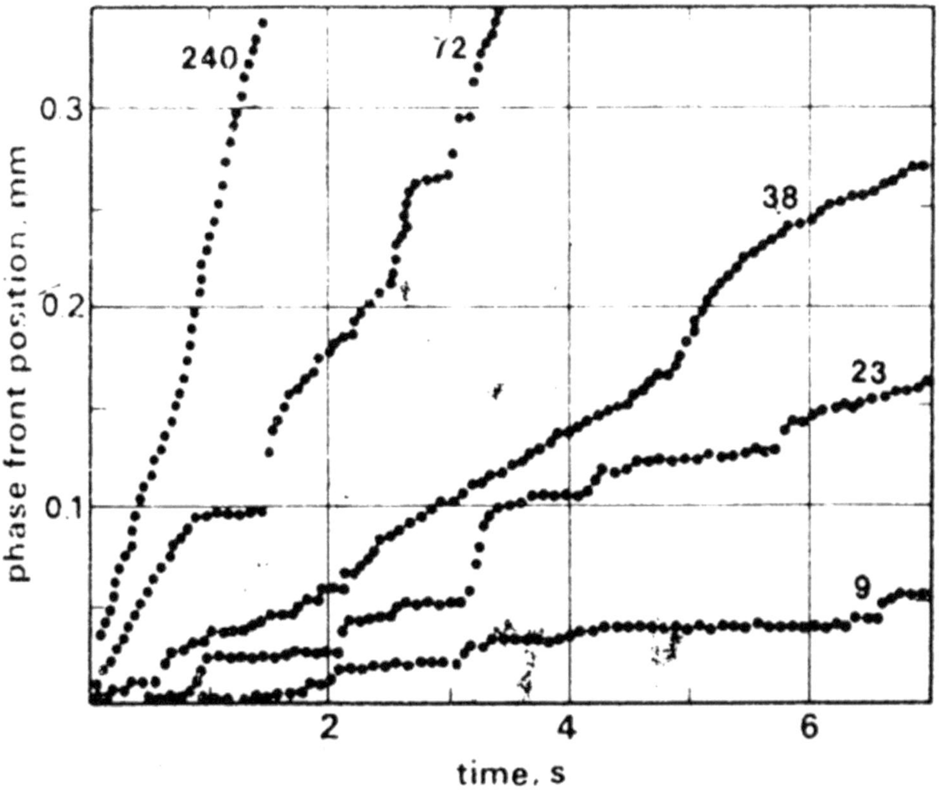

Figure 2. The position of phase boundary as a function of time for $NaNbO_3$ single crystal for the various values of average velocity v_a (the values in μm/s given against the curves) [12].

The non-equilibrium variational principles are just the analogue and generalization of the variational principles in analytical dynamics. The principle of least dissipation of energy, the Gauss's principle of least constraint and the Hamiltonian principle etc., in non-equilibrium thermodynamics play the fundamental roles as those in analytical dynamics. They describe the characteristics of stationary states or determine the real path of a non-equilibrium process.

For the basic characteristics of non-equilibrium processes is the dissipation of energy, the dissipation function φ is defined as

$$\varphi = \sigma_s - \pi \tag{11}$$

where σ_s is the rate of local entropy production and π is the external power supply (per unit volume and temperature). After the rather lengthy deducing and utilizing the thermodynamic Gauss's principle of least constraint which makes the system choose a real path [13], the evolution with time t of the deviation from a given non-equilibrium stationary state, $\xi(t)=\{\xi_i(t)\}$, was obtained in two cases. If no external power supply,

$$\xi_i(t) = \xi_i(0) e^{\frac{2S_i}{R_i}t} \tag{12}$$

where $\xi_i(0)$ is the initial value of $\xi_i(t)$, S_i is the coefficient between the linear variation in the thermodynamic force $\chi_{Ti}^{(1)}(\xi)$ and the deviation of the extensive pseudo-thermodynamic variable from a given non-equilibrium stationary state ξ_i, R_i is the coefficient between the linear variation in the dissipative force $\chi_{Di}^{(1)}(\xi)$ and the time-derivative of the deviation of the extensive pseudo-thermodynamic variable from a given non-equilibrium stationary state ξ_i. Equation (12) define the real path with the addition that R_i should be a suitable value R_i^*. If the external power supply exists, similarly the evolution of deviation ξ_i can be obtained

$$\xi_i(t) = \xi_i(0) e^{\frac{2(S_i - V_i)}{R_i}t} \tag{13}$$

where V_i is the coefficient between the linear variation in the force related to the external power $\chi_{Ei}^{(1)}$ and the deviation of the extensive pseudo-thermodynamic variable from a given non-equilibrium stationary state ξ_i. If the coefficients V_i, R_i assume the suitable values V_i^{**}, R_i^{**}, the system choose a real path.

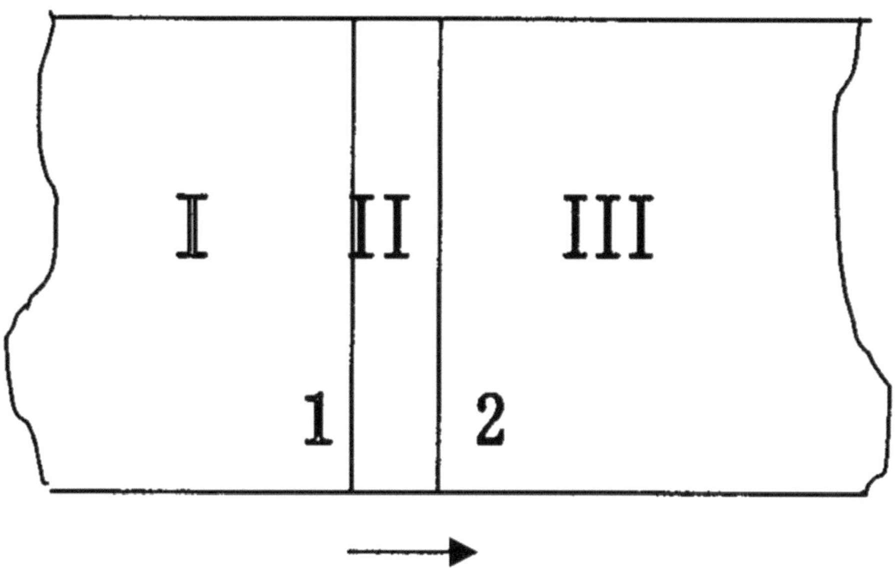

Figure 3. Three types of regions and their interfaces in the ferroelectric-paraelectric system in which a first-order phase transition is occurring [13].

Both the real paths in the two cases reveal that the deviations decrease exponentially when the system regresses to the stationary states. Stationary states are a king of attractors to non-equilibrium states. The decreases are steep. So the regressions are quick. It should be noted that we are interested in calculating the change in the generalized displacements during a macroscopically small time interval. In other words, we are concerned with the determination of the path of an irreversible process which is described in terms of a finite difference equation. In the limit as the time interval is allowed to approach zero, we obtain the variational equation of thermodynamic path.

So, if the irreversible process is not quick enough, it can be regarded as the one that consists of a series of stationary states. The ferroelectric phase transitions are usually achieved by the quasi-static heating or cooling in the experiments. So, the processes are not quick enough to make the states deviate from the corresponding stationary states in all the time. In Figure 3, three types of regions and their interfaces are marked with I, II, III, 1, 2 respectively. The region III where the phase transition will occur is in equilibrium and has no dissipation. In the region I where the phase transition has occurred, there is no external power supply, and in the region II (i.e. the paraelectric-ferroelectric interface as a region with finite thickness instead of a geometrical plane) where the phase transition is occurring, there exists the external power supply, i.e. the latent heat (per unit volume and temperature). According to the former analysis in two cases, we may conclude that they are in stationary states except for the very narrow intervals of time after the sudden lose of phase stability.

3. Thermo-electric coupling

In the paraelectric-ferroelectric interface dynamics induced by the latent heat transfer [6,7], the normal velocity of interface v_n was obtained

$$v_n = \frac{1}{l\rho}\left[k_{fer}\left(\nabla T\right)_{fer} - k_{par}\left(\nabla T\right)_{par}\right]\cdot \boldsymbol{n} \tag{14}$$

where l is the latent heat (per unit mass), ρ is the density of metastable phase (paraelectric phase), k_{fer} is the thermal conductivity coefficient of ferroelectric phase, k_{par} is the thermal conductivity coefficient of paraelectric phase, $(\nabla T)_{fer}$ is the temperature gradient in ferro-electric phase part, $(\nabla T)_{par}$ is the temperature gradient in paraelectric phase part, n is the unit vector in normal direction and directs from the ferroelectric phase part to the paraelectric phase part. The temperature gradients can be studied from the point of view that a ferroelectric phase transition is a stationary, thermo-electric coupled transport process [8].

3.1. Local entropy production

In the thermo-electric coupling case, the Gibbs equation was given as the following [8]

$$Tds = du - \boldsymbol{E} \cdot d\boldsymbol{D} - \sum_i \mu_i dn_i \tag{15}$$

where T, E, D is the temperature, the electric field intensity and the electric displacement within a random small volume, respectively; s, u, μ_i, n_i is the entropy density, the internal energy density, the chemical potential and the molar quantity density in the small volume, respectively. And there, it was assumed that the crystal system is mechanically-free (i.e. no force is exerted on it). Differentiating Equation (15) and using the following relations

$$\frac{\partial u}{\partial t} + \nabla \cdot \boldsymbol{J}_u = 0 \tag{16}$$

$$\frac{\partial \boldsymbol{D}}{\partial t} = \frac{\partial \left(\varepsilon_0 \boldsymbol{E} + \boldsymbol{P} \right)}{\partial t} = \boldsymbol{J}_P \tag{17}$$

$$\frac{\partial n_i}{\partial t} + \nabla \cdot \boldsymbol{J}_{ni} = 0 \tag{18}$$

we have

$$\frac{\partial s}{\partial t} = -\frac{1}{T} \nabla \cdot \boldsymbol{J}_u - \frac{\boldsymbol{E}}{T} \cdot \boldsymbol{J}_P + \sum_i \frac{\mu_i}{T} \nabla \cdot \boldsymbol{J}_{ni} \tag{19}$$

where J_u, J_P, J_{ni} is the energy flux, the polarization current and the matter flux; P is the polarization. J_u should consist of three parts: the energy flux caused by the heat conduction, the energy flux caused by the charge transport and the energy flux caused by the matter transport

$$\boldsymbol{J}_u = \boldsymbol{J}_q + \varphi \boldsymbol{J}_P + \sum_i \mu_i \boldsymbol{J}_{ni} \tag{20}$$

where J_q is the heat flux, φ is the electrical potential and satisfies

$$\boldsymbol{E} = -\nabla \varphi \tag{21}$$

Then we deduce the following

$$\frac{\partial s}{\partial t} = -\nabla \cdot \left(\frac{\boldsymbol{J}_q + \varphi \boldsymbol{J}_P}{T} \right) + \boldsymbol{J}_q \cdot \nabla \left(\frac{1}{T} \right) + \boldsymbol{J}_P \cdot \nabla \left(\frac{\varphi}{T} \right) - \sum_i \frac{\boldsymbol{J}_{ni}}{T} \cdot \nabla \mu_i \tag{22}$$

If we define a entropy flux J_s and a rate of local entropy production σ_s as

$$\boldsymbol{J}_s = \frac{\boldsymbol{J}_q + \varphi \boldsymbol{J}_P}{T} \tag{23}$$

$$\sigma_s = \boldsymbol{J}_q \cdot \nabla \left(\frac{1}{T} \right) + \boldsymbol{J}_P \cdot \nabla \left(\frac{\varphi}{T} \right) - \sum_i \frac{\boldsymbol{J}_{ni}}{T} \cdot \nabla \mu_i \tag{24}$$

Equation (22) can be written as

$$\frac{\partial s}{\partial t} + \nabla \cdot \boldsymbol{J}_s = \sigma_s \tag{25}$$

This is the local entropy balance equation. We know, the system is in the crystalline states before and after a phase transition so that there is no diffusion of any kind of particles in the system. So, J_{ni}=0. The local entropy production can be reduced as

$$\sigma_s = \boldsymbol{J}_q \cdot \nabla \left(\frac{1}{T} \right) + \boldsymbol{J}_P \cdot \nabla \left(\frac{\varphi}{T} \right) \tag{26}$$

We know the existence of ferroics is due to the molecular field. It is an internal field. So we must take it into account. Here, the electric field should be the sum of the external electric field E_e and the internal electric field E_i

$$\boldsymbol{E} = \boldsymbol{E}_e + \boldsymbol{E}_i \tag{27}$$

Correspondingly, there are the external electrical potential φ_e and the internal electrical potential φ_i and they satisfy

$$\boldsymbol{E}_e = -\nabla \varphi_e \tag{28}$$

$$\boldsymbol{E}_i = -\nabla \varphi_i \tag{29}$$

If the external electric field is not applied, φ_e can be a random constant. There is no harm in letting the constant equal zero. Then the entropy production equals

$$\sigma_s = \boldsymbol{J}_q \cdot \nabla\left(\frac{1}{T}\right) + \boldsymbol{J}_P \cdot \nabla\left(\frac{\varphi_i}{T}\right) \tag{30}$$

According to the crystal structures of ferroelectrics [2], we know the polarization current J_P originates from the displacement or ordering of ions in ferroelectrics. We may consider it as the transport of charges influenced by the internal electric field.

3.2. Description of phase transitions and verification of interface dynamics

Assume the external electric field is not applied. Here are the thermodynamic fluxex J_q, J_P and the corresponding thermodynamic forces X_q, X_P

$$\boldsymbol{X}_q = \nabla\left(\frac{1}{T}\right) \tag{31}$$

$$\boldsymbol{X}_P = \nabla\left(\frac{\varphi_i}{T}\right) \tag{32}$$

J_i can be expanded linearly with $X_j (i,\ j = q,\ P)$

$$\boldsymbol{J}_q = \boldsymbol{L}_{qq} \cdot \boldsymbol{X}_q + \boldsymbol{L}_{qP} \cdot \boldsymbol{X}_P \tag{33}$$

$$\boldsymbol{J}_P = \boldsymbol{L}_{Pq} \cdot \boldsymbol{X}_q + \boldsymbol{L}_{PP} \cdot \boldsymbol{X}_P \tag{34}$$

where L_{qq}, L_{qP}, L_{Pq} and L_{PP} are the transport coefficients, which are four second-order tensors. They should satisfy the generalized Onsager relations [14]
$L_{qq} = L_{qq}^T$, $L_{qP} = L_{Pq}^T$,

$$\boldsymbol{L}_{PP} = \boldsymbol{L}_{PP}^T \tag{35}$$

Because to a first-order ferroelectric phase transition the electric displacement changes suddenly and so does the internal electrical potential, the force X_P of the region where the phase transition is occurring can be regarded as a large constant roughly in the characteristic times of phase transition (i.e. the times in which the interface keeps rest). For the ferroelectric phase transition may be regarded as a stationary states process, the principle of minimum entropy production must be satisfied [15].

According to Equations (30)-(35), we have

$$\sigma_s = \boldsymbol{L}_{qq} : \boldsymbol{X}_q \boldsymbol{X}_q + 2\boldsymbol{L}_{qP} : \boldsymbol{X}_q \boldsymbol{X}_P + \boldsymbol{L}_{PP} : \boldsymbol{X}_P \boldsymbol{X}_P \tag{36}$$

If there is no any restriction on X_q and X_P, according to the conditions on which the entropy production is a minimum

$$\left(\frac{\partial \sigma_s}{\partial \boldsymbol{X}_q} \right)_{\boldsymbol{X}_P} = 2\boldsymbol{L}_{qq} \cdot \boldsymbol{X}_q + 2\boldsymbol{L}_{qP} \cdot \boldsymbol{X}_P = 2\boldsymbol{J}_q = 0 \tag{37}$$

$$\left(\frac{\partial \sigma_s}{\partial \boldsymbol{X}_P} \right)_{\boldsymbol{X}_q} = 2\boldsymbol{L}_{Pq} \cdot \boldsymbol{X}_q + 2\boldsymbol{L}_{PP} \cdot \boldsymbol{X}_P = 2\boldsymbol{J}_P = 0 \tag{38}$$

We know the stationary states are equilibrium ones actually. If we let X_q (or X_P) be a constant, according to Equation (38) (or (37)) we know J_P (or J_q) which is corresponded to another force X_P (or X_q) should be zero.

Then, a first-order ferroelectric phase transition can be described by the second paradigm. Since the force X_P of the region where the phase transition is occurring is a large constant, the flux J_q of this region should be zero (but $J_P \neq 0$). This states clearly that the pure heat conduction and the heat conduction induced by the thermo-electric coupling cancel out each other so as to release or absorb the latent heat. It is certain that the latent heat passes through the region where the phase transition has occurred (at the outside of the region where the phase transition is occurring) and exchanges itself with the thermal bath. Accompanied with the change of the surface's temperature and the unceasing jerky moving of the region where the phase transition is occurring, a constant temperature gradient is kept in the region where the phase transition has occurred, i.e. the force X_q is a constant. So, the flux $J_P=0$ (but $J_q \neq 0$). This states clearly that the electric displacement of the region where the phase transition has occurred will not change but keep the value at Curie temperature or zero until the phase transition finishes. Differently, the region where the phase transition will occur should be described by the first paradigm for there is no restriction on the two forces X_P, X_q. The states of this region are equilibrium ones. So the temperature gradient ∇T should be zero.

Considering that $(\nabla T)_{par} \approx 0$ for the region where the phase transition will occur (i.e. the paraelectric phase part) can be regarded as an equilibrium system, we modify Equation (14) as

$$v_n = \frac{k_{fer} \left(\nabla T \right)_{fer} \cdot \boldsymbol{n}}{l\rho} \tag{39}$$

In order to compare it with the experiments, we make use of the following values which are about $PbTiO_3$ crystal: ρ =7.1g/cm^3 [16], l =900cal/mol [17], k_{fer} =8.8 × 10^{5}erg/cmsK [18]. The value of the velocity of the interface's fast motion, which has been measured in the experiments, is 0.5mm/s [11]. According to Equation (39), we calculate the corresponding temperature gradient to be 57.35K/cm. However, in [19] it is reported that the experimental temperature gradient varies from 1.5 to 3.5K/mm while the experimental velocity of the interface's motion varies from 732 to 843μm/s. Considering the model is rather rough, we may conclude that the theory coincides with the experiments.

3.3. Relation between latent heat and spontaneous polarization

In the experiments, the latent heat and the spotaneous polarization are measured often for first-order ferroelectric phase transitions. So in the follows, we will establish the relation between latent heat and spontaneous polarization in the realm of non-equilibrium thermodynamics.

All the quantities of the region where the phase transition has occurred are marked with the superscript "I"; all the quantities of the region where the phase transition is occurring are marked with the superscript "II"; and all the quantities of the region where the phase transition will occur are marked with the superscript "III". Let's consider the heating processes of phase transition firstly. In the region where the phase transition has occurred,

$$\boldsymbol{J}_q^{I-para} = \boldsymbol{L}_{qq}^{I-para} \cdot \boldsymbol{X}_q^{I-para} = l\rho \boldsymbol{v}_a \tag{40}$$

where we have ignored the difference between the mass density of ferroelectric phase and that of paraelectric phase (almost the same) and denote them as ρ, v_a is the average velocity of interface. In the region where the phase transition is occurring,

$$\boldsymbol{J}_q^{II} = \boldsymbol{L}_{qq}^{II} \cdot \boldsymbol{X}_q^{II} + \boldsymbol{L}_{qP}^{II} \cdot \boldsymbol{X}_P^{II} = 0 \tag{41}$$

$$\boldsymbol{J}_P^{II} = \boldsymbol{L}_{Pq}^{II} \cdot \boldsymbol{X}_q^{II} + \boldsymbol{L}_{PP}^{II} \cdot \boldsymbol{X}_P^{II} \tag{42}$$

The heat which is transferred to the region where the phase transition is occurring is absorbed as the latent heat because the pure heat conduction and the heat conduction induced by the thermo-electric coupling cancel out each other. So,

$$l\rho \boldsymbol{v}_a = \boldsymbol{L}_{qq}^{II} \cdot \boldsymbol{X}_q^{II} \tag{43}$$

According to Eqations (41)-(43), we work out

$$\boldsymbol{J}_P^{II} = l\rho \left[\boldsymbol{L}_{Pq}^{II} \cdot \left(\boldsymbol{L}_{qq}^{II} \right)^{-1} - \boldsymbol{L}_{PP}^{II} \cdot \left(\boldsymbol{L}_{qP}^{II} \right)^{-1} \right] \cdot \boldsymbol{v}_a \tag{44}$$

where the superscript "-1" means reverse. While

$$\int \boldsymbol{J}_P^{II} \mathrm{d}t = -\boldsymbol{D}^{spon} = \varepsilon_0 \nabla \varphi_i^{III} - \boldsymbol{P}^{spon-III} = -\boldsymbol{P}^{spon-III} \tag{45}$$

where we utilized the boundary condition of D and considered the region where the phase transition will occur is in equilibrium, the superscript "spon" means spontaneous. So,

$$\boldsymbol{P}^{spon} = \boldsymbol{P}^{spon-III} = -\int \boldsymbol{J}_P^{II} \mathrm{d}t \tag{46}$$

The relation between latent heat and spontaneous polarization are obtained. In the cooling processes of phase transition,

$$-l\rho \boldsymbol{v}_a = \boldsymbol{L}_{qq}^{I-ferr} \cdot \boldsymbol{X}_q^{I-ferr} + \boldsymbol{L}_{qP}^{I-ferr} \cdot \boldsymbol{X}_P^{I-ferr} = \boldsymbol{L}_{qq}^{II} \cdot \boldsymbol{X}_q^{II} \tag{47}$$

$$\int \boldsymbol{J}_P^{II'} \mathrm{d}t = \boldsymbol{D}^{spon-I} = -\varepsilon_0 \nabla \varphi_i^{I} + \boldsymbol{P}^{spon-I} \tag{48}$$

Repeating the above steps, we obtain

$$\boldsymbol{P}^{spon-I} = \int \boldsymbol{J}_P^{II'} \mathrm{d}t + \varepsilon_0 \nabla \varphi_i^{I} \tag{49}$$

With

$$\boldsymbol{J}_P^{II'} = -l\rho \left[\boldsymbol{L}_{Pq}^{II} \cdot \left(\boldsymbol{L}_{qq}^{II} \right)^{-1} - \boldsymbol{L}_{PP}^{II} \cdot \left(\boldsymbol{L}_{qP}^{II} \right)^{-1} \right] \cdot \boldsymbol{v}_a = -\boldsymbol{J}_P^{II} \tag{50}$$

Then we find that $P^{spon-III}$ (serves as the equilibrium polarization) is not equal P^{spon-I} (serves as the non-equilibrium polarization). For the region where the phase transition will occur is stressed [9], there is some difference between $P^{spon}\left(=P^{spon-III}\right)$ and the equilibrium spontaneous polarization without the affects of stress $P^{spon\prime}$ because of the piezoelectric effect.

4. Thermo-electro-mechanical coupling

The comprehensive thermo-electro-mechanical coupling may be found in ferroelectric phase transition processes. Because there exists not only the change of polarization but also the changes of the system's volume and shape when a ferroelectric phase transition occurs in it, the mechanics can not be ignored even if it is mechanically-free, i.e. no external force is exerted on it. To a first-order ferroelectric phase transition, it occurs at the surface layer of system firstly, then in the inner part. So, the stress may be found in the system.

Since one aspect of the nature of ferroelectric phase transitions is the thermo-electro-mechanical coupling, we take the mechanics into account on the basis of Section 3, where only the thermo-electric coupling has been considered. This may lead to a complete description in the sense of continuum physics.

4.1. Deformation mechanics

For a continuum, the momentum equation in differential form can be written as

$$\nabla\cdot\boldsymbol{\sigma}+\rho\boldsymbol{f}=\rho\boldsymbol{a} \tag{51}$$

where σ, f, a, ρ is the stress, the volume force exerted on unit mass, the acceleration and the mass density, respectively. Let $k=\frac{1}{2}v^2$ be the local kinetic energy density (per unit mass), with v is the velocity. Then

$$\frac{\mathrm{d}k}{\mathrm{d}t}=\boldsymbol{v}\cdot\boldsymbol{a} \tag{52}$$

where t is the time. In terms of

$$\nabla\cdot(v\cdot\sigma)=v\cdot(\nabla\cdot\sigma)+(\nabla v):\sigma \tag{53}$$

$$\boldsymbol{\sigma}:\left(\nabla\boldsymbol{v}\right)=\boldsymbol{\sigma}:\boldsymbol{d} \tag{54}$$

we can deduce the following balance equation of mechanical energy basing on Equations (51) (52)

$$\rho \frac{\mathrm{d}k}{\mathrm{d}t} - \nabla \cdot (v \cdot \sigma) = \rho v \cdot f - \sigma : d \tag{55}$$

which is in differential form and in Lagrangian form. Or in Eulerian form

$$\frac{\partial(\rho k)}{\partial t} + \nabla \cdot (\rho k v - v \cdot \sigma) = \rho v \cdot f - \sigma : d \tag{56}$$

where $d = \frac{1}{2}(\nabla v + v \nabla) = \frac{1}{2}[\nabla v + (\nabla v)^T]$ is the rate of deformation or strain rate (the superscript "T" means transposition).

To a ferroelectric phase transition, f and σ may be the nominal volume force and stress, which are the embodiments of the actions of thermo-electro-mechanical coupling and are two internal fields. Generally, they are the sums of real and nominal volume force or stress

$$\boldsymbol{f} = \boldsymbol{f}^{real} + \boldsymbol{f}^{nom} \tag{57}$$

$$\boldsymbol{\sigma} = \boldsymbol{\sigma}^{real} + \boldsymbol{\sigma}^{nom} \tag{58}$$

The nominal volume force and stress are not zero until the eigen (or free) deformation of system finishes in phase transitions. If they are zero, the eigen (or free) deformation finishes.

4.2. Local entropy production and description of phase transitions

The Gibbs equation was given as the following [9]

$$T\mathrm{d}s + \sum_i \mu_i \mathrm{d}n_i + \frac{1}{\rho} E \cdot \mathrm{d}D + f \cdot \mathrm{d}r + \frac{1}{\rho} \nabla \cdot (v \cdot \sigma)\mathrm{d}t = \mathrm{d}u + \mathrm{d}k \tag{59}$$

where T, E, D is the local temperature, the local electric field intensity and the local electric displacement, respectively; s, u, n_i, μ_i is the local entropy density (per unit mass), the local internal energy density (per unit mass), the local molar quantity density (per unit mass) and the chemical potential, respectively; r is the displacement vector. If the external electric field is not applied, the quantity E is the internal electric field E_i only [8].

Make the material derivative of Equation (59) with t, then obtain

$$\rho \frac{\mathrm{d}s}{\mathrm{d}t} + \frac{\rho}{T} \sum_i \mu_i \frac{\mathrm{d}n_i}{\mathrm{d}t} + \frac{1}{T} E_i \cdot \frac{\mathrm{d}D}{\mathrm{d}t} + \frac{\rho}{T} f \cdot \frac{\mathrm{d}r}{\mathrm{d}t} + \frac{1}{T} \nabla \cdot (v \cdot \sigma) = \frac{\rho}{T} \frac{\mathrm{d}e}{\mathrm{d}t} \tag{60}$$

where e is the total energy with $e=u+k$. We know, the system is in the crystalline states before and after a phase transition so that there is no diffusion of any kind of particles in the system. So, n_i as the local molar quantity (per unit mass) does not change with t, i.e. $\frac{dn_i}{dt}=0$. $\frac{dD}{dt}$ stands for the polarization current J_P, while $\frac{dr}{dt}$ stands for the velocity v, $E_i=-\nabla\varphi_i$ (φ_i stands for the internal electric potential).

After the lengthy and troublesome deduction [9], the local entropy balance equation in Lagrangian form can be obtained

$$\rho\frac{\mathrm{d}s}{\mathrm{d}t}=-\nabla\cdot\boldsymbol{J}_s+\sigma_s \tag{61}$$

with the entropy flux J_s

$$J_s=\frac{J_q^{diff}+\varphi_i J_P-v\cdot\sigma}{T} \tag{62}$$

and the rate of local entropy production σ_s

$$\sigma_s=J_q^{diff}\cdot\nabla\left(\frac{1}{T}\right)+J_P\cdot\nabla\left(\frac{\varphi_i}{T}\right)-\sigma:\left[v\nabla\left(\frac{1}{T}\right)\right]-\rho f\cdot\frac{v}{T} \tag{63}$$

where J_q^{diff} is the diffusion of heat, i.e. heat conduction.

Here are the thermodynamic fluxes $J_q^{diff}(=J_q)$, J_P, $\sigma(=J_\sigma)$, $\rho f(=J_f)$ and the corresponding thermodynamic forces X_q, X_P, X_σ, X_f

$$\boldsymbol{X}_q=\nabla\left(\frac{1}{T}\right) \tag{64}$$

$$\boldsymbol{X}_P=\nabla\left(\frac{\varphi_i}{T}\right) \tag{65}$$

$$\boldsymbol{X}_\sigma=-\boldsymbol{v}\nabla\left(\frac{1}{T}\right) \tag{66}$$

$$\boldsymbol{X}_f = -\frac{\boldsymbol{v}}{T} \tag{67}$$

$J_i(i=q,\ P,\ \sigma,\ f)$ can be expanded linearly with $X_j(j=q,\ P,\ \sigma,\ f)$

$$\boldsymbol{J}_q = \boldsymbol{L}_{qq}\cdot\boldsymbol{X}_q + \boldsymbol{L}_{qP}\cdot\boldsymbol{X}_P + \boldsymbol{L}_{q\sigma}:\boldsymbol{X}_\sigma + \boldsymbol{L}_{qf}\cdot\boldsymbol{X}_f \tag{68}$$

$$\boldsymbol{J}_P = \boldsymbol{L}_{Pq}\cdot\boldsymbol{X}_q + \boldsymbol{L}_{PP}\cdot\boldsymbol{X}_P + \boldsymbol{L}_{P\sigma}:\boldsymbol{X}_\sigma + \boldsymbol{L}_{Pf}\cdot\boldsymbol{X}_f \tag{69}$$

$$\boldsymbol{J}_\sigma = \boldsymbol{L}_{\sigma q}\cdot\boldsymbol{X}_q + \boldsymbol{L}_{\sigma P}\cdot\boldsymbol{X}_P + \boldsymbol{L}_{\sigma\sigma}:\boldsymbol{X}_\sigma + \boldsymbol{L}_{\sigma f}\cdot\boldsymbol{X}_f \tag{70}$$

$$\boldsymbol{J}_f = \boldsymbol{L}_{fq}\cdot\boldsymbol{X}_q + \boldsymbol{L}_{fP}\cdot\boldsymbol{X}_P + \boldsymbol{L}_{f\sigma}:\boldsymbol{X}_\sigma + \boldsymbol{L}_{ff}\cdot\boldsymbol{X}_f \tag{71}$$

where L_{qq}, L_{PP}, L_{ff}, L_{qP}, L_{Pq}, L_{qf}, L_{fq}, L_{Pf}, L_{fP} are nine second-order tensors, $L_{q\sigma}$, $L_{\sigma q}$, $L_{P\sigma}$, $L_{\sigma P}$, $L_{\sigma f}$, $L_{f\sigma}$ are six third-order tensors, $L_{\sigma\sigma}$ is a fourth-order tensor. They should satisfy the generalized Onsager relations [14]

$L_{\alpha\alpha} = L_{\alpha\alpha}^T$

$$\left(\alpha = q,\ P,\ \sigma,\ f\right) \tag{72}$$

$L_{\alpha\beta} = L_{\beta\alpha}^T$

$$\left(\alpha,\ \beta = q,\ P,\ \sigma, f,\ \text{and}\ \alpha \neq \beta\right) \tag{73}$$

So the rate of local entropy production can be written as

$$\begin{aligned}\sigma_s &= \boldsymbol{L}_{qq}:\boldsymbol{X}_q\boldsymbol{X}_q + \boldsymbol{L}_{qP}\cdot\cdot\boldsymbol{X}_P\boldsymbol{X}_q + \boldsymbol{L}_{q\sigma}\cdots\boldsymbol{X}_\sigma^T\boldsymbol{X}_q + \boldsymbol{L}_{qf}\cdot\cdot\boldsymbol{X}_f\boldsymbol{X}_q\\ &+ \boldsymbol{L}_{Pq}\cdot\cdot\boldsymbol{X}_q\boldsymbol{X}_P + \boldsymbol{L}_{PP}:\boldsymbol{X}_P\boldsymbol{X}_P + \boldsymbol{L}_{P\sigma}\cdots\boldsymbol{X}_\sigma^T\boldsymbol{X}_P + \boldsymbol{L}_{Pf}\cdot\cdot\boldsymbol{X}_f\boldsymbol{X}_P\\ &+ \boldsymbol{L}_{\sigma q}\cdots\boldsymbol{X}_q\boldsymbol{X}_\sigma^T + \boldsymbol{L}_{\sigma P}\cdots\boldsymbol{X}_P\boldsymbol{X}_\sigma^T + \boldsymbol{L}_{\sigma\sigma}\cdots\cdot\boldsymbol{X}_\sigma^T\boldsymbol{X}_\sigma^T + \boldsymbol{L}_{\sigma f}\cdots\boldsymbol{X}_f\boldsymbol{X}_\sigma^T\\ &+ \boldsymbol{L}_{fq}\cdot\cdot\boldsymbol{X}_q\boldsymbol{X}_f + \boldsymbol{L}_{fP}\cdot\cdot\boldsymbol{X}_P\boldsymbol{X}_f + \boldsymbol{L}_{f\sigma}\cdots\boldsymbol{X}_\sigma^T\boldsymbol{X}_f + \boldsymbol{L}_{ff}:\boldsymbol{X}_f\boldsymbol{X}_f\end{aligned} \tag{74}$$

According to the condition on which the local entropy production is a minimum, from Equation (74) we can deduce the following

$$\begin{aligned}\left(\frac{\partial\sigma_s}{\partial \boldsymbol{X}_q}\right)_{\boldsymbol{X}_P,\boldsymbol{X}_\sigma,\boldsymbol{X}_f} &= 2\boldsymbol{L}_{qq}\cdot\boldsymbol{X}_q+\boldsymbol{L}_{qP}\cdot\boldsymbol{X}_P+\boldsymbol{L}_{q\sigma}:\boldsymbol{X}_\sigma+\boldsymbol{L}_{qf}\cdot\boldsymbol{X}_f\\ &+\boldsymbol{L}_{Pq}^T\cdot\boldsymbol{X}_P+\boldsymbol{L}_{\sigma q}^T:\boldsymbol{X}_\sigma+\boldsymbol{L}_{fq}^T\cdot\boldsymbol{X}_f\\ &=2\left(\boldsymbol{L}_{qq}\cdot\boldsymbol{X}_q+\boldsymbol{L}_{qP}\cdot\boldsymbol{X}_P+\boldsymbol{L}_{q\sigma}:\boldsymbol{X}_\sigma+\boldsymbol{L}_{qf}\cdot\boldsymbol{X}_f\right)\\ &=2\boldsymbol{J}_q\\ &=0\end{aligned} \tag{75}$$

$$\begin{aligned}\left(\frac{\partial\sigma_s}{\partial \boldsymbol{X}_P}\right)_{\boldsymbol{X}_q,\boldsymbol{X}_\sigma,\boldsymbol{X}_f} &= \boldsymbol{L}_{qP}^T\cdot\boldsymbol{X}_q+\boldsymbol{L}_{Pq}\cdot\boldsymbol{X}_q+2\boldsymbol{L}_{PP}\cdot\boldsymbol{X}_P+\boldsymbol{L}_{P\sigma}:\boldsymbol{X}_\sigma\\ &+\boldsymbol{L}_{Pf}\cdot\boldsymbol{X}_f+\boldsymbol{L}_{\sigma P}^T:\boldsymbol{X}_\sigma+\boldsymbol{L}_{fP}^T\cdot\boldsymbol{X}_f\\ &=2\left(\boldsymbol{L}_{Pq}\cdot\boldsymbol{X}_q+\boldsymbol{L}_{PP}\cdot\boldsymbol{X}_P+\boldsymbol{L}_{P\sigma}:\boldsymbol{X}_\sigma+\boldsymbol{L}_{Pf}\cdot\boldsymbol{X}_f\right)\\ &=2\boldsymbol{J}_P\\ &=0\end{aligned} \tag{76}$$

$$\begin{aligned}\left(\frac{\partial\sigma_s}{\partial \boldsymbol{X}_\sigma}\right)_{\boldsymbol{X}_q,\boldsymbol{X}_P,\boldsymbol{X}_f} &= \boldsymbol{L}_{q\sigma}^T\cdot\boldsymbol{X}_q+\boldsymbol{L}_{P\sigma}^T\cdot\boldsymbol{X}_P+\boldsymbol{L}_{\sigma q}\cdot\boldsymbol{X}_q+\boldsymbol{L}_{\sigma P}\cdot\boldsymbol{X}_P\\ &+2\boldsymbol{L}_{\sigma\sigma}:\boldsymbol{X}_\sigma+\boldsymbol{L}_{\sigma f}\cdot\boldsymbol{X}_f+\boldsymbol{L}_{f\sigma}^T\cdot\boldsymbol{X}_f\\ &=2\left(\boldsymbol{L}_{\sigma q}\cdot\boldsymbol{X}_q+\boldsymbol{L}_{\sigma P}\cdot\boldsymbol{X}_P+\boldsymbol{L}_{\sigma\sigma}:\boldsymbol{X}_\sigma+\boldsymbol{L}_{\sigma f}\cdot\boldsymbol{X}_f\right)\\ &=2\boldsymbol{J}_\sigma\\ &=0\end{aligned} \tag{77}$$

$$\begin{aligned}\left(\frac{\partial\sigma_s}{\partial \boldsymbol{X}_f}\right)_{\boldsymbol{X}_q,\boldsymbol{X}_P,\boldsymbol{X}_\sigma} &= \boldsymbol{L}_{qf}^T\cdot\boldsymbol{X}_q+\boldsymbol{L}_{Pf}^T\cdot\boldsymbol{X}_P+\boldsymbol{L}_{\sigma f}^T:\boldsymbol{X}_\sigma+\boldsymbol{L}_{fq}\cdot\boldsymbol{X}_q\\ &+\boldsymbol{L}_{fP}\cdot\boldsymbol{X}_P+\boldsymbol{L}_{f\sigma}:\boldsymbol{X}_\sigma+2\boldsymbol{L}_{ff}\cdot\boldsymbol{X}_f\\ &=2\left(\boldsymbol{L}_{fq}\cdot\boldsymbol{X}_q+\boldsymbol{L}_{fP}\cdot\boldsymbol{X}_P+\boldsymbol{L}_{f\sigma}:\boldsymbol{X}_\sigma+\boldsymbol{L}_{ff}\cdot\boldsymbol{X}_f\right)\\ &=2\boldsymbol{J}_f\\ &=0\end{aligned} \tag{78}$$

This reveals that if the k forces among those are kept constant, i.e. $X_i = const(i=1,\ 2,\ \cdots,\ k,\ k<4)$, the fluxes corresponding to the left $4-k$ forces are zero. Of course, if there are no restrictions on $X_i (i=1,\ 2,\ 3,\ 4)$, all the flues are zero. (For convenience, we have modified the superscripts $q,\ P,\ \sigma,\ f$ to be 1,2,3,4).

We may describe a ferroelectric phase transition by using the two paradigms above similarly as we have done in Section 3. To a first-order ferroelectric phase transition, the forces X_P, X_σ, X_f of the region where the phase transition is occurring can be regarded as three large constants roughly in the characteristic times of phase transition (i.e. the times in which the interface keeps rest) because the electric displacement, the volume and the shape change suddenly. So, the flux J_q^{diff} of the region should be zero (but $J_P \neq 0$, $\sigma \neq 0$, $\rho f \neq 0$). This states clearly that the pure heat conduction and the heat conduction induced by the thermo-electric coupling and the thermo-mechanical coupling cancel out each other so as to release or absorb the latent heat. The phase transition occurs at the surface layer firstly, which is mechanically-free. So, when the phase transition occurs in this region, the flux σ maybe the nominal stress σ^{nom} only, which does work to realize the transformation from internal energy to kinetic energy. When the phase transition occurs in the inner part, the flux σ should be the sum of σ^{real} and σ^{nom} because the sudden changes of the inner part's volume and shape have to overcome the bound of outer part then σ^{real} arises. The region where the phase transition is occurring, i.e. the phase boundary is accompanied with the real stress σ^{real} usually, which does work to realize the transformation from kinetic energy to internal energy. This has been predicted and described with a propagating stress wave [20].

It is certain that the latent heat passes through the region where the phase transition has occurred (at the outside of the region where the phase transition is occurring) and exchange itself with the thermal bath. For $\pm l\rho v_a = J_q^{diff} = -\kappa \cdot \nabla T$, a constant temperature gradient ∇T is kept in the region where the phase transition has occurrd, i.e. the force X_q at every site is a constant (which does not change with the time but may vary with the position). So, the fluxes $J_P = \sigma = \rho f = 0$ (but the flux $J_q^{diff} \neq 0$). This states clearly that the electric displacement D will not change but keep the value at Curie temperature or zero until the phase transition finishes and $\sigma^{real} = -\sigma^{nom}$ in this region. Because the electric displacement D and the strain (or deformation) are all determined by the crystal structure of system, $J_P = 0$ reveals that D of this region does not change so does not the crystal structure then does not the strain (or deformation). According to [20], we know the region where the phase transition has occurred is unstressed, i.e. $\sigma^{real} = 0$, then $\sigma^{nom} = 0$. This reveals that the eigen (or free) strain (or deformation) of system induced by the thermo-electro-mechanical coupling of phase transition is complete and the change of it terminates before the phase transition finishes. The two deductions coincide with each other. σ^{real} may relaxes via the free surface.

The region where the phase transition will occur should be in equilibrium because there are no restrictions on the forces X_q, X_P, X_σ, X_f. Whereas, according to [20], the region is stressed, i.e. $\sigma^{real} \neq 0$. To the heating process of phase transition, this may lead to a change of the spontaneous polarization of this region because of the electro-mechanical coupling (piezo-electric effect).

An immediate result of the above irreversible thermodynamic description is that the action at a distance, which is the kind of heat transfer at phase transitions, is removed absolutely. The latent heat is transferred within a finite time so the occurrence of phase transition in the inner part is delayed. (Of course, another cause is the stress, just see Section 5) In other words, the various parts absorb or release the latent heat at the various times. The action at a distance does not affect the phase transition necessarily.

5. Irreversibility: Thermal hysteresis and occurrences of domain structure

5.1. Thermal hysteresis

The "thermal hysteresis" of first-order ferroelectric phase transitions is an irreversible phenomenon obviously. But it was treated by using the equilibrium thermodynamics for ferroelectric phase transitions, the well-known Landau-Devonshire theory [2]. So, there is an inherent contradiction in this case. The system in which a first-order ferroelectric phase transition occurs is heterogeneous. The occurrences of phase transition in different parts are not at the same time. The phase transition occurs at the surface layer then in the inner part of system. According to the description above, we know a constant temperature gradient is kept in the region where the phase transition has occurred. The temperature of surface layer, which is usually regarded as the temperature of the whole system in experiments, must be higher (or lower) than the Curie temperature. This may lead to the thermal hysteresis.

No doubt that the shape and the area of surface can greatly affect the above processes. We may conclude that the thermal hysteresis can be reduced if the system has a larger specific surface and, the thermal hysteresis can be neglected if a finite system has an extremely-large specific surface. So, the thermal hysteresis is not an intrinsic property of the system.

The region where the phase transition will occur can be regarded as an equilibrium system for there are no restrictions on the forces X_q, X_P, X_σ, X_f. In other words, the forces and the corresponding fluxes are zero in this region. To a system where a second-order ferroelectric phase transition occurs, the case is somewhat like that of the region where a first-order ferroelectric phase transition will occur. The spontaneous polarization, the volume and the shape of system are continuous at the Curie temperature and change with the infinitesimal magnitudes. This means X_q, X_P, X_σ, X_f and J_q^{diff}, J_P, σ, ρf can be arbitrary infinitesimal magnitudes. The second-order phase transition occurs in every part of the system simultaneously, i.e. there is no the co-existence of two phases (ferroelectric and paraelectric). So, there is no the latent heat and stress. The thermal hysteresis disappears.

The region where a first-order ferroelectric phase transition will occur is stressed. This reveals that the occurrences of phase transition in the inner part have to overcome the bound of outer part, where the phase transition occurs earlier. This may lead to the delay of phase transition in the inner part.

5.2. Occurrences of domain structure

Though the rationalization of the existence of domain structures can be explained by the equilibrium thermodynamics, the evolving characteristics of domain occurrences in ferroelectrics can not be explained by it, but can be explained by the non-equilibrium thermodynamics.

In the region where the phase transition is occurring, the thermodynamic forces X_P (= $\nabla\left(\frac{\varphi_i}{T}\right)$), X_σ (= $-v\nabla\left(\frac{1}{T}\right)$), X_f (= $-\frac{v}{T}$) can be regarded as three large constants in the characteristic times of transition and the thermodynamic flux $J_q^{diff}=0$ (but $J_P\neq0,\ \sigma\neq0,\ \rho f\neq0$). The local entropy production (cf. Equation (63)) reduces to

$$\sigma_s = \boldsymbol{J_P}\cdot\nabla\left(\frac{\varphi_i}{T}\right) - \quad :\left[\boldsymbol{v}\nabla\left(\frac{1}{T}\right)\right] - \rho\boldsymbol{f}\cdot\frac{\boldsymbol{v}}{T} \tag{79}$$

Now, we are facing a set of complicated fields of T, v and φ_i, respectively. Assume that the phase transition front is denoted by S. The points included in S stand for the locations where the transition is occurring. Because the transition occurs along all directions from the outer part to the inner part, we may infer that the orientations of $\nabla\left(\frac{\varphi_i}{T}\right)$, or of $-v\nabla\left(\frac{1}{T}\right)$ and or of $-\frac{v}{T}$ vary continuously such that they are differently oriented at different locations.

There are always several (at least two) symmetry equivalent orientations in the prototype phase (in most cases it is the high temperature phase), which are the possible orientations for spontaneous polarization (or spontaneous deformation or spontaneous displacement). Therefore, the spontaneous polarization, the spontaneous deformation and the spontaneous displacement must take an appropriate orientation respectively to ensure σ_s is a positive minimum when the system transforms from the prototype (paraelectric) phase to the ferroelectric (low temperature) phase. The underlying reasons are that

$$\boldsymbol{J_P} = \frac{\mathrm{d}\boldsymbol{D}}{\mathrm{d}t} = \frac{\mathrm{d}\left(\varepsilon_0\boldsymbol{E}_i+\boldsymbol{P}\right)}{\mathrm{d}t} \tag{80}$$

$$= \boldsymbol{L}:\varepsilon \tag{81}$$

$$\rho\boldsymbol{f} = \rho\frac{\mathrm{d}\boldsymbol{v}}{\mathrm{d}t} - \nabla\cdot \tag{82}$$

where L , ε, and ε_0 are the modulus of rigidity, the strain and the permittivity of vacuum, respectively. Therefore, P, ε at different locations will be differently oriented. The domain structures in ferroelectrics thus occur.

http://dx.doi.org/10.5772/52089

It seems that the picture of domain occurrences for first-order ferroelectric phase transition systems should disappear when we face second-order ferroelectric phase transition systems. This is true if the transition processes proceed infinitely slowly as expounded by the equilibrium thermodynamics. But any actual process proceeds with finite rate, so it is irreversible. Then the above picture revives.

In [21], the domain occurrences in ferromagnetics can be described parallelly by analogy. And the case of ferroelastic domain occurrences is a reduced, simpler one compared with that of ferroelectrics or ferromagnetics.

It is well known that the Landau theory or the Curie principle tells us how to determine the symmetry change at a phase transition. A concise statement is as follows [22]: for a crystal undergoing a phase transition with a space-group symmetry reduction from G_0 to G, whereas G determines the symmetry of transition parameter (or vice versa), it is the symmetry operations lost in going from G_0 to G that determine the domain structure in the low-symmetry phase. The ferroic phase transitions are the ones accompanied by a change of point group symmetry [23]. Therefore, the substitution of "point group" for the "space group" in the above statement will be adequate for ferroic phase transitions. From the above statement, the domain structure is a manifestation of the symmetry operations lost at the phase transition. In our treatment of the domain occurrences in ferroics, we took into account the finiteness of system (i.e. existence of surface) and the irreversibility of process (asymmetry of time). The finiteness of system make the thermodynamic forces such as $\nabla\left(\frac{\varphi_i}{T}\right)$, $-v\nabla\left(\frac{1}{T}\right)$, $-\frac{v}{T}$ have infinite space symmetry. The infinite space symmetry, combined with the asymmetry of time, reproduces the symmetry operations lost at the phase transition in the ferroic phase. It can be viewed as an embodiment of time-space symmetry.

After all, for the domain structures can exist in equilibrium systems, they are the equilibrium structures but not the dissipative ones, for the latter can only exist in systems far from equilibrium [24].

6. Concluding remarks

In order to overcome the shortcoming of Landau-Devonshire theory, the non-equilibrium thermodynamics was applied to study the ferroelectric phase transitions. The essence of transitions is the thermo-electro-mechanical coupling. Moreover, the irreversibility, namely thermal hysteresis and domain occurrences can be explained well in the realm of non-equilibrium thermodynamics.

The non-equilibrium thermodynamic approach utilized here is the linear thermodynamic one actully. In order to get the more adequate approaches, we should pay attention to the new developments of non-equilibrium thermodynamics. The thermodynamics with internal variables [25] and the extended (irreversible) thermodynamics [26] are two current ones. They all expand the fundamental variables spaces to describe the irreversible processes more adequately. Whereas, the relevant theoretical processing must be more complicated undoubtly. This situation needs very much effort.

Acknowledgements

The work is supported by the Natural Science Foundation Program of Shandong Province, China (Grant No. ZR2011AM019).

Author details

Shu-Tao Ai*

School of Science, Linyi University, Linyi, People's Republic of China

References

[1] Grindlay, I. An Introduction to the Phenomenological Theory of Ferroelectricity. Oxford: Pergamon; (1970).

[2] Lines, M E, & Glass, A M. Principles and Applications of Ferroelectrics and Related Materials. Oxford: Clarendon; (1977).

[3] Tilley, D R, & Zeks, B. Landau Theory of Phase Transitions in Thick Films. Solid State Communications (1984)., 49(8), 823-828.

[4] Scott, J F, Duiker, H M, Beale, P D, Pouligny, B, Dimmler, K, Parris, M, Butler, D, & Eaton, S. Properties of Ceramic KNO_3 Thin-Film Memories. Physica B+C (1988).

[5] Zhong, W L. Physics of Ferroelectrics (in Chinese). Beijing: Science Press; (1996).

[6] Gordon, A. Finite-Size Effects in Dynamics of Paraelectric-Ferroelectric Interfaces Induced by Latent Heat Transfer. Physics Letters A (2001).

[7] Gordon, A, Dorfman, S, & Fuks, D. Temperature-Induced Motion of Interface Boundaries in Confined Ferroelectrics. Philosophical Magazine B (2002)., 82(1), 63-71.

[8] Ai, S T. Paraelectric-Ferroelectric Interface Dynamics Induced by Latent Heat Transfer and Irreversible Thermodynamics of Ferroelectric Phase Transitions. Ferroelectrics (2006)., 345(1), 59-66.

[9] Ai, S T. Mechanical-Thermal-Electric Coupling and Irreversibility of Ferroelectric Phase Transitions. Ferroelectrics (2007)., 350(1), 81-92.

[10] Yufatova, S M, Sindeyev, Y G, Garilyatchenko, V G, & Fesenko, E G. Different Kinetics Types of Phase Transformations in Lead Titanate. Ferroelectrics (1980). , 26(1), 809-812.

[11] Dec, J. Jerky Phase Front Motion in $PbTiO_3$ Crystal. Journal of Physics C (1988). , 21(7), 1257-1263.

[12] Dec, J, & Yurkevich, J. The Antiferroelectric Phase Boundary as a Kink. Ferroelectrics (1990). , 110(1), 77-83.

[13] Ai, S T, Xu, C T, Wang, Y L, Zhang, S Y, Ning, X F, & Noll, E. Comparison of and Comments on Two Thermodynamic Approaches (Reversible and Irreversible) to Ferroelectric Phase Transitions. Phase Transitions (2008). , 81(5), 479-490.

[14] Ai, S T, Zhang, S Y, Jiang, J S, & Wang, C C. On Transport Coefficients of Ferroelectrics and Onsager Reciprocal Relations. Ferroelectrics (2011). , 423(1), 54-62.

[15] Lavenda, B H. Thermodynamics of Irreversible Processes. London and Basingstoke: Macmillan; (1978).

[16] Chewasatn, S, & Milne, S T. Synthesis and Characterization of $PbTiO_3$ and Ca and Mn Modified $PbTiO_3$ Fibres Produced by Extrusion of Diol Based Gels. Journal of Material Science (1994). , 29(14), 3621-3629.

[17] Nomura, S, & Sawada, S. Dielectric Properties of Lead-Strontium Titanate. Journal of the Physical Society of Japan (1955). , 10(2), 108-111.

[18] Mante, A, & Volger, H. J. The Thermal Conductivity of $PbTiO_3$ in the Neighbourhood of Its Ferroelectric Transition Temperature. Physics Letters A (1967). , 24(3), 139-140.

[19] Dec, J. The Phase Boundary as a Kink. Ferroelectrics (1989). , 89(1), 193-200.

[20] Gordon, A. Propagation of Solitary Stress Wave at First-Order Ferroelectric Phase Transitions. Physics Letters A (1991).

[21] Ai, S T, Zhang, S Y, Ning, X F, Wang, Y L, & Xu, C T. Non-Equilibrium Thermodynamic Explanation of Domain Occurrences in Ferroics. Ferroelectrics Letters Section (2010). , 37(2), 30-34.

[22] Janovec, V. A Symmetry Approach to Domain Structures. Ferroelectrics (1976). , 12(1), 43-53.

[23] Wadhawan, V K. Ferroelasticity and Related Properties of Crystals. Phase Transitions (1982). , 3(1), 3-103.

[24] Glansdoff, P, & Prigogine, I. Thermodynamic Theory of Structure, Stability and Fluctuation. New York: Wiley-Interscience; (1971).

[25] Maugin, G A, & Muschik, W. Thermodynamics with Internal Variables (I) General Concepts, (II) Applications. Journal of Non-Equilibrium Thermodynamics (1994).

[26] Jou, D, Casas-vázquez, J, & Lebon, G. Extended Irreversible Thermodynamics (3rd edition). Berlin: Springer; (2001).

Chapter 3

$Pb(Mg_{1/3}Nb_{2/3})O_3$ (PMN) Relaxor: Dipole Glass or Nano-Domain Ferroelectric?

Desheng Fu, Hiroki Taniguchi, Mitsuru Itoh and Shigeo Mori

Additional information is available at the end of the chapter

http://dx.doi.org/10.5772/52139

1. Introduction

In sharp contrast to normal ferroelectric (for example $BaTiO_3$), relaxors show unusually large dielectric constant over a large temperature range (~100 K) (Fig. 1 (a)) [1,2]. Such large dielectric response is strongly dependent on the frequency. Its origin has been the focus of interest in the solid-state physics. Unlike the dielectric anomaly in $BaTiO_3$, which is associated with a ferroelectric phase transition, the maximum of dielectric response in relaxor does not indicate the occurrence of a ferroelectric phase transition. Such huge dielectric response suggests that local polarization might occur in the crystal. This was envisioned by Burns and Dacol from the deviation from linearity of refractive index $n(T)$ (Fig. 1(c)) around the so-called Burns temperature (T_{Burns}) [3] because the deviation Δn is proportional to polarization P_s. The local polarization is suggested to occur in a nano-region, and is generally called as polar nano-region (PNR). The existence of PNR in relaxor is well confirmed from the neutron scattering measurements [4] and transition electron microscopy (TEM) observation. However, it is still unknown how PNRs contribute to the large dielectric response. More recently, it is also suggested that a strong coupling between zone-center and zone-boundary soft-modes may play a key role in understanding the relaxor behaviors [5]. Clearly, the question "what is the origin of the giant dielectric constant over a broad temperature range?"is still unclear [6,7].

Another longstanding issue on relaxors is how PNRs interact at low temperature. There are two acceptable models: (1) dipole glass model [2,8-11], and (2) random-field model [12,13]. In spherical random-bond–random-field (SRBRF) model, Pirc and Blinc assumed that PNRs are spherical and interact randomly and proposed a frozen dipole glass state for relaxor (Fig. 2(a)) [10,11]. It predicts that the scaled third-order nonlinear susceptibility $a_3=-\varepsilon_3/\varepsilon_1^4$ will shows a nearly divergent behavior at the freezing temperature T_f of the spherical glass phase. In sharp contrast, the random field model of Fisch [13], which assumes non-random two-spin exchange, predicts a ferroelectric or ferroelectric domain state in relaxor (Fig. 2(b)). This random Potts field model also predicts a broadening specific heat peak for the glass phase and shows that the latent heat at ferroelectric transition T_c is so small that it may be difficult to be detected, which reasonably explains the data reported by Moriya et al. [14].

Combining our recent results [15] from the electrical polarization, Raman scattering, TEM measurements and those reported in the literature, here, we propose a physical picture to understand the dielectric behaviors of $Pb(Mg_{1/3}Nb_{2/3})O_3$ (PMN) relaxor.

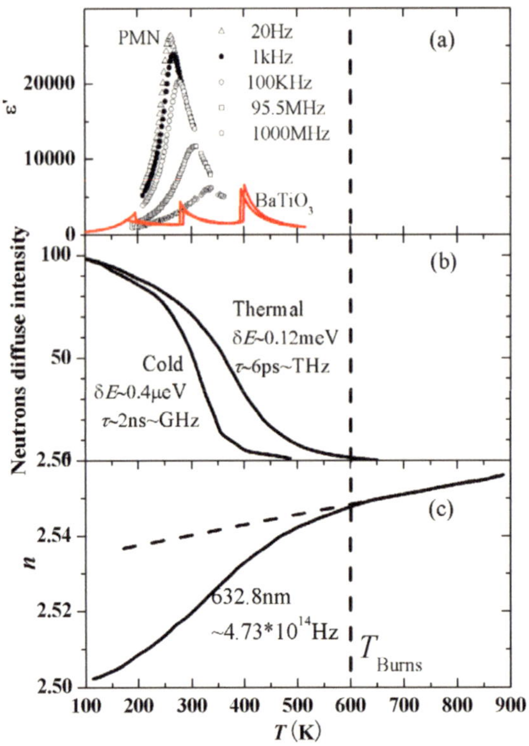

Figure 1. Temperature dependence of (a) linear dielectric constant [9], (b) cold [35] & thermal [4] neutrons diffuse intensities, and (c) refractive index [3] for PMN relaxor crystal.

2. Multiple inhomogeneities in relaxors

PMN is a prototypical relaxor with $A(B',B'')O_3$ perovskite structure (Fig. 3), in which B-sites are occupied by two kinds of heterovalent cations. Such chemical inhomogeneity is a common feature of relaxor crystals. Although it remains an average centrosymmetric cubic structure down to 5 K [16], local structural inhomogeneity has been detected in PMN relaxor. In addition to PNR mentioned above, chemically ordering region (COR) [17-19] with size of several nm has been observed in PMN crystal by TEM. It should be noticed that PNR and COR belong to different symmetry groups and are considered to have non-centrosymmetry of$R3m$ and centrosymmetryof$Fm\bar{3}m$, respectively. Therefore, there is spontaneous polarization P_s along the <111>$_c$ direction of pseudocubic structure in PNRs [20,21], but none of P_s exists in CORs. In addition to chemical and structural inhomogeneities, we will show that PMN relaxor also has inhomogeneity of ferroelectric domain structure. Multiple inhomogeneities are thus considered to play a crucial role in inducing the intriguing behaviors in relaxors.

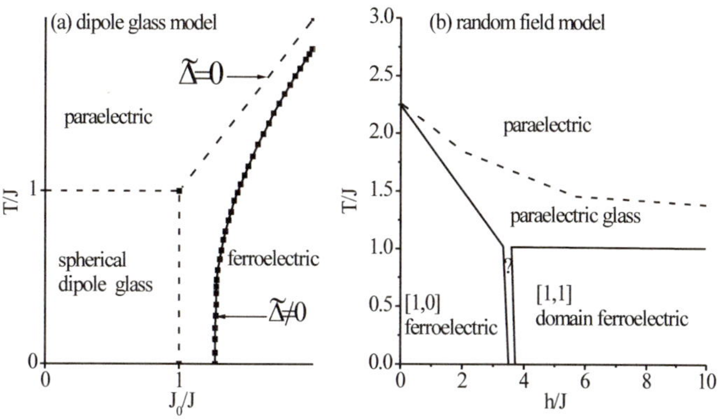

Figure 2. Two phase diagrams proposed for relaxor by (a) a dipole glass model [10], and (b) a random field model [13].

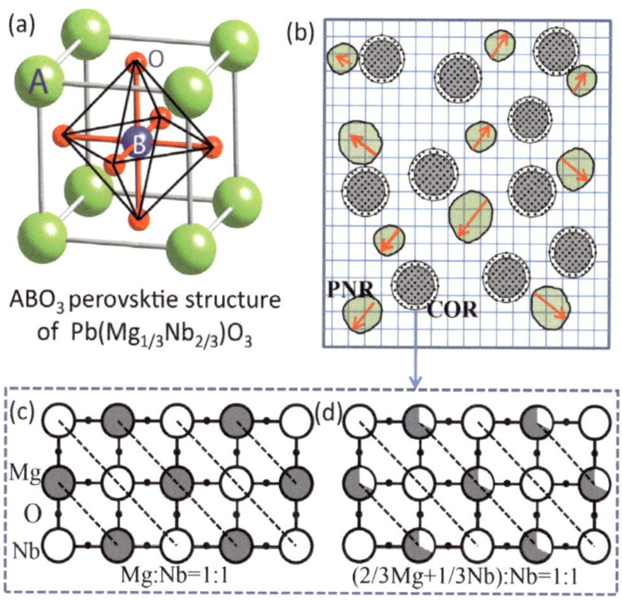

Figure 3. (a) ABO_3perovskite structure. (b) Model for relaxor structure. PNR and COR represent the polar nano-region and chemically order region, respectively. (c) &(d) show two models of atom arrangement for COR. To maintain the electric neutrality, a Nb-rich layer is required for case (c).

3. Evolution of the electrical polarization and origin of the huge dielectric responses

In order to understand the nature of the huge dielectric response and the ground state of the electrical polarization in PMN, it is essential to know the polarization hysteresis of all states including virgin state in PMN crystal. Although there are many reports on the polarization hysteresis of PMN crystal, there is a lack of understanding of the polarization hysteresis of the virgin state. In our polarization measurements, in order to access the virgin state of the crystal at a temperature, it was firstly annealed at 360 K and then cooled to the desired temperature for the measurements.

Figure 4 shows the *D-E* hysteresis at three typical temperatures observed for $(110)_c$-cut PMN crystal A*t* T=360 K that is greatly higher than the freezing temperature T_f=224 K assumed for PMN crystal [9],there is no remnant polarization within the experimental time scale of τ~10 ms (one cycle of the *D–E* loop) and the polarization is history-independent, indicating that the crystal is macroscopically paraelectric at this temperature. When temperature is lower than room temperature (for example, T=250 K), remnant polarization was observed but it generally disappears after removing the electric field.

Upon further cooling to temperatures lower than ~220 K,PMN shows polarization hysteresis similar to that of normal ferroelectric [22]. Fig. 4(a) shows an example of the characteristic hysteresis loop in this temperature range. In the virgin state as indicated by the thick red line, it appears that there is no remnant polarization in the crystal at zero electric field. However, as increasing the electric field, we can see the gradual growth of the polarization. This is a characteristic behavior of the polarization reversal (switching) in ferroelectric. When the applied field is larger than the coercive field E_c, ferroelectric domains are aligned along the direction of the electric field, leading to a stable macroscopic polarization in the crystal. This is evident from the fact that the remnant polarization is identical to the saturation polarization. These results clearly indicate that PMN is a ferroelectric rather than a dipole glass at temperature lower than 220 K.

There are many reports on the electric-field induced phase transition in PMN. On the basis of the change of dielectric constant under the application of a DC electric field, Colla et al. proposed an *E-T* phase diagram for PMN [8], which suggests a phase transition from a glass phase to a ferroelectric phase at a critical field of E_t=1. 5kV/cm in the temperature range of 160 K - 200 K. However, from our polarization results, we cannot find such a critical field except the coercive field E_c. If we assume that E_c is the critical field, then its value completely disagrees with the reported value. We observed that E_c increases rapidly with lowering the temperature, for example, E_c can reach a value of ~11 kV/cm at 180 K, which is about 7 times of E_t. Moreover, E_c is strongly dependent on frequency (Fig. 4b), and has an exponential form($1/f \propto \exp(\alpha/E_c)$) (Fig. 4c), which will result in an undefined critical field if we assume E_c=E_t because E_c is dependent on the measurement time-scale.

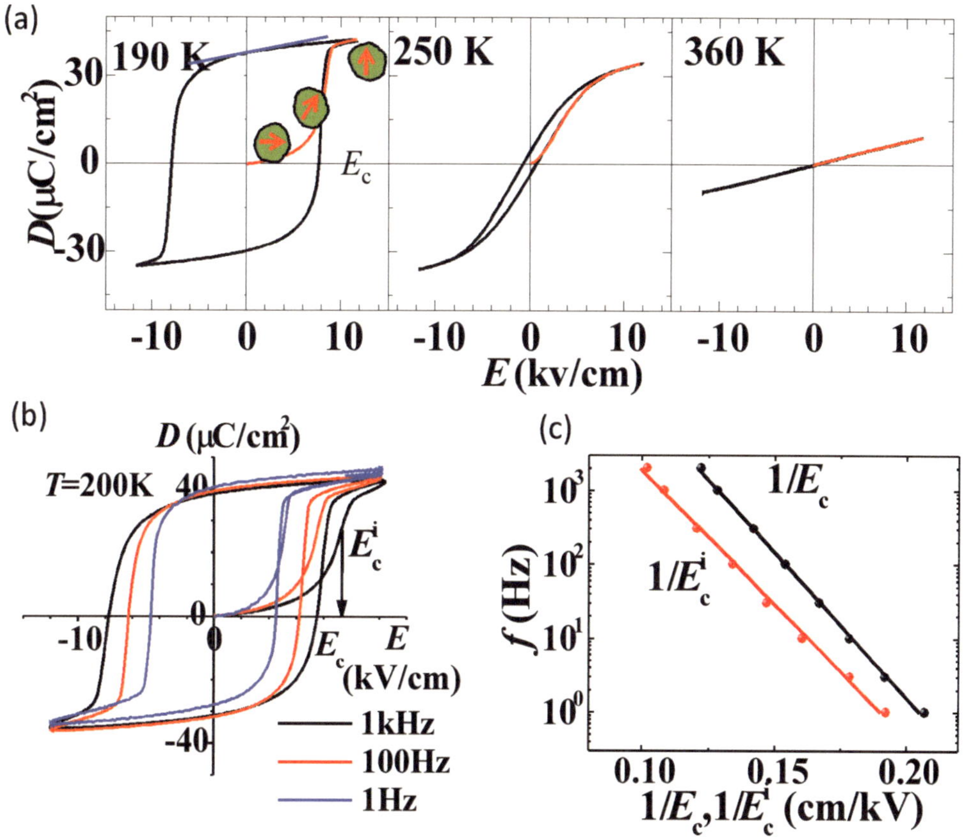

Figure 4. (a) Polarization hysteresis in PMN crystal at 190 K, 250 K, 360K. (b) Frequency dependence of polarization hysteresis. (c) Relationship between frequency and coercive field determined from the peak of switching currents. Superscript *i* denotes the value observed for the virgin state.

As discussed in following, ferroelectric micro-domain and soft-mode behaviors have also been observed at zero field in PMN in our measurements. Also, lowering of symmetry of local structure at zero field was also revealed around 210 K by a NMR study [21]. All these results direct to the fact of occurrence of a ferroelectric state at zero field in PMN crystal. We therefore consider that it is more rational to attribute E_c to the coercive field required for the domain switching rather than the critical field for a field-induced phase transition. In fact, the exponential relationship of coercive field with frequency is well-known as Merz's law ($f=1/\tau \propto \exp(-\alpha/E_c)$, τ=switching time, α=activation field) [23-28] in the normal ferroelectrics (for example, $BaTiO_3$, TGS) [23,24]. Using this relationship, the activation field is estimated to be 83. 5 kV/cm for PMN crystal at 200 K (Fig. 4c), which is one order of magnitude greater than that of the $BaTiO_3$ crystal. This indicates that the domain switching in PMN become more and more difficult as lowering temperatures. For example, when an electric field of 1 kV/cm is applied to the crystal at 200 K, an unpractical time of 2. 3×10^{29}s (~7. 3×10^{21} years) is required for the domain switching. Actually, it is impossible to observe the spontaneous polarization by this weak field within a limit time at this temperature.

The high-resolution data of the polarization obtained by a 14-bit oscilloscope allow us to calculate the linear and nonlinear dielectric susceptibilities (defined by the expansion $P=\varepsilon_0(\chi_1 E +\chi_3 E^3+))$ directly from the *D-E* hysteresis by differentiating the polarization with respective to the electric field. The calculated results are summarized in Fig. 5 for various electric fields in the virgin state and the zero electric field after the polarization reversal. These results allow us to have a deep insight into the nature of abnormal dielectric behaviors and the phase transition in PMN relaxor. Fig. 5(a) shows the linear dielectric response for the virgin state in various electric fields. When comparing the response obtained at zero field with that obtained by LCR meter at the frequency corresponding to the sampling rate used in *D-E* hysteresis measurements, one might find that they both behavior in the same way with the temperature. As mentioned above that the polarization response under the electric field in the virgin state is essentially due to the polarization reorientation, we therefore can reasonably attribute the dielectric anomaly usually observed in PMN relaxor to the polarization reorientation. This indicates that the reorientation of the PNRs dominates the huge dielectric response in PMN relaxor. Fig. 5 (a) also shows that the peak of dielectric response shifts to lower temperature at a higher electric field. This means that the activation field required for domain switching increases with lowering the temperature.

The nonlinear dielectric susceptibility ε_{3v} and its scaled value a_{3v} for the virgin state are given in Fig. 5(c) and (d), respectively. ε_{3v}shows a broad peak around 255 K, which is in good agreement with those obtained by Levstik et al. using a lock-in to wave analyzer technique for various frequencies [9]. Levstik et al. attributed this behavior to the freezing of dipole glass at T_f=224 K in PMN. However, this picture is inconsistent with the results shown in the above polarization measurements, which indicates that PMN relaxor is ferroelectric but not glass at $T<T_f$. Such dipole glass picture is also excluded by the results of the scaled susceptibility a_3 shown in Fig. 5(d), and those reported in the previous studies [29, 30]. We can see that there is no divergent behavior of a_3 in PMN relaxor. This result again suggests that there is no freezing of dipole glass in PMN as predicted for the dipole glass model. The observed anomaly of nonlinear dielectric susceptibility around 255 K is more consistent with the phase diagram of the random field model proposed by Fisch [13], in which a glass phase occurs between the paraelectric phase and the ferroelectric phase in relaxors. Such anomaly of nonlinear dielectric susceptibility may be a manifestation of spherical dipole glass with random interaction used in SRBRF model, and its nature requires further theoretical investigations.

In the random field model, the ferroelectric phase transition is suggested to be smeared due to the quenched random fields, but it may be visible if the random fields are overcome by an external electric field [12]. This ferroelectric phase transition has been convincingly shown by the sharp peak of the linear dielectric susceptibility ε_{1p} (Fig. 5(b)) and the anomaly of its nonlinear components ε_{3p} and a_{3p} (solid circles in Fig. 5(c) and (d)) when ferroelectric domains are aligned by an external field. The sharp peak of linear susceptibility indicates that the ferroelectric phase transition occurs at $T=T_c$=225 K, around which Curie-Weise law was observed. The value of Curie constant is estimated to be C = 2. 05 *10^5 K, which is characteristic of that of the displacive-type phase transition, suggesting a soft mode-driven phase transition in this system. This conclusion is supported by the occurrence of soft–mode in the crystal observed by neutron [31] and Raman scattering measurements [32].

http://dx.doi.org/10.5772/52139

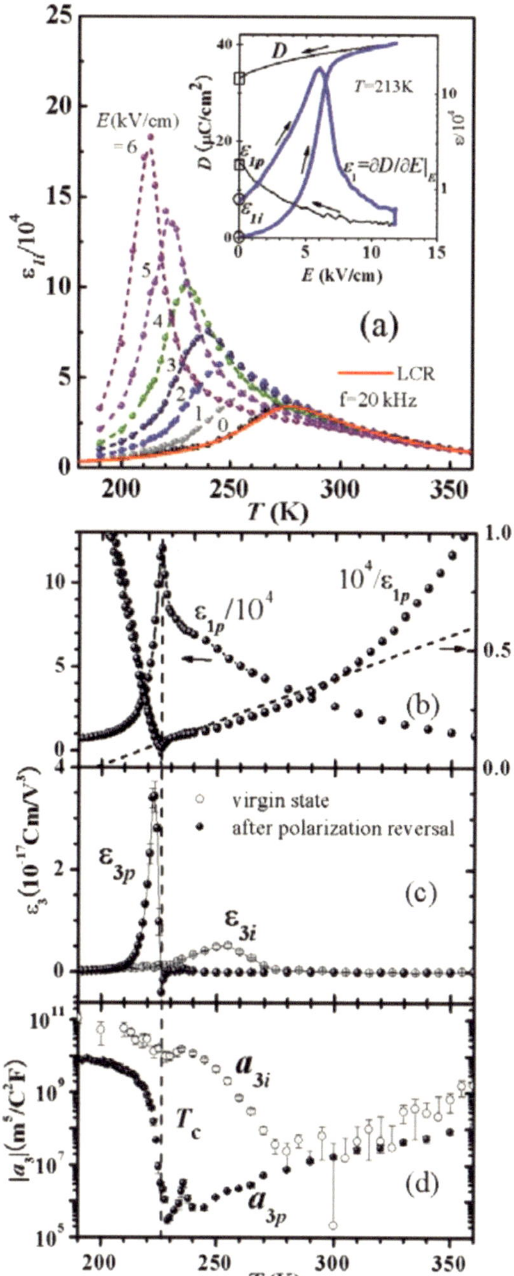

Figure 5. Dielectric responses in PMN crystal. (a) Chang of linear dielectric responses (ε_1=$\partial D/\partial E|_E$) with the electric field for the virgin state. Inset shows an example of the polarization and dielectric responses, in which the thick lines indicate the virgin state. Dielectric constant obtained by LCR impedance measurements at an ac level of 1 V/cm (red solid line) is also shown for comparison. (b) Linear dielectric response (ε_{1p}=$\partial D/\partial E|_{E=0}$) at zero field after the polarization reversal (indicated by the square in inset of (a)), and its inverse. (c) Nonlinear dielectric constant(ε_3=$\partial^3 D/\partial^3 E|_{E=0}$) at zero field for the virgin state, and the state after the polarization reversal. (d) The corresponding a_3=$\varepsilon_3/\varepsilon_1{}^4|_{E=0}$ for these two states.

Here, we can see that there are two characteristic temperatures in relaxors: Burns temperature T_{Burns} and ferroelectric phase transition temperature T_c. At T_{Burns}, local polarizations (PNRs) begin to occur. Before the ferroelectric transition occurs, PNRs are dynamic, and more importantly, interactions among them are random. Consequently, the existence of PNRs between T_c and T_{Burns} can be considered as precursor phenomenon in a phase transition [33]. Actually, such precursor phenomenon has also been observed in the normal ferroelectric $BaTiO_3$ at temperatures far above T_c[34]. A difference between $BaTiO_3$ and PMN relaxor is that the temperate region of precursor existence in PMN is greatly lager than in $BaTiO_3$. To probe the precursor behavior, one has to consider the time scale used. For example, we cannot detect a spontaneous polarization at ms scale by *D-E* measurements for PMN at room temperature, but at the probe time scale of 2 ns, cold neutron high-flux backscattering spectrometer can detect PNRs up to ~400K [35] (Fig. 1b). In contrast, due to a shorter probe time scale of ~6 ps [4], thermal neutron scattering can probe PNRs up to 600 K, which is close to T_{Burns} determined by the optical measurements that have the shortest probe time scale.

4. Soft mode behaviors in PMN relaxor

In the displacive-type ferroelectrics, soft-modes should occur in the lattice dynamics of the crystal. Actually, in a study of neutron inelastic scattering, a FE soft mode was revealed to recovers, i. e., becomes underdamped, below 220 K, and from there its energy squared $(\hbar\omega_s)^2$ increases linearly with decreasing T as for normal FEs below T_c(see also Fig. 6(e)) [31]. This has long been a puzzle for PMN relaxor: how can this be, since it has been thought that PMN remains cubic to at least 5 K [16]? However, such soft-mode behavior is exactly consistent with our results from polarization measurements shown in previous section, which show a ferroelectric phase transition at T_c~225 K. Our Raman scattering measurements also support the occurrence of FE softmode in PMN relaxor [32].

In Raman scattering studies for relaxors, the multiple inhomogeneities due to the coexistence of different symmetry regions such as the PNR and COR has been a tremendous barrier to clarify the dynamical aspect of relaxor behavior in PMN. In particular, the intense temperature-independent peak at 45 cm^{-1} (indicated by ↓ in Fig. 6(a)), which stems from the COR with $Fm\bar{3}m$ [32], always precludes a detailed investigation of low-wave number spectra of PMN crystal. Our angular dependence of the Raman spectra together with the results from the Raman tensor calculations clearly indicate that the strong F_{2g} mode located at 45 cm^{-1} can be eliminated by choosing a crossed Nicols configuration with the polarization direction of the incident laser along <110>$_c$ direction (see right panel in Fig. 6(c)).

Such special configuration allows us to observe the other low-wave number modes easily. Fig. 6 (d) shows the spectra obtained by this configuration. At the lowest temperature, a well-defined mode can be seen from the spectrum, which softens as increasing the temperature, indicating the occurrence of FE soft mode in PMN relaxor. Due to the multiple inhomogeneities of the system, the shape of soft-mode of PMN relaxor is not as sharp as that observed in normal displacive-type ferroelectrics. However, we still can estimate its frequency reliably from the careful spectrum analysis. Its temperature dependence is shown in Fig. 6(e) (indicated by ●) in comparison with the results obtained by neutron inelastic scattering (○) [31].

http://dx.doi.org/10.5772/52139

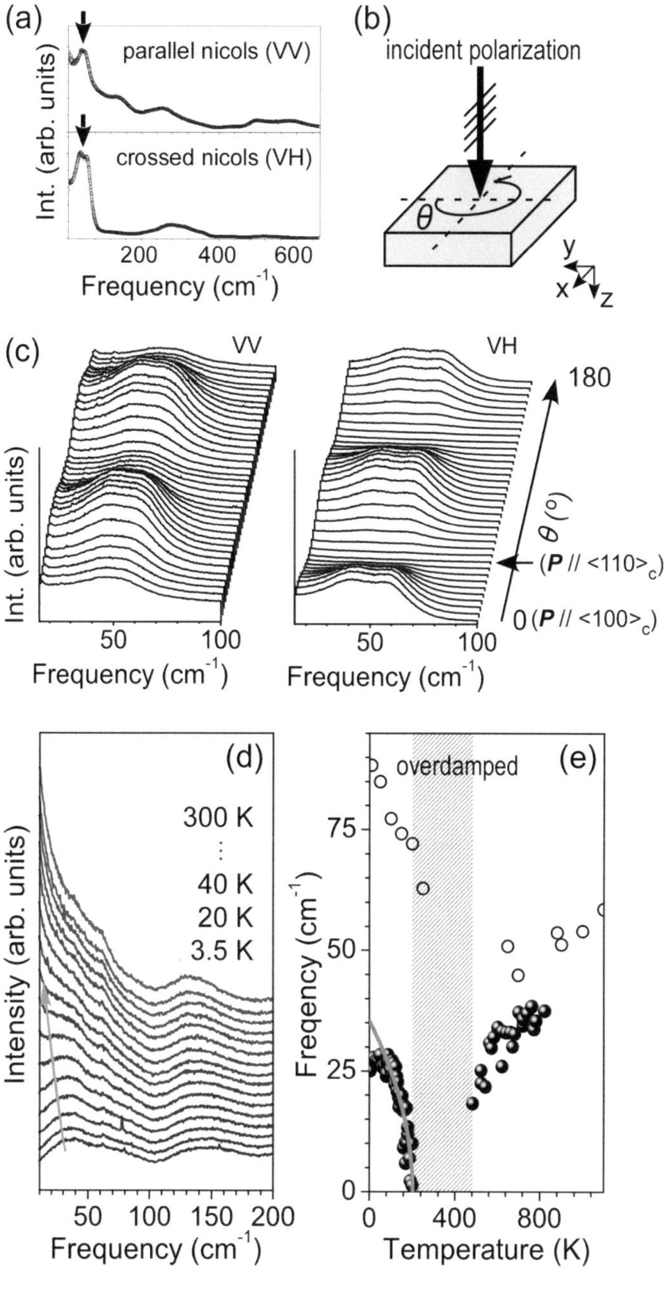

Figure 6. a) Room-temperature Raman spectra in PMN observed by the parallel (upper panel) and crossed (bottom panel) Nicols configurations, with the polarization direction of the incident laser parallel to <100>$_c$(***P*** // <100>$_c$). (b) Scattering configurations used in measurements on the angular dependence of the Raman spectra. (c) Angular dependence of the low-wave number Raman spectra obtained at room temperature. (d) Temperature dependence of Raman spectra observed by the crossed Nicols configuration with the polarization direction of the incident laser along <110>c direction (***P*** // <110>$_c$). (e) Soft mode wave numbers obtained by Raman scattering (●) and by neutron inelastic scattering by Wakimoto et al. (○) [31].

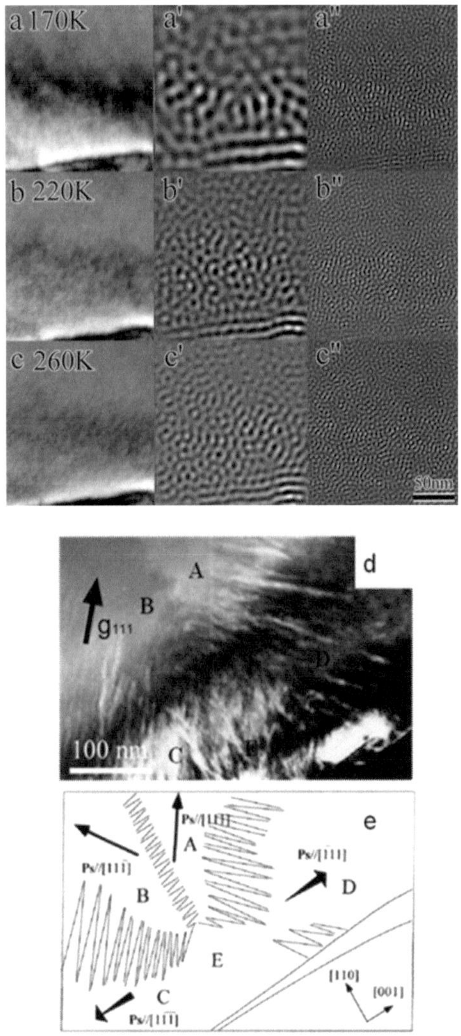

Figure 7. a)-(c) Temperature variation of TEM images observed for PMN relaxor. (a')-(c') & (a")-(c") show images of PNR and COR derived from (a)-(c). (d)Micrometric domain structure observed in the ferroelectric phase at 130K. (e) Schematic domain patterns shown in (d). Arrows and lines indicate the polarization directions and domain boundaries, respectively.

We find that the soft-mode exhibits softening towards T_c on heating and follows the conventional Curie–Weiss law (solid line) over a large temperature region. However, upon further heating, the soft mode becomes over damped in a temperature region extending over ~200 K, which does not allow us to estimate the frequency of the mode. At temperatures above 480 K, the soft mode recovers the under damped oscillation and hardens as the temperature increases. These phenomena are very similar to those revealed by neutron inelastic scattering [31]. A major difference is that the wave number of the soft mode in the present study is significantly lower than that observed by the neutron inelastic scattering. This can be reasonably understood by the splitting of the soft mode due to the lowering of symmetry as demonstrated in the NMR study [21]. According to previous results, the local symmetry in the PNR changes from cubic to rhombohedral. Therefore, the soft mode can be assumed to split from the F_{1u} mode to the A_1 and E modes. Generally, the A_1 mode is higher in wave number than the E mode due to the depolarization field effect.

In a short summary, we may say that the polarization in PMN is induced by the soft mode. This interpretation is essentially consistent with the results described in the polarization measurements, and the results obtained in previous neutron studies [36, 37], in which the crystallographic structure of PNR is attributed to the displacement pattern of the soft mode. The results of the Raman study also support that a ferroelectric state exists in PMN even at the zero-bias field.

5. Ferroelectric domain structures observed by TEM

In order to understand the microstructures of COR and PNR together with the domain structures and its evolution with temperature in the ferroelectric phase of PMN relaxor, we have carried out a detailed TEM observation. The typical results are summarized in Fig. 7. As shown in Fig. 7 (a")-(c"), COR was found to be spherical shape and has size less than 5 nm. It is very stable and remains unchanged within the temperature range of 130 K-675K. In the TEM observation, large amount of CORs were found to distribute in the PMN crystal. In a previous HRTEM study, its volume fraction has been estimated to be ~1/3 of the crystal [18]. CORs are thus considered to be the intense sources of the strong random fields.

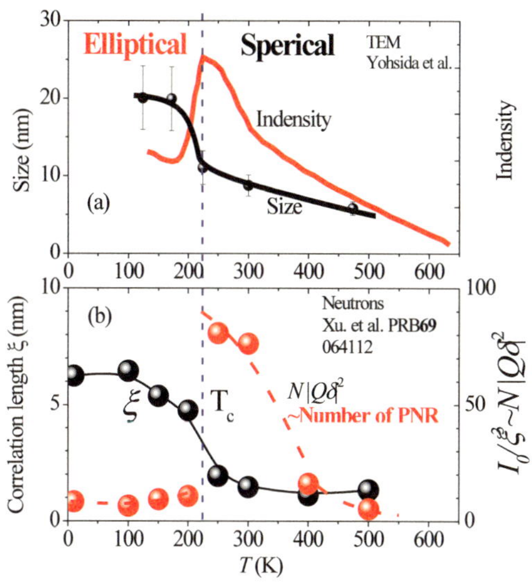

Figure 8. Temperature variation of PNR observed by (a) TEM [38]and (b) Neutron scattering [4]. In (b), ξ is the correlation length, and I_0 is the integrated diffuse scattering intensity and can be written to $N\xi^3|Q\delta|^2$, where N is the total number of PNR, δ is the average displacement of atoms within the PNR, and Q is the wave vector.

In contrast to CORs, PNRs exhibit remarkable change with temperature. As shown in Fig. 7(a')-(c') and Fig. 8, PNRs with size of several nm were found to occur in the crystal for $T<T_{Burns}$. These spherical PNRs show continuous growth as lowering the temperature. However, PNRs change from spherical shape to elliptical shape around T_c. Associating with the change in shape, its intensity was also found to drop rapidly. These results are consistent with those derived from neutron scattering (Fig. 8(b)) [4]. Neutron scattering study by Xu et al. [4] shows that the "correlation length" ξ, which is a direct measure of the length scale of the PNR, increases on cooling and changes remarkably around T_c. At the same time, the number of PNR increases on cooling from high temperatures and then drops dramatically at around T_c, remaining roughly constant below T_c.

Associating with the change in the number and the shape of PNR, micrometric ferroelectric domains were found to occur for $T<T_c$. It is because of growing into macroscopic domain that the number of PNR drop sharply in the ferroelectric phase. Figs. 7(d) and (e) show the structure of a FE micrometric domain in the FE phase of 130 Kand its schematic patterns, respectively. The micrometric domains are formed in the crystal with the spontaneous P_s along the <111>$_c$ direction. In comparison with the domain structure of normal ferroelectric such as $BaTiO_3$, domain size is relatively small and the domain boundaries blur in PMN relaxor.

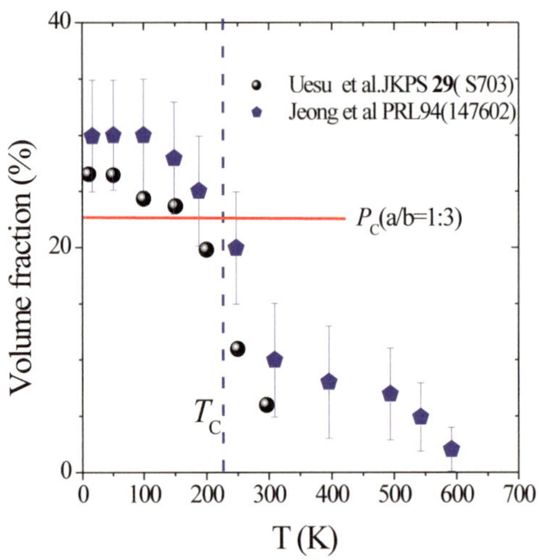

Figure 9. Volume fraction of PNR estimated from neutron scattering measurements [37,39]. Solid line denotes the threshold of percolation for elliptical shape [40].

The occurrence of micrometric domains, the soft-mode observed by Raman scattering, and the macroscopic polarization all direct to the same conclusion: PMN is essentially ferroelectric but not dipole-glass at $T<T_c$ although it exhibits some unique characteristic properties including broadening soft-mode, smearing domain wall, and very large activation field required for domain switching. Our TEM measurements clearly indicate that interactions among PNRs for $T<T_c$ are not random, but cooperative, which is different from the picture expected in the SRBRF model [10]. It is due to such non-random interaction, PNRs team up

http://dx.doi.org/10.5772/52139

together to form micrometric domain in the FE phase. Thus, our TEM observations support the random field model suggested by Fisch [13]. It should be emphasized that PNRs cannot merge together completely due to the blocking of the intense CORs in the crystal (Fig. 10).

Here, we made a discussion on the volume fraction of PNRs in PMN crystal. Neutron scattering technique has been used to estimate the volume fraction of PNRs. Fig. 9 replots two results reported by Jeong et al. [37] and Uesu et al. [39], respectively. Both studies indicate that PNRs occupy a volume fraction > 25% at the lowest measurement temperature. This volume fraction is larger than the threshold of 22% to form a percolated ferroelectric state with an ellipsoidal-shape [40], supporting again the picture of a ferroelectric state in PMN relaxor.

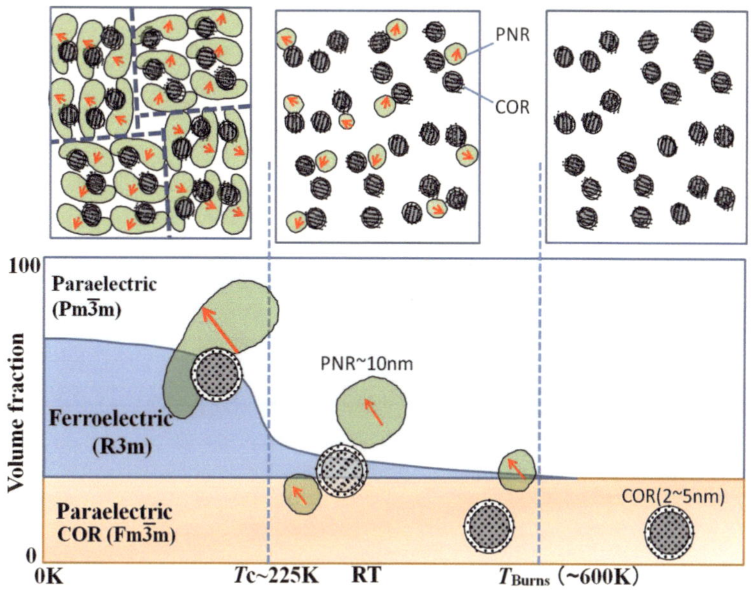

Figure 10. A physics picture of structure evolution in PMN relaxor.

6. A physics picture of relaxor

In summary, we propose a physics picture for relaxors. Figure 10 schematically shows a model of structure evolution in PMN relaxor. Since COR has been observed at $T>T_{Burns}$, it can be considered that there is a coexistence of paraelectric phase of COR with $Fm\bar{3}m$ symmetry and paraelectric $Pm\bar{3}m$phasein this high temperature. Upon cooling, spherical PNRs occur from paraelectric $Pm\bar{3}m$ phase for $T<T_{Burns}$. Both number and size of PNR increase as lowering the temperature. Around room temperature, PNRs grow to a size of about 10 nm. For $T<T_c$ (~225K), neighboring PNRs merge together to form elliptical shape with anisotropy, associating with the reduction of its number. Due to the blocking by the intense CORs, individual PNR merely grows to a size of about 20 nm at $T<T_c$. However, PNRs with elliptical

shape tend to team up together to form a larger domain at $T<T_c$ at which a ferroelectric state occurs. The existence of a multi-scale inhomogeneity of ferroelectric domain structure provides a key point to understand the huge electromechanical coupling effects in relaxors and piezoelectrics with morphotropic phase boundary (MPB). This also gives idea to design new material with domain structure to have large elastic deformation with the application of an electric or magnetic field [41].

Acknowledgements

We thank Mr. M. Yoshida, Prof. N. Yamamoto and Prof. Shin-ya Koshihara of Tokyo Institute of Technology for their contributions in this work. We acknowledge the support of a Grant-in-Aid for Scientific Research, MEXT, Japan.

Author details

Desheng Fu[1*], Hiroki Taniguchi[2], Mitsuru Itoh[2] and Shigeo Mori[3]

*Address all correspondence to: ddsfu@ipc. shizuoka. ac. jp

1 Division of Global Research Leaders, Shizuoka University, Johoku, Naka-ku, Hamamatsu, Japan

2 Materials and Structures Laboratory, Tokyo Institute of Technology, Nagatsuta, Yokohama, Japan

3 Department of Materials Science, Osaka Prefecture University, Sakai, Osaka, Japan

References

[1] Smolenskii, G. A., Agranovskaia, A. I. *Sov. Phys. Tech. Phys.* (1958), 3, 1380.

[2] Cross, L. E. Relaxor ferroelectrics: An overview. *Ferroelectrics* (1994), 151, 305-320.

[3] Burns, G., Dacol, F. H. Glassy polarization behavior in ferroelectric compounds $Pb(Mg_{1/3}Nb_{2/3})O_3$ and $Pb(Zn_{1/3}Nb_{2/3})O_3$. *Solid State Commun.* (1983), 48, 853-856.

[4] Xu, G., Shirane, G., Copley, J. R. D. , Gehring, P. M. Neutron elastic diffuse scattering study of $Pb(Mg_{1/3}Nb_{2/3})O_3$. *Phys. Rev. B* (2004), 69, 064112.

[5] Swainson, I. P., Stock, C., Gehring, P. M., Xu, G., Hirota, K., Qiu, Y., Luo, H., Zhao, X., Li, J.-F., Viehland, D., Soft phonon columns on the edge of the Brillouin zone in the relaxor $PbMg_{1/3}Nb_{2/3}O_3$. *Phys. Rev. B* (2009), 79, 224301.

[6] Vugmeister, B. E. Polarization dynamics and formation of polar nanoregions in relaxor ferroelectrics. *Phys. Rev. B* (2006), 73, 174117.

[7] Grinberg, I., Juhas, P., Davies, P. K., Rappe, A. M. Relationship between local structure and relaxor behavior in perovskite oxides. *Phys. Rev. Lett.* (2007), 99, 267603.

[8] Colla, E.V., Koroleva, E. Yu., Okuneva, N. M., Vakhrushev, S. B. Long-time relaxation of the dielectric response in lead magnoniobate. *Phys. Rev. Lett.* (1995), 74, 1681-1684.

[9] Levstik, A., Kutnjak, Z., Filipič, C., Pirc, R. Glassy freezing in relaxor ferroelectric lead magnesium niobate. *Phys. Rev. B* (1998), 57, 11204-11211.

[10] Pirc, R., Blinc, R. Spherical random-bond–random-field model of relaxor ferroelectrics. *Phys. Rev. B* (1999), 60, 13470-13478.

[11] Bobnar, V., Kutnjak, Z., Pirc, R., Blinc, R., Levstik, A. Crossover from glassy to inhomogeneous-ferroelectric nonlinear dielectric response in relaxor ferroelectrics. *Phys. Rev. Lett.* (2000), 84, 5892-5895.

[12] Westphal, V., Kleemann, W., Glinchuk, M. D. Diffuse phase transitions and random-field-induced domain states of the "relaxor" ferroelectric $PbMg_{1/3}Nb_{2/3}O_3$. *Phys. Rev. Lett.* (1992), 68, 847.

[13] Fisch, R. Random-field models for relaxor ferroelectric behavior. *Phys. Rev. B* 2003; 67, 094110.

[14] Moriya, Y., Kawaji, H., Tojo, T., Atake, T. Specific-heat anomaly caused by ferroelectric nanoregions in $Pb(Mg_{1/3}Nb_{2/3})O_3$ and $Pb(Mg_{1/3}Ta_{2/3})O_3$ relaxors. *Phys. Rev. Lett.* (2003), 90, 205901.

[15] Fu, D., Taniguchi, H., Itoh, M., Koshihara, S., Yamamoto, N., and Mori S. Relaxor $Pb(Mg_{1/3}Nb_{2/3})$ O_3: A ferroelectric with multiple inhomogeneities. *Phys. Rev. Lett.* (2009), 103, 207601.

[16] Mathan, N. de, Husson, E., Calvarn, G., Gavarri, J. R., Hewat, A. W., Morell, A. A structural model for the relaxor $PbMg_{1/3}Nb_{2/3}O_3$ at 5 K. *J. Phys. Condens. Matter* (1991), 3, 8159.

[17] Hilton, A. D., Barber, D. J., Randall, C.A., Shrout, T. R. On short range ordering in the perovskite lead magnesium niobate. *J. Mater. Sci.* (1990), 25, 3461-3466.

[18] Boulesteix, C., Varnier, F., Llebaria, A., Husson, E. Numerical determination of the local ordering of $PbMg_{1/3}Nb_{2/3}O_3$ (PMN) from high resolution electron microscopy images. *J. Solid State Chem.* (1994), 108, 141-147.

[19] Chen, J., Chan, H. M., Harmer, M. P. Ordering structure and dielectric properties of undoped and La/Na-doped $Pb(Mg_{1/3}Nb_{2/3})O_3$. *J. Am. Ceram. Soc.* (1989), 72, 593-598.

[20] Jeong, I. K., Darling, T. W., Lee, J. K., Proffen, Th., Heffner, R. H., Park, J. S., Hong, K. S., Dmowski, W., Egami, T. Direct observation of the formation of polar nanoregions

in $Pb(Mg_{1/3}Nb_{2/3})O_3$ using neutron pair distribution function analysis. *Phys. Rev. Lett.* (2005), 94, 147602.

[21] Blinc, R., Laguta, V. V., Zalar, B. Field cooled and zero field cooled ^{207}Pb NMR and the local structure of relaxor $PbMg_{1/3}Nb_{2/3}O_3$. *Phys. Rev. Lett.* (2003), 91, 247601.

[22] Wieder, H. H. Electrical behavior of barium titanatge single crystals at low temperatures. *Phys. Rev.* (1955), 99, 1161-1165.

[23] Wieder, H. H. Activation field and coercivity of ferroelectric barium titanate. *J. Appl. Phys.* (1957), 28, 367-369.

[24] Wieder, H. H. Ferroelectric polarization reversal in rochelle salt. *Phys. Rev.* (1958), 110, 29-36.

[25] Merz, W. J. Domain formation and domain wall motions in ferroelectric $BaTiO_3$ single crystals. *Phys. Rev.* (1954), 95, 690-698.

[26] Fatuzzo, E., Merz, W. J. Switching mechanism in triglycine sulfate and other ferroelectrics. *Phys. Rev.* 1959; 116, 61-68.

[27] Fatuzzo, E. Theoretical Considerations on the Switching transient in ferroelectrics. *Phys. Rev.* (1962), 127, 1999-2005.

[28] Shin, Y-H., Grinberg, I., Chen, I-W., Rappe, A. M. Nucleation and growth mechanism of ferroelectric domain-wall motion. *Nature* (2007), 449, 881-884.

[29] Glazounov, A. E., Tagantsev, A. K. Phenomenological model of dynamic nonlinear response of relaxor ferroelectrics. *Phys. Rev. Lett.* (2000), 85, 2192-2195.

[30] Dec, J., Miga, S., Kleemann, W., Dkhil, B. Nonlinear dielectric properties of PMN relaxor crystals within Landau-Ginzburg-Devonshire approximation. *Ferroelectrics* (2008), 363, 141-149.

[31] Wakimoto, S., Stock, C., Birgeneau, R. J., Ye, Z.-G., Chen, W., Buyers, W. J. L., Gehring, P. M., Shirane, G. Ferroelectric ordering in the relaxor $Pb(Mg_{1/3}Nb_{2/3})O_3$ as evidenced by low-temperature phonon anomalies. *Phys. Rev. B* (2002), 65, 172105.

[32] Taniguchi, H., Itoh, M., Fu, D. Raman scattering study of the soft mode in $Pb(Mg_{1/3}Nb_{2/3})O_3$. *J. Raman Spectrosc.* (2010), 42, 706-714.

[33] Bussmann-Holder, A., Beige, H., Völkel, G. Precursor effects, broken local symmetry, and coexistence of order-disorder and displacive dynamics in perovskite ferroelectrics. *Phys. Rev. B* (2009), 79, 184111.

[34] Tai, Z. R., Namikawa, K., Sawada, A., Kishimoto, M., Tanaka, M., Lu, P., Nagashima, K., Maruyama, H., Ando, M. Picosecond view of microscopic-scale polarization clusters in paraelectric $BaTiO_3$. *Phys. Rev. Lett.* (2004), 93, 087601.

[35] Gehring, P. M., Hiraka, H., Stock, C., Lee, S.-H., Chen, W., Ye, Z.-G., Vakhrushev, S. B., Chowdhuri, Z. Reassessment of the Burns temperature and its relationship to the

diffuse scattering, lattice dynamics, and thermal expansion in relaxor $Pb(Mg_{1/3}Nb_{2/3})O_3$. *Phys. Rev. B* (2009), 79, 224109.

[36] Hirota, K., Ye, Z. G., Wakimoto, S., Gehring, P. M., Shirane, G. Neutron diffuse scattering from polar nanoregions in the relaxor $Pb(Mg_{1/3}Nb_{2/3})O_3$. *Phys. Rev. B* (2002), 65, 104105.

[37] Jeong, I. K., Darling, T. W., Lee, J. K., Proffen, T., Heffner, R. H., Park, J. S., Hong, K. S., Dmowski, W., Egami, T. Direct observation of the formation of polar nanoregions in $Pb(Mg_{1/3}Nb_{2/3})O_3$ using neutron pair distribution function analysis. *Phys. Rev. Lett.* (2005), 94, 147602.

[38] Yoshida, M. Study of $Pb(Mg_{1/3}Nb_{2/3})O_3$ ferroelectric relaxor by Transmission Electron Microscopy, *Undergraduate thesis*, Tokyo Institute of Technology, Tokyo, (1996).

[39] Uesu, Y., Tazawa, H., Fujishiro, K., Yamada, Y. Neutron scattering and nonlinear-optical studies on the phase transition of ferroelectric relaxor $Pb(Mg_{1/3}Nb_{2/3})O_3$. *J. Korean Phys. Soc.* (1996), 29, S703-S705.

[40] Garboczi, E. I., Snyder, K. A., Douglas, J. F. Geometrical percolation threshold of overlapping ellipsoids. *Phys. Rev. E* (1995), 52, 819-828.

[41] "Ferroelectrics: Nanoregions team together", *NPG Asia Materials*, Feb. 1, 2010, doi: 10.1038/asiamat.2010.17.

Chapter 4

Self Assembled Nanoscale Relaxor Ferroelectrics

Ashok Kumar, Margarita Correa, Nora Ortega,
Salini Kumari and R. S. Katiyar

Additional information is available at the end of the chapter

http://dx.doi.org/10.5772/54298

1. Introduction

Worldwide research on relaxor ferroelectric (RFE) has been carried out since 1950s. There are several schools in the world who have defined the evolution and origin of relaxor properties in the ferroelectric materials in their own way. One common consensus among the scientists is the presence of polar nano regions (PNRs) i.e. self assembled domains of short range ordering typically of less than 20-50 nm in the ferroelectric relaxor materials which causes the dielectric dispersion near the phase transition temperature. Another common approach has been also believed among the relaxor ferroelectric scientists i.e. the existence of random field at nanoscale. Random field model is considered on the experimental and theoretical facts driven from the different dielectric dispersion response under zero field cooled (ZFC) and field cooled (FC) among the relaxors.

Overall relaxor ferroelectrics have been divided into two main categories such as (i) classical relaxors (only short range ordering), the most common example is $PbMg_{1/3}Nb_{2/3}O_3$ (PMN), (ii) Semi-classical relaxor ferroelectrics (a combination of short and long range ordering). In the latter case, the relaxor properties can be arises due to compositional inhomogeneities, artificially induced strain, growth conditions (temperature, pressure, medium, etc.), and due to different ionic radii mismatched based chemical pressure in the matrix. The local domains (PNRs) reorientation induces polar-strain coupling which makes RFE the potentially high piezoelectric coefficient materials widely used in Micro/Nanoelectromechanical system (MEMS/NEMS). The basic features of the RFE over a wide range of temperatures and frequencies are as follows (i) dispersive and diffuse phase transition, (ii) partially disordered structure, (iii) existence of polar nano-ordered regions, etc. over a wide range of temperatures and frequencies. Physical and functional properties of relaxors are very different from normal ferroelectrics due to the presence of self assembled ordered regions in the configurational disorder matrix.

This article deals with the historical development of the relaxor ferroelectrics, microstructural origin of RFE, strain mediated conversion of normal ferroelectric to relaxor ferroelectrics, superlattice relaxors, lead free classical relaxor ferroelectrics, distinction of classical relaxors and semiclassical ferroelectric relaxors based on the polarization study, and their potential applications in microelectronic industry.

2. History

Relaxor ferroelectric materials were discovered in complex perovskites by Smolenskii [1] in early fifties of twentieth centuries. Since then a vast scientific communities have been working on relaxor materials and related phenomena but the plethora of mesoscopic and microscopic heterogeneities over a range of lengths and timescales made difficult to systematic observations [2]. The microscopic image and the dielectric response of relaxor are qualitatively different for that of normal ferroelectrics. The universal signature of RFE is as follows [3 - 18]: (i) the occurrence of broad frequency-dependent peak in the real and imaginary part of the temperature dependent dielectric susceptibility (χ') or permittivity (ε') which shifts to higher temperatures with increasing frequency, G. A. Smolenskii and A. I. Agranokskaya, Soviet Physics Solid State 1, 1429 (1959), (ii) Curie Weiss law is observed at temperatures far above dielectric maxima temperature (T_m), (iii) slim hysteresis loops because polar domains are nanosized and randomly oriented, (iv) existence of polar nano domains far above T_m, (v) no structural phase transition (overall structure remains pseudo-cubic on decreasing temperature but rhombohedral-type distortion occurs in crystal at local level) across T_m in relaxors in contrast with the normal ferroelectric in which phase transition implies a macroscopic symmetry change,(vii) history dependent functional properties and skin effects (surface behavior is quite different than the interior of the system) (vi) in most cases, relaxor crystal showed no optical birefringence (either far below the freezing temperature and/or under external electric field).

As we know ferroelectric and related phenomenon are synonyms of ferromagnetism, antiferromagnetism, spin glass, and random field model which have been originally studied for magnetic materials. Relaxors ferroelectrics possess a random field state, as initially proposed by Westphal, Kleemann and Glinchuk [16-18]. Near Burns temperature, small polar nano regions start originating in different directions inside the crystal at mesoscopic scale; however, these polar nano regions are not static in nature at Burns temperature. At low temperature these polar nanoregions (PNRs) become static and developed a more defined regions called random field, the nature of these fields are short range order. The evidence of the existence of nanosized polar domains at temperatures well above the dynamic transition temperature comes for experimental observations. The presence of these polar nanoregions (PNRs) was established by measuring properties that depend on square of the polarization (P^2) and by direct imaging with TEM studies. The first evidences came from a report by Burns and Dacol [5] who studied the electro-optical effect in a PMN crystal. In case of normal FE, the temperature dependence of the refraction index *(n)* show a linear decrease of *n* from the paraelectric phase down to phase transition temperature (T_c), below T_c *n* deviates from linearity. The deviation is proportional to the P^2 and it increases as the polarization changes with temperature. Burns and Dacol observed deviation from linear *n(T)* in relaxor PMN crystal well above $T_{m,}$ the onset of the deviation was observed at 620 K, almost 350 K above T_m. The temperature in which onset of deviation from linearity of *n(T)* occurs (in any relaxor) had been identified in the literature as the Burns temperature (T_B or T_d).

Electrostriction is a property which depends on P^2. Thermal strain in a cubic perovskite has two components, one due to the linear coefficient of thermal expansion, α, and another due to electrostriction accounted by the electrostrictive coefficients Q_{ijkl}. Measures on the temperature dependence of thermal expansion in RFE crystals have shown that the contribution due to electrostriction vanishes only above T_B where P=0.

The existence of PNRs well above T_m and the growth of these domains with decreasing temperature have been demonstrated by TEM [8 - 12]. Transmission electron studies also account for the B-cation order in complex perovskites [11 - 12]. Ordering of the B-site cations occurs if there is sufficiently large interaction energy between neighboring cations.

3. Theory of relaxor ferroelectricity

The general formula of perovskite having A and B-site cations with different charge states can be written as $A'_xA''_{1-x}B'_yB''_{1-y}O_3$. The randomness occupancy of A′,A″ and B′,B″ in A, B site respectively, depends on the ionic sizes, distribution of cations ordering at sub lattices and charge of cations. If the charges of cations at B-sites are same it is unlikely to have the polar randomness at nanoscale. Long range order (LRO) is defined as a continuous and ordered distribution of the B cations on the nearest neighbor sites. The short coherence length of LRO occurs when the size of the ordered domains are in a range of 20 to 800 Å in diameter. The long coherence range of LRO occur when the size of the ordered domains are much greater than 1000 Å. Randall et al. made a classification of complex lead perovskite and their solid solutions based on B-cation order studied by TEM and respective dielectric, X-rays and optical properties [11]-[12]. This classification divides the complex lead perovskites into three subgroups; random occupation or disordered, nanoscale or short coherent long-range order and long coherent long-range order of B-site cations.

Diffuse phase transition behavior is characteristic of the disordered structures. In these structures, random lattice disorder introduces dipolar impurities and defects that influence the static and dynamic properties of these materials. The presence of the dipolar entities on a lattice site of the highly polarizable FE structure, induced dipoles in a region determined by the correlation length (r_c). The correlation length is a measure of the extent of dipoles that respond in a correlated manner. In normal FE, r_c is larger than the lattice parameter *(a)* and it is strongly temperature dependent. On decreasing the temperature, a faster increase of r_c promotes growing of polar domains yielding a static cooperative long-range ordered FE state at $T < T_c$. This is not the case for RFE where a small correlation length of dipoles leads to formation of polar nanodomains frustrating the establishment of long-range FE state. Therefore, the dipolar nanoregions form a dipolar-glass like or relaxor state at low temperature with some correlation among nanodomains.

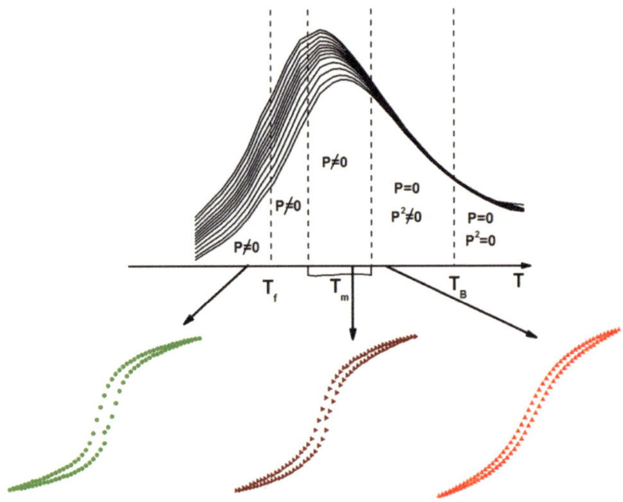

Figure 1. Temperature evolution of dielectric constant showing the characteristic temperatures in RFE. Representative hysteresis loops for each temperature interval are showed below.

These PNRs are dynamic and they experience a slowing down of their fluctuations at $T \leq T_m$. The dielectric relaxation does not fit the classical Debye relaxation model; instead there is a distribution of relaxation times related to the sizes of the nanodomains. The temperature dependence of dielectric constant as shown in Fig. 1 indentifies the main temperatures associated with relaxor ferroelectric behavior. These temperatures are T_B, T_m and T_f, we already defined T_B and T_m but the definition of T_f follow from the fit of the frequency dispersion of T_m with the Vogel-Fulcher law [15]. The dynamics of polar nanoregions does not follow Arrhenius type temperature dependence; instead nice fit of the frequency dispersion for each relaxor system is obtained with the Vogel-Fulcher law:

$$f = f_0 exp\left(\frac{-E_a}{k_B(T_m - T_f)}\right) \tag{1}$$

where f_0 is the attempt frequency which is related to the cut-off frequency of the distribution of relaxation times, E_a is the activation barrier to dipole reorientation, T_m is the dielectric maxima temperature, and T_f is the freezing temperature where polarization fluctuations "frozen-in".

The temperature dependence of the dielectric constant below T_B and in the vicinity of T_m is normally fitted by the empirical power law [19]:

$$\frac{1}{\varepsilon(\omega,\ T)} = \left(\frac{1}{\varepsilon_{max}(\omega,\ T)}\right)\left(1 + \frac{(T - T_{max})^{\gamma}}{2\delta^{2}}\right) \tag{2}$$

where γ and δ are parameters describing the degree of relaxation and diffuseness of the transition respectively. The parameter γ varies between 1 and 2, where values closer to 1 indicate normal ferroelectric behavior whereas values close to 2 indicate good relaxor behavior. Above T_B the inverse of dielectric permittivity is fitted by the Curie-Weiss law:

$$\chi = \frac{C}{T - \Theta} \tag{3}$$

where C and Θ are the Curie constant and Curie temperature respectively.

Last sixty years of extensive research in the field of relaxor ferroelectrics, several models have been proposed to explain the unusual dielectric behavior of these materials. Some of these models are: statistical composition fluctuations [20, 21], superparaelectric model [22], dipolar glass model [23], random field model [24], spherical random field model [25] etc. These models can explain much of the experimental observed facts but a clear understanding of the relaxor nature or even a comprehensive theory is not available yet. Despite the absence of a comprehensive theory of relaxor ferroelectricity, the literature agree to define relaxor materials in terms of the existence of polar nano regions as motioned above, these ordered regions exist in a disordered environment.

4. Strain induced relaxor phenomenon

The influence of epitaxial clamping on ferroelectric properties of thin films to rigid substrates has been largely studied and successfully explained in a Landau-Ginzburg-Devonshire (LGD) framework [26-30]. However, when the same treatment was applied to RFE, discrepancies among the predicted values and those observed in experiments were found [30]. For instance, a downward shift in T_m of PMN films in the presence of compressive in-plane strain was found, contrary to the expected upward shift. Catalan [30] et al. developed a model to analyze the influence of epitaxial strain on RFE. This model is based on LGD theory but assuming a quadratic dependence on (T-T_m), rather than linear, for the first coefficient of the free energy. Catalan's model shows that the shift in T_m for relaxors thin films do not depend on the sign of the epitaxial mismatch strain, but rather on the thermal expansion mismatch between substrate and film. Reference [30] also provided a list of compounds which exhibited shift in T_m when they were grown in film forms. We have already observed in-plane strain in our $Pb(Sc_{0.5}Nb_{0.25}Ta_{0.25})O_3$(PSNT)/$La_{0.5}Sr_{0.5}CoO_3$ (LSCO)/MgO heterostructure, which may have caused the measured shift of T_m with respect to bulk values [31-33].

We have calculated shift in the peak position of the dielectric maxima of PSNT films compared to bulk values following the model developed by Catalan et.al. The model is as follow: In LGD formalism, the thermodynamical potential of a perovskite dielectric thin films is described as [27-28].

$$\Delta G=\left(\frac{1}{2}\alpha - Q_{i3}X_i\right)P_3^2+\frac{1}{4}\beta P_3^2-\frac{1}{2}s_{ij}\left(X_iX_j\right) \tag{4}$$

where α and β correspond to linear and nonlinear terms of the inverse permittivity, and s_{ij}, X_i, and Q_{i3} are the elastic compliances, the stress tensor, and the electrostriction tensor in Voigt notation; P_3 is out of the plane polarization. Strain gradients across the films are neglected. The inverse permittivity is the second derivative of the free energy with respect to the polarization:

$$\chi_3'=\frac{\partial^2 \Delta G}{\partial P^2}=\alpha - 4Q_{13}\frac{Y}{1-v}x_m+3\beta P_3^2 \tag{5}$$

where x_m, Y, and v are the mismatch strain, Young's modulus, and Poisson's ratio of the film respectively. For conventional ferroelectrics, α is usually expanded in a power series of $(T-T_c)$:

$$\alpha=\frac{1}{C\varepsilon_0}\left(T-T_c\right)+f\left(T-T_c\right)^2 \tag{6}$$

which directly leads to Curie-Weiss behavior, where C is the Curie constant. Substituting α into the above equation showing that the strain shifts the critical temperature for a ferroelectric by:

$$\Delta T_c=4C\varepsilon_0 Q_{13}\frac{Y}{1-v}x_m \tag{7}$$

Since Q_{13} is always negative, it has been observed that the shift in the critical temperature depends on the sign of in-plain strain: T_c decreases for the tensile strain and it increases for compressive strain. Equation 7 is not adequate to explain the shifts in the dielectric maximum temperature for the relaxor ferroelectric materials

In case of PSNT films, we observed a quadratic temperature dependence of the permittivity:

$$\frac{1}{\varepsilon(\omega,T)}-\frac{1}{\varepsilon_m(\omega,T)}=\frac{1}{2\varepsilon_m\varepsilon_0}\left(T-T_m\right)^2 \tag{8}$$

Since the inverse of the dielectric constant is the second derivative of free energy with respect to polarization (Equation 5), the measured dielectric constant can be related to the appropriate coefficients in the LGD thermodynamical potential. For P=0 and x_m=0 (in the absence of strain), χ correlates with the first coefficient of the Landau equation, i.e.

$$\chi(T,\ f)=\alpha(T,\ f)=\frac{1}{2\varepsilon_m\varepsilon_0}(T - T_m)^2 \qquad (9)$$

We substituted the values of α in Equation 5 and calculated the dielectric maxima temperature under strain $\left(T_m^{'}\right)$. To calculate $T_m^{'}$, we differentiated the inverse of permittivity with respect to temperature at $T_m^{'}$.

$$\frac{\partial \chi^{'}}{\partial T_m^{'}}=0=\frac{1}{\varepsilon_0\varepsilon_m\delta^2}\left(T_m^{'} - T_m\right) - 4Q_{13}\frac{Y}{1 - \nu}\frac{\partial x_m}{\partial T} \qquad (10)$$

where it is assumed that Q_{13}, Y, and ν do not change substantially [34] with temperature in the dielectric maxima regions; only the mismatch strain changes due to differential thermal expansion:

$$\frac{\partial x_m}{\partial T}=\frac{\partial}{\partial T}\left[\frac{\left(a_{film}\left[1 - \lambda_{substrate}(T - T_0)\right] - a_{bulk}\right)}{a_{bulk}}\right] \qquad (11)$$

where $\lambda_{substrate}$ is the coefficient of thermal expansion of the substrate, a_{film} and a_{bulk} are the lattice constant of the film and the bulk respectively. Using Equations, 9, 10, and 11 the expected shift in T_m is:

$$\Delta T_m=4\delta^2\varepsilon_m\varepsilon_0Q_{13}\frac{Y}{1 - \nu}\frac{a_{film}}{a_{bulk}}\lambda_{substrate} \qquad (12)$$

We have used the above equation to calculate the shift in the position of dielectric maxima of PSNT films. Due to lack of experimental mechanical data in literature for PSNT films or ceramics, we have used the standard Q_{13} for PST thin films [34] and Y value for the PMN single crystal [35] other values we got from our experimental observation i.e. (δ, ε_m, a_{film}, a_{bulk}). For MgO substrate, $\lambda_{substrate} = 1.2\text{x}10^{-5}\ K^{-1}$. Putting all these data in Equation 12 we got a shift in the dielectric maximum temperature (T_m) of the same magnitude order than those of experimental observation.

5. Experimental observation of strain induced relaxor phenomenon

Fig. 2 shows the dielectric behavior and the microstructures of the PSNT thin films and their bulk matrix [31-33]. It has been shown that the two-dimensional (2D) clamping of films by the substrate may change profoundly the physical properties of ferroelectric heterostructure with respect to the bulk material. We observed a transfer and relaxing of epitaxial strain between the layers due to in-plane oriented heterostructure. However the mismatch strain due to different thermal expansion properties of each layer provokes a shift of 62 K in the temperature of the dielectric maxima.

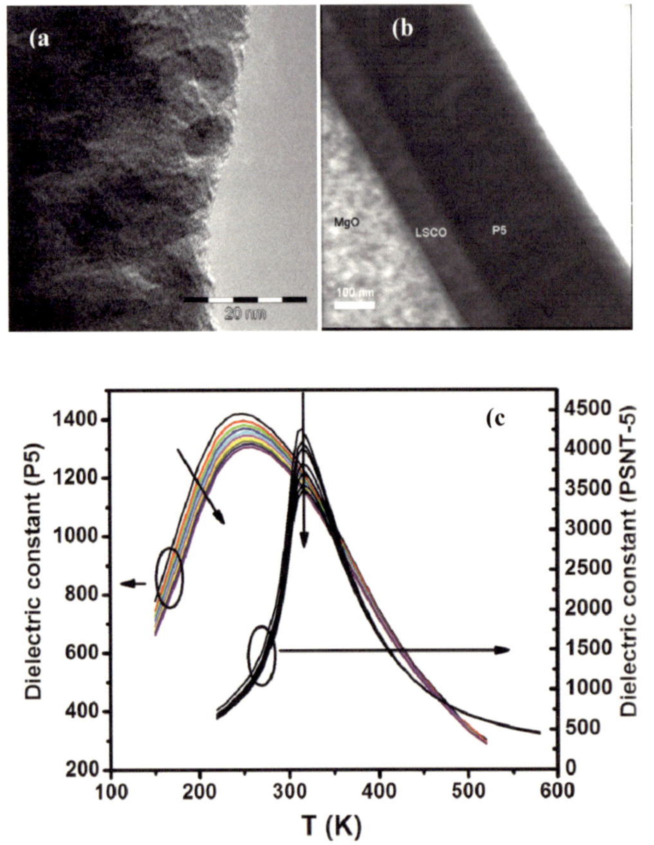

Figure 2. Microstructure of the PSNT bulk nanoceramics (a) and the thin films (b) grown on the LSCO coated MgO substrate, and a drastic shift in dielectric maxima temperature (c) in PSNT thin film with respect same compositions of their bulk counterpart. The microstructures of film and ceramic support the existence of strain induced relaxor phenomenon, as their microstructures- property correlation indicate the presence of some ordered polar nano-regions in the thin films whereas absence of such effect in bulk. Figure 2(a) was adapted from Ref. [31] and reproduced with permission (© 2008 John Wiley and Sons).

Transmission electron microscopy (TEM) images of PSNT in bulk (Fig. 2(a)) and thin film (Fig. 2(b)) form are shown in Fig. 2. The interlayer of about 20 nm between MgO and LSCO and the fibril structures along a single direction in the film confirm the in-plane strain state of the film. TEM image of bulk shows well ordered nanoregions in the grain matrix that meet a basic ingredient to produces relaxor behavior: the existence of nano-ordered regions surrounding by a disordered structure, however, it is unable to produce it. It has been observed from the neutron diffraction data that if the ordered nanoregions are in the range of ~ 5-10 nm, then it is capable of yielding frequency dispersion in the dielectric spectra [30]. But the relaxor state also depends on the coherency with what the dipoles respond to the probing field (and frequency). They can establish long range or short range coherent response. PSNT ceramic was incapable to produce the frequency dispersion behavior, but in thin films form the in plane strain and the breaking of long range ordering response, induced dielectric dispersion and shifting in dielectric maxima temperature towards lower temperature side. We have observed from the temperature evolution of the Raman spectra of bulk PSNT ceramic [31-33], the competitions between ordered domains of short and long-range order due to Nb- and Ta-rich regions respectively. The average disorder arrangement of $Sc^{3+}/Nb^{5+}/Ta^{5+}$ ions in the B-site octahedra leads to the observed diffuse phase transition. In film only short range ordering in B'/B'' ions was developed within the disordered matrix. Although bulk matrix exhibited nano-ordered regions its average size must be higher than film. Stress effects change the ionic positions somehow favoring short-range ordering. These microstructural differences trigger the different observed dielectric responses. The microstructure-property relation of PSNT thin films and ceramics, one can build conclusions that perovskite with similar compositions have well defined relaxor behavior than their bulk matrix.

6. Birelaxors

We have discussed the origin and functional properties of relaxor, in nutshell the basic characteristic of relaxors are the existence of ordered PNRs in a disordered matrix [37-42]. Similarly, relaxor ferromagnets (synonyms: "Mictomagnets") are also known in the literature since the 1970s. A mictomagnet is described as having magnetic clusters (superparamagnetism) which form a spin glass and have a tendency to form short-range ordering [40, 41]. The materials hold both ferroelectric and magnetic orders at nanoscale are called "birelaxors". Birelaxors are suppose to possess only short range order over the entire temperature range with broken symmetry such as "global spatial inversion" and "time reversal symmetry" at nanoscale. Nanoscale broken symmetries and length scale coherency make birelaxors difficult to investigate microscopically; it is advisable to use optical tools for proper probing. Birelaxors do not usually show linear ME coupling ($a_{ij}\ P_i\ M_j$) but may have large nonlinear terms such as bP^2M^2. Since linear coupling is not allowed, local strain-mediated PNR-PNR, PNR-MNR (magnetic nanoregions), and MNR-MNR interactions may provide very strong ME effects [43-44].

A single-phase perovskite $PbZr_{0.53}Ti_{0.47}O_3$ (PZT) and $PbFe_{2/3}W_{1/3}O_3$ (PFW), (40% PZT-60 % PFW) solid solutions thin film grown by pulsed laser deposition system have shown interesting birelaxor properties. Raman spectroscopy, dielectric spectroscopy and temperature-dependent zero- field magnetic susceptibility indicate the presence of both ferroic orders at nanoscale [45].

Dipolar glass and spin glass properties are confirmed from the dielectric and magnetic response of the PZT-PFW system, the micro Raman spectra of this system also revealed the presence of polar nanoregions. Dielectric constant and tangent loss show the dielectric dispersion near the dielectric maxima temperature that confirms near-room-temperature relaxor behavior. Magnetic irreversibility is defined as ($M_{irr}= M_{FC}-M_{ZFC}$) and it represents the degree of spin glass behavior. PZT-PFW demonstrates special features in M_{irr} data at 100 Oe indicate that irreversibility persists above 220 K up to 4th or 5th order of the magnitude, T_{irr} (inset Fig. 3 (b)). The evidence of ZFC and FC splitting at low field ~100 Oe and even for high 1 kOe field suggests and supports glass-like behavior with competition between long-range ordering and short-range order. The behavior of the ZFC cusp is the same for the entire range of field (>100 Oe), with a shift in blocking transition temperature (T_B) (maximum value of magnetization in ZFC cusp) to lower temperature, i.e. from 48 K to 29.8 K with increasing magnetic field. The ZFC cusps disappeared in the FC process, suggesting that competing forces were stabilized by the field- cooled process. Magnetic irreversibilities were found over a wide range of temperature with a sharp cusp in the ZFC data, suggesting the presence of MNRs with spin-glass-like behavior. The detailed fabrication process, the complete characterization process and the functional properties are reported in reference [43]. This is not only the unique system to have birelaxor properties, however, Levstik et al also observed the birelaxor properties in $0.8Pb(Fe_{1/2}Nb_{1/2})O_3$–$0.2Pb(Mg_{1/2}W_{1/2})O_3$ thin films [38].

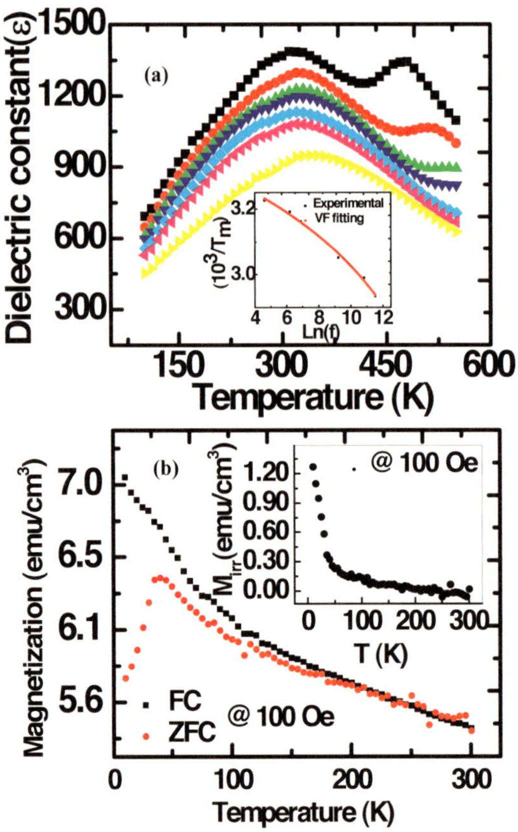

Figure 3. (a) Dielectric response and Vogel-Fulcher fitting (inset), (b) Zero field cooled and field cool magnetic response of PZT-PFW thin films indicate the presence of PNRs and MNRs at nanoscale. Adapted from Ref. [45] and reproduced with permission (© 2011 AIP).

7. Superlattice relaxors and high energy density capacitors

Material scientists are looking for a system or some novel materials that possess high power density and high energy density or both. Relaxors demonstrate high dielectric constant, low loss, non linear polarization under external electric field, high bipolar density, nano dipoles (polar nano regions (PNRs) and moderate dielectric saturation, these properties support their potential candidature for the high power as well as high energy devices. At present high-k dielectric (dielectric constant less than 100, with linear dielectric) dominates in the high energy, high power density capacitor market. The high-k dielectrics show very high electric breakdown strength (> 3 MV/cm to 12 MV/cm) but their dielectric constant is relatively very low which in turns offer 1-2 J/cm^3 volume energy density. Recently, polymer ferroelectric, antiferroelectric, and relaxor ferroelectric have shown better potential and high energy density compare to the existing linear dielectrics [46-51].

$BaTiO_3/Ba_{0.30}Sr_{0.70}TiO_3$ (BT/BST) superlattices (SLs) with a constant modulation period of Λ= 80 Å were grown on (001) MgO substrate by pulsed laser deposition. The modulation periodicity Λ/2 was precisely maintained by controlling the number of laser shots; the total thickness of each SL film was ~6,000 Å = 0.6 μm. An excimer laser (KrF, 248 nm) with a laser energy density of 1.5 J/cm^2, pulse repetition rate of 10 Hz, substrate temperature 830 °C, and oxygen pressure at 200 mTorr was used for SL growth. The detailed growth and characterization techniques are presented elsewhere [52].

Dielectric responses of BT/BST SLs display the similar response as for normal relaxor ferroelectric which can be seen in Figure 4. It follows the non linear Vogel- Fulcher relationship (inset Figure. 4 (a)), frequency dispersion near and below the dielectric maxima temperature (T_m), merger of frequency far above the T_m, shift in T_m towards higher temperature side (about 50-60 K) with increase in frequencies, low dielectric loss, and about 60-70% dielectric saturation.

BT/BST SLs demonstrate a "in-built" field in as grown samples at low probe frequency (<1 kHz), whereas it becomes more symmetric and centered with increase in probe frequency system (>1 kHz) that ruled out the effect of any space charge and interfacial polarization. Energy density were calculated for the polarization-electric field (P-E) loops provide ~ 12.24 J/cm^3 energy density within the experimental limit, but extrapolation of this data in the energy density as function of applied field graph for different frequencies (see inset in Figure 4) suggests huge potential in the system such as it can hold and release more than 40 J/cm^3 energy density.

Experimental limitations restrict to proof the extrapolated data, however the current density versus applied electric field indicates exceptionally high breakdown field (5.8 to 6.0 MV/cm) and low current density (~ 10-25 mA/cm^2) near the breakdown voltage. Both direct and indirect measurements of the energy density indicate that it has ability to store very high energy density storage capacity (~ 46 J/cm^3).

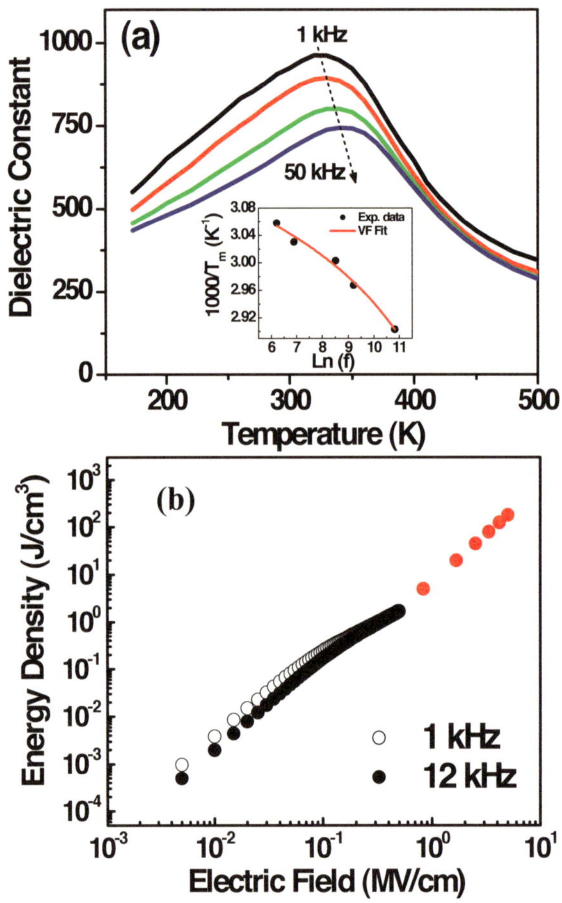

Figure 4. (a) Dielectric responses of BT/BST superlattice relaxors as function of temperature over wide range of frequencies, nonlinear Vogel-Fulcher relationship (inset), (b) Energy density capacity as function of applied electric filed, red dots show the extrapolated data. Adapted from Ref. [46] and reproduced with permission (© 2012 IOP).

The above experimental facts suggest the relaxor nature of BT/BST ferroelectric superlattices. It also indicates their potential to store and fast release of energy density which is comparable to that of high-k (<100) dielectrics, hopefully in the coming years it might be suitable for high power energy applications. These functional properties of relaxor superlattices make it plausible high energy density dielectrics capable of both high power and energy density applications.

http://dx.doi.org/10.5772/54298

8. Conclusions

Extensive studies on the relaxor ferroelectrics suggest the presence of localized nano size ordered regions in the disordered matrix or static random field are responsible for the dielectric dispersion near T_m. The nano regions and their coherence length are also critical for the relaxor behavior. A normal ferroelectric with diffuse phase transition system can be turn to relaxor ferroeletrics in their thin film forms under the suitable applications of strain, utilizing highly lattice mismatch substrate, growth conditions (thermal, oxygen partial pressure, atmosphere), etc.. It also indicates that the defects, oxygen vacancies, ordering of cations at A and B site of perovskite, tensile or comprehensive strain across the interface, etc. originate the ordered nano regions in films mainly responsible for dielectric dispersion. Materials with same compositions can have normal ferroelectric in bulk form whereas become relaxor in the thin film. Birelaxors hold both ferroics orders at nano scale with only short range ordering (SRO). These nanocale SRO make it potential candidates for non linear biquadratic magneto-electric coupling suitable for magnetic field sensors and non volatile memory applications. BT/BST superlattice has shown very high dielectric constant, high breakdown field, relaxor, and high energy density functional properties. A Relaxor superlattice with high functional properties is capable for both high power and energy density applications.

Acknowledgement

This work was partially supported by NSF-EFRI-RPI-1038272 and DOE-DE-FG02-08ER46526 grants.

Author details

Ashok Kumar[1,2], Margarita Correa[2], Nora Ortega[2], Salini Kumari[2] and R. S. Katiyar[2]

*Address all correspondence to: ashok553@gmail.com

1 National Physical Laboratory, Council of Scientific and Industrial Research (CSIR), New Delhi, India

2 Department of Physics and Institute for Functional Nanomaterials, University of Puerto Rico, San Juan, Puerto Rico

References

[1] G. A. Smolenskii and A. I. Agranokskaya, Soviet Physics Solid State 1, 1429 (1959)

[2] R. E. Cohen, Nature, 441, 941 (2006)

[3] Eric Cross, Ferroelectrics, 151, 305, (1994)

[4] G. A. Samara, J. Phys.: Condens. Matter 15, R367, (2003)

[5] G. Burns and F. H. Dacol, Solid State Communications, 48, 853, (1983)

[6] G. A. Samara, Solid State Physics, vol. 56, Edited by H. Ehrenreich, and R. Spaepen, New York: Academic, New York (2001)

[7] B. Krause, J. M. Cowley, and J. Wheatly, Acta Cryst. A35, 1015 (1979)

[8] C.A. Randall, D. J. Barber, and R. W. Whatmore, J. Micro. 145, 235 (1987)

[9] J. Chen, H. Chen, and M. A. Harmer, J. Am. Ceram. Soc. 72, 593 (1989)

[10] I. M. Brunskill, H. schmid, and P. Tissot, Ferroelectrics 37, 547 (1984)

[11] C. A. Randall, A. S. Bhalla, T. R. Shrout, and L. E. Cross, J. Mater. Res., 5, 829 (1990)

[12] C. A. Randall and A. S. Bhalla, Japanese Journal of Applied Physics, 29, 327 (1990)

[13] Ferroelectrics and Related Materials, G. A. Smolenskii (Ed.), V. A. Bokov, V. A. Isupov, N. N. Krainik, R. E. Pasynkov and A. I. Sokolov, Gordon & Breach Science Publisher, New York (1984)

[14] Z.-G Ye, Key Engineering Materials, 155-156, 81 (1998)

[15] H. Vogel. Z. Phys. 22, 645 (1921). G. Fulcher. J. Am. Ceram. Soc. 8, 339 (1925)

[16] V.Westphal,W. Kleeman, and M.D. Glinchuk, Phys. Rev. Lett. 68, 847 (1992).

[17] R. Pirc and R. Blinc, Phys. Rev. B. 60, 13470 (1999).

[18] R. Blinc, J. Dolinsek, A. Gregorovic, B. Zalar, C. Filipic, Z. Kutnjak, A. Levstik, and R. Pirc, Phys. Rev. Lett. 83, 424 (1999).

[19] H.T. Martirena and J.C. Burfoot. Ferroelectrics, 7, 151 (1974)

[20] G. A. Smolenskii, V. A. Isupov, A. I. Agranovskaya and S. N. Popov, Soviet Phys.-Solid State. 2, 2584 (1961).

[21] G. A. Smolenskii, J. Phys. Soc. Jpn. 28, Suppl. 26 (1970).

[22] L.E. Cross, Ferroelectrics 76, 241 (1987).

[23] D. Viehland, J. F. Li, S. J. Jang, and L. E. Cross and M. Wuttig. Phys. Rev. B. 43, 8316 (1991).

[24] R. Fisch. Phys. Rev. B. 67, 094110 (2003).

[25] R. Pirc and R. Blinc, Phys. Rev. B. 60, 13470 (1999)

[26] F. Devonshire, Philos. Mag. 40, 1040, (1949); 42, 1065, (1951)

[27] G.A. Rossetti Jr., L.E. Cross, and K. Kushida, Appl. Phys. Lett., 59, 2504 (1991)

[28] N.A. Perstev, A.G. Zembilgotov, and A. K. Tagantsev, Phys. Rev. Lett., 80, 1988 (1998)

[29] M. Dawber, K.M. Rabe, and J.F. Scott, Reviews of Modern Physics, 77, 1083, (2005)

[30] G. Catalan, M. H. Corbett, R. M. Bowman, and J. M. Gregg, J. App. Phys., 91, [4] 2295 (2002).

[31] Margarita Correa, A. Kumar and R.S. Katiyar, J. Am. Cer. Soc. 91 (6) 1788 (2008).

[32] Margarita Correa, Ashok Kumar, and R. S. Katiyar, Applied Physic Letters, 91, 082905 (2007).

[33] Margarita Correa, Ashok Kumar, and R. S. Katiyar, Integrated Ferroelectrics, 100, 297 (2008)

[34] K. Brinkman, A. Tagantsev, P. Muralt, and N. Setter, Jpn. J. App. Phys., 45, [9B] 7288 (2006)

[35] Q.M. Zhang and J. Zhao, Appl. Phys. Lett., 71, 1649 (1997)

[36] D. L. Orauttapong, J. Toulouse, J. L. Robertson, and Z. G. Ye, Physical Rev. B , 64, 212101 (2001).

[37] A. Kumar, N. M. Murari, and R. S. Katiyar, Appl. Phys. Lett. 90, 162903 (2007).

[38] A. Levstik, V. Bobnar, C. Filipič, J. Holc, M. Kosec, R. Blinc, Z. Trontelj, and Z. Jagličič, Appl. Phys. Lett. 91, 012905, (2007).

[39] V. V. Shvartsman, S. Bedanta, P. Borisov, W. Kleemann, A. Tkach, and P. M. Vilarinho, Phys. Rev. Lett. 101, 165704 (2008).

[40] G. A. Samara, in Solid State Physics, edited by H. Ehrenreich and R. Spaepen Academic, New York, 2001, Vol. 56, p. 239.

[41] J. Dho, W. S. Kim, and N. H. Hur, Phys. Rev. Lett. 89, 027202 (2002).

[42] R. Pirc and R. Blinc, Phys. Rev. B 76, 020101R (2007).

[43] A. Kumar, G. L. Sharma, R. S. Katiyar, R. Pirc, R. Blinc, and J. F. Scott, J. Phys.: Condens. Matter 21, 382204 (2009).

[44] R. Pirc, R. Blinc, and J. F. Scott, Phys. Rev. B 79, 214114 (2009).

[45] A. Kumar, G. L. Sharma, R. S. Katiyar, Appl. Phys. Lett. 99, 042907 (2011).

[46] J. H. Tortai, N. Bonifaci, A. Denat, J. Appl. Phys. 97, 053304 (2005).

[47] M. Rabuffi, G. Picci, IEEE Trans. Plasma Sci. 30, 1939 (2002).

[48] X. Hao, J. Zhai, and, X. Yao, J. Am. Ceram. Soc., 92, 1133 (2009).

[49] J. Parui, S. B. Krupannidhi, Appl. Phys. Lett., 92, 192901 (2008).

[50] R. M. Wallace, G. Wilk, MRS Bulletin. 27, 186 (2002).

[51] G. D. Wilk, R. M. Wallace, J. M. Anthony, J. Appl. Phys. 89, 5243 (2001).

[52] N. Ortega, A Kumar, J. F. Scott, Douglas B. Chrisey, M. Tomazawa, Shalini Kumari, D. G. B. Diestra and R. S. Katiyar. J. Phys.: Condens. Matter. 24 445901 (2012).

Chapter 5

Relaxor Behaviour in Ferroelectric Ceramics

A. Peláiz-Barranco, F. Calderón-Piñar,
O. García-Zaldívar and Y. González-Abreu

Additional information is available at the end of the chapter

http://dx.doi.org/10.5772/52149

1. Introduction

Ferroelectric materials are commonly characterized by high dielectric permittivity values [1]. Usually, for the well-known 'normal' ferroelectrics the temperature of the maximum real dielectric permittivity (T_m) corresponds to the ferroelectric-paraelectric (FE-PE) phase transition temperature (T_C) [2]. On the other hand, there are some kinds of ferroelectrics, so-called relaxor ferroelectrics, which have received special attention in the last years because of the observed intriguing and extraordinary dielectric properties [3-25], which remain not clearly understood nowadays. For instance, some remarkable characteristics of the dielectric response of relaxor materials can be summarized as follows: i- they are characterized by wide peaks in the temperature dependence of the dielectric permittivity, ii- the temperature of the corresponding maximum for the real (ε') and imaginary (ε'') component of the dielectric permittivity (T_m and $T_{\varepsilon''max}$, respectively) appears at different values, showing a frequency dependent behaviour, and iii- the Curie-Weiss law is not fulfilled for temperatures around T_m. So that, the temperature of the maximum real dielectric permittivity, which depends on the measurement frequency, cannot be associated with a FE-PE phase transition.

Lead zirconatetitanate (PZT) system is a typical ferroelectric perovskite showing 'normal' FE-PE phase transition [1]. Nevertheless, the partial substitution by different elements, such as lanthanum, contributes to enhance such relaxor characteristics [11-15]. In fact, for some lanthanum concentrations, the distortion of the crystalline lattice in the PZT system due to ions displacement could promote the formation of the so-called polar nanoregions (PNRs). Another interesting relaxor ferroelectric perovskite is known as PZN-PT-BT [16-19]. The $Pb(Zn_{1/3}Nb_{2/3})O_3$ ferroelectric material (PZN) belongs to the relaxor ferroelectrics family, receiving special attention for its technical importance [20]. However, its preparation usually needs the addition of $BaTiO_3$ (BT) and $PbTiO_3$ (PT) to obtain pure phases [16,21].

On the other hand, a considerable number of compositions from the Aurivillius family exhibit a relaxor ferroelectric behaviour [22-26]. The Aurivillius compounds are layered bismuth $[Bi_2O_2]^{2+}[A_{n-1}B_nO_{3n+1}]^{2-}$, where the sites *A* and *B* can be occupied for different elements. These are formed by the regular stacking of Bi_2O_2 slabs and perovskite-like blocks $A_{n-1}B_nO_{3n+1}$. These materials have received great attention due to their large remanent polarization, lead-free nature, relatively low processing temperatures, high Curie temperatures and excellent piezoelectric properties, which made them good candidates for high-temperature piezoelectric applications and memory storage [27]. The origin of the relaxor behaviour for these materials have been associated to a positional disorder of cations on *A* or *B* sites of the perovskite blocks that delay the evolution of long-rage polar ordering [28].

Several models have been proposed to explain the dielectric behaviour of relaxor ferroelectrics. The basic ideas have been related to the dynamics and formation of the polar nanoregions (PNRs). In this way, Smolenskii proposed the existence of compositional fluctuations on the nanometer scale taking into account a statistical distribution for the phase transition temperature [29]. On the other hand, Cross extended the Smolenskii's theory to a superparaelectric model associating the relaxor behaviour to a thermally activated ensemble of superparaelectric clusters [30]. Viehland et al. have showed that cooperative interactions among these superparaelectric clusters could produce a glass-like freezing behaviour, commonly exhibited in spin-glass systems [31]. Later, Qian and Bursill [32] analyzed the possible influence of random electric fields on the formation and dynamics of the polar clusters, which can be originated from nano-scaled chemical defects. They have also proposed that the relaxor behaviour can be associated to a dipolar moment in an anisotropic double-well potential, taking into account only two characteristic relaxation times. According to this model, the dispersive behaviour is produced by changes in the clusters size and the correlation length (defined as the distance, above which such PNRs become non-interactive regions) as a function of the temperature, which provides a distribution function for the activation energy. However, despite their very attractive physical properties, the identification of the nature of the dielectric response in relaxors systems still remains open and requires additional theoretical and experimental information, which can be very interesting to contribute to the explanation of the origin of the observed anomalies.

The present chapter shows the studies carried out on several relaxor ferroelectric ceramic materials, which have been developed by the present authors. Lanthanum modified lead zirconatetitanate ceramics will be evaluated considering different La^{3+} concentrations and Zr/Ti ratios. It will be discussed the dynamical behaviour of the PNRs taking into account a relaxation model, which considers a distribution function for the relaxation times. The influence of *A* or *B* vacancies on the relaxor behaviour will be also analyzed considering the decoupling effects of these defects in the Pb-O-Ti/Zr bounding. For the PZN-PT-BT system, the relaxor behaviour will be explained considering the presence of local compositional fluctuation on a macroscopic scale. Finally, the relaxor behaviour of $Sr_{0.50}Ba_{0.50}Bi_2Nb_2O_9$ ferroelectric ceramics, which belong to the Aurivillius family, will be discussed considering a positional disorder of cations on *A*or *B* sites of the perovskite blocks.

http://dx.doi.org/10.5772/52149

2. Lanthanum modified Lead Zirconate Titanate (PLZT)

2.1. Relaxor behaviour and coexistence of AFE and FE phases

Several researches concerning PZT properties have shown that very good properties can be obtained by suitable selection of the Zr/Ti ratio and the substitution of a small amount of isovalent or heterovalent elements for the Pb or (Zr,Ti) sublattices [1]. Especial attention receives the Zr-rich PZT-type ceramics, which exhibit interesting dielectric and pyroelectric characteristics, suitable for pyroelectric detectors, energy converters, imaging systems, etc [1,33-34]. They show orthorhombic antiferroelectric (AFE-O), low temperature rhombohedral ferroelectric (FR-LT), high temperature rhombohedral ferroelectric (FR-HT) and cubic paraelectric (PC) phase sequence [1].

One of the most common additives to the PZT ceramics is the lanthanum cation, which substitute the Pb-ions in the A-site of the perovskite structure [1]. The lanthanum-modified lead zirconatetitanate $Pb_{1-x}La_x(Zr_{1-y}Ti_y)_{1-x/4}O_3$ (PLZT) ferroelectric ceramics show excellent properties to be considered for practical applications, especially PLZT x/65/35, x/70/30 and x/80/20 compositions due to the relaxor ferroelectric behaviour [35-36]. For Zr-rich PZT ceramics, which are close to the antiferroelectric-ferroelectric (AFE-FE) phase boundary, the FE phase is only marginally stable over the AFE [33]. In this way, as the FE state is disrupted by lanthanum modification, the AFE state is stabilized [33].

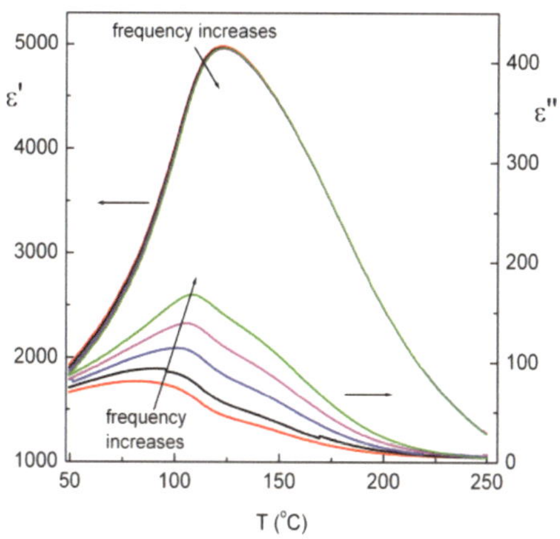

Figure 1. Temperature dependence of the real (ε') and imaginary (ε'') parts of the dielectric permittivity at several frequencies for PLZT 6/80/20 ferroelectric ceramic system.

Figure 1 shows the temperature dependence of the real (ε') and imaginary (ε'') parts of the dielectric permittivity, at several frequencies, for PLZT 6/80/20 ferroelectric ceramic system. Typical characteristics of relaxor ferroelectric behaviour are observed. The frequency dependence of ε' is very weak but the frequency dependence of ε'' clearly reflects the relaxor nature of the system. The asymmetrical shape of the curves could suggest the existence of more than one contribution. X-ray diffraction analysis for the PLZT 6/80/20 ceramic samples have previously shown a mixture of FE and AFE phases at room temperature [13]. Polarized light microscopy studies were performed on the ceramic system in a wide temperature range. Aggregates of individual regions were observed, which were associated to FE and AFE phases. As the temperature increased, a gradual change was observed, suggesting a transition to a single phase [13].

Figure 2 shows the hysteresis loops obtained for the PLZT 6/80/20 ceramics for two temperatures, as example of the obtained behaviour in a wide temperature range. Around 70ºC a double-loop-like behaviour was observed suggesting an induced FE-AFE transformation.The FE state is disrupted by the lanthanum modification and the AFE state is stabilized, which may be a consequence of the short-range nature of the interaction between the AFE sublattices. The double-loop remains for higher temperature and around 120ºC the loop disappears suggesting a transition to a PE phase. Then, it is possible to consider the coexistence of two phase transitions: a FE-AFE phase transition observed around 70ºC and an AFE-PE phase transition around 120ºC. Note, that the maximum of ε' is observed around 120ºC. Negative values for T_C were found, which has confirmed that around 120ºC an AFE-PE transition take places [13]. Thus, the relaxor ferroelectric behaviour for this system could be associated to the coexistence of FE and AFE phases in the studied material.

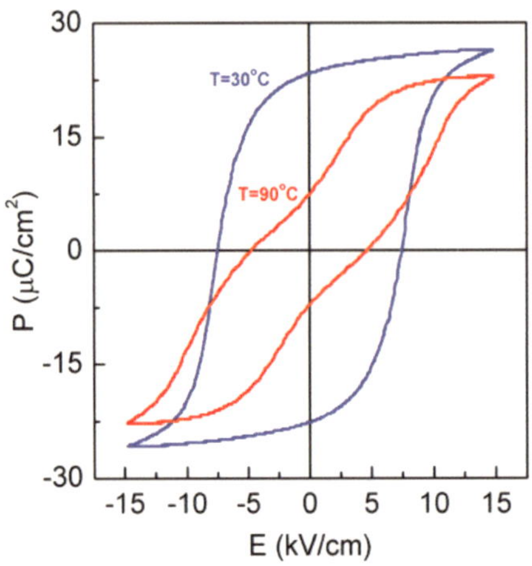

Figure 2. Hysteresis loops for PLZT 6/80/20 ferroelectric ceramic system.

2.2. A relaxation model

The nature of the relaxor behaviour is determined by the existence of PNRs, which possess different relaxation times [32,37]. The relaxation time (τ) represents the time response of such PNRs or polarization mechanisms to change with the applied electric field. However, this process does not occur instantaneously. Indeed, there exists certain inertia, which is the cause of the pronounced dielectric relaxation in relaxor ferroelectrics. The polar nanoregions appear below a certain temperature, the so-called the Burns' temperature (T_B) [38], which is typically hundreds degrees above the temperature of the maximum real dielectric permittivity (T_m). On cooling, the number, size (some of them) and the interaction of the PNRs increase [39]. The increase of such interactions promotes the freezing of some regions around certain temperatures below T_m, known as the freezing temperature (T_F). Two fundamental polarization mechanisms have been reported, which have been associated with the dynamics of the PNRs; (i) dipole reorientation [37,40] and (ii) domain wall vibrations [32]. Both mechanisms have a characteristic time response, which depends on the temperature and size of the PNRs. If the contributions of an ensemble of these regions are considered, the macroscopic polarization function, at fixed temperature, can be expressed by [41]:

$$P(t) = P_{01}e^{-\frac{t}{\tau_1}} + P_{02}e^{-\frac{t}{\tau_2}} + P_{03}e^{-\frac{t}{\tau_3}} + \ldots\ldots = \sum_i P_{0i}e^{-\frac{t}{\tau_i}} \tag{1}$$

where $\tau_i$is the relaxation time of the ith PNR and P_{0i} takes the form of equation (2), in analogy with the Debye single relaxation time model, which takes into account a distribution function of relaxation times [41]:

$$P_{0i} = \frac{\varepsilon_s - \varepsilon_\infty}{\tau_i} g(\tau_i) \tag{2}$$

$$g(\tau_i) = \frac{2\sigma}{\pi} \frac{1}{4(\ln\frac{\tau_i}{\tau_0})^2 + \sigma^2} \tag{3}$$

In these relations $g(\tau_i)$ is a distribution function for the logarithms of the relaxation times, which has been assumed to be a Lorentz rather than a Gaussian distribution function; τ_0 and σ are the mean relaxation time and the Standard Deviation, respectively. According to the Debye model, the frequency dependence of the complex dielectric permittivity can be expressed as:

$$\varepsilon^*(\omega) = \varepsilon' - i\varepsilon'' = \varepsilon_\infty + \int P(t)e^{-i\omega t}dt \tag{4}$$

where ε_s and ε_∞ are the low (static) and high (optical) dielectric permittivity, respectively, ω the measurement frequency, and P(t) the decay polarization function. Substituting equations (1), (2) and (3) into equation (4), the real and imaginary component of the dielectric permittivity for the multi-relaxation times approximation can be obtained and expressed as in equations (5) and (6). It is important to point out that equations (5) and (6) have been derived from the discrete expression (1), taking into account the values of the relaxation times (τ_i) close to each other [41].

$$\varepsilon'(\omega,T) = \varepsilon_\infty + (\varepsilon_s - \varepsilon_\infty)\frac{2\sigma}{\pi}\int_{-\infty}^{+\infty}\frac{1}{(4z^2+\sigma^2)(1+\omega^2\tau_0^2\exp(2z))}dz \quad (5)$$

$$\varepsilon''(\omega,T) = (\varepsilon_s - \varepsilon_\infty)\frac{2\sigma}{\pi}\int_{-\infty}^{+\infty}\frac{\omega\tau_0\exp(z)}{(4z^2+\sigma^2)(1+\omega^2\tau_0^2\exp(2z)}dz \quad (6)$$

It has been considered in equations (5) and (6) that $z = \ln\tau/\tau_0$. By using the experimental results of $\varepsilon'(\omega,T)$ and $\varepsilon''(\omega,T)$ for two frequencies, the temperature dependence of dielectric parameters, such as, τ_0, ε_s and σ can be obtained as a solution of the equations system. After that, by using the theoretical results of $\tau_0(T)$, $\varepsilon_s(T)$ and $\sigma(T)$, the theoretical dependences of $\varepsilon'(\omega,T)$ and $\varepsilon''(\omega,T)$ can be obtained for the studied frequency and temperature ranges. The parameter ε_∞ has been considered negligible because of the high values of the dielectric permittivity in ferroelectric systems [1].

Figures 3 and 4 show the temperature dependence of the real (ε') and imaginary (ε'') components of the dielectric permittivity (symbols), at several frequencies, for the studied PLZT 10/80/20 composition. A relaxor characteristic behaviour can be observed. The maximum real dielectric permittivity decreases, while its corresponding temperature (T_m) increases, with the increase of the measurement frequency. There is not any peak for ε'' due to the low temperature range where the maximum of the real part of the dielectric permittivity appears, i.e. the maximum for ε'' should appear below the room temperature.

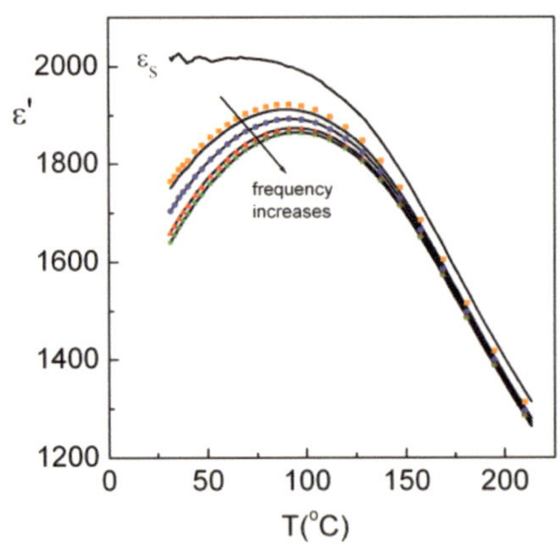

Figure 3. Temperature dependence of the real part (ε') of the dielectric permittivity at several frequencies for PLZT 10/80/20 ferroelectric ceramic system. The experimental values are represented by solid points and the theoretical results by solid lines. It has been included the theoretical temperature dependence of the static dielectric permittivity (ε_S).

http://dx.doi.org/10.5772/52149

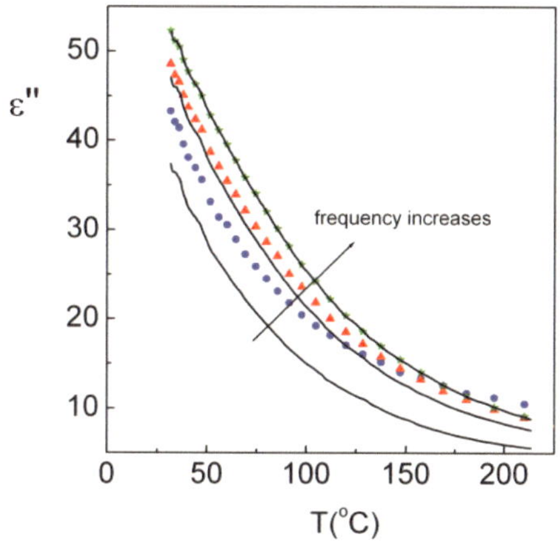

Figure 4. Temperature dependence of the imaginary part (ε'') of the dielectric permittivity at several frequencies for PLZT 10/80/20 ferroelectric ceramic system. The experimental values are represented by solid points and the theoretical results by solid lines.

The number of PNRs able to follow the applied external electric field switching decreases with the increase of the frequency, so that only such inertial-less regions contribute to the dielectric permittivity (those regions whose activation energy is close to thermal energy k_BT, being $k_B$the Boltzmann constant). Therefore, due to the cooperative nature, the real component of the dielectric permittivity should decrease with the increase of the measurement frequency.

The maximum for ε' in relaxor systems is not related to a crystallographic transition. Indeed, such a maximum corresponds to rapid changes of the fraction of the frozen Polar Regions [39]. Hence, the shift up to higher temperatures of the maximum for ε' with the increase of the frequency is a direct consequence of the delay in dielectric response of the frozen regions.

By using equations (5) and (6) and the experimental data of Figures 3 and 4, the temperature dependences of ε_s, σ and τ_0 were obtained. The results were obtained by numerical methods because there was no analytical solution for the equation system. The theoretical curves for ε' and ε'' were obtained for all the studied frequency range, and shown in the same Figures 3 and 4 as solid lines, for the studied system. The temperature dependence of the static dielectric permittivity (ε_s) is also shown in Figure 3. A good agreement between experimental and theoretical results can be observed. It is important to point out that the deviation between the experimental and theoretical results, observed at low frequencies for the temperature dependence of ε'', can be associated with the contribution of the electric conductivity to the dielectric response, which has not been considered in the proposed model.

Figure 5 shows the temperature dependence of $\ln\tau_0$ and σ for the studied composition. As can be seen, $\ln\tau_0$ increases with the increase of the temperature, passes through a maximum and then decreases for higher temperatures. Similar results have been previously reported by Lin et al [41]. According to the model, the logarithmic mean relaxation time ($\ln\tau_0$) should have has approximately the same tendency as that of σ(T). Both parameters increases when the temperature decreases which can be understood as a consequence of the increased inertia of the dipolar clusters as they become more correlated with each other upon cooling. This observation also could reflect a dynamic change in the well potential for the shifting ions. However, the reduction in $\ln\tau_0(T)$ at lower temperatures possibly could reflect a freezing phenomenon of some clusters that are saturated in correlation, which leads to the frustration of cooperative interactions and, hence, leaves only relatively unstrained or smaller clusters available to couple with the electric field. Also, one could speculate that such clusters are the regions within the domain boundaries between the frozen regions.

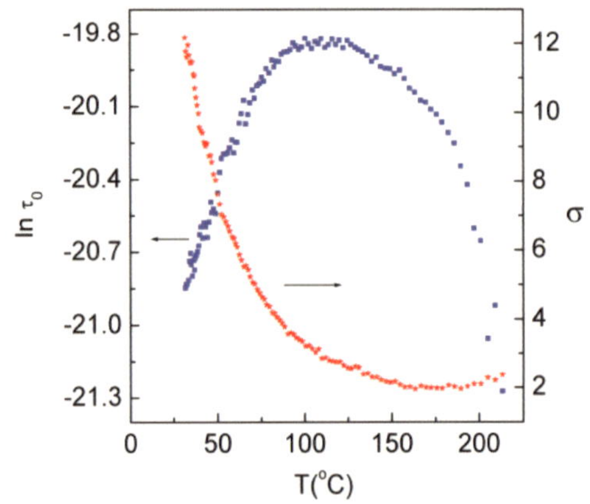

Figure 5. Temperature dependence of the main relaxation time (logarithmic representation, $\ln\tau_0$) and its standard deviation (σ), for PLZT 10/80/20 ferroelectric ceramic system.

In this way, the size and the interaction between PNRs increase on cooling from the high temperature region and its contribution to the dielectric permittivity becomes negligible below the freezing temperature (T_F) [39]. In this temperature range ($T < T_F$), the applied electric field is not strong enough to break such interactions and only the smallest regions can switch with the electric field, i.e. that is to say, there is a frustration of the cooperative effect. For temperatures above T_F, there are fluctuations between equivalent polarization states [30], leading to a decrease of the macroscopic polarization upon heating [42]. The applied electric field cannot reorient the ferroelectric dipoles because of the thermal fluctuation. The potential barrier between equivalent polarization states decreases with the increase of the temperature and the thermal energy promotes the spontaneous switching of the dipoles, even when an electric field is applied. The contribution to the dielectric permittivity is due to those PNRs, which can switch with the applied electric field. Thus, for temperatures slightly above T_F all the PNRs could contribute to the dielectric permittivity and a maximum value of the mean relaxation time could be expected. Therefore, the temperature corresponding to the maximum of $\ln\tau_0$ can be related to the freezing temperature. The standard deviation (σ), which can be interpreted as the correlation between the PNRs [32,41], decreases upon heating and it is relatively small at high temperatures. When the temperature increases the thermal energy is high enough to break down the interaction between the PNRs.

So that, a decrement of the correlation between the PNRs is promoted by an increase of the temperature and the Standard deviation decreases up to a relatively constant value around the Burns' temperature (T_B) [38,43], where the PNRs completely disappear.

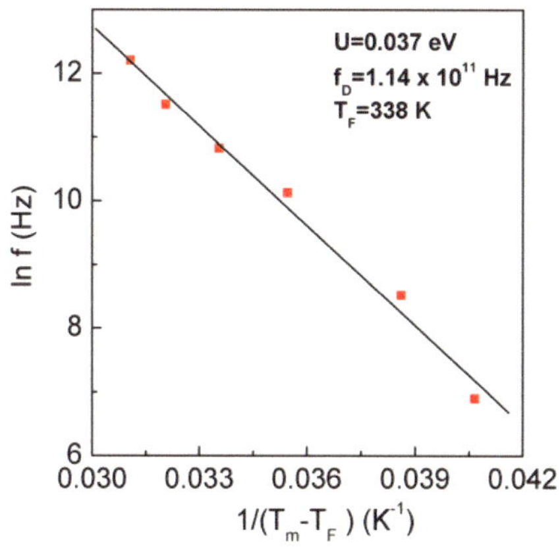

Figure 6. Temperature dependence of the frequency (lnf versus $1/(T_m–T_F)$ curve) for PLZT 10/80/20 ferroelectric ceramic system. Solid lines represent the fitting by using the Vogel–Fulcher relation.

The freezing temperature (T_F) was determined by using the Vogel–Fulcher relation [44], as expressed by equation (7):

$$f = f_D e^{-\frac{U}{k_B(T_m - T_F)}} \tag{7}$$

where f_D is the Debye frequency and U is the activation energy. Figure 6 show the temperature dependence of the frequency (lnf versus $1/(T_m–T_F)$ curve) and the fitting by using equation (7). In order to maintain the standard representation, the temperature in Figure 6 was expressed in Kelvin. The fitting parameters U, f_D and T_F have been included in the figure. The results are in agreement with previously reported results in the literature [42]. The disagreement between the T_F value, which has been determined by using the Vogel-Fulcherrelation, and the temperature corresponding to the maximum value of $\ln\tau_0$ could be associated to the diffusivity of the phase transition in the studied system.

2.3. Relaxor behaviour and vacancies.

Relaxor ferroelectrics are formed by temperature dependent Polar Nanometric Regions (PNRs), which possess different volumes [45] and orientations for the polarization [30,37,46-47]. The PNRs appear at elevated temperatures (in the paraelectric state), at the so-called Burns' temperature (T_B) [38,48-49], due to short-range interactions, which establish a local polarization thermally fluctuating between equivalents polar states.

The disorder in the arrangement of different ions on the crystallographic equivalent sites is the common feature of relaxors [50]. This is associated, in general, with a complex perovskite of the type A(B'B")O_3, where the *B* sites of the structure are occupied by different cations. Nevertheless, in lead zirconate titanate based ceramics (PZT) the disorder due to the arrangement of the isovalent cations Zr^{4+} and Ti^{4+}, in the *B* site of the structure, does not lead to relaxor behaviour for any Zr/Ti ratio. The lanthanum modification on this system, above certain La^{3+} concentrations, provides a relaxor ferroelectric state [11-12]. Electrical neutrality can be achieved, stoichiometrically, considering the vacancies formation either in the *A*-site (Pb^{2+}) or in the *B*-site (Zr^{4+}, Ti^{4+}), or on both. The distortion of the crystalline lattice in the PZT system could promote the formation of the PNRs. In this section, the results concerning the influence of the *A* or *B* vacancies defects in the relaxor ferroelectric behaviour of lanthanum modified PZT ceramic samples will be presented. The studied compositions will be $Pb_{0.85}La_{0.10}(Zr_{0.60}Ti_{0.40})O_3$ and $Pb_{0.90}La_{0.10}(Zr_{0.60}Ti_{0.40})_{0.975}O_3$, labelled as PLZT10-VA and PLZT10-VB, respectively.

The temperature dependence of the real dielectric permittivity (ε') and the dielectric losses (tan δ) are shown in the Figures 7 and 8, respectively, for several frequencies. Typical characteristics of relaxor ferroelectrics are exhibited for both samples, i.e. wide peaks for ε', the maximum real dielectric permittivity shifts to higher temperatures with the increase of the frequency, and the maximum dielectric losses temperature appears at temperatures below T_m. On the other hand, it can be observed that the dielectric response is highly affected by the type of compensation. The maximum values of the real dielectric permittivity in the PLZT10-VA sample are lower than those obtained for the PLZT-10VB sample, and appear at lower temperatures. The PLZT10-VA sample also exhibits a larger temperature shift of T_m (ΔT=15°C), from 1 kHz to 1 MHz, than that of the PLZT10-VB sample (ΔT=9°C).

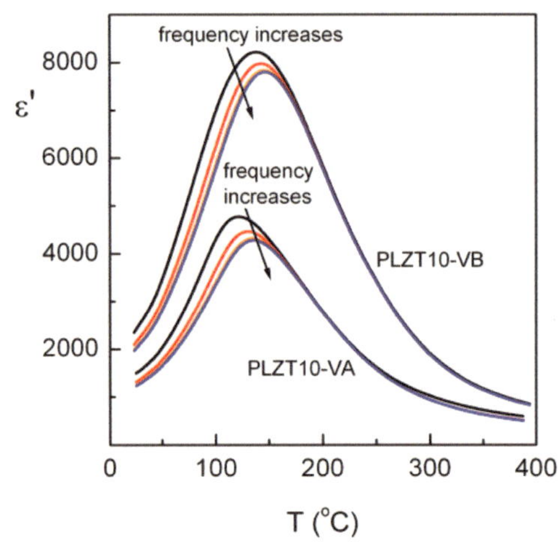

Figure 7. Temperature dependence of the real (ε') component of the dielectric permittivity at several frequencies, for PLZT10-VA and PLZT10-VB ceramic samples.

http://dx.doi.org/10.5772/52149

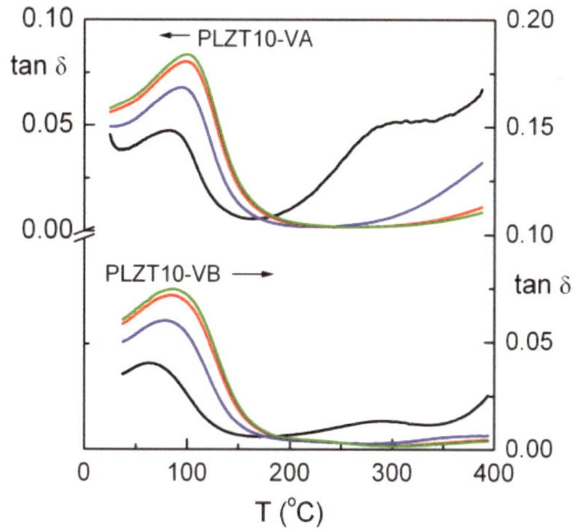

Figure 8. Temperature dependence of the dielectric losses (tan δ) at several frequencies for PLZT10-VA and PLZT10-VB ceramic systems.

It has been previously commented that, in relaxors, the maximum real dielectric permittivity is not a consequence of crystallographic changes but it is associated with a rapid change in the volume fraction of frozen polar nanoregions [39]. During cooling, the PNRs grow in size and number [39,43] as well as the interactions among them increases. Thus, a decrement of its mobility is expected; the thermal energy (k_BT) is not enough to switch the polar state of the PNRs (freezing phenomenon). While the regions grow, with the decreasing of the temperature, the thermal energy k_BT decreases. The freezing process begins at high temperatures and finish at low temperatures, below T_m, in the so called freezing temperature (T_F), where all the PNRs are frozen.

The dielectric and ferroelectric response in relaxor ferroelectrics strongly depend of the temperature dependence of the freezing process (dynamic of the PNRs) as well as on the interactions between external electric field and the dipole moment of the PNRs. When a weak alternating electric field is applied, only the unfrozen regions and the frozen ones with activation energies close to k_BT could switch with the applied alternating electric field. Therefore, only these regions contribute to the dielectric permittivity. However, not all these regions can switch at any excitation frequency; the number of PNRs able to follow the applied external electric field switching decreases with the increase of the frequency [14]. The response also depends on the relaxation time of each region [14, 41]. Thus, the dielectric response in relaxors depends of the number of regions that can contribute at each temperature and the distribution of the corresponding relaxation times.

The Pb^{2+} ions (*A*-sites of the perovskite structure) establish the long-range order in PZT based ceramics due to strong coupling of Pb-O-Ti/Zr bonding [51]. The coupling defines the height of the energy barrier between polar states via Ti/Zr-hopping [51]. The Pb^{2+} substitution by La^{3+} ions, above certain level of La^{3+} ions concentration, weakens this bonding resulting in small and broadly distributed energy barriers [51], which disrupts the long-range order and promotes the PNRs formations. An additional contribution, which plays a fundamental role in the relaxor behaviour, can be originated from the *A*- or *B*-sites vacancies formations. There are two important differences between the *A*- or *B*-site compensation, which can explain the different dielectric behaviour observed in the studied samples: *i*) for the same lanthanum concentration the number of lead vacancies is twice the Zr/Ti vacancies, if it is considered *A*- or *B*-site compensation, respectively, i.e. $2La^{3+} \rightarrow 1V_{Pb}$ and $4La^{3+} \rightarrow 1V_{Zr/Ti}$; *ii*) the *A*-site vacancies generate large inhomogeneous electric fields, which reduce the barriers between energy minima for different polarization directions [20], and promote the decoupling of the Pb-O-Ti/Zr bonding, making the corresponding Ti/Zr-hopping easier due to the lack of lead ions. For the case of the *B*-site vacancies, all the Ti/Zr ions are well coupled to the lead or the lanthanum ions.

From this point of view, it could be pointed out that for the PLZT10-VA sample the defects concentration, which could promote the PNRs formation, is higher than that for the PLZT10-VB sample. On the other hand, the thermal energy, which could change the nanoregions polarization state, is smaller for the PLZT10-VA sample as a consequence of the smaller hopping barrier. Considering these analyses, it could be explained the lower T_m values of the real part of the dielectric permittivity for the PLZT10-VA. The observed results are in agreement with previous reports for lead based systems with and without Pb^{2+} vacancies [52].

The larger temperature shift of T_m, which was analyzed for the PLZT10-VA, could be related to a higher temperature dependence of the freezing process around T_m, i.e. higher temperature dependence in the change of the volume fraction of frozen polar nanoregions around T_m. The number of PNRs, which can follow the electric field switching, decreases with the increase of the frequency. Thus, the maximum value of ε' decreases when the frequency increases, and also shifts to higher temperatures.

From the Figure 8, it is observed that the maximum values of the dielectric losses, for both samples, are observed in the same temperature range. No remarkable differences exist as can be expected from the difference observed in the T_m values between both systems. This could suggest similar dipolar dynamic in the temperature range for both ceramics.

Figure 9 shows the temperature dependence of the remanent polarization (P_R) for both samples. The PLZT10-VB sample exhibits higher values, which is in agreement with the previous discussion concerning the hopping barrier. The decoupling introduced by *A*-site vacancies affects the total dipolar moment of the system and, consequently, the macroscopic polarization is affected. Thus, the magnitude of the polarization decreases with the increase of the *A*-site vacancies concentration.

On the other hand, the remanent polarization for both samples shows an anomaly in the same temperature region (70–85°C), which could be associated with the 'offset' of the freezing process [14, 53]. This result suggests that the freezing temperature (T_F) could be the same for both samples or, at less the dynamic of PNRs is the same in both systems for that temperature range. Furthermore, the anomaly is observed in the same region where the maximum values of the dielectric losses have been observed, which could suggest a relation between the freezing phenomenon and the temperature evolution of dielectric losses.

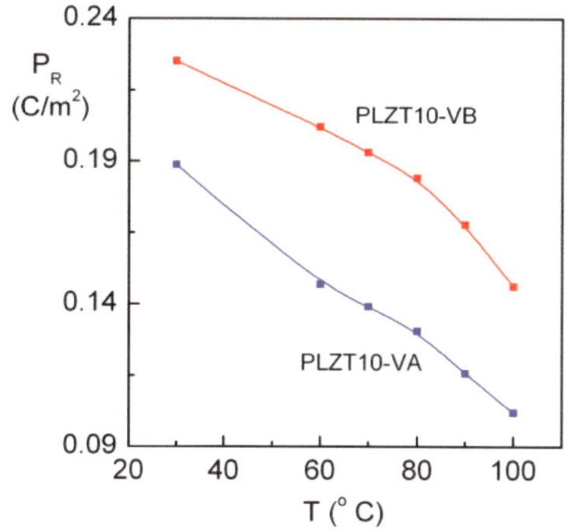

Figure 9. Temperature dependence of remnant polarization for the PLZT10-VA and PLZT10-VB samples. The lines between dots are only a guide to the eyes.

3. Other typical relaxor ferroelectric perovskites

Relaxor-based ferroelectric materials $PbMg_{1/3}Nb_{2/3}O_3$–$PbTiO_3$ (PMN–PT) and $PbZn_{1/3}Nb_{2/3}O_3$–$PbTiO_3$ (PZN–PT) have been extensively studied due to their high electromechanical coupling factor and piezoelectric coefficient [3,8,16-21,43-44,54-57].

It is known that the large dielectric permittivity of perovskite relaxor ferroelectrics, such as $(PbMg_{1/3}Nb_{2/3}O_3)_{1-x}(PbTiO_3)_x$, comes from ferroelectric like polar nanoregions [30]. Upon lowering the temperature these nanoregions grow slightly [43] but do not form long-range ferroelectric order. The temperature dependence of the dielectric permittivity shows a frequency-dependent peak as the polar nanoregion orientations undergo some sort of glassy kinetic freezing. Analyses of the phase and microstructure evolution for these materials have showed that the increase in the sintering temperature caused the weakening of the dielectric relaxor behaviour [54]. It has been related to compositional fluctuation and the release of the internal stresses, leading to the decrease in the short-range *B*-site order.

The PNRs freezing process in $PbMg_{1/3}Nb_{2/3}O_3$ has been analyzed considering both polarization and strain operating on subtly different timescales and length scales [57]. The strain fields are considered as relatively weak but longer ranging, while the dipole interactions tend to be short range and relatively strong. It has been discussed that the elastic shear strain fields is partly suppressed by cation disordering and screened by the presence of antiphase boundaries with their own distinctive strain and elastic properties, which provide a mechanism for suppressing longer ranging correlations of strain fields.

On the other hand, the barium modified PZN-PT system has shown excellent piezoelectric properties [58]. The preparation of PZN system usually needs the addition of $BaTiO_3$ (BT) to obtain pure phases. Therefore, the ternary system PZN-PT-BT has received wider attention, showing high values of the dielectric permittivity and the pyroelectric coefficient [21,59]. This material shows lower diffuseness of the phase transition and weaker frequency dispersion of the dielectric response than that of the PZN-BT system [16]. It has been shown that it could be interesting try to decrease the transition temperature and to grow the value of the figure of merit in order to obtain better materials for practical applications [17-18,21,60-61].

The present authors have developed several researches concerning $(Pb_{0.8}Ba_{0.2})[(Zn_{1/3}Nb_{2/3})_{0.7}Ti_{0.3}]O_3$ (PZN-PT-BT) system [19,62-63]. A Positive Temperature Coefficient of Resistivity (PTCR) effect has been studied. This effect has found extensive applications as thermal fuses, thermistors, safety circuits and other overload protection devices [64]. It has been discussed that the PTCR behaviour primarily arise from the Schottky barrier formed at grain boundary regions, which act as effective electron traps of the available electrons from the oxygen vacancies in the ceramic, increasing the Schottky barrier height of the material. Excellent properties to be used in dielectric and pyroelectric applications have been also reported [63].

Figure 10 shows the temperature dependence of the real part (ε') of the dielectric permittivity for the $(Pb_{0.8}Ba_{0.2})[(Zn_{1/3}Nb_{2/3})_{0.7}Ti_{0.3}]O_3$ ferroelectric ceramic system at several frequencies. Typical characteristics of relaxor ferroelectrics are observed. There is a strong dispersion of the maximum of ε' and its corresponding temperature shift towards higher temperatures when the frequency increases. On the other hand, the system did not follow a Curie–Weiss–like behaviour above T_m. For relaxor ferroelectrics, some of the dipoles are frozen during the time scale of measurement. The fraction of frozen dipoles in itself is a function of the temperature, so Curie-Weiss's law is no longer valid.

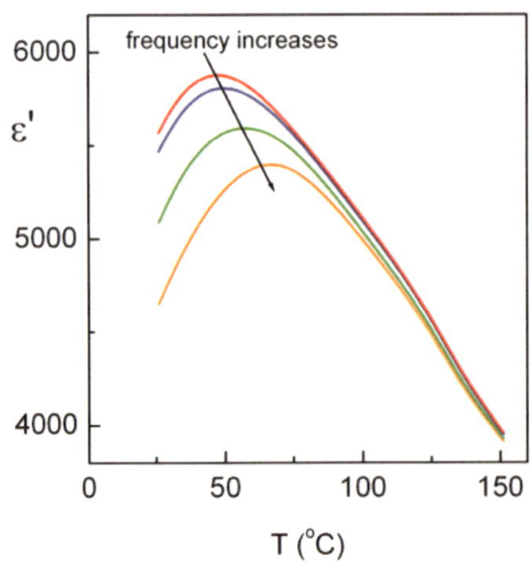

Figure 10. Temperature dependence of the real (ε') part of the dielectric permittivity at several frequencies for $(Pb_{0.8}Ba_{0.2})[(Zn_{1/3}Nb_{2/3})_{0.7}Ti_{0.3}]O_3$ ferroelectric ceramic system.

For $Pb(B'_{1/3}B''_{2/3})O_3$ type pervoskites, it has been reported a particular microstructure where the B-site 1:1 short-range-ordered nanodomains (rich in B′ ions) are embedded in a matrix rich in B″ ions, which promotes a compositional inhomogeneity [65]. The nonstoichiometric ordering induces strong charge effects; i.e., the ordered domains have a negative charge with respect to the disordered matrix. Thus, the charge imbalance inhibits the growth of domains and nanodomains are obtained in a disordered matrix with polar micro-regions.

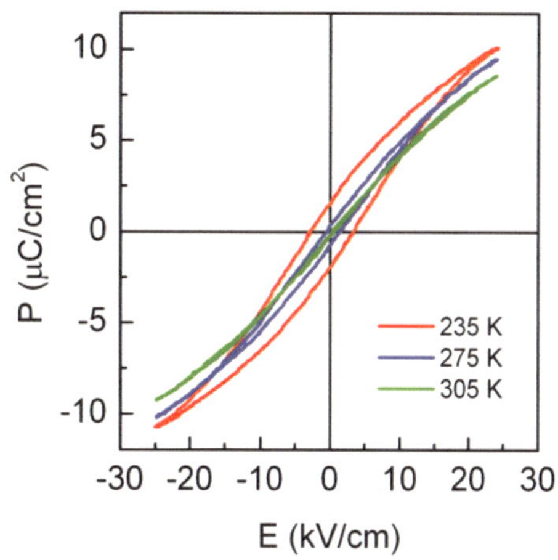

Figure 11. Hysteresis loops for $(Pb_{0.8}Ba_{0.2})[(Zn_{1/3}Nb_{2/3})_{0.7}Ti_{0.3}]O_3$ ferroelectric ceramic system.

Figure 11 shows the hysteresis loops at several temperatures for the studied system. Slim P-E loops and small remanent polarization values are observed. The slim-loop nature suggests that most of the aligned dipole moments switch back to a randomly oriented state upon removal of the field. It could be interpreted in terms of correlated polar nanodomains embedded in a paraelectric matrix [44]. For relaxor ferroelectric materials, there is a micro- to macro-domain transition [66]. In the absence of any external field, the domain structure of relaxor ferroelectrics contains randomly oriented micropolar regions. When an electric field is applied, the micro-domains are oriented along the field direction and the macro-domains appear. The micro- to macro-domain transition has been confirmed by 'in-situ' switching by means of an electron beam inducing local stresses to align the domains [66].

4. Lead free Aurivillius relaxor ferroelectrics

The temperature dependence of the real (ε') and imaginary (ε'') parts of the dielectric permittivity, at several frequencies, for the $Sr_{0.5}Ba_{0.5}Bi_2Nb_2O_9$ sample are showed in the Figure 12. The maximum of ε' decreases with the increase of the frequency; its corresponding temperature (T_m) shifts with the frequency, showing a high frequency dispersion. For the imaginary part (ε'') the maximum values are observed at lower temperatures than those the observed for ε' and the corresponding temperature again shows a significant frequency dispersion. These characteristics are typical of relaxor ferroelectric materials. On the other hand, an abrupt increase of ε'' is observed in the higher temperature zone, which is more clear in the low frequency range. This behaviour could be associated to the conductivity losses.

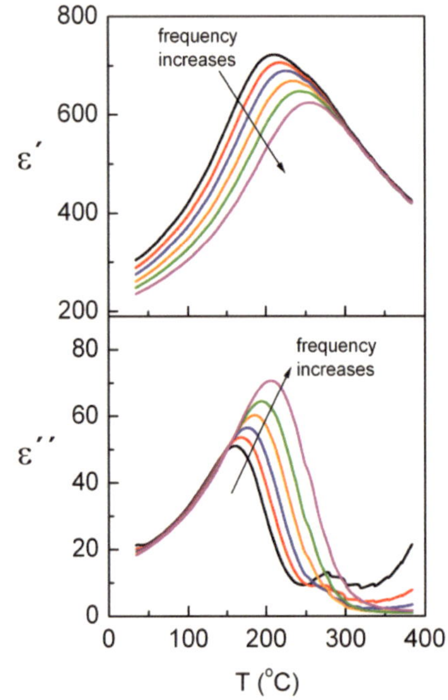

Figure 12. Temperature dependence of the real (ε') and imaginary (ε'') parts of the dielectric permittivity, at several frequencies, for the $Sr_{0.5}Ba_{0.5}Bi_2Nb_2O_9$.

The origin of the relaxor behaviour for this material can be explained from a positional disorder of cations on *A* or *B* sites of the perovskite blocks that delay the evolution of long-range polar ordering [28]. Previous studies in the $SrBi_2Nb_2O_9$ system have showed that the incorporation of barium to this system conduces to a relaxor behaviour in the ferroelectric-paraelectric phase transition [28,67] and the increases of barium content enhance the degree of the frequency dispersion of the dielectric parameters [67]. This behaviour can be explained considering that the barium ions can substitute the strontium ions in the *A* site of the perovskite block but enter in the $Bi_2O_2^{2+}$ layers, conducting to an inhomogeneous distribution of barium and local charge imbalance in the layered structure. The incorporation of barium ions to bismuth sites takes care to reduce the constrain existent between the perovskite blocks and the layered structure [68].

The relaxor behaviour is usually characterized by using the variation of T_m with frequency (ΔT), degree of the frequency dispersion, and the critical exponent (δ) obtained from the following law [67]:

$$\frac{1}{\varepsilon'} - \frac{1}{\varepsilon'_{\max}} = \frac{\left(T - T_m\right)^{\delta}}{C} \tag{8}$$

whereε'_{max} is the maximum value for the real part of the dielectric permittivity, and δ and C are constants. For an ideal relaxor ferroelectric δ =2, while for a normal ferroelectric δ =1 and the system follows the Curie-Weiss law.

For the studied material, ΔT=50°C between 1 kHz and 1 MHz, which is higher than that of the obtained values for the PMN and PLZT 8/65/35 systems [67]. Thus, the studied material shows higher frequency dispersion. However, ΔT is lower than that of the $BaBi_2Nb_2O_9$ system (BBN), which is in agreement with previous reports where ΔT decreases with the decrease of the barium concentration [67].

The Figure 13 shows the dependency of the log $(1/\varepsilon'-1/\varepsilon'_{max})$ vs. log $(T-T_m)$ at 1 MHz for the study ceramic sample. The solid points represent the experimental data and the line the fitting, which was carried out by using the equation 8. The value obtained for δ parameter is 1.71, which is in agreement with other reports for the studied system [67].

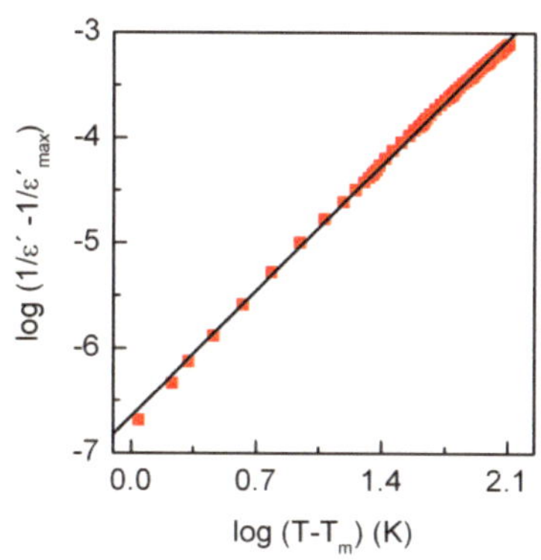

Figure 13. Dependence of the log $(1/\varepsilon'-1/\varepsilon''_{max})$ vs. log $(T-T_m)$ at 1 MHz for the $Sr_{0.5}Ba_{0.5}Bi_2Nb_2O_9$.

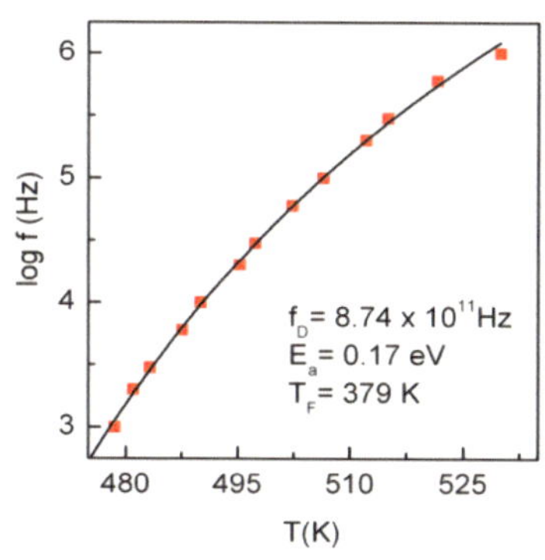

Figure 14. Dependence of the log f with T_m for the $Sr_{0.5}Ba_{0.5}Bi_2Nb_2O_9$ considering a frequency range from 1 kHz to 1 MHz. The fitting was carried out by using the Vogel-Fulcher law.

Figure 14 shows the dependence of log f with T_m for the studied material. The experimental data was fitted by using the Vogel-Fulcher law (equation 7). The fitting parameters have been included in the figure. The difference between T_m at 1 kHz and T_F is about 100 K, a higher value considering the studies carried out in PMN type relaxors [69]. On the other hand, the BBN system has showed a difference higher than 300 K, which is agreement with its strong shift of T_m [28, 69].

5. Conclusions

The relaxor behaviour has been discussed in ferroelectric perovskites-related structures. The influence of the coexistence of ferroelectric and antiferroelectric phases on the relaxor behaviour of lanthanum modified lead zirconate titanate ceramics has been analyzed. A relaxation model to evaluate the dynamical behaviour of the polar nanoregions in a relaxor ferroelectric PLZT system has been presented and the results have been discussed considering the correlation between the polar nanoregions and the freezing temperature. On the other hand, the influence of *A* or *B* vacancies on the relaxor behaviour for PLZT materials have been studied considering the decoupling effects of these defects in the Pb-O-Ti/Zr bounding and in terms of the dynamic of PNRs. Other typical relaxor ferroelectrics were analyzed too. For the PZN-PT-BT system, the relaxor behaviour was discussed considering the presence of local compositional fluctuation on a macroscopic scale. The $Sr_{0.50}Ba_{0.50}Bi_2Nb_2O_9$ ferroelectric ceramic, which is a typical relaxor from the Aurivillius family, was analyzed considering a positional disorder of cations on A or B sites of the perovskite blocks, which delay the evolution of long-rage polar ordering.

Acknowledgement

The authors wish to thank to TWAS for financial support (RG/PHYS/LA No. 99-050, 02-225 and 05-043) and to ICTP for financial support of Latin-American Network of Ferroelectric Materials (NET-43). Thanks to FAPESP, CNPq and FAPEMIG Brazilian agencies. Thanks to CONACyT, Mexico. Dr. A. Peláiz-Barranco wishes to thank the Royal Society. M.Sc. O. García-Zaldívar wishes to thanks to Red de Macrouniversidades / 2007 and to CLAF/ICTP fellowships NS 32/08.

Author details

A. Peláiz-Barranco, F. Calderón-Piñar, O. García-Zaldívar and Y. González-Abreu

*Address all correspondence to: pelaiz@fisica.uh.cu

Physics Faculty–Institute of Science and Technology of Materials, Havana University, San Lázaro y L, Vedado, La Habana, Cuba

http://dx.doi.org/10.5772/52149

References

[1] Xu Y. Ferroelectric Materials and Their Applications. Netherland: Elsevier Science Publishers; 1991.

[2] Strukov BA., Levanyuk AP. Ferroelectric Phenomena in Crystals. Berlin: Springer-Verlag; 1998.

[3] Tai CW., Baba-Kishi KZ. Relationship between dielectric properties and structural long-range order in $(x)Pb(In_{1/2}Nb_{1/2})O_3$:$(1 - x)Pb(Mg_{1/3}Nb_{2/3})O_3$ relaxor ceramics. Acta Materialia 2006;54; 5631-5640.

[4] Huang S., Sun L., Feng Ch., Chen L. Relaxor behaviour of layer structured $SrBi_{1.65}La_{0.35}Nb_2O_9$. J. Appl. Phys. 2006;99; 076104.

[5] Ko JH., Kim DH., Kojima S. Correlation between the dynamics of polar nanoregions and temperature evolution of central peaks in $Pb[(Zn_{1/3}Nb_{2/3})_{0.91}Ti_{0.09}]O_3$ ferroelectric relaxors. Appl. Phys. Lett. 2007;90; 112904.

[6] Kumar P., Singh S., Juneja JK., Prakash Ch., Raina KK. Dielectric behaviour of La substituted BPZT ceramics. Physica B 2009;404(16) 2126–2129.

[7] Craciun F. Strong variation of electrostrictive coupling near an intermediate temperature of relaxor ferroelectrics. Physical Review B 2010;81; 184111.

[8] Ko JH., Kim DH., Kojima S. Low-frequency central peaks in relaxor ferroelectric PZN-9%PT single crystals. Ferroelectrics 2007; 347(1) 25-29.

[9] Samara GA. The relaxational properties of compositionally disordered ABO_3 perovskites. J. Phys.: Condens. Matter 2003;15(9) R367.

[10] Shvartsman VV., Lupascu DC. Lead-free relaxor ferroelectrics. J. Am. Ceram. Soc. 2012;95(1) 1–26.

[11] Dai X., Xu Z., Li JF., Viehland D. Effects of lanthanum modification on rombohedral $Pb(Zr_{1-x}Ti_x)O_3$ ceramics: Part I. Transformation from normal to relaxor ferroelectric behaviours. J. Mat. Res. 1996;11(3) 618-625.

[12] Dai X., Xu Z., Li JF., Viehland D. Effects of lanthanum modification on rombohedral $Pb(Zr_{1-x}Ti_x)O_3$ ceramics: Part II. Relaxor behaviour versus enhanced antiferroelectric stability. J. Mat. Res. 1996;11(3) 626-638.

[13] Peláiz-Barranco A., Mendoza ME., Calderón-Piñar F., García-Zaldívar O., López-Noda R., de los Santos-Guerra J., Eiras JA., Features on phase transitions in lanthanum-modified lead zirconate titanate ferroelectric ceramics. Sol. Sta. Comm. 2007;144(10-11) 425-428.

[14] García-Zaldívar O., Peláiz-Barranco A., Calderón-Piñar F., Fundora-Cruz A., Guerra JDS., Hall DA., Mendoza ME.. Modelling the dielectric response of lanthanum modi-

fied lead zirconate titanate ferroelectric ceramics - an approach to the phase transitions in relaxor ferroelectrics. J. Phys.: Cond. Matt. 2008;20; 445230.

[15] Peláiz Barranco A., Calderón Piñar F., Pérez Martínez O. Normal-Diffuse Transition in (Pb,La)$Zr_{0.53}Ti_{0.47}O_3$ Ferroelectric Ceramics. Phys. Sta. Sol. (b) 2000;220(1) 591-595.

[16] Fan HQ., Kong LB., Zhang LY., Yao X. Structure-property relationships in lead zinc niobate based ferroelectric ceramics. J. Appl. Phys. 1998;83; 1625-1630.

[17] Zhu WZ., Yan M., Effect of Mn-doping on the morphotropic phase boundary of PZN-BT-PT system. J. Mater. Sci. Lett. 2001;20(16) 1527-1529.

[18] Zhu WZ., Yan M., Kholkin AL., Mantas PQ., Baptista JL., Effect of tungsten doping on the dielectric response of PZN–PT–BT ceramics with the morphotropic phase boundary composition. J. Eur. Ceram. Soc. 2002;22(3) 375-381.

[19] Peláiz-Barranco A., González-Carmenate I., Calderón-Piñar F., Relaxor behaviour in PZN-PT-BT ferroelectric ceramics. Sol. Sta. Comm. 2005;134; 519-522.

[20] Haertling HG. Ferroelectric Ceramics: History and Technology. J. Am. Ceram. Soc. 1999;82(4) 797-818.

[21] Halliyal A., Kumar U., Newnham RE., Cross LE. Stabilization of the perovskite phase and dielectric properties of ceramics in the $Pb(Zn_{1/3}Nb_{2/3})O_3$-$BaTiO_3$ system. Am. Ceram. Soc. Bull. 1987;66(4) 671-676.

[22] Jennet DB., Marchet P., El Maaoui M., Mercurio JP., From ferroelectric to relaxor behaviour in the Aurivillius-type $Bi_{4-x}Ba_xTi_{3-x}Nb_xO_{12}$ ($0<x<1.4$) solid solutions. Mat. Lett. 2005;59; 376-382.

[23] Karthik C., Ravishankar N., Maglione M., Vondermuhll R., Etourneau J., Varma KBR. Relaxor behaviour of $K_{0.5}La_{0.5}Bi_2Ta_2O_9$ ceramics. Sol. Sta. Comm. 2006;139; 268–272.

[24] González-Abreu Y., Peláiz-Barranco A., Araújo EB., Franco Júnior A. Dielectric relaxation and relaxor behaviour in bi-layered perovskite. Appl. Phys. Lett. 2009;94(26) 262903.

[25] Zhang H., Yan H., Reece MJ. High temperature lead-free relaxor ferroelectric: intergrowth Aurivillius phase $BaBi_2Nb_2O_9$–$Bi_4Ti_3O_{12}$ ceramics. J. Appl. Phys. 2010;107; 104111.

[26] Chakrabarti A., Bera J. Effect of La-substitution on the structure and dielectric properties of $BaBi_4Ti_4O_{15}$ ceramics. J. Alloys Comp. 2010;505; 668–674.

[27] Rabe KM., Ahn CA. Triscone JM., editors. Physics of Ferroelectrics. Berlin Heidelberg, New York: Springer; 2007.

[28] Miranda C., Costa MEV., Avdeev M., Kholkin AL., Baptista JL., Relaxor properties of Ba-based layered perovskites. J. Eur. Ceram. Soc. 2001;21; 1303-1306.

[29] Smolienskii GA. Physical phenomena in ferroelectric with diffused phase transition. J. Phys. Soc. Japan 1970;28,Supplement; 26-30.

[30] Cross LE. Relaxor ferroelectrics. Ferroelectrics 1987;76; 241-267.

[31] Viehland D., Jang SJ., Cross LE., Wutting M. Freezing of the polarization fluctuations in lead magnesium niobate relaxors. J. Appl. Phys. 1990;68; 2916-2921.

[32] Quian H., Bursill LA., Phenomenological theory of the dielectric response of lead magnesium niobate and lead scandium tantalate. Int. J. Mod Phys. B 1996;10; 2007-2025.

[33] Ishchuk VM., Baumer VN., Sobolev VL., The influence of the coexistence of ferroelectric and antiferroelectric states on the lead lanthanum zirconate titanate crystal structure. J. Phys. Condens. Matter 2005;17(19) L177-L182.

[34] Peláiz-Barranco A., Hall DA. Influence of composition and pressure on the electric field-induced antiferroelectric to ferroelectric phase transformations in lanthanum modified lead zirconate titanate ceramics. IEEE Trans. Ultras. Ferroel. Freq. Cont. 2009;56(9) 1785-1791.

[35] Jullian Ch., Li JF., Viehland D. Comparisons of polarization switching in "hard," "soft," and relaxor ferroelectrics. J. Appl. Phys. 2005;95(8) 4316-4318.

[36] Peláiz-Barranco A., García-Zaldívar O., Calderón-Piñar F., López-Noda R., Fuentes-Betancourt J. A multi-Debye relaxation model for relaxor ferroelectrics showing diffuse phase transition. Phys. Sta. Sol. (b) 2005;242(9) 1864-1867.

[37] Bovtun V., Petzelt J., Porokhonskyy V., Kamba S., Yakimenko Y. Structure of the dielectric spectrum of relaxor ferroelectrics. J. Eur. Ceram. Soc. 2001;21(10-11) 1307-1311.

[38] Burns G., Dacol FH., Crystalline ferroelectrics with glassy polarization behaviour. Phys. Rev. B 1983;28; 2527-2530.

[39] Cheng ZY., Katiyar RS., Yao X., Bhalla AS. Temperature dependence of the dielectric constant of relaxor ferroelectrics. Phys. Rev. B 1998;57; 8166-8177.

[40] Bokov AA. Recent advances in diffuse ferroelectric phase transitions. Ferroelectrics 1992;131; 49-55.

[41] Lin HT., Van Aken DC., Huebner W. Modelling the dielectric response and relaxation spectra of relaxor ferroelectrics. J. Am. Ceram. Soc. 1999;82(10) 2698-2704.

[42] Dal-Young K., Jong-Jin Ch., Hyoun-Ee K. Birefringence study of the freezing mechanism of lanthanum-modified lead zirconate titanate relaxor ferroelectrics. J. Appl. Phys. 2003;93; 1176-1179.

[43] Xu G., Shirane G., Copley JRD., Gehring PM. Neutron elastic diffuse scattering study of $Pb(Mg_{1/3}Nb_{2/3})O_3$. Phys. Rev. B 2004;69; 064112.

[44] Viehland D., Li JF., Jang SJ., Cross LE., Wuttig M. Dipolar-glass model for lead magnesium niobate. Phys. Rev. B 1991;43; 8316-8320.

[45] Xu G., Gehring PM., Shirane G. Coexistence and competition of local- and long-range polar orders in a ferroelectric relaxor. Phys. Rev B 2006;74; 104110.

[46] Kirillov VV., Isupov VA. Relaxation polarization of $PbMg_{1/3}Nb_{2/3}O_3$ (PMN) a ferroelectric with a diffused phase transition. Ferroelectrics 1973;5; 3-9.

[47] Isupov VA. Nature of physical phenomena in ferroelectric relaxors. Phys. Sol. Sta. 2003;45(6) 1107-1111.

[48] Pirc R., Blinc R. Vogel-Fulcher freezing in relaxor ferroelectrics. Phys. Rev. B 2007;76; 020101R.

[49] Jiménez R., Jiménez B., Carreaud J., Kiat JM., Dkhil B., Holc J., Kosec M., Algueró M. Transition between the ferroelectric and relaxor states in $0.8Pb(Mg_{1/3}Nb_{2/3})O_3$-$0.2PbTiO_3$ ceramics. Phys. Rev. B 2006;74; 184106.

[50] Bokov AA., Ye Z-G. Recent progress in relaxor ferroelectrics with perovskite structure. J. Mat. Sci. 2006;4(1) 31-52.

[51] Chen IW., Li P., Wang Y. Structural origin of relaxor perovskites. J. Phys. Chem. Sol. 1996;57(10); 1525-1536.

[52] Bellaiche L., Íñiguez J., Cockayne E., Burton BP. Effects of vacancies on the properties of disordered ferroelectrics: A first-principles study. Phys. Rev. B 2007;75; 014111.

[53] Shen M., Han J., Cao W. Electric-field-induced dielectric anomalies in C-oriented $0.955Pb\ Zn_{1/3}Nb_{2/3}O_3$-$0.045PbTiO_3$ single crystals. Appl. Phys. Lett. 2003;83(4) 731-733.

[54] Wu N-N., Hou Y-D, Wang Ch., Zhu M-K, Song X-M, Hui Y. Effect of sintering temperature on dielectric relaxation and Raman scattering of $0.65Pb(Mg_{1/3}Nb_{2/3})O_3$–$0.35PbTiO_3$ system. J. Appl. Phys. 2009;105; 084107.

[55] Park SE., Shrout TR. Ultrahigh strain and piezoelectric behaviour in relaxor based ferroelectric single crystals. J. Appl. Phys. 1997;82(4) 1804-1811.

[56] Park SE., Shrout TR. Characteristics of relaxor-based piezoelectric single crystals for ultrasonic transducers. IEEE Trans. Ultrason. Ferroelectr. Freq. Control 1997;44(5) 1140-1147.

[57] Carpenter MA., Bryson JFJ., Catalan G., Zhang SJ., Donnelly NJ. Elastic and inelastic relaxations in the relaxor ferroelectric $Pb(Mg_{1/3}Nb_{2/3})O_3$: II. Strain–order parameter coupling and dynamic softening mechanisms. J. Phys.: Condens. Matter 2012;24; 045902.

[58] Uchino K. High electromechanical coupling piezoelectrics: relaxor and normal ferroelectric solid solutions. Sol. Sta. Ion. 1998;108(1-4) 43-52.

[59] Deb KK., Hill MD., Roth RS., Kelly JF. Dielectric and pyroelectric properties of doped lead zinc niobate (PZN) ceramic materials. Am. Cer. Bull. 1992;71; 349-352.

[60] Krumin A. Specific solid state features of transparent ferroelectric ceramics. Ferroelectrics 1986;69(1) 1-16.

[61] Xia F., Yao X. Piezoelectric and dielectric properties of PZN-BT-PZT solid solutions. J. Mat. Sci. 1999;34; 3341-3343.

[62] Peláiz Barranco A., González Carmenate I., Calderón Piñar F., Torres García E. AC behaviour and PTCR effect in PZN-PT-BT ferroelectric ceramics. Sol. Sta. Comm. 2004;132; 431-435.

[63] Peláiz Barranco A., Gonzalez Carmenate I., Calderón Piñar F. Dielectric and pyroelectric behaviour of $(Pb_{0.8}Ba_{0.2})[(Zn_{1/3}Nb_{2/3})_{0.7}Ti_{0.3}]O_3$ ferroelectric ceramics. Rev. Cub. Fis. 2009;26(2B) 238-241.

[64] Chang H-Yi., Liu K-S., Lin IN. Modification of PTCR behaviour of $(Sr_{0.2}Ba_{0.8})TiO_3$ materials by post-heat treatment after microwave sintering. J. Eur. Ceram. Soc. 1996;16(1) 63-70.

[65] Chen J., Chan H., Harmer MP. Ordering structure and dielectric properties of undoped and La/Na-doped $Pb(Mg_{1/3}Nb_{2/3})O_3$. J. Am. Ceram. Soc. 1989;72(4) 593-598.

[66] Randall CA., Barber DJ., Whatmore RW., Groves P. Short-range order phenomena in lead-based perovskites. Ferroelectrics 1987;76(1) 277-282.

[67] Huang S., Feng Ch., Chen L., Wang Q. Relaxor Behaviour of $Sr_{1-x}Ba_xBi_2Nb_2O_9$ Ceramics. J. Am. Ceram. Soc. 2006;89(1) 328-331.

[68] Haluska MS., Misture ST. Crystal structure refinements of the three-layer Aurivillius ceramics $Bi_2Sr_{2-x}A_xNb_2TiO_{12}$ (A=Ca,Ba, x=0,0.5,1) using combined X-ray and neutron powder diffraction. J. Solid State Chem. 2004;177; 1965-1975.

[69] Kholkin AL., Avdeev M., Costa MEV., Baptista JL. Dielectric relaxation in Ba-based layered perovskites. Appl. Phys. Letters 2001;79(5) 662-664.

Chapter 6

Electronic Band Structures and Phase Transitions of Ferroelectric and Multiferroic Oxides

Zhigao Hu, Junhao Chu, Yawei Li, Kai Jiang and Ziqiang Zhu

Additional information is available at the end of the chapter

http://dx.doi.org/10.5772/52189

1. Introduction

Perovskite ferroelectric (FE) materials have attracted considerable attention for a wide range of applications, such as dynamic random access memories (DRAM), microwave tunable phase shifters and second harmonic generators (SHGs). [1–3] Moreover, materials that have coupled electric, magnetic, and structural order parameters that result in simultaneous ferroelectricity, ferromagnetism, and ferroelasticity are known as multiferroics. [4–6] These multiferroics materials have attracted a lot of attention in recent years because they can potentially offer a whole range of new applications, including nonvolatile ferroelectric memories, novel multiple state memories, and devices based on magnetoeletric effects. Although there are some reports on the electrical and magnetic properties of perovskite-type ferroelectric and multiferroics materials, optical properties and electronic transitions have not been well investigated up to now. On the other hand, phase transition is one of the important characteristics for the ferroelectric/multiferroics system. As we know, the phase transition is strongly related to the structural variation, which certainly can result in the electronic band modifications. Therefore, one can study the phase transition of the above material systems by the corresponding spectral response behavior at different temperatures.

Among these materials, barium strontium titanate (BST) has been considered to be one of the most promising candidates for devices due to its excellent dielectric properties of high dielectric functions, low leakage current and an adjustable Curie temperature T_C through variation of the composition between barium titanate (BT) and strontium titanate (ST). However, the limited figure of merit at high frequency microwave region restricts the BST practical applications. In order to improve the physical properties of the BST materials, introducing small compositions of dopants has been used for several decades. Many experimental and theoretical studies have been performed on the dielectric properties of BST in the ferroelectric state by adding dopants such as Magnesium [7, 8], Aluminum [9], Manganese (Mn) [10–12], Samarium [13], and different rare earth [14], In particular, the Mn doping BST shows some advantages in reducing the dielectric loss, enhancing the resistivity, and increasing dielectric tunability. It can significantly improve the dielectric properties, which makes it a potential candidate for microwave elements. For example, the Mn doping can cause the variation of the oxygen vacancy, which is the crucial role in modifying the dielectric loss mechanism. On the other hand, the doping of Mn can also reduce the dielectric constant peak and broaden the dielectric phase transition temperature range, which results in a smaller temperature coefficient of capacitance in BSMT materials. [10] Therefore, it is important to further investigate the physical properties of BSMT materials in order to develop the potential applications.

Meanwhile, many research groups have focused on the doping effect on the fabrications and dielectric properties for strontium barium niobate $SrBi_2Nb_2O_9$ (SBN) materials. [15–20] The substitution of Ca ions in the Sr site for the SBN ceramic induced the T_C increasing, which is useful for the application in high-temperature resonators. [15, 18] However, the substitution of some rare earth ions such as La^{3+} or Pr^{3+} for Bi^{3+} in the Bi_2O_2 layers can result in a shift for the T_C to lower temperature. [16, 20, 21] It is found that the behavior of the Nd-doped SBN ceramic tends to change from a normal ferroelectrics to a relaxor type ferroelectrics owing to the introduction of Nd ions in the Bi_2O_2 layers. [16, 17, 20, 22] Up to now, a detailed understanding of the lattice dynamic properties and the phase transition behavior of Nd-doped SBN ceramics are still lacking. Raman spectroscopy is a sensitive technique for investigating the structure modifications and lattice vibration modes, which can give the information on the changes of lattice vibrations and the occupying positions of doping ions. Thus, it is a powerful tool for the detection of phase transition in the doping-related ferroelectric materials. [22, 23]

The optical properties such as the dielectric functions provide an important insight on dielectric and ferroelectric behaviors of the material and play an important role in design, optimization, and evaluation of optoelectronic devices. [24–27] In addition, the doping of Mn or Nd can induce more defects in the lattice structure, which can affect its electronic band structures and optoelectronic properties. Hence, the doping composition dependence of optical properties for BST and SBN ceramics is technically important for practical optoelectronic device development. Compared to film structure, the optical properties of bulk material (single crystals and dense ceramics) are not affected by interface layer, stress from clamping by the substrate, non-stoichiometry and lattice mismatch between film and substrate. Hence, it is desirable to carry out a delicate investigation regarding the optical properties of the BSMT and SBNN ceramics. Note that spectroscopic ellipsometry (SE), Raman scattering, and transmittance spectra are potentially valuable techniques for the studies of ferroelectric materials due to their high sensitivity of local structure and symmetry. Compared with the other techniques, they can provide dielectric functions of the materials. SE and transmittance spectra can provide optical band gap and optical conductivity, whereas Raman spectra can provide Raman-active modes of the materials. [28–31]

On the other hand, bismuth ferrite ($BiFeO_3$, BFO) is known to be the only perovskite material that exhibits multiferroic at room temperature (RT). At RT, it is a rhombohedrally distorted ferroelectric perovskite with the space group *R3c* and a Curie temperature (T_C) of about 1100 K. [32–36] Since the physical properties of BFO films are related to their domain structure and phase states, which is sensitive to the applied stress, composition, and fabrication condition for BFO materials. PLD technique has the ability to exceed the solubility of magnetic impurity and to permit high quality film grown at low substrate temperature. Recent studies of photoconductivity, [35] photovoltaic effect, [37] and low open circuit voltage in a working solar device, [38] illustrate the potential of polar oxides as the active photovoltaic material.

In spite of the promising properties, there are no systematical reports focused on the optical properties of BFO films. In order to make BFO useful in actual electrical and optoelectronic devices, the physical properties, especially for electronic band structure and optical response behavior, need to be further clarified.

The objectives of the chapter will tentatively answer the interesting questions: (1) Is there an effective method to directly analyze electronic structure of FE materials by optical spectroscopy? (2) What kind of temperature dependence have FE oxides from band-to-band transitions? (3) Can spectral response at high-temperature be used to judge phase transition? Correspondingly, this chapter is arranged in the following way. In Sec. 2, detailed growths of $Ba_{0.4}Sr_{0.6-x}Mn_xTiO_3$ (BSMT), $SrBi_{2-x}Nd_xNb_2O_9$ (SBNN) ceramics and BFO films are described; In Sec. 3, solid state spectroscopic techniques are introduced; In Sec. 4, electronic band structures of BSMT ceramics are presented; In Sec. 5, phase transitions of SBNN ceramics are derived; In Sec. 6, temperature effects on electronic transitions of BFO films have been discussed; In Sec. 7, the main results and remarks are summarized.

2. Experimental

2.1. Fabrications of FE ceramics

The ceramics based on $Ba_{0.4}Sr_{0.6-x}Mn_xTiO_3$ (with x = 1, 2, 5 and 10%) specimens were prepared by the conventional solid-state reaction sintering. High purity $BaCO_3$ (99.8%), $SrCO_3$ (99.0%), TiO_2 (99.9%), and $MnCO_3$ were used as the starting materials. Weighted powers were mixed by ball milling with zirconia media in the ethanol as a solvent for 24 h and then dried at 110 °C for 12 h. After drying, the powders were calcined at 1200 °C for 4 h, and then remilled for 24 h to reduce the particle size for sintering. The calcined powders were added with 8 wt.% polyvinyl alcohol (PVA) as a binder. The granulated powders were pressed into discs in diameter of 10 mm and thickness of 1.0 mm. The green pellets were kept at 550 °C for 6 h to remove the solvent and binder, followed by sintering at 1400 °C for 4 h. More details of the preparation process can be found in Ref. [39]. On the other hand, the SBNN (x=0, 0.05, 0.1, and 0.2) ceramics were prepared by a similar method, and $SrCO_3$, Bi_2O_3, Nb_2O_5, and Nd_2O_3 were used as the starting materials. Details of the fabrication process for the ceramics can be found elsewhere. [17, 40]

2.2. Depositions of BFO films

The $BiFeO_3$ films were deposited on c-sapphire substrates by the PLD technique. The $BiFeO_3$ targets with a diameter of 3 cm were prepared through a conventional solid state reaction method using reagent-grade Bi_2O_3 (99.9%) and Fe_2O_3 (99.9%) powders. Weighed powders were mixed for 24 h by ball milling with zirconia media in ethanol and then dried at 100 °C for 12 h. The dried powders were calcined at about 680 °C in air for 6 h to form the desired phase, and followed by sintering at about 830 °C for 2 h. Before the deposition of the BFO films, c-sapphire substrates need to be cleaned in pure ethanol with an ultrasonic bath to remove physisorbed organic molecules from the surfaces, followed by rinsing several times with de-ionized water. Then the substrates were dried in a pure nitrogen stream before the film deposition. A pulsed Nd:YAG (yttrium aluminum garnet) laser (532 nm wavelength, 5 ns duration) operated with an energy of 60 mJ/pulse and repetition rate of 10 Hz was used as the ablation source. The films were deposited immediately after the target was

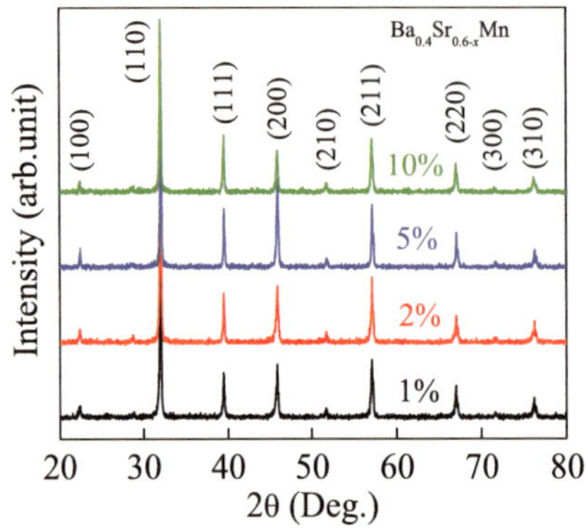

Figure 1. The XRD patterns of the BSMT ceramics with the Mn composition of 1, 2, 5 and 10%, respectively.

preablated in order to remove any surface contaminants. The distance between the target and the substrate was kept at 3 cm. The deposition time was set to about 30 min. Finally, the films were annealed at 600 °C in air atmosphere by a rapid thermal annealing process. A detailed preparation of the films can be found in Ref. [6].

3. Optical spectroscopy

The ellipsometric measurements were carried out in the photon energy range of 0.7-4.2 eV (300-1700 nm) with a spectral resolution of 5 nm by near-infrared-ultraviolet (NIR-UV) SE (SC630UVN by Shanghai Sanco Instrument, Co., Ltd.). The measurements were performed under the incident angle of 67° for all the ceramics corresponding to the experimental optimization near the Brewster angle of the BSMT. Raman scattering experiments were carried out using a Jobin-Yvon LabRAM HR 800 UV micro-Raman spectrometer, excited by a 632.8 nm He-Ne laser with a spectral resolution of 0.5 cm^{-1}. Temperature dependent measurements from 80 to 873 K were performed using the Linkam THMSE 600 heating stage, and the set-point stability is of better than 0.5 K. The normal-incident transmittance spectra were recorded using a double beam ultraviolet-infrared spectrophotometer (PerkinElmer Lambda 950) at the photon energy from 0.5 to 6.5 eV (190-2650 nm) with a spectral resolution of 2 nm. The samples at 5.3-300 K were mounted into an optical cryostat (Janis SHI-4-1) for variable temperature experiments. [41]

4. Electronic band structures of BSMT ceramics

The XRD patterns of the BSMT ceramics with different Mn composition are shown in Fig. 1 and no secondary phase appears within the detection limit of the XRD. Besides the strongest (110) peak, some weaker peaks (100), (111), (200), (210), (211), (220) can be also observed, which indicate that the ceramics are polycrystalline with single perovskite phase. The diffraction patterns are fitted by the Gaussian lineshape analysis to extract the peak positions and full width at half maximum (FWHM). The lattice constant *a* of the BSMT ceramics, which can be estimated from the (110) diffraction peak, is calculated to be about 3.954 Å. [39] The ionic radius of Mn^{2+} (1.27 Å) is smaller than that of Sr^{2+} (1.44 Å) and Ba^{2+} (1.61 Å), and is larger than that of Ti^{4+} (0.61 Å), which can be attributed to the change of the lattice constant. When the Mn composition is below 5%, the (110) diffraction peak positions shift from smaller angles to larger angles and the lattice constant slightly decreases, which can be

http://dx.doi.org/10.5772/52189

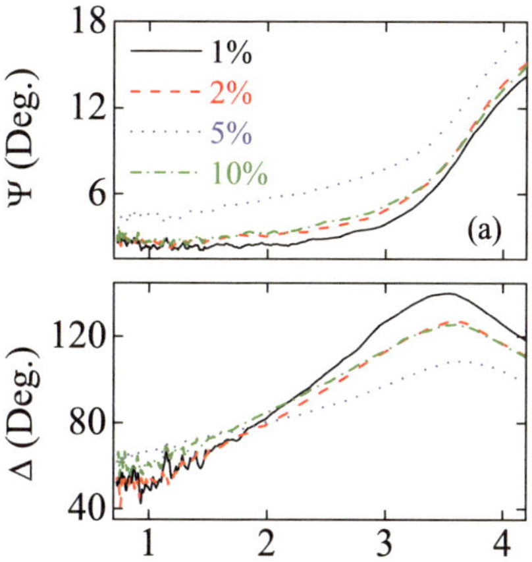

Figure 2. Experimental ellipsometric data (a) Ψ and (b) Δ for the BSMT ceramics from near-infrared to ultraviolet photon energy region at the incident angle of 67°. (Figure reproduced with permission from [39]. Copyright 2012, Springer.)

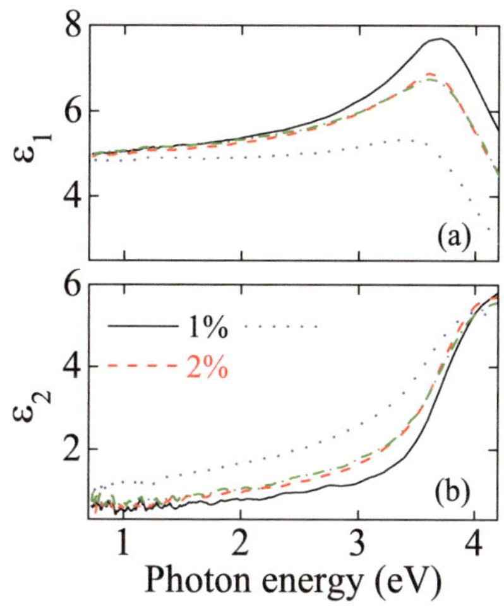

Figure 3. The (a) real and (b) imaginary parts of the NIR-UV dielectric functions for the BSMT ceramics with different Mn composition. (Figure reproduced with permission from [39]. Copyright 2012, Springer.)

ascribed to the fact that Mn occupies the A site together with Ba^{2+} and Sr^{2+}. In order to reduce distortion of the crystal lattice, the Mn mainly substitutes Sr^{2+} because the differences in ionic radius between Mn ion and Ba ion is larger than that between Mn and Sr. On the other hand, when the Mn composition is larger than 5%, the excess Mn can also substitute the Ti site, which results in the increase of the lattice constant and the smaller shift of the peak position. It can be concluded that the Mn ions substitute Sr sites of the BST lattice at first, then occupy Ti sites when the Mn composition is beyond 5%.

The experimental ellipsometric spectra of Ψ and Δ recorded at an incident angle of 67° for the BSMT ceramics are depicted in Figs. 2(a) and (b), respectively. The observed changes in the Ψ and Δ data for different Mn composition may be attributed to the lattice distortion and variation in atomic coordinate. Because the sample is bulk material with a thickness of several millimeters, the dielectric functions of the BSMT ceramics can be directly calculated

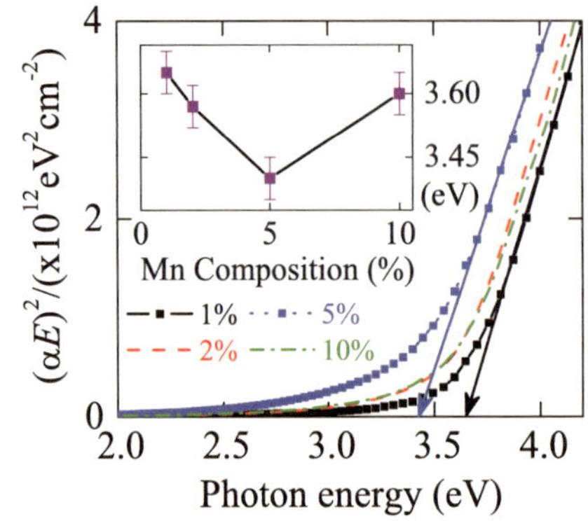

Figure 4. Absorption coefficient *vs.* incident photon energy near the optical band gap of the BSMT ceramics. The insert is the optical band gap E_g with the different Mn composition. (Figure reproduced with permission from [39]. Copyright 2012, Springer.)

according to the ellipsometric spectra Figs. 3(a) and (b) show the real (ε_1) and imaginary (ε_2) parts of the dielectric functions in the photon energy range of 0.7-4.2 eV, respectively. The evolution of ε with the photon energy is a typical optical response behavior of ferroelectric and/or semiconductors. The optical band gap (E_g) of the BSMT ceramics is one of the important optical behaviors, which is calculated by considering a direct transition from the VB to the CB when the photon energy falls on the materials. In the BSMT system, the VB is mainly composed of the O 2*p* orbital and the CB is mainly composed of the Ti 3*d* orbital. It should be noted that because of the splitting of the Ti 3*d* conduction bands into t_{2g} and e_g subbands, the lowest CB arises from the threefold degenerate Ti 3*d* t_{2g} orbital, which has lower energy than the twofold degenerate Ti 3*d* e_g orbital. [42] The absorption coefficient related E_g of the BSMT ceramics can be determined according to the Tauc's law: $(\alpha E)^2 = A(E - E_g)$, where A is a constant, α and E are the absorption coefficient and incident photon energy, respectively. For the allowed direct transition, the straight line between $(\alpha E)^2$ and E will provide the value of the band gap, which is extrapolated by the linear portion of the plot to $(\alpha E)^2$=0, as seen in Fig. 4. The E_g is estimated to 3.65, 3.57, 3.40 and 3.60 eV corresponding to x=1, 2, 5 and 10% for the BSMT ceramics, respectively, as shown in the inset of Fig. 4. The results suggest that the band gap of the BSMT ceramics decreases and then increases with increasing Mn composition.

As we know, the optical band gap can be affected by some factors such as grain size, oxygen vacancy, stress and amorphous nature of the materials. [2] The decreasing trend of the band gap with the Mn composition below 5% can be attributed to the increase of the grain size and the smaller lattice constant, which are caused by the Mn introduction in the A sites. When the Mn composition is 10%, there is a sharp increase in the optical band gap because the excess Mn will substitute the Ti site at the Mn composition of 10%, causing the increase of the oxygen vacancies. In addition, the creation of an oxygen vacancy which is associated with the generation of free charge carriers can be described as the following: $MnO(-TiO_2) \rightarrow Mn''_{Ti} + V_O^{\cdot\cdot} + O_O$; $O_O \rightarrow V_O^{\cdot\cdot} + 2e^- + 1/2O_2$, where $V_O^{\cdot\cdot}$ represents the doubly charged oxygen vacancy, O_O is an oxygen ion at its normal site, e^- is the free electronic charge generated through the vacancy formation. [9, 43] The heavy doping blocks the lowest states in the CB and the effective band-gap increases, which is known as the Burstein-Moss (BM) effect. [44] When a large number of vacancy-related charge carriers are

http://dx.doi.org/10.5772/52189

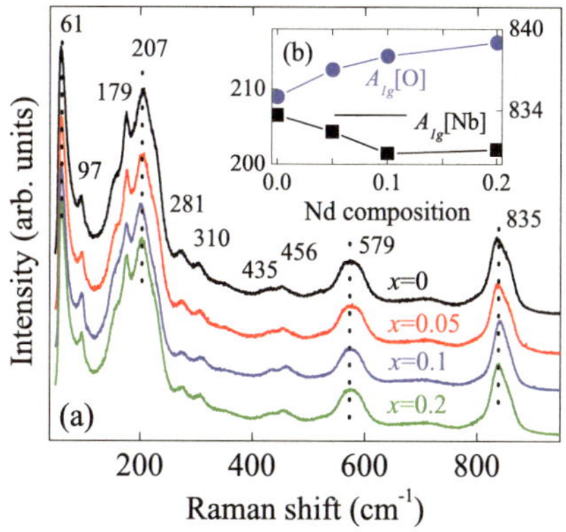

Figure 5. Raman scattering spectra of the SBNN ceramics with different Nd composition (x) recorded at RT. The dashed lines clearly indicate some Raman-active phonon modes. The inset shows the peak frequency variation of the A_{1g}[Nb] phonon mode at about 207 cm^{-1} and the A_{1g}[O] phonon mode at about 836 cm^{-1} as a function of Nd composition. (Figure reproduced with permission from [46]. Copyright 2012, Wiley.)

generated in the system, the lowest lying states in the CB are filled and the vertical distance needed for the optical transition increases. [44] Hence, it can be concluded that, when the Mn composition is 10%, the sharply increase in the band gap may be attributed to the BM shift caused by electrons generated by oxygen vacancies. Besides, the grain size for the ceramic decreases induced by the heavy doping with the Mn composition of 10%, which may also result in a sharply increase of the optical band gap. It can be concluded that the difference of the optical band gap could be due to the dopant composition, the oxygen vacancies, and the crystallinity of the BSMT ceramics.

5. Phase transitions of SBNN ceramics

The general formula of bismuth layer structure ferroelectrics (BLSFs) is given as $(Bi_2O_2)^{2+}$ $(A_{m-1}B_mO_{3m+1})^{2-}$, where A and B are the two types of cations that enter the perovskite unit, and m is the the number of perovskite unit cell between bismuth oxide layers. [20] SBN, which is known to be m=2 member of BLSFs family, has been regarded as a promising ferroelectric material due to low dielectric constants and excellent fatigue resistance. [40] Fig. 5(a) shows the Raman spectra of the SBNN ceramics with different Nd compositions at RT in the spectral range of 50 – 950 cm^{-1}. The Raman selection rules allow 18 phonon modes $(4A_{1g} + 2B_{1g} + 6B_{2g} + 6B_{3g})$ for SBN ceramics at RT. [45] However, less than 10 phonon modes are observed because of the possible overlap of the same symmetry vibration or the weak feature of some vibration bands. [46, 47] According to the assignment of $SrBi_2Nb_2O_9$ single crystal, [45] the Raman phonon modes at about 61, 207 and 835 cm^{-1} can be assigned to the A_{1g} phonon mode, the vibrations at about 179 and 579 cm^{-1} can be assigned to the E_g phonon mode. However, the assignment of other phonon modes are still not clear now. The internal vibrations of NbO_6 octahedra occur in the high-frequency mode region above 200 cm^{-1} because the intragroup binding energy within the NbO_6 octahedra is much larger than the intergroup or crystal binding energy. [15] The composition dependence of the frequencies for two typical phonon modes is illustrated in Fig. 5(b). Note that the A_{1g}[Nb] phonon mode at 207 cm^{-1}, which arises from the distortion of NbO_6 octahedra, generally decreases with the Nd composition, whereas the A_{1g}[O] phonon mode at 835 cm^{-1} mode corresponding to the symmetric Nb–O stretching vibration, increases with the introduction of Nd ions.

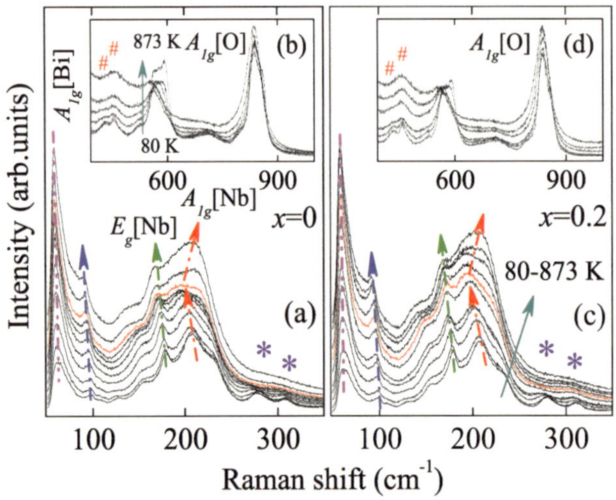

Figure 6. Temperature dependence of the SBNN ceramics with the composition of (a)(b) x=0 and (c)(d) x=0.2. The solid arrows indicate the temperature increasing from 80 K to 873 K, and the dash arrows show the shift of the frequency for the phonon modes with the temperature. The symbol asterisk (∗) and pound sign (#) indicate the two weak E_g phonon modes in the range of 281-310 cm^{-1} and 435-456 cm^{-1}, respectively. (Figure reproduced with permission from [46]. Copyright 2012, Wiley.)

In order to further understand the effect of Nd^{3+} ion substitution on the phonon modes, Fig. 6 presents the temperature dependence of the Raman spectra for the SBNN ceramics with two Nd compositions of x=0 and 0.2 in the temperature range from 80 to 873 K. It suggests that the intensities for all the phonon modes increase with the temperature except for the phonon mode at about 579 cm^{-1}, whose peak has been overwhelmed at high temperature. The broadening band can be assigned to a rigid sublattice mode, in which all the positive and negative ion displacements are equal and opposite. [48] It is found that a strong broadening peak can be observed at 80 K due to the combined effects of two modes splitting from the E_g character mode. However, with increasing the temperature, the frequency and intensity of the mode present a decreasing trend. Because the mode is assigned to the asymmetric Nb−O vibration, it can be concluded that the NbO_6 octahedra is sensitive to the temperature. On the other band, the phonon modes at about 281 and 310 cm^{-1} (labeled by ∗), which are associated with the O−Nb−O bending, become more difficult to be distinguished as the temperature increases and disappear at high temperature. Similar phenomena can be observed for the phonon modes in the range of 435-456 cm^{-1} (labeled by #). The band at about 456 cm^{-1}, which is described to a Ti−O torsional mode, has been assigned as the E_g character and splits into two phonon modes centered at 435 and 456 cm^{-1} at lower temperature. As pointed out by Graves *et al.*, [45] it can be ascribed to the fact that the several E_g phonon modes split into the B_{2g} and B_{3g} phonon modes during the tetragonal to orthorhombic transition. Moreover, the splitting of the phonon modes reveals the structural changes in the SBNN ceramics with the temperature.

Considering that the phase transition temperature is related to the distortion extent of the NbO_6 octahedra for the SBNN ceramics, the temperature dependence of the Raman shift for the A_{1g}[Nb] phonon mode is plotted in Fig. 7(a)-(d). For all the SBNN ceramics, the decrease of the A_{1g}[Nb] phonon mode can be observed as the temperature is increased. Note that an obviously anomalous vibration occurs around the phase transition temperatures: the Raman shift sharply increases with increasing the temperature. In addition, the temperature

http://dx.doi.org/10.5772/52189

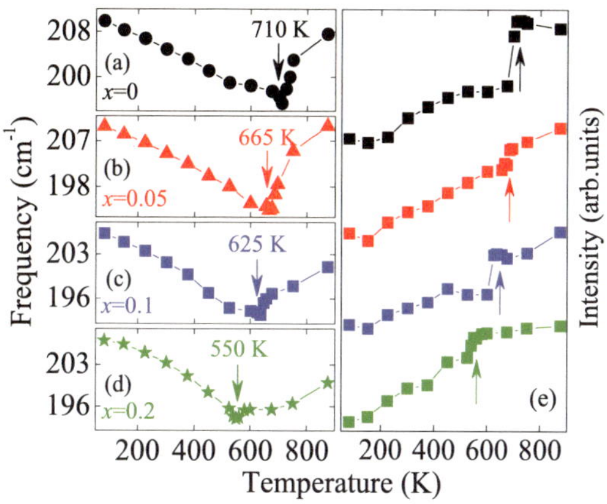

Figure 7. (a)-(d) Raman shift and (e) intensities of the A_{1g}[Nb] phonon mode as a function of the temperature for the SBNN ceramics. The arrows indicate that the anomalous vibration occurs around the ferroelectric to paraelectric phase transition temperatures. (Figure reproduced with permission from [46]. Copyright 2012, Wiley.)

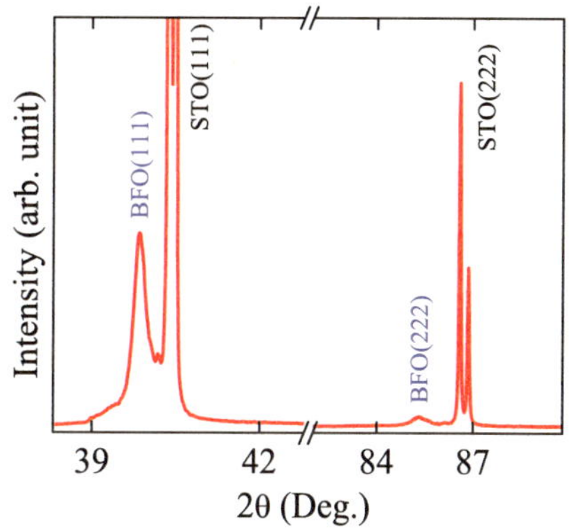

Figure 8. The XRD patterns of the BFO films deposited on STO (111) substrate. (Figure reproduced with permission from [5]. Copyright 2010, American Institute of Physics.)

of the anomalous points is different for the four ceramics: 710 K for x=0, 665 K for x=0.05, 625 K for x=0.1 and 550 K for x=0.2. The different anomalous points can be ascribed to the Nd^{3+} incorporation in the Bi_2O_2 layers. In addition, the phase transition from the orthorhombic to the tetragonal phase may occur at the temperature when the frequency of the A_{1g}[Nb] phonon mode shows the anomalous behavior. Therefore, the sharp change in the temperature dependence of both wavenumber and intensity of the A_{1g}[Nb] phonon mode was successfully applied to probe the phase transition of the SBNN ceramics.

6. Temperature effects on electronic transitions of BFO films

BFO film with the nominal thickness of about 330 nm was prepared on STO (111) substrate by pulsed laser deposition. [5] Fig. 8 shows the XRD pattern of the BFO film and there is no impurity phase. As can be seen, the film is well crystallized with the rhombohedral phase and presents a (111) single crystalline orientation. According to the known Scherrer's equation, the grain size from the (111) diffraction peak was evaluated to about 32 nm. A three-phase layered structure (air/film/substrate) was constructed to

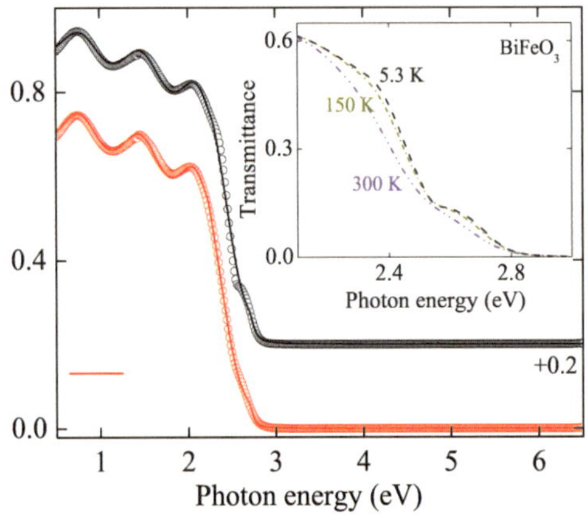

Figure 9. Experimental (dotted curves) and fitting (solid curves) transmittance spectra at temperatures of 300 and 5.3 K, respectively. The inset shows the enlarged band gap region of the BFO film at temperatures of 300, 200, and 5.3 K, respectively. (Figure reproduced with permission from [5]. Copyright 2010, American Institute of Physics.)

simulate the transmittance spectra of the BFO film. It should be emphasized that the normal-incident transmittance spectra cannot be sensitive to the thinner surface rough layer, which could be several nanometers and much less than the film thickness. Therefore, the surface rough layer can be reasonably neglected owing to a slight contribution in the evaluation of the optical properties. The optical constants of the BFO film can be expressed using four Tauc-Lorentz (TL) oscillators. [50] As an example, the experimental and fitting transmittance spectra of the BFO film at 300 and 5.3 K are shown in Fig. 9 with the dotted and solid curves, respectively. Note that the symmetrical interference period indicates that the film is of good uniformity and crystallization. From Fig. 9, it can be observed that the absorption edge remarkably shift toward the lower energy with increasing the temperature, suggesting that the OBG of the film has a negative temperature coefficient. Note that the shift at high temperature region (100-300 K) is larger than that at low temperature region (5.3-100 K). This is because the quantities of the conduction band downward and the valence band upward are different under the distinct temperature regions. Especially, two broadening shoulder structures appear and the intensities become much stronger with decreasing the temperature. The similar phenomena have been observed at 2.5 eV when the temperature decreases to about 4 K, which represents the onset of the optical absorption. [51] Furthermore, the shoulders are simply low-lying features of the electronic structure or evidence for excitonic character.

Based on the theoretical calculations and experimental observations, the four energy bands can be uniquely assigned to the following electronic transitions: (1) on-site Fe^{3+} d to d crystal field transition; (2) majority channel Fe 3d to O 2p charge transfer excitation; (3) minority channel dipole-allowed O p to Fe d charge transfer excitation; and (4) strong hybridized majority channel O p and Fe d to Bi p state excitation, respectively. [51–55] Within the experimental error bars, the energy positions shift toward the higher energy at the temperature of 5.3 K except for the second excitation, which can be attributed to the energy band variations. Nevertheless, the origin of the abnormal shift for the second excitation is unclear in the present work. Under the influence of the tetrahedral crystal field, the Fe 3d orbital states split into t_{2g} and e_g state and the t_{2g} state strongly hybrided with the O p orbital. [52] With decreasing the temperature, the t_{2g} and e_g states can be located at different level in

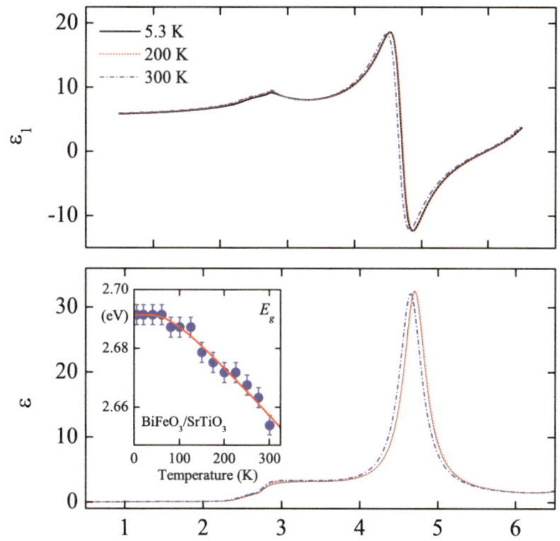

Figure 10. The dielectric functions of the BFO films in the photon energy range of 0.5-6.5 eV at 300, 200, and 5.3 K, respectively. The inset shows the temperature dependence of the E_g (dotted curve) and Bose-Einstein model fitting result (solid curve). (Figure reproduced with permission from [5]. Copyright 2010, American Institute of Physics.)

the energy space, which can affect the electronic excited ability of the Bi, Fe, and O states. On the other hand, the evaluated optical constants of the BFO film is presented in Fig. 10. The real and imaginary parts of dielectric functions increase with the temperature at the lower photon energy while decrease with further increasing the photon energy. With decreasing the temperature, both the ε_1 and ε_2 shift toward the higher energy. The phenomena are related to the modification of the electronic structure such as the fundamental band gap absorption under the lower temperature. From the inset of Fig. 10, it can be found that the E_g value increases from 2.65 ± 0.01 to 2.69 ± 0.01 eV, corresponding to decreasing the temperature from 300 to 5.3 K, which indicates that the total shift value of the E_g is about 40 meV. The observed decrease in the E_g with the temperature can be described using the Bose-Einstein model. It is widely recognized that the electron-phonon interaction and the lattice thermal expansion are responsible for the shrinkage in the optical band gap with the temperature.

7. Conclusion and remarks

In summary, electronic band structures and phase transitions of perovskite (ABO_3)-type Manganese (Mn) doped $Ba_{0.4}Sr_{0.6}TiO_3$ ceramics, $SrBi_{2-x}Nd_xNb_2O_9$, and $BiFeO_3$ materials have been investigated from infrared to ultraviolet transmittance, SE and temperature-dependent Raman scattering techniques. The interband electronic transitions and dielectric functions of these materials could be readily obtained in the wider photon energy region. Meanwhile, the phase transition temperature can be easily determined by phonon scattering measurements, indicating that the present solid state spectroscopy is useful to further clarify the physical phenomena for perovskite-type ferroelectric/multiferroics oxides.

It should be emphasized that optical properties and their related phenomena, such as phase transition and domain status have not been well investigated. The following factors must be addressed: (1) Growth of high-quality perovskite FE materials (crystal, ceramic and film); (2) Theoretical model and explanations; (3) Improved experimental methods. As for solid state spectroscopic technique, however, it can uniquely discover the electronic band structure of FE system. One can think that the electronic transition will be changed during the phase transition process owing to the crystal structure variation. Thus, we can check the status by recording the spectral response. Evidently, some optical setup at elevated temperatures are necessary because most of FE materials have the high T_C. In our research group, Transmittane/Reflectance, SE, and Raman systems from LHe temperature to about 800 K have been developed. Our next goal is to characterize the physical information of some typical FE materials in a wider spectral and temperature ranges, which play an important role in clarifying the structure transitions and the intrinsic origin.

Acknowledgments

The authors would like to thank Dr. Wenwu Li for valuable discussions. This work was financially supported by the Major State Basic Research Development Program of China (Grant Nos. 2011CB922200 and 2013CB922300), the Natural Science Foundation of China (Grant Nos. 11074076, 60906046 and 61106122), the Program of New Century Excellent Talents, MOE (Grant No. NCET-08-0192) and PCSIRT, the Projects of Science and Technology Commission of Shanghai Municipality (Grant Nos. 11520701300, 10DJ1400201 and 10SG28), and the Program for Professor of Special Appointment (Eastern Scholar) at Shanghai Institutions of Higher Learning.

Author details

Zhigao Hu[1,*], Junhao Chu[1,2,*],
Yawei Li[1], Kai Jiang[1] and Ziqiang Zhu[1]

1 Key Laboratory of Polar Materials and Devices, Ministry of Education, Department of Electronic Engineering, East China Normal University, Shanghai, People's Republic of China
2 National Laboratory for Infrared Physics, Shanghai Institute of Technical Physics, Chinese Academy of Sciences, Shanghai, People's Republic of China

References

[1] Nagaraj, B.; Sawhney, T.; Perusse, S.; Aggarwal, S.; Ramesh, R.; Kaushik, V. S.; Zafar, S.; Jones, R. E.; Lee, J. -H.; Balu, V. & Lee, J. (1999). (Ba,Sr)TiO_3 thin films with conducting perovskite electrodes for dynamic random access memory applications. *Appl. Phys. Lett.*, 74, 21, 3194-3196

[2] Roy, S. C.; Sharma, G. L. & Bhatnagar, M. C. (2007). Large blue shift in the optical band-gap of sol-gel derived $Ba_{0.5}Sr_{0.5}TiO_3$ thin films. *Solid State Commun.*, 141, 5, 243-247

[3] Kozyrev, A. B.; Kanareykin, A. D.; Nenasheva, E. A.; Osadchy, V. N. & Kosmin, D. M. (2009). Observation of an anomalous correlation between permittivity and tunability of a doped (Ba,Sr)TiO_3 ferroelectric ceramic developed for microwave applications. *Appl. Phys. Lett.*, 95, 01, 012908(1-3)

[4] Catalan, G. & Scott, J. F. (2009). Physics and applications of bismuth ferrite. *Adv. Mater.*, 21, 24, 2463-2485

[5] Li, W. W.; Zhu, J. J.; Wu, J. D.; Gan, J.; Hu, Z. G.; Zhu, M. & Chu, J. H. (2010). Temperature dependence of electronic transitions and optical properties in multiferroic $BiFeO_3$ nanocrystalline film determined from transmittance spectra. *Appl. Phys. Lett.*, 97, 12, 121102(1-3)

[6] Jiang, K.; Zhu, J. J.; Wu, J. D.; Sun, J.; Hu, Z. G. & Chu, J. H. (2011). Influences of oxygen pressure on optical properties and interband electronic transitions in multiferroic bismuth ferrite nanocrystalline films grown by pulsed laser deposition. *ACS Appl. Mater. Interfaces*, 03, 12, 4844-4852

[7] Su, B. & Button, T. W. (2004). Microstructure and dielectric properties of Mg-doped barium strontium titanate ceramics. *J. Appl. Phys.*, 95, 03, 1382-1385

[8] Xu, S.; Qu, Y. & Zhang, C. (2009). Effect of Mg^{2+} content on the dielectric properties of $Ba_{0.65-x}Sr_{0.35}Mg_xTiO_3$ ceramics. *J. Appl. Phys.*, 106, 01, 014107(1-5)

[9] Kim, S. S. & Park, C. (1999). Leakage current behaviors of acceptor- and donor-doped $(Ba_{0.5}Sr_{0.5})TiO_3$ thin films. *Appl. Phys. Lett.*, 75, 17, 2554-2556

[10] Yuan, Z.; Lin, Y.; Weaver, J.; Chen, X.; Chen, C. L.; Subramanyam, G.; Jiang, J. C. & Meletis, E. I. (2005). Large dielectric tunability and microwave properties of Mn-doped $(Ba,Sr)TiO_3$ thin films. *Appl. Phys. Lett.*, 87, 15, 152901(1-3)

[11] Zhang, J. J.; Zhai, J. W. & Yao, X. (2009). Dielectric tunable properties of low-loss $Ba_{0.4}Sr_{0.6}Ti_{1-y}Mn_yO_3$ceramics. *Scripta Mater.*, 61, 07, 764-767

[12] Liu, M.; Ma, C.; Collins, G.; Liu, J.; Chen, C.; Shui, L.; Wang, H.; Dai, C.; Lin, Y.; He, J.; Jiang, J.; Meletis, E. I. & Zhang, Q. (2010). Microwave dielectric properties with optimized Mn-doped $Ba_{0.6}Sr_{0.4}TiO_3$ highly epitaxial thin films. *Cryst. Growth Des.*, 10, 10, 4221-4223

[13] Li, Y. L. & Qu, Y. F. (2009). Dielectric properties and substitution mechanism of samarium-doped $Ba_{0.68}Sr_{0.32}TiO_3$ ceramics. *Mater. Res. Bull.*, 44, 01, 82-85

[14] Zhang, J. J.; Zhai, J. W.; Chou, X. J. & Yao, X. (2008). Influence of rare-earth addition on microstructure and dielectric behavior of $Ba_{0.6}Sr_{0.4}TiO_3$ ceramics. *Mater. Chem. Phys.*, 111, 02, 409-413

[15] Huang, S. M.; Li, Y. C.; Feng, C. D.; Gu, M. & Liu, X. L. (2008). Dielectric and structural properties of layer-structured $Sr_{1-x}Ca_xBi_2Nb_2O_9$. *J. Am. Ceram. Soc.*, 91, 09, 2933-2937

[16] Huang, S. M.; Feng, C. D.; Chen, L. D. & Wen, X. W. (2005). Dielectric properties of $SrBi_{2-x}Pr_xNb_2O_9$ ceramics (x=0, 0.04 and 0.2). *Solid State Commun.*, 133, 6, 375-379

[17] Sun, L.; Feng, C. D.; Chen, L. D. & Huang, S. M. (2007). Effect of substitution of Nd^3 for Bi^3 on the dielectric properties and structures of $SrBi_{2-x}Nd_xNb_2O_9$ bismuth layer-structured ceramics. *J. Appl. Phys.*, 101, 08, 084102(1-5)

[18] Wu, Y.; Forbess, M. J.; Seraji, S.; Limmer, S. J.; Chou, T. P.; Nguyen, C. & Cao, G. (2001). Doping effect in layer structured $SrBi_2Nb_2O_9$ ferroelectrics. *J. Appl. Phys.*, 90, 10, 5296-5302

[19] Liu, G. Z.; Wang, C.; Gu, H. S. & Lu, H. B. (2007). Raman scattering study of La-doped $SrBi_2Nb_2O_9$ ceramics. *J. Phys. D: Appl. Phys.*, 40, 24, 7817-7820

[20] Fang, P.; Fan, H.; Li, J. & Liang, F. (2010). Lanthanum induced larger polarization and dielectric relaxation in Aurivillius phase $SrBi_{2-x}La_xNb_2O_9$ ferroelectric ceramics. *J. Appl. Phys.*, 107, 06, 064104(1-4)

[21] Verma, M.; Sreenivas, K. & Gupta, V. (2009). Influence of La doping on structural and dielectric properties of $SrBi_2Nb_2O_9$ ceramics. *J. Appl. Phys.*, 105, 02, 024511(1-6)

[22] Speghini, A.; Bettinelli, M.; Caldiño, U.; Ramírez, M. O.; Jaque, D.; Bausá, L. E. & Solé, J. G. (2006). Phase transition in $Sr_xBa_{1-x}Nb_2O_6$ ferroelectric crystals probed by Raman spectroscopy. *J. Phys. D: Appl. Phys.*, 39, 23, 4930-4934

[23] Tenne, D. A.; Bruchhausen, A.; Kimura, N. D. L.; Fainstein, A.; Katiyar, R. S.; A. Cantarero,; Soukiassian, A.; Vaithyanathan, V.; Haeni, J. H.; Tian, W.; Schlom, D. G.; Choi, K. J.; Kim, D. M.; Eom, C. B.; Sun, H. P.; Pan, X. Q.; Li, Y. L.; Chen, L. Q.; Jia, Q. X.; Nakhmanson, S. M.; Rabe, K. M. & Xi, X. X. (2006). Probing nanoscale ferroelectricity by ultraviolet Raman spectroscopy. *Science*, 313, 5793, 1614-1616

[24] Zhu, J. J.; Li, W. W.; Xu, G. S.; Jiang, K.; Hu, Z. G. & Chu, J. H. (2011). A phenomenalogical model of electronic band structure in ferroelectric $Pb(In_{1/2}Nb_{1/2})O_3$-$Pb(Mg_{1/3}Nb_{2/3})O_3$-$PbTiO_3$ single crystals around morphotropic phase boundary determined by temperature-dependent transmittance spectra. *Acta Materialia*, 59, 17, 6684-6690

[25] Zhang, W. J.; Li, W. W.; Chen, X. G.; Hu, Z. G.; Liu, W.; Wang, G. S.; Dong, X.L. & Chu, J.H. (2011). Phonon mode and phase transition behaviors of $(1-x)PbSc_{1/2}Ta_{1/2}O_3$-$xPbHfO_3$ relaxor ferroelectric ceramics determined by temperature-dependent Raman spectra. *Appl. Phys. Lett.*, 99, 04, 041902(1-3)

[26] Zhu, J. J.; Li, W. W.; Xu, G.S.; Jiang, K.; Hu, Z. G.; Zhu, M. & Chu, J. H. (2011). Abnormal temperature dependence of interband electronic transitions in relaxor-based ferroelectric $(1-x)Pb(Mg_{1/3}Nb_{2/3})O_3$-$xPbTiO_3$ (x=0.24 and 0.31) single crystals. *Appl. Phys. Lett.*, 98, 09, 091913(1-3)

[27] Chen, X.; Jiang, K.; Hu, Z. G.; Chen, X. F.; Wang, G. S.; Dong, X. L. & Chu, J. H. (2012). Abnormal electronic transition variations of lanthanum-modified lead zironate stannate titanate ceramics near morphotropic phase boundary: A spectroscopic evidence. *Appl. Phys. Lett.*, 101, 01, 011914(1-5)

[28] Liao, Y. Y.; Li, Y. W.; Hu, Z. G. & Chu, J. H. (2012). Temperature dependent phonon Raman scattering of highly *a*-axis oriented $CoFe_2O_4$ inverse spinel ferromagnetic films grown by pulsed laser deposition. *Appl. Phys. Lett.*, 100, 07, 071905(1-4)

[29] Li, W. W.; Yu, Q.; Liang, J. R.; Jiang, K.; Hu, Z. G.; Liu, J.; Chen, H. D.; & Chu, J. H. (2011). Intrinsic evolutions of optical functions, band gap, and higher-energy electronic transitions in VO_2 film near the metal-insulator transition region. *Appl. Phys. Lett.*, 99, 24, 241903(1-3)

[30] Han, M. J.; Jiang, K.; Zhang, J. Z.; Li, Y. W.; Hu, Z. G. & Chu, J. H. (2011). Temperature dependent phonon evolutions and optical properties of highly *c*-axis oriented $CuGaO_2$ semiconductor films grown by the sol-gel method. *Appl. Phys. Lett.*, 99, 13, 131104(1-3)

[31] Yu, W. L.; Jiang, K.; Wu, J. D.; Gan, J.; Zhu, M.; Hu, Z. G. & Chu, J. H. (2011). Electronic structures and excitonic transitions in nanocrystalline iron-doped tin dioxide diluted magnetic semiconductor films: an optical spectroscopic study. *Phys. Chem. Chem. Phys.*, 13, 13, 6211-6222

[32] Wang, J.; Neaton, J. B.; Zheng, H.; Nagarajan, V.; Ogale, S. B.; Liu, B.; Viehland, D.; Vaithyanathan, V.; Schlom, D. G.; Waghmare, U. V.; Spaldin, N. A.; Rabe, K. M.; Wuttig, M. & Ramesh, R. (2003). Epitaxial $BiFeO_3$ multiferroic thin film heterostructures. *Science*, 299, 1719-1722

[33] Singh, M. K. & Katiyar, R. S. (2011). Phonon anomalies near the magnetic phase transitions in $BiFeO_3$ thin films with rhombohedral R_{3C} symmetry. *J. Appl. Phys.*, 109, 07, 07D916(1-3)

[34] Choi, S. G.; Yi, H. T.; Cheong, S. W.; Hilfiker, J. N.; France, R. & Norman, A. G. (2011). Optical anisotropy and charge-transfer transition energies in $BiFeO_3$ from 1.0 to 5.5 eV. *Phys. Rev. B*, 83, 10, 100101(R)(1-4)

[35] Basu, S. R.; Martin, L. W.; Chu, Y. H.; Gajek, M.; Ramesh, R.; Rai, R. C.; Xu, X. & Musfeldt, J. L. (2008). Influence of sample processing parameters on thermal boundary conductance value in an Al/AlN system. *Appl. Phys. Lett.* 92, 09, 091905(1-3)

[36] Balke, N.; Choudhury, S.; Jesse, S.; Huijben, M.; Chu, Y. H.; Baddorf, A. P.; Chen, L. Q.; Ramesh, R. & Kalinin, S. V. (2009). Deterministic control of ferroelastic switching in multiferroic materials. *Nature Nanotechnology*, 4, 12, 868-875

[37] Yang, S. Y.; Martin, L. W.; Byrnes, S. J.; Conry, T. E.; Basu, S. R.; Paran, D.; Reichertz, L.; Ihlefeld, J.; Adamo, C.; Melville, A.; Chu, Y. H.; Yang, C. H.; Musfeldt, J. L.; Schlom, D. G.; Ager, I. J. W. & Ramesh, R. (2009). Photovoltaic effects in $BiFeO_3$. *Appl. Phys. Lett.*, 95, 06, 062909(1-3)

[38] Choi, T.; Lee, S.; Choi, Y. J.; Kiryukhin, V. & Cheong, S.-W. (2009). Switchable ferroelectric diode and photovoltaic effect in $BiFeO_3$. *Science*, 324, 5923, 63-66.

[39] Jiang, K.; Zhang, J. Z.; Yu, W. L.; Hu, Z. G. & Chu, J. H. (2012). Manganese doping effects on interband electronic transitions, lattice vibrations, and dielectric functions of perovskite-type $Ba_{0.4}Sr_{0.6}TiO_3$ ferroelectric ceramics. *Appl. Phys. A-Mater. Sci. Process.*, 106, 04, 877-884

[40] Zhu, M.; Sun, L.; Li, W. W.; Yu, W. L.; Li, Y. W.; Hu, Z. G. & Chu, J. H. (2010). Lattice vibrations and dielectric functions of ferroelectric $SrBi_{2-x},Nd_xNb_2O_9$ bismuth

layer-structured ceramics determined by infrared reflectance spectra. *Mater. Res. Bull.*, 45, 11, 1654-1658

[41] Yu, W. L.; Li, W. W.; Wu, J. D.; Sun, J.; Zhu, J. J.; Zhu, M.; Hu, Z. G. & Chu, J. H. (2010). Far-infrared-ultraviolet dielectric function, lattice vibration, and photoluminescence properties of diluted magnetic semiconductor $Sn_{1-x}Mn_xO_2/c$-sapphire nanocrystalline films. *J. Phys. Chem. C*, 114 , 18, 8593-8600

[42] Kohiki, S.; Arai, M.; Yoshikawa, H.; Fukushima, S.; Oku, M. & Waseda, Y. (2000). Energy-loss structure in core-level photoemission satellites of $SrTiO_3$, $SrTiO_3$:La, and $SrTiO_3$:Nb. *Phys. Rev. B*, 62, 12, 7964-7969

[43] Warren, W. L.; Vanheusden, K.; Dimos, D.; Pike, G. E. & Tuttle, B. A. (1996). Oxygen vacancy motion in perovskite oxides. *J. Am. Ceram. Soc.*, 79, 02, 536-538

[44] Burstein, E. (1954). Anomalous optical absorption limit in InSb. *Phys. Rev.*, 93, 03, 632-633

[45] Graves, P. R.; Hua, G.; Myhra, S. & Thompson, J. G. (1995). The Raman modes of the Aurivillius phases: Temperature and polarization dependence. *J. Solid State Chem.*, 114, 01, 112-122

[46] Jiang, K.; Chen, X. G.; Li, W. W.; Zhan, Z. N.; Sun, L.; Hu, Z. G. & Chu, J. H. (2012). Doping effect on the phase transition temperature in ferroelectric $SrBi_{2-x}Nd_xNb_2O_9$ layer-structured ceramics: A micro-Raman scattering study. *J. Raman Spectrosc.*, 43, 04, 583-587

[47] Liang, K.; Qi, Y. J. & Lu, C. J. (2009). Temperature-dependent Raman scattering in ferroelectric $Bi_{4-x}Nd_xTi_3O_{12}$ (x=0, 0.5, 0.85) single crystals. *J. Raman Spectrosc.*, 40, 12, 2088-2091

[48] Dobal, P. S. & Katiyar, R. S. (2002). Studies on ferroelectric perovskites and Bi-layered compounds using micro-Raman spectroscopy. *J. Raman Spectrosc.*, 33, 06, 405-423

[49] Kitaev, Y. E.; Aroyo, M. I. & Mato, J. M. P. (2007). Site symmetry approach to phase transitions in perovskite-related ferroelectric compounds. *Phys. Rev. B*, 75, 06, 064110(1-11)

[50] Jellison, G. E. Jr. & Modine, F. A. (1996). Parameterization of the optical functions of amorphous materials in the interband region. *Appl. Phys. Lett.*, 69, 03, 371-373. Erratum: "Parameterization of the optical functions of amorphous materials in the interband region". 69, 03, 2137

[51] Basu, S. R.; Martin, L. W.; Chu, Y. H.; Gajek, M.; Ramesh, R.; Rai, R. C.; Xu, X. & Musfeldt, J. L. (2008). Photoconductivity in $BiFeO_3$ thin films. *Appl. Phys. Lett.*, 92, 09, 091905(1-3)

[52] Chen, P.; Podraza, N. J.; Xu, X. S.; Melville, A.; Vlahos, E.; Gopalan, V.; Ramesh, R.; Schlom, D. G. & Musfeldt, J. L. (2010). Optical properties of quasi-tetragonal $BiFeO_3$ thin films. *Appl. Phys. Lett.*, 96, 13, 131907(1-3)

[53] Xu, X. S.; Brinzari, T. V.; Lee, S.; Chu, Y. H.; Martin, L. W.; Kumar, A.; McGill, S.; Rai, R. C.; Ramesh, R.; Gopalan, V.; Cheong, S. W. & Musfeldt, J. L. (2009). Optical properties and magnetochromism in multiferroic $BiFeO_3$. *Phys. Rev. B*, 79, 13, 134425(1-4)

[54] Ramirez, M. O.; Kumar, A.; Denev, S. A.; Podraza, N. J.; Xu, X. S.; Rai, R. C.; Chu, Y. H.; Seidel, J.; Martin, L. W.; Yang, S. -Y.; Saiz, E.; Ihlefeld, J. F.; Lee, S.; Klug, J.; Cheong, S. W.; Bedzyk, M. J.; Auciello, O.; Schlom, D. G.; Ramesh, R.; Orenstein, J.; Musfeldt, J. L. & Gopalan, V. (2009). Magnon sidebands and spin-charge coupling in bismuth ferrite probed by nonlinear optical spectroscopy. *Phys. Rev. B*, 79, 22, 224106(1-9)

[55] Ramirez, M. O.; Kumar, A.; Denev, S. A.; Chu, Y. H.; Seidel, J.; Martin, L. W.; Yang, S. -Y.; Rai, R. C.; Xue, X. S.; Ihlefeld, J. F.; Podraza, N. J.; Saiz, E.; Lee, S.; Klug, J.; Cheong, S. W.; Bedzyk, M. J.; Auciello, O.; Schlom, D. G.; Orenstein, J.; Ramesh, R.; Musfeldt, J. L.; Litvinchuk, A. P. & Gopalan, V. (2009). Spin-charge-lattice coupling through resonant multimagnon excitations in multiferroic $BiFeO_3$. *Appl. Phys. Lett.*, 94, 16, 161905(1-3)

Chapter 7

Phase Diagram of The Ternary BaO-Bi_2O_3-B_2O_3 System: New Compounds and Glass Ceramics Characterisation

Martun Hovhannisyan

Additional information is available at the end of the chapter

http://dx.doi.org/10.5772/52405

1. Introduction

Restriction and an interdiction on use of toxic materials in electronics products since 2006 have promoted an intensification of development new ecologically friendly materials (glasses, glass ceramics, ceramics) with attractive properties. It has stimulated new lead (cadmium) free systems with good glass forming abilities investigations and new stoiciometric and eutectic points revealing and characterization. Alkaline-earth bismuth borate ternary systems were a good candidate for this purpose, because the binary Bi_2O_3-B_2O_3 system have propensity for glass formation and set of binary compounds and eutectics [1 - 3]. Furthermore, bismuth borate single crystals and glass ceramics have nonlinear optical (NLO) properties and other attractive properties [4 - 7]. Both these factors are reasons for further study of binary and ternary bismuth borate systems, and the glasses which they form.

The phase diagram of the Bi_2O_3–B_2O_3 system was first determined by Levin &Daniel in 1962 [2] and five crystalline compounds, $Bi_{24}B_{12}O_{39}$, $Bi_4B_2O_9$, $Bi_3B_5O_{12}$, BiB_3O_6 and $Bi_2B_8O_{15}$, were identified. Later Pottier revealed a sixth compound, $BiBO_3$ (bismuth orthoborate) [8], which was missing in the original phase diagram [2]. There are no doubts about the existence of $BiBO_3$ now: Becker with co-workers have confirmed existence of bismuth orthoborate [5, 9, 10] and its transparent colourless single crystals of $BiBO_3$ have recently been grown from the melt and characterized by Becker & Froehlich [10]. Monophase samples of both crystalline $BiBO_3$ modifications were obtained by crystallisation below 550°C of bismuth borate glasses with 50-57 mol% B_2O_3 [10]. However, these authors did not correct the phase diagram, and did not determine the melting point of $BiBO_3$ or the eutectic composition between $BiBO_3$ and $Bi_3B_5O_{12}$. The compound $BiBO_3$ and this eutectic point are clearly given on the Zargarova & Kasumova's version of the B_2O_3–Bi_2O_3 phase diagram, without indication of their melting points and the eutectic composition [11].

Kargin with co-authors [12] by DTA and X-ray analysis have studied conditions of metastable phases formation at system Bi_2O_3-B_2O_3 melts crystallization. They have confirmed existence of metastable $BiBO_3$ compound and for the first time have specified on congruent character of its melting. Authors also establish formation of a metastable phase of $5Bi_2O_3 \bullet 3B_2O_3$ composition. Both compounds together with initial Bi_2O_3 and B_2O_3 are present on the metastable state diagram of the Bi_2O_3-B_2O_3 system constructed by them.

Presence of five compounds on the known Bi_2O_3-B_2O_3 phase diagram has naturally led to formation of five eutectics compositions containing (mol % B_2O_3): 19.14 (622°C), 44.4 (646 °C), 73.5 (698°C), 76.6 (695°C) and 81.04 (709°C). There is an area of phase separation traditional for borate systems, observed for compositions containing 81-100 mol % B_2O_3 [2]. Though according to [12], the area of stable phase separation is stretched to 58-95 mol % B_2O_3.

Interest to ternary alkali free bismuth borate systems M_xO_y-Bi_2O_3-B_2O_3 (M=Zn,Sr,Ca,Ba) studies has amplified recently. Various research groups (Russian, Canadian, Armenian) worked in this area during 1990-2009 and revealed a number of ternary compounds, determined their structure, optical and nonlinear optical properties. Thus, three ternary zinc bismuth borate compounds have been revealed in the ZnO-Bi_2O_3-B_2O_3 system. At first Zargarova& Kasumova have revealed $ZnBi_4B_2O_{10}$ and $ZnBiBO_4$ compounds [11]. Later Barbier with co-authors by solid-state reaction have synthesized third melilite type $ZnBi_2B_2O_7$ compound with large SHG (four time higher as KDP) [13].

Barbier & Cranswick at first two novel noncentrosymmetric $MBi_2B_2O_7$ or $MBi_2O(BO_3)_2$ (M=Ca, Sr) compounds have synthesized by solid-state reactions in air at temperatures in the 600–700°C range [14]. Their crystal structures have been determined and refined using powder neutron diffraction data. $CaBi_2B_2O_7$ compound has SHG response two time higher as KDP [14]. However, authors didn't pay attention for both compounds melting behavior.

Egorisheva with co-authors have studied phase relation in the CaO-Bi_2O_3-B_2O_3 system and constructs the 600 °C (subsolidus) section of its phase diagram [15]. A new ternary compound of composition $CaBi_2B_4O_{10}$ was identified and the existence of $CaBi_2B_2O_7$ ternary compound was comfirmed. Both compounds had incongruent melting at 700 and 783 °C respectively and liquidus temperature about 900-930 °C.

Kargin with co-workers have studied phase relation in the SrO-Bi_2O_3-B_2O_3 system in subsolidus at 600 °C [16]. Two new ternary compound of $Sr_7Bi_8B_{18}O_{46}$ and $SrBiBO_4$ compositions were identified. Both compounds had incongruent melting at 760 and 820 °C without indication liquidus temperature. However, later Barbier et el. have discribe new novel centrosymmetric borate $SrBi_2OB_4O_9$ ($SrBi_2B_4O_{10}$) forming in the SrO–Bi_2O_3–B_2O_3 system [17], thereby having substituted under doubt existence of previously reported $Sr_7Bi_8B_{18}O_{46}$ compound [16].

The uniqueness of the BaO-Bi_2O_3-B_2O_3 system is shown by the available sets of compounds and eutectics both in the binary Bi_2O_3–B_2O_3 and BaO–B_2O_3 systems. Seven compounds are known in the BaO-B_2O_3 system. Four congruent melting binary compounds $Ba_3B_2O_6$, BaB_2O_4, BaB_4O_7, BaB_8O_{13} with melting points(m.p.) 1383, 1105, 910, 889°C accordingly were found by Levin & McMurdie [18, 19]. Further, Green and Wahler have found out new congruent melted at 890°C $Ba_2B_5O_{17}$ compound at the ternary BaO-B_2O_3-Al_2O_3 system investigation [20].

Hubner confirmed an existence of the congruent melted $Ba_2B_5O_{17}$ compound with m.p. 890 °C, and revealed two new compounds $Ba_4B_2O_7$, $Ba_2B_2O_5$ [21]. However, all scientists and researches have used the melting diagram of the BaO-B_2O_3 system created by Levin & McMurdie up to now, without the indication in it the specific areas of existence of new compounds and eutectic points among them [18].

Both these factors were the reason of the BaO-B_2O_3 system phase diagram correction made by Hovhannisyan R.M. [22]. Author has revealed fields of $Ba_2B_5O_{17}$ and BaB_4O_7 compounds crystallisation and new eutectic points which are absent on the diagram constructed by Levin & McMurdie [18]. Six binary eutectic compositions containing 31.5, 37.5, 63.5, 68.5, 76.0, 83.4 mol % B_2O_3 with melting points 1025, 915, 905, 895, 869 and 878°C accordingly were on the diagram after correction.

Area of two immiscible liquids established by Levin & McMurdie [35] in the BaO-B_2O_3 in an interval of 1.5 to 30wt. % BaO content, it has been confirmed in the subsequent by other authors. However, the temperature of the liquation couple, which are 1150, 1180, 1256 °C according to [23] and 1539°C according to [24] is discussed till now.

There are no full version of the phase diagram of the BaO-Bi_2O_3 system till now [1]. It is very complex system, which is very critical to atmosphere and pressure at experiment carrying out [25 - 28]. Two low melted eutectic areas (740-790°C) clear observed on phase diagram studied in air or oxigen in high bismuth content region around 5-7 mol%BaO and 25-30 mol %BaO [26 - 28].

All research groups payd special attention to BaO-Bi_2O_3-B_2O_3 system studies and new ternary compounds revealing and characterisation. Barbier *et al.* have studied seven compositions in the ternary BaO-B_2O_3-Bi_2O_3 system by solid state synthesis at temperatures below 650°C and $BaBiBO_4$, or $BaBi(BO_3)O$, a novel borate compound, has been made and chracterisised [29]. Above 650°C it decays with bismuth borate glass formation. A powder sample of $BaBiBO_4$ had a second harmonic signal with a NLO efficiency equal to five times that of KDP.

Practically in parallel, Egorysheva with co-workers have been investigated phase equilibrium in the Bi_2O_3-BaB_2O_4-B_2O_3 system by X-ray analysis and DTA [30, 31]. Studies were spent by the samples solid state synthesis in closed Pt crucibles in muffle furnaces at the temperature range 500-750 ^{0}C, that corresponds to sub-solidus area. The synthesis duration (with intermediate cakes regrinding) were 6-16 days. They confirmed presence of $BaBiBO_4$ and have revealed three new compounds: $BaBiB_{11}O_{19}$, $BaBi_2B_4O_{10}$, $Ba_3BiB_3O_9$. $BaBiB_{11}O_{19}$, $BaBi_2B_4O_{10}$ have congruent melting at 830 and 730 °C respectively and $BaBiBO_4$ melt incongruently at 780^{0}C. $Ba_3BiB_3O_9$ undergoes a phase transition at 850°C and exist up to 885°C, were decompose in the solid state [31].

Recently single crystals of $BaBi_2B_4O_{10}$ composition were grown by cooling of a melt with the stoichiometric composition with cooling rate 0.5 K/h [32]. They have once again confirmed existence of $BaBi_2B_4O_{10}$ stoichiometric compound earlier obtained by solid state synthesis.

In 1972 Elwell with co-workers investigated the $BaO–B_2O_3–Bi_2O_3$ system by hot stage microscopy and a new ternary eutectic composition, $23.4BaO•62.4Bi_2O_3•14.2B_2O_3$ (wt%), with a low liquidus temperature of 600°C, was revealed for ferrite spinel growth [33].

Using different melts cooling rates Hovhannisyan, M. with co-authors at first have determined large glass-forming field in the $BaO-Bi_2O_3-B_2O_3$ system, which includes all eutectics in the binary $Bi_2O_3-B_2O_3$, $BaO-B_2O_3$ and $BaO-Bi_2O_3$ systems and covers majority of the concentration triangles, reaching up to 90 mol% Bi_2O_3. [34, 35].

The methodology based on glass samples investigation was more effective at $BaO-Bi_2O_3-B_2O_3$ system phase diagram construction, than a traditional technique based on solid state sintered samples studies. Because DTA curves of glasses, to the contrary DTA curves of solid state sintered samples, indicates their all characteristics temperatures, includes exothermal effects of glass crystallizations and endothermic effects of formed crystalline phases melting. It has allowed us to reveal two new $BaBi_2B_2O_7$ and $BaBi_{10}B_6O_{25}$ congruent melted at 725 and 690°C respectively compounds in the $BaO-Bi_2O_3-B_2O_3$ system [34 - 36].

However, our further studies of glasses and glass ceramics in this system have shown necessity of glass forming diagram correction and phase diagram construction in the ternary $BaO-Bi_2O_3-B_2O_3$ system and present these data to scientific communitty. Another aim of this work is both known and novel stoichiometric ternary barium bismuth borates compounds characterisation in glassy, glass ceramic and ceramic states for further practical application.

2. Experimental

About three hundred samples of various binary and ternary compositions have been synthesized and tested in $BaO- Bi_2O_3-B_2O_3$ system. Compositions were prepared from "chemically pure" grade $BaCO_3$, H_3BO_3 and Bi_2O_3 at 2.5-5.0 mol % intervals. The most part of samples has been obtained as glasses by various cooling rates depending on melts glass forming abilities: as bulk glass plates with thickness 6,5 ÷7mm by casting on metallic plate (up to 10 $^K/s$), as monolithic glass plates with thickness up to 3mm by casting between two steel plates(~10^2 $^K/s$), and glass tapes samples with thickness 30–400 μm through super cooling method (10^3÷10^4 $^K/s$). Glass formation was determined visually or by x-ray analysis. The glass melting was performed at 800–1200°C for 15–20 min with a 20–50 g batch in a 20–50 ml uncovered quartz glass or corundum crucible, using an air atmosphere and a "Superterm 17/08" electric furnace. Compositions in the $BaO-B_2O_3$ system were melted in a 25 or 50 ml uncovered Pt crucibles at 1400–1500°C for 30 min with a 20–50 g batch. The chemical composition of some glasses was determined by traditional chemical analysis, and the results indicate a good compatibility between the calculated and analytical amounts of B_2O_3, BaO and Bi_2O_3. SiO_2 contamination from quartz glass crucibles did not exceed 2 wt%, and alumina contamination did not exceed 0 5–1 wt%, according to the chemical analysis data.

Samples of compositions laying outside of a glass formation field or having high melting temperature, have been obtained by solid-phase synthesis. Mixes (15-20 g) were carefully frayed in an agate mortar, pressed as tablets, located on platinum plates and passed the thermal treatment in "Naber"firm electric muffles. After regrinding powders were tested by DTA and X-ray methods. The synthesized samples of binary barium borate system compositions containing 60 mol% and more of BaO and also compositions containing over 90mol % B_2O_3 had very low chemical resistance and were hydrolyzed on air at room temperature. In this connection the synthesized samples were kept in a dryer at 200°C.

DTA and X-ray diffraction data of glass and crystallized glass samples have been used for phase diagram construction in the ternary BaO- Bi_2O_3 -B_2O_3 system. The DTA analysis (pure Al_2O_3 crucible, powder samples weight ~600 mg, heating rates 10 K/min) on Q-1500 type derivatograph were carried out. Glass transition -T_g, crystallization peaks -T_{cr}, melting -T_m and liquidus -T_L temperatures have been determined from DTA curves. Reproducibility of temperatures effects on DTA curves from melting to melting was ±10K. The accuracy of temperature measurement is ±5 K.

Thermal expansion coefficient (TEC) and glass transition temperature (T_g) measurements were made on a DKV-4A type vertical quartz dilatometer with a heating rate of 3K/min. Glass samples in the size of 4×4×50 millimeters have been prepared for TEC measurement. The dilatometer was graduated by the quartz glass and sapphire standards. The TEC measurement accuracy is $\pm(3\div4)\bullet10^{-7}K^{-1}$, T_g ±5 °C.

X-ray patterns were obtained on a DRON-3 type diffractometer (powder method, CuKα–radiation, Ni-filter). Samples for glass crystallization were prepared with glass powder pressed in the form of tablets. Crystallization process was done in the electrical muffles of "Naber" firm by a single-stage heat treatment. This was done within 6-12 hours around a temperature at which the maximum exothermal effects on glasses by DTA were observed.

Crystalline phases of binary and ternary compounds formed both at glasses crystallization and at solid-phase synthesis have been identified by using JCPDS-ICDD PDF-2 release 2008 database [43].

Computerized methodic of ferroelectric hysteresis test and measurement of ferroelectric properties such as coercive field and remanent polarization at wide temperature (up to 250°C and frequency (10-5000Hz) ranges was used. Methodic based on the well known Sawyer – Tower's [44] modified scheme, which is allowing to compensate phase shifts concerned with dielectric losses and conductivity. The desired frequency signal from waveform generator is amplifying by high voltage amplifier and applying to sample. The signals, from the measuring circuit output, proportional to applied field and spontaneous polarization are passing throw high impedance conditioning amplifiers, converting by ADC and operating and analyzing in PC. The technique allows to perform tests of synthesized glass ceramics obtained by means of controlling crystallization of thin (above 30 micrometer thick) monolithic tape (film) specimens by applying up to 300kV/cm field to our thin samples (~50 micrometer thick) and obtain hysteresis loops for wide diversity of hard FE materials.

3. Results

3.1. Glass forming and phase diagrams of the BaO-Bi_2O_3-B_2O_3 system

The traditional method of phase diagram construction based on solid-phase sintered samples investigation takes long time and is not effective. The glass samples investigation technique is progressive, because the DTA curves have registered all processes taking place in glass samples, including the processes of glass crystallizations, quantity of crystal phases and temperature intervals of their formation and melting. However, inadequate amount of glass samples restrict their use during phase diagram construction. The super-cooling method promotes the mentioned problem solving and open new possibilities for phase diagrams constructions.

Hovhannisyan R.M. with co-workes successfully developed this direction last time and have constructed phase diagrams in binary and ternary alkiline-eath bismuth borate, barium boron titanate, barium aluminum boron titanate, barium gallium borate, yttrium aluminum borate, yttrium gallium borate, lantanum gallium borate, zinc tellurium molibdate and other systems [34 - 42].

3.1.1. Glass forming diagram of the BaO-Bi_2O_3-B_2O_3 system

Figure 1 shows the experimental data on glass formation in the BaO-Bi_2O_3-B_2O_3 system obtained by different authors from 1958 to 2007 [45 - 49]. For defining the glass forming ability of the pointed system, the authors of the mentioned works used different amounts of melt, glass melting crucibles, temperature–time melting regimes, and technological methods of melt cooling. Imaoka & Yamazaki studied glass formation by melts cooling on air Glasses were melted at temperatures below 1200 °C in gold-palladium or platinum-rhodium crucibles (Fig.1.1) [45]. Janakirama-Rao glass formation studied by melts cooling on air. Glasses melted in platinum crucibles at 600- 1400 °C with 0.5-1.0 h melts exposition and its cooling in air (Fig. 1.2) [46]. Izumitani [47] experiments spent in 10g crucibles at 1100-1350 °C with melts cooling on air (Fig. 1.3). Milyukov with co-authors glass formation studied by melts casting in steel mold. Glasses melted in platinum crucibles at 600-1400 °C with melts stirring by Pt stirrer for 1h (Fig. 1.4) [48]. Kawanaka & Matusita glass formation studied by silica rod stirred melts pouring into preheated to 250-300°C carbon mold (Fig. 1.5) [49].

Authors used different weights of glass forming melts, melting crucibles, temperature-time of melting regimes and technological methods of melts cooling. Obtained data are difficultly comparable and remote from two basic criteria promoting glass formation: liquidus temperature and speeds of melts cooling.

http://dx.doi.org/10.5772/52405

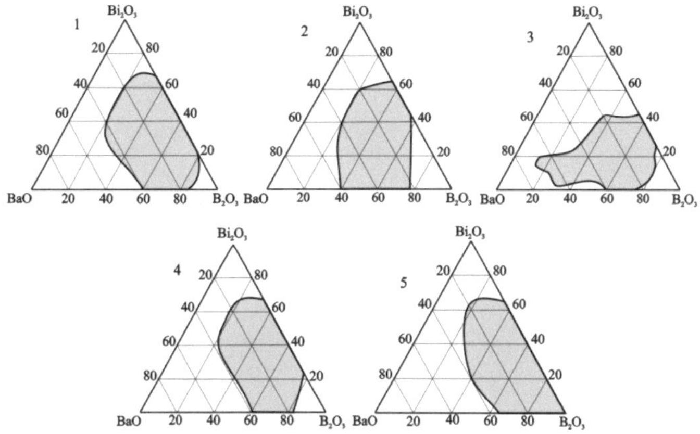

Figure 1. Glass forming regions in the BaO-Bi$_2$O$_3$-B$_2$O$_3$ system according to the data of the authors: 1- [45], 2-[46], 3-[47], 4-[48], 5-[49].

Figure 2 shows corrected glass formation diagram in the BaO-Bi$_2$O$_3$-B$_2$O$_3$ system based on phase diagrams of the BaO–B$_2$O$_3$, BaO–Bi$_2$O$_3$, and B$_2$O$_3$–Bi$_2$O$_3$ binary systems and controllable melt cooling rates. Using the term "diagram," but not the glass formation region, we take into account the interrelation between the phase diagram and the glass forming ability of the system.

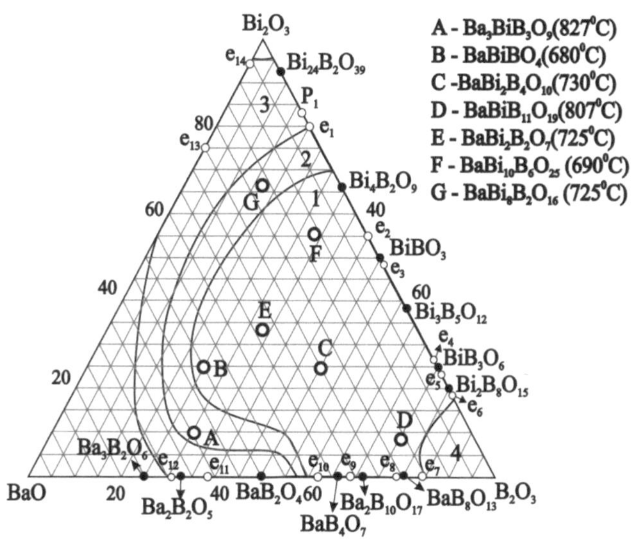

Figure 2. Glass forming diagram in the BaO-Bi$_2$O$_3$-B$_2$O$_3$ system depending of melts cooling rates: 1-up to 10 K/s; 2~10^{2}K/s; 3-(10^3-10^4) K/s; 4- stable phase separation region.

Area of glass compositions with low crystallization ability and stable glass formation in the studied ternary BaO-Bi$_2$O$_3$ -B$_2$O$_3$ system have been determined at melts cooling rate ~ 10 K/s

(Fig.2-1). It included binary $Bi_4B_2O_9$, $BiBO_3$, $Bi_3B_5O_{12}$, BiB_3O_6, $Bi_2B_8O_{15}$, $Ba_2B_{10}O_{17}$, BaB_8O_{13} compounds in the BaO-Bi_2O_3 and BaO-B_2O_3 systems and five ternary $BaBiB_{11}O_{19}$(D), $BaBi_2B_4O_{10}$(C), $BaBi_2B_2O_7$ (E), $BaBi_{10}B_6O_{25}$ (F), and $BaBiBO_4$ (B) compounds. However, we didn't comfirm presence of $Ba_3BiB_3O_9$ (A) compound in area of stable glasses at melts cooling rate ~ 10 K/s, which was reported earlier [34, 35].

Increasing of melts cooling speed up to ~10^2 K/s has led to glass formation area expansion (Fig.2-2). This cooling rate is enough for monolithic glass plates with thickness up to 3mm fabrication by melts casting between two steel plates (Fig.2-2). The glass plates of compositions correspondings to $Ba_3BiB_3O_9$ (A) and supposed $BaBi_8B_2O_{16}$ (G) compounds have been obtained by this way.

Super cooling technique constructed by our group allowed to expand the borders of glass formation in studied system under high melts cooling rates equal to (10^3-10^4) K/s (Fig.2-3). Determined glassforming area include compositions content: 80 – 95 mol% Bi_2O_3 in the binary B_2O_3–Bi_2O_3 system; 43-70 mol% BaO in the binary BaO–B_2O_3 system, including BaB_2O_4 composition. Area of glass formation from both these areas moves to 55-95 mol% content compositions in the binary BaO–Bi_2O_3 system (Fig.2-3).

Traditional for borate systems a stable phase separation region was also observed for high B_2O_3 content compositions contents more than 84 -87 mol%B_2O_3 (Fig.2-4).

3.1.2. Phase diagram of the BaO-Bi_2O_3-B_2O_3 system

Our investigation of the ternary BaO-B_2O_3-Bi_2O_3 system have purposefully been directed on construction of the phase diagram through first of all glass forming diagram construction and revealing both new compounds and eutectic compositions. Constructed by us glass forming diagram (Fig.2) practically occupies the most part of the BaO-B_2O_3-Bi_2O_3 concentration triangles. It has allowed to use synthesized glasses as initial compositions at phase diagram construction. It was basic difference of our methodology from technologies used by other authors. Phase equilibriums reached at isothermal sections construction do not allow to have a full picture of processes in cases of the solid state synthesized samples investigations. Whereas at glass samples studies we determine not only characteristic points of glasses(T_g and T_s) by DTA, but also quantity of crystal phases, temperatures of their crystallization and then temperatures of their melting. It has allowed us to reveal new stoichiometric compositions which have been lost by other research groups at isothermal sections construction by traditional methods. In some cases we also in parallel used samples obtained by solid state synthesis for comparison with their glassy analogues or in those cases, when their obtaining in the glassy form was impossible.

http://dx.doi.org/10.5772/52405

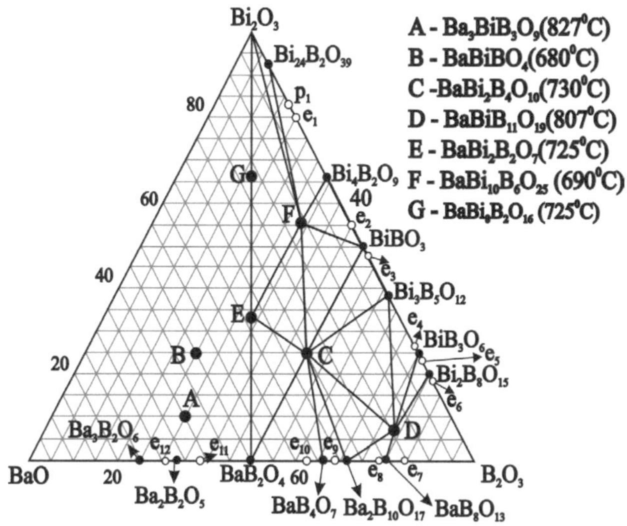

Figure 3. The BaO-B_2O_3-Bi_2O_3 system triangulation

Experimental data concerning phase diagrams of binary systems Bi_2O_3-B_2O_3, BaO-B_2O_3, BaO-Bi_2O_3 and pseudo-binary sections in the BaO-B_2O_3-Bi_2O_3 system have allowed us to estimate fields of primary crystallization of co-existing phases and divided all concentration triangle into elementary triangles, i.e. to make triangulation.

The triangulation scheme in the BaO-B_2O_3-Bi_2O_3 system is presented on Fig. 3. By means of a triangulation all concentration triangle is divided into following elementary triangles: $BiBO_3$-F-C, $BiBO_3$-F-$Bi_4B_2O_9$, $BiBO_3$-C-$Bi_3B_5O_{12}$, F-E-C, C-BaB_2O_4-E, C-$Bi_3B_5O_{12}$-D, $Bi_4B_2O_9$-F-Bi_2O_3, E-Bi_2O_3-F, C-BaB_2O_4-$Ba_2B_{10}O_{17}$, $Ba_2B_{10}O_{17}$-D-C, $Ba_2B_{10}O_{17}$-D-B_2O_3, $Bi_3B_5O_{12}$-D-$Bi_2B_8O_{15}$, D-$Bi_2B_8O_{15}$-B_2O_3.

3.1.2.1. Phase diagram of the binary Bi_2O_3–B_2O_3 system

First of all we have attempt to finished phase diagram construction in area of compositions around of $BiBO_3$ compound. Compositions containing 45–65 mol% B_2O_3 in the Bi_2O_3–B_2O_3 system were tested to determine the melting point of $BiBO_3$ and to determine the eutectic composition between $BiBO_3$ and $Bi_3B_5O_{12}$. The compositions used to correct the B_2O_3–Bi_2O_3 phase diagram were prepared by solid state synthesis at 520°C, with steps of 0.5–1.0 mol% B_2O_3 over the interval 45–55 mol% B_2O_3. As a result, the eutectic composition, 48.5Bi2O3•51.5B2O3 (mol%), between $BiBO_3$ and $Bi_3B_5O_{12}$, was determined, and its melting point was measured by DTA as 665±5°C (Fig. 4). It was also found that $BiBO_3$ melts congruently at 685±5°C.

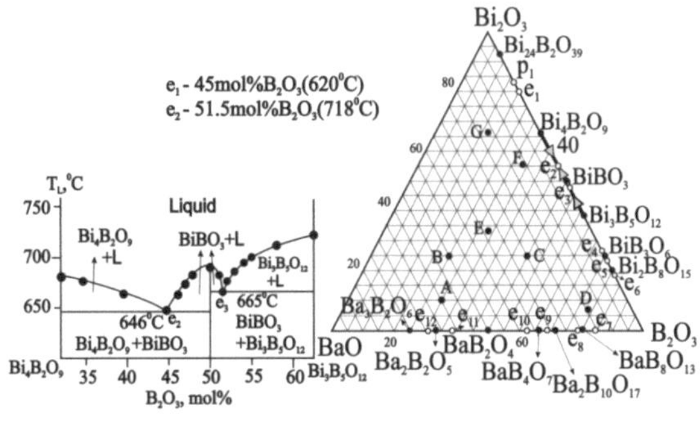

Figure 4. Corrected phase diagram of the Bi_2O_3-B_2O_3 system in the interval 30–65 mol% B_2O_3.

3.1.2.2. Phase diagram of the pseudo-binary $BiBO_3$–BaB_2O_4 system

$BaBi_2B_4O_{10}$ is a congruently melting compound, with a melting point of 730°C, and it occupies the central area of the $BiBO_3$–BaB_2O_4 pseudo-binary system (Fig. 5). This system forms two simple pseudo-binary eutectics, E_1 at 15 mol% BaB_2O_4, with a melting point of 620°C, and E_2 at 60 mol% BaB_2O_4, with a melting point of 718°C.

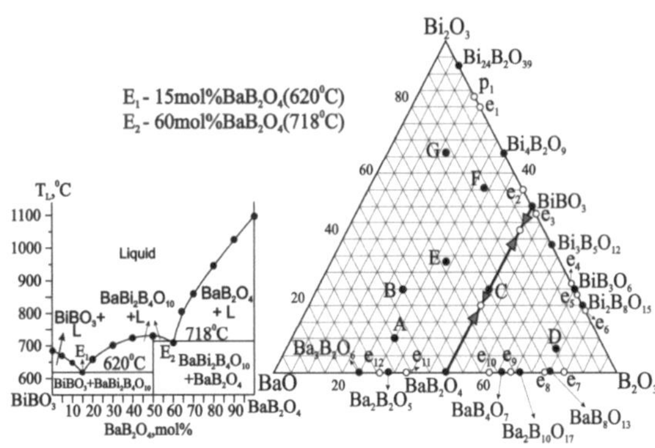

Figure 5. Phase diagram of the pseudo-binary system $BiBO_3$-BaB_2O_4

3.1.2.3. Phase diagram of the pseudo-binary $Bi_4B_2O_9$–$BaBi_2B_2O_7$ system

The introduction of 12 mol% $BaBi_2B_2O_7$ in the pseudo-binary system $Bi_4B_2O_9$-$BaBi_2B_2O_7$ reduced the melting point of initial $Bi_4B_2O_9$, and resulted in the formation of a simple pseudo-binary eutectic, E_3, with melting point 605°C (Fig. 6). A maximum of the liquidus with melting point of 690°C is seen at 33.33 mol% $BaBi_2B_2O_7$, which indicates the formation of the new congruently melting ternary compound $BaBi_{10}B_6O_{25}$ (11.11BaO•55.55Bi_2O_3•33.33B_2O_3). Further increase of the $BaBi_2B_2O_7$ content (49 mol%) leads to a second pseudo-binary eutectic, E_4, with melting point 660°C. Increasing of the liquidus temperature is observed in the post eutectic region of composition, with a maximum at 725°C. It corresponds to the formation of the new congruently melting ternary compound $BaBi_2B_2O_7$ (Fig. 6).

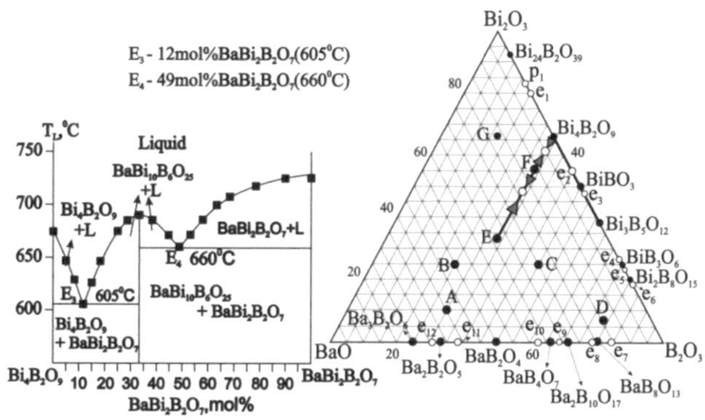

Figure 6. Phase diagram of the pseudo-binary system Bi$_4$B$_2$O$_9$-BaBi$_2$B$_2$O$_7$.

Two new crystalline ternary compounds, BaBi$_2$B$_2$O$_7$ and BaBi$_{10}$B$_6$O$_{25}$, were revealed by crystallisation at the same glass composition. Both compounds, BaBi$_2$B$_2$O$_7$ and BaBi$_{10}$B$_6$O$_{25}$, melt congruently at 725±5°C and 690±5°C, respectively. The X-ray characteristics of BaBi$_2$B$_2$O$_7$ and BaBi$_{10}$B$_6$O$_{25}$ were determined and are given in Tables 1 and 2.

No.	d_{exp}	I/I_0	hkl	No.	d_{exp}	I/I_0	hkl	No.	d_{exp}	I/I_0	hkl
1	6.23	9	101	20	2.15	25	140	39	1.349	7	125
2	5.02	9	111	21	2.12	5	232	40	1.33	3	543
3	4.80	5	201	22	2.06	24	123	41	1.28	8	035
4	4.29	5	020	23	2.01	6	523	42	1.24	6	560
5	4.11	14	120	24	1.977	25	241	43	1.217	5	263
6	3.88	6	300	25	1.84	15	142	44	1.21	9	843
7	3.67	4	021	26	1.826	7	133	45	1.206	13	271
8	3.59	26	301	27	1.786	6	004	46	1.19	14	145
9	3.56	50	121	28	1.729	52	114	47	1.173	12	245
10	3.52	23	220	29	1.679	23	250	48	1.17	4	1010
11	3.19	100	112	30	1.636	34	251	49	1.14	6	126
12	3.12	8	221	31	1.63	5	532	50	1.11	6	662
13	3.05	9	202	32	1.57	4	052	51	1.10	6	326
14	2.91	43	030	33	1.556	10	243	52	1.09	4	180
15	2.696	90	122	34	1.522	8	034	53	1.042	6	146
16	2.51	12	222	35	1.488	23	632	54	1.021	9	065
17	2.376	21	003	36	1.458	6	060	55	1.018	9	943
18	2.31	5	421	37	1.428	10	811	56	1.01	5	274
19	2.254	22	113	38	1.373	11	821				

Table 1. X-ray characteristics of the new ternary compound BaBi$_2$B$_2$O$_7$, synthesized at the same glass composition crystallization (640°C, 20 h).

Single crystals of $BaBi_{10}B_6O_{25}$ were obtained by cooling of a melt with the stoichiometric composition. Glass powder of composition $11.11BaO \bullet 55.55Bi_2O_3 \bullet 33.33B_2O_3$ (mol%) was heated in a quartz glass ampoule up to 750°C at a rate 10 K/min. After 2 h exposition at high temperature, the melt was cooled at a rate 0.5 K/h. Single crystals with sizes up to 1.66×0.38×0.19 mm^3 were grown.

No.	d_{exp}	I/I_0	hkl	No.	d_{exp}	I/I_0	hkl	No.	d_{exp}	I/I_0	hkl
1	9.21	3.0	012	23	2.91	31	0110	45	2.01	5.2	321
2	6.26	3.0	101	24	2.8	6	221	46	1.98	23.7	323
3	6.02	3.0	005	25	2.7	75.9	206	47	1.92	3.0	1214
4	5.01	7.3	006	26	2.64	3.4	045	48	1.88	3.0	0314
5	4.89	10.8	104	27	2.57	3.0	2.07	49	1.86	3.0	162
6	4.63	6.5	024	28	2.53	4.3	046	50	1.84	15.1	254
7	4.19	4.3	025	29	2.52	5.2	144	51	1.83	3.0	1412
8	4.18	4.3	122	30	2.49	6.9	230	52	1.82	3.0	164
9	4.11	10	115	31	2.47	9.5	1210	53	1.81	3.0	255
10	3.92	6.0	030	32	2.45	4.7	232	54	1.79	3.4	165
11	3.80	3.4	032	33	2.38	16	0310	55	1.77	3.0	328
12	3.65	10.8	033	34	2.35	4.3	050	56	1.75	3.4	1413
13	3.56	39.7	107	35	2.34	5.2	051	57	1.73	55.2	341
14	3.51	9.5	125	36	2.33	3.9	1012	58	1.71	4.3	343
15	3.41	3.4	117	37	2.31	3.4	1212	59	1.69	3.0	069
16	3.38	7.3	130	38	2.25	17.2	228	60	1.68	16.4	070
17	3.33	8.6	131	39	2.21	7.8	150	61	1.65	3.0	259
18	3.27	9.1	132	40	2.19	3.0	055	62	1.64	31.0	074
19	3.18	100	133	41	2.15	21.2	237	63	1.61	2.9	400
20	3.07	14.7	203	42	2.09	3.4	312	64	1.6	6.0	402
21	3.04	9.2	212	43	2.06	21.6	304	65	1.55	6.5	421
22	2.94	14.2	040	44	2.03	3.0	314	66	1.52	9.5	177

Table 2. X-ray characteristics of the new ternary $BaBi_{10}B_6O_{25}$ single crystals.

The x-ray powder diffraction patterns of $BaBi_2B_2O_7$ and $BaBi_{10}B_6O_{25}$ could be indexed on an orthorhombic cell with lattice parameters as follows:

- for $BaBi_2B_2O_7$ a=11.818 Å, b=8.753 Å, c=7.146 Å, cell volume V=739.203 Å^3, Z=4;
- for $BaBi_{10}B_6O_{25}$ a=6.434 Å, b=11.763 Å, c=29.998 Å, cell volume V=2270.34 Å^3, Z=8.

3.1.2.4. *Phase diagram of the pseudo-binary $BiBO_3$–$BaBi_{10}B_6O_{25}$ system*

$BiBO_3$– $BaBi_{10}B_6O_{25}$ is a very important system (Fig.7). Initial $BiBO_3$ has a melting point of 685°C. The second maximum in the liquidus curve (Fig.7) of 690°C is connected with the formation of the new ternary compound $BaBi_{10}B_6O_{25}$. There is a simple pseudo-binary eutectic E_5 between these two compounds at 54 mol%$BaBi_{10}B_6O_{25}$, with a melting point of 595°C.

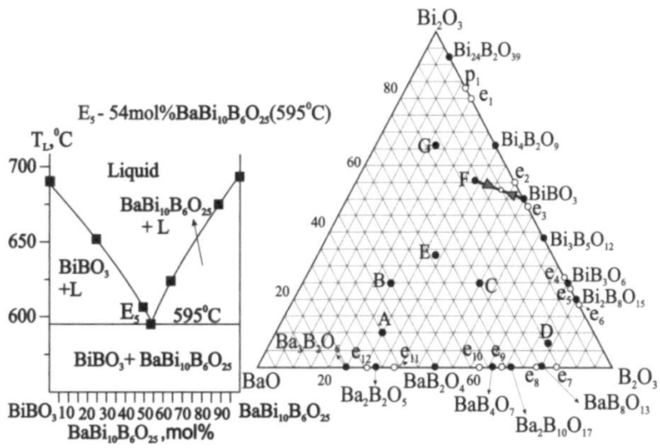

Figure 7. Phase diagram of the pseudo-binary system $BiBO_3$- $BaBi_{10}B_6O_{25}$.

3.1.2.5. *Phase diagram of the pseudo-binary $BaBi_{10}B_6O_{25}$ - $BaBi_2B_4O_{10}$ system*

The $BaBi_{10}B_6O_{25}$–$BaBi_2B_4O_{10}$ system confirms the presence of the new congruently melting ternary compound $BaBi_{10}B_6O_{25}$, with a melting point of 690°C (Fig.8). $BaBi_2B_4O_{10}$ melts congruently at 730°C. There is a simple pseudo-binary eutectic E_6 between these two compounds at 28 mol % $BaBi_2B_4O_{10}$, with a melting point of 660°C.

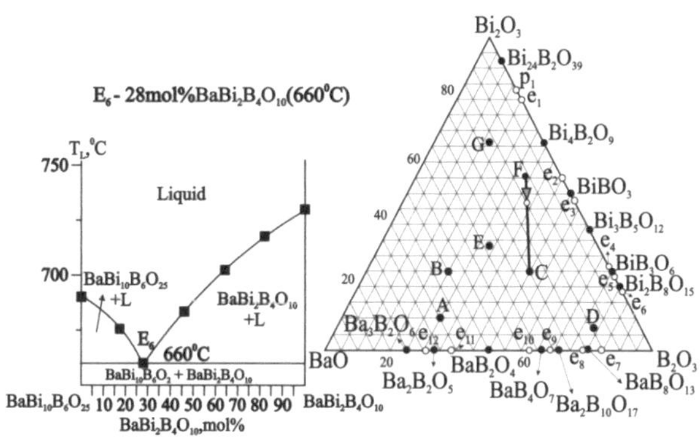

Figure 8. Phase diagrams of the pseudo-binary system $BaBi_{10}B_6O_{25}$ -$BaBi_2B_4O_{10}$.

3.1.2.6. *Phase diagram of the pseudo-binary $BaBi_2B_4O_{10}$ – 50BaO•50Bi_2O_3 section*

This pseudo-binary section consists of two ternary compounds $BaBi_2B_4O_{10}$, $BaBi_2B_2O_7$ and two eutectics E_7, E_8 dividing fields primary crystallisations these compounds. Initial composition is $BaBi_2B_4O_{10}$ (Fig.9). The introduction of 20 mol% 50%BaO•50%Bi_2O_3 in the pseudo-binary system $BaBi_2B_4O_{10}$ – $50BaO \bullet 50Bi_2O_3$ reduced the melting point of initial $BaBi_2B_4O_{10}$, and resulted in the formation of a simple pseudo-binary eutectic, E_7, with melting point 680°C (Fig.9). A maximum of the liquidus with melting point of 725°C is seen at 33.33 mol% of $50\%BaO \bullet 50\%Bi_2O_3$, which indicates the formation of the new congruently melting ternary compound $BaBi_2B_2O_7$. Further increase of the $50\%BaO \bullet 50\%Bi_2O_3$ content (52.5 mol%) leads to a second pseudo-binary eutectic, E_8, with melting point 700°C. Increasing of the liquidus temperature is observed in the post eutectic region of composition (more than 52.5mol% of $50BaO \bullet 50Bi_2O_3$). Un identified phase is in the post eutectic (E_8) region of composition (Fig.9).

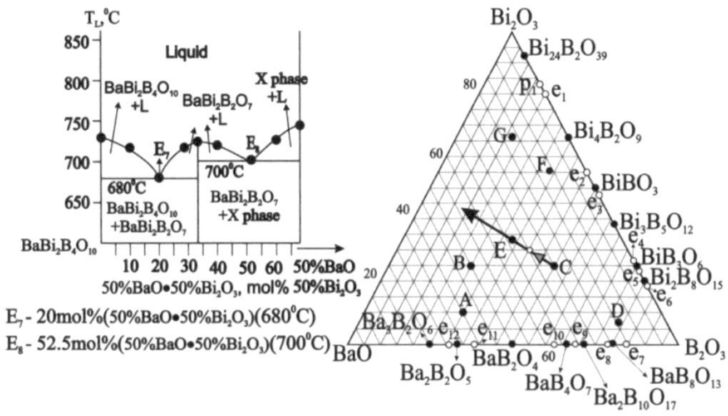

Figure 9. Phase diagram of the pseudo-binary section $BaBi_2B_4O_{10}$ – $50BaO \bullet 50Bi_2O_3$.

3.1.2.7. Phase diagram of the pseudo-binary BaB_2O_4–Bi_2O_3 section

BaB_2O_4 - Bi_2O_3 section consist of two pseudo-binary BaB_2O_4- $BaBi_2B_2O_7$, $BaBi_2B_2O_7$- Bi_2O_3 systems (Fig.10). There are two eutectics: E_9 between BaB_2O_4 and $BaBi_2B_2O_7$, E_{10} between $BaBi_2B_2O_7$ and $BaBi_8B_2O_{16}$, and peritectic point P_1 between $BaBi_8B_2O_{16}$ and Bi_2O_3 (Fig.10). The introduction of 26 mol% Bi_2O_3 in the pseudo-binary system BaB_2O_4 – Bi_2O_3 sharp reduced the melting point of initial BaB_2O_4 on 445^{0}C, and resulted in the formation of a simple eutectic, E_9, with melting point 685°C (Fig.10). A maximum of the liquidus with melting point of 725°C is seen at 33 33 mol% of Bi_2O_3, which indicates the formation of the new congruently melting ternary compound $BaBi_2B_2O_7$.

Further increase of the Bi_2O_3 content (42 mol%) leads to a second eutectic, E_{10}, formation with melting point 690°C. Increasing of the liquidus temperature is observed in the post eutectic region of composition (more than 42mol% of Bi_2O_3) and formation of new incongruent melted at 725 °C $BaBi_8B_2O_{16}$ ternary compound (Fig.10). It is very difficult determined of $BaBi_8B_2O_{16}$ X-ray characteristics, because they very closed to Bi_2O_3 characteristics.

Constructed by us this section's diagram essentially differs from that constructed by Russian researches Egorisheva & Kargin [30] because they could not find out two new compounds revealed by us: congruent melted at 725°C $BaBi_2B_2O_7$ and incongruent melted at 725°C $BaBi_8B_2O_{16}$(Fig.10). At repeated, even more detailed studies they also could not find out these compounds [31].

http://dx.doi.org/10.5772/52405

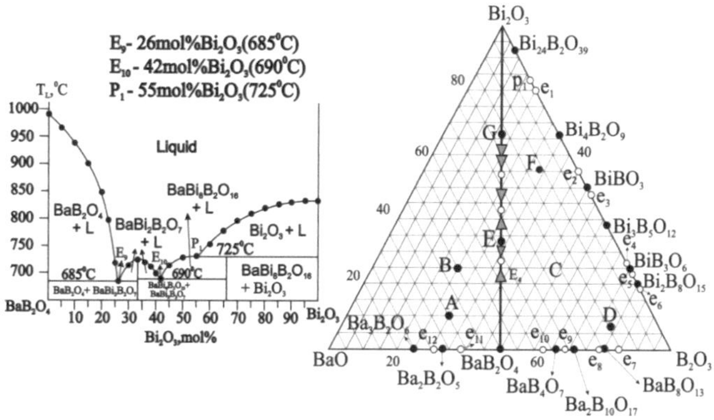

Figure 10. Phase diagram of the $BaB_2O_4 - Bi_2O_3$ section.

3.1.2.8. *Phase diagram of the pseudo-binary $Bi_3B_5O_{12} - BaBi_2B_4O_{10}$ system*

It is very simple system with pseudo-binary eutectic E_{11} between two congruent melted $Bi_3B_5O_{12}$ and $BaBi_2B_4O_{10}$ compounds. Eutectic E_{11} content 38 mol% of $BaBi_2B_4O_{10}$ and has melting point 680°C (Fig.11).

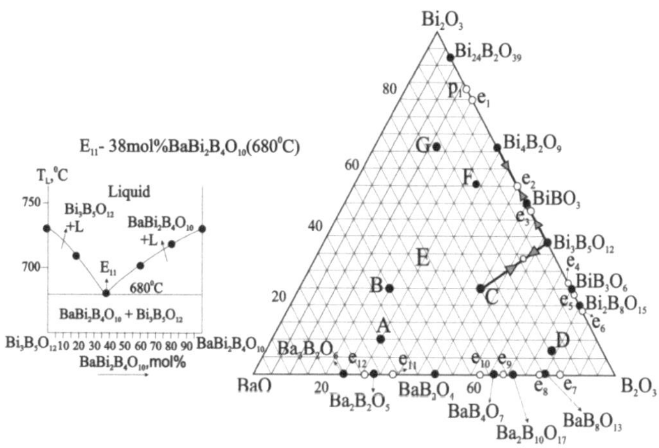

Figure 11. Phase diagram of the pseudo-binary system $Bi_3B_5O_{12} - BaBi_2B_4O_{10}$

3.1.2.9. *Phase diagram of the pseudo-binary $BaBi_2B_4O_{10} - BaBiB_{11}O_{19}$ system*

Initial $BaBi_2B_4O_{10}$ has a melting point of 730°C. The second maximum in the liquidus curve (Fig.12) of 807°C is connected with the formation of the ternary compound $BaBiB_{11}O_{19}$. There is a simple pseudo-binary eutectic, E_{12}, between these two compounds at 65.5 mol% Ba-$Bi_2B_4O_{10}$, with a melting point of 695°C.

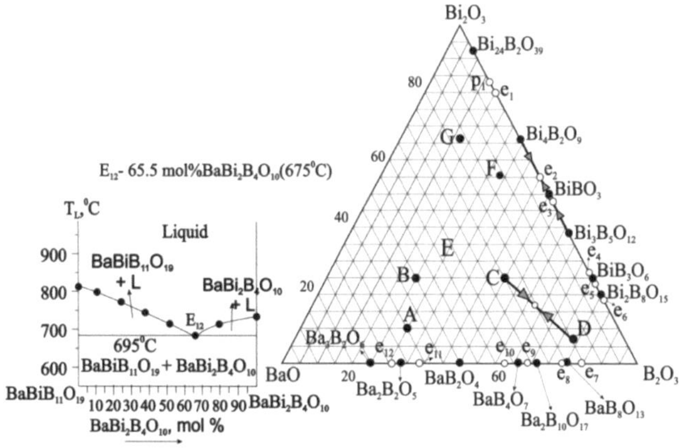

Figure 12. Phase diagram of the pseudo-binary system $BaBi_2B_4O_{10}$ – $BaBiB_{11}O_{19}$.

3.1.2.10. Phase diagram of the pseudo-binary $Bi_3B_5O_{12}$ – $BaBiB_{11}O_{19}$ system

Pseudo-binary system $Bi_3B_5O_{12}$ -$BaBiB_{11}O_{19}$ has simple eutectic E_{13} formed between two congruent melted $BaBiB_{11}O_{19}$ and $Bi_3B_5O_{12}$ compounds. According to DTA eutectic E_{13} has melting point 705°C and content 28 mol% $BaBiB_{11}O_{19}$ (Fig.13).

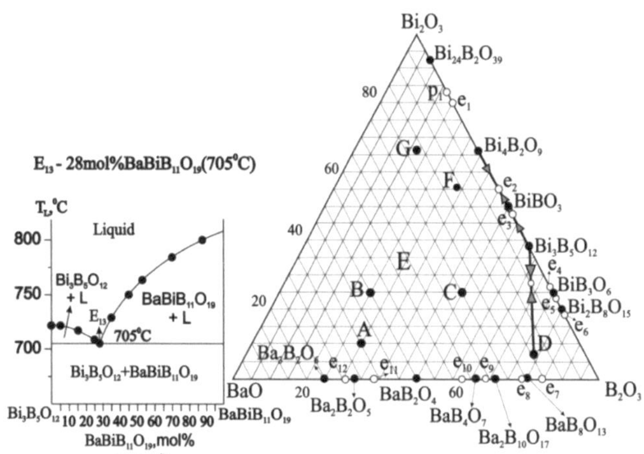

Figure 13. Phase diagram of the pseudo-binary system $Bi_3B_5O_{12}$ - $BaBiB_{11}O_{19}$.

3.1.2.11. Phase diagram of the pseudo-binary $BaBiB_{11}O_{19}$ – $Ba_2B_{10}O_{17}$ system

Pseudo-binary system $BaBiB_{11}O_{19}$ – $Ba_2B_{10}O_{17}$ has simple eutectic E_{14} formed between two congruent melted compounds $Ba_2B_{10}O_{17}$ and $BaBiB_{11}O_{19}$. According to DTA eutectic E_{14} has melting point 780°C and content 26 mol% $Ba_2B_{10}O_{17}$(Fig.14).

http://dx.doi.org/10.5772/52405

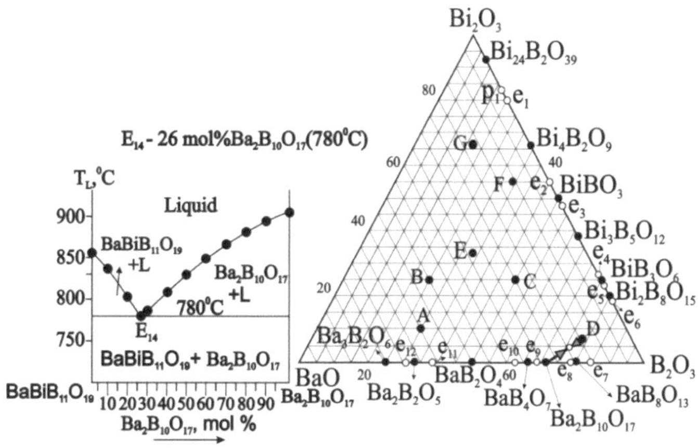

Figure 14. Phase diagram of the pseudo-binary system $BaBiB_{11}O_{19} – Ba_2B_{10}O_{17}$.

3.1.2.12. *Phase diagram of the pseudo-binary $BaBi_2B_4O_{10} – Ba_2B_{10}O_{17}$ system*

It is very simple system with eutectic E_{15} formed between two congruent melted $Ba_2B_{10}O_{17}$ and $BaBi_2B_4O_{10}$ compounds. Pseudo-binary eutectic E_{15} content 24 mol% of $Ba_2B_{10}O_{17}$ and has melting point 710°C (Fig. 15).

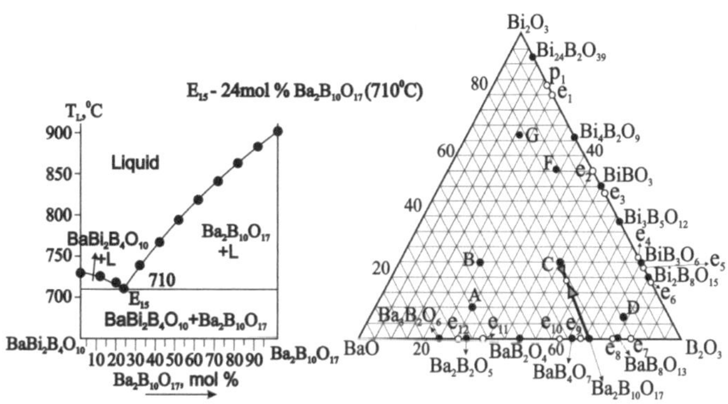

Figure 15. Phase diagram of the pseudo-binary system $BaBi_2B_4O_{10} – Ba_2B_{10}O_{17}$.

3.1.2.13. *Phase diagram of the pseudo-binary $BaBi_2B_4O_{10} – BaB_4O_7$ system*

The same picture is observe for $BaBi_2B_4O_{10} – BaB_4O_7$ system: simple eutectic E_{16} is formed between two congruent melted compounds. Pseudo-binary eutectic E_{16} content 24 mol% of BaB_4O_7 and has melting point 715°C (Fig. 16).

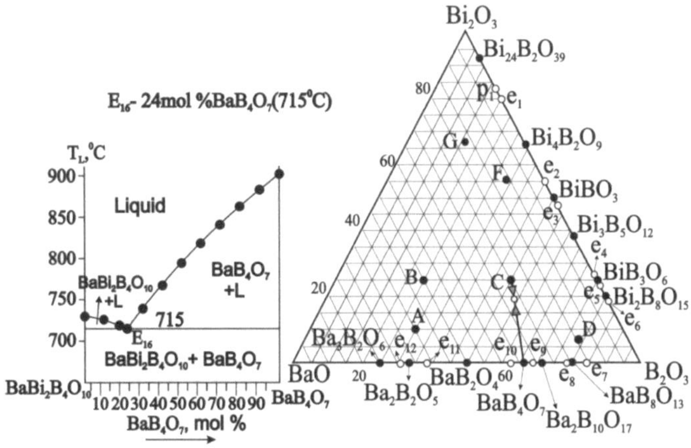

Figure 16. Phase diagram of the pseudo-binary system $BaBi_2B_4O_{10} - BaB_4O_7$.

3.1.2.14. Phase diagram of the pseudo-binary $BaBiB_{11}O_{19} - BaB_8O_{13}$ system

It is simple system with pseudo-binary eutectic E_{17} between two congruent melted $BaBiB_{11}O_{19}$ and BaB_8O_{13} compounds. Eutectic E_{17} content 38 mol% of BaB_8O_{13} and has melting point 770°C (Fig. 17).

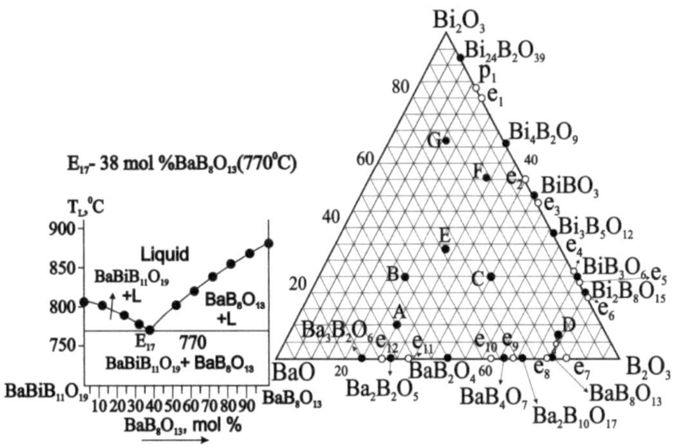

Figure 17. Phase diagram of the pseudo-binary system $BaBiB_{11}O_{19} - BaB_8O_{13}$.

3.1.2.15. Phase diagram of the $BaO-Bi_2O_3-B_2O_3$ ternary system

As result of huge work under project the phase diagram in the ternary $BaO-B_2O_3-Bi_2O_3$ system has been constructed for the first time and presented on Fig.18. Three new compounds $BaBi_2B_2O_7$, $BaBi_{10}B_6O_{25}$ and $BaBi_8B_2O_{16}$ have been revealed and characterized.

Fields of binary bismuth and barium borates as well as all ternary barium bismuth borates compounds crystallizations have been determined and outlined and sixteen ternary eutectic points E_1-E_{16} have been revealed as result of phase diagram construction (Fig. 18, table 3). The phase diagram evidently represents interaction of binary and ternary compounds taking place in the pseudo-ternary systems. The ternary eutectic E_1 with m.p 590°C has been determined among $BiBO_3$, F and $Bi_4B_2O_9$ compounds; ternary eutectic E_2 with m.p. 585°C has been formed among $BiBO_3$, F and C compounds; ternary eutectic E_3 with m.p. 640°C has been formed among F, E and C compounds; ternary eutectic E_4 with m.p. 622°C has been formed among C, BaB_2O_4 and E compounds; ternary eutectic E_5 with m.p. 610°C has been formed among $BiBO_3$, C and $Bi_3B_5O_{12}$ compounds; ternary eutectic E_6 with m.p. 675°C has been formed among C, $Bi_3B_5O_{12}$ and D compounds; ternary eutectic E_7 with m.p. 680°C has been formed among $Bi_3B_5O_{12}$, D and BiB_3O_6 compounds; ternary eutectic E_8 with m.p. 675°C has been formed among BiB_3O_6, D and $Bi_2B_8O_{15}$ compounds; ternary eutectic E_9 with m.p. 680°C has been formed among $Bi_2B_8O_{15}$, D and B_2O_3 compounds; ternary eutectic E_{10} with m.p. 730°C has been formed among BaB_8O_{13}, D and B_2O_3 compounds; ternary eutectic E_{11} with m.p. 750 °C has been formed among $Ba_2B_{10}O_{17}$-D- BaB_8O_{13} compounds; ternary eutectic E_{12} with m.p. 680°C has been formed among C, $Ba_2B_{10}O_{17}$ and D compounds; ternary eutectic E_{13} with m.p. 690°C has been formed among $Ba_2B_{10}O_{17}$, C and BaB_4O_7 compounds; ternary eutectic E_{14} with m.p. 700°C has been formed among BaB_4O_7, C and BaB_2O_4 compounds; ternary eutectic E_{15} with m.p. 645 °C has been formed among G, F and C compounds; ternary eutectic E_{16} with m.p. 615°C has been formed among $Bi_{24}B_2O_{39}$, F and $Bi_4B_2O_9$ compounds (Fig18, Table 3).

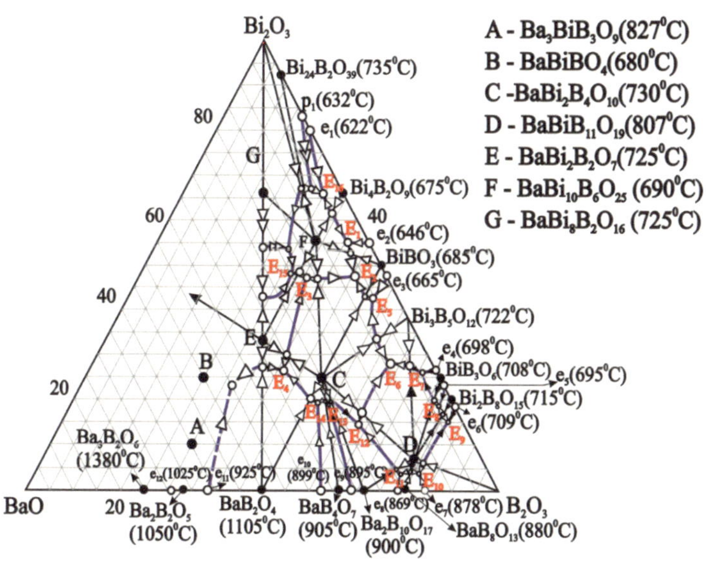

Figure 18. Phase diagram of the BaO-Bi$_2$O$_3$-B$_2$O$_3$ system

Point	Composition, mol%			T_m, °C
	BaO	B_2O_3	Bi_2O_3	
E_1	4.5	40.5	55	590
E_2	7.3	45.1	47.6	585
E_3	15	38	47	640
E_4	32.4	41.3	26.3	622
E_5	5.4	52	42.6	610
E_6	9	63	28	675
E_7	3	71	26	680
E_8	3.5	78	18.5	675
E_9	3.2	81.8	15	680
E_{10}	15	81.2	3.8	730
E_{11}	18.9	77.5	3.6	750
E_{12}	22.4	63.2	14.4	680
E_{13}	26.6	54.7	18.7	690
E_{14}	28.8	52	19.2	700
E_{15}	20	32	48	645
E_{16}	4.5	30	65.5	615

Table 3. The temperature and compositions for ternary eutectic points in the $BaO-Bi_2O_3-B_2O_3$ system

3.2. DTA and X-ray characterisation of ternary stoichiometric glasses and glass ceramics from the $BaO-Bi_2O_3-B_2O_3$ system

The glasses corresponding to known sixth stoichiometric compounds in the $BaO-Bi_2O_3-B_2O_3$ system examined in the present study and following glass compositions (mol%) have been melted: $14.28BaO•7.14Bi_2O_3•78.57B_2O_3$ ($BaBiB_{11}O_{19}$), $25BaO•25Bi_2O_3•50B_2O_3$($BaBi_2B_4O_{10}$), $33.33BaO•33.33Bi_2O_3•33.33B_2O_3$($BaBi_2B_2O_7$), $11.11BaO•55.55Bi_2O_3•33.33B_2O_3$ ($BaBi_{10}B_6O_{25}$), $50BaO•$ $25Bi_2O_3•$ $25B_2O_3$ ($BaBiBO_4$) and $60BaO•10Bi_2O_3•30B_2O_3$ ($Ba_3BiB_3O_9$). These glasses DTA curves are shown in Fig. 19, giving the peaks due to the glass transition, crystallization, melting, and liquidus temperatures. The glass characteristics points T_g (glass transition), T_s(glass softening), T_c (peak of exothermal effects connected with crystalline phases crystallizations) and T_m (minimum of endothermic effects associated with these phases melting) observed on DTA curves (Fig. 19, curves 1-6) of all tested powder samples summarized on table 4.

http://dx.doi.org/10.5772/52405

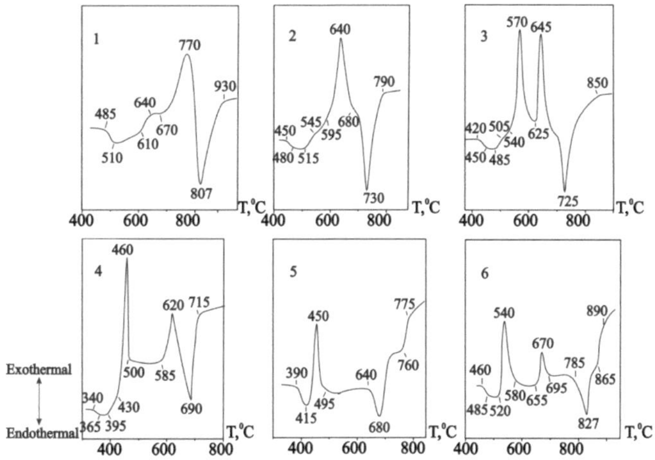

Figure 19. DTA curves (heating rate 10K/min) of glasses corresponding to ternary compounds in the $BaO-Bi_2O_3-B_2O_3$ system: 1-$BaBiB_{11}O_{19}$, 2-$BaBi_2B_4O_{10}$, 3-$BaBi_2B_2O_7$, 4-$BaBi_{10}B_6O_{25}$,5-$BaBiBO_4$, 6-$Ba_3BiB_3O_9$.

##	Glass compositions, correspoding to stoichometric compounds	Dilatometric characteristics			DTA characteristics			
		TEC(α_{20-300})•10^7K^{-1}	T_g,°C	T_s,°C	T_g, °C	T_{cr}, °C	T_m, °C	T_L, °C
1	$BaBiB_{11}O_{19}$ (glass)	72	498	535	485	640; 770	807	807
	$BaBiB_{11}O_{19}$ (615°C 24h)	49.8						
2	$BaBi_2B_4O_{10}$ (glass)	96	445	475	450	545; 640	730	730
	$BaBi_2B_4O_{10}$ (640°C 24h)	77.9						
3	$BaBi_2B_2O_7$ (glass)	108	415	455	420	570; 645	725	725
	$BaBi_2B_2O_7$ (640°C 24h)	96						
4	$BaBi_{10}B_6O_{25}$ (glass)	99	350	380	340	460; 620	690	690
	$BaBi_{10}B_6O_{25}$(590°C 24h)	97						
5	$BaBiBO_4$ (glass)	120	400	450	390	450	680	760
	$BaBiBO_4$ (570°C 24h)	110.8						
6	$Ba_3BiB_3O_9$ (glass)	127	460	490	460	540; 670	827	865
	$Ba_3BiB_3O_9$ (690°C 24h)	109.8						

Table 4. Chemical compositions, DTA (glass transition -T_g, crystallization peak -T_{cr}, melting -T_m, liquidus- T_L) and dilatometric characteristics (glass transition temperature -T_g, softening point - T_S, thermal expansion coefficient -TEC) of $BaO-Bi_2O_3-B_2O_3$system glasses and crystallaised glasses.

Two exothermic effects were observed on DTA curve of $14.28BaO•7.14Bi_2O_3•78.57B_2O_3$ (mol %) glass composition: first weak effect at 640°C and second strong effect at 770°C(Fig.19, curve1). The melting temperature (T_m) is equal to 807°C and corresponding to Egorisheva and Kargin's data [30]. X-ray patterns of this glass crystallization products show one Ba-$BiB_{11}O_{19}$ crystalline phase presence [30], which formed at powder samples crystallization in an temperature interval 640-770°C (Fig.20, curve1). It is possible to assume, that weak exothermic effect at 640°C apparently is connected with pre-crystallisation fluctuations taking place in glass matrix [50]. Diffuse character of second exothermic effect at 770°C testifies about dominating surface crystallisation of the given glass particles.

One sharp exothermic effect at 640°C and sharp endothermic effect at 730°C were observed on DTA curve of $BaBi_2B_4O_{10}$ glass composition (Fig. 19, curve2). The melting temperature (T_m) is equal to 730°C and corresponding to Egorisheva's data [30]. X-ray diffraction patterns of this glass crystallization products show one $BaBi_2B_4O_{10}$ crystalline phase crystallization[30], which formed at glass powder samples crystallization at temperature 640°C (Fig. 20, curve2) and its melting. T_m is equal to 730°C and corresponding to Egorisheva's data [30]. Hardly visible pre-crystallisation fluctuation exothermal effect is observed also at 545°C (Fig. 19, curve2).

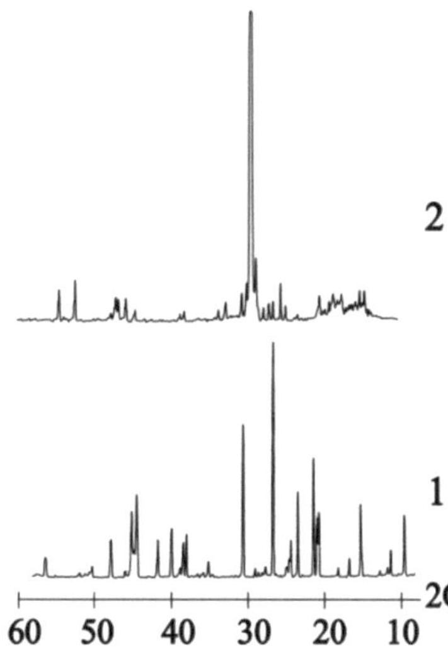

Figure 20. XRD-patterns of the crystallized glasses corresponding to ternary $BaBiB_{11}O_{19}$ (1- 760 °C 24h,cooling in the muffle) and $BaBi_2B_4O_{10}$ (2- 640°C 24h,cooling in the muffle) compounds

On the DTA curves of stoichiometric $BaBi_2B_2O_7$ and $BaBi_{10}B_6O_{25}$ glass compositions observed two exothermal effects at 570 and 645°C for $BaBi_2B_2O_7$ and at 460 and 620°C for $BaBi_{10}B_6O_{25}$ (Fig. 19, curves 3,4). But both compositions have one endothermic effect of melting at 725 and 690°C respectively for $BaBi_2B_2O_7$ and $BaBi_{10}B_6O_{25}$ testifying to one formed crystalline phase melting (Fig. 19, curves 3,4). X-ray data of these samples confirmed monophase crystallizations in each samples (Fig. 21, curves 1,2; Fig.22, curves 1,2).

According to [34] the X-ray powder diffraction patterns of formed $BaBi_2B_2O_7$ crystalline phase at stoichiometric glass composition (33.33BaO•33.33Bi_2O_3• 33.33B_2O_3 mol%) at second exothermic peaks temperature (640°C 24h) was indexed on an orthorhombic cell with following lattice parameters: a=11.818Å, b=8.753 Å, c=7.146Å, cell volume V=739.203 Å, Z=4 (Fig. 21, curve2). XRD-patterns of products of same glass crystallization at 570°C 24h keeps all diffraction lines of its analogue obtained at 640°C 24h (Fig. 21, curve1). Difference is observed only in sharp increasing of intensity (I/I_o) of [030] diffraction line from 4 to 43 at high temperature crystallization. That leads to reorientation of crystal structure, decreasing [030] diffraction line and accompanied with occurrence of the second exothermal effect on DTA curve (Fig. 19, curve 3).

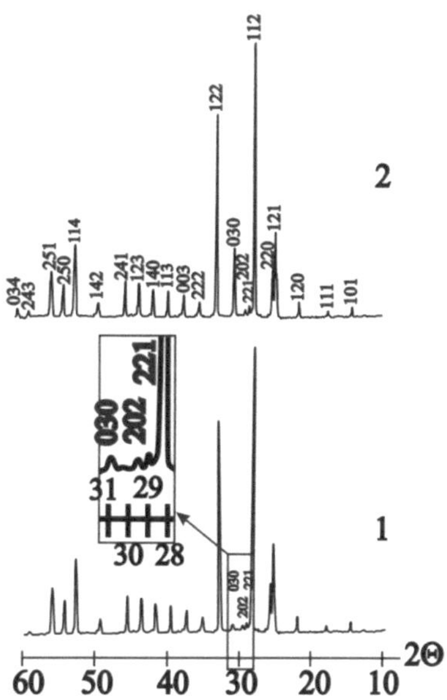

Figure 21. XRD-patterns of the crystallized glasses corresponding to ternary $BaBi_2B_2O_7$ composition:

1-570°C 24h, cooling in the muffle; 2-640°C 24h, casting in the cold water

X-ray powder diffraction patterns of $BaBi_{10}B_6O_{25}$ single crystals was indexed on an orthorhombic cell with following lattice parameters: a=6.434Å, b=11.763 Å, c=29.998Å, cell volume V=2270.34 Å, Z=8 [34]. XRD-patterns of products of same compositions (11.11BaO• $55.55Bi_2O_3$• $33.33B_2O_3$ mol%) glass crystallization at both exothermal effects (420°C 24h and 620°C 24h) have all diffraction lines of the $BaBi_{10}B_6O_{25}$ single crystals (Fig. 22, curves1-3). Naturally, $BaBi_{10}B_6O_{25}$ single crystal has well generated planes and clear observed diffraction lines on XRD-patterns in contrast to crystalline phases formed at same composition glasses crystallization. However, the most intensive diffraction line (I/I_o=100) of single crystals is [133], whereas products of glass crystallizations have [203] strongest diffraction line and [133] diffraction line became 5-10 times less (Fig. 22, curves1-3). Now it is difficult to us only on the basis of XRD-patterns analysis of glass crystallizations products to assume the nature of the second exothermal effect at 620°C on DTA curve of $BaBi_{10}B_6O_{25}$ of glass composition (Fig.19, curve 4). Their XRD-patterns are identical each other and to single crystals, but contain slightly quantity of not indexed reflexes, which are absent in X-ray powder diffraction patterns of $BaBi_{10}B_6O_{25}$ single crystals (Fig. 22, curves1-3).

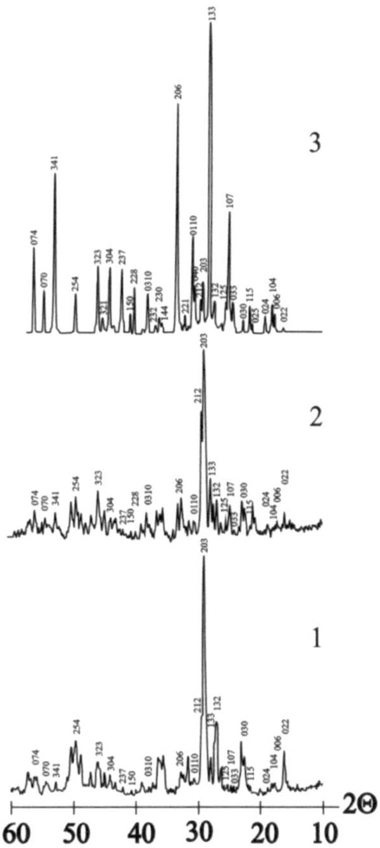

Figure 22. XRD-patterns of the crystallized glasses corresponding to ternary $BaBi_{10}B_6O_{25}$ composition:(1-460°C 24h,cooling in the muffle; 2-620°C 24h, casting in the cold water) and $BaBi_{10}B_6O_{25}$ single crystals (3).

The DTA curve of 50BaO• $25Bi_2O_3$• $25B_2O_3$ mol% ($BaBiBO_4$) glass composition contain exothermal effect of glass crystallization at 450°C and endothermic effect of this crystalline phase melting at 680°C (Fig. 19, curv.5). X-ray diffraction patterns of this glass crystallization products show one $BaBiBO_4$ crystalline phase formation at glass powder samples crystallization at temperature interval 450-640°C(Fig. 23, curve 1), which completely correspond to Barbier with co-authors data [29]. A second endothermic effect within the interval of 745-775°C with minimum at 760°C is associated with $BaBiBO_4$ incongruent melting (Fig. 19, curve 5).

We have revealed also, that the crystalline $BaBiBO_4$ compound is melted incongruently at 680°C with the melt and crystalline $BaBiO_3$ formation (Fig. 23, curve 2). The $BaBiO_3$ crystalline phase was observed on XRD-patterns of thermal treated at 720 °C and fast freeze in cold water products and identified according to X-ray database [43, fail # 01-074-7523]. The dissolution of this $BaBiO_3$ phase in a melt leads to the appearance on a DTA curve the second endothermic effect in an interval 745-775°C (Fig. 19, curve 5). Above 775°C we have glass-forming $BaBiBO_4$ composition melt without presence of any crystalline phase.

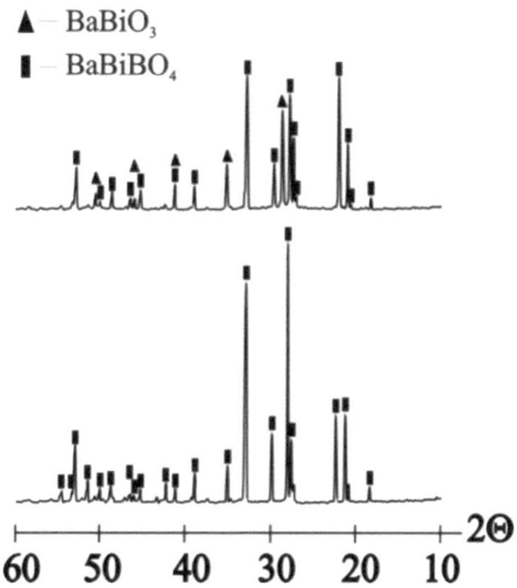

Figure 23. XRD-patterns of the crystallized glasses corresponding to ternary $BaBiBO_4$ composition:

1-450°C 24h, cooling in the muffle; 2-720°C 3h, casting in the cold water

Two exothermal effects of glass crystallization at 540 and 670°C and one endothermic effect of crystalline phase melting at 827°C are seen on the DTA curve of the $60BaO\bullet10Bi_2O_3\bullet30B_2O_3$ mol% ($Ba_3BiB_3O_9$) glass composition (Fig. 19, curve 6). X-ray diffraction patterns of this glass crystallization products at 540 and 670°C show one $Ba_3BiB_3O_9$ crystalline phase formation (Fig. 24, curves 1,2) at glass powder samples crystallization at 540 and 670°C and fully correspond to Egorisheva with co-authors data, which synthesized for the first time and have describe $Ba_3BiB_3O_9$ compound [31]. However, we didn't indicate polymorphic transition of $Ba_3BiB_3O_9$ at 850°C as reported in [31]. Presence of second endothermic effect within the interval of 840-890°C with minimum at 865°C is associated with $Ba_3BiB_3O_9$ incongruent melting (Fig. 19, curve 6). We have revealed that the crystalline $Ba_3BiB_3O_9$ compound is melted incongruently at 827°C with the glass forming melt and crystalline phase formation (Fig. 24, curve 3). The $Ba_2B_2O_5$ crystalline phase was observed in amorphous matrix on XRD-patterns of thermal treated at 830 °C and fast freeze in cold water products and identified according to X-ray database [43, fail # 024-0087]. For clear $Ba_2B_2O_5$ observation on XRD-patterns the preliminary crystallized at 670°C 24h sample have been exposed at 830°C 3 h (Fig. 23, curve3). Dissolution of this $Ba_2B_2O_5$ phase in a melt leads to the appearance on a DTA curve the second endothermic effect in an interval 840-890°C (Fig. 19, curve 6). Above 890°C we have glass-forming $Ba_3BiB_3O_9$ composition melt without presence of any crystalline phase at cooling rate 10^2 K/s.

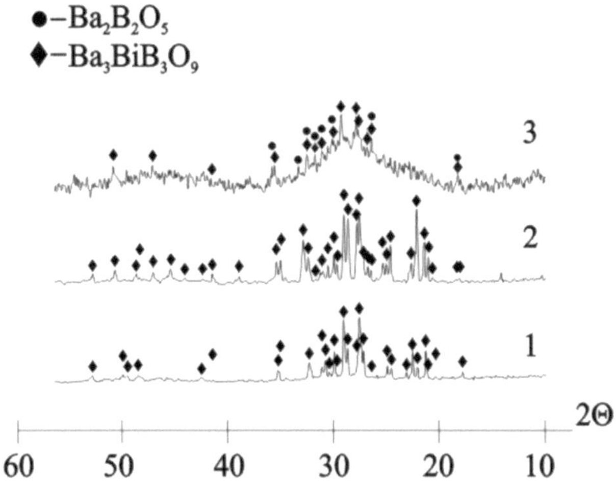

Figure 24. XRD-patterns of the crystallized glasses corresponding to ternary $6BaBi3BO_4$ composition:1.540°C 24h, cooling in the muffle; 2-670°C 24h, casting in the cold water; 3-670°C 24h+830°C 3 h, casting in the cold water

3.3. TEC study of the stoichiometric compositions glasses in the BaO-Bi_2O_3-B_2O_3 system

The isolines diagram of BaO-Bi_2O_3-B_2O_3 system glasses TEC values is given on Fig. 25. It is clear observed common regularity, that the increase of barium and bismuth oxides amounts in glasses of binary BaO-Bi_2O_3 and Bi_2O_3-B_2O_3 systems leads to increase TEC of glasses. The same tendency is observed for glasses of ternary system: joint presence of BaO and Bi_2O_3 and increase their amounts leads to increase glasses TEC values from 70 to 127•$10^{-7}K^{-1}$ (Fig. 25).

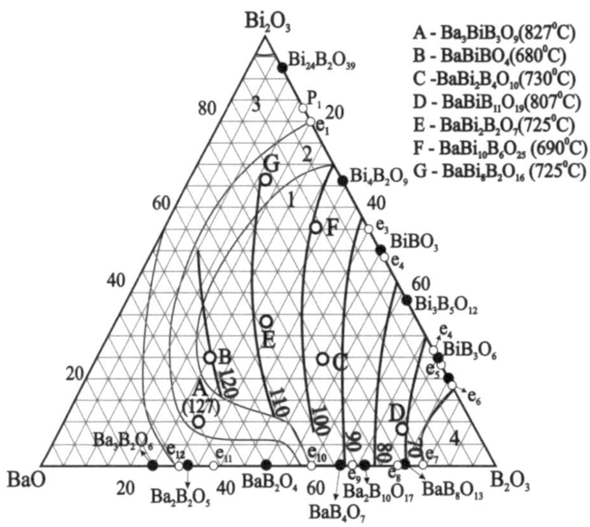

Figure 25. BaO-Bi_2O_3-B_2O_3 system's glasses TEC (α_{20-300}•$10^{-7}K^{-1}$) values isolines

The high boron content glass composition corresponds to $BaBiB_{11}O_{19}$ (14.28BaO•7.14Bi_2O_3•78.57B_2O_3 mol %) have TEC=72•$10^{-7}K^{-1}$ and T_g=498°C calculated from dilatometric curve (Table 4). Reduction the B_2O_3 amount together with increasing of BaO and Bi_2O_3 amounts in glass compositions leads to increase TEC and reduction Tg values: for glass composition 25BaO•25Bi_2O_3•50B_2O_3 mol % ($BaBi_2B_4O_{10}$) TEC=96•$10^{-7}K^{-1}$ and T_g= 445°C; 33.3BaO•33.3 Bi_2O_3•33.3B_2O_3 mol %($BaBi_2B_2O_7$) TEC=108•$10^{-7}K^{-1}$ and T_g=415°C; 11.1BaO•55.5Bi_2O_3•33.3B_2O_3 mol %($BaBi_{10}B_6O_{25}$) TEC=97•$10^{-7}K^{-1}$ and T_g=350°C; 16.67BaO•66.67Bi_2O_3•16.67B_2O_3 mol%($BaBi_8B_2O_{16}$) TEC=110•$10^{-7}K^{-1}$ and T_g=415°C. However, for 50BaO•25 Bi_2O_3•25B_2O_3 mol % ($BaBiBO_4$) and 60BaO•10Bi_2O_3•30B_2O_3 mol % ($Ba_3BiB_3O_9$) glass compositions simultaneous increase both TEC and T_g values were observed: TEC=120•$10^{-7}K^{-1}$ and T_g=400°C; TEC=127•10^{-7} K^{-1} and T_g=460°C respectively for $BaBiBO_4$ and $Ba_3BiB_3O_9$ (Fig. 25, table4).

TEC values of crystallized glasses corresponding to the ternary barium bismuth borates given in Table 4. Crystallized barium bismuth borate glass samples have TEC values lower, than initial glasses and equals to: 49•$10^{-7}K^{-1}$ for $BaBiB_{11}O_{19}$ sample (750°C 24h), 78•$10^{-7}K^{-1}$ for $BaBi_2B_4O_{10}$ sample (630°C 24 h), 96•10^{-7} K^{-1} for $BaBi_2B_2O_7$ sample (640°C 24h), 97•$10^{-7}K^{-1}$ for $BaBi_{10}B_6O_{25}$ sample (610°C 24h), 110•10^{-7} K^{-1} for $BaBiBO_4$ sample (450°C24h) and 109•10^{-7} K^{-1} for $Ba_3BiB_3O_9$ sample (690°C 24h). The same tendency, as well as for their glassy analogues, is observed for crystallized glass samples: increase of barium and bismuth oxides amounts in ternary compounds leads to their TEC values increase.

4. Ferroelectric properties of new ternary $BaBi_2B_2O_7$ and $BaBi_{10}B_6O_{25}$ stoichometric compositions glass ceramics.

The ferroelectric (polarization - electric field) hysteresis, is a defining property of ferroelectric materials. Thus, the most widely studied characteristics of ferroelectric hysteresis were those of interest for this particular application: the value of the switchable polarization (the difference between the positive and negative remanent polarization, $P_R - (-P_R)$, dependence of the coercive field *Ec* on sample thickness, decrease of remanent or switchable polarization with number of switching cycles, polarization imprint, endurance, retention [51].

Electric field induced polarization (P) and remanent polarization(P_r) were measured at room temperature for $BaBi_2B_2O_7$ and $BaBi_{10}B_6O_{25}$ glass tape samples crystallized using various regimes (Fig. 26).

a. $BaBi_2B_2O_7$ glass tape sample of 0.07 mm in thickness crystallized at 450°C 24h, $2P_r$ = 0.15 μC/cm²;

b. $BaBi_{10}B_6O_{25}$ glass tape sample of 0.06 mm in thickness crystallized at 380 °C 12h, $2P_r$ = 0.32 μC/cm²;

c. $BaBi_{10}B_6O_{25}$ glass tape sample of 0.06 mm in thickness crystallized at 410°C 12h, $2P_r$ = 0.62 μC/cm²;

d. $BaBi_{10}B_6O_{25}$ glass tape sample of 0.05 mm in thickness crystallized at 410°C 24h, $2P_r = 0.9$ μC/cm^2

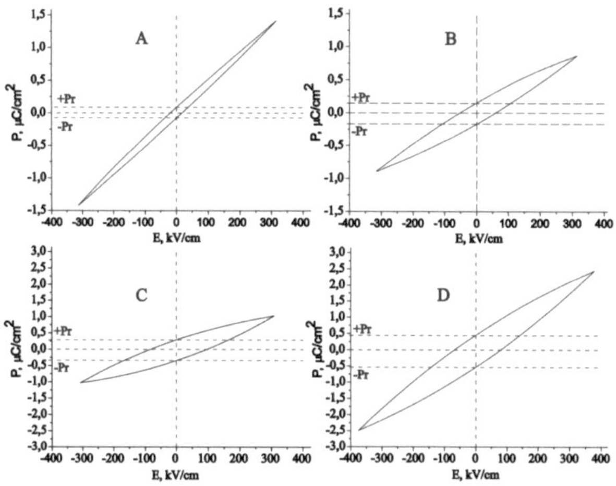

Figure 26. Dependence of polarization (P) on electric field (E) for crystallized stoichiometric glass compositions:

Linear *P–E* curves are observed up to fields of 40-120 kV/cm for all measured samples with thickness 0.05-0.07mm. The polarization becomes nonlinear with increasing of applied electric field, and at 140-380 kV/cm the remanent polarization $2P_r$ values were found 0.15 μC/cm^2 for the $BaBi_2B_2O_7$ (Fig. 26, A). The remanent polarization $2P_r$ value for $BaBi_{10}B_6O_{25}$ crestallized glass tape samples encreasing with termal treatment temperature from 0,32 to 0, 64 μC/cm^2 (Fig. 26, B & C) and time (Fig. 26, D). The highest remanent polarization value ($2P_r$=0.9 μC/cm^2) has $BaBi_{10}B_6O_{25}$ glass tape sample crystallized at 410°C 24h (Fig.26, D). According to obtained results it is possible to conclude that samples are ferroelectrics.

5. Discussion

The pricipial diference of our methodology from traditional is a glass samples using as initial testing substance for phase diagram of very complex ternary BaO-Bi_2O_3-B_2O_3 system construction. It is a very effective method, due possibility to indicate temperature intervals of all processes taking place in glass samples: glass transition, crystallization, quantity of formed crystalline phases and their melting. Whereas, samples prepared by traditional solid phase synthesis are less informative and often lose a lot of information. Super cooling technique created by our group allowed us both to expand borders of glass formation and to have enough quantity samples for DTA and X-ray investigations and BaO-Bi_2O_3-B_2O_3 system phase diagram construction (Fig.2).

The region of stable glasses includes the binary compounds BaB_4O_7, $Ba_2B_{10}O_{17}$, BaB_8O_{13}, $Bi_4B_2O_9$, $BiBO_3$, $Bi_3B_5O_{12}$, BiB_3O_6, and $Bi_2B_8O_{15}$ in the BaO–B_2O_3 and Bi_2O_3–B_2O_3 systems (Fig.

2). Binary BaB_4O_7, $Ba_2B_{10}O_{17}$, and BaB_8O_{13} barium borates have melting temperatures (Tm) of 910, 905, and 890°C and can be found between the eutectics e_7, e_8, e_9, and e_{10} with Tm = 878, 869, 895, and 899°C, respectively. The transition to a crystallization field of barium metaborate is accompanied by a sharp increase of liquidus temperature (T_L) and a decrease of the glass forming ability of the melts. The final compound, forming a stable glass, contains ~43 mol %BaO and has T_L = 950°C. Compounds BaB_2O_4 and $Ba_2B_2O_5$, having higher Tm, which are 1095 and 1050°C, respectively [18, 21, 22], are found in the region of the compounds obtained in the form of glasses with a cooling rate of (10^3–10^4) K/s. Glass formation in the BaO–B_2O_3 binary system is limited by the eutectic e_{12} with Tm = 1025°C (Fig. 2) because of the sharp increase of the liquidus temperature during the transition to a field of crystallization of the $Ba_3B_2O_6$ compound (Tm = 1383°C). BaB_2O_4 (Tm = 1095°C) is the dominating compound in the system and does not form stable glasses. Its considerable crystallization field narrows the region of stable glasses in the ternary system, which is only restricted by compounds with T_L ~ 950°C (Fig. 2).

Binary bismuth borates $Bi_4B_2O_9$, $BiBO_3$, $Bi_3B_5O_{12}$, BiB_3O_6, and $Bi_2B_8O_{15}$ have Tm 675, 685, 722, 708, and 715°C can be found between the eutectics e_1, e_2, e_3, e_4, e_5, e_6 with a Tm of 622, 646, 665, 698, 695, and 709°C, respectively [2, 34]. The region of stable glasses in the Bi_2O_3–B_2O_3 system is limited by a compound containing ~70%mol Bi_2O_3 and having T_L = 670°C. Compounds that are found in the range of 70–80mol% Bi_2O_3 (before the e_1 point) are obtained in the form of glasses during cooling at a rate of ~10^2 K/s. During the transition to the crystallization field of $Bi_{24}B_2O_3$ and Bi_2O_3, T_L increases to 825°C (Tm of Bi_2O_3). Glasses in this part of the system are obtained during melt cooling at a rate of (10^3–10^4) K/s. These compounds have a low liquidus temperature; however, the structure factor essentially influences their glass forming ability, not allowing glass formation at low melt cooling rates.

Six ternary compounds are known in the BaO–Bi_2O_3–B_2O_3 system: $Ba_3BiB_3O_9$, $BaBiBO_4$, $BaBi_2B_4O_{10}$, and $BaBiB_{11}O_{19}$ synthesized by Barbie and Egorysheva in 2005–2006, and $BaBi_2B_2O_7$ and $BaBi_{10}B_6O_{25}$ revealed by our research group in 2008–2009. $BaBi_2B_4O_{10}$, $BaBiB_{11}O_{19}$, $BaBi_2B_2O_7$, and $BaBi_{10}B_6O_{25}$ melt congruently at 730, 807, 725, and 695°C respectively, and $BaBiBO_4$ melts incongruently at 680°C and has T_L = 760 °C. All these five ternary compounds along with the eutectics formed between each other and with binary barium and bismuth borates form a "plateau" with low T_L, which is responsible for the formation of the region of stable glasses in the ternary system BaO–Bi_2O_3–B_2O_3.

Compounds joining low temperature eutectic e_1 (622°C) [2] in the Bi_2O_3–B_2O_3 binary system and the eutectics e_{13} (~790°C) and e_{14} (~750°C) in the BaO–Bi_2O_3 binary system 26 - 28] form glasses only at higher cooling rates of their melts (10^3–10^4) K/s. Glass formation in the BaO–Bi_2O_3 binary system stops at 45 mol% BaO content (T_L ~930°C) [26 - 28]. Along with the factor of the liquidus temperature [52], a considerable contribution to the glass formation of the pointed compositions is made by the structural factor of the melt. The combination of the structural factors of the melt and the liquidus temperature is also considerable during the transition to the vitreous state of the compositions, which are found in the crystallization fields of BaB_2O_4 and $Ba_2B_2O_5$, where they show the tendency towards glass formation only at high rates of melt cooling.

There are very stable congruent melted binary BaB_2O_4 and ternary $BaBi_2B_4O_{10}$, $BaBiB_{11}O_{19}$, and $BaBi_{10}B_6O_{25}$ compounds in the studied ternary BaO-Bi_2O_3-B_2O_3 system. They have dominating positions in ternary diagram and occupied the biggest part of it (Fig.18). Mutual influence of these compounds and other binary and ternary compounds (BaB_4O_7, $Ba_2B_{10}O_{17}$, BaB_8O_{13}, $Bi_4B_2O_9$, $BiBO_3$, $Bi_3B_5O_{12}$, BiB_3O_6, $Bi_2B_8O_{15}$, $BaBi_2B_2O_7$, and $BaBi_8B_2O_{16}$) lead to formation of sixteen revealed at present time ternary eutectics (Fig.18& Table 3), which have essential influence on liquidus temperature decrease and to assist in glass formation. Ternary $BaBi_2B_4O_{10}$ compound forms eight eutectics with binary and ternary compounds, its neighbors: E_4(622°C), E_3(640°C), E_2(585°C), E_5(610°C), E_6(675°C), E_{12}(680°C), E_{13} (690°C), and E_{14} (700°C) (Fig.18& Table 3). $BaBiB_{11}O_{19}$ compound forms seven eutectics with its neighbors: E_6 (675°C), E_7 (680°C), E_8 (675°C), E_9 (680°C), E_{10} (730°C), E_{11} (750°C),and E_{12} (680°C) (Fig.18& Table 3). $BaBi_{10}B_6O_{25}$ compound forms five eutectics with its neighbors: E_1 (590°C), E_2 (585°C), E_3 (640°C), E_{15} (645°C), and E_{16} (615°C) (Fig.18& Table 3). Determined ternary eutectics together with binary eutectics e_1, e_2, e_3, e_4, e_5, and e_6 of Bi_2O_3-B_2O_3 system have allowed to outline the fields of binary $Bi_4B_2O_9$, $BiBO_3$, $Bi_3B_5O_{12}$, BiB_3O_6, and $Bi_2B_8O_{15}$ bismuth borates crystallisation, as well as together with binary eutectics e_7, e_8, e_9, and e_{10} of BaO-B_2O_3 system have allowed to outline the fields of binary BaB_4O_7, $Ba_2B_{10}O_{17}$, BaB_8O_{13} barium borates and partly BaB_2O_4 crystallization on the BaO-Bi_2O_3-B_2O_3 system phase diagram (Fig.18).

The clear correlation between glass forming and phase diagrams has been observed in studied system. The glass melting temperature and level of glass formation depending on the cooling rate of the studied melts are in good conformity with boundary curves and eutectic points (Fig.2& 18).

The phase diagram of the well known binary Bi_2O_3–B_2O_3 system has been corrected in the interval between the $Bi_4B_2O_9$ and $Bi_3B_5O_{12}$ compounds. The eutectic composition, $48.5Bi_2O_3 \bullet 51.5B_2O_3$ (mol%), between $BiBO_3$ and $Bi_3B_5O_{12}$, with m.p. 665±5°C, has been determined. It is shown that the compound $BiBO_3$ is congruently melting with a m.p. of 685±5°C.

The next unexpected results were obtained at phase diagram construction: two new ternary $BaBi_2B_2O_7$ and $BaBi_{10}B_6O_{25}$ compounds have been revealed at the same glass compositions crystallisation. X-ray characteristics of the new ternary compound $BaBi_2B_2O_7$, synthesized at the $33.33BaO \bullet 33.33Bi_2O_3 \bullet 33.33B_2O_3$ (mol%) glass composition crystallization at 640°C, 20 h. The x-ray powder diffraction patterns of $BaBi_2B_2O_7$ could be indexed on an orthorhombic cell with lattice parameters as follows: $BaBi_2B_2O_7$ a=11.818 Å, b=8.753 Å, c=7.146 Å, cell volume V=739.203 Å^3, Z=4.

Single crystals of $BaBi_{10}B_6O_{25}$ were obtained by cooling of a melt with the stoichiometric composition. Glass powder of composition $11.11BaO \bullet 55.55Bi_2O_3 \bullet 33.33B_2O_3$ (mol%) was heated in a quartz glass ampoule up to 750°C at a rate 10 K/min. After 2 h at high temperature, the melt was cooled at a rate 0.5 K/h. Single crystals with sizes up to 1.66×0.38×0.19 mm^3 were grown. The x-ray powder diffraction patterns of $BaBi_{10}B_6O_{25}$ could be indexed on an orthorhombic cell with lattice parameters as follows: a=6.434 Å, b=11.763 Å, c=29.998 Å, cell volume V=2270.34 Å^3, Z=8.

http://dx.doi.org/10.5772/52405

Common regularities of bulk glass samples TEC changes in studied BaO-Bi_2O_3-B_2O_3 system have been determined: the increase of barium and bismuth oxides amounts in glasses of binary BaO-Bi_2O_3 and Bi_2O_3-B_2O_3 systems leads to increase TEC of glasses. The same tendency is observed for glasses of ternary BaO-Bi_2O_3-B_2O_3system: joint presence of BaO and Bi_2O_3 and increase their amounts leads to increase glasses TEC values from 70 to $127 \bullet 10^{-7} K^{-1}$ (Fig. 25).

Crystallized barium bismuth borate glass samples have TEC values lower, than initial glasses and equals to: $49 \bullet 10^{-7} K^{-1}$ for $BaBiB_{11}O_{19}$ sample (750°C 24h), $78 \bullet 10^{-7} K^{-1}$ for $BaBi_2B_4O_{10}$ sample (630°C 24 h), $96 \bullet 10^{-7} K^{-1}$ for $BaBi_2B_2O_7$ sample (640°C 24h), $97 \bullet 10^{-7} K^{-1}$ for $BaBi_{10}B_6O_{25}$ sample (610°C 24h), $110 \bullet 10^{-7} K^{-1}$ for $BaBiBO_4$ sample (450°C24h) and $109 \bullet 10^{-7} K^{-1}$ for $Ba_3BiB_3O_9$ sample (690°C 24h). The same tendency, as well as for their glassy analogues, is observed for crystallized glass samples: increase of barium and bismuth oxides amounts in ternary compounds leads to their TEC values increase.

Electric field induced polarization (P) and remanent polarization (P_r) were measured at room temperature for $BaBi_2B_2O_7$ and $BaBi_{10}B_6O_{25}$ glass tape samples crystallized at various regimes. All tested samples shown loop of hysteresis.

Linear *P–E* curves are observed up to fields of 40-120 kV/cm for all measured samples with thickness 0.05-0.07mm. The polarization becomes nonlinear with an increase of applied electric field, and at 140-400 kV/cm the remanent polarization $2P_r$ values were found 0.15 μC/cm² for the $BaBi_2B_2O_7$ (Fig.26, A), and 0.32- 0.9 μC/cm² for the $BaBi_{10}B_6O_{25}$ (Fig.26, B-D), crystallized glass tape samples. According to obtained results it is possible to conclude that all tested samples are ferroelectrics.

5. Conclusion

Effective way of new system investigation and new compounds and characteristic points revealing via simultaneous glass forming and phase diagrams construction have been shown. Phase diagram of the ternary BaO-Bi_2O_3-B_2O_3 system have been constructed for the first time us result of fourteen pseudo-binary systems and sections phase diagrams investigations.

The phase diagram of the well known binary Bi_2O_3–B_2O_3 system has been corrected in the interval between the $Bi_4B_2O_9$ and $Bi_3B_5O_{12}$ compounds. The eutectic composition, $48.5Bi_2O_3 \bullet 51.5B_2O_3$ (mol%), between $BiBO_3$ and $Bi_3B_5O_{12}$, with m.p. 665±5°C, has been determined. It is shown that the compound $BiBO_3$ is congruently melting with a m.p. of 685±5°C.

Two new ternary $BaBi_2B_2O_7$ and $BaBi_{10}B_6O_{25}$ compounds have been revealed at the same glass compositions crystallisation. The new ternary compound $BaBi_2B_2O_7$ synthesized at the $33.33BaO \bullet 33.33Bi_2O_3 \bullet 33.33B_2O_3$ (mol%) glass composition crystallization at 640°C, 20 h. The x-ray powder diffraction patterns of $BaBi_2B_2O_7$ could be indexed on an orthorhombic cell with lattice parameters as follows: $BaBi_2B_2O_7$ a=11.818 Å, b=8.753 Å, c=7.146 Å, cell volume V=739.203 Å³, Z=4.

Single crystals of $BaBi_{10}B_6O_{25}$ were obtained by cooling of a melt with the stoichiometric composition. Glass powder of composition $11.11BaO \bullet 55.55Bi_2O_3 \bullet 33.33B_2O_3$ (mol%) was heated in a quartz glass ampoule up to 750°C at a rate 10 K/min. After 2 h exposition at high temperature, the melt was cooled at a rate 0.5 K/h. Single crystals with sizes up to 1.66×0.38×0.19 mm^3 were grown. The x-ray powder diffraction patterns of $BaBi_{10}B_6O_{25}$ could be indexed on an orthorhombic cell with lattice parameters as follows: a=6.434 Å, b=11.763 Å, c=29.998 Å, cell volume V=2270.34 $Å^3$, Z=8.

Ternary $BaBi_2B_4O_{10}$, $BaBiB_{11}O_{19}$, and $BaBi_{10}B_6O_{25}$ compounds have dominating positions in ternary diagram and occupied the biggest part of it. Mutual influence of these and other binary and ternary compounds (BaB_4O_7, $Ba_2B_{10}O_{17}$, BaB_8O_{13}, $Bi_4B_2O_9$, $BiBO_3$, $Bi_3B_5O_{12}$, BiB_3O_6, $Bi_2B_8O_{15}$, $BaBi_2B_2O_7$, and $BaBi_8B_2O_{16}$) lead to formation of sixteen ternary eutectics, which have essential influence on liquidus temperature decrease and to assist in glass formation.

The clear correlation between glass forming and phase diagrams has been observed: glass melting temperature and level of glass formation depending on the cooling rate of the studied melts are in good conformity with boundary curves and eutectic points.

Common regularities of bulk glass samples TEC changes in studied BaO-Bi_2O_3-B_2O_3 system have been determined: the increase of barium and bismuth oxides amounts in glasses of binary BaO-Bi_2O_3 and Bi_2O_3-B_2O_3 systems and their joint amounts increasing in ternary compositions leads to increase glasses TEC values from 70 to $127 \bullet 10^{-7} K^{-1}$.

Crystallized barium bismuth borate glass samples have TEC values lower, than initial glasses. Increase of barium and bismuth oxides amounts in ternary compounds leads to their TEC values increasing from 49 to $109 \bullet 10^{-7} K^{-1}$.

Electric field induced polarization (P) and remanent polarization (P_r) were measured at room temperature for $BaBi_2B_2O_7$ and $BaBi_{10}B_6O_{25}$ glass tape samples crystallized at various regimes. The remanent polarization $2P_r$ values were found 0.15 μC/cm^2 for the $BaBi_2B_2O_7$, and 0.32- 0.9 μC/cm^2 for the $BaBi_{10}B_6O_{25}$ crystallized glass tape samples. According to obtained results it is possible to conclude that all tested samples are ferroelectrics.

Acknowledgements

This work was supported by the International Science and Technology Center (Projects # A-1591). Author is very gratefull to all project team and to Dr. Rafael Hovhannisyan , Dr. Nikolay Knyazyan and Prof. Heli Jantunen for effective cooperation and fruitful discussions.

Author details

Martun Hovhannisyan*

Address all correspondence to: eni_arm@yahoo.com

Address all correspondence to: raf.hovhannisyan@yahoo.com

http://dx.doi.org/10.5772/52405

Address all correspondence to: martun_h@yahoo.com

«ENI» Institute of Electronic Materials LTD, Armenia

References

[1] The American Ceramic Society & National Institute of Standards and Technology [ACerS & NIST]. (2004). Phase Equilibria Diagrams. *CD-ROM Database, Version 3.0,* 157498215.

[2] Levin, E. M., & Mc Daniel, C. L. (1962). The System Bi_2O_3-B_2O_3. *J. Am.Cer.Soc*, 45(8), 355-360.

[3] Masurin, O. V., et al. (1979). Properties of Glasses and Glass-Forming Melts. Handbook., In Russian, v.III, part 2, L.,, Nauka Press.

[4] Muehlberg, M., Burianek, M., Edongue, H., & Poetsch, C. (2002). $Bi_4B_20_9$ crystal growth and some new attractive properties. *J. Cryst. Growth*, 237, 740-744.

[5] Becker, P., & Bohaty, L. (2003). Linear electro-optic properties of bismuth triborate, BiB_3O_6 (BIBO). *Phys. Chem. Glasses*, 44, 212-214.

[6] Ihara, R., Honma, T., Benino, Y., Fujiwara, T., & Komatsu, T. (2004). Second-order optical nonlinearities of metastable BiBO3 phases in crystallized glasses. *Optical Materials*, 27, 403-408.

[7] Oprea, I. I. (2005). Optical Properties of Borate Glass- Ceramic. *Dissertation zur Erlangung des Grades Doktor der Naturwissenschaften, Osnabruck University, Germany.*

[8] Pottier , M. J. (1974). Mise en evidence d'un compose BiBO3 et de son polymorphisme par spectroscopie vibrationnelle. *Bull. Soc. Chim. Belg.*, 83, 235-238.

[9] Becker, P., Liebertz, J., & Bohatý, L. (1999). Top-seeded growth of bismuth triborate, BiB_3O_6. *J.Cryst.Growth*, 203, 149-155.

[10] Becker, P., & Froehlich, R. (2004). Crystal growth and crystal structure of metastable bismuth ortoborate $BiBO_3$. *Z. Naturforsch. B*, 59, 256-258.

[11] Zargarova, M. I., & Kasumova , M. F. (2004). Liquidus surface of the system ZnO - Bi_2O_3-B_2O_3 projection. *Zh. Neorg. Mater*, In Russian, 26(8), 1678-1681.

[12] Kargin, Yu. F., Zhereb, V. P., & Egorysheva, A. V. (2002). Phase diagram of metastable stats of the Bi_2O_3-B_2O_3 system. *Zh. Neorg. Khim*, 47(8), 1362-1364, In Russian.

[13] Barbier, J., Penin, N., & Cranswick, L. M. (2005). Melilite-Type Borates $Bi_2ZnB_2O_7$ and $CaBiGaB_2O_7$. *Chem. Mater.*, 17, 3130-3136.

[14] Barbier, J., & Cranswick, L. M. (2006). The Non-Centrosymmetric Borate Oxides, $MBi_2B_2O_7$(M=Ca,Sr). *Solid State Chem*, 179, 3958-3964.

[15] Egorisheva, A. V., Skorikov, V. M., & Volodin, V. D. (2008). Calcium Bismuth Borates in the CaO-Bi_2O_3-B_2O_3 system. Zh.Nerg.Mater. In Russian)., 44(1), 76-81.

[16] Kargin, Yu. F., Ivicheva, S. N., Komova, M. G., & Krut'ko, V. A. (2008). Subsolidus Phase Relation in the SrO-Bi_2O_3-B_2O_3 system. *Zh.Nerg.Khim*, 53(3), 474-478, In Russian.

[17] Barbier, J., Davis, L. J. M., Goward, G. R., & Cranswick, L. M. D. (2009). Ab initio structure determination of $SrBi_2OB_4O_9$ by powder X-ray/neutron diffraction and NMR spectroscopy. *Powder Diffr*, 24(1), 35-40.

[18] Levin, E. M., & Mc Murdie, H. F. (1949). The System BaO-B2O3. *J. Res.Nat.Bur.Standards*, 42(2), 131-138.

[19] Levin, E. M., & Ugrinic, G. M. (1953). The System Barium Oxide-Boric Oxide-Silica. *J. Res. Nat.Bur. Stand*, 51, 37-56.

[20] Green, C. H., & Wahler, R. L. (1969). Crystallization of Barium Aluminum Borate Glasses-Rhodium as a Specific Catalyst. *In:Kinetics of reactions in ionic systems. Proc. Intern. Symp. 18-23 IV, New York*, 545-553.

[21] Hubner, K. H. (1969). Ueber die Borate $2BaO \bullet 5B_2O_3$, tief-$BaO \bullet B_2O_3$, $2BaO \bullet B_2O_3$ und $4BaO \bullet B_2O_3$. *Neues Jahrb. Mineral., Monatsch*, 335-343.

[22] Hovhannisyan, R. M. (2006). Binary alkaline-earth borates: phase diagrams correction and low thermal expansion of crystallized stoichiometric glass compositions. *Phys.Chem.Glasses: Eur. J.Glass Sci. Technol. B*, 47(4), 460-465.

[23] Crichton, S. N., & Tomozawa, M. (1997). Prediction of phase separation in binary borate glasses. *J. Non-Crystalline Solids*, 215, 244-251.

[24] Hageman, V. B. M., & Oonk, H. A. J. (1987). Liquid immiscibility in the B_2O_3-MgO, B_2O_3-CaO, B_2O_3-SrO and B_2O_3-BaO systems. *Physics and Chemistry of Glasses*, 28, 183-187.

[25] Klinkova, L. A., Nikolaychik, V. I., Barkovskiy, N. V., & Fedotov, V. K. (1999). Phase relations in the system Ba-Bi-O (20-80 mol% $BiO_{1.5}$) at p_{O2}=0.01, 0,21 and 1 atmosphere. Zh.Neorg. Khim In Russian)., 44(11), 1774-1782.

[26] Mueller, K., Meen, J. K., & Elthon, D. (1999). System BaO-Bi_2O_3 in O_2. *In ACerS & NIST, 2004.*

[27] Shevchuk, A. V., Skorikov, V. M., Kargin, Yu. F., & Konstantinov, V. V. (1985). The system BaO-Bi_2O_3. *Zh.Neorg. Khim*, 30(6), 1519-1522, In Russian.

[28] Witler, D., & Roth, R. S. (2004). System BaO-Bi2O3 in Air,. *in Phase Equilibria Diagrams, CD-ROM Database, A CerS-NITS.*

[29] Barbier, J., Penin, N., Denoyer, A., & Cranswic, L. M. D. (2005). $BaBiBO_4$, a novel noncentrosymmetric borate oxide. *Solid State Science*, 7, 1055-1061.

[30] Egorysheva, A.V, & Kargin Yu, F. (2006). Phase equilibrium in the sub-solidus parts of the Bi_2O_3-BaB_2O_4-B_2O_3 system. *Zh.Neorg. Khim ,In Russian*, 51(7), 1185-1189.

[31] Egorysheva, A.V, Skorikov, V.M, Volodin, V.D, Myslitskiy, O.E, & Kargin, Yu, F. (2006). Phase equilibrium in the BaO-B_2O_3-Bi_2O_3 system. *Zh.Neorg. Khim ,In Russian*, 51(12), 2078-2082.

[32] Bubnova, R. S., Krivovichev, S. V., Filatov, S. K., Egorysheva, A. V., & Kargin, Y. F. (2007). Preparation, crystal structure and thermal expansion if a new bismuth barium borate, $BaBi_2B_4O_{10}$. *Journal of Solid State Chemistry*, 180, 596-603.

[33] Elwell, D., Morris, A. W., & Neate, B. W. (1972). The flux system BaO-B_2O_3-Bi_2O_3. *J. of Crystal Growth*, 16, 67-70.

[34] Hovhannisyan, M. R., et al. (2009). A study of the phase and glass forming diagrams of the BaO-Bi_2O_3-B_2O_3 system. *Phys. Chem.Glasses: Eur. J.Glass Sci. Technol. B*, 50(6), 323-328.

[35] Hovhannisyan, M. R. (2009). Development of low melting sealing glasses on the bismuth borate and tellur molibdate systems basis. PhD Thesis. *State Engineering University of Armenia, Yerevan.*

[36] Hovhannisyan, M. R. (2008). Synthesis of new $BaBi_{10}B_6O_{25}$ compond in BaO-Bi_2O_3-B_2O_3 system. *Chemical J. of Armenia*, 62(2), 28-29.

[37] Hovhannisyan, R. M., et al. (2008). Mutual influence of barium borates, titanates and boron titanates on phase diagram and glass formation in the BaO-TiO2-B2O3 system. *Phys.Chem.Glasses: Eur. J.Glass Sci. Technol. B*, 49(2), 63-67.

[38] Hovhannisyan, R. M. (2010). Melts supercooling-effective way of inorganic systems glassforming and phase diagrams construction. *Proceedings of the II International conference of inorganic chemistry and chemical technology, 13-14th Sept., Yerevan*, 37-39, 978-9-99412-450-3.

[39] Hovhannisyan, R. M. (2011). New generation highly effective electro-and light-active nonlinear oriented tape glass ceramic on the basis of stoichiometric three-coordinated borates. *Final Technical Report of ISTC project # A-1486, Moscow.*

[40] Hovhannisyan, R. M., et al. (2011). Phase diagram, crystallization behavior and ferroelectric properties of stoichiometric glass ceramics in the BaO-TiO_2-B_2O_3 system. *in: Ferroelectrics- Physical Effects, InTech publishing, Croatia*, 49-76.

[41] Hovhannisyan, R. M. (2012). New Lead & Alkali Free Low Melting Sealing Glass Frits & Ceramic Fillers. *Final Technical Report of ISTC project # A-1591, Moscow.*

[42] Barseghyan, A. H., Hovhannisyan , R. M., Petrosyan , B. V., Aleksanyan , H. A., & Toroyan , V. P. (2012). Glass formation and crystalline phases in the ternary CaO-Bi_2O_3-Bi_2O_3 and SrO-Bi_2O_3-B_2O_3 systems. *Phys.Chem.Glasses: Eur. J.Glass Sci. Technol. B, in press.*

[43] International Center for Diffraction Data [ICDD]. (2008). Powder Diffraction Fails, PDF-2 release database,. *Pennsylvania, USA*, 1084-3116.

[44] Sawyer, C. B., & Tower, C. H. (1930). Rochelle Salt as a Dielectric. *Phys. Rev*, 35(3), 269-273.

[45] Imaoka, M., & Yamazaki, T. (1972). Glass-formation ranges of ternary systems. *III. Borates containing b-group elements, Rep. Inst. Ind. Sci. Univ. Tokyo*, 22(3), 173-212.

[46] Janakirama-Rao, Bh. V. (1965). Unusual properties and structure of glasses in the system BaO-B_2O_3-SrO, BaO-B_2O_3-Bi_2O_3, BaO-B_2O_3-ZnO, BaO-B_2O_3-PbO, Compt. *Rend. VII Congr. Intern. Du Verre, Bruxelles*, 1(104-104).

[47] Izumitani, T. (1958). Fundamental studies on new optical glasses,. *Rep. Governm. Ind. Res. Inst. Osaka* [311], 127.

[48] Kawanaka, Y, & Matusita, K. (2007). Various properties of bismuth oxide glasses for PbO free new solder glasses. *Proc.XXIth Intern.Congr.on Glass (in CD-ROM), Strasbourg* [I23].

[49] Milyukov, E. M., Vilchinskaya, N. N., & Makarova, T. M. (1982). Optical constants and some other characteristics of glasses in the systems BaO-Bi_2O_3 and La_2O_3-Bi_2O_3-B_2O_3. Fizika i Khimiya Stekla in Russian)., 8(3), 347-349.

[50] Sarkisov, P. D. (1997). The Directed Glass Crystallization- Basis for Development of Multipurpose Glass Ceramics,. *Mendeleev Russian Chemical Technological University Press, Moscow(In Russian.*

[51] Damjanovic, D. (2005). Hysteresis in Piezoelectric and Ferroelectric Materials,. In: The Science of Hysteresis, Mayergoyz, I & Bertotti, G Elsevier, 0-12480-874-3 , 3, 337-465.

[52] Rawson, H. (1967). Inorganic glass-forming systems. *Academic Press, London & New York*, 312.

Chapter 8

Optical Properties of Ferroelectrics and Measurement Procedures

A.K. Bain and Prem Chand

Additional information is available at the end of the chapter

http://dx.doi.org/10.5772/52098

1. Introduction

It is well known that the optical properties of ferroelectric materials find wide ranging applications in laser devices. Particularly in the recent years, there has been tremendous interest in the investigation of the nonlinear optical properties of ferroelectric thin films [1-5] for planar waveguide and integrated –optic devices. A new class of thin film waveguides has been developed using $BaTiO_3$ thin films deposited on MgO substrates [6]. Barium strontium titanate $Ba_{1-x}Sr_xTiO_3$ (BST) is one of the most interesting thin film ferroelectric materials due to its high dielectric constant, composition dependent Curie temperature and high optical nonlinearity. The composition dependent T_c enables a maximum infrared response to be obtained at room temperature. The BST thin films in the paraelectric phase, have characteristics such as good chemical and thermal stability and good insulating properties, due to this nature they are often considered the most suitable capacitor dielectrics for successful fabrication of high density Giga bit (Gbit) scale dynamic random access memories (DRAMs). Compositionally graded ferroelectric films have exhibited properties not previously observed in conventional ferroelectric materials. The most notable property of the graded ferroelectric devices or graded Functionally Devices (GFDs) is the large DC polarization offset they develop when driven by an alternating electric field. Such GFDs can find applications as tunable multilayer capacitors, waveguide phase shifters and filters [7]. Recently, BST thin films were used in the formation of graded ferroelectric devices by depositing successive layers of BST with different Ba/Sr ratios [8].

In our work, the Barium strontium titanate ($Ba_{0.05}Sr_{0.95}TiO_3$) ferroelectric thin films were prepared on single crystal [001] MgO substrates using the pulsed laser deposition method. The refractive index of BST ($Ba_{0.05}Sr_{0.95}TiO_3$) thin films is determined in the wavelength range between 1450-1580 nm at the room temperature. The dispersion curve is found to decrease gradually with increasing wavelength. The average value of the refractive index is found to be 1.985 in the wavelength range between 1450-1580 nm which is important for optoelectronic device applications [9].

1: Present address: Director, PM College of Engineering, Kami Road, Sonepat-131001.

2: Present Address: 99 Cyril Road, Small Heath, Birmingham B10 0ST.

Lithium heptagermanate $Li_2Ge_7O_{15}$ (LGO) is regarded as a weak ferroelectric and its curie point T_c is 283.5K [10,11]. Due to its intermediate behaviour between order-disorder and displacive types in a conventional grouping of ferroelectric materials LGO remains a subject of interest from both the theoretical and the application point of view. The paraelectric phase above T_c is orthorhombic $D^{14}{}_{2h}$ ~ pbcn and below T_c the ferroelectric phase is $C^5{}_{2v}$ ~ $pbc2_1$ with four formula units in a unit cell in both the phases. Below T_c LGO shows dielectric hysteresis loop and the permittivity shows a sharp peak at T_c [10-12]. Below T_c the spontaneous polarization appears along the c-axis. Many interesting physical properties of LGO such as birefringence [13], elastic behaviour [14], thermal expansion [11], dielectric susceptibility [12,15, 16] and photoluminescence [17] exhibit strong anomalies around T_c. The optical properties, however vary only to such a small degree that the transition could not be detected with the aid of a standard polarization microscope [13]. Employing a high resolution polarization device, Kaminsky and HaussÜhl [13] studied the birefringence in LGO near T_c and observed anomalies at the phase transition.

The study of piezo-optic dispersion of LGO (un-irradiated and x-irradiated) in the visible region of the spectrum of light at room temperature (RT=298 K) shows an optical zone/ window in between 5400Å and 6200Å with an enhanced piezo-optical behavior [18]. The temperature dependence of the photoelastic coefficients of the ferroelectric crystals $Li_2Ge_7O_{15}$ (both un-irradiated and x-irradiated) in a cooling and a heating cycle between room temperature and 273K shows an interesting observation including the lowering of the T_c under uniaxial stress contrary to the increase of T_c under hydrostatic pressure and observation of thermal photoelastic hysteresis similar to dielectric behavior [19]. The study of fluorescence spectra of the crystals $Li_2Ge_7O_{15}:Cr^{3+}$ in the temperature interval 77-320 K shows the sharply decrease of intensities of the R_1 and R_2 lines (corresponding to the Cr^{3+} ions of types I and II) during cooling process near the temperature T_c = 283.5 K[16].

The present chapter includes optical properties of the ferroelectric BST thin films and the Lithium heptagermanate ($Li_2Ge_7O_{15}$) single crystals, fabrication methods, measurement procedures of the refractive index of BST thin films on MgO substrates, the fluorescence spectra and the photoelastic coefficients of LGO single crystals (un-irradiated and x-irradiated) at different wave lengths and temperatures around the phase transition temperature T_c. The potential of these materials for practical applications in the opto-electronic devices will also be discussed.

1.1. Optical property of Barium strontium titanate thin films

The Barium strontium titanate ($Ba_{0.05}Sr_{0.95}TiO_3$) ferroelectric thin films were prepared on single crystal MgO substrates using the pulsed laser deposition (PLD) method at a substrate temperature of 780 °C and then annealed at 650 °C for 55 min. The x-ray diffraction (XRD) analysis revealed that the films are oriented with [001] parallel to the substrate [001] axis and thus normal to the plane of the films [9]. The films were grown to a thickness of 430 nm.

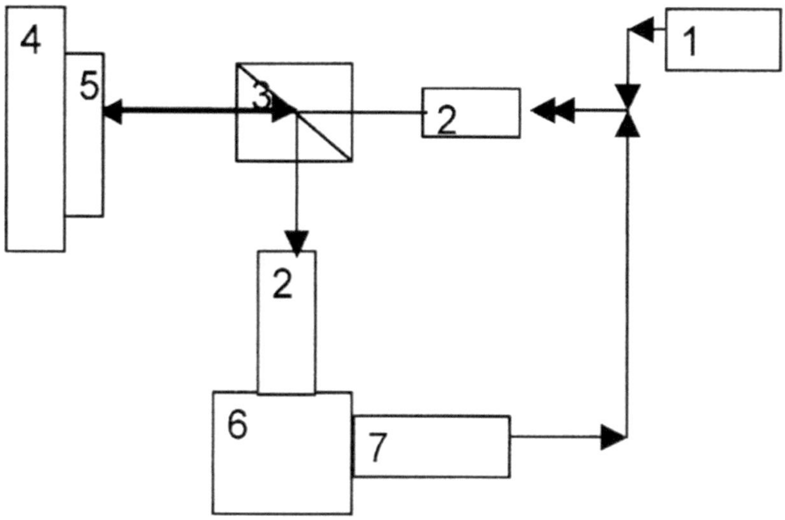

Figure 1. Schematic diagram of the experimental setup for the measurement of refractive index of the $Ba_{0.05}Sr_{0.95}TiO_3$ thin films at room temperature. He-Ne Laser for alignment (1), Lenses (2), Polarizer (3), sample holder (4), BST sample (5), Agilent light wave measurement system (6) and Tunable Laser (7).

Figure 1 shows the schematic diagram of the experimental setup for the measurement of the refractive index of $Ba_{0.05}Sr_{0.95}TiO_3$ thin films on MgO substrates through a reflection method. The He-Ne laser beam is used as a source of light to setup the alignment of the reflected beam of light from the samples to the detector. The incident beam is allowed to pass through a polarizer onto the sample. The reflected light is then passed through the same polarizing beam splitter oriented at 45° relative to the incident light and finally allowed to fall on the detector that is at 90° to the reflected/incident beam of light. The reflectivity measurement of the black metal, mirror, MgO substrate and $Ba_{0.05}Sr_{0.95}TiO_3$ thin films were carried using the Agilent 8164A Light Wave Measurement system in the wavelength region of 1450-1580 nm at room temperature.

The refractive index of substrate MgO is taken to be 1.7 [9]. The reflectivity of $Ba_{0.05}Sr_{0.95}TiO_3$ film is then normalized with respect to the mirror. The value of refractive index is derived from model described in ref. [20]. The fitting is done with the calculated data of the reflectivity of $Ba_{0.05}Sr_{0.95}TiO_3$ in the wavelength range between 1450-1580 nm. Figure 2 shows the refractive index of $Ba_{0.05}Sr_{0.95}TiO_3$ thin films as a function of wavelength at room temperature. The refractive index of $Ba_{0.05}Sr_{0.95}TiO_3$ with $Ba_xSr_{1-x}TiO^3$ (BST) and other materials of ferroelectric thin films at different wavelengths are presented in table 1.

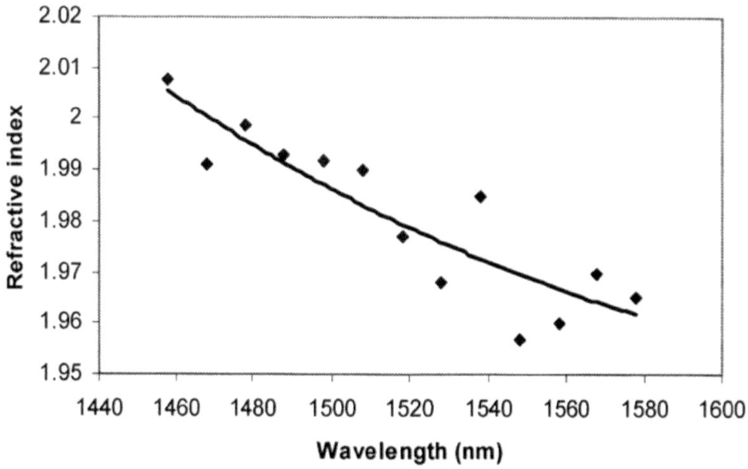

Figure 2. The variation of refractive index of $Ba_{0.05}Sr_{0.95}TiO_3$ thin films as a function of wavelength.

Sample Name	Refractive index	Wavelength	Remarks
$Ba_xSr_{1-x}TiO3$(x=0%)	2.37	600 nm (RT)	Ref. [21]
$Ba_xSr_{1-x}TiO3$(x=30%)	2.42	600 nm (RT)	Ref. [21]
$Ba_xSr_{1-x}TiO3$(x=50%)	2.37	600 nm (RT)	Ref. [21]
$Ba_xSr_{1-x}TiO3$(x=70%)	2.34	600 nm (RT)	Ref. [21]
$Ba_xSr_{1-x}TiO3$(x=50%)	2.45	470 nm (RT)	Ref. [22]
$Ba_xSr_{1-x}TiO3$(x=80%)	2.27	430 nm (RT)	Ref. [23]
$PbZr_{1-x}Ti_xO3$ (PZT)	2.87	400 nm (300 ^{0}C)	Ref. [24]
$PbZr_{1-x}Ti_xO3$ (PZT)	2.82	400 nm (50 ^{0}C)	Ref. [24]
$PbZr_{1-x}Ti_xO3$ (PZT)	2.66	500 nm (300 ^{0}C)	Ref. [24]
$PbZr_{1-x}Ti_xO3$ (PZT)	2.67	500 nm (50 ^{0}C)	Ref. [24]
$PbTiO_3$	3.10	470 nm (RT)	Ref. [25]
$PbTiO_3$	2.80	490 nm (RT)	Ref. [25]
$PbTiO_3$	2.75	550 nm (RT)	Ref. [25]
$Sr_xBa_{1-x}Nb_2O_6$ (x=61%)	2.35	400 nm (RT)	Ref. [26]

Table 1. The refractive index of BST and other ferroelectric thin films.

As shown in Figure 2, the dispersion curve decreases gradually with increasing wavelength. The average value of the refractive index is found to be ~ 1.985 in the wavelength range of 1450-1580 nm which is important for optoelectronic device (optical waveguide) applications. The variation of refractive index is attributed predominantly to the changes of electronic structure associated with the larger lattice parameter and variations in atomic co-ordination [27] that is local relaxations.

1.2. Growth and structure of $Li_2Ge_7O_{15}$ Crystals

Single crystals of $Li_2Ge_7O_{15}$ are grown in an ambient atmosphere by Czochralski method from stoichiometric melt, employing a resistance heated furnace. Stoichiometric mixture of powdered Li_2CO_3 and GeO_2 in the ratio of 1.03 and 7.0 respectively was heated at 1100 K for 24 hours to complete the solid state reaction for the raw material for the crystal growth. The crystals were grown by rotating the seed at the rate of 50 rpm with a pulling rate of 1.2 mm/hour. The cooling rate of temperature in the process of growth was 0.8-1.2 K/hour. The crystals grown were colourless, fully transparent and of optical quality. The crystal axes were determined by x-ray and optical methods.

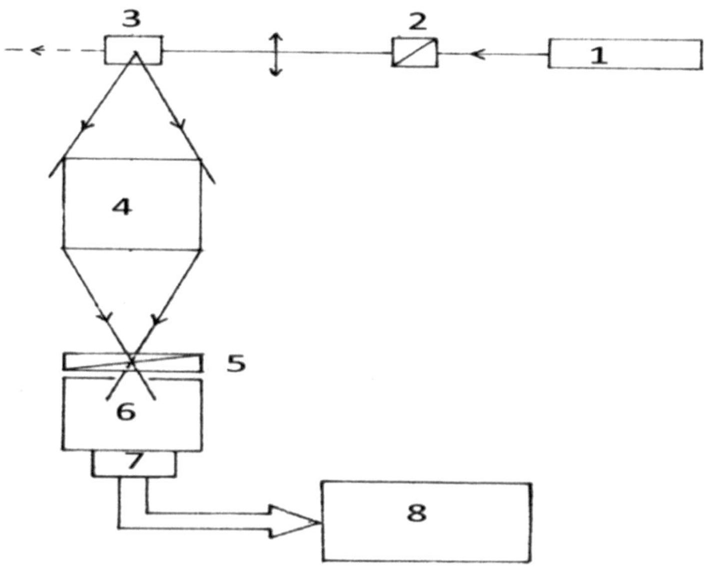

Figure 3. Schematic diagram of the experimental setup for the measurement of fluorescence spectra of the crystals $Li_2Ge_7O_{15}$:Cr^{3+}. He-Ne Laser for radiation (1), Glen prism (2), crystal sample (3), condenser (4), polarizer (5), spectrograph (6) Multichannel analyzer(7) and computer (8).

The desired impurities such as Cr^{+3}, Mn^{+2}, Bi^{+2}, Cu^{2+} and Eu^{+2} etc are also introduced in desired concentration by mixing the appropriate amount of the desired anion salt in the growth mixture. The crystal structure of LGO above T_c is orthorhombic (psedohexagonal) with the space group $D^{14}{}_{2h}$ (Pbcn). The cell parameters are a: 7.406 Å, b: 16.696 Å, c: 9.610 Å, Z = 4 and b~√3c. Below T_c a small value of spontaneous polarization occurs along c-axis and the ferroelectric phase belongs to $C^5{}_{2v}$ ($Pbc2_1$) space group. The crystal structure contains strongly packed layers of GeO_4 tetrahedra linked by GeO_6-octahedra to form a three dimensionally bridged frame work in which Li atoms occupy the positions in the vacant channels extending three dimensionally [14, 28, 29]. The size of the unit cell (Z = 4) does not change at the phase transition and ferroelectric phase transition is associated with a relaxational mode as well as the soft phonon [30]. Activation of the pure crystals with impurity ions will demand charge compensating mechanism through additional defects in the pure lattice.

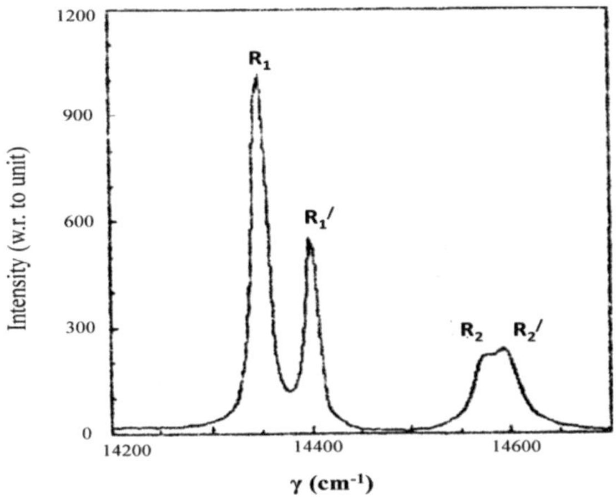

Figure 4. Fluorescence Spectra of the crystals $Li_2Ge_7O_{15}:Cr^{3+}$ at T=77 K.

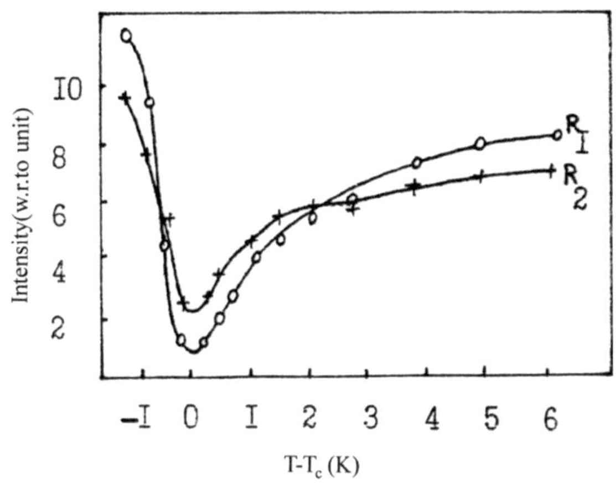

Figure 5. Temperature dependence of intensity of R_1 and R_2 Lines of fluorescence in $Li_2Ge_7O_{15}:Cr^{3+}$ crystals near the phase transition temperature T_c.

1.3. Study of fluorescence spectra of $Li_2Ge_7O_{15}:Cr^{3+}$ crystals

The fluorescence spectra of the crystals $Li_2Ge_7O_{15}:Cr^{3+}$ were studied in the temperature interval 77-320 K including the phase transition temperature $T_c = 283.5$ K. The experimental set to record the fluorescence spectra of the crystals $Li_2Ge_7O_{15}:Cr^{3+}$ is shown in fig.3. A laser with the pair of mode (λ_1=510.6 nm, λ_2=578.2 nm) was used as a source of excitation of the crystal sample. The recording of fluorescence spectra were carried out by the optical multichannel analyzer in combination with the polychromator. The radiation beam was initially polarized with glen prism. The plane Polaroid was used as an analyzer that was placed before the input aperture of the polychromator.

The fluorescence spectra consist of narrow intensity lines referred to R_1 and R_2 with frequencies γ_1~14348 cm^{-1} and γ_2 ~14572 cm^{-1}. These lines split further into two components each $R_1{}^{/}$ and $R_2{}^{/}$ respectively at lowering the temperature towards 77 K. Besides this, a wide long wavelength region/zone is observed in the spectra. It may be related with the effect of electron-photon interaction. It is known that Cr^{3+} doping ions in the structure of $Li_2Ge_7O_{15}:Cr^{3+}$ crystals substitute the Ge^{4+} host ions within oxygen octahedral (GeO_6) complexes [31-36]. The optical spectra of Cr^{3+} ions shows the existence of two types of Cr^{3+} centre (type I and II with different values of effective g-factor) as observed in EPR (Electron Paramagnetic Resonance) spectra of Cr^{3+} ions in ferroelectric phase of the crystals $Li_2Ge_7O_{15}:Cr^{3+}$ [32, 33]. Two pair of R lines $4A_2$ - ^2E (at T=77 K, its positions are R_1=14348 cm^{-1}, $R_1{}^{/}$=14402 cm^{-1}, R_2=14572 cm^{-1} and $R_2{}^{/}$=14593 cm^{-1}) are observed at low temperature region (T<190 K) in the optical spectra of the crystals $Li_2Ge_7O_{15}:Cr^{3+}$ as shown in Fig. 4. Actually the two different types of Cr^{3+} centers (R and R$^{/}$) with pretty different positions below $\bar{E}$ and above $2\bar{A}$ levels of the excited E^2 level are duplicate [37, 38] and conform to the EPR observations.

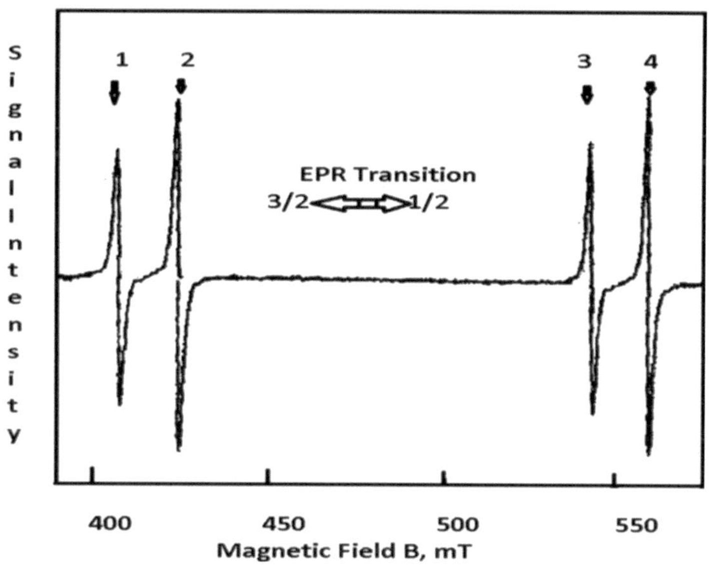

Figure 6. Part of EPR spectrum of Cr^{3+} doped LGO crystal in an arbitrary orientation at RT. The four EPR signals are attributed to four distinct Cr^{3+} sites per unit cell of LGO.

The intensity of fluorescence of the R_1 and R_2 lines of the crystals $Li_2Ge_7O_{15}:Cr^{3+}$ were studied near the phase transition temperature T_c at the direction E⊥[001]. It is observed that the intensity of R_1 and R_2 lines are decreased sharply near the phase transition temperature T_c but at high temperature region (T>T_c) the intensity again increases as shown in figure 5. Such nature of suppression of R_1 and R_2 lines was not observed previously and it may be related with the mechanism of interaction of excitation spectra of light in the crystals $Li_2Ge_7O_{15}$:Cr^{3+} near the phase transition temperature T_c[17].

The Crystals doped with chromium impurity give EPR (Electron Paramagnetic Resonance) signals characteristics of the trivalent chromium ions [Fig.6]. It is known that impurity Cr^{3+} ions substitute the Ge^{4+} host ions within oxygen octahedral in the basic structure of (LGO) crystal [31-36]. Incorporation of tri-positive chromium ions into GeO_6-octahedra changes the local symmetry of the lattice site from monoclinic C_2 group to triclinic C_1 group. The local symmetry lowering is attributed to the effect of the additional Li^+ defect required for compensating the charge misfit of Cr^{3+} ion at the Ge^{4+} site. Taking into account a weak coupling of lithium ions with the germanium – oxygen lattice framework, the interstitial Li^+ is considered to be the most probable charge compensating defect, located within the structural cavity near the octahedral CrO_6 complex (Fig.7). Subsequent measurements of optical spectra have confirmed the model of Cr^{3+}– Li^+ pair centers in the LGO crystal structure [32, 33].

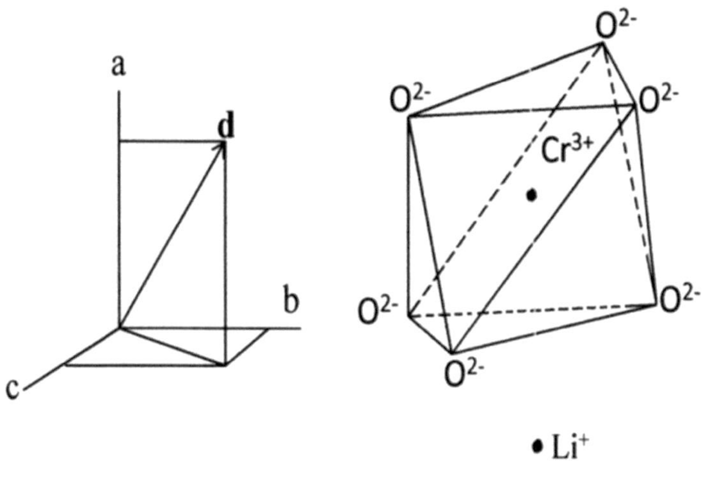

Figure 7. Physical model of Cr^{3+} centers in $Li_2Ge_7O_{15}$:Cr^{3+} crystals and its dipole moment d.

The available data make it possible to assume that electric dipole moments of Cr^{3+}– Li^+ pairs are directed along the crystal axis "a" of the crystal. Interstitial Li^+ ions locally break the symmetry axis C_2 of the sites within the oxygen octahedral complexes [34]. As a result, there are two equivalent configurations of the pair centers which are conjugated by broken C_2 axis and have dipole moments with opposite orientations. It may be assumed that pair centers can reorient due to thermal activation. Reorientation of the pair centers should be accompanied by: i) shortening of the configuration life time and ii) switching of defect dipole moments [35]. This is reflected in the typical temperature dependence of the imaginary part of dielectric permittivity of chromium doped LGO single crystals [35] along the a-axis of the crystal shown in fig.8.

http://dx.doi.org/10.5772/52098

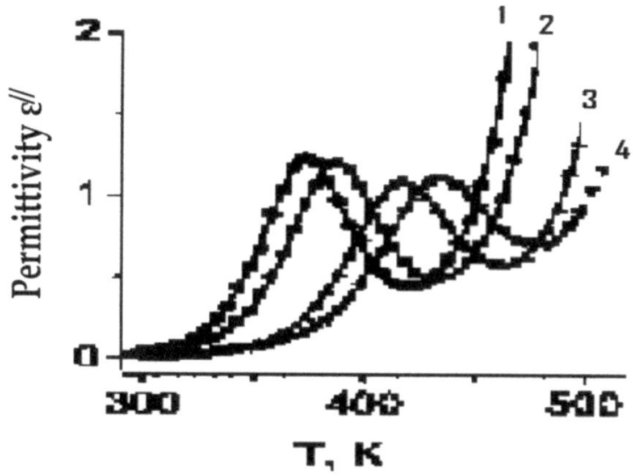

Figure 8. Temperature dependence of Imaginary part ε''of permittivity ε=ε'-jε'' of LGO:Cr^{3+}, measured along crystal a -axis at the following frequencies : 0.5 kHz (1); 1 kHz (2); 5 kHz (3); 10 kHz (4). Distinct peak is observed in ε'' in the temperature range 350-450 K and the peak is observed to shift to higher temperature for higher frequencies. In contrast no such peak is observed for the real part ε' of the permittivity. The typical behavior is observed only along crystal a –axis which incidentally coincides with the proposed Cr^{3+}- Li^{+} centers.

1.4. Principle of Photoelasticity

If a rectangular parallelepiped with edges parallel to x[100], y[010] and z[001] axes is stressed along z-axis and observation is made along y-axis, as shown in Fig.9, then the path retardation δ_{zy} introduced per unit length due the stress introduced birefringence is given by

$$\delta_{zy} = (\Delta n_z - \Delta n_x) = C_{zy} P_{zz} \tag{1}$$

where Δn_z and Δn_x are the changes in the corresponding refractive indices, $(\Delta n_z - \Delta n_x)$ is the corresponding stress induced birefringence, P_{zz} is the stress along z-axis and C_{zy} is a constant called the Brewster constant or the relative photoelastic coefficient. In general the Brewster constant is related to the stress optical and strain optical tensors of forth rank [39] and is a measure of the stress induced (piezo-optic) birefringence. It is conveniently expressed in the unit of 10^{-13} cm^2/dyne per cm thickness along the direction of observation is called a Brewster [39].

1.5. Measurement procedure of photoelastic constants

To study the piezo-optical birefringence the experimental set up consists of a source of light (S), a lens (L) to render the rays parallel, a polarizer (P), an analyzer Polaroid (A), a Babinet compensator (B) and a detector (D), as shown in Fig.10. The P and A combination are adjusted for optimal rejection of light. The sample with stressing arrangement and a Babinet compensator are placed between P and A. A monochromator and a gas flow temperature controlling device are used to obtain the piezo-optic coefficients (C_λ) at different wavelengths and temperature. The subscript λ in the symbol C_λ denotes that the piezo-optic coefficient depends on the wavelength of light used to measure it. The experiments are carried out for different wavelengths using white light and a monochromator and the monochromatic sodium yellow light. An appropriate stress along a desired direction of the sample is applied with the help of a stressing apparatus comprising a mechanical lever and load.

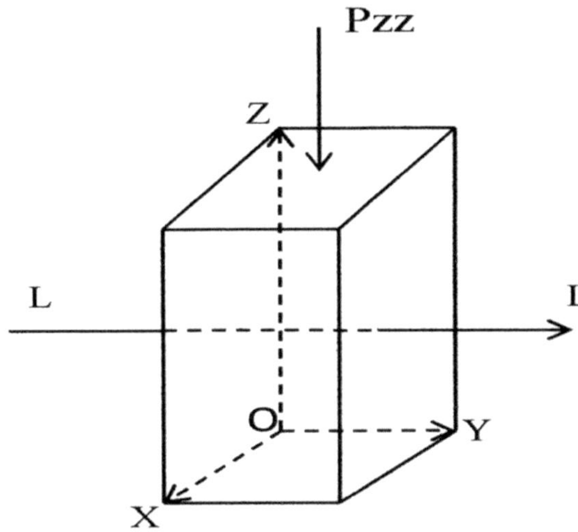

Figure 9. A solid under a linear stress of stress-optical measurements (P_{zz} is the applied stress and LL is the direction of light propagation and observation).

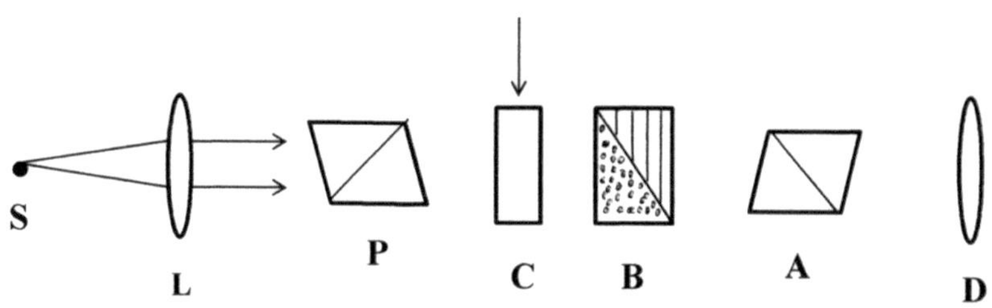

Figure 10. A schematic diagram of the experimental setup for the measurement of photoelastic constants of the crystals at room temperature. Source of light (S), Lense (L), Polarizer (P), Crystals (C) under stress, Babinet Compensator (B), Analyzer (A) and Detector (D).

To start with, the Babinet compensator is calibrated and the fringe width is determined for different wavelengths of light in the visible region. The crystal specimen is placed on the stressing system so that the stress could be applied along vertical axis and observation made along horizontal axis. A load on the crystal shifts the fringe in the Babinet compensator and this shift is a measure of the piezo-optic behavior. The piezo-optic coefficients (C_λ) are now calculated using the calibration of the Babinet compensator. The experiment is repeated for other orientations of the crystals and the results are obtained.

http://dx.doi.org/10.5772/52098

1.6. Piezo-optic Dispersion of $Li_2Ge_7O_{15}$ Crystals

The experimental procedure for the piezo-optic measurements is described in section 1.5. The polished optical quality samples worked out to dimensions i) 5.9 mm, 9.4 mm and 5.0 mm; ii) 3.17 mm, 5.88 mm and 6.7 mm, along the crystallographic a, b and c axes respectively. The stress was applied with an effective load of ~23 kg in each case [40].

The values of C_λ thus obtained at different wavelengths are given in Table 2 and the results are plotted in Fig. 11. Here C_{pq} is the piezo-optic coefficient with the stress direction being p and observation direction being q. The results show an interesting piezo-optic behavior. A survey of literature indicates that the piezo-optic behavior of materials studied till now shows a reduction of C_λ with increasing wavelength in the visible region [39]. In the present case, C_λ decreases with wavelength up to a certain wavelength as in other normal materials and then suddenly shows a peak and later on the usual behavior of reduction in the values of piezo-optic coefficients is observed.

		Wavelengths				
Obs.	C_{pq}	4358**Å**	4880**Å**	5390**Å**	5890**Å**	6140**Å**
1	C_{xy}	4.024	3.819	3.722	4.328	3.677
2	C_{xz}	5.243	4.895	4.770	5.552	4.451
3	C_{yx}	4.084	3.525	3.092	3.562	2.913
4	C_{yz}	4.353	4.118	3.946	4.261	3.866
5	C_{zy}	4.179	2.814	3.177	3.713	3.172
6	C_{zx}	3.312	2.991	2.650	4.190	2.618

Table 2. Stress optical coefficients c_{pq} (in Brewster) for $Li_2Ge_70_{15}$ at different wave lengths.

To the best knowledge of the authors this behavior is unique to the LGO crystals. For the sake of convenience we denote C_λ measured at λ = 5890 Å as C_{5890} and so on. The results show that sometimes the value of C_{5890} is even higher than that at C_{4400}, the value of piezo-optic coefficient obtained at the lowest wavelength studied here. This is the case with C_{xy}, C_{zx} and C_{xz}. For other orientations the value is lower than that at 4400 Å. Further, C_λ is found to have increased to more than 50% in the case of stress along [001] and observation along [100]. Also, it is interesting to note that the value of C_{6140}, is less than that of C_{5390}, in tune with usual observation of piezo-optic dispersion. Thus one can see an "optical window" in between 5400 Å and 6200 Å. The height of this optical window is different for various orientations, though the width seems approximately the same. The maximum height of about 1.5 Brewster was found for C_{zx} followed by C_{xz} with about 0.9 Brewster. It should be noted here that z-axis is the ferroelectric axis for LGO. It is also interesting to note that the change in height is more in the former while the actual value of C_λ, is less compared to that of the latter. The percentage dispersion also is different for various orientations. It is very high, as high as 25% for C_{zy}, while it is just 10% for C_{xy}.

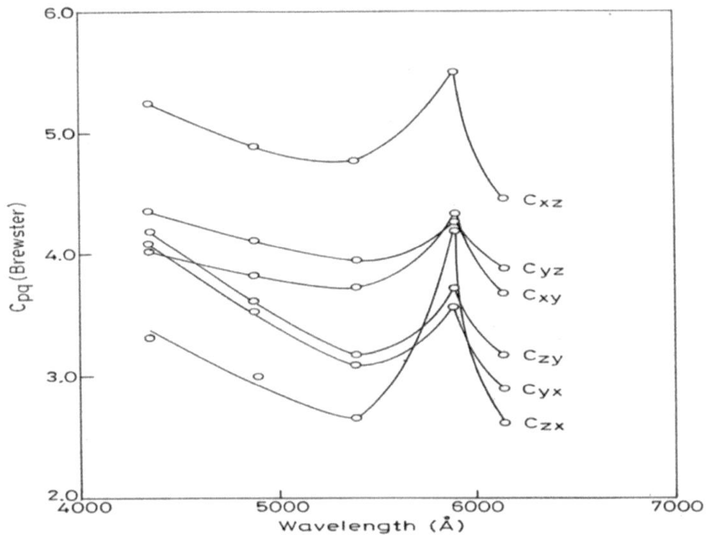

Figure 11. Stress optical dispersion of $Li_2Ge_7O_{15}$ crystals with wavelength at room temperature (298 K).

Figure 12 shows the variation of $C_{zx}(\lambda)$ at the temperatures ranging from 298K to 283K on cooling process of the sample LGO. It is clear from the figure that the distinct peak of $C_{zx}(\lambda)$ appears only at the sodium yellow wavelength of 5890 Å for the whole range of temperatures (298 K–283 K) investigated. It is also interesting to note that a temperature anomaly is also observed around 283 K. LGO undergoes a second order phase transition at 283.5 K from the high temperature paraelectric phase to the low temperature ferroelectric phase. So this anomaly is related to this phase transition of the LGO crystal.

The observed peculiarity of piezo-optic behavior could be due to many factors, viz., i) anomalous behavior of refractive index or birefringence ii) anomalous ferroelastic transformation at some stage of loading iii) shift of absorption edge due to loading. The following have been done to identify the reasons for this peculiar behaviour.

Birefringence dispersion has been investigated in the visible region and no anomalies in its behavior has been observed. This rules out the first of the reasons mentioned. The reason due to ferroelastic behavior also is ruled out since the effect would be uniform over all the wavelengths investigated. It was not possible to investigate the effect of load on the absorption edge. Hence an indirect experiment has been performed. If there is a shift in the absorption edge due to loading the sample, the peak observed now at sodium yellow light would shift with load. No clear shift of the peak could be observed within the experimental limits. Another interesting experiment was done to identify the source of the anomaly. It is well known that T_c of LGO changes under uniaxial stress. The measurements were made near T_c under different stress (loads). Although T_c was found to shift a little with load the dispersion peak did not show any discernible shift. No particular reason could be established as to why a dispersion peak appears around sodium yellow region. Another interesting work in this

http://dx.doi.org/10.5772/52098

direction is on $Gd_2(Mo0_4)_3$ — where an anomalous peak was recorded in spontaneous birefringence at 334.7 nm [41], an observation made for the first time.

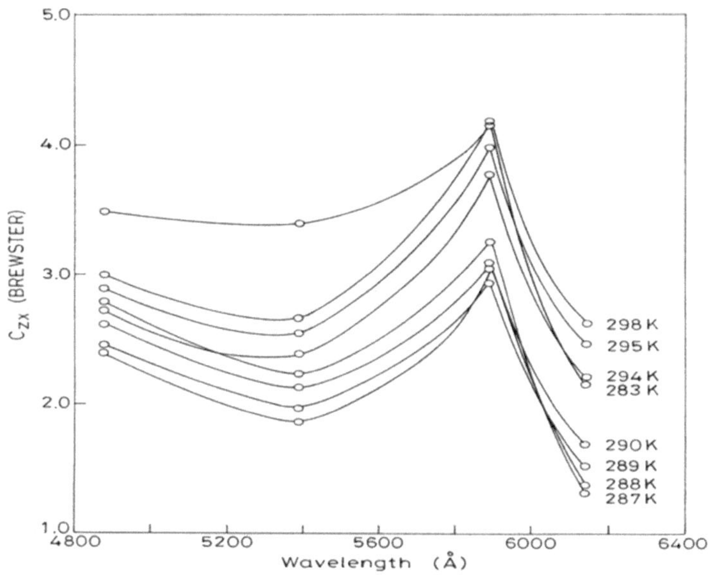

Figure 12. The variation of $C_{zx}(\lambda)$ at the temperatures ranging from 298 K to 283 K on cooling process of the sample $Li_2Ge_70_{15}$.

It is well known that the photoelasticity in crystals arises due to change in number of oscillators, effective electric field due to strain and the polarisability of the ions. In the present case, as the wavelength approaches around 5400 Å, the ionic polarisability seems to be changing enormously. There is no optical dispersion data available on LGO. We have conducted an experiment on transmission spectra of LGO along x, y and z-axes, which shows a strong absorption around 5400 Å. The observed anomaly in the piezo-optic dispersion may be attributed to the absorption edge falling in this region. This explanation needs further investigation in this direction. It is also known that the strain optical dispersion arises due to the shift in absorption frequencies and a change in the oscillator strength caused by the physical strain in the crystal.

1.7. Irradiation Effect on Piezo-optic Dispersion of $Li_2Ge_7O_{15}$ Crystals

The ferroelectric single crystals $Li_2Ge_7O_{15}$ was irradiated by x-ray for one hour and the experimental processes described in section 1.5 were repeated for the crystal (irradiated) LGO in order to understand the radiation effect on piezo-optical birefringence dispersion [18]. The values of C_λ of the crystal (irradiated) LGO thus obtained at different wavelengths are given in Table 3 and the results are plotted in Fig. 13.

		Wavelengths				
Obs.	C'_{pq}	4358**Å**	4880**Å**	5390**Å**	5890**Å**	6140**Å**
1	C'_{xy}	4.08	3.87	3.72	4.33	3.73
2	C'_{xz}	5.35	5.00	4.88	5.59	4.55
3	C'_{yx}	4.02	3.47	3.01	3.50	2.83
4	C'_{yz}	4.39	4.19	4.01	4.26	3.90
5	C'_{zx}	4.63	4.46	4.41	4.66	4.29
6	C'_{zy}	3.71	3.26	2.97	3.43	2.72

Table 3. Stress Optical Coefficients C_{pq} (in Brewsters) for $Li_2Ge_70_{15}$ (irradiated) at different wavelengths.

Some interesting results are obtained in the case of irradiated crystal LGO. The peak value of C'_{zx} has decreased about 18% and that of C'_{zy} has increased about 25% at the wave length λ= 5890 Å. Also, it is interesting to note that the value of C_{6140}, is less than that of C_{5390} for the un-irradiated and irradiated sample of LGO crystal, in tune with usual observation of piezo-optic dispersion.

Irradiation of crystals can change physical properties of the crystals. Irradiation brings about many effects in the crystal such as creating defects, internal stress and electric fields etc. These irradiation effects in turn are supposed to affect the physical properties of the irradiated crystal as compared to un-irradiated crystal. While there was no appreciable change in the lattice parameters, a significant drop in the value of dielectric constant and tan δ was observed upon x-irradiation of ferroelectric glycine phosphate. An appreciable shift in the phase transition temperature towards the lower temperature was observed. These changes are attributed to the defects produced in it by irradiation [42]. The studies of triglycine sulphate (TGS) showed that very small doses of x-irradiation can give large changes of the ferroelectric properties. The direct evidence of domain clamping by defects was obtained from optical studies. With increasing dosage the dielectric constant peak and polarization curve broaden and move to lower temperatures. In our present studies, the x-irradiation is believed to produce internal stress and electric fields inside the crystals LGO due to defects that can change the values of piezo-optic constants [43].

1.8. Piezo-optic Birefringence in $Li_2Ge_7O_{15}$ Crystals

The temperature dependence of the photoelastic coefficients of the ferroelectric crystals $Li_2Ge_7O_{15}$ in a cooling and heating cycle between 298 K and 273 K was carried out with the experimental procedure described in section 1.5 [19]. A special arrangement was made to vary the temperature of the sample. The temperature was recorded with a digital temperature indicator and a thermocouple sensor in contact with the sample.

http://dx.doi.org/10.5772/52098

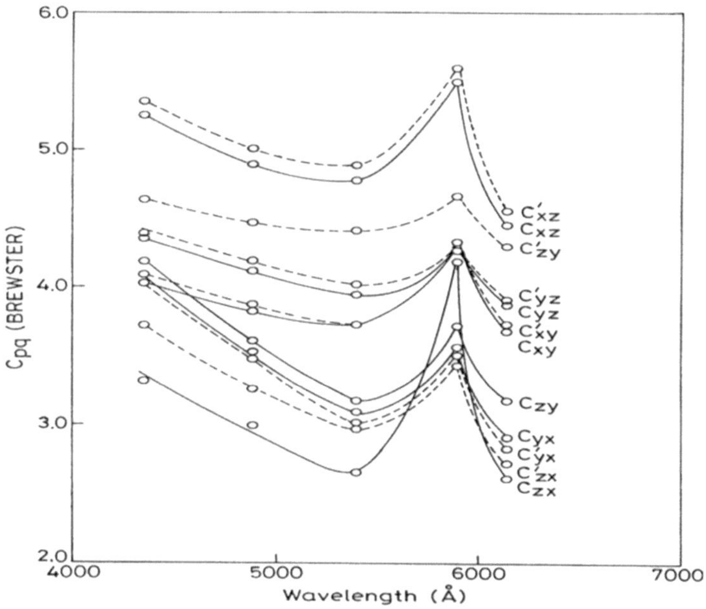

Figure 13. Stress optical dispersion of $Li_2Ge_70_{15}$ crystals (un-irradiated and irradiated) with wavelength at room temperature (298 K).

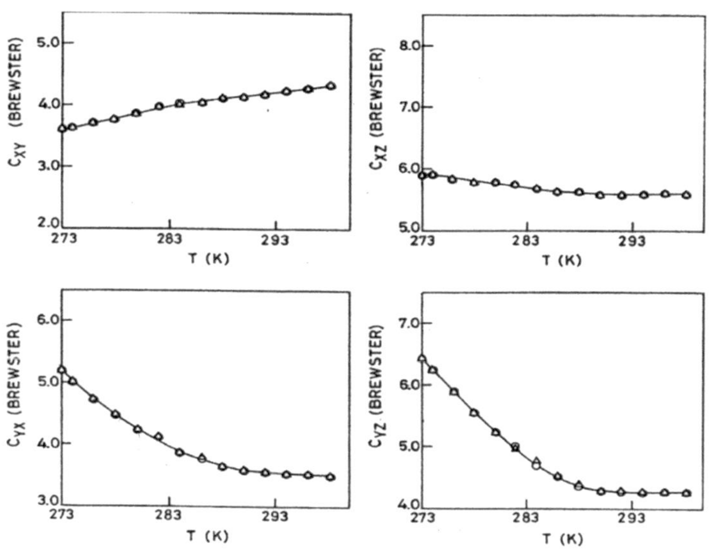

Figure 14. Temperature dependence of the piezo-optic coefficients C_{xy}, C_{xz}, C_{yz} and C_{yx} of the crystals LGO in a cooling (0) and heating (Δ) cycle.

The temperature dependence of piezo-optic coefficients C_{pq} of the crystals $Li_2Ge_7O_{15}$ between 298 K and 273 K were determined and are shown in Fig. 14 and Fig. 15. The values of C_{pq} at

291 K and 278 K were reported in paper [44] and it was observed that there were large changes in the values of C_{zy} and C_{yz} at 278 K and 291 K as compared to other components and C_{zy} did not show a peak in its temperature dependence between 291K and 278 K.

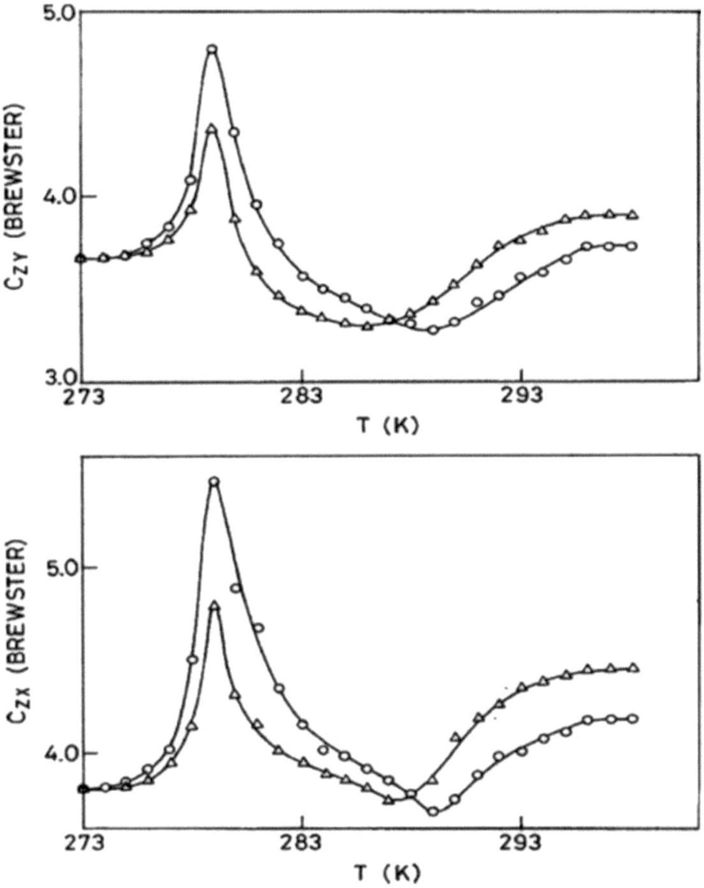

Figure 15. Anomalous temperature dependence of the piezo-optic coefficients C_{zx} and C_{zy} of the crystals LGO in a cooling (0) and heating (Δ) cycle.

Here in contrast we observed a peak in the temperature dependence of both C_{zy} and C_{zx} at 279 K. The temperature dependence of C_{pq} are quite interesting, for example the piezo-optic coefficients C_{yz}, C_{yx} and C_{xz} have negative temperature derivatives but C_{xy} has a positive temperature derivative. In complete contrast both C_{zy} and C_{zx} have both positive and negative temperature derivatives at different temperature intervals between 298 K and 273 K (Table: 4). Besides a clear thermal hysteresis is observed in C_{zy} and C_{zx} in a complete cooling and heating cycle (Fig. 15) whereas no discernible hysteresis is observed in rest of the piezo-optic coefficients (Fig. 14). The two distinct anomalies in the temperature dependence of C_{zy} and C_{zx} are characterized by a valley at T_m (~289 K) and a peak at T_c (~279 K). Anomalous temperature dependence of C_{zx} at different wave lengths is also shown in Fig. 16. The temperature dependence of the dielectric permittivity along the c-axis of LGO shows a sharp peak at T_c (283.5 K) and the Curie-Weiss law holds only for a narrow range of temperature ($T_c \pm 4$ K) [11,15, 16]. The peak for piezo-optic coefficient is attributed to the paraelectric to ferroelectric phase transition of LGO at T_c. To check the curie-Weiss law like dependence near T_c the following relation is used.

$$C^T_{pq} - C^0_{pq} = K_{pq} / (T - T_c) \tag{2}$$

Where C^T_{pq} and C^0_{pq} denote the value of the corresponding piezo-optic coefficients at temperature T and 273 K respectively and K_{pq} is a constant. Fig. 17 shows the $(C^T_{pq} - C^0_{pq})^{-1}$ vs $(T-T_c)$ curve for C_{zx} and C_{zy}. It is clear from these curves that like dielectric constant the relation fits well only within a narrow range of temperature near $T_c(T_c \pm 4$ K). The solid lines denote the theoretical curves with the following values K_{zx} = 1.05; K_{zy} = 0.92 for $T > T_c$, K_{zx} = −0.40; K_{zy} = −0.34 for $T < T_c$ and T_c = 279 K.

C_{pq}	Value of Derivative (Brewster/K)	Temperature Range Ratio
C_{zx}	0.013	296K-289K -1.69
	-0.022	289K-283K
	-0.090	282K-279K -2.0
	0.181	279K-276K
	~0	276K-273K
C_{xz}	-0.003	293K-273K
C_{zy}	0.020	296K-289K -0.75
	-0.015	289K-283K
	-0.095	282K-279K -1.9
	0.179	279K-276K
	~0	276K-273K
C_{yz}	-0.026	293K-273K
C_{xy}	0.007	293K-273K
C_{yz}	-0.023	293K-273K

Table 4. The temperature derivative $[dl_nC_{pq}/dT]$ of the piezo-optic coefficients of $Li_2Ge_7O_{15}$.

Furthermore the magnitudes of the ratio of the temperature derivatives below and above T_m and T_c are given in Table 4 and we can see that the ratio near T_c comes out to be about 2. Therefore it satisfies the law of two for the ratio of such derivatives of quantities which are coupled with the spontaneous polarization in second order ferroelectric phase transition such as in the case of triglycine sulphate [45] and LGO. Therefore the peak around T_c is [13, 15, 16] attributed to the paraelectric to ferroelectric phase transition of LGO. The smallness of K_{pq} and the applicability of relation (2) above only in a narrow range of temperature suggest that LGO may be an improper ferroelectric. The law of two does not hold for the ratio at T_m (Table 4). Therefore this anomaly is not related to the spontaneous polarization.

From the behaviour that only C_{zx} and C_{zy} show anomalous it is obvious that birefringence (Δn_z–Δn_y) and (Δn_z–Δn_x) show steep increase around T_c and below T_c show a $(T–T_c)^{1/2}$ behaviour correlated to the spontaneous polarization which is parallel to the z-axis (crystallographic c-axis). From the behaviour of C_{xy} and C_{yx} which do not show any temperature anomaly we may say that only n_z is responsible for the anomaly in accordance with the behaviour of the dielectric properties where only ε_{33} is strongly affected by the phase transition. These observations are in accordance with the results of Faraday effect and birefringence in LGO [13].

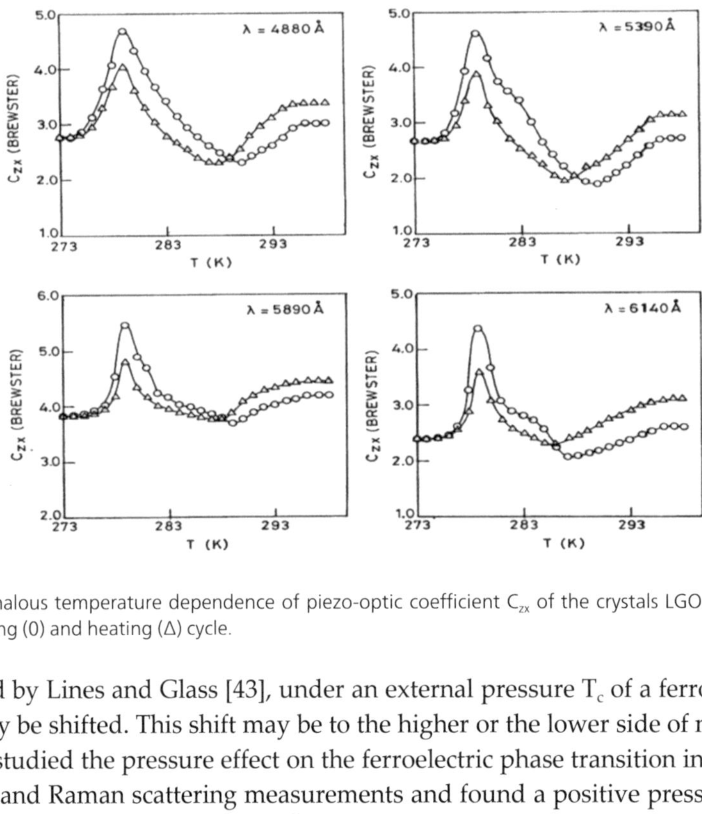

Figure 16. Anomalous temperature dependence of piezo-optic coefficient C_{zx} of the crystals LGO at different wave lengths in a cooling (0) and heating (Δ) cycle.

As mentioned by Lines and Glass [43], under an external pressure T_c of a ferroelectric phase transition may be shifted. This shift may be to the higher or the lower side of normal T_c. Wada et al. [46] studied the pressure effect on the ferroelectric phase transition in LGO through the dielectric and Raman scattering measurements and found a positive pressure coefficient dT_c/dp = 14.6 K/kbar. Preu and HaussÜhl [12] studied the dependences of dielectric constants on hydrostatic and uniaxial pressure as well as temperature. They observed a shift of T_c at a rate of 14.02 K/kbar for the hydrostatic pressure and ~7 K/kbar for the uniaxial pressure. In the present case the position of the peak of C_{zy} is found to depend on the stress applied. If the peak position is believed to represent the T_c, it appears to shift to the lower side under the uniaxial stress. To see whether T_c shifts linearly with uniaxial stress similar to the earlier observations [12, 46], we used different stresses within the elastic limits of LGO for

C_{zx} and found a linear relationship (Fig. 18). However, a negative stress coefficient $dT_c/dp \sim -22$ K/kbar is obtained in this case which agrees only in magnitude with the hydrostatic pressure coefficient. The linear curve (Fig. 18) extrapolates to a $T_c = 281.5$ K in the unstressed state instead of 283.5K as determined by dielectric measurements [11, 15, 16]. This may be due to a non linear dependence of shift of T_c under stress near 283.5 K.

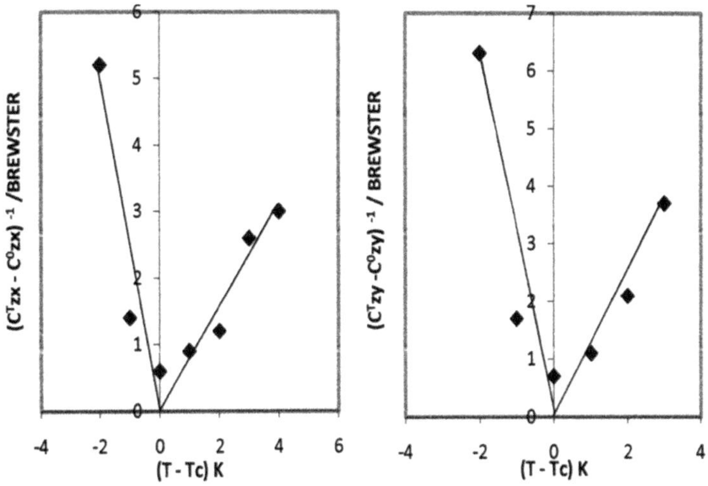

Figure 17. Plots of $(C^T_{pq} - C^o_{pq})^{-1}$ vs $(T-T_c)$ curve for C_{zx} and C_{zy}.

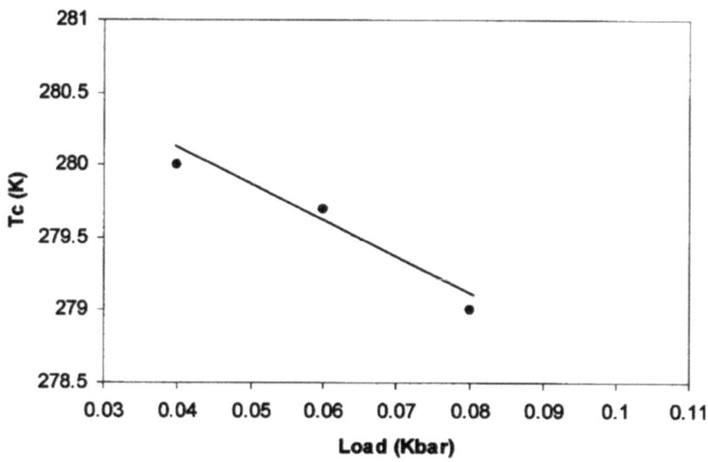

Figure 18. The stress dependence of the shift of T_c for C_{zx}.

Now we turn to the anomaly around T_m. Morioka et al. [47] proposed that there is an interaction between the soft phonon mode and a relaxational mode in the paraelectric phase in the temperature interval 300 K to T_c. The critical slowing down of the relaxational mode near T_c is expected to cause the increase of the fluctuation of the spatially homogeneous polarization and thereby the increase of the fluctuation of the hyperpolarizability with $k_c = 0$. Wada et al. [48] measured the soft phonon mode with the help of their newly designed FR-IR spectrometer and proposed that as T_c is approached from above soft phonon mode becomes over damped and transforms to a relaxational mode.

On the other hand there may exist a relaxational mode with an independent degree of freedom as well as the soft phonon mode and the character of the softening transfers from the phonon to the relaxational mode. This is an important problem in determining the dynamics of the peculiar ferroelectric phase transition of LGO, where both the dielectric critical slowing down characteristic of the order-disorder phase transition and the soft phonon mode characteristic of the displacive phase transition are observed [11, 14]. In the light of the above discussion we may say that the change up to T_m is caused by the softening of mode and the softening character transforms to the relaxations mode near T_m causing a change in the trend below T_m and near T_c the relaxational mode becomes dominant. The valley around T_m is perhaps caused by the interplay between the competitive relaxational mode and the soft phonon mode. It has been observed that softening of the velocity and rise of the damping of acoustic phonon occur in the paraelectric phase of LGO even quite far from T_c, i.e.$(T-T_c) \sim 30$ K and the effect is attributed to the fluctuation induced contributions [49].

Obs.	Cpq	Paraelectric (PE) phase (RT)	At $T_c = 279$ K
1	C_{xy}	4.38	3.85
2	C_{xz}	5.55	5.85
3	C_{yx}	3.60	4.46
4	C_{yz}	4.26	5.50
5	C_{zy}	3.71	4.83
6	C_{zx}	4.19	5.45

Table 5. Stress optical coefficients c_{pq} (in Brewster's) of $Li_2Ge_70_{15}$ at RT=298 K and at $T_c = 279$ K.

Another interesting aspect is the observation of a significant thermal photoelastic hysteresis (Fig. 15). Although the peak position does not shift in the heating cycle the values of the photoelastic constants get reduced significantly in the heating cycle as compared to the corresponding values in a cooling cycle. A similar kind of hysteresis was observed in the dielectric behaviour of LGO and the appearance of the dielectric hysteresis is attributed to the internal space charge (electrets state) effects which produce an internal electric field in LGO on heating from the ferroelectric phase [15-17]. It was possible to compensate the internal electric field effects in dielectric measurements by an external electric field [15-17]. It is suspected that the photoelastic hysteresis also occurs due to similar effects. Although it was not possible to try to compensate the electric field effects in the present investigation, it is possible to attempt experiment under the simultaneous application of a suitable electric field and stress along z-direction.

Obs	C_{pq}	Rochelle (RS)	SaltKDP	ADP	Remarks
1	C_{xz}	3.74	0.28	1.25	Ref. [50] for RS
2	C_{yz}	4.29	0.28	1.25	a- polar axis
3	C_{yx}	3.56	1.04	4.30	Ref. [51] for KDP
4	C_{zx}	0.85	1.54	3.50	Ref. [52] for ADP
5	C_{zy}	2.61	1.54	3.50	
6	C_{xy}	3.04	1.04	4.30	

Table 6. Piezo-optic coefficients c_{pq} (in Brewsters) for some ferroelectric crystals in their paraelectric (PE) phases.

The Stress optical coefficients C_{pq} of the crystals $Li_2Ge_7O_{15}$ at paraelectric phase (RT = 298 K) and at T_c = 279 K are presented in Table 5. It is important to compare the values of C_{pq} for $Li_2Ge_7O_{15}$ with other ferroelectric crystals given in Table 6 particularly with Rochelle-salt (RS) which belongs to the orthorhombic class like LGO [44]. The values of C_{pq} are significantly higher for LGO as compared to these ferroelectric systems. So, the large photoelastic coefficients and the other properties like good mechanical strength, a transition temperature close to room temperature and stability in ambient environment favour LGO as a potential candidate for photoelastic applications.

The EPR (Electron Paramagnetic Resonance) spectroscopy of the transition metal ion doped crystals of LGO (Mn^{2+}, Cr^{3+}) has also been studied both in Paraelectric (PE) and ferroelectric (FE) phases in the temperature interval from 298 K to 279 K during cooling and heating cycles [17, 36, 53]. It is observed that on approaching T_c in a cooling cycle, the EPR lines are slightly shifted to the high field direction and undergo substantial broadening. At the temperature T_c (≈ 283.4 K), the EPR lines are splitted into two components which are shifted to the higher and lower field directions progressively as a result of cooling the sample below T_c as shown in Fig.19.

During heating cycle (i.e. approaching T_c from below), the phenomena occurred were just opposite to the above processes observed in the cooling cycle. However, the EPR line width (peak to peak ΔH_{pp}) for H ∥ c, H⊥a was found to decrease to about one third of its value at T_c in a heating cycle as compared to its value in the cooling cycle. The shape of the EPR resonance lines far from T_c has a dominant Lorentzian character (a Lorentzian line shape) but very near to T_c, the line shape has been described mainly by Gaussian form of distribution (a Gaussian line shape). All the peculiarities observed are attributed to the PE ↔ FE phase transition of the LGO crystals. The line width reduction near T_c is attributed to the internal space charge (electret state) effects which produce an internal electric field inside the crystals on heating process from the ferroelectric phase. This observation is similar to the photoelastic hysteresis behavior of the crystals LGO near T_c.

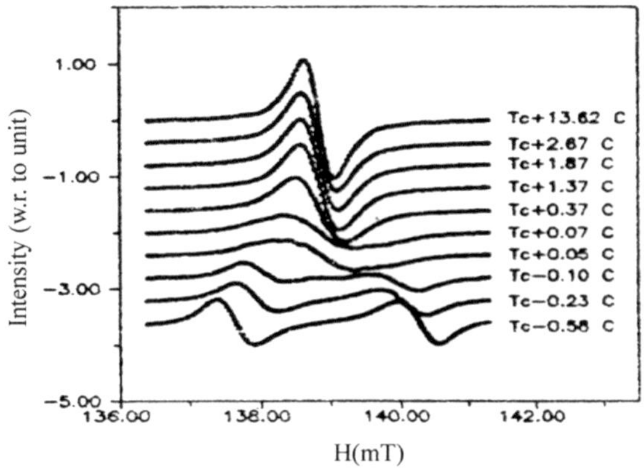

Figure 19. Temperature dependence of EPR lines of $Li_2Ge_7O_{15}$:Cr^{+3} crystals for |M|= ½ ↔3/2, H ∥ a, H⊥c near T_c during cooling process.

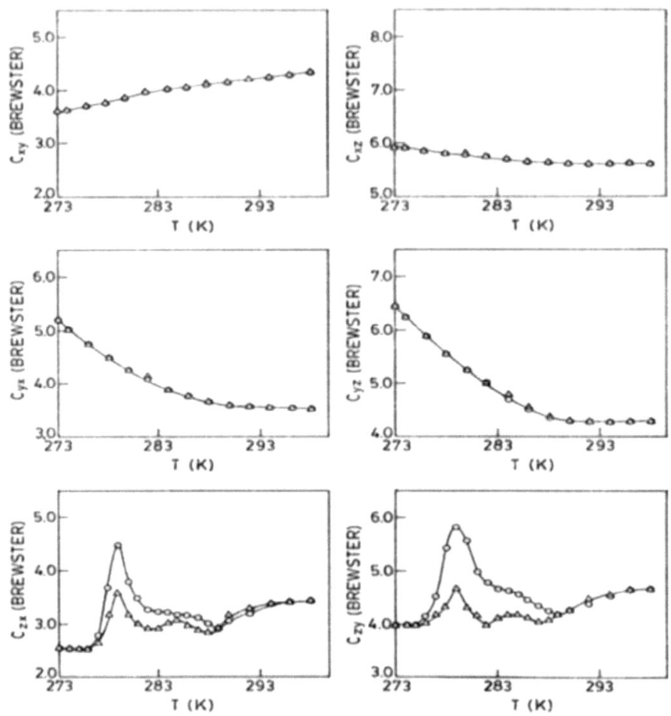

Figure 20. Temperature dependence of photoelastic coefficients C_{xy}, C_{xz}, C_{yz}, C_{yx}, C_{zx} and C_{zy} of the crystal (x-irradiated) LGO in a cooling (0) and heating (Δ) cycle.

http://dx.doi.org/10.5772/52098

1.9. Irradiation Effect on piezo-optic Birefringence in $Li_2Ge_7O_{15}$ Crystals

The photoelastic coefficients C_{pq} of the ferroelectric crystals $Li_2Ge_7O_{15}$ (x-irradiated) in a cooling and heating cycle between 298 K and 273 K was carried out with the experimental procedure described in section 1.5 and are shown in Fig. 20 [54]. The results show an interesting photoelastic behaviour.

Peaks are observed in the temperature dependence of the photoelastic coefficients C_{zy} and C_{zx} at temperature ~ 279 K in a complete cooling and heating cycle whereas no discernible hysteresis is observed in rest of the photoelastic coefficients. Anomalous temperature dependence of C_{zx} of the crystal (x-irradiated) LGO at different wave lengths are shown in Fig.21.

It is observed that the peak value of C_{zy} has increased about 25% and that of C_{zx} has decreased about 18% at the wave length λ=5890 Å during cooling process of the crystal (Fig.15 and Fig.20). The peak value of C_{zx} of the crystal (un-irradiated and x-irradiated) LGO thus obtained at different wave lengths (Fig.16 and Fig.21) are given in Table 7 and the results are plotted in Fig.22.

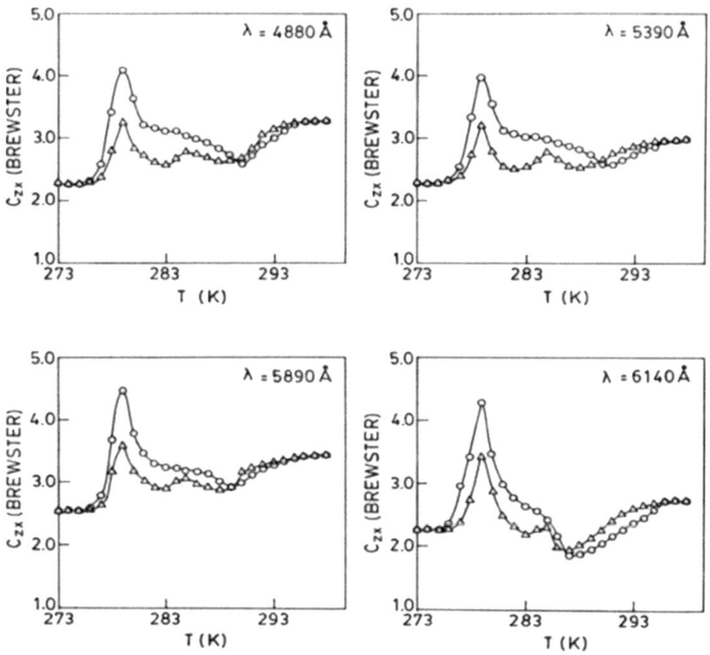

Figure 21. Temperature dependence of photoelastic coefficient C_{zx} of the crystal (x-irradiated) LGO at different wave lengths in a cooling (0) and heating (Δ) cycle.

It has been observed that the changes in the value of photoelastic coefficients C_{zy} and C_{zx} of the crystal (x-irradiated) LGO in a cooling and heating cycle occur only if the crystal is stressed along the polar axis (z-axis). It is known that the irradiation of crystals can change physical properties of the crystals.

Wave lengths	C_{zx} (un-irradiated)		C_{zx} (x-irradiated)	
(Å)	Cooling	Heating	Cooling	Heating
4880	4.8	4.0	4.05	3.3
5390	4.7	3.9	3.95	3.2
5890	5.6	4.8	4.6	3.7
6140	4.5	3.6	4.3	3.4

Table 7. The peak value of C_{zx} (in Brewster) for the Crystal (un-irradiated and x-irradiated) LGO at different wave lengths in the cooling and heating cycles.

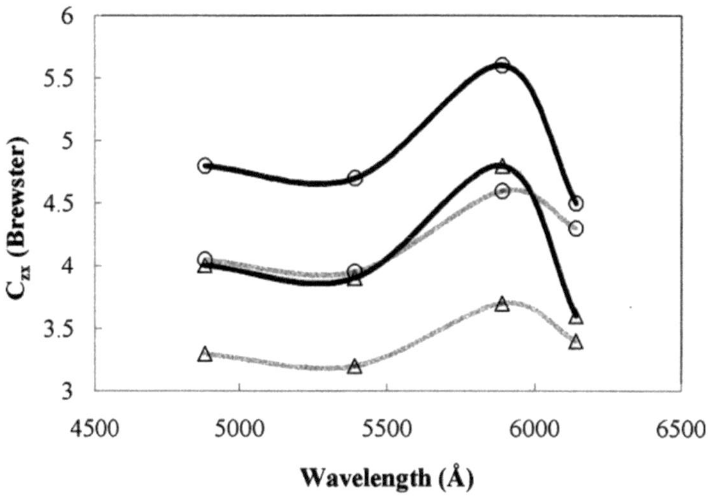

Figure 22. The peak value of C_{zx} for the un-irradiated (black colour) and x-irradiated (ash colour) crystal LGO at different wave lengths in a cooling (0) and heating (Δ) cycle.

Irradiation brings about many effects in the crystal such as creating defects, internal stress and electric fields etc [43]. In our present studies, the x-irradiation is believed to produce internal stress and electric fields inside the crystals $Li_2Ge_7O_{15}$ due to defects that can change the values of photoelastic coefficients.

2. Summary

It is known that the Barium strontium titanate $Ba_{1-x}Sr_xTiO_3$(BST) is one of the most interesting thin film ferroelectric materials due to its high dielectric constant, composition dependent curie temperature and high optical nonlinearity. The wavelength dependence of refractive index of BST ($Ba_{0.05}Sr_{0.95}TiO_3$) thin films has shown a nonlinear dependence in the 1450-1580 nm wavelength range at room temperature as described in section 1.1. The dispersion curve decreases gradually with increasing wavelength. The average value of the refrac-

http://dx.doi.org/10.5772/52098

tive index is found to be 1.985 in the 1450-1580 nm wavelength range which is considered to be important for optoelectronic device applications.

The study of fluorescence spectra of the crystals $Li_2Ge_7O_{15}:Cr^{3+}$ in the temperature interval 77-320 K shows the sharply decrease of intensities of the R_1 and R_2 lines (corresponding to the Cr^{3+} ions of types I and II) during cooling process near the temperature T_c = 283.5 K as described in section 1.3. Such nature of suppression of R_1 and R_2 lines was not observed previously and it may be related with the mechanism of interaction of excitation spectra of light in the crystals $Li_2Ge_7O_{15}:Cr^{3+}$ at the temperature T_c. The doping of chromium in LGO is believed to create Cr^{3+}- Li^{+} defect pairs in the host LGO lattice at Ge^{4+} sites creating dipoles in two conjugate directions. The EPR, optical, dielectric and fluorescence studies conform each other and pose more scope for further studies.

The high optical quality, good mechanical strength and stability in ambient environment and large photoelastic coefficients in comparison with other ferroelectric crystals like Rochelle-salt, KDP and ADP favour the crystals LGO as a potential candidate for photoelastic applications. The piezo-optic dispersion of the crystals (un-irradiated and x-irradiated) LGO in the visible region of the spectrum of light at room temperature (298 K) have been described in sections 1.6 and 1.7. It shows an "optical zone or optical window" in between the wavelengths 5400 Å and 6200 Å with an enhanced piezo-optical behavior. This peculiar optical window can have a technical importance for example this window region can act as an optical switch for acousto-optical devices. From the studies undertaken it may be concluded that LGO is an attractive acousto-optic material which deserves further probe. It may be possible to understand the observed behavior if extensive piezo-optic and refractive index data become available over an extended range of wavelengths.

The temperature dependence of the photoelastic coefficients of the crystals (un-irradiated and x-irradiated) LGO in a cooling and heating cycle between room temperature (298 K) and 273 K have shown an interesting observations: lowering of the T_c under uniaxial stress contrary to the increase of T_c under hydrostatic pressure and observation of thermal photoelastic hysteresis similar to dielectric hysteresis behavior as described in sections 1.8 and 1.9. In our studies, the x-irradiation is believed to produce internal stress and electric fields inside the crystals LGO due to defects that can change the values of photoelastic coefficients, as described in sections 1.7 and 1.9.

Author details

A.K. Bain* and Prem Chand

Department of Physics Indian Institute of Technology Kanpur, Kanpur-208016, INDIA

References

[1] Lepingard, F. A., Kingston, J. J., & Fork, D. K. (1996). Second harmonic generation in $LiTaO_3$ thin films by modal dispersion and quasi phase matching. *Appl. Phys. Lett.*, 68, 3695-3697.

[2] Hewing, G. H., & Jain, K. (1983). Frequency doubling in a $LiNbO_3$ thin film deposited on sapphire. *J. Appl. Phys.*, 54, 57-61.

[3] Lee, S. H., Kim, D. W., Noh, T. W., Lee, J. H., Lim, M. J., & Lee, S. D. (1996). Non-linear optical properties of epitaxial $LiNbO_3$ film prepared by pulsed laser deposition. *J. Korean Phys. Soc.*, 29, S628-S631.

[4] Moon, S. E., Back, S. B., Kwun, S. I., Lee, Y. S., Noh, T. W., Song, T. K., & Yoon, J. G. (2000). Orientational Dependence of Electro-optic Properties of $SrBi_2Ta_2O_9$ Ferroelectric Thin Films. Jpn. J. Appl. Phys., 39, 5916-5917.

[5] Reitze, D. H., Haton, E., Ramesh, R., Etemad, S., Leaird, D. E., & Sands, T. (1993). Electrical and Electro-optic Properties of Single Crystalline, Epitaxial Thin Films Grown on Silicon Substrates. *Appl. Phys. Lett.*, 63(5), 596-598.

[6] ORNL's. *Thin film Waveguide and the Information Highway by Carolyn Krause.*

[7] Varadam, V. K., Varadam, V. V., Selmi, F., Ounaise, Z., & Jose, K. A. (1994). Multilayer Tunable Ferroelectrci Materiasl and Thin Films. SPIE-Int. *Soc. Opt. Eng.*, 2189, 433-447.

[8] Mohammed, M. S., Naik, R., Mantese, J. V., Schubring, N. W., Micheli, A. L., & Catalan, A. B. (1996). Microstructure and ferroelectric properties of fine-grained $Ba_xSr_{1-x}TiO_3$ thin films prepared by metalorganic decomposition. *J. Mater. Res.*, 11, 2588-2593.

[9] Bain, A. K., Jackson, T. J., Koutsonas, Y., Cryan, M., Yu, S., Hill, M., Varrazza, R., Rorison, J., & Lancaster, M. J. (2007). Optical Properties of Barium Strontium Titanate (BST) Ferroelectric Thin Films. *Ferroelectric Letters*, 34, 149-154.

[10] Wada, M., Swada, A., & Ishibashi, Y. (1981). Ferroelectricity and Soft Mode in $Li_2Ge_7O_{15}$. *J. Phys. Soc. Jpn.*, 50(6), 1811-1812.

[11] Wada, M., & Ishibashi, Y. (1983). Ferroelectric Phase Transition in $Li_2Ge_7O_{15}$. *J.Phys. Soc. Jpn .*, 52(1), 193-199.

[12] Preu, P., & Hauss Ühl , S. (1982). Dielectric Properties and Phase Transition in $Li_2Ge_7O_{15}$. *Solid State Commun*, 41(8), 627-630.

[13] Kaminsky, W., & Hauss, Ühl. S. (1990). Faraday effect aid birefringence in orthorhombic $Li_2Ge_7O_{15}$ near the ferroelectric phase transition. *Ferroelectrics Lett.*, 11(3), 63-67.

[14] Hauss, Ühl. S., Wallrafen, F., Recker, K., & Eckstein, J. (1980). Growth, Elastic Properties and Phase Transition of Orthorombic $Li_2Ge_7O_{15}$. *Z.krist.*, 153, 329-337.

[15] Kudzin, A., Yu, Volnyanskii M. D., & Bain, A. K. (1994). Temperature Hysteresis of the Permittivity of $Li_2Ge_7O_{15}$. *Phys. Solid State*, 36(2), 228-230.

[16] Kudzin, A., Yu, Volnyanskii M. D., & Bain, A. K. (1995). Influence of Space Charges on Ferroelectric Property of Weak Ferroelectric $Li_2Ge_7O_{15}$. *Ferroelectrics*, 164(1), 319-322.

[17] Bain, A.K. (1994). Study of Pecularities of Ferroelectric Phase Transition of the Crystals $Li_2Ge_7O_{15}$ (Ph.D. Thesis). *Dniepropetrovsk State University, Ukraine*, 128.

[18] Bain, A. K., Chand, P., Rao, K. V., Yamaguchi, T., & Wada, M. (2008). Irradiation Effect on Piezo-optic Dispersion of $Li_2Ge_7O_{15}$ Crystals. *Ferroelectrics*, 377(1), 86-91.

[19] Bain, A. K., Chand, P., Rao, K. V., Yamaguchi, T., & Wada, M. (2009). Anomalous Temperature Dependence of Piezo-optic Birefrigence in $Li_2Ge_7O_{15}$ Crystals. *Ferroelectrics*, 386(1), 152-160.

[20] O'Mahony, M.J. (1988). Semiconductor laser optical amplifers for use in future fiber systems. *J. Light wave Technology*, 6(4), 531-544.

[21] Miranda, F. A., Vankeuls, F. W., Romanofsky, R. R., Mueller, C. H., Alterovitz, S., & Subramanyam, G. (2002). Ferroelectric thin-film based technology for frequency and phase agile microwave communication applications. *Integrated Ferroelectrics*, 42, 131-149.

[22] Tian, H. Y., Luo, W. G., Pu, X. H., Qiu, P. S., He, X. Y., & Ding, A. L. (2001). Synthesis and characteristics of strontium-barium titanate graded thin films at low temperature using a sol-gel technique. *Solid State Communications*, 117, 315-319.

[23] Pontes, F. M., Leite, E. R., Pontes, D. S. L., Longo, E., Santos, E. M. S., Mergulhao, S., Pizani, P. S., Lanciotti Jr, F., Boschi, T. M., & Varela, J. A. (2002). Ferroelectric and optical properties of $Ba_{0.8}Sr_{0.2}TiO_3$ thin film. *J. Appl. Phys.*, 91(9), 5972-5978.

[24] Deineka, A., Jastrabik, L., Suchaneck, G., & Gerlach, G. (2002). Optical Properties of Self-Polarized PZT Ferroelectric Films. *Ferroelectrics*, 273, 155-160.

[25] Gu, Y. Z., Gu, D. H., & Gan, F. X. (2001). Optical Nonlinearity in $PbTiO_3$ Thin Film Deposited on $Al2O_3$ with RF-Sputtering System. *Nonlinear Optics*, 28(4), 283-289.

[26] Fernandez, F. E., Gonzalez, Y., Liu, H., Martinez, A., Rodriguez, V., & Jia, W. (2002). Structure, morphology, and properties of strontium barium niobate thin films grown by pulsed laser deposition. *Integrated Ferroelectrics*, 42, 219-233.

[27] Wohlecke, M., Marrello, V., & Onton, A. (1977). Refractive index of $BaTiO_3$ and $SrTiO_3$ films. *J. Appl. Phys.*, 48, 1748-1750.

[28] Wada, M., Swada, A., & Ishibashi, Y. (1984). The Oscillator Strength of the Soft Mode in $Li_2Ge_7O_{15}$. *J. Phys. Soc. Jpn.*, 53(10), 3319-3320.

[29] Iwata, Y., Shibuya, I., Wada, M., Sawada, A., & Ishibashi, Y. (1987). Neutron Diffraction study of structural Phase Transition in Ferroelectric $Li_2Ge_7O_{15}$. *J. Phys. Soc. Jpn.*, 56(7), 2420-2427.

[30] Wada, M. (1988). Soft Mode Spectroscopy Study of Ferroelectric Phase Transition in $Li_2Ge_7O_{15}$. *Ind. J. Pure and Appl. Phys.*, 26, 68-71.

[31] Galeev, A. A., Hasanova, N. M., Bykov, A. B., Vinokurov, B. M., Nizamutdinov, N. M., & Bulka, G. R. (1990). EPR of Cr^{3+} and Fe^{3+} in $Li_2Ge_7O_{15}$ single crystal. In: Spectroscopy, a crystal chemistry and a realstructure of minerals and their analogues. *Kazan' state university*, 77-87, in Russian).

[32] Basun, S. A., Kaplyanski, A. A., & Feofilov, S. P. (1994). Dipolar centres in $Li_2Ge_7O_{15}$ crystal activated with ($3d^3$) ions: a microstructure and spectroscopic effects of an internal and external electric field. *Solid State Phys.*, 36(11), 3429-3449.

[33] Kaplyanski, A. A., Basun, S. A., & Feofilov, S. P. (1995). Ferroelectric transition induced dipole moments in probe ions in $Li_2Ge_7O_{15}$ crystals doped with Mn^{4+} and Cr^{3+} . *Ferroelectrics*, 169, 245-248.

[34] Iwata, Y., Koizumi, H., Koyano, N., Shibuya, I., & Niizeki, N. (1973). Crystal structure determination of ferroelectric phase of $5PbO_3GeO_2$. *J. Phys. Soc. Jap.*, 35(1), 314-315.

[35] Volnianskii, M. D., Trubitsyn, M. P., & Obaidat, Yahia A. H. (2007). EPR and dielectric spectroscopy of reorienting Cr^{3+}- Li^{+} pair centres in $Li_2Ge_7O_{15}$ Crystal. *Condensed Matter Physics*, 10(49), 75-78.

[36] Trubitsyn, M. P., Volnianskii, M. D., & Busoul, I. A. (1998). EPR study of the ferroelectric phase transition in $Li_2Ge_7O_{15}$:Cr^{3+} crystal. *Solid State Phys.*, 40, 6, 1102-1105.

[37] Powell, R.C. (1968). Temperature Dependence of the Widths and Positions of the R Lines in $Li_2Ge_7O_{15}$:Cr^{3+} . *J. Appl. Phys.*, 39, 4517-4521.

[38] Powell, R.C. (1968). Energy Transfer between Chromium Ions in Nonequivalent Sites in $Li_2Ge_7O_{15.}$ *Phys. Rev.*, 173, 358-366.

[39] Narsimhamurty, T.S. (1981). Photoelastic and Electro-optic properties of Crystals. *New York: Plenum Press.*, 514.

[40] Bain, A. K., Chand, P., Rao, K. V., Yamaguchi, T., & Wada, M. (2008). Piezo-optic Dispersion of $Li_2Ge_7O_{15}$ Crystals. *Ferroelectrics*, 366(1), 16-21.

[41] Saito, K., Ashahi, T., Takahashi, N., Hignao, M., Kamiya, I., Sato, Y., Okubo, K., & Kobayashi, J. (1994). Optical activity of $Gd_2(MoO_4)_{3.}$ *Ferroelectrics*, 152(1), 231-236.

[42] Vanishri, S., & Bhat, H. L. (2005). Irradiation Effects on Ferroelectric Glycine Phosphite Single Crystal. *Ferroelectrics*, 323(1), 151-156.

[43] Lines, M. E., & Glass, A. M. (2004). Principles and Applications of Ferroelectrics and Related Materials. *Oxford: Clarendon press*, 680.

[44] Bain, A. K., Chand, P., Rao, K. V., Yamaguchi, T., & Wada, M. (1998). Determination of the Photoelastic Coefficients in Lithium heptagermanate Crystals. *Ferroelectrics*, 209(1), 553-559.

[45] Hauss, Ühl. S., & Albers, J. (1977). Elastic and thermoelastic constants of triglycine sulphate (TGS) in the paraelectric phase. *Ferroelectrics*, 15(1), 73-75.

[46] Wada, M., Orihara, H., Midorikawa, M., Swada, A., & Ishibashi, Y. (1981). Pressure Effect on the Ferroelectric Phase Transition in $Li_2Ge_7O_{15}$. *J. Phys. Soc. Jpn.*, 50(9), 2785-2786.

[47] Morioka, Y., Wada, M., & Swada, A. (1988). Hyper-Raman Study of Ferroelectric Phase Transition of $Li_2Ge_7O_{15}$. *J. Phys. Soc. Jpn.*, 57(9), 3198-3203.

[48] Wada, M., Shirawachi, K., & Nishizawa, S. (1991). A Fourier Transform Infrared Spectrometer with a Composite Interferometer for Soft Mode Studies. *Jpn. J. Appl. Phys.*, 30(5), 1122-1126.

[49] Sinii, I. G., Fedoseev, A. I., & Volnyanskii, M. D. (1990). Relaxation- and fluctuation-induced damping of hypersound in the presence of dispersion. *Sov. Phys. Solid State*, 32(10), 1817-1818.

[50] Narasimamurty, T. S. (1969). Piezooptical constants of Rochelle salt crystals. *Phys. Rev.*, 186, 945-948.

[51] Veerabhadra, Rao K., & Narasimamurty, T. S. (1975). Photoelastic behavior of KDP. *J. Mat. Sci.*, 10(6), 1019-1021.

[52] Narasimamurty, T. S., Veerabhadra, Rao. K., & Petterson, H. E. (1973). Photoelastic constants of ADP. *J. Mat. Sci.*, 8(4), 577-580.

[53] Trubitsyn, M. P., Kudzin, A., Yu, Volnyanski. M. D., & Bain, A. K. (1992). Critical Broadening of EPR lines near the Ferroelectric Phase Transition in $Li_2Ge_7O_{15}$:Mn^{2+} . Sov. Phys. Solid State , 34(6), 929-932.

[54] Bain, A. K., & Chand, P. (2011). Irradiation Effect on Photoelastic Coefficients in Ferroelectric $Li_2Ge_7O_{15}$ Crystals. *Integrated Ferroelectrics*, 124, 10-18.

Chapter 9

Ferroelectric Domain Imaging Multiferroic Films Using Piezoresponse Force Microscopy

Hongyang Zhao, Hideo Kimura, Qiwen Yao,
Lei Guo, Zhenxiang Cheng and Xiaolin Wang

Additional information is available at the end of the chapter

http://dx.doi.org/10.5772/52519

1. Introduction

Recently, multiferroic materials with the magnetoelectric coupling of ferroelectric (or anti-ferroelectric) properties and ferromagnetic (or antiferromagnetic) properties have attracted a lot of attention.[1-4] Among them, $BiFeO_3$(BFO) and $YMnO_3$has been intensively studied. For such ABO_3perovskite structured ferroelectric materials, they usually show antiferromagnetic order because the same B site magnetic element except $BiMnO_3$ is ferromagnet. While for the $A_2BB'O_6$ double perovskite oxides, the combination between B and B′ give rise to a ferromagnetic coupling. They are also expected to be multiferroic materials. Several bismuth-based double perovskite oxides ($BiBB'O_6$) have aroused great interest like Bi_2NiMnO_6, La_2NiMnO_6, $BiFeO_3$-$BiCrO_3$. But as we know, few researches are focused on Bi_2FeMnO_6. We believe it is particular interesting to investigate Bi_2FeMnO_6 (BFM). The origin of the ferromagnetism in these compounds has been discussed in many reports. According to Goodenough-Kanamori's (GK)rules, many ferromagnets have been designed through the coupling of two B site ions with and without e_g electrons.In BFM, the 180 degree -Fe^{3+}-O-Mn^{3+}- bonds is quasistatic, partly because the strong Jahn-Teller uniaxial strain in an octahedral site.Because the complication of the double perovskite system, there are still some questions about the violation of GK rules in some cases and the origin of the ferromagnetism or antiferromagnetism.The other problem is the bad ferroelectric properties. In order to characterization of their ferroelectric/piezoelectric properties, scanning probe microscopy (SPM) techniques were used.

Multiferroic materials can be classified into two categories: one is single-phase materials; the other is multilayer or composite hetero-structures that contain more than one ferroic phases [5]. The most desirable multiferroic material is the intrinsic single-phase material, which is still rarely produced although significant advancements had been made recently. Therefore, it is essential to broaden the searching field for new candidates of multiferroics. This work focuses on both typesmultiferroic materials: Single-phase Bi_2FeMnO_6 and multilayered $YMnO_3/SnTiO_3$. Scanning probe microscopy (SPM) techniques were used for ferroelectric domain imaging of these multiferroic materials.

As we know, most piezoelectric/ferroelectric materials with good performances are based on the perovskite-type oxides of ABO_3, among them, the most extensively studied ones should be PbTiO3 and $Pb(ZrTi)O_3$ based materials due to their high dielectric constants and good piezoelectric/ferroelectric properties. However, these materials containing lead which leads to environmental problems. Thus, it is desirable to develop new lead-free piezoelectric materials to replace PZT based piezoelectrics for environmental protection. Through first principle calculations, $SnTiO_3$ containing Sn^{2+} ions was estimated to have excellent ferroelectric properties [6, 7]. The calculated results indicated that the spontaneous polarization and piezoelectric coefficients of $SnTiO_3$ is comparable with those of $PbTiO_3$. Moreover, the most stable structure is tetragonal perovskite with a=b=3.80Å and c=4.09Å [6, 7]. The metastable $SnTiO_3$ phase is very difficult to obtain, a good way to stabilize the $SnTiO_3$ phase is to mix it with other compounds. Several studies have been focused on the synthesis of Sn-doped ceramics in $BaTiO_3$ system, such as $(Ba_{0.6}Sr_{0.4})(Ti_{1-x}Sn_x)O_3$ [8], $(Ba_{1-x-y}Ca_xSn_y)TiO_3$ [9]. However, it is difficult to obtain Sn^{2+} by traditional bulk synthesis techniques. On the other hand, it is easier to fabricate $SnTiO_3$ using pulsed laser deposition method (PLD). Our interest was to find out whether the $SnTiO_3$ phase can be stabilized through the layer-by-layer deposition using PLD, and whether it is possible to obtain such metastable materials in non-equilibrium conditions.

The hexagonal manganite $YMnO_3$, which shows an antiferromagnetic transition at $T_N = 75$ K, and a ferroelectric transition $T_C = 913$ K, is one of the rare existing multiferroics [10-12]. It was chosen as the basic composition to sandwich $SnTiO_{3+x}$ phase and to form YST multilayer film. In this work, we reported the effect of substrate orientation and layer numbers of $YMnO_3$ and $SnTiO_{3+x}$ on the piezoelectric/ferroelectric and magnetic properties in the designed layered YST system. It was hoped that our method would serve as a model system to introduce the use of PLD techniques in the growth of multilayer multiferroic materials and metastable materials. Such techniques could be promising for device design and the searching in fabricating new materials.

2. SPM system

SPM has emerged as a powerful tool for high-resolution characterization of ferroelectrics for the first time providing an opportunity for non-destructive visualization of ferroelectric domain structures at the nanoscale. [13]These nanostructures can be used for studying the intrinsic size effects in ferroelectrics as well as for addressing such technologically important issues as processing damage, interfacial strain, grain size, aspect ratio effect, edge effect, domain pinning and imprint. [14] The system composes of a conducting tip in contact with the dielectric surface on a conductive substrate can be considered as a capacitor. SPM is a well-established field offering multiple opportunities for new discoveries and breakthroughs. [15-17]

Piezoelectric materials provide an additional response to the applied ac electric field due to the converse piezoelectric effect: $\Delta l=d_{33}V$, whereΔl is the displacement, d_{33} is the effective longitudinal piezoelectric coefficient. [18, 19] If follows that both electrostatic and piezoelectric signals are linear with the applied voltage and thus contribute to the measured PFM response. These measurements are referred to as out-of-plane (OP) measurement. Measurements of local hysteresis loops are of great importance in inhomogeneous or polycrystalline ferroelectrics because they are able to quantify polarization switching on a scale significantly smaller than the grain size or inhomogeneity variation. Macroscopically, the switching occurs via the nucleation and growth of a large number of reverse domains in the situation where the applied electric field is uniform. Therefore, the d_{33} hysteresis reflects the switching averaged over the entire sample under the electrode. [20, 21] In the PFM experimental conditions, the electric field is strongly localized and inhomogeneous; therefore, the polarization switching starts with the nucleation of a single domain just under the tip.

3. Material designation and characterization

Multiferroic materials have been attracting considerable attention especially in the last ten years. [22-25] One of the most appealing aspects of multiferroics is their magnetoelectric coupling. Among them, the most studied is the provskite $BiFeO_3$(BFO) with room temperature multiferroic properties and $YMnO_3$. Based on BFO and $YMnO_3$, we designed new multiferroics and fabricated the films using pulsed laser deposition (PLD) method. The material system includes double perovskitemultiferroic Bi_2FeMnO_6 (BFM) and multilayered $YMnO_3$/ $SnTiO_{3+x}$ (YST). Compared to BFO, the magnetic properties of BFM and YST were greatly improved. However, until now there is no report about their ferroelectric properties because the difficulty of obtaining well-shaped polarization hysteresis loops. It is important to study ferroelectric properties because the possible coupling between ferroelectric and antiferromagnetic domains and its lead-free nature. As is well known, the ferroelectric property is mainly determined by the domain structures and domain wall motions. The most exciting recent developments in the field of multiferroics and the most promise for future discoveries are in interfacial phenomena. [26] The interfaces include those that emerge spontaneously and those that are artificially engineered. Therefore, the domain structure and polarization switching were studied in these three new multiferroic films using piezoresponse force microscopy (PFM).

The emerging technique of PFM is proved to be a powerful tool to study piezoelectric and ferroelectric materials in such cases and extensive contributions have been published In PFM, the tip contacts with the sample surface and the deformation (expansion or contraction of the sample) is detected as a tip deflection. The local piezoresponse hysteresis loop and information on local ferroelectric behavior can be obtained because the strong coupling between polarization and electromechanical response in ferroelectric materials. In the present study, we attempts to use PFM to study the ferroelectric/piezoelectric properties in films of BFM and YST. PFM response was measured with a conducting tip (Rh-coated Si cantilever, k~1.6 N m^{-1}) by an SII Nanotechnology E-sweep AFM. PFM responses were measured as a function of applied DC bias (V_{dc}) with a small ac voltage applied to the bottom electrode (substrate) in the contact mode, and the resulting piezoelectric deformations transmitted to the cantilever were detected from the global deflection signal using a lock-in amplifier.

3.1. Characterization of BFM film

Magnetism and ferroelectricity exclude each other in single phase multiferroics. It is difficult for designing multiferroics with good magnetic and ferroelectric properties. Our interest is to design new candidate multiferroics based on $BiFeO_3$. $BiFeO_3$ is a well-known multiferroic material with antiferromagntic Neel temperature of 643K, which can be synthesized in a moderate condition. [22, 23, 25] In contrast, $BiMnO_3$ is ferromagnetic with Tc = 110K and it needs high-pressure synthesis. [27, 28] The possible magnetoelectric coupling has motivated a lot of studies on the ferroelectric and antiferromagnetic domains. The ferroelectric domain structures of BFO in the form of ceramics, single crystals and thin films have been intensively studied using PFM, TEM and other techniques.

Single phase BFM ceramics could be synthesized by conventional solid state method as the target. For BFM ceramics, the starting materials of Bi_2O_3, Fe_3O_4, $MnCO_3$ were weighed according to the molecular mole ratio with 10 mol% extra Bi_2O_3. They were mixed, pressed into pellets and sintered at 800 °C for 3 h. Then the ceramics were crushed, ground, pressed into pellets and sintered again at 880 °C for 1 h. BFM films were deposited on $SrTiO_3$(STO) substrate by pulsed laser deposition (PLD) method at 650°C with 500 ~ 600mTorr dynamic oxygen. [29, 30]

The structure of BFM was calculated [31] and it is connected with the magnetic configurations as shown in Figure 1. It has three possible space groups of Pm3m, R3 and C2 and the magnetic configurations were presumedfor each structure symmetry to be G-type antiferromagnetic (G-AFM) and ferromagnetic (FM) structures. The most stable structure of BFM is monoclinic with C2 space group. Mn tends to show 3+ valence which will induce a large distortion because it is Jahn-Teller ion. The valence of Mn and Fe has been studied in the former work. [30, 32]

Figure 2 shows the results of the PFM images of BFM film. Several features could be observed: firstly, the obvious contrast could be seen and the grains in PFM and topography are correspondence; secondly, the existence of contrasts on both OP and IP indicates multiple orientations of domains; thirdly, the IP contrast is not so clear as OP contrast, that is to say, the suppression of the in-plane response for heterostructures suggesting a constrained ferroelectric domain-orientation along the OP direction. In the former paper, the ferroelectric domain switching and typical butterfly loops were observed. [32]

http://dx.doi.org/10.5772/52519

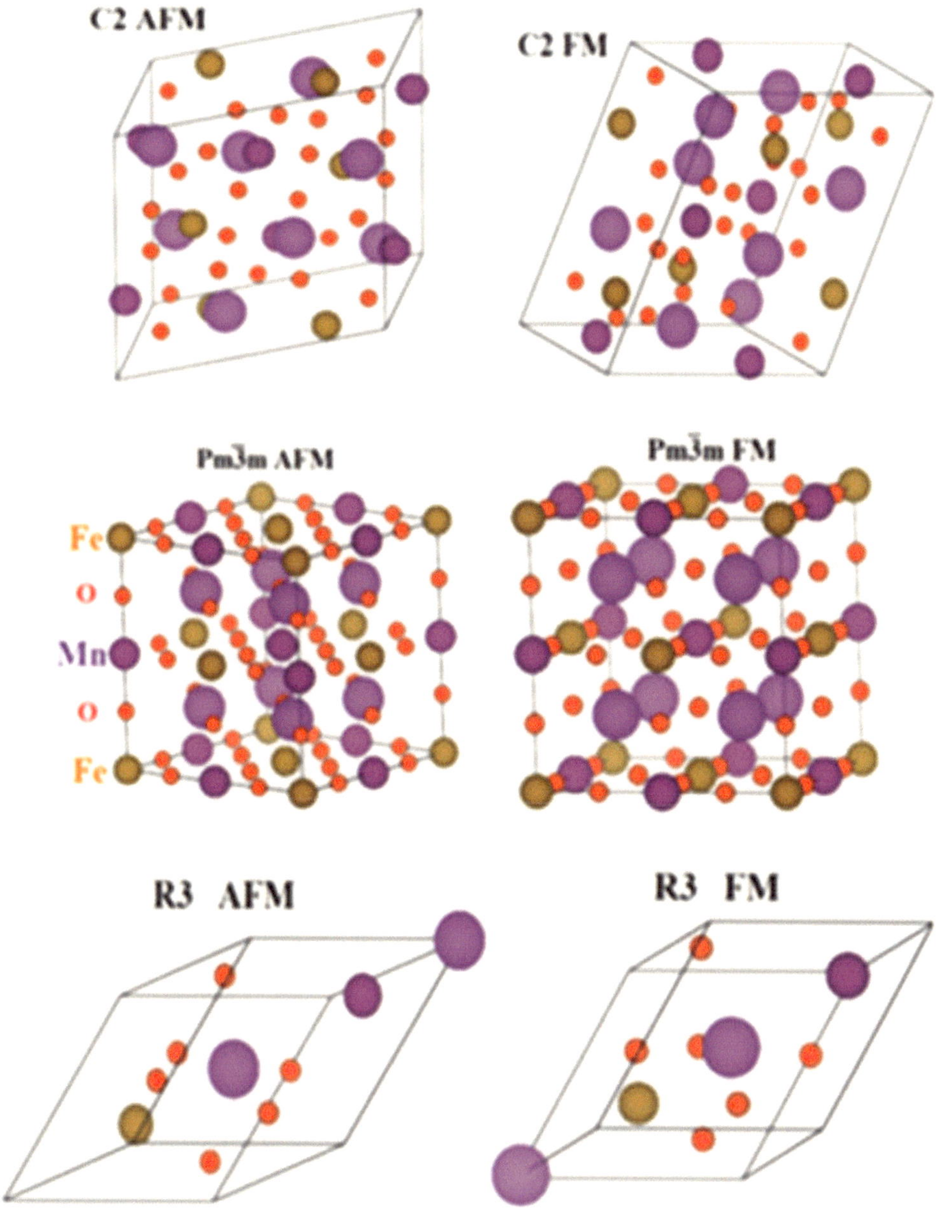

Figure 1. Calculated six structures of BFM.

The domains in BFO films have been analyzed in detail in Ref [19]: the bigger spontaneous ferroelectric domains were observed in BFO than in other ferroelectrics without multiferroic properties; the domains were irregular but the the model was predicted and consistent with the experimental results. According to the present results of BFM, further study is needed to obtain the domain morphology and do the calculation.

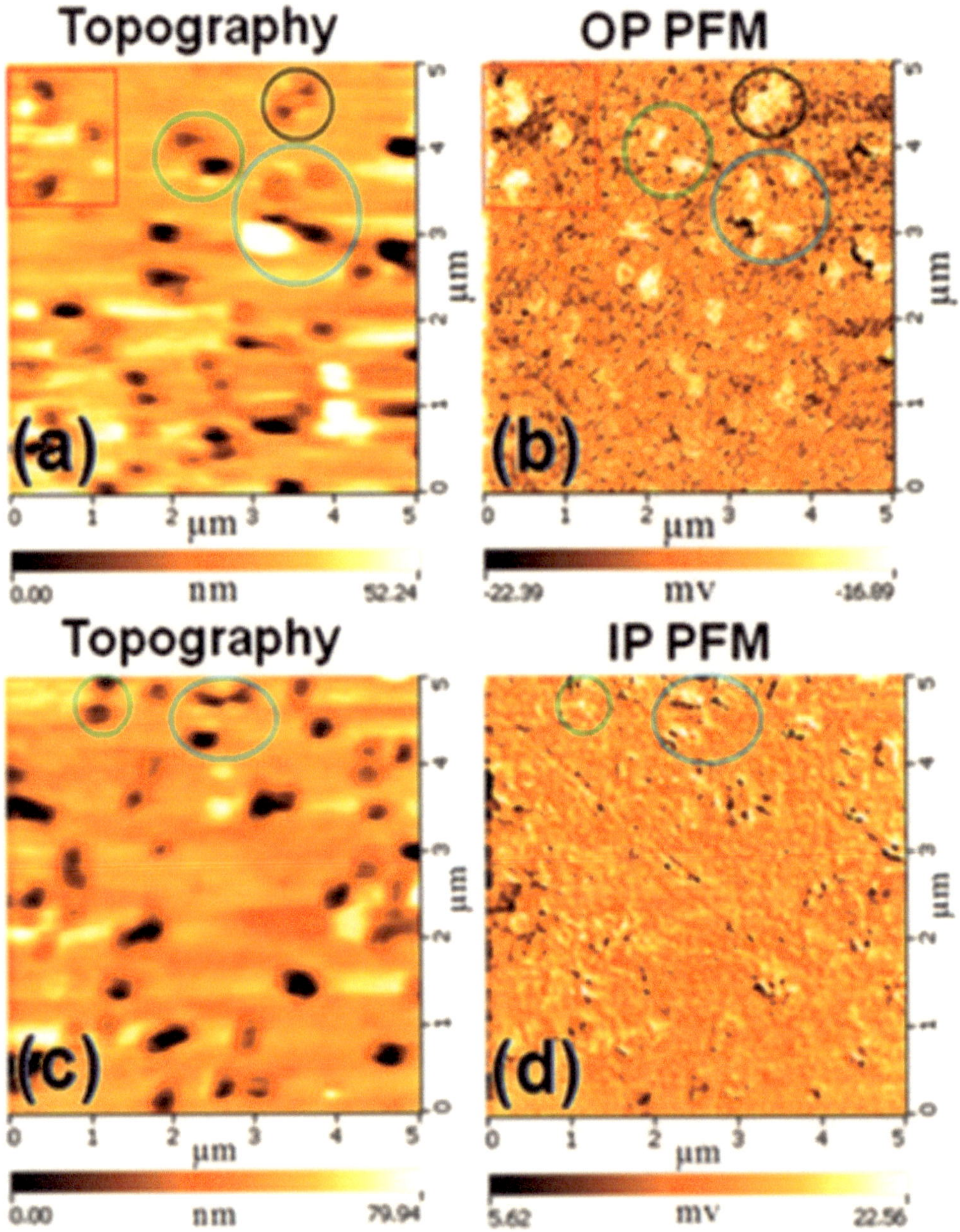

Figure 2. PFM images of BFM film

http://dx.doi.org/10.5772/52519

3.2. Characterization of YST film

Polycrystalline $YMnO_3$ and TiO_2 ceramics were synthesized by conventional solid state method as the targets. On addition the SnO and SnO_2 commercial targets were also used at the same time in our process. The deposition of these films includes the following major steps. YST films were deposited on (100) and (110) Nb: $SrTiO_3$ (STO) substrate using a pulsed laser deposition (PLD) system at 700°C with 10^{-1} ~10^{-5}Torr dynamic oxygen. The targets were alternately switched constantly and the films were obtained in a layer-by-layer growth mode. After deposition, the films were annealed in the same condition for 15 minutes at 700°C and then cooled to room temperature. [33]

In YST multilayered films, one layer is defined to be comprised of two sub-layers: (1) $YMnO_3$ and (2) $SnTiO_{3+x}$. The films deposited on (100), (111) and (110) STO are expressed as YST100, YST111 and YST110, respectively. The film YST110-4 denotes the films deposited on (110) STO with four layers, as shown clearly in Figure 3. The total deposition time for $YMnO_3$ and $SnTiO_{3+x}$ is the same of 40 and 20 minutes, respectively.

XRD patterns for the five films of YST100-2, YST100-4 were shown in Figure 4. The peaks were identified using the XRD results of $SnTiO_3$ (◎), $YMnO_3$ (Y) (shown in Figure 9, $FeTiO_3$ (▽) [34, 35]. $SnTiO_3$ is metastable and it showed two combined phases, one is $FeTiO_3$, and the other is the good ferroelectric phase which has tetragonal structure. XRD patterns of $SnTiO_3$ (◎) were shown in Figure 5 in supplementary materials, which is obtained using the calculated data[6, 7]. The symbol of "?" represents the phase which we cannot identify so far, it could be a peak from the (111) TiO_2 phase.

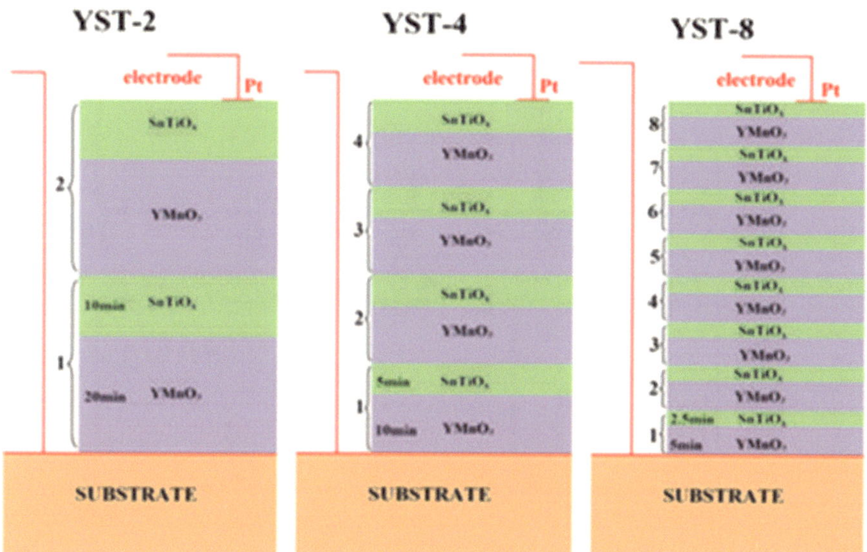

Figure 3. Schematic explanation of films with two, four and eight layers, each layer contains two sublayers of $YMnO_3$ and $SnTiO_3$.

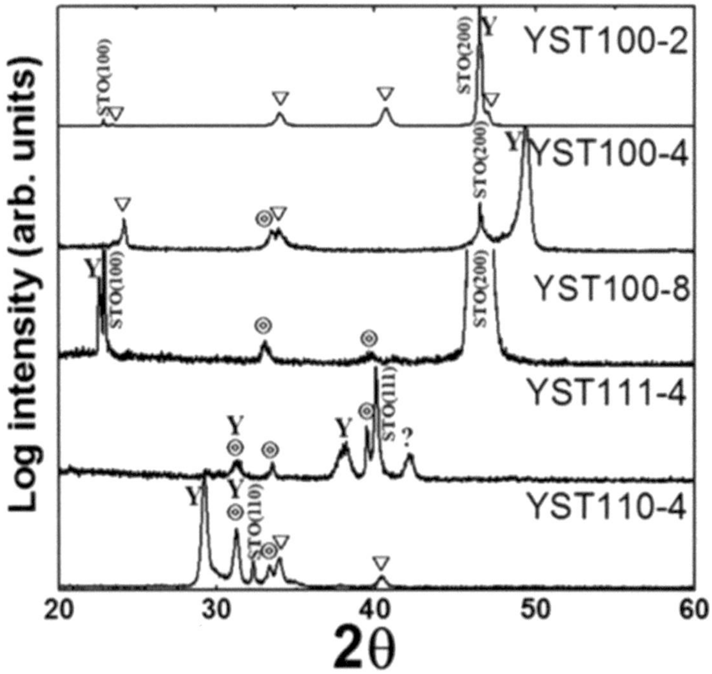

Figure 4. XRD patterns for the five films of YST100-2, YST100-4. [31]

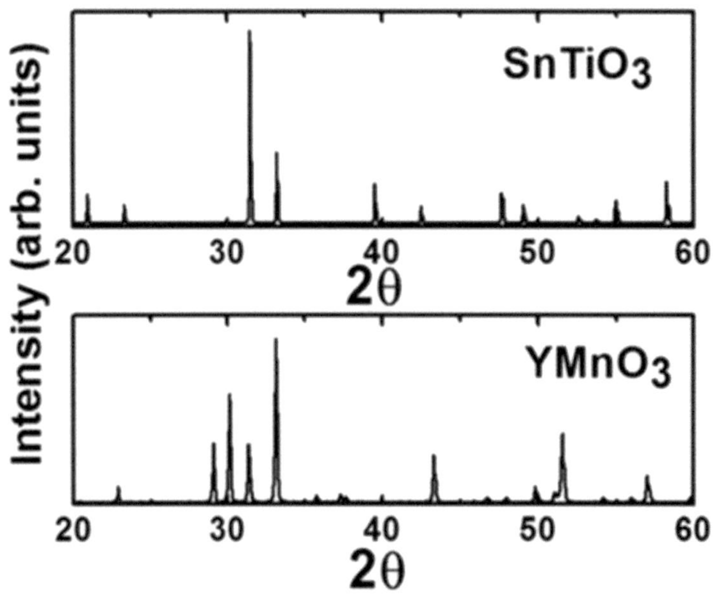

Figure 5. XRD patterns of calculated $SnTiO_3$ and $YMnO_3$ targets. [33]

Many reports about the $YMnO_3$ films can be found but most of them without ferroelectric characterization provided. This is mainly ascribed to the difficulty in obtaining a satisfactory ferroelectric measurement result. Figure 6 shows the electrical polarization hysteresis loops (P-E loops) of YST110-4 film. The ferroelectric type hysteresis loop was observed. It is obvious that the P-E loop is significantly different for different layer-number films, as well as for different substrate orientations. For the same STO orientation, the P-E loops were improved as the layer number increase, and the film fabricated on (110) STO shows improved properties compared to the film on (100) STO with the same layer number. As for the same deposition time for the four samples, it is found that more layers of $SnTiO_3$ in the YST system shows much better ferroelectric properties. It is suggested that the $SnTiO_3$ phase can exist only in ultra-thin thickness and can be stabilized by $YMnO_3$ sub-layers. That is to say, $SnTiO_3$ is indeed a ferroelectric materials but it is a great challenge [36] to obtain stable single phase $SnTiO_3$ films that is with good ferroelectric performance (it will be tried in our future works). Although the electric charge in the interface may affect its ferroelectric properties, the improved properties of YST110-4 proved that its effect was not significant. While the P-E loops were not improved in YST111-4, all the observations showed the anisotropy of the films.

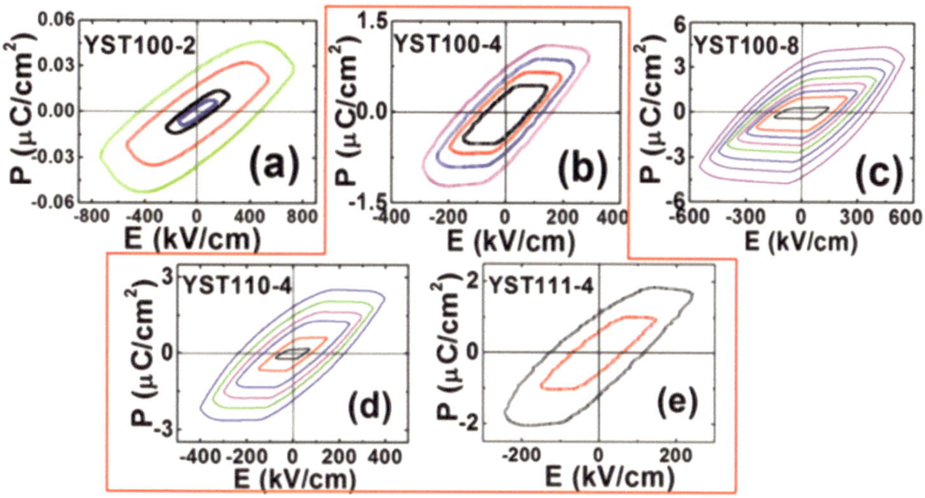

Figure 6. Ferroelectric polarization loop of YST films. [31]

Although well developed P-E loops were observed for YST100-8 and YST110-4, the loops still show the rounded features, indicating that contributions to the hysteresis loop from movable charges was significant. In addition, the P-E loops of the four films have revealed that the polarization is rather small (just a few $\mu C/cm^2$) and they do not exhibit saturation. The fatigue properties were shown in Figure 7 for YST110-4. The polarization measurement was at 21V, the switching voltage was 14V and the frequency was 5k Hz. The remnant polarization decreased about 40% after 10^7 read/write cycles for YST110-4. From all the comparisons, YST110-4 sample shows the comparable ferroelectric properties with YST100-8.

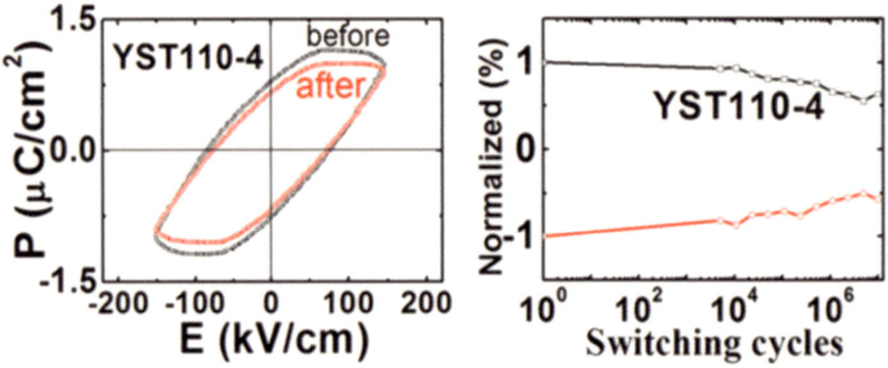

Figure 7. Fatigue measurement of YST110-4 film. [31]

Figure 8 shows TEM results of the YST110-4 film. It clearly shows four layers and one layer which is close to the substrate seems have some reaction with the substrate. The other three layers are homogeneous and we could see the domain structures through this image. The surface of the film was studied using AFM as shown in Figure 9. As expected, Figure 10 displays contrast over the polarized square after poled by positive and negative 10 V voltage, due to the different phases of the PFM response for the up and down domains. The obvious change of the contrast in YST110-4 film confirms that the polarization reversal is indeed possible and that the film is ferroelectric at room temperature.

Figure 8. TEM image of YST110-4 film.

http://dx.doi.org/10.5772/52519

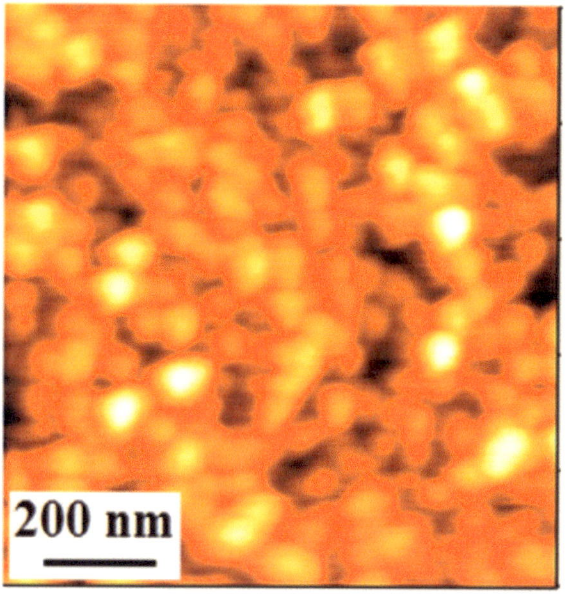

Figure 9. AFM topography image of YST110-4 film.

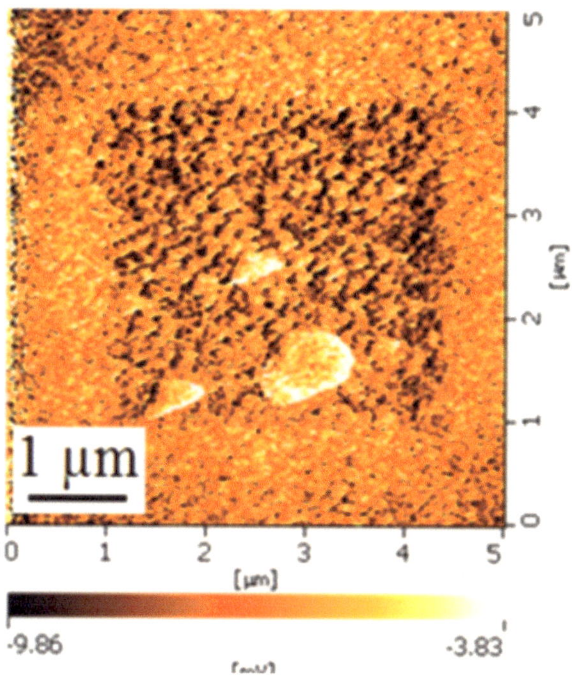

Figure 10. PFM image of YST110-4 film after poled using ± 10 V voltage

As seen in Figure 11, when a direct current voltage up to 10 V was applied on the samples, the sample exhibited "butterfly" loop. The loops were not symmetrical due to the asymmetry of the upper (tip) and bottom (substrate) electrodes. In addition, the substrate may also affect the d_{33}. According to the equation $d_{33}=\Delta l/V$, where Δl is the displacement, the effective d_{33} could be calculated. At the voltage of 10 V, both samples show the maximum effective $d_{33}$of about 6.21 pm/V for YST110-4 sample.

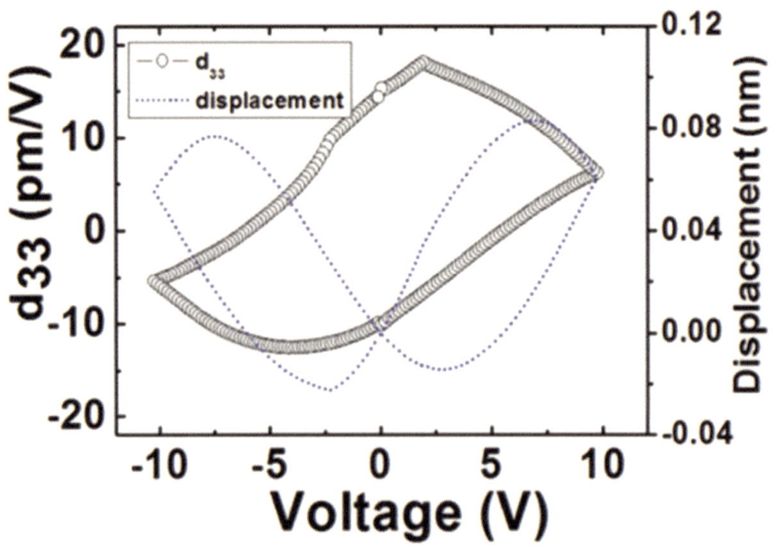

Figure 11. Butterfly and local piezoresponse hysteresis loops of YST110-4 film.

4. Conclusions

The novel multiferroic films of BFM and multilayered YST were successfully produced using PLD method. Both of them were characterized using PFM method. The special or excellent properties can often be found in the metastable materials. Through calculation, s4ix structures are presumed in BFM with two magnetic configurations of G-AFM and FM. In YST films, the metastable $SnTiO_3$ phase was obtained. The improved ferroelectric properties were observed through increasing the layer numbers and more $SnTiO_3$ phase was stabilized, moreover, the change in contrast after bias is applied indicates a change in polarization direction and hence ferroelectric switching. Although it is a great challenge in obtaining the $SnTiO_3$ thin films, we believe that through the optimization of fabrication process and conditions, the single phase $SnTiO_3$ or multilayer films, or composite materials containing $SnTiO_3$ could become a new generation lead-free piezoelectric/ferroelectric material. For a thorough understanding of the mechanisms of the films, knowledge of the domain structures is a prerequisite, which is of crucial importance to increase and tune the functionality of multiferroic films.

http://dx.doi.org/10.5772/52519

Acknowledgements

The authors gratefully acknowledge Dr. Minoru Osada, Dr. Kazuya Terabe in NIMS and Prof. H. R. Zeng in Shanghai Institute of Ceramics for their experimental help and discussions. Part of this work was supported by JST ALCA and MEXT GRENE.

Author details

Hongyang Zhao[1,2*], Hideo Kimura[1], Qiwen Yao[1], Lei Guo[1], Zhenxiang Cheng[3] and Xiaolin Wang[3]

*Address all correspondence to: zhao.hongyang@nims.go.jp

1 National Institute for Materials Science, Sengen 1-2-1, Tsukuba, Japan

2 Shanghai Institute of Ceramics, Chinese Academy of Sciences, Shanghai, China

3 Institute for Superconducting and Electronics Materials, University of Wollongong, Innovation Campus, Fairy Meadow, Australia

References

[1] Hill N. A., Why are there so few magnetic ferroelectrics? J. Phys. Chem. B 2000; 104: 6694.

[2] Fiebig M., T. Lottermoser, D. Frohlich, A. V. Goltsev, R. V. Pisarev, Observation of cou-pled magnetic and electric domains.Nature 2002; 419: 818.

[3] Eerenstein W., Mathur N. D. and Scott J.F., Multiferroic and magnetoelectric materials. Nature 2006; 442: 759-765.

[4] Singh M. K., Prellier W., Singh M. P., Katiyar R. S., Scott J. F., Spin-glass transition in single-crystal BiFeO3. Phys. Rev. B 2008; 77: 144403.

[5] J. Das, Y.Y. Song, M.Z. Wu, Electric-field control of ferromagnetic resonance in monolithic $BaFe_{12}O_{19}$–$Ba_{0.5}Sr_{0.5}TiO_3$heterostructures. J. Appl. Phys. 2010; 108: 043911.

[6] Konishi Y., Ohasawa M., Yonezawa Y., Tanimura Y., Chikyow T., Wakisaka T., Koinuma H., Miyamoto A., Kubo M., Sasata K., Mater. Res. Soc. Symp. Proc. 2003; 748 U3.13.1.

[7] Uratani Y., Shishidou T., Oguchi T., First-Principles Study of Lead-Free Piezoelectric $SnTiO_3$.Jpn. J. Appl. Phys. 2008; 47 (9), 7735-7739.

[8] Ha J. Y., Lin L. W., Jeong D. Y., Yoon S. J., Choi J. W., Improved Figure of Merit of (Ba,Sr)TiO3-Based Ceramics by Sn Substitution.Jpn. J. Appl. Phys. 2009; 48 011402.

[9] Suzuki S., Takeda T., Ando A., Takagi H., Ferroelectric phase transition in Sn2+ ions doped (Ba,Ca)TiO3 ceramics.Appl. Phys. Lett. 2010; 96: 132903.

[10] Marin L.W., Crane S.P., Chu Y-H., Holcomb M.B., Gajek M., Huijben M., Yang C-H., Balke N., Ramesh R., Multiferroics and magnetoelectrics: thin films andnanostructures.J. Phys.: Condens. Matter 2008; 20: 434220.

[11] Yakel H.L., Koehler W.C., Bertaut E.F., Forrat E.F., On the crystal structure of the manganese(III) trioxides of the heavy lanthanides and yttrium. ActaCrystallogr. 1963; 16: 957-962.

[12] Bertaut E. F., Pauthenet R., Mercier M., Proprietesmagnetiqueset structures du manganite d'yttrium. Phys. Lett. 1963; 7: 110-111.

[13] Gruverman A, Kholkin A, Nanoscale Ferroelectrics: Processing, Characterization and Future Trends. Rep. Prog. Phys. 2006; 69: 2443-2474.

[14] Catalan G., Noheda B, McAneney J, Sinnamon L J, Gregg J M Strain gradients in epitaxial ferroelectrics. Phys. Rev. B 2005; 72: 0201102.

[15] Muralt P. J. Micromech. Ferroelectric thin films for micro-sensors and actuatorsMicroeng. 2000;10: 136.

[16] Terabe K, Nakamura M, Takekawa S. Kitamura K, Higuchi S. Gotoh Y, Cho Y Microscale to nanoscale ferroelectric domain and surface engineering of a near-stoichiometric $LiNbO_3$ crystal. Appl. Phys. Lett. 2003; 82: 433-435.

[17] Fong D D, Stephenson G B, Streigger S K, Eastman J A, Auciello O, Fuoss P H, Thompson C Ferroelectricity in ultrathin perovskite films Science. 2004; 304: 1650.

[18] Kholkin A L, Bdikin I K, Shvartsman V V, Orlova A, Kiselev D, Bogomolov V, MRS Proc. E 2005; 838: O7.6.

[19] Catalan G., Bea H., Fusil S., Bibes M., Paruch P., Barthelemy A., Scott J. F., Fractal Dimension and Size Scaling of Domains in Thin Films of Multiferroic $BiFeO_3$. Phy. Rev. Lett. 2008; 100 027602.

[20] Kalinin S V, Gruverman A and Bonnell D A Quantitative analysis of nanoscale switching in SrBi2Ta2O9 thin films by piezoresponse force microscopy Appl. Phys. Lett. 2004; 85: 795-797

[21] Tybell T, Paruch P, Giamarchi T, Triscone J-M Domain wall creep in epitaxial ferroelectric $Pb(Zr_{0.2}Ti_{0.8})O_3$ thin films 2002 Phys. Rev. Lett. 89 097601

[22] Catalan G., Scott J. F., Adv. Mater. Physics and Applications of Bismuth Ferrite.2009; 21, 2643.

[23] Wang J., Neaton B. J., Zheng H., Nagarajan V., Ogale S. B., Liu B., Viehland D., Vaithyanathan V., Schlom D. G.,. Waghmare U. V, Spaldi N. A.n, Rabe K. M., Wutting

M., Ramesh R., Epitaxial BiFeO3 Multiferroic Thin Film Heterostructures. Science 2003; 299: 1719-1722.

[24] Kimura T., Goto T., Shintani H., Ishizaka K., Arima T., Tokura Y., Magnetic Control of Ferroelectric Polarization. Nature 2003; 55: 426.

[25] Cheng Z. X., Wang X. L., Dou S. X., Ozawa K., Kimura H., Improved ferroelectric properties in multiferroic $BiFeO_3$ thin films through La and Nbcodoping. Phys. Rev. B 2008; 77: 092101.

[26] Spaldin N. A., Cheong S. W., Ramesh R.,Multiferroics: Past,present, and future. Physics Today 2010; 63: 38-43.

[27] T. Atou, H. Chiba, K. Ohoyama, Y. Yamaguchi, Y. Syono, Structure determination of ferromagnetic perovskite $BiMnO_3$. J. Solid State Chem. 1999: 145(2) 639-642.

[28] T. Kimura, S. Kawamoto, I. Yamada, M. Azuma, M. Takano, Y. Tokura, Magnetocapacitance effect in multiferroic $BiMnO_3$, Phys. Rev. B 2003: 67 (R), 180401-180404.

[29] Zhao, H.Y.; Kimura, H.; Cheng, Z.X.; Wang, X.L. &Nishida, T. Room temperature multiferroic properties of Nd:BiFeO3/Bi2FeMnO6 bilayered films.Appl. Phys. Lett. 2009; 95(23)232904

[30] Zhao, H.Y.; Kimura, H.; Cheng, Z.X.; Wang, X.L.; Ozawa, K. &Nishida, T. Magnetic characterization of Bi_2FeMnO_6 film grown on (100) $SrTiO_3$ substrate Phys. Status Solidi RRL 2010; 4(11) 314.

[31] Bi, L.; Taussig, A.R.; Kim, H-S.; Wang, L.; Dionne, G.F.; Bono, D.; Persson, K.; Ceder, G.&Ross, C.A. Structural, magnetic, and optical properties of $BiFeO_3$ and Bi_2FeMnO_6 epitaxial thin films: An experimental and first-principles study. Phys. Rev. B 2008; 78 (10) 104106.

[32] ZhaoH Y, Kimura H.; Cheng Z.X.; Wang X.L., New multiferroic materials: Bi_2FeMnO_6. Ferroelectrics-Material Aspects. InTech; 2011.

[33] Zhao H Y, Kimura H.; Cheng Z.X.; Wang X.L., Yao Q W, Osada M, Li B W, Nishida T A new multiferroicheterostructure of YMnO3/$SnTiO_3$+x. Scrip. Mater. 2011; 65: 618-621

[34] Harrison R.J., Redfern S.A.T., Smith R.I., Thermodynamics of the R(3)over-bar to R(3)over-barc phase transition in the ilmenite-hematite solid solution.Am. Mineral. 2000; 85: 1694-1705.

[35] Leinenweber K., Utsmi W., Tsuchida Y., Yagi T., Kuita K., Unquenchable High-Pressure Perovskite Polymorphs Of $MnSnO_3$ And $FeTiO_3$. Phys. Chem. Miner. 1991; 18: 244.

[36] Venkatesan S., Daumont C., Kooi B.J., Noheda B., J Hossoneff Th.M. De, Nanoscale domain evolution in thin films of multiferroic $TbMnO_3$. Phys. Rev. B 2009; 80: 214111.

Chapter 10

Gelcasting of Ferroelectric Ceramics: Doping Effect and Further Development

Dong Guo and Kai Cai

Additional information is available at the end of the chapter

http://dx.doi.org/10.5772/53995

1. Introduction

Ferroelectric ceramics may be seen as the most important type of ferroelectric materials, which have been used in a wide spectrum of electrical and microelectronic devices, including underwater transducers, micro-pumps and valves, ultrasonic motors, thermal sensors, probes for medical imaging and non-destructive testing, and accelerometers, etc (Cross LE, 1996; Setter et al., 2000).

Ferroelectric ceramics used in the devices have various shapes. A certain shape formation technique is required to make ceramics with the desired shapes. Dry pressing is the most commonly used ceramic forming technique. In this technique, dry powders containing organic binder are filled into a solid mold, then dry ceramic green bodies with the shape of the mold cavity are formed under mechanical or hydraulic compacting presses selected for the necessary force and powder fill depth. The pressure is around several tens of MPa or higher. Therefore, large ceramic parts require a much higher compacting force. If the ceramic parts are unable to have pressure transmit suitably for a uniform pressed density then isostatic pressing may be used. One of the most serious disadvantages of dry pressing lies in the difficulty in fabricating high quality large and complex-shaped ceramics or ceramics with a fine structure, which are required for various devices.

To resolve the problems associated with the conventional dry pressing, new wet forming techniques, such as gelcasting (Omatete et al., 1991), electrophoretic casting (Biesheuvel et al., 1999), hydrolysis assisted solidification (Novak et al., 2002), and direct coagulation casting (Graule et al., 1996), etc., are becoming increasingly attractive for advanced ceramic materials. Since in these techniques the ceramic powders are dispersed in a liquid medium and thoroughly mixed, wet forming techniques have the advantages of reducing some structure defects that are difficult to remove in dry pressed ceramic parts. Among the many wet forming techniques, aqueous gelcasting represents the latest improvements. In the technique, a high solids loading slurry obtained by dispersing the ceramic powders in the pre-mixed solution containing monomer and cross-linker is cast in a mold of the desired shape. When heated, the monomer and cross-linker polymerize to form a three-dimensional network structure, thus the slurry is solidified *in situ* and green bodies of the desired shape are obtained, which consists mainly of ceramic powders with a low polymer content. Gelcasting may be seen as a milestone in fabricating complex-shaped ceramic parts since it has initiated a new branch of research in ceramic processing due to its intriguing properties of near-net-shape forming, high green strength, and low binder concentration, etc. As a pressure free method, it can also be used in fabricating large ceramic components that have simple shapes. For example, to make a ceramic disc with a diameter of 20 cm by dry pressing, a load of hundreds of tons is required. In contrast, such a ceramic disc can be easily made by gelcasting via a simple ring-shaped mold at much less cost. Furthermore, gelcasting may be developed to fabricate complex-shaped (Cai et al., 2003) or fine-structured (Guo et al., 2003) ceramic parts that are rather difficult or even impossible to be formed by the conventional dry pressing method.

Since its invention in 1991 gelcasting has been widely used for the fabrication of structural ceramics, including Al_2O_3 (Young et al., 1991), SiC (Zhou et al., 2000), and Si_3N_4 (Huang et al., 2000), etc. Later, it was applied to a number of functional ceramics in a number of ways including $LaGaO_3$ (Zha et al., 2001), ZnO (Bell et al., 2004), and $Sr_{0.5}Ba_{0.5}Nb_2O_6$ (Chen et al., 2006), etc. The first journal report about the application of gelcasting to piezoelectric ceramics appeared at 2002 (Guo et al., 2003). Compared to structural ceramics, much less research about gelcasting of functional ceramics has been conducted so far.

Among the many ferroelectric ceramic materials, lead zirconate titanate (PZT) is the most widely used one owing to their superior piezoelectric, pyroelectric and dielectric properties. A fascinating feature of multicomponent ferroelectric ceramics is that their electrical properties can be modified by doping with acceptors and donors (Shaw et al., 2000). As a result, a series of PZT materials with tailored properties are commercially available. Unfortunately, this feature also leads to the problem of high sensitivity of the electrical properties of PZT to composition. In addition, PZT is commonly used with a composition close to the morphotropic phase boundary (MPB) at a Zr/Ti ratio of about 52/48, where properties such as piezoelectric coefficients, dielectric permittivity, and coupling factors are maximized and thus may be more sensitive to the composition (Noheda et al., 2006). On the other hand, since shape formation in gelcasting is achieved through *in situ* polymerization, organic additives are used in the premix solution. Also, addition of commercial surfactants that may have a complicated composition is indispensable, because successful fabrication of ceramics via gelcasting or other colloidal methods requires to prepare high solids loading ceramic slurry with still a low viscosity. Consequently, in order to apply gelcasting to the formation of ferroelectric ceramics such as PZT, it is necessary to remove the possible influence of the impurities introduced by the various additives on the electrical performance of the final products. This makes the problems more complicated than that of structural ceramics.

http://dx.doi.org/10.5772/53995

2. Colloidal chemistry and rheological properties of PZT suspensions

We first give a short introduction the gelcasting technique. The details are out of the focus of this chapter, and interested readers can refer to other papers. According to the liquid medium used, there are two types of gelcasting systems: aqueous and nonaqueous systems. In aqueous gelcasting deionized water is used as the medium, and acrylamide ($C_2H_3CONH_2$) may be seen as a prototype monomer. Generally, N,N-methylenebisacrylamide ($(C_2H_3CONH)_2CH_2$, MBAM) N,N,N_0,N_0-Tetramethylethylenediamine (TEMED) and $(NH_4)_2S_2O_8$ are used as the cross-linker, the catalyst and the initiator, respectively. The gelcasting process of PZT is similar to that of previous studies. The flowchart of the gelcasting is shown in Figure 1. First, the PZT powders are added in the premix solution containing AM and MBAM and thoroughly mixed. After addition of initiator and catalyst, the slurry is de-aired in vacuum, then the slurry is cast into the mold with desired shape and heated around 60-80°C in a oven for several hours for polymerizatoin and drying. After demolding the green ceramic body is obtained.

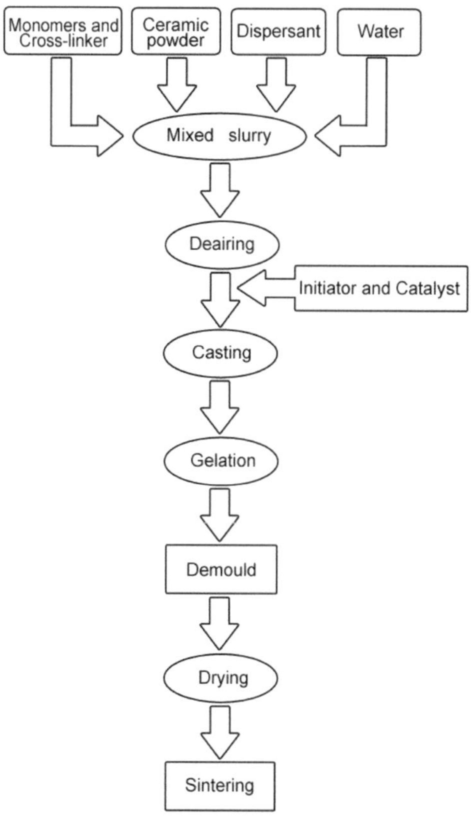

Figure 1. The flowchart of gelcasting

The colloidal and rheological properties of the PZT suspension are important issues that should be addressed first. Homogenous dispersion of the powder in the premix solution and stability of the suspension are determined mainly by attractive and repulsive forces between the particles in the system. The former generally arises from the van der Waals forces while the latter can arise from electrostatic repulsion or steric repulsion of surfactant materials absorbed on the particle surfaces (Israelachvili, 1992). Magnitude of the van der Waals forces is mainly determined by the nature of the particle surface and the solvent. While the repulsive forces can be modified over a wide range by dispersants.

Commercial Name	TAC	JN281	SP2	SGA
Producer	Beijing Chemical Reagent Co.	Beijing Pinbao Chemical Co.	Beijing Tongfang Chemical Co.	Beijing Tongfang Chemical Co.
Composition	Triammonium citrate	Poly (acrylic acid), NH_4^+ salt solution	Poly (acrylic acid-co-maleic acid), Na^+ salt solution	Poly (acrylic acid), Na^+ salt solution

Table 1. Composition and producer of the dispersants

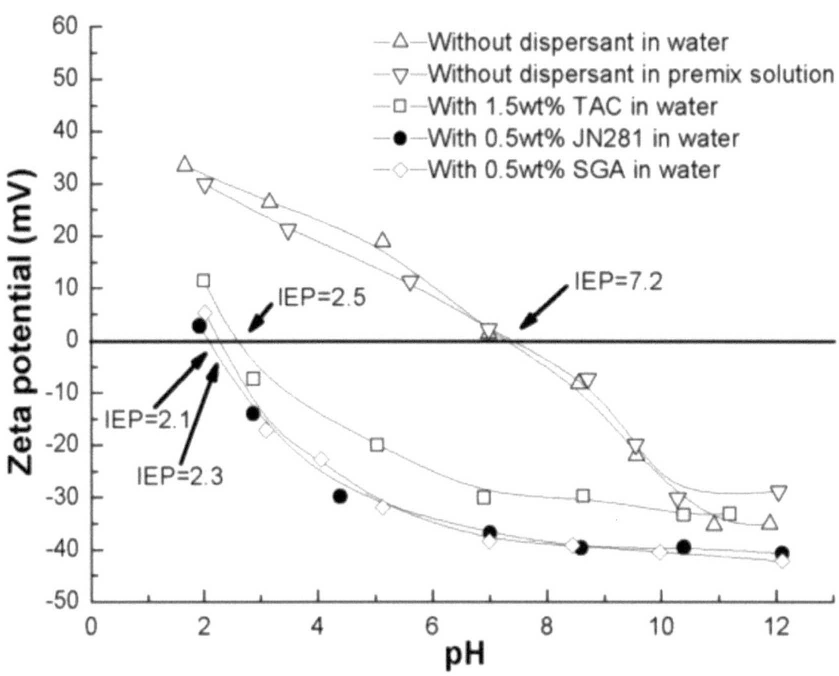

Figure 2. Effect of dispersants on the zeta potentials of the PZT suspensions at different pH values.

Polyelectrolyte dispersants are well known to be effective for various ceramic slurries due to both electrostatic and steric repulsions of the macromolecules. Here we show the effects of typical polyelectrolyte dispersants and a widely used organic surfactant triammonium citrate (TAC). The details are listed in Table 1. Electrostatic repulsion is dependent on the zeta potential (ζ) of the powders. The higher this value with the same polarity, the stronger the electrostatic repulsion between the particles. When close to the isoelectric point (IEP), the particles tend to flocculate. Zeta potentials of various PZT aqueous suspensions (0.06 vol % solids) at different pH values are shown in Figure 2. The zeta potential of pure PZT suspension changes from 33 mV at pH = 1.7 to - 35.1 mV at pH = 11.9 with an IEP at about pH = 7.2, suggesting that neutral environment is disadvantageous for good dispersion. Addition of monomer only slightly decreases the relative value of the potential and has little effect on IEP, indicating that the uncharged AM molecules screen the charge of the PZT particles. With the addition of TAC, JN281 and SGA the IEP is moved to pH = 2.5, 2.1 and 2.3, respectively. In the range of neutral environment to pH = 12, the zeta potential is almost constant. The higher absolute potential values of suspensions with JN281 and SGA than that with TAC imply that polyelectrolyte dispersants are more effective as far as the electrostatic repulsion is concerned.

Figure 3 shows the viscosity as a function of the shear rate for different PZT slurries solids loading. Adjustment of pH values to either acid or basic conditions has little effect, while addition of dispersants can greatly decrease the viscosity. The high viscosities at the beginning indicate a 'Bingham' type behavior, which is followed by a shear thinning at low shear rates. Shear-thinning behavior can be attributed to a certain kind of rearrangement of the relative spatial disposition of the particles. For concentrated suspensions of hard solid particles in Newtonian liquids, a flow-induced layered structure has been verified (Ackerson, 1990). Such a structure can provide a low resistance of the particle movement between different layers under the shear flow. For the systems containing dispersants, when the shear rates increase to a critical value (γ_c) a shear-thickening behavior appears, indicating that the flow-induced structure is destroyed. It is clear that the polyelectrolyte dispersants are much more effective than TAC. In addition, PZT slurries in the premix solution have almost the same viscosity values as those in the pure water, suggesting that addition of AM (15 wt.%) has little effect on the viscosity. These and the zeta potential results indicate that the polyelectrolyte dispersants work by both electrostatic and steric stabilization mechanism.

Generally, there is a complex nonlinear relationship between viscosity and solids volume fraction, which is closely related to many factors, including the continuous phase viscosity, particle-size distribution, and particle shape, etc. Influence of solids loading on the apparent viscosity is shown in Figure 4. At low solids loading the slurries show a low viscosity. The continuous addition of particles finally three-dimensional contact throughout the suspension, making flow impossible. The particular solid phase volume at which this happens is called the maximum packing fraction Φ_m (Barnes & Hutton, 1989). Before adding dispersant the PZT suspension has a measurable viscosity at a solids loading slightly smaller than 32 vol.%. Adding a little more PZT powders leads to a 'solidified' slurry whose viscosity is impossible to be measured by the normal rheometer. The Φ_m is thus determined to be 32 vol.%. Adding 0.8 and 1.5 wt.% of TAC increase Φ_m to about 47 and 53 vol.%, respectively. The lower Φ_m for a TAC concentration of 2.2 wt.% indicates that excess dispersant is harmful.

The relationship between the viscosity and the solids volume fraction for monodispersed suspension can be explained by the Krieger–Dougherty (K–D) model:

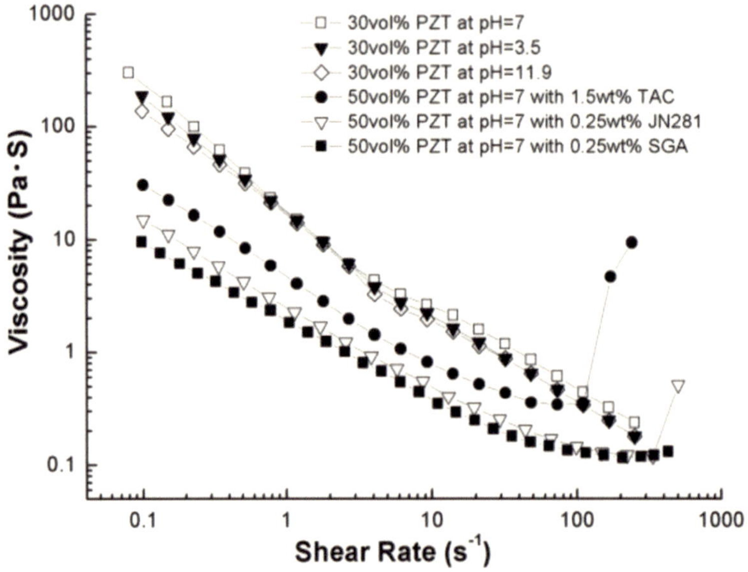

Figure 3. Influence of dispersants and pH value on the viscosity of the PZT suspensions

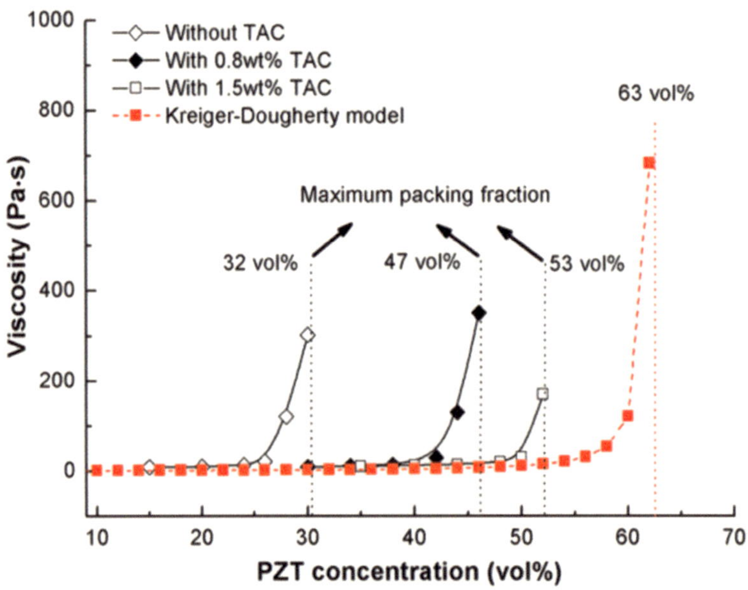

Figure 4. Influence of solids loading on the viscosities of the PZT suspensions

http://dx.doi.org/10.5772/53995

$$\eta = \eta_o(1 - \phi / \phi)^{-[\eta]\phi_m} \tag{1}$$

where η and η_o are the viscosity of the suspension and the solvent, respectively. The true volume fraction of the powder in the suspension is represented by f. The intrinsic viscosity [η] is a function of particle geometry; a value of 2.5 is suitable for spherical particles. A maximum packing fraction of 0.63±0.002 is suitable for random close packing at low shear rate. The K–D model curve with [η] = 2.5 and Φ_m=0.63 is plotted in the Figure. Although the experimental curves are somewhat deviated from the K–D model, they still show a similar shape: low viscosity at low solids loading and sharp increase at high solids loading. The discrepancy may partly be ascribed to the inhomogeneity of the particle size. In addition, particle flocculation will lead to a lower Φ_m because the flocs themselves are not close-packed (Starov et al., 2002) and they can trap part of the liquid phase, thus, leading to higher 'effective phase volume' and viscosity than those of the primary particles. The results clearly show the remarkable effect of dispersants in getting low viscosity ceramic slurry.

3. Electrical characterization and analysis of the doping effects

3.1. Electrical characterization for identifying the doping type

The piezoelectricity of PZT type materials originates from the displacement of Zr and Ti sublattices and the electrical properties of the materials can be dramastically affected by doping atoms. There are primarily two types of dopants for PZT, i.e. the donor type (soft type), and the acceptor type (hard type) (Jaffe et al., 1971). The former is mainly caused by substitution of higher valence ions for the A site Pb or B site Zr and Ti, and correspondingly higher piezoelectric coefficient (d_{33}), planar electromechanical coupling factor (K_p), relative permittivity (ε_r), loss tangent ($tg\delta$), P_r values and a lower mechanical quality factor (Q_m) value are obtained. The latter, which is caused by substitution of lower valence ions for the A or B site atoms, has contrary effects.

Sample	Average d_{33}	ε_r	tgδ	K_p
Dry pressed	469 pC/N	1430	0.0311	0.621
G-TAC	468 pC/N	1369	0.0325	0.633
G-JN281	479 pC/N	1352	0.0401	0.675
G-SP2	403 pC/N	1285	0.0250	0.591
G-SGA	330 pC/N	1244	0.0202	0.541

Table 2. Comparison of densities and some electrical parameters of dry pressed and gelcast PZT samples.

Some electrical parameters of different soft-doped PZT samples are compared in Table 2, where G-TAC, G-JN281, G-SP2 and G-SGA represent the best available gelcast PZT samples with TAC, JN281, SP2 and SGA as the dispersants, respectively. The ε and $tg\delta$ are 1 kHz data under room temperature. The dry pressed sample is obtained under a pressure of ~ 80MPa. G-TAC shows similar electrical properties with those of the dry pressed one. This indicates that the organic species used in gelcasting, including the dispersant TAC, the monomer and the cross-linker, etc., have almost no doping effects. Compared to the dry pressed sample, G-SP2 and G-SGA show higher Q_m and decreased d_{33}, K_p, ε, $tg\delta$ and P_r values, while change of these parameters for G-JN281 is to the contrary. Thus, the data in reveal that SP2 and SGA induced evident 'hard' doped characteristics, while JN281 induced 'soft' doped characteristics. Except for the dispersant species, the samples were prepared under the same conditions. Hence, the different properties should be mainly attributed to the dispersants used. A 'fingerpint' of hard doping effect is the increased Q_m (Damjanovic, 1998). Since Na^+ is the main metal cation in SP2 and SGA, we neglect other difference and check the correlation between Na concentration (inspected by X-ray Fluorescence) and Q_m. As shown in Figure 5, Q_m roughly shows an increasing trend with Na mole fraction except for G-JN281. Then we intentionally introduced Na into the G-JN281 by adding NaOH to the corresponding slurry to increase its pH value to ~13. As expected, a higher Q_m is indeed obtained after adding Na to G-JN281. These reveal that the Na^+ ion introduced by the dispersants has a stronge doping effect for gelcast PZT. From the valence and diameter (1.02Å) of Na^+, we deduce that it should substitute for the A site Pb^{2+} (1.18Å) as an acceptor dopant.

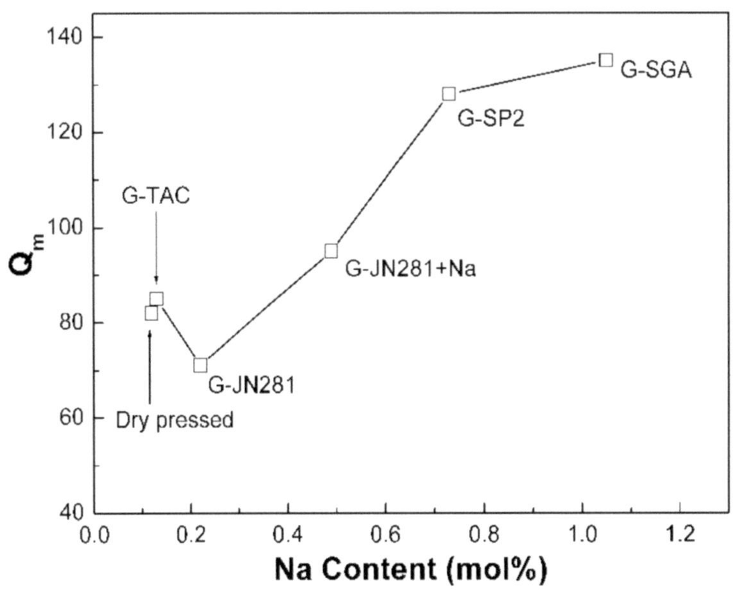

Figure 5. Illustration of the relationship between Na concentration (mole percentage in total metal elements) and Q_m of various PZT samples.

3.2. Complex impedance spectra

Figure 6(a), 6(b), 6(c) and 6(d) show the complex impedance spectra in the temperature range of 300~540 °C of the dry pressed sample, G-JN281, G-JN281+Na and G-SP2, respectively. The spectra of G-TAC is very similar with those of the dry pressed sample and thus not shown. The complex impedance spectra of all samples show only one Cole-Cole semicircle crossing the origin, which can be assigned with an equivalent circuit composed of a simple parallel RC element with a impedance that can be expressed as equation (2):

$$Z = \frac{R}{1+(i\omega RC)^{1-n}}, \tag{2}$$

where the parameter n characterizes the distribution width of the relaxation times around a mean value τ_0=RC (West et al., 1997). The semicircles of the samples show no obvious depression, i.e. the depression angle β (=nπ/2) between the real axis and the line from the high frequency intercept to the centre of the circle is close to 0. Thus, n is close to 0, indicating a debye-like behavior with a single relaxation time (West, 1997; Cao, 1990). Then based on equation (2) the following relationship about the real (Z′) and imaginary (Z″) parts of the impedance can be derived:

$$(Z' - R/2)^2 + Z''^2 = (R/2)^2. \tag{3}$$

The resistance R derived from the diameter of the semicircle, and the capacitance C can be derived from the angular frequency ω at the peak of the circular arc based on the relationship ωRC=1. The C values of the samples are in the order of 10^{-10}-10^{-9} F, which can be associated with the intrinsic ferroelectric bulk (or grain) response.

Using the R values, the logarithm of conductivity of different samples as a function of reciprocal temperature is plotted in Figure 6(e). The conductivity σ of the ferroelectric bulk phase shows an Arrhenius type behavior that can be described by equation (4):

$$\sigma = \sigma_o \exp(\frac{-E_s}{kT}), \tag{4}$$

where σ_0, E_g, k and T are the pre-exponential factor, the activation energy, the Boltzmann's constant and the absolute temperature, respectively. The logarithm of equation (4) gives equation (5):

$$\ln\sigma_T = \ln\sigma_o - \frac{E_s}{k}(\frac{1}{T}) = -2.30259 \log\rho_{T,} \tag{5}$$

where ρ_T is the resistivity. Based on the sample geometry, the E_g values associated with the intrinsic bulk phase conduction of the samples were derived by using equation (5). The data are compared in Table 3.

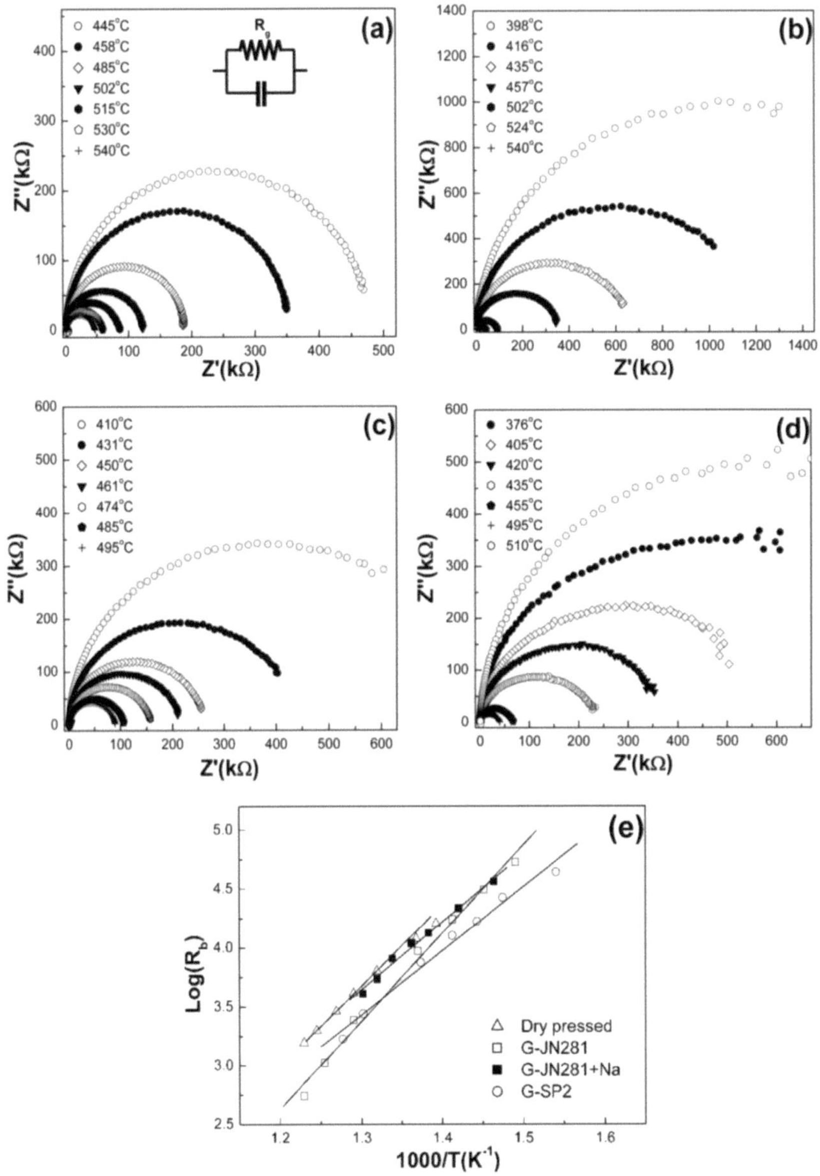

Figure 6. Complex impedance spectra of different PZT samples. (a) Dry pressed. (b). G-JN281. (c) G-JN281+Na. (d) G-SP2. (e) Arrhenius plot for the ferroelectric bulk resistivity of the samples derived from the impedance spectra.

Sample	Dry pressed	G-JN281	G-JN281+Na	G-SP2	G-SGA
E_g	1.24 eV	1.48 eV	1.14 eV	1.08 eV	1.07 eV

Table 3. Comparison of the conductivity activation energies of different PZT samples.

http://dx.doi.org/10.5772/53995

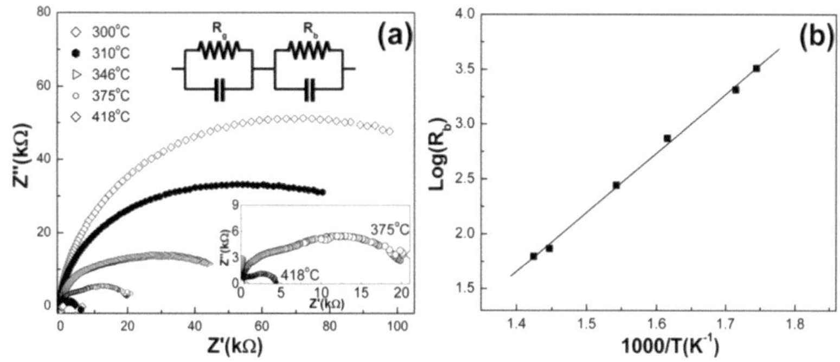

Figure 7. (a) Complex impedance spectra of G-SGA. The inset shows the magnified 375_0C and 418_0C curves. (b) Arrhenius plot of the ferroelectric bulk phase resistivity of the sample derived from the grain response of its impedance spectra.

The spectra of G-SGA shown in Figure 7(a), particularly the high temperature curves shown in the inset, consist of two semicircles, which can be assigned with an equivalent circuit composed of two parallel RC elements. This indicates the much different 'electrical microstructure'[14] of G-SGA, which has the highest Na concentration. The capacitance obtained by fitting the first semicircle can also be attributed to the ferroelectric bulk response, while the second semicircle may be attributed to the grain boundary response due to the higher capacitance. Actually, one semicircle spectrum may also appear if different responses in a sample overlap as a result of their similar relaxation time. A closer check of Figure 6(d) indicates that the one semicircle impedance spectra of G-SP2 have a slightly higher depression angle than other spectra in Figure 6, consistent with its higher Na concentration (see Figure 5). These imply that the grain boundary response is gradually magnified with increasing Na concentration. The Arrhenius plot of the conductivity of the bulk phase response of G-SGA is shown in Figure 7 (b). The derived E_g value is also listed in Table 3.

The loss of PbO through its volatility causes oxygen vacancies in PZT, leading to a p-type conductivity (Barranco et al., 1999). The acceptor Na^+ incorporated by the dispersants can substitute for Pb^{2+}, which can be expressed as following:

$$Na_2O=2Na_{pb}'+O_o+V_o \tag{6}$$

Thus, Na^+ ions replace Pb^{2+} ions and more oxygen vacancies are simultaneously created for charge compensation. Oxygen vacancies are the only lattice defects in the perovskite oxides that have a significant mobility, and the conductivity should be improved by the oxygen vacancy conduction mechanism via hopping of atoms in the oxygen octahedral network (Raymond & Smyth, 1996). This well explains the lower E_g data of G-SP2 and G-SGA than that of the dry pressed sample (Table 3), revealing again the hard doping effect of the Na-containing dispersants. The relatively higher activation energy of G-JN281 implies the presence of a certain donor-type impurity in G-JN281. Addition of NaOH into G-JN281 leads to a lower E_g value and further confirms the doping effect of Na.

3.3. Ferroelectric hysteresis loops

The ferroelectric (polarization-electric field) hysteresis loops of the PZT samples are shown in Figure 8. In Figure 8(a), G-SP2 and G-SGA show remnant polarization (P_r) of 32.7 μC/cm² and 16.7 μC/cm², respectively. Both values are smaller than the value of 39.8 μC/cm² of the dry pressed sample. The difference can also be interpreted by increased oxygen vacancies due to doping of Na. As afore mentioned, oxygen vacancies can move in the oxygen octahedral network. This may lead to a low stability of the $2Na_{pb}'$-V_o defect dipoles. The defect dipoles tend to orient themselves along the polarization direction, resulting in stabilized ferroelectric domains. A stabilized domain wall structure in turn give rise to more difficult poling and depoling (switching) and a smaller P_r (Warren, et al., 1996; Lambeck & Jonker, 1978). The round open loop of G-SP2 implies a higher leakage current. The hysteresis loops of G-JN281 and G-JN281+Na in Figure 8 (b) also clearly demonstrates the effect of Na: decreased P_r and a loop with a rounder shape.

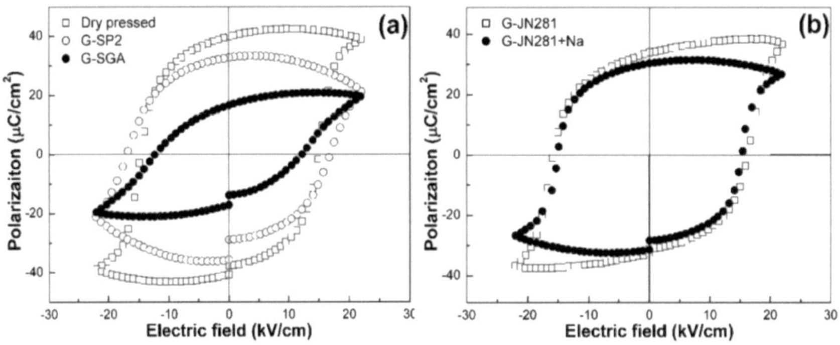

Figure 8. Ferroelectric hysteresis loops of the PZT samples.

4. Microstructural characterization

The XRD spectra of All samples (Figure 9) show typical perovskite structures (Soares et al., 2000; Hammer et al., 1998). The XRD patterns of G-TAC, G-SP2 and G-SGA are very similar to that of the dry pressed sample. While the peaks of the XRD pattern of G-JN281 shift very slightly to higher diffraction angles, indicating a contracted lattice cell. Considering the soft doped characteristics of G-JN281, the change in its XRD pattern may again be attributed to a certain unknown donor impurity ion introduced by the dispersant.

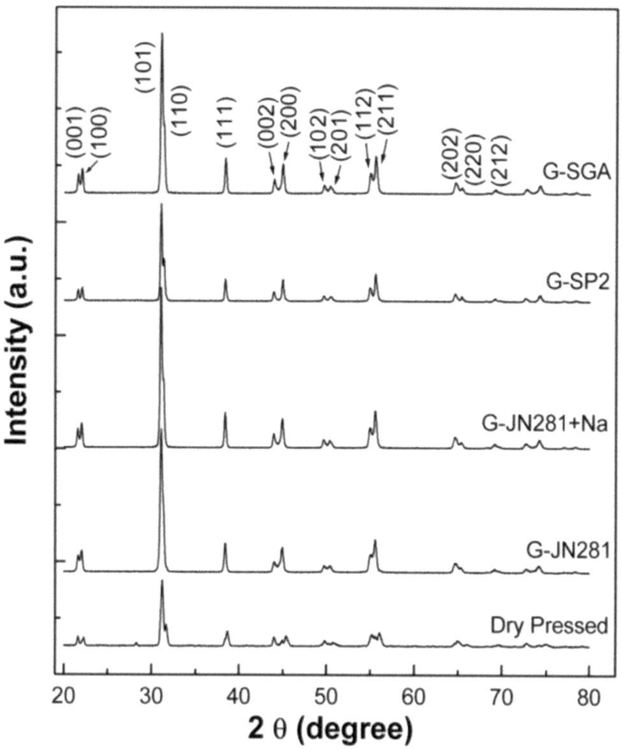

Figure 9. XRD patterns of different PZT samples.

As shown in Figure 10, although sintered under the same procedure, the samples exhibit rather different morphologies. SEM image of G-JN281+Na is not illustrated since it is very similar to that of G-JN281. Dry pressed sample shows an intergranular fracture surface and relative uniform grains with a diameter of 3~5 μm. G-JN281 shows a morphology very similar to that of the dry pressed one. G-SP2 also shows a basically intergranular fracture surface, but the grains are much larger with a diameter of 6~10 μm. Much different from other samples, G-SGA shows a transgranular fracture surface with the largest grains with a diameter of about 12 μm. Such a transgranular growth may result from chemical inhomogeneity and presence of intergranular phases or expanded grain boundary region. This is consistent with the two semicircle impedance spectra of G-SGA shown in Figure 7 (a).

In summary, electrical and structural characterization indicate that the Na^+ ion, which is the main cation in many widely used commercial dispersants (Xu et al., 1996; Tomasik et al., 2003), shows detrimental hard doping effects, leading to deteriorated electrical performance. Also, the impurity species introduced by dispersants or other additives may have a complicated influence on the electrical properties and microstructure of gelcast PZT samples. The results indicate that the specific doping effect, e.g. the change in electrical performance by the additives during processing, is a critical issue that should be paid special attention when applying gelcasting to the formation of electronic ceramics.

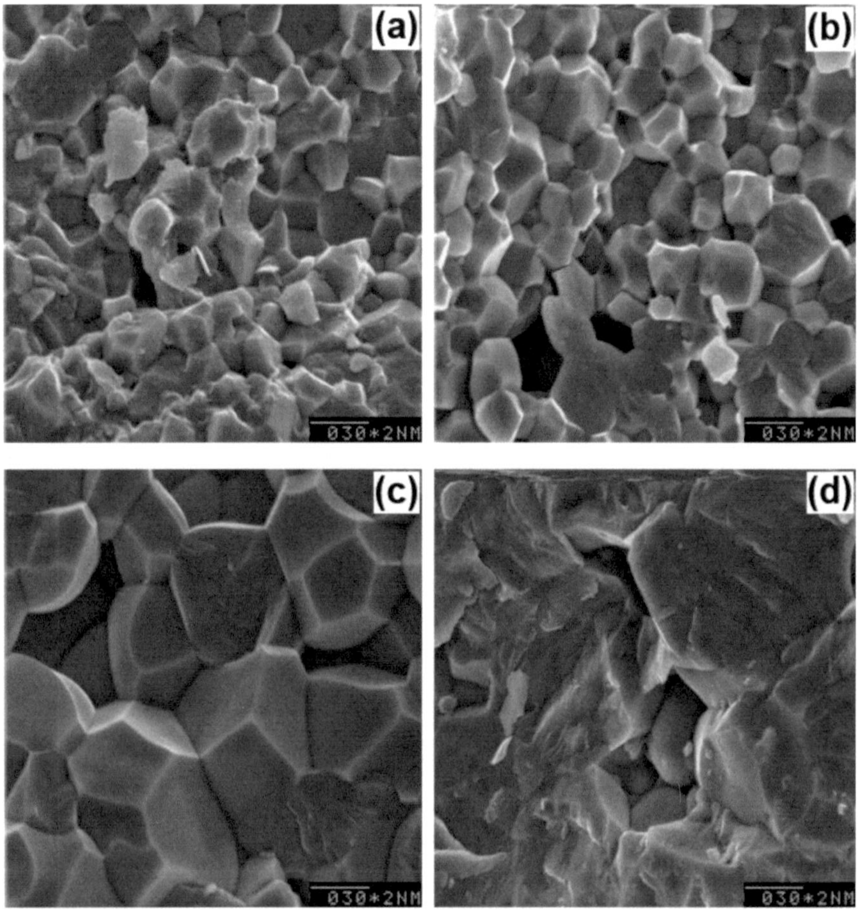

Figure 10. SEM images of the fracture surfaces of different PZT samples. (a) Dry pressed sample; (b) G-JN281; (c) G-SP2; (d) G-SGA.

5. Development of gelcasting for special-shaped PZT ceramics

Application of gelcasting to ferroelectric ceramics is not a mechanical copy of the technique to a different powder material. In addition to the demonstrated doping effect of the necessary additives, there are many other specific issues deserving further investigation. The many devices require the ferroelectric ceramic components to have various specific shapes (Scott, 2007). The shapes of the gelcast bodies are formed during gelation and drying process with the confinement of the molds. The two most critical issues about shape formation that have attracted more and more attention are probably the fracture (or crack growth) and deformation of the ceramics. No doubt that the former should be avoided, however, what is interesting is that the latter may even be used to form shapes that are difficult to formed by the molds. The two issues are briefly discussed in the following.

http://dx.doi.org/10.5772/53995

5.1. Factors affecting fracture and crack growth

Fracture or crack growth results from the competition of the strength of the gelcast body and the stress developed during gelation and drying. Many factors can affect fracture or crack growth, including the monomer concentration, monomer/cross-linker ratio, initiator concentration, initiator/catalyst ratio, gelation temperature and drying condition (humidity), etc (Ma et al., 2006). It is easy to be understood that too low a monomer concentration is insufficient to keep the three dimensional polymer network structure. A suitable monomer/cross-linker ratio is necessary to keep a strong gelcast body with still controllable stresses. Compared to other factors, initiator and catalyst seem to be more critical in controlling the gelcast body strength and stresses, since they can determine the 'joints' of the gel network. As shown in Figure 11. A slight decrease of the $(NH_4)_2S_2O_8$ concentration from the normal level causes a seriously broken green body. In addition to use of optimized premix solution, drying is a complicated process that should be carried out under contolled humidity (Barati et al., 2003).

Fracture or crack growth may also appear after sintering (Zheng et al., 2008), even if the shapes of the ceramics are well preserved after drying. During sintering the macromolecular gel network is destroyed and the stresses developed due to material densification may give rise to fracture or cracks in the ceramic body with decreased strength.

Figure 11. A fractured PZT-4 type green body caused by a lower initiator concentration

5.2. Deformation controllable gelcasting

Factors that affect crack growth discussed in the forgoing section also affect deformation. In most cases, deformation should be kept as smaller as possible. However, well controlled deformation can also be used to fabricate certain devices that use specially-shaped PZT as the active components. Here we show two examples: spherical PZT shell vibrator and PZT minitube.

As an effective non-invasive surgical tool, High intensity Focused sound (HIFU) has been used for the treatment of human tissues such as liver, kidney, breast, uterus and pancreas, and it is receiving growing interest (Wan, et al., 2008; Davies et al., 1998). As shown in Figure 12, some therapeutic HIFU transducers require to use spherical piezoelectric PZT shells as the active components (Saletes et al., 2011). Conventionally, such a spherical plate is fabricated by grinding both sides of a cylindrical plate. The mechanical method is time-consuming and wastes a lot of PZT material. Gelcasting of flat PZT plates generally uses a cylindrical ring mold and a flat bottom plate. During drying, the upper surface of the gelcast body is exposed to atmosphere while the other side is still berried until the whole body solidifies. Then we found an interesting phenomenon: the upper part of the gelcast body might dry and shrink first and a spherical plate was formed. By using suitable premix solution and under well controlled temperature and humidity, the plate edge bends up and a rather ideal spherically shaped PZT green plate can be formed. After sintering a spherical PZT plate was formed, which only needs to be grinded slightly to produce a homogenous spherical vibrator. Furthermore, because a lower humidity and a higher temperature can give rise to a larger deformation, the curvature radius can be controlled in a certain range under well controlled experimental conditions. So far the smallest focal length obtained for a plate with a diameter of 10 cm without further machining is around 15 cm. A PZT-8 plate with a diameter of 12 cm is shown in Figure 13.

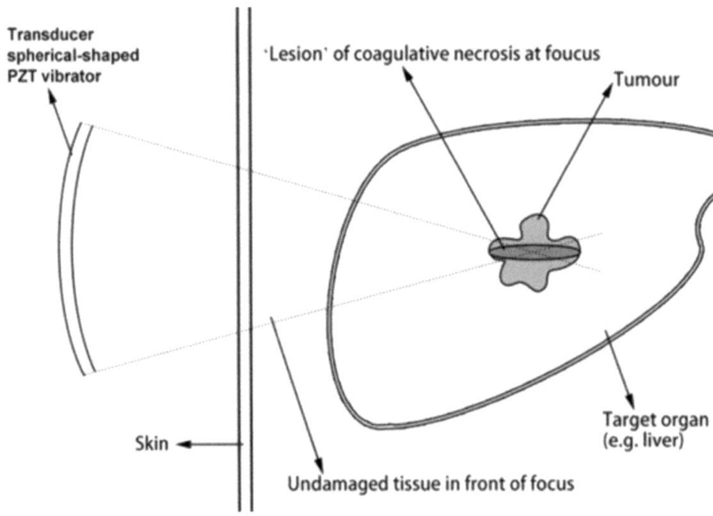

Figure 12. Schematic illustration of the therapeutic effect of a HIFU transducer

5.3. Fabrication of hollow spherical PZT shell

Thin-walled hollow sphere Omnidirectional Transducer has been used in hydrophones for many years (Li et al., 2010), which uses a hollow spherical PZT shell as the active component. Hollow spherical PZT shell is generally fabricated by Cold isostatic pressing, which requires complicated facilities. We show here that gelcasting can be developed to fabricate such Hollow spherical PZT shells by using a ball-shaped removable polymeric mold. First, the polymeric ball mold is fixed in a normal metal mold with a large spherical cavity, then the aqueous PZT premix slurry is poured into the mold. After gelation and drying, the whole mold containing the PZT green body and polymeric ball is removed by a careful thermal treatment. The most critical step in this method is thermal treatment, which solidifies the PZT slurry and melt the polymeric ball in due time. A PZT-4 hollow spherical shell with a focal length of ~ 200 mm, a diameter of ~ 120 mm and a wall thickness of 2 mm is also show in Figure 13.

Figure 13. Illustration of special-shaped gelcast PZT ceramics: spherical PZT plate and hollow spherical PZT shell.

In addition to the above mentioned advanced techniques, other gelcasting based ceramic fabrication or processing techniques have been developed in the authors' group, including spatter laser drilling techniques (Guo, 2004), and Rapid Prototyping of PZT bodies by combining gelcasting and Selective Laser Sintering (Guo, 2003), etc.

6. Conclusions

Comparison of various electrical properties and microstructures of various gelcast soft PZT samples with those of the dry pressed PZT one suggests that the Na^+ ion, which is the main cation in many widely used commercial dispersants, shows detrimental hard doping effects, leading to deteriorated electrical performance. The conclusion may be transferable to other doping ions like K^+, which is also contained in many commercial dispersants. Also due to the doping ions introduced by dispersants, the performance of the glecast PZT sample may also be improved as well. Considering that dispersants are indispensible in getting concentrated low viscosity ceramic slurries, the possible doping effect of metal ions or impurities introduced by the dispersants or other additives should be generally considered when applying gelcasting to forming multicomponent electronic ceramic materials whose electrical properties are sensitive to the composition.

In addition, we also demonstrate some advanced gelcasting techniques, including deformation controllable gelcasting of spherical PZT disc and gelcasting based hollow spherical PZT fabrication technique, etc. Application of gelcasting to PZT is not a mechanical copy of the technique to a different powder material. The results show that successful application of gelcasting to ferroelectric ceramics is not a mechanical copy of the technique to a different powder material.

Author details

Dong Guo[1] and Kai Cai[2]

1 Institute of Acoustics, Chinese Academy of Sciences, Beijing, China

2 Beijing Center for Chemical and Physical Analysis, Beijing Municipal Science and Technology Research Institute, Beijing, China

References

[1] Crsos, LE. (1996). Ferroelectric materials for electromechanical transducer applications. *Materials Chemistry and Physics*, 43, 108-115

[2] Setter, N.; Waser, R. (2000). Electroceramic materials. *Acta Materialia*, 48, 151-178

[3] Omatete, O.O.; Janney, M.A.; Strehlow, R.A. (1991). Gelcasting: a new ceramic forming process. *American Ceramic Society Bulletin*, 70, 1641-1649

[4] Biesheuvel, PM.; Verweij, H. (1999). Theory of Cast Formation in Electrophoretic Deposition. *Journal of the Amarican Ceramic Society*, 82, 1451-1455

[5] Novak, S.; Kosmac, T.; Krnel, K.; Drazic, G. (2002). Principles of the hydrolysis assisted solidification (HAS) process for forming ceramic bodies from aqueous suspension. *Journal of the European Ceramic Society*, 22, 289-295

[6] Graule, TJ.; Gauckler, LJ.; Baader, FH. (1996). Direct coagulation casting–a new green shaping technique.PT.1. processing principles. *Industrial Ceramics*, 16, 31-34

[7] Cai, K.; Guo, D.; Huang, Y.; Yang, JL. (2003). Solid freeform fabrication of alumina ceramic parts through a lost mould method. *Journal of the European Ceramic Society*, 23, 921-925

[8] Guo, D.; Li, LT.; Gui, ZL.; Nan, CW. (2003). B-Solid state materials for advanced technology. *Materials Science and Engineering*, 99, 25-28

[9] Young, A.C.; Omatete, O.O.; Janney, M. A.; Menchhofer, P.A.(1991). Gelcasting of Alumina. *Am.Ceram. Soc*, 74, 612-618

[10] Zhou, LJ.; Huang, Y.; Xie, ZP.; (2000). Gelcasting of concentrated aqueous silicon carbide suspension. *Journal of the European Ceramic Society*, 20, 85-90

[11] Huang, Y.; Ma, LG.; Tang, Q.; Yang, JL.; Xie, ZP.; Xu, XL. (2000). Surface oxidation to improve water-based gelcasting of silicon nitride. *Journal of Materials Science*, 35, 3519-3524

[12] Zha, SW.; Xia, CR.; Fang, XH.; Wang, HB.; Peng, DK.; Meng, GY. (2001). Processing and electrical properties of doped-$LaGaO_3$ by gelcasting. *Ceramics Intrernational*, 27, 649-654

[13] Bell, NS.; Voigt, JA.; Tuttle, BA. (2004). Colloidal processing of chemically prepared zinc oxide varistors. Part II: Near-net-shape forming and fired electrical properties. *Journal of Materials Research*, 19, 1341-1347

[14] Chen, W.; Kinemuchi, Y.; Watari, K.; Tamura, T.; Miwa, K. (2006). Preparation of Grain-Oriented $Sr_{0.5}Ba_{0.5}Nb_2O_6$ Ferroelectric Ceramics by Magnetic Alignment. *Journal o f the American Ceramic Society* , 89, 381-384

[15] Guo, D.; Cai, K.; Li, LT.; Gui, ZL. (2003). Application of gelcasting to the fabrication of piezoelectric ceramic parts. *Journal of the European Ceramic Society*, 23, 1131-1137

[16] Shaw, TM.; Trolier-McKinstry, S; McIntyre, PC.; (2000). The Properties of Ferroelectric Films at small dimensions. *Annual Review of Materials Science*, 30, 263-298

[17] Noheda, B.; Cox, DE. (2006). Bridging phases at the morphotropic boundaries of lead oxide solid solutions. *Phase Transitions*, 79, 5-20

[18] Israelachvili, J.; (1992). *Intermolecular & Surface Forces, 2nd ed.*

[19] Ackerson, B. J. (1990). Shear induced order and shear processing of model hard sphere suspensions. *J. Rheol*, 34, 553–590

[20] Barnes, H.A. (1989). Shear-thickening (dilatancy) in suspensions of nonaggregating solid particles dispersed in Newtonian liquids. *J. Rheol*, 33, 329–366

[21] Barnes, H.A.; Hutton,J.F.; Walers, K.; (1989). *An Introduction to Rheology, Elsevier, Oxford.*

[22] Starov, V.; Zhdanov, V.; Meireles M. et al. (2002) Viscosity of concentrated suspensions: influence of cluster formation. *Advances in Colloid and Interface Science* , 96, 279-293

[23] Jaffe, B.; Cook, W. (1971). *Piezoelectric Ceramics*. Academic Press,London/New York.

[24] Damjanovic, D. (1998). Ferroelectric, dielectric and piezoelectric properties of ferroelectric thin films and ceramics. Reports on Progress in Physics,61, 1267.

[25] West, A. R.; Sinclair, D. C. and Hirose, N. (1997). *J. Electroceram*, 65, 71

[26] Cao, W. Q.; and Gerhardt, R. (1990). *Solid State Ionics*, 42, 213

[27] Barranco, A. P.; Pinar, F. C.; Martinez, O. P. ; Guerra, J. D. ; and Carmenate,I. G. (1999). *J. Eur. Ceram. Soc*, 19, 2677

[28] Raymond, M. V.; Smyth, D. M. (1996). Defects and charge transport in perovskite ferroelectrics. *J. Phys. Chem. Solids*, 57, 1507-1511

[29] Warren, W.L.; Pike, G.E.; Vanheusden,K.; Dimos, D.; Tuttle, B. A.; Robertson,J. (1996). Defect-dipole alignment and tetragonal strain in ferroelectrics. *J. Appl. Phys*, 79, 9250-9257

[30] Lambeck, P. V.; Jonker,G. H. (1978). *Ferroelectrics*, 22, 729

[31] Soares, MR.; Senos, AMR.; Mantas, PQ. (2000). Phase coexistence region and dielectric properties of PZT ceramics. *J. Eur. Ceram. Soc*, 20, 321-334

[32] Hammer, M.; Monty, C.; Endriss, A. (1998). Correlation between surface texture and chemical composition in undoped, hard, and soft piezoelectric PZT ceramics. *Journal of the American Ceramic Society*, 81, 721-724

[33] Xu, ZH.; Ducker, W.; Israelachvili, J. (1996). *Langmuir*, 12, 2263-2270

[34] Tomasik, P.; Schilling, CH.; Jankowiak, R.; Kim, JC. (2003). The role of organic dispersants in aqueous alumina suspensions. *Journal of the European Ceramic Society*, 23, 913-919

[35] Scott, J. F. (2007). *Science*, 315, 954-959

[36] Ma, LG.; Huang, Y.; Yang, JL.; Le, HR.; Sun, Y. (2006). Control of the inner stresses in ceramic green bodies formed by gelcasting. *Ceramics International*, 32, 93-98

[37] Barati, A.; Kokabi, M.; Famili, N. (2003). Modeling of liquid desiccant drying method for gelcast ceramic parts. *Ceramics International*, 29, 199-207

[38] Zheng, ZP.; Zhou, DX.; Gong, SP. (2008). Studies of drying and sintering characteristics of gelcast $BaTiO_3$-based ceramic parts. *Ceramics International*, 34, 551-555

[39] Wan, Yayun.; Ebbini, Emad S. (2008). Transactions on Ultrasonics Ferroelectrics and frequency Control. *IEEE* 55, 1705-1718

[40] Davies, BL.; Chauhan, S.; Lowe, M. (1998). International Conference on Medical Image Computing and Computer-Assisted Intervention. *Lecture Notesin Computer Science*, 1496, 386-396

[41] Saletes, Izella.; Gilles, Bruno.; Bera, Jean-Christophe. (2011). Promoting inertial cavitation by nonlinear frequency mixing in a bifrequency focused ultrasound beam. *Ultrasonics*,51, 94-101

[42] Li, XF.; Peng, XL.; Lee, KY. (2010). The static response of functionally graded radially polarized piezoelectric spherical shells as sensors and actuators. *Smart Materials & Structures*, 19, 3

[43] Guo, D.; Li, LT.; Cai, K.; Gui, ZL.; Nan, CW. (2004). Rapid Prototyping of Piezoelectric Ceramics via Selective Laser Sintering and Gelcasting. *Journal of the American Ceramic Society*, 87, 17-22

[44] Guo, D.; Cai, K.; Huang, Y.; Li, LT. (2003). A novel anti-spatter and anti-crack laser drilling technique: application to ceramics. *Applied Physics A:materials Science & Processing* , 76, 1121-1124

Chapter 11

Electronic Ferroelectricity in II-VI Semiconductor ZnO

Akira Onodera and Masaki Takesada

Additional information is available at the end of the chapter

http://dx.doi.org/10.5772/52304

1. Introduction

II-VI semiconductor Zinc oxide (ZnO) is a well-known electronic material [1-3]. Because of large piezoelectric constant and electromechanical-coupling constant, ZnO has been applied to ultrasonic transducer, SAW filter, gas sensor *etc* [3-5]. In addition, ZnO has been used widely as pigment and UV cut cosmetics extensively, so it is a safe material for our living environment, compared with heavy metals used in semiconducting process and materials science. Recently, ZnO has been studied as a material for the solar cell, transparent conductors formed on liquid crystal displays, and the blue laser [6]. Then, *p*-type ZnO was found by Joseph *et al* [7].

Ferroelectricity is recognized to appear mainly by the delicate balance between a long-range dipole-dipole interaction and a short-range interaction. When electrons should be well localized in dielectrics, the electronic distribution in the unit cell is determined by atomic positions of constituent ions. Slight distortion of electronic distribution due to structural changes in dielectrics gives a rise of dipole moments. From this point of view, the ferroelectric phase transition should be understood as a structural phase transition from the paraelectric phase with high symmetry to the ferroelectric phase with low symmetry. The atomic displacements are generally 0.01~0.1 Å, which are small compared with Bohr radius (0.53 Å). Therefore ferroelectrics are considered as a group of materials which are sensitive to structural changes. In this sense, the ferroelectric phase transition is classified into one of the structural phase transitions.

The basic Hamiltonian is generally given as

H= H(phonon) + H(electron) + H(electron-phonon),

where H(phonon) and H(electron) are due to motions of ion cores and valence electrons, and H(electron-phonon) presents interactions between ions and valence electrons. In dielectric materials, the contribution of electron systems is usually omitted since the band gap is generally wide. However, it should be necessary to consider the electron-phonon interaction in the case of ferroelectric semiconductors, since the correlation energy of dipole-dipole interaction (about 0.2 eV) is comparable with band-gap energy in some narrow-gap and wide-gap semiconductors. For this type of compounds, the electronic contribution due to bond charges and conduction electrons should play a key role and must be taken into account for understanding the nature of ferroelectricity. For long time, the importance of electronic contribution has been pointed out in the field of ferroelectrics, since the simple superposition of electronic polarizability does not hold in many ferroelectric substances. Recently, the novel ferroelectricity was discovered in narrow-gap and wide-gap semiconductors such as $Pb_{1-x}Ge_xTe$ [8], $Cd_{1-x}Zn_xTe$ [9] and $Zn_{1-x}Li_xO$ [10].The appearance of ferroelectricity is primarily due to electronic origin in $Zn_{1-x}Li_xO$. Although the ferroelectric phase transition accompanies with structural distortions in usual ferroelectrics, only small structural changes of the order of 10^{-3} Å are observed in $Zn_{1-x}Li_xO$. The change in *d-p* hybridization caused by Li-substitution is responsible for the novel ferroelectricity and dielectric properties. In this chapter, we summarized recent works mainly onZnO.

2. Zinc oxide

The crystal Structure of ZnO is wurtzite-type ($P6_3mc$) as shown in Fig. 1, which belongs to hexagonal system. It does not have the center of symmetry, and is polar along the *c*-axis. Although it has been pointed out that the structure of pure ZnO has the possibility to exhibit ferroelectricity, the polarization switching does not observed until its melting point (1975°C) because of large activation energy accompanied by dipole switching process. The lattice constants are a = 3.249858 Å, c= 5.206619 Å at room temperature (298 K) [11]. Zn and O ions are bonded tetrahedrally and form ZnO_4 groups. This tetrahedron is not perfect: the apical bond length of Zn-O is 1.992 Å (parallel to the *c*-axis) and the basal one is 1.973 Å. Bond length along the *c*-axis is longer than other three bonds by 0.96%: the ZnO_4tetrahedra distort along the *c*-axis, which result in dipole moment. According to recent first-principles studies, it was reported that the hybridization between the Zn 3*d*-electron and the O 2*p*-electron plays an important role for dielectric properties of ZnO [12, 13].The energy gap E_g is 3.44 eV [14]. Although stoichiometric ZnO is an insulator (intrinsic semiconductor), it exhibits *n*-type conductivity because of excess Zn atoms. The resistivity is about 300 Ωcm, which, however, changes drastically by doping various impurities. By doping of trivalent ions, such as Al^{3+}, In^{3+}, Ga^{3+}, it reduces to the order of 10^{-4}Ωcm and shows the conductivity near that of metals [15].

Although *p*-type conductivity is expected by doping of monovalent Li^+ ions, ZnO becomes an insulator and the resistivity increases to as much as 10^{10}Ωcm [16-18]. It is remarkable that the change in resistivity reaches over the order of 10^{14}, which covers from metal-like to insulating region. It suggests that the physical nature in this material changes drastically by a little amount of dopant.

http://dx.doi.org/10.5772/52304

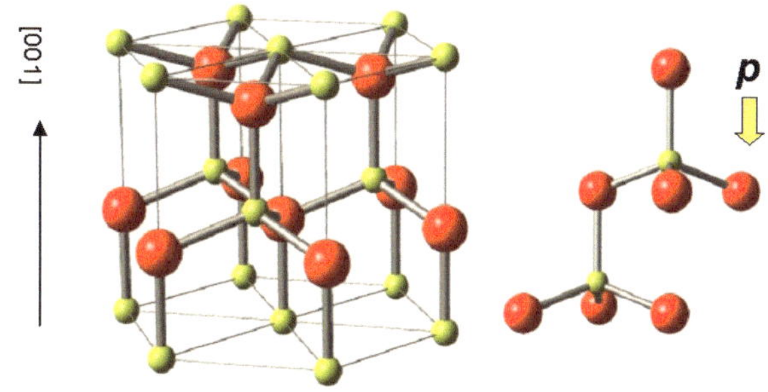

Figure 1. Crystal structure of ZnO (Wurtzite structure).

3. Electronic structure

Electronic structure of ZnO has been studied by LMTO (linear muffin-tin orbital) method [19] and by LAPW (linearized augmented-plane-wave) method using LDA (local density approximation). Band structure and the density of states is shown in Figs. 2 and 3 [20]. Around -17 eV, there are two bands originating from the O 2*s*-states. The narrow bands between -6 and -4 eV consist mainly of the Zn 3*d*-orbitals, and moderately dispersive bands from -4 to 0 eV consist mainly of the O 2*p*-orbitals. Figure 4 shows significant *d*-*p* hybridization. The 3*d* derived bands split into two groups, leading to double-peak structure in DOS (density of states). The lower peak is characterized by a strong *d*-*p* hybridization. The sharp upper peak between -4.8 and -4.2 eV has strong Zn 3*d* character and the hybridization with the O 2*p*-state is very small. Band gap is 0.77 eV. Discrepancy with the experimental value is larger than that of LMTO. It is due to the LDA gap error [21].

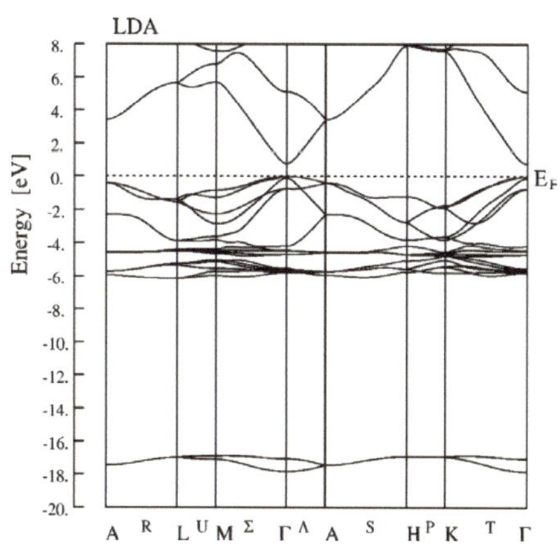

Figure 2. Band structure of ZnO using LAPW.

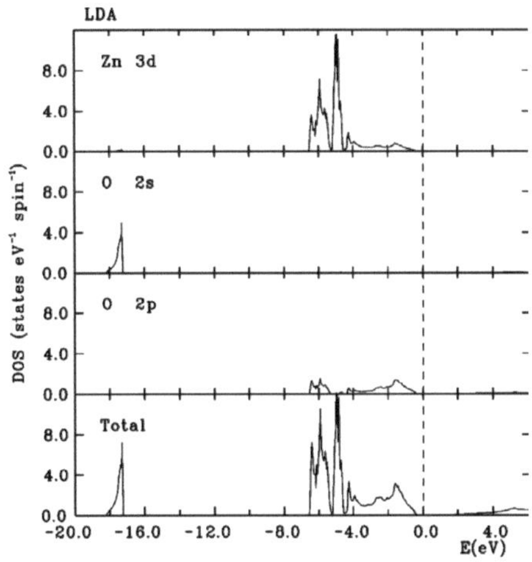

Figure 3. Density of states of ZnO using LAPW.

4. Ferroelectricity in binary semiconductors

Recently, the novel ferroelectricity was found in Li-doped ZnO [10, 22, 23], although pure ZnO has not shown any evidence of ferroelectricity. Generally representative ferroelectrics have complicate crystal structures such as Rochelle salt ($KNaC_4H_4O_6$ $4H_2O$), TGS (tri-glycine sulfate, $(NH_3CH_2COOH)_3H_2SO_4$) and $BaTiO_3$. Since ZnO is a simple binary compound, it is very convenient to study the microscopic mechanism and its electronic contribution for the appearance of ferroelectricity. Besides ZnO, IV-VI narrow-gap semiconductor $Pb_{1-x}Ge_xTe$ and II-VI wide-gap semiconductor $Cd_{1-x}Zn_xTe$, are known as materials of binary crystals accompanying ferroelectricity.

4.1. IV-VI Narrow-gap semiconductor

Among IV-VI semiconductor, $Pb_{1-x}Ge_xTe$ has been investigated extensively about its ferroelectricity [8]. $Pb_{1-x}Ge_xTe$ has a NaCl (rock-salt) type structure at room temperature. Although any ferroelectric activities have not been observed in pure PbTe and GeTe crystals, the ferroelectric phase transition is induced only in solid solution $Pb_{1-x}Ge_xTe$. In the low temperature phase of $Pb_{1-x}Ge_xTe$, the crystal becomes rhombohedral and exhibits ferroelectric activity. In the case of $x = 0.003$, it shows a large dielectric anomaly of the order of 10^3 at T = 100 K as shown in Fig. 5 and the softening of TO mode was observed.

The energy gap of $Pb_{1-x}Ge_xTe$ is no more than 0.3 eV, which is 3000 K in temperature. This value is comparable to the energy of the Lorentz field of dielectrics, $(4\pi/3)P$. Therefore, conduction electrons can couple strongly with phonons in this solid solution. The electron-phonon interaction decreases the frequency of TO phonon mode which results in the ferroelectric phase transition. In $Pb_{1-x}Ge_xTe$, doped Ge ions displace from the center position of Pb ion by 0.8 Å and behave as *off-center ions* because of the ionic size-mismatch between Pb ion (the ionic radius: 1.2 Å) and Ge ion (the ionic radius: 0.73 Å) [24]. The direction of displacement is the eight equivalent [111] directions, which is spatially vacant and polar in the rock-salt structure. Above T_c, the ions shift toward any one of these directions at random, but its displacement would be ordered along the trigonal axis below T_c. The ordering of the *off-center ion* triggers the softening of TO phonon mode and induces rhombohedral distortion in the way such as $\cdots Pb^{2+}$-$Te^{2-}\cdots Pb^{2+}$-$Te^{2-}\cdots Pb^{2+}$-$Te^{2-}\cdots$or $\cdots Te^{2-}$-$Pb^{2+}\cdots Te^{2-}$- $Pb^{2+}\cdots$ Te^{2-}-$Pb^{2+}\cdots$. This structural ordering generates spontaneous polarization. However, in $Pb_{1-x}Ge_xTe$, the *D-E* hysteresis loop has not been observed, while it is a direct evidence of the ferroelectricity. Whereas traditional dielectric measurements are performed on a parallel-plate capacitor, it is difficult to measure dielectric constant of such lossy $Pb_{1-x}Ge_xTe$ because of current leaks. Therefore dielectric constant is determined by the *C-V* measurement or the optical reflectiveity, which are commonly used in semiconductors. The large dielectric anomaly and soft mode of $Pb_{1-x}Ge_xTe$ suggest the ferroelectric activity.

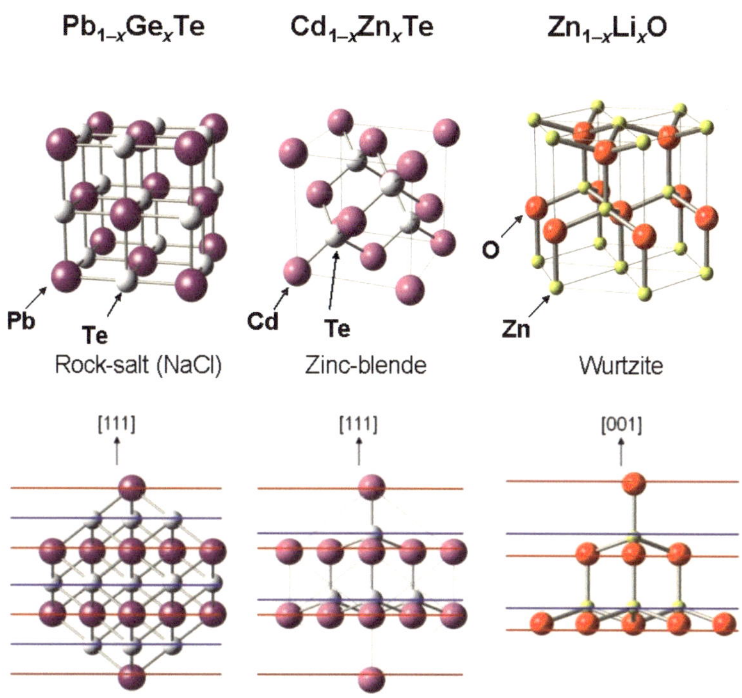

Figure 4. Crystal structures of $Pb_{1-x}Ge_xTe$, $Cd_{1-x}Zn_xTe$ and $Zn_{1-x}Li_xO$. The lower figures are plots along the polar [111] directions for $Pb_{1-x}Ge_xTe$ and $Cd_{1-x}Zn_xT$, and polar [0 0 1] direction for $Zn_{1-x}Li_xO$.

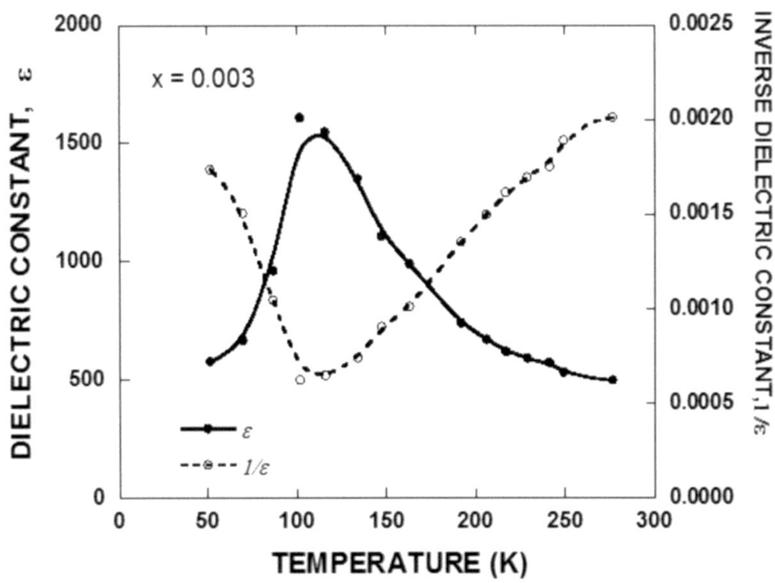

Figure 5. Temperature dependence of dielectric constant in $Pb_{1-x}Ge_xTe$ (x=0.003).

4.2. II-VI Wide-gap semiconductor $Cd_{1-x}Zn_xTe$

Among many wide-gap semiconductors, the ferroelectricity of $Cd_{1-x}Zn_xTe$ was discovered by R. Weil *et al* in 1988 [9, 25]. The E_g of $Cd_{1-x}Zn_xTe$ is 1.53 eV. It is a wide-gap semiconductor unlike the narrow-gap IV-VI $Pb_{1-x}Ge_xTe$. The crystal structure is cubic zinc-blende-type. Cd(Zn) ion is surrounded by four Te ions tetrahedrally. The center of symmetry in this compound does not exist. When x=0.1, the crystal exhibits a dielectric anomaly at 393 K as shoen in Fig. 6. The peak value of dielectric constant (ε_{peak}) is only 50, which is smaller by about two orders than that of $BaTiO_3$ (~14000) and that of $Pb_{1-x}Ge_xTe$ (~1000). Because of the large E_g and the resistivity (~kΩ), a ferroelectric *D-E* hysteresis loop was successfully observed in the low-temperature phase. The direction of spontaneous polarization is along the apex of a tetrahedron, [111], which is reported as P_s= 0.0035 μC/cm^2 in their first paper, and 5 μC/cm^2 in the second paper. Because the ionic radius of Zn ion (0.83 Å) is smaller than that of Cd ion (1.03 Å), Zn ion also locates at the *off-center* position which deviates from the center of tetrahedra toward the apex. According to the result of EXAFS (X-ray absorption fine structure), it shifts by 0.04 Å [26]. It is considered that the ordering of *off-center ions* causes rhombohedral distortion of the cubic lattice by about 0.01 Å, and induces ferroelectricity.

In $Cd_{1-x}Zn_xTe$, the soft mode has not been observed and the dielectric anomaly is small. These dielectric properties are different from those found in $Pb_{1-x}Ge_xTe$, which are summarized in Table 1. These facts are consistent each other when one consider the LST (Lyddane-Sachs-Teller) relation ($\omega_{LO}{}^2/\omega_{TO}{}^2 = \varepsilon/\varepsilon_\infty$). In $Pb_{1-x}Ge_xTe$, the existence of soft mode is responsible for large dielectric anomaly. Although the behavior of *off-center ions* plays an important role in this ferroelectricity like $Pb_{1-x}Ge_xTe$, the occurrence of phase transition seems to be driven in different way from that of $Pb_{1-x}Ge_xTe$, because of the different dielectric properties described above. Furthermore, the band gap is much larger than that of the narrow-gap ferroelectric semiconductors, which suggests that the electron-phonon coupling is not so strong in $Cd_{1-x}Zn_xTe$ solid solutions.

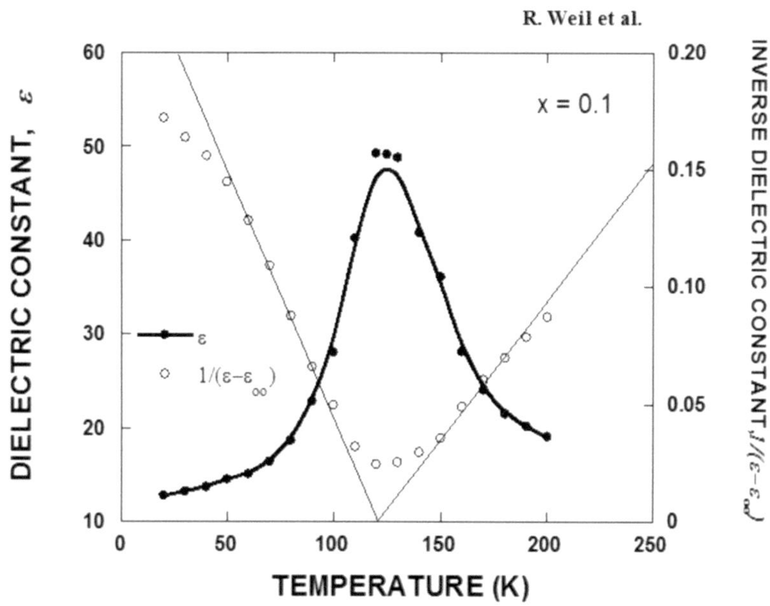

Figure 6. Temperature dependence of dielectric constant in $Cd_{1-x}Zn_xTe$ (x=0.1).

4.3. $Zn_{1-x}Li_xO$

Dielectric constants of $Zn_{1-x}Li_xO$ ceramics with x=0.09 show an anomaly at 470 K (T_c) (Fig. 7), although a high purity of ZnO does not show any anomaly from 20 K to 700 K [27]. The peak value of dielectric anomaly is 21 (x=0.1), which is the same order with $Cd_{1-x}Zn_xTe$ ($\varepsilon \sim 50$), but much smaller than ordinal ferroelectrics by 2~4 orders. This means that the dipole-dipole correlation accompanied with this ferroelectric phase transition is not so large in $Zn_{1-x}Li_xO$. In the measurement of *D-E* hysteresis loop, a small and clear hysteresis curve was observed. The spontaneous polarization varies by samples from 0.05 μC/cm² to 0.59 μC/cm². This is due to that the preferred orientation of the *c*-axis, the direction of P_s of the samples varies depending on samples. Naturally, powder of ZnO has egg-shaped grains which are elongated along the *c*-axis due to the anisotropy of elastic constants. Therefore, the orientation of the ceramic sample depends on the condition of the pressing process in the sample preparation. After the correction of the preferred orientation by using X-ray diffraction, the value of spontaneous polarizations of each sample converges to 0.9 μC/cm² as shown in Fig. 8 [28, 29]. The transition temperature, T_c, depends on the Li concentration. The phase diagram between T_c and x is shown in Fig. 9. As temperature increases, T_c becomes higher,which reminds us the phase diagram of quantum ferroelectrics such as $KTa_{1-x}Nb_xO_3$ and $Sr_{1-x}Ca_xTiO_3$.

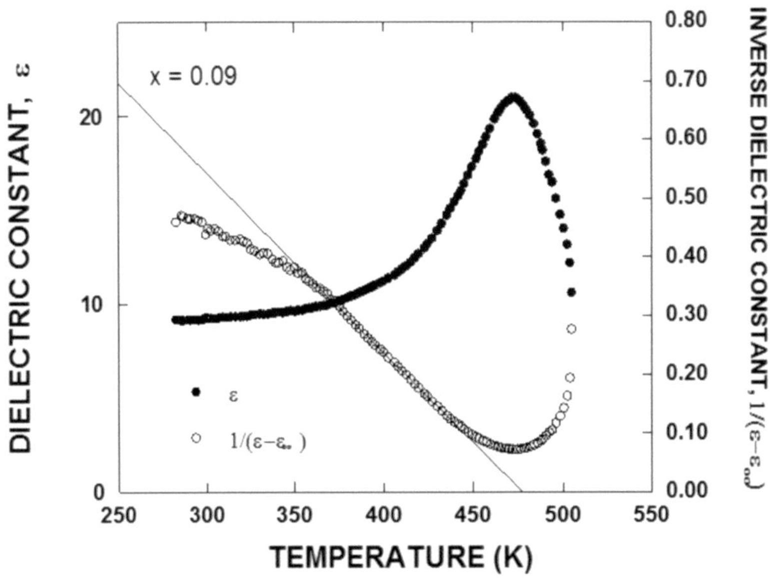

Figure 7. Temperature dependence of dielectric constant of $Zn_{1-x}Li_xO$ ceramics (x= 0.09).

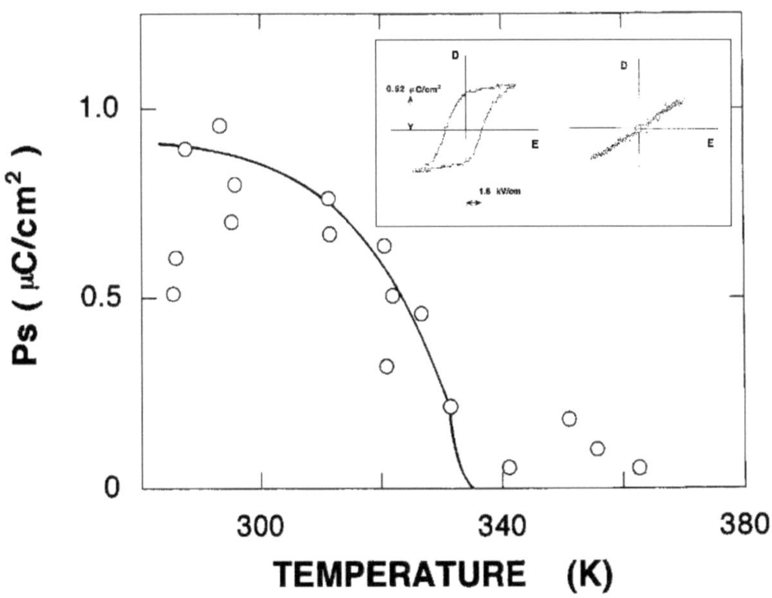

Figure 8. Temperature dependence of spontaneous polarization of $Zn_{1-x}Li_xO$ ceramics(x = 0.06). The inset is *D-E*hyteresis loop observed below (left) and above T_c (right).

http://dx.doi.org/10.5772/52304

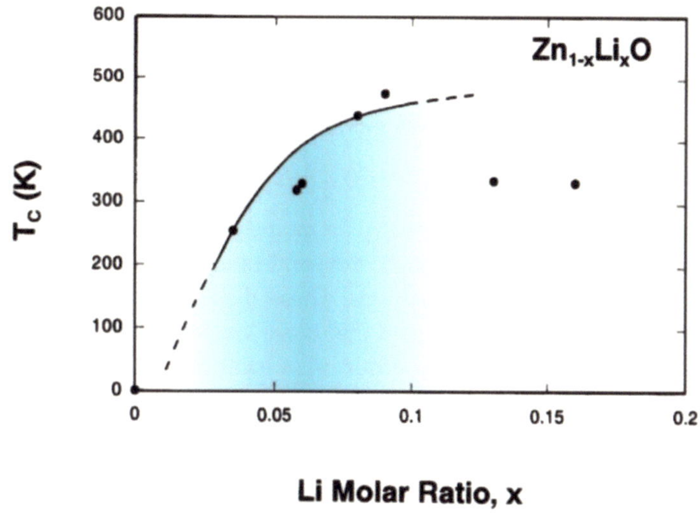

Figure 9. The T_c-x phase diagram of $Zn_{1-x}Li_xO$.

Specific heat anomaly is small at the ferroelectric-paraelectric phase transition temperature. The transition entropy ΔS is almost 0. If Li ions which substitute Zn ions occupy the off-center position and behave an order-disorder motion, ΔS must be about $R\ln 2$. If the nature of the transition is displacive type, $\Delta S \doteq 0$. The observed thermal behavior suggests displacive nature, but, however, no soft mode was observed in Raman scattering measurement [30]. The small dielectric anomaly and the absence of soft mode are consistent each other, when we consider the LST-relation. Relaxation mode corresponding to an order-disorder motion has not been found also by Raman scattering [31]. These peculiar dielectric properties of $Zn_{1-x}Li_xO$ is resemble to those of $Cd_{1-x}Zn_xTe$. These evidences suggests a new type of ferroelectric phase transition.

	$Pb_{1-x}Ge_xTe$	$Cd_{1-x}Zn_xTe$	$Zn_{1-x}Li_xO$
Crystal Structure of Paraelectric Phase	Rock-salt (NaCl) (($Fm3m$))	Zinc-blende (($F\bar{4}3m$))	Wurtzite (($P6_3mc$))
Ferroelectric Phase	Rhombohedral (($R3m$))	Rhombohedral (($R3m$))	Wurtzite (($P6_3mc$))
T_c (K) (x=0.1)	200	390	470
E_g (eV)	0.3	1.53	3.2
ρ (Ωm)	10	10^3	10^{10}
P_s (μC/cm²)	-	5	0.9
Soft mode	○	×	×
ε_{peak}	10^3	50	21

Table 1. Dielectric properties of binary ferroelectric semiconductors.

5. Electronic ferroelectricity in ZnO

It is considered that the replacement of host Zn atoms by substitutional Li atoms plays an primary role for the appearance of ferroelectricity in ZnO. The problem is the effect of Li-doping. Here we consider the following two models [32].

i. Structural size-mismatch model

Pure ZnO is polar along the *c*-axis and has dipole moments in crystals. Because of the size-mismatch between Zn ion (the ionic radius: 0.74 Å) and Li ion (the ionic radius: 0.60 Å), substituted Li ions displace from the Zn positions. These displacements may force to induce extra dipole moments locally. It is considered that these local dipoles couple with dipoles of the mother ZnO crystal and the structural ordering of local dipoles triggers to induce a ferroelectric phase transition.

ii. Electronic model

According to the first-principle study by Corso *et al* [10], the hybridization between Zn 3*d* electron and O 2*p* electron plays an important role in dielectric properties of ZnO. As the Li atom has no *d*-electrons, it is considered that the partial replacement of Zn ions by Li ions changes the nature of d the *d-p* hybridization. Therefore this doping is considered to induce local extra dipole moments. These extra local dipoles couple with parent dipoles and induce a new type of ferroelectricity.

To clarify which model is effective in the appearance of ferroelectricity in ZnO, the following structural and dielectric measurements were performed.

5.1. Rietveld analysis of structural changes in $Zn_{1-x}Li_xO$ ceramics

Structural changes associated with Li-substitution were studied by X-ray powder diffraction [33]. Ceramic samples of $Zn_{1-x}Li_xO$ prepared by using a SPS (spark plasma sintering) method were used in this experiment. The nominal value of x is 0.1 for the first sample. The T_c of the sample was 470 K which is determined by dielectric constant measurements. The systematic check of the absence of possible reflections (*h-k*=3*n* and *l*= odd for (*hkl*), n = integer) suggests that the space group of the ferroelectric phase still remains $P6_3mc$, which is same as pure ZnO. The obtained pattern was analyzed by the Rietveld method including the Li concentration, x, as a refinable parameter. All parameters refined are shown in Table 2. The lattice constants are a = 3.2487(1) and c = 5.2050(1) Å at 293 K. The actual Li concentration x is determined to be 0.09. The final discrepancy factors were R_{wp}=7.4%, R_p=5.4% and R_I=5.9%. The Zn-O bond lengths are 1.988_7 Å along the *c*-axis and 1.973_5Å in the basal plane, respectively. This means that the bond length along the *c*-axis shows a decrease by an amount of 0.003 Å by Li-substitution in Li-doped ZnO. This lattice distortion is the order of 10^{-3} Å only, while the displacement of Ti ions is the order of 0.1 Å in well-known ferroelectric $BaTiO_3$. In this phase transition, the crystal symmetry does not change associated with Li-substitution and structural change is considerably small compared with the structural phase transition in typical ferroelectrics.

Atom	*x*	*y*	*z*	*B*	Occupation
Zn	1/3	2/3	0.3821(3)	0.52	0.91
Li	1/3	2/3	03821(3)	0.31	0.09
O	1/3	2/3	0	0.40	1.0

Table 2. Positional (*x,y,z*) and isotropic thermal (*B*) parameters in $Zn_{1-x}Li_xO$ at 293K.

5.2. Effect of Be and Mg dopants on T_c

The effect of dopants on T_c was studied based on the two models mentioned above. If the ionic size-mismatch between the substituted and the host ions is important for ferroelectricity, the introduction of Be ions should be more effective than Li and Mg doping. The ionic radii of Li^+, Be^{2+} and Mg^{2+} ions are 0.60 Å, 0.3 Å and 0.65 Å, respectively. If the changes in electronic configuration are important, the Mg^{2+} ion ($1s^2 2s^2 2p^6$) should play a different role from the isoelectronic Li^+ and Be^{2+} ions ($1s^2$). Dielectric constant measurements were done on Mg- and Be-doped ZnO ceramics [32]. A dielectric anomaly of $Zn_{0.9}Be_{0.1}O$ was found at 496 K, which is very similar to those observed in $Zn_{1-x}Li_xO$. However, $Zn_{1-x}Mg_xO$ samples show a clear decrease of T_c. The dielectric anomaly in 30% Mg-doped ZnO found at 345 K is 150 K lower than the Li-doped one. The concentration dependence of T_c of Be- and Mg-doped ZnO is summarized in Fig. 10. The series of dielectric measurements show that the introduction of Mg^{2+} suppresses T_c, while isoelectronic Be^{2+} shows almost the same T_c.

It is considered that the appearance of ferroelectricity in ZnO is primarily due to electronic origin. The change in *d-p* hybridization caused by Li-substitution is responsible for this novel ferroelectricity and dielectric properties.

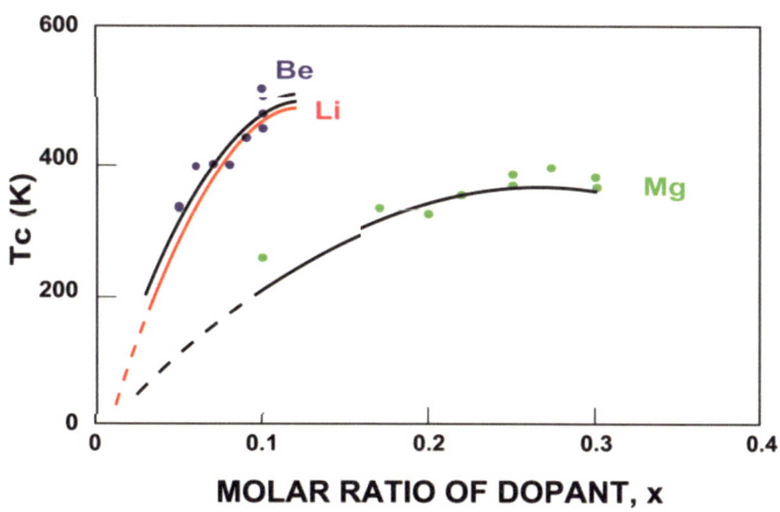

Figure 10. T_c *vs.* molar concentrations(*x*)of dopants, Li, Be and Mg in ZnO.

In the next section, we shall study the change in electronic distribution, especially in the nature of *d-p* hybridization by Li-doping directly by X-ray diffraction, and discuss new electronic ferroelectricity in ZnO.

6. X-ray study of electronic density distribution

X-ray diffraction measurements were performed for single crystals of pure ZnO and $Zn_{1-x}Li_xO$ at 297 K and 19 K in order to investigate the changes in the crystal structure and the electronic density distribution by Li-substitution in detail. Single crystal of pure ZnO was prepared by the hydrothermal method [34]. The content of excess Zn ions of obtained single crystal was 1.7 ppm and color was light yellowish (Fig. 11). In this hydrothermal method, it is rather difficult to add a large amount of Li ions into single crystal. In order to dope Li ions, several *c*-plate samples (0.16 mm thick) were annealed at an atmosphere of Li ions at 920 K for 24 hours. The light yellowish color of as-grown single crystal became transparent after this doping as shown in Fig. 12. Li concentration x was measured by chemical analysis and it is confirmed that $x = 0.082\sim0.086$.

Figure 11. Single crystal of as grown ZnO.

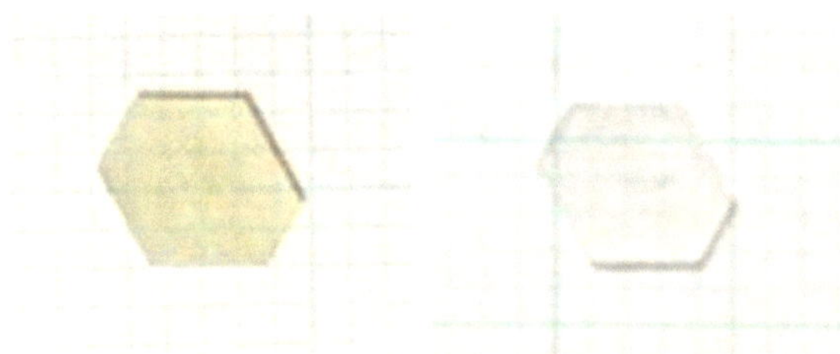

Figure 12. Single crystal of pure (left) and Li-doped ZnO (right).

Table 3 shows an experimental data of X-ray diffraction performed at room temperature (293 K) and low temperature (19 K) by using a He-gas closed-cycle cryostat (Cryogenic RC-110) mounted on a Huber off-center four-circle diffractometer to reduce the effect of thermal vibration. The space group is confirmed to be $P6_3mc$ for both pure and Li-doped crystals. The Lorentz and polarization, thermal diffuse scattering (TDS) (the elastic constants used in TDS collection are C_{11}=20.96, C_{33}=21.09, C_{12}=12.10, C_{13}=10.51, C_{44}=4.243 [10^{11}dyn/cm^2], absorption (spherical) and extinction (anisotropic, type I) corrections were made. Crystal structures were refined using a full-matrix least squares program (RADIEL) [35].

	ZnO		$Zn_{1-x}Li_xO$	
Temperature	293 K	19 K	293 K	19 K
X-ray Radiation	AgKα	AgKα	MoKα	AgKα
$(\sin\theta/\lambda)_{max}$	1.36	1.36	1.00	1.27
Number of Reflections (|F$_o$| "/> 3σ |F$_o$|)	3157	2533	1296	2377
μ*R*	1.91	2.84	7.67	3.96
R(*F*) (%)	2.61	3.84	3.05	3.62
$R_W(F)$ (%)	3.38	4.86	3.90	4.54
S	1.00	1.06	1.09	1.01

Table 3. Experimental data.

Accurate electron densities of single crystals of paraelectricZnO and ferroelectric Li-doped ZnO at 19 K were analyzed by the maximum-entropy method (MEM) [36]. The MEM analyses were performed using the MEED (maximum-entropy electron density) program [37].

6.1. Crystal structures of single crystals of ZnO and Li-doped ZnO at 293 K

The final positional parameters and thermal factors at 293 K are given with their estimated standard deviation in Table 4 [38]. The final discrepancy factors are $R(F)$=2.61 %, $R_w(F)$=3.38 % for ZnO, and $R(F)$=3.05 %, $R_w(F)$=3.90 % for $Zn_{1-x}Li_xO$. The lattice constants *a* and *c* of $Zn_{1-x}Li_xO$ become smaller than those of pure ZnO by 0.002 ~ 0.003 Å. Lattice distortion in $Zn_{1-x}Li_xO$ is only the order of 10^{-3} Å along the polar axis, which is consistent with that of ceramic sample measured by thermal expansion [39]. These changes are regarded as the changes associated with ferroelectric phase transition because Li-doped ZnO is ferroelectric phase and pure ZnO is paraelectric phase. In $BaTiO_3$, the lattice constants change by 0.02 Å. In ZnO, changes in the lattice constants are smaller than that of $BaTiO_3$ by one order. The thermal factors U_{33} of Zn and O atoms in $Zn_{1-x}Li_xO$ are smaller than U_{11} by 25-35 %, although those values in pure ZnO are almost same. This implies that the thermal vibration in $Zn_{1-x}Li_xO$ is suppressed along the *c*-axis. But in ZnO, the effect from bonding electron is also included in the thermal factors since strong covalence exists. The effects of thermal vibration and bonding electron cannot be distinguished at 297 K. The Li concentration *x*, the positional parameter *u* and isotropic thermal factor *U* of Li are refined after all the other parameters are determined. Two cases are assumed for the Li position: the first is the interstitial position and the second is host Zn position. The discrepancy factor became larger when Li locates at interstitial Li position than Zn position. Therefore, present results support that Li substitutes Zn. The precise refinement shows that Li locates at the off-center position from the Zn position by 0.02 Å in $Zn_{1-x}Li_xO$ along the *c*-axis.

			ZnO (a=3.2489(1) Å, c=5.2049(3) Å) at 293 K				
	x	*y*	*z*	U_{11}	U_{33}	U_{eq}	Occupation
Zn	1/3	2/3	0.3815(1)	0.0086(1)	0.0088(1)	0.0087(1)	1.0
O	1/3	2/3	0	0.0085(1)	0.0088(1)	0.0086(1)	1.0
			$Zn_{1-x}Li_xO$ (a=3.2467(3) Å, c=5.2032(1) Å) at 293 K				
	x	*y*	*z*	U_{11}	U_{33}	U_{eq}	Occupation
Zn	1/3	2/3	0.3804(3)	0.0075(1)	0.0055(1)	0.0068(1)	0.914
Li	1/3	2/3	0.376(36)			0.0036(48)	0.086
O	1/3	2/3	0	0.0088(3)	0.0057(41)	0.0078(3)	0.957

Table 4. Crystal structure of ZnO and $Zn_{1-x}Li_xO$ at 293 K.(The form of thermal factors are $\exp[-2\pi^2(U_{11}a^{*2}h^2 + U_{22}b^{*2}k^2 + U_{33}c^{*2}l^2 + 2U_{12}a^*b^*hk + 2U_{23}b^*c^*kl + 2U_{31}c^*a^*lh)]$ for anisotropic (Zn and O) and $\exp(-8\pi^2U_{eq}\sin^2\theta/\lambda)$ for isotropic (Li) atoms.)

The bond lengths and angles of ZnO_4 group are shown in Table 5. The Zn-O bond length does not change in the basal plane, but becomes short along the *c*-axis by an amount of 0.007 Å by Li-substitution.

	Bond length		Bond Angle	
	Apical	Basal	Apical	Basal
ZnO	1.986(1) Å	1.975(1) Å	108.20(3)°	110.71(3)°
$Zn_{1-x}Li_xO$	1.979(1) Å	1.975(1) Å	108.37(4)°	110.55(4)°

Table 5. Bond lengths and angles in ZnO and $Zn_{1-x}Li_xO$ at 293 K.

The distributions of electronic density around Zn and O atoms at 293 K are obtained by using Fourier analysis (Fig. 13). The difference Fourier maps between observed distribution ϱ_{obs} and calculated distribution ϱ_{cal} in the (110) plane are shown in Fig. 14. Since the bonding electron is not distributed spherically in ZnO, the section of the bonding electron appears in the difference map. It is seen that there are large negative distributions (blue region) around Zn atom along the *c*-axis in $Zn_{1-x}Li_xO$. This suggests that the core electrons disappear from the Zn atom. The broadening of map due to anharmonic thermal vibrations is also appreciated. The positive electronic density is observed near the O atom. It corresponds to the anti-bonding orbital of O 2*p*-electron.

http://dx.doi.org/10.5772/52304

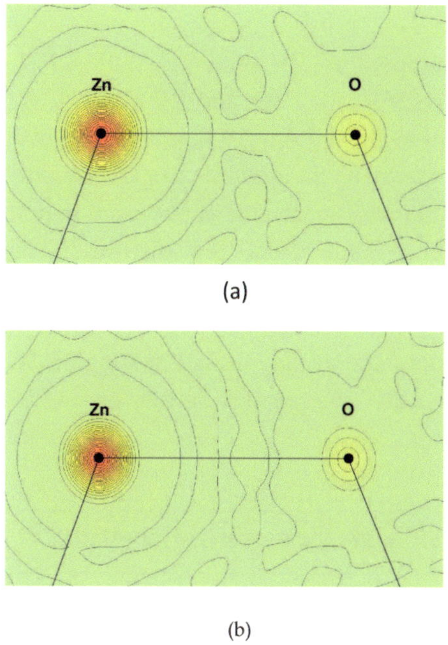

Figure 13. The charge density maps of (a) ZnO and (b) $Zn_{1-x}Li_xO$ at 293 K in the (110) plane obtained by Fourier analysis.

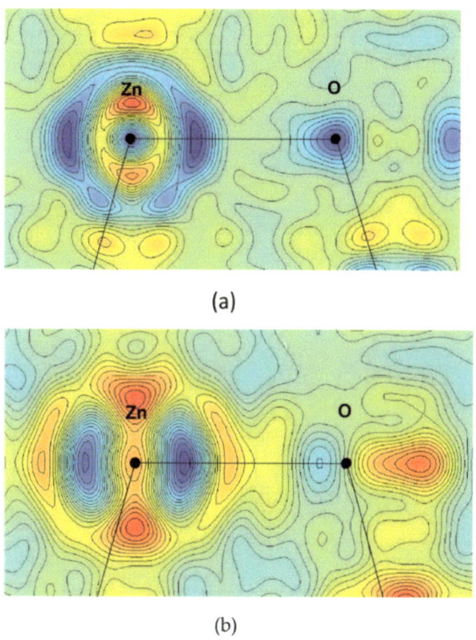

Figure 14. The difference fourier maps of charge densities of (a) ZnO and (b) $Zn_{1-x}Li_xO$ at 293 K in the (110) plane with a contour increment of 0.2 $e/Å^3$. Bluish cold color means negative charge density and reddish warm region is positive charge density.

6.2. Crystal structures at 19 K

Crystal structures at low temperature are shown in Table 6 [38]. The final discrepancy factors are $R(F)$=3.84 %, $R_w(F)$=4.86 % for ZnO, and $R(F)$=3.62 %, $R_w(F)$=4.54 % for $Zn_{1-x}Li_xO$. Li ion shifts from the Zn position by 0.08 Å at 19 K, which is four times larger than that at r. t.. The Zn-O bond lengths decrease by an amount of 0.002 Å along both basal and apical axes in $Zn_{1-x}Li_xO$, while their bond angles are almost the same in both crystals (Table 7 and Fig. 15). The Fourier and difference maps of electronic distribution are shown in Figs. 16 and 17. Comparing with the result of 293 K, bonding electrons are clearly observed around the center of Zn-O bond in both crystals. In $Zn_{1-x}Li_xO$, negative distribution is observed around Zn atom, whose shape corresponds to Zn-$3d_z{}^2$-orbital. It suggests the disappearance of $3d$-electrons from Zn atom in $Zn_{1-x}Li_xO$. The positive density near O atoms corresponds to the $2p$-orbital of O atom.

ZnO (a=3.2465(8) Å, c=5.2030(19) Å) at 19 K							
	x	y	z	U_{11}	U_{33}	Occupation	
Zn	1/3	2/3	0.3812(1)	0.0033(1)	0.0032(1)	1.0	
O	1/3	2/3	0	0.0041(1)	0.0048(2)	1.0	
$Zn_{1-x}Li_xO$ (a=3.2436(5) Å, c=5.1983(30) Å) at 19 K							
	x	y	z	U_{11}	U_{33}	U_{eq}	Occupation
Zn	1/3	2/3	0.3811(2)	0.0032(1)	0.0026(1)	0.0030(1)	0.914
Li	1/3	2/3	0.366(12)	-	-	0.0036(48)	0.086
O	1/3	2/3	0	0.0053(3)	0.0043(3)	0.0050(2)	0.957

Table 6. Crystal structure of ZnO and $Zn_{1-x}Li_xO$ at 19 K.(The form of thermal factors are $\exp[-2\pi^2(U_{11}a^{*2}h^2 + U_{22}b^{*2}k^2 + U_{33}c^{*2}l^2 + 2U_{12}a^*b^*hk + 2U_{23}b^*c^*kl + 2U_{31}c^*a^*lh)]$ for anisotropic (Zn and O) and $\exp(-8\pi^2U_{eq}\sin^2\theta/\lambda)$ for isotropic (Li) atoms.)

	0.957		**0.957**	
	Apical	Basal	Apical	Basal
ZnO	1.983(2) Å	1.974(1) Å	108.25(4)°	110.66(4)°
$Zn_{1-x}Li_xO$	1.981(2) Å	1.972(1) Å	108.27(3)°	110.65(3)°

Table 7. Bond lengths and angles in ZnO and $Zn_{1-x}Li_xO$ at 19 K.

http://dx.doi.org/10.5772/52304

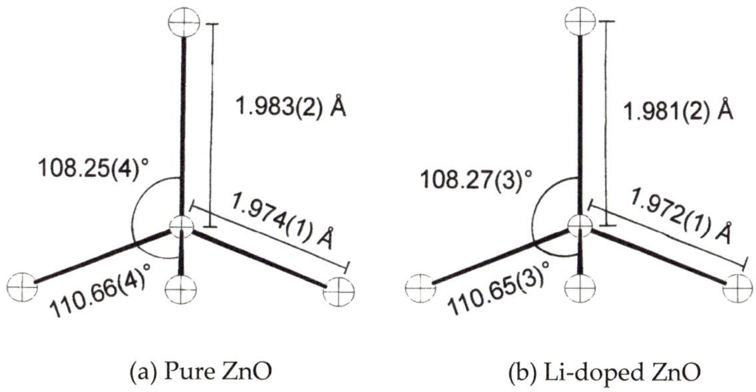

Figure 15. Structural Changes of ZnO_4 group at 19 K. The shift of Li ion is 0.08 Å from the Zn position.

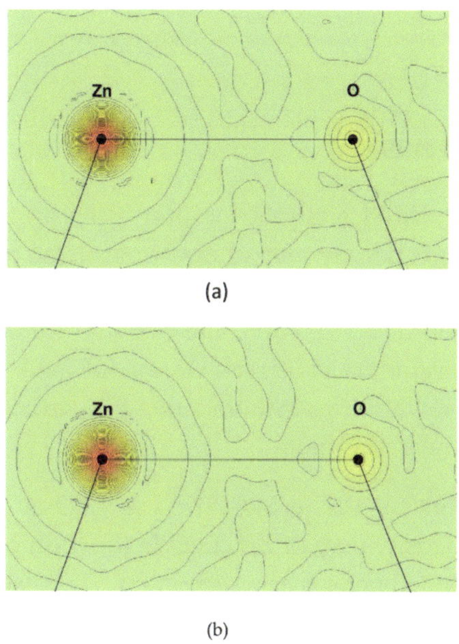

Figure 16. The charge density maps of (a) ZnO and (b) $Zn_{1-x}Li_xO$ at 19 K in the (110) plane obtained by Fourier analysis.

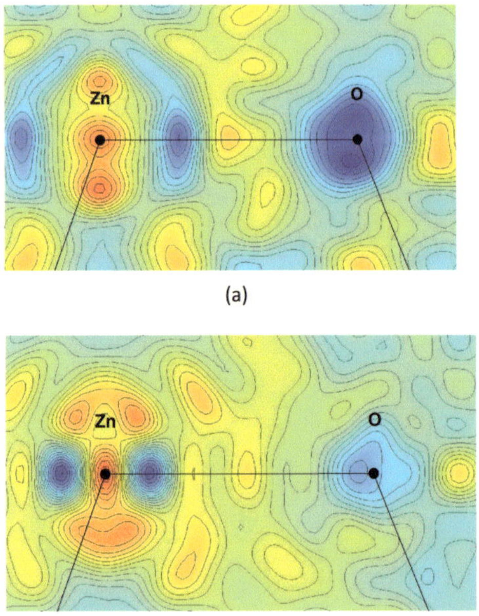

(b)

Figure 17. The difference maps of charge densities of (a) ZnO and (b) $Zn_{1-x}Li_xO$ at 19 K in the (110) plane with a contour increment of 0.2 e/ Å ^{3}. Bluish cold color means negative charge density and reddish warm color region is positive charge density.

6.3. Bond electron densities at 19 K

The discrepancy factors of MEM analysis for pure ZnO at 19 K are $R(F)$=1.20 % and $R_w(F)$=1.39 % [40]. The electron density map of ZnO in the (110) plane is shown in Fig. 18(b). The covalent character of Zn-O bonds is clearly observed. Charge densities at the center of Zn-O bonds are 0.58 e/ Å ^{3} for the apical bond, and 0.56 e/Å ^{3} for the basal one. The density of bonding electron of the apical bond is larger than that of the basal by 3.6 %. The electron density of O atom distorts toward the Zn atom along the c-axis. The electron density of Zn atom elongates to the third nearest O atom.

The final discrepancy factors for the ferroelectric $Zn_{1-x}Li_xO$ at 19 K are $R(F)$=0.87 % and $R_w(F)$=0.85%. The electron density map of $Zn_{1-x}Li_xO$ in the (110) plane is shown in Fig. 18(c). The covalent character of Zn-O bonds is also seen in $Zn_{1-x}Li_xO$. Charge densities of the Zn-O at the saddlepoint are 0.46 e/ Å ^{3} for the apical bond, and 0.49 e/ Å ^{3} for the basal one. Each value is smaller than that of pure ZnO. It is considered that this may be due to the decrease of total charge by Li-substitution. The density of bonding electron of the basal bond is larger than that of the apical by 6.5 %, on the contrary to the case of ZnO. The electron density of O atom is distorted anisotropically, similar to the pure ZnO, but the direction is opposite. The extension of Zn atom toward the third nearest O atom, which is observed in ZnO, was not detected in $Zn_{1-x}Li_xO$.

http://dx.doi.org/10.5772/52304

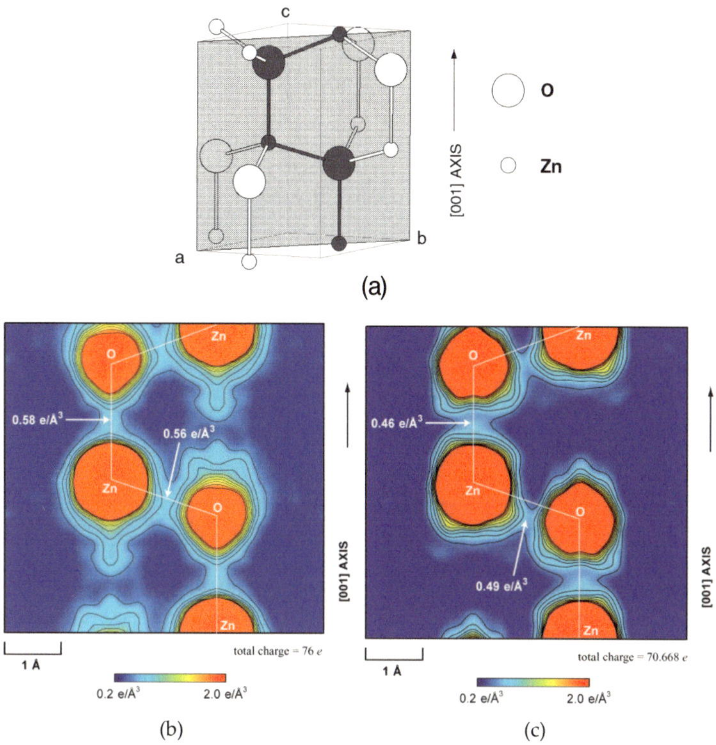

Figure 18. The electronic Distribution by MEM method. (a) The (110) plane of ZnO. Shaded atoms are included in the plane.(b)The MEM charge density map of ZnO at 19 K in the (110) plane. (c) The MEM charge density map of $Zn_{0.914}Li_{0.086}O$ at 19 K in the (110) plane. Contours are drawn from 0.4 e/ Å^3 at 0.2 e/ Å^3 intervals.

6.4. Difference between Li-doped ZnO and pure ZnO

The difference of the charge densities between two crystals is calculated by subtracting the MEM charge densities of pure ZnO from those of $Zn_{1-x}Li_xO$ as in Fig. 19 [40]. The values of MEM charge densities of pure ZnO are normalized by multiplying 0.930, because the total charges of two crystals are different. The positive and negative peaks were observed around the O atom. This is due to the shift of O atom from the host position opposite to the Zn atom along the *c*-axis in $Zn_{1-x}Li_xO$. Four negative peaks are observed around the Zn atom. These peaks correspond to the Zn 3*d*-orbitals and suggest that the 3*d*-electrons disappear from the Zn site, compared with wave functions of Zn 3*d*- and O 2*p*-orbitals of ZnO obtained by DV-Xα calculation in Fig. 20. This result is the same as that calculated by the Fourier synthesis previously. The bonding region around the apical and the basal Zn-O bonds is covered by positive electron distribution. This implies that the Zn 3*d*-electrons transfer to the bonding electrons in $Zn_{1-x}Li_xO$.

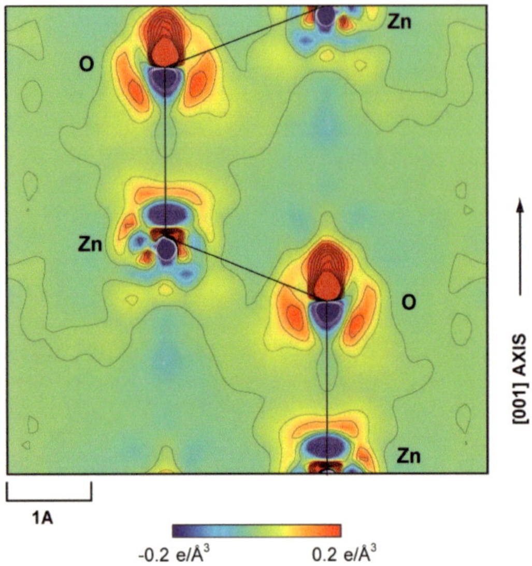

Figure 19. The difference map between MEM charge densities of $Zn_{1-x}Li_xO$ and ZnO at 19 K in the (110) plane. Contours are drawn from -5.0 e/ Å^3 to 5.0 e/ Å^3 at 0.5 e/ Å^3 intervals. Shaded area eith reddish color indicates positive charge.

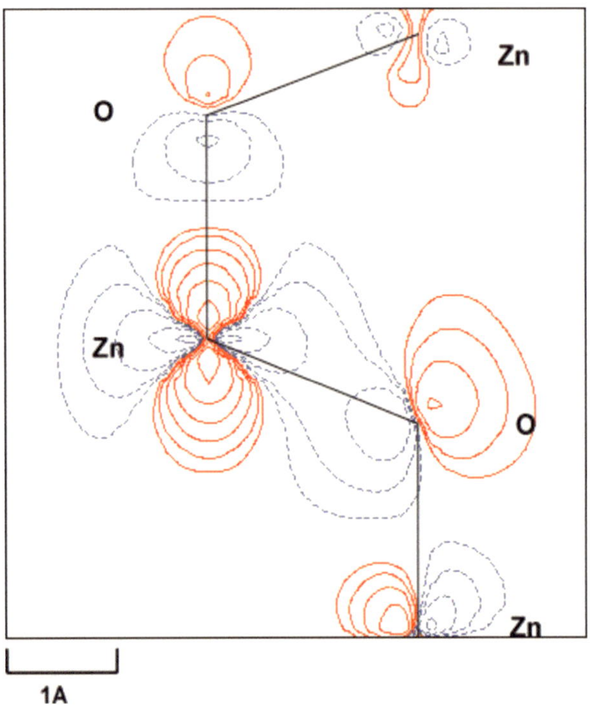

Figure 20. Wave functions of Zn 3*d*- and O 2*p*-orbitals of ZnO obtained by DV-Xα calculation. Red lines indicate the positive and Blue lines indicate the negative area.

In Li-doped ZnO, negative peaks around Zn atom corresponding to $3d_z^2$ orbital were observed in difference Fourier map, and also in the MEM difference map between the ferroelectric phase and the paraelectric phase. These results may be related to the existence of localized *d*-holes in Li-doped ZnO. It is considered that $3d_z^2$ electrons transfer to bonding region and play a role for contribution for the covalence observed from the result of the MEM analysis. Furthermore, the antibonding orbital of 2*p* electron is observed in difference Fourier map of Li-doped ZnO. It is suggested that the interaction between *d*-holes and *p*-electrons should be closely related to the appearance of the ferroelectricity in Li-doped ZnO.

7. Discussion

The ferroelectric phase transition accompanies with structural distortions in usual ferroelectrics. Portengen, Ostreich and Sham reported the theory of electronic ferroelectricity [41], which examined possibilities of electronic ferroelectricity, based on the spinless Falikov-Kimball (FK) model [42] with a *k*-dependence hybridization in Hartree-Fock approximation. The FK model introduces two types of electrons, itinerant *d*-electrons and localized *f*-electrons. The valence transition is driven by on-site Coulomb repulsion between the *d*-electrons and *f*-electrons. They found that the Coulomb interaction between itinerant *d*-electrons and the localized *f*-electrons give rise to an excitonic$\langle d^+ f \rangle$ expectation value, which breaks the center of symmetry of the crystal and leads to electronic ferroelectricity in mixed-valent compounds. In this electronic model, the transition involves a change in the electronic structure rather than the structural one. The estimated spontaneous polarization is of the order of 10 $\mu C/cm^2$, which is comparable to those in displacive type ferroelectrics such as $BaTiO_3$.

In the case of Li-doped ZnO, the *d-p* hybridization should be changed by Li-doping. The *d*-holes and itinerant electrons should be closely related to the appearance of Li-doped ZnO. It should be further detailed studies would be necessary whether this proposed theory is applicable for the electronic ferroelectricity found in ZnO or not.

Recently Glinchuk *et al* propose the mechanism of impurities induced ferroelectricity in nonperovskite semiconductor matrices due to indirect dipole interaction via free carriers [43]. They showed that the ferroelectricity in Li-doped ZnO might appear due to indirect interaction of dipoles, formed by off-center impurities, *via* free charge carriers, namely, the Ruderman–Kittel–Kasuya–Yosida (RKKY)-like indirect interaction of impurity dipoles *via* free charge carriers. They estimated that the typical semiconducting concentration of the carriers like 10^{17} cm^{-3} is sufficient for the realization of the ferroelectricity. The finite conductivity does not mean complete destruction of possible ferroelectric order and shows rather many interesting effects. In this theory, they have succeeded to obtain the spontaneous polarization and the phase diagram of T_c*vs*. impurity molar ratio (x) as shown in Fig. 21.

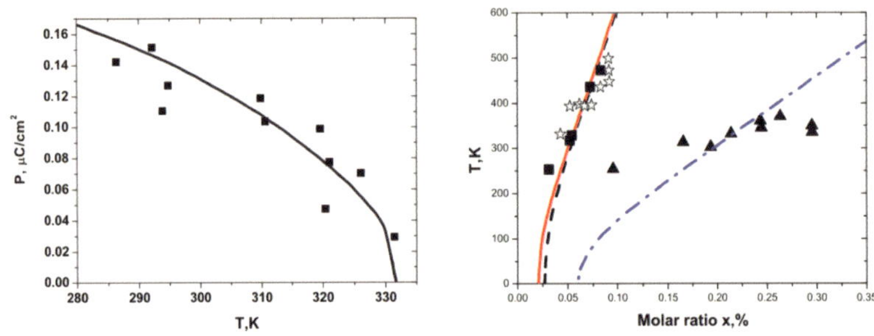

Figure 21. The calculated spontaneous polarization and the phase diagram of T_c vs. impurity molar ratio (*x*) after Glinchuk et al [36].

Experimental results suggest the existence of localized *d*-holes and *p*-electrons in Li-doped ZnO. We consider that the interaction between *d*-holes and *p*-electrons may be related to the appearance of the ferroelectricity in Li-doped ZnO. However, there are many proposals for this peculir novel ferroelectricity in ZnO. Scott and Zubko have pointed out the possibility of a classic electret mechanism for Li-doped ZnO[44]. Tagantsev discussed a Landau theory, where a crystal on cooling from a state with polar symmetry exhibits a maximum of dielectric permittivity and *D-E* hysteresis loops [45]. He proposed that these ferroelectric like phenomena corresponds to the case for Li-doped ZnO. Furthermore, the multiferroic behavior has been reported in impurity-doped ZnO [46]. At present, further detailed experiments should be expected to clarify the nature of Li-doped ZnO.

8. Conclusion

Dielectric properties and crystal structures and electron density distributions studied on pure ZnO and $Zn_{1-x}Li_xO$ by the precise X-ray diffraction are reviewed in comparison with ferroelectric semiconductors $Pb_{1-x}Ge_xTe$ and $Cd_{1-x}Zn_xTe$. It is considered that the appearance of ferroelectricity in ZnO is primarily due to electronic origin. The change in *d-p* hybridization caused by Li-substitution is considered to be responsible for this novel ferroelectricity and dielectric properties. Although the ferroelectric phase transition accompanies with structural distortions in usual ferroelectrics, the structural changes observed in $Zn_{1-x}Li_xO$ are the order of 10^{-3} Å. The clear change of Zn 3*d*-electron is observed. It suggests Zn 3*d*-electron of $Zn_{1-x}Li_xO$ transfers from Zn atom to the bonding. The positive electronic density is observed near the O atom. It corresponds to the antibonding orbital of O 2*p*-electron. These results suggest the existence of localized *d*-holes and *p*-electrons in Li-doped ZnO. It may be probable that the interaction between *d*-holes and *p*-electrons may be related to the appearance of the ferroelectricity in Li-doped ZnO.

For long time, the importance of electronic contribution has been pointed out in the field of ferroelectrics, since the simple superposition of electronic polarizability does not hold in many ferroelectric substances. Many efforts have been done in vain because of complexity of crystal structures of ferroelectrics. As the crystal structure of ZnO is very simple, the electronic contribution could be observed rather easily. It is considered that this result is the first example to discuss the electronic contribution for ferroelectricity.

Author details

Akira Onodera and Masaki Takesada

Department of Physics, Faculty of Science, Hokkaido University, Sapporo, Japan

References

[1] Klingshirn C F, Meyer B K, Waag A, Hoffmann A and Geurts J. Zinc Oxide From Fundamental Properties Towards Novel Applications. Springer, 2010.

[2] Yao T (ed.). ZnO Its Most Up-to-date Technology and Application, Perspectives (in japanese). CMC Books, 2007.

[3] Heiland G. Mollwo E and Stockmann F. Solid State Phys 1959; 81: 193.

[4] Campbell C. Surface Acoustic Wave Devices and Their Signal Processing Application. San Diego: Academic Press; 1989.

[5] Hirshwald W. Current Topics Mater Sci 1981; 7, 143.

[6] Tsukazaki A, Ohtomo A, Onuma T, Ohtani M, Makino T, Sumiya M, Ohtani K, Chichibu S F, Fuke S, Segawa Y, Ohno H, Koinuma H and Kawasaki M. Nature Materials 2005; 4: 42.

[7] Joseph M, Tabata H and Kawai T. Jpn J Appl Phys 1999; 38: L1205.

[8] Bilz H, Bussmann-Holder A, Jantsch W and Vogel P. Dynamical Properties of IV-VI

[9] Compounds. Berlin: Springer-Verlag; 1983.

[10] Weil R, Nkum R, Muranevich E and Benguigui L. Phys Rev Lett 1989; 62: 2744.

[11] Onodera A, Tamaki N, Kawamura Y, Sawada T and Yamashita H. Jpn. J Appl Phys 1996; 35: 5160.

[12] Abrahams C S and Bernstein L J. Acta Cryst 1969; B25: 1233.

[13] Corso D A, Posternak M, Resta R and Baldereshi A. Phys Rev 1994; B50: 10715.

[14] Zakharov O, Rudio A, Blase X, Cohen L M and Louie G S. Phys Rev1994; B50: 10780.

[15] Massida S, Resta R, Posternak M and Baldereschi A. Phys Rev 1995; B50.

[16] Osikiri M and Aryasetiawan F. J Phys Soc Jpn 2000; 69: 2123.

[17] Minami T, Nanto H and Tanaka S. Jpn J Appl Phys 1984; 23: L280.

[18] Laudise A R, Kolb D E and Caporaso J A. J Am Ceram Soc 1964; 47: 9.

[19] Kolb D E and Laudise A R. J Am Ceram Soc 1966; 49: 302.

[20] Yang CK and Dy S K. Solid State Commun 1993; 88: 491.

[21] Usuda M, Hamada N, Kotani T and Schilfgaarde M. Phys Rev 2001; B66: 075205.

[22] Zhang B S, Wei S -H and Zunger A. Phys Rev 2001; B63: 075205.

[23] Tamaki N, Onodera A, Sawada T and Yamashita H. J Kor Phys 1966; 29: 668.

[24] Onodera A, Tamaki N, Jin K and Yamashita H. Jpn J Appl Phys 1997; 36: 6008.

[25] Islam T Q and Bunker A B. Phys Rev Lett 1987; 59: 2701.

[26] Benguigui L, Weil R, Muranevich E, Chack A and Fredj E. J Appl Phys 1993; 74:513.

[27] Terauchi H, Yoneda Y, Kasatani H, Sakaue K, Koshiba T, Murakami S, Kuroiwa Y, Noda Y, Sugai S, Nakashima S and Maeda H. Jpn J Appl Phys 1993; 32: 728.

[28] Onodera A, Yoshio K, Satoh H, Yamashita H and Sakagami N. Jpn. J. Phys. Phys. 1998; 37; 5315.

[29] Onodera A, Tamaki N, Satoh H, Yamashita H and Sakai A. Ceramic Transactions 1999; 100; 77.

[30] Onodera A and Satoh H. Frontiers in Science and Technology – Ferroelectrics Vol. 1, Stefan University Press, 2002; 93.

[31] Islam E, Sakai A and Onodera A. J Phys Soc Jpn 2001; 70: 576.

[32] Kagami D, Takesada M, Onodera A and Satoh H. J Kor Phys 2011; 59: 2532.

[33] Hagino S, Yoshio K, Yamazaki T, Satoh H, Matsuki K and Onodera A. Ferroelectrics 2001; 264: 235.

[34] Onodera A, Yoshio K, Satoh H, Tamaki T, Takama T, Fujita M and Yamashita H. Ferroelectrics 1999; 230: 465.

[35] Sakagami N. J Cryst Growth 1990; 99: 905.

[36] Coppens P, Guru Row N T, Leung P, Stevens D E, Becker J P and Yang W Y. Acta Crystallogr 1979; A35: 63.

[37] Sakata M and Sato M. Acta Crystallogr 1990; A46: 263.

[38] Kumazawa S, Kubota Y, Tanaka M, Sakata M and Ishibashi Y. J Appl Cryst 1993; 26: 453.

[39] Yoshio K, Onodera A, Satoh H, Sakagami N and Yamashita H. Ferroelectrics 2001; 264; 133.

[40] Onodera A, Tamaki N, Satoh H and Yamashita H. Ferroelectrics 1998; 217: 9.

[41] Yoshio K, Onodera A, Satoh H, Sakagami N and Yamashita H. Ferroelectrics 2002; 270; 357.

[42] Portengen T, Ostreich Th and Sham J L. Phys Rev 1996; B54: 17452.

[43] Falicov M L and Kimball C J. Phys Rev Lett 1969; 22: 997.

[44] Glinchuk D M, Kirichenko V E, Stephanovich A V and Zaulychny Y B. Appl Phys 2009; 105: 104101.

[45] Scott F J and Zubko P. IEEE Xplore 2005; ISE-12 (12th International Symposium on Electrets): 113-116.

[46] Tagantsev K A. Appl Phys Let 2008; 93: 202905.

[47] Yang C Y, Zhong F C, Wang H X, He B, Wei Q S, Zeng F and Pan F. Appl Phys 2008; 104: 064102.

Chapter 12

Doping-Induced Ferroelectric Phase Transition and Ultraviolet-Illumination Effect in a Quantum Paraelectric Material Studied by Coherent Phonon Spectroscopy

Toshiro Kohmoto

Additional information is available at the end of the chapter

http://dx.doi.org/10.5772/45744

1. Introduction

There has been significant interest in a quantum paraelectric material strontium titanate ($SrTiO_3$), and its lattice dynamics and unusual dielectric character have been extensively studied. In low temperatures, its dielectric constant increases up to about 30 000. The dielectric constant increases extraordinarily with decreasing temperature while the paraelectric phase is stabilized by quantum fluctuations without any ferroelectric phase transition even below the classical Curie temperature T_c=37 K [1]. In $SrTiO_3$, a ferroelectric transition is easily induced by a weak perturbation such as an uniaxial stress [2], an isotopic substitution of oxygen 18 for oxygen 16 [3], and an impurity doping [4,5].

$SrTiO_3$ has a perovskite structure as shown in Fig. 1(a) and is known to undergo a structural phase transition at T_c=105 K [6]. The cubic (O_h) structure above T_c, where all phonon modes are Raman forbidden, changes into the tetragonal (D_{4h}) structure below T_c, where Raman-allowed modes of symmetries A_{1g} and E_g appear [7]. The phase transition is due to the collapse of the Γ_{25} mode at the R point of the high-temperature cubic Brillouin zone. Below T_c, the R point becomes the Γ point of the D_{4h} phase. The phase transition is characterized by the softening of phonons at the R point and concomitant doubling of the unit cell.

The distortion consists of an out-of-phase rotation of adjacent oxygen octahedra in the (100) planes [6]. The order parameter for the phase transition is inferred to be the angle of rotation of the oxygen octahedra. Only a small rotation of the oxygen octahedra is involved for the transition. The rotation angle for the oxygen octahedra varies from ~2° of arc near 0 K down to zero at T_c=105 K; the transition is second order. At liquid nitrogen temperature, the rotation angle is about 1. 4° and the linear displacement of the oxygen ions about their high-temperature equilibrium positions is less than 0.003 nm. This oxygen octahedron motion can be described as a rotation only as a first approximation; the oxygen ions actually remain on the faces of each cube and therefore increase in separation from the titanium. Because the (100) planes are equivalent in the cubic phase, the distortion produces domains below *Tc* in which the [100], [010], or [001] axis becomes the unique tetragonal *c* axis.

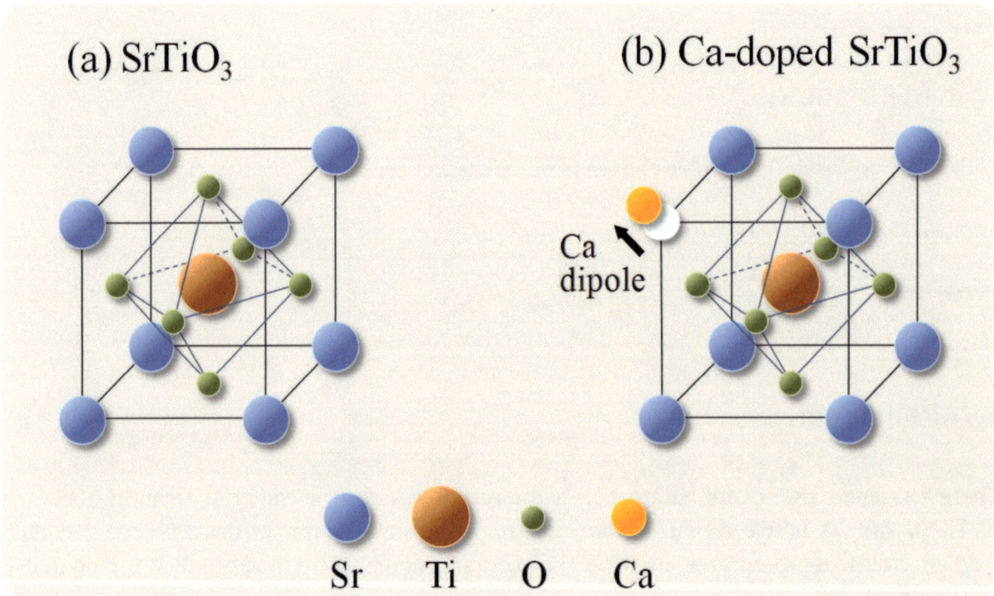

Figure 1. (a) Perovskite structure in $SrTiO_3$. (b) Doped Ca ions are substituted for the Sr ions in Ca-doped $SrTiO_3$.

According to the measurement of dielectric constants, $Sr_{1-x}Ca_xTiO_3$ undergoes a ferroelectric transition above the critical Ca concentration x_c=0. 0018, where doped Ca ions are substituted for the Sr ions [8–11] as shown in Fig. 1(b). The cubic structure above the structural phase-transition temperature T_{c1} changes in to the tetragonal structure below T_{c1} and into the rhombohedral structure below the ferroelectric transition temperature T_{c2}. The structural phase transition at T_{c1} is also antiferrodistortive as in pure $SrTiO_3$ at 105 K [8]. As in the case of impurity systems, Li-doped $KTaO_3$ and Nb-doped $KTaO_3$ [4], off-centered impurity ions are supposed. Their polarized dipole moments show a ferroelectric instability below the ferroelectric transition temperature [4]. In the case of Ca-doped$SrTiO_3$, a spontaneous polarization occurs along [110] directions in the *c* plane, where the tetragonal (D_{4h}) symmetry is lowered to C_{2v} [8]. With decreasing temperature a ferroelectric ordering process dominates, that is, due to the thermal growth of the polarization clouds surrounding the off-center Ca^{2+} dipoles [8]. The system behaves like a super paraelectric as the ferroelectric nano-ordered regions contain disordered cluster like regions. The investigation by x-ray and neutron diffractions and first-principles calculations [12] suggests that polar instabilities originating from the off-center displacements of

Ca^{2+} ions are not likely to directly polarize the host matrix by an electrostatic mechanism. Instead, the possible role of random fields in inducing the presence of disordered polar clusters was suggested, which is similar to polar nanoregions in relaxor materials.

Recently, a gigantic change in the dielectric constant by an ultraviolet (UV) illumination was discovered [13,14], and a deeper interest has been taken in $SrTiO_3$ again. The origin of the giant dielectric constants, however, has not yet been clarified. In Ca-doped $SrTiO_3$, it was reported that a UV illumination causes a ferroelectric peak shift of the dielectric constant toward the lower temperature side [11]. In several ferroelectric materials such as $BaTiO_3$ [15], SbSI [16], and oxygen-isotope-substituted $SrTiO_3$ [17], the T_c reduction under a UV illumination has been observed.

The optical information on the dielectric response is usually obtained from the experiments of Raman scattering or infrared spectroscopy. The usefulness of the investigation of low-frequency dielectric response by observing coherent phonons have also been demonstrated by the time-resolved study of the dynamics of phonons [18] and phonon polaritons [19]. At low frequencies this technique is very sensitive and provides a very good signal-to-noise ratio as compared to the conventional frequency-domain techniques while at higher frequencies a better performance will be achieved by using the conventional techniques. Therefore the coherent phonon spectroscopy and the conventional frequency-domain techniques can be considered to be complementary methods for the investigation of the dielectric response.

The observed signal of the Raman scattering [20] in $SrTiO_3$ is very weak because the distortion from cubic structure in the low-temperature phase is very small. The intensity of the first-order Raman signal is of the same order of magnitude with many second-order Raman signals, and then a background-free signal of the first-order Raman scattering cannot be observed.

Under a UV illumination, $SrTiO_3$ and Ca-doped $SrTiO_3$ show a broadband luminescence in the visible region originated from a relaxed excited state [21]. The coherent phonon spectroscopy is not sensitive to the luminescence and a powerful technique to investigate UV-illumination effects in paraelectric materials as compared to the Raman-scattering measurement because in Raman-scattering experiments, it is not easy to separate Raman-scattering signals from the luminescence.

In the present study, ultrafast polarization spectroscopy is used to observe the coherent optical phonons in pure and Ca-doped $SrTiO_3$, which are generated by femtosecond optical pulses through the process of impulsive stimulated Raman scattering [22,23]. Time-dependent linear birefringence induced by the generated coherent phonons is detected as a change of the polarization of probe pulses. High detection sensitivity of $\sim 10^{-5}$ in polarization change has been achieved in our detection system. Damped oscillations of coherent phonons in $SrTiO_3$ were observed below the structural phase-transition temperature (T_c=105 K), and temperature dependences of the phonon frequency and the relaxation rate are measured [24]. The mechanism of the phonon relaxation is discussed by using a population decay model, in which an optical phonon decays into two acoustic phonons due to an harmonic phonon-phonon coupling.

The doping-induced ferroelectric phase transition in Ca-doped $SrTiO_3$ is investigated by observing the birefringence and coherent phonons [25]. In the birefringence measurement, the structural and the ferroelectric phase-transition temperatures are examined. In the observation of coherent phonons, the soft phonon modes related to the structural (T_{c1}=180 K) and the ferroelectric (T_{c2}=28 K) phase transitions are studied, and their frequencies are obtained from the observed coherent phonon signals. The behavior of the softening toward each phase-transition temperature and the UV-illumination effect on the ferroelectric transition are discussed. In addition to the ferroelectric phase transition at T_{c2}=28 K, another structural deformation at 25 K is found. A shift of the ferroelectric phase-transition temperature under the UV illumination and a decrease in the phonon frequencies after the UV illumination are found. We show the approach in the time domain is very useful for the study of the soft phonon modes and their UV-illumination effect in dielectric materials.

2. Experiment

The experiments are performed on single crystals of pure $SrTiO_3$ and Ca-doped $SrTiO_3$ with the Ca concentration of x=0.011. $SrTiO_3$ was obtained commercially (MTI Corporation) and Ca-doped $SrTiO_3$ was grown by the floating zone method [11]. The thickness of the sample is 1. 0 and 0.5 mm for the pure and Ca-doped crystals, respectively. In Ca-doped$SrTiO_3$, the structural phase-transition ($O_h \rightarrow D_{4h}$) temperature, T_{c1}=180 K, is obtained from the result of the birefringence measurement in section 3. The ferroelectric phase-transition temperature, T_{c2}=28 K, was determined by the measurement of dielectric constants [11]. The value of x was determined by the empirical relation of Bednorz and Müller [5].

Schematic diagram of the experiment of polarization spectroscopy is shown in Fig. 2. The change in optical anisotoropy (birefringence) is detected by a polarimeter as the change in the polarization of the probe light (ellipticity). In the birefringence measurement, the birefringence generated by the lattice deformation is detected by a continuous-wave (cw) probe beam with no pump beam. In the coherent phonon spectroscopy, the transient birefringence due to the coherent phonons generated by a pump pulse is detected by a probe pulse.

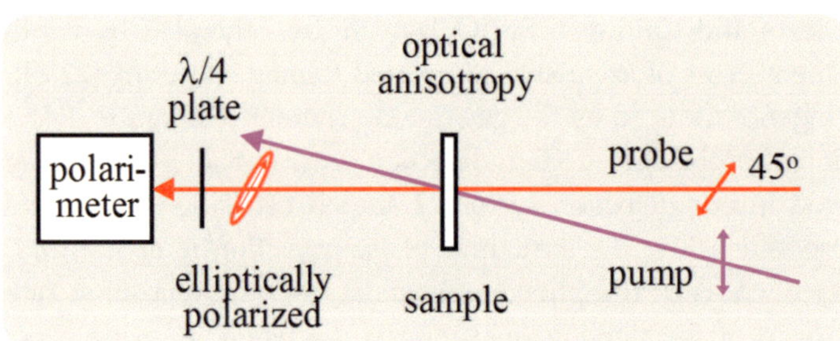

Figure 2. Schematic diagram of the experiment of polarization spectroscopy.

2.1. Birefringence measurement

In the birefringence measurement, the linearly polarized probe beam is provided by a Nd:YAG laser (532 nm, cw) and is perpendicular to the (001) surface of the sample.

The construction of the polarimeter is shown in Fig. 3. The polarimeter [26,27] detects the rotation of polarization plane of a light beam. A linearly-polarized beam is split by a polarized beam splitter (PBS) and incident on the two photodiodes (PD) whose photocurrents are subtracted at a resistor (R). When the polarized beam splitter is mounted at an angle of 45° to the plane of polarization of the light beam, the two photocurrents cancel. If the plane of polarization rotates, the two currents do not cancel and the voltage appears at the resistor.

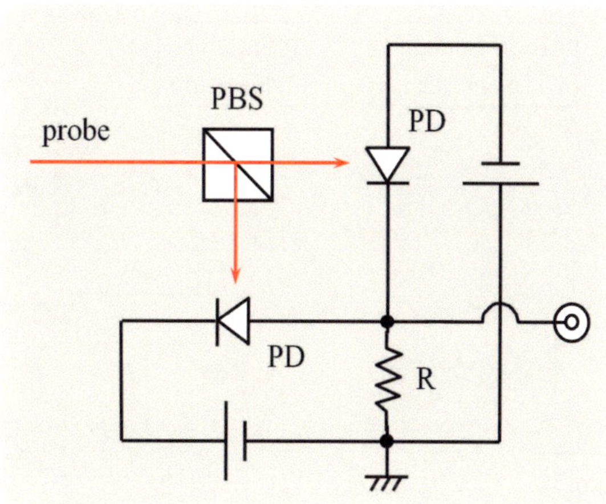

Figure 3. Construction of the polarimeter.

In the present experiment, the birefringence generated by the lattice deformation is detected as the change in polarization of the probe beam using a quarterwave plate and the polarimeter. The birefringence generated in the sample changes the linear polarization before transmission to an elliptical polarization after transmission. The linearly-polarized probe beam is considered to be a superposition of two circularly-polarized components which have the opposite polarizations and the same intensities. The generated birefringence destroys the intensity balance between the two components. The two circularly-polarized beams are transformed by the quaterwave plate to two linearly-polarized beams whose polarizations are crossed each other, and the unbalance of circular polarization is transformed to the unbalance of linear polarization or the rotation of polarization plane. This rotation is detected by the polarimeter as the signal of the lattice deformation.

2.2. Observation of coherent phonons

Coherent phonons are observed by ultrafast polarization spectroscopy with the pump-probe technique. The experimental setup for coherent phonon spectroscopy is shown in Fig. 4. Coherent phonons are generated by femtosecond optical pulses through the process of impulsive stimulated Raman scattering [22,23], and are detected by monitoring the time-dependent anisotropy of refractive index induced by the pump pulse. The pump pulse is provided by a Ti: sapphire regenerative amplifier whose wavelength, pulse energy, and pulse width at the sample are 790 nm, 2 μJ, and 0.2 ps, respectively. The probe pulse is provided by an optical parametric amplifier whose wavelength, pulse energy, and pulse width are 690 nm, 0. 1 μJ, and 0.2 ps, respectively. The repetition rate of the pulses is 1kHz. The linearly polarized pump and probe beams are nearly collinear and perpendicular to the (001) surface of the sample, and are focused on the sample in a temperature-controlled refrigerator. The waist size of the beams at the sample is about 0.5 mm.

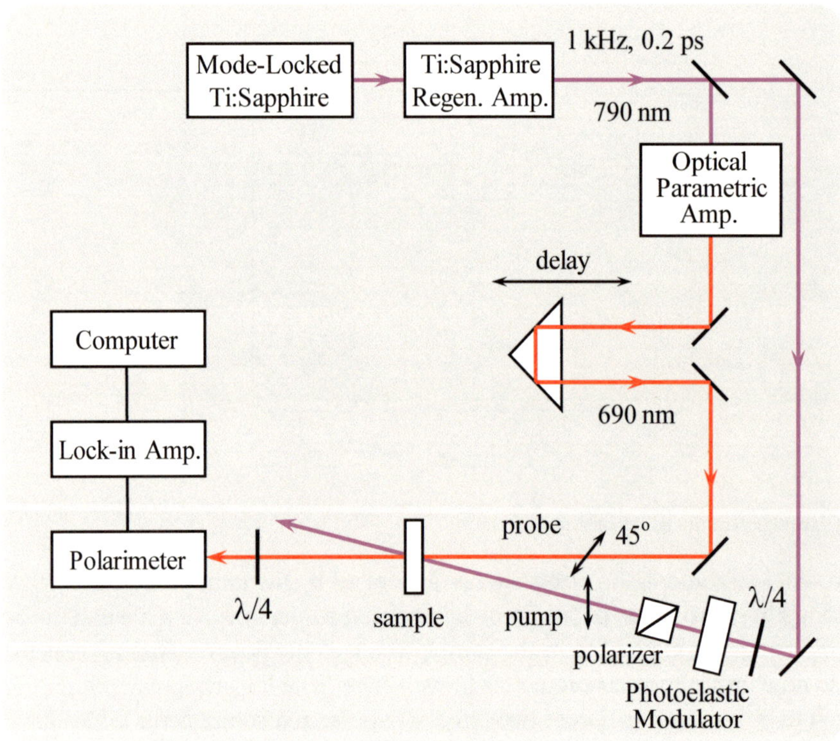

Figure 4. Experimental setup for coherent phonon spectroscopy.

The induced anisotropy of refractive index is detected by the polarimeter with a quarter-wave plate as the polarization change in the probe pulse. The plane of polarization of the probe pulse is tilted by 45° from that of the pump pulse. The two different wavelengths for the pump and probe pulses and pump-cut filters are used to eliminate the leak of the pump light from the input of the polarimeter. The time evolution of the signal is observed by changing the optical delay between the pump and probe pulses. The pump pulse is switched on and off shot by shot by using a photoelastic modulator, a quarter-wave plate, and a polarizer, and the output from the polarimeter is lock-in detected to improve the signal-to-noise ratio.

The source of UV illumination is provided by the second harmonics (380 nm, 3. 3 eV) of the output from another mode-locked Ti: sapphire laser, whose energy is larger than the optical band gap of $SrTiO_3$ (3.2 eV). Since the repetition rate of the UV pulses is 80 MHz, this UV illumination can be considered to be continuous in the present experiment, where the UV-illumination effect appearing more than one minute after is studied.

3. Birefringence measurement

Figure 5 shows the temperature dependences of the change in birefringence in $SrTiO_3$ and Ca-doped $SrTiO_3$ between 4.5 and 250 K, where the polarization plane of the probe light is along the [110]and [100]axes. In $SrTiO_3$, a change in birefringence appears below T_c=105 K, which is the temperature of the structural phase transition, and is increased as the temperature is decreased. In Ca-doped $SrTiO_3$, large changes in birefringence come out at T_{c1}=180 K, which is the temperature of the structural phase transition, and at T_{c2}=28 K, which is the temperature of the ferroelectric phase transition. The change in birefringence is increased as the temperature is decreased from T_{c1} to T_{c2}, as well as the case in $SrTiO_3$, and shows a increase in the gradient for both axes around T_{c2}. Below T_{c2}, another kind of lattice distortion is added, and the peaks of Δn due to the competition between the two kinds of lattice distortion appear. In both $SrTiO_3$ and Ca-doped$SrTiO_3$, a cusp on the birefringence curve appears around T_c=105 K and T_{c1}=180 K, respectively, because of the fluctuation associated with the second-order structural phase transition [28].

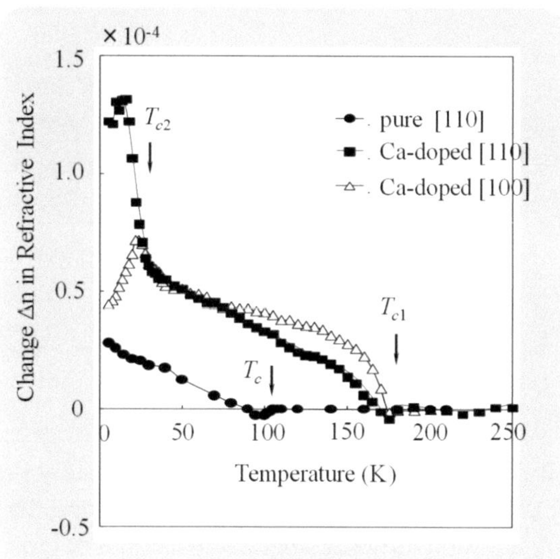

Figure 5. Temperature dependences of the change in birefringence between 4. 5 and 250 K in $SrTiO_3$ and Ca-doped $SrTiO_3$. The polarization plane of the probe light for $SrTiO_3$ is along the [110] axis (solid circles). That for Ca-doped $SrTiO_3$ is along the [110] axis (solid squares) and along the [100] axis (open triangles).

4. Coherent phonon spectroscopy in $SrTiO_3$

4.1. Angular dependence of the coherent phonon signal

Figure 6 shows the transient birefringence in $SrTiO_3$ observed at 6 K for the 0° pumping, where the polarization direction of the pump pulse is parallel to the [100] axis of the crystal. Vertical axis is the ellipticity η in electric-field amplitude of the transmitted probe pulse. At zero delay, a large signal due to the optical Kerr effect, whose width is determined by the pulse width, appears. After that, damped oscillations of coherent phonons are observed as shown in Fig. 6(b), where the vertical axis is enlarged by 50 from that of Fig. 6(a). The change of the polarization for the oscillation amplitude of the coherent phonon signal is 4×10^{-4} of the electric-field amplitudeof the probe pulse, which corresponds to the change $\Delta n=7\times10^{-8}$ of the refractive index. In our detection system, polarization change of ~10^{-5} in the electric field amplitude can be detected.

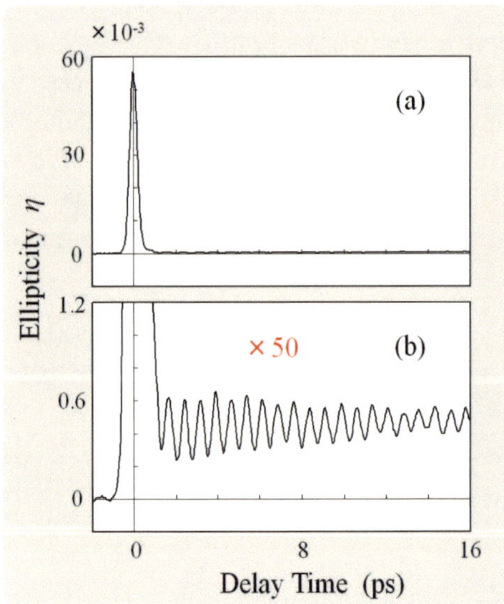

Figure 6. Transient birefringence in $SrTiO_3$ observed at 6 K for the 0° pumping. The vertical axis is the ellipticity η in electric-field amplitude of the transmitted probe pulse. (a) A large signal due to the optical Kerr effect appears at zero delay. (b) Damped oscillations of coherent phonons follow after the Kerr signal. The vertical axis of (b) is enlarged by 50 from that of (a).

Angular dependence of the coherent phonon signal in $SrTiO_3$ observed at 6 K is shown in Fig. 7(a), where the angle between the [100] axis of the crystal and the polarization direction of the pump pulse is 0°, 15°, and 45°. The angle between the polarization directions of the pump and probe pulses is fixed to 45°. The 0.7 ps period signal for the 0° pumping disappears for the 45°pumping, where the 2.3 ps period signal appears. The Fourier transform of the coherent phonon signals in Fig. 7(a) is shown in Fig. 7(b). Oscillation frequency of the signal for the 0° pumping is 1.35 THz, and that for the 45° pumping is 0.4 THz. For other

http://dx.doi.org/10.5772/45744

pumping angles both frequency components coexist in the coherent phonon signal. From the oscillation frequencies the 1.35 THz component is considered to correspond to the A_{1g} mode, and the 0.4 THz component to the E_g mode [7].

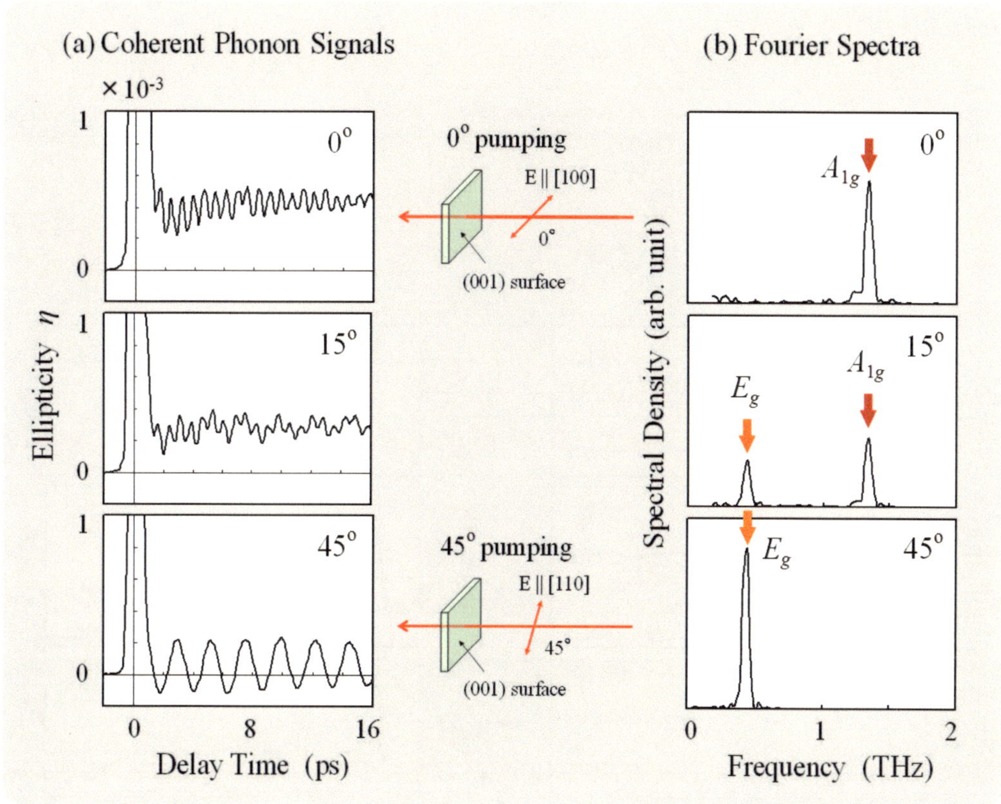

Figure 7. (a) Coherent phonon signals in $SrTiO_3$ observed at 6 K for the 0°, 15°, and 45° pumping, where the angle between the [100] axis of the crystal and the polarization direction of the pump pulse is changed. (b) Fourier transform of the coherent phonon signals in (a).

In addition to the oscillation signal there exists a dc component. The dc component has a maximum amplitude for the 0° pumping and minimum amplitude for the 45° pumping. However, the creation mechanism is not clear at present. In the following we pay attention to the oscillation component.

4.2. Temperature dependence of the coherent phonon signal

The temperature dependence of the coherent phonon signal in $SrTiO_3$ observed for the 0° pumping, which corresponds to the A_{1g} mode, is shown in Fig. 8(a). The oscillation period and the relaxation time of coherent phonons at 10 K are 0.7 and 12 ps. As the temperature is increased, the oscillation period becomes longer and the relaxation time becomes shorter. At

T_c=105 K, the phase transition point, the oscillation disappears. Above T_c no signal of coherent phonons is observed. The temperature dependence of the coherent phonon signal in $SrTiO_3$observed for the 45° pumping, which corresponds to the E_g mode, is shown in Fig. 8(b). The oscillation period and the relaxation time of coherent phonons at 10 K are 2.3 and 45 ps. Similar behavior to that of the A_{1g} mode was observed as the temperature was increased.

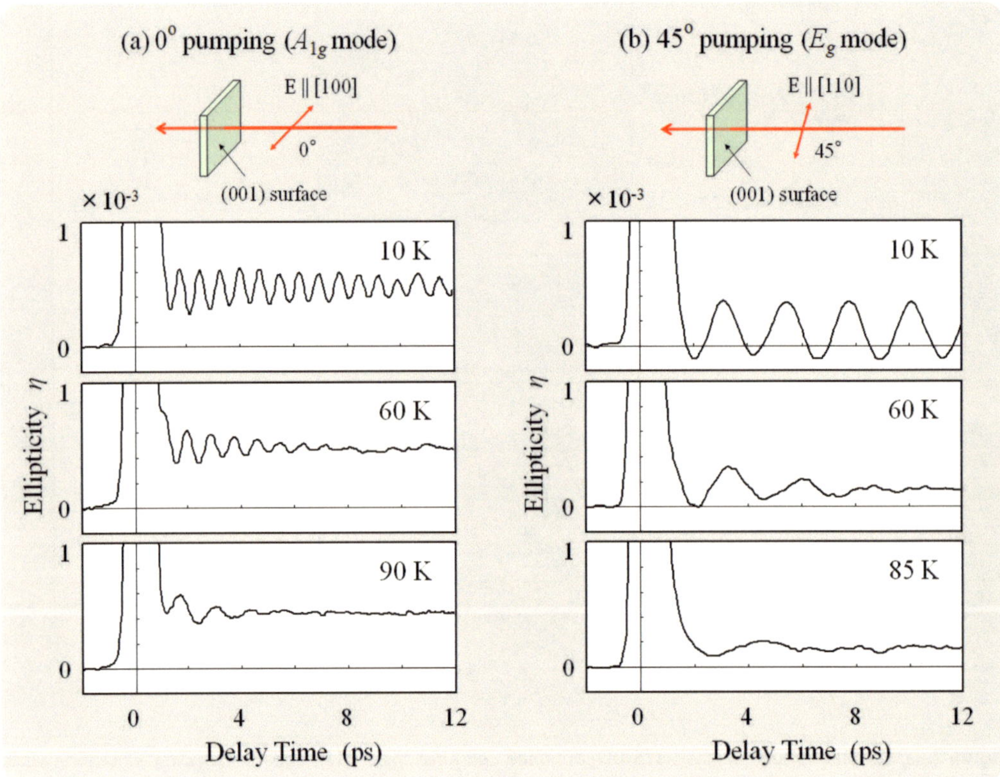

Figure 8. Temperature dependence of the coherent phonon signals in $SrTiO_3$ observed (a) for the 0° pumping which corresponds to the A_{1g} mode and (b) for the 45° pumping which corresponds to the E_g mode.

4.3. Phonon frequencies

The observed coherent phonon signal $S(t)$ is expressed well by the damped oscillation

$$S(t) = Ae^{-\gamma t} \sin \omega t \, , \tag{1}$$

where ω is the oscillation frequency and γ is the relaxation rate. This sine-type function is expected for phonons induced by impulsive stimulated Raman scattering [22,23]. The temperature dependence of the oscillation frequency in $SrTiO_3$ obtained from the observed co-

http://dx.doi.org/10.5772/45744

herent phonon signal below T_c is shown in Fig. 9. The diamonds are the oscillation frequency for the A_{1g} mode, and the squares are that for the E_g mode. As the temperature is increased from 6 K, the oscillation frequencies decrease and approach zero at the phase transition temperature T_c for both modes. This result is consistent with the temperature dependence of phonon frequency observed by Raman scattering [7]. The solid curves describe a temperature dependence of the form $\omega \propto (T_c - T)^n$. The experimental results for the temperature region between 50 K and T_c are explained well by n=0.4 for both modes, while those below 40 K deviate from that form.

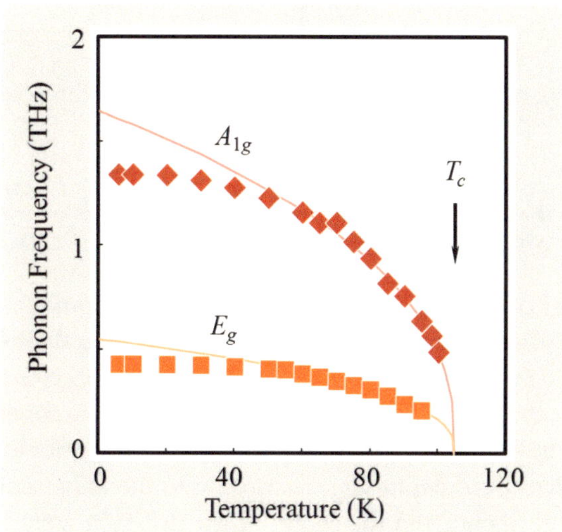

Figure 9. Temperature dependence of the oscillation frequency in $SrTiO_3$ obtained from the coherent phonon signal below T_c. The diamonds are the oscillation frequency for the A_{1g} mode, and the squares are that for the E_g mode. The solid curves describe a temperature dependence of the form $\omega \propto (T_c - T)^n$, where T_c=105 K and n=0.4 for both modes.

The intensity of the first-order Raman-scattering signal in $SrTiO_3$ is very weak and is of the same order of magnitude as many second-order Raman-scattering signals because the distortion from the cubic structure in the low-temperature phase is very small. Then observation of a background-free signal of first-order Raman scattering is not easy, and information on the relaxation, or the spectral width, is not given in a study of Raman scattering [20]. By the present method of coherent phonon spectroscopy in the time domain, on the other hand, background-free damped oscillations can be observed directly, and the oscillation frequency and the relaxation rate can be obtained accurately.

4.4. Relaxation rates

The temperature dependence of the relaxation rate in $SrTiO_3$ obtained from the observed coherent phonon signal below T_c is shown in Fig. 10. The diamonds in Fig. 10(a) are the relaxation rate for the A_{1g} mode and the squares in Fig. 10(b) are that for the E_g mode. The relaxation rates increase as the temperature is increased.

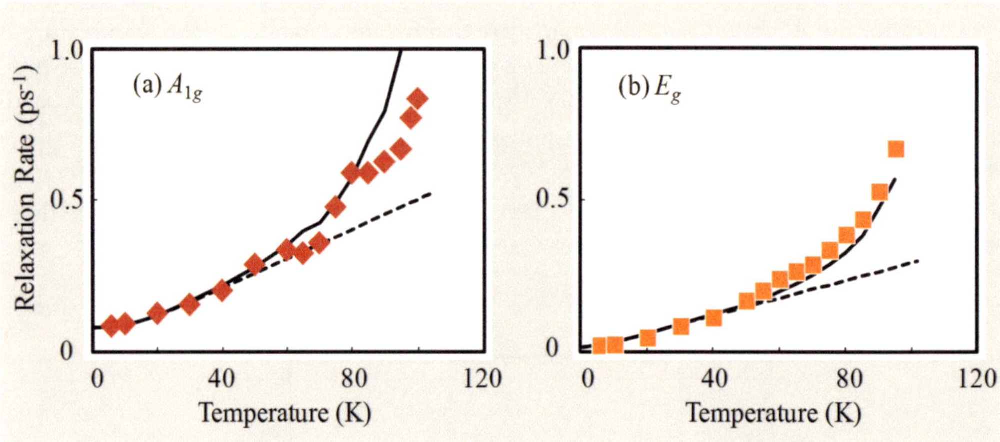

Figure 10. Temperature dependence of the relaxation rate in $SrTiO_3$ obtained from the coherent phonon signal below T_c. (a) The diamonds are the relaxation rate for the A_{1g} mode, and (b) the squares are that for the E_g mode. The solid curves show the theoretical curves including the frequency change obtained from Eq. (2).

In general, relaxation of coherent phonons is determined by population decay (inelastic scattering) and pure dephasing (elastic scattering). In metals, pure dephasing due to electron-phonon scattering, which depends on the hot electron density, contributes to the phonon relaxation [29]. In dielectric crystals, the relaxation process of the coherent phononis considered to be dominated by the population decay due to the anharmonic phonon-phonon coupling [30-32], rather than pure dephasing. According to the anharmonic decay model [30], the relaxation of optical phonons in the center of the Brillouin zone is considered to occur through two types of decay process, the down-conversion and up-conversion processes. In a down-conversion process, the initial ω_0 phonon with wave vector $k\cong 0$ decays into two lower-energy phonons ω_i and ω_j, with opposite wave vectors k and $-k$, which belong to the i branch and the j branch of the phonon. Energy and wave-vector conservation is given by $\omega_0=\omega_{ik}+\omega_{j-k}$. In an up-conversion process, the initial excitation is scattered by a thermal phonon (ω_{ik}) into a phonon of higher energy (ω_{jk}), where $\omega_0+\omega_{ik}=\omega_{jk}$. The down-conversion process can be realized either for $i=j$ (overtone channel), or for $i\neq j$ (combination channel), depending on the phonon band structure of the material, while the up-conversion process contains only the combination channel and has no overtone channel. The combination channel is less likely, because three frequencies of phonons and three phonon branches have to be concerned and stringent limitations are imposed by the energy and wave-vector conservation. The overtone channel, on the other hand, is more likely because two (an optical and an acoustic) phonon branches are concerned, and the energy and wave-vector conservation are necessarily satisfied by two acoustic phonons with the same frequency and opposite wave vectors, if the frequency maximum of the acoustic branch is higher than half the frequency of the initial optical phonon.

Here we consider the down-conversion process in which an optical phonon decays into two acoustic phonons with half the frequency of the optical phonon and with opposite wave vec-

http://dx.doi.org/10.5772/45744

tors. Schematic diagram of this down-conversion process is shown in Fig. 11. The temperature dependence of the relaxation rate γ of the coherent phonon is given by [30,31]

$$\gamma = \gamma_0 \left(1 + \frac{2}{\exp\left[\hbar(\omega_0 / 2) / k_B T\right] - 1}\right) \tag{2}$$

where ω_0 is the frequency of the optical phonon, and k_B is the Boltzmann constant.

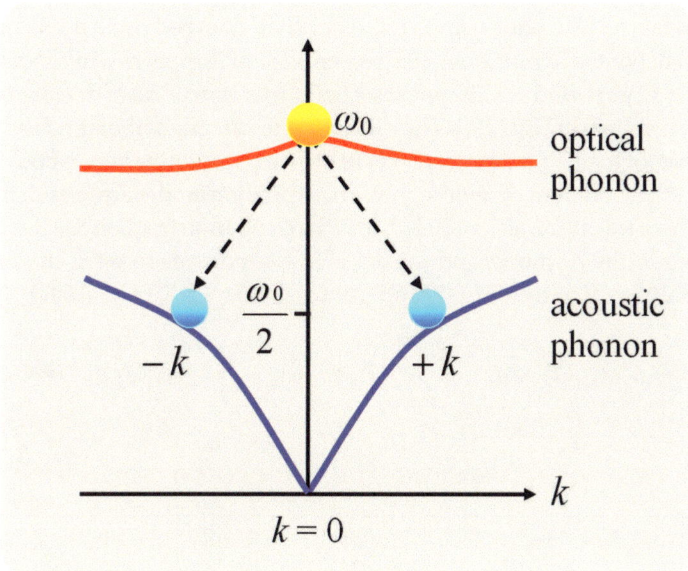

Figure 11. Schematic diagram of the down-conversion processe (overtone channel) in the an harmonic decay model of optical phonons. The initial ω_0 optical phonon with wave vector $k \cong 0$ decays into two $\omega_0 / 2$ acoustic phonons, with opposite wave vectors k and $-k$.

In ordinary materials the temperature dependence of the phonon frequency is small, and a theoretical curve with a fixed value of phonon frequency fits the experimental data well. In $SrTiO_3$, however, the phonon frequencies are changed greatly as the temperature is increased; thus a frequency change has to be considered. The solid curves in Fig. 10 show the theoretical curves including the frequency change obtained from Eq. (2) with γ_0=8. 0×10^{10} s^{-1} for the A_{1g} mode and 1.5×10^{10} s^{-1} for the E_g mode, where the observed phonon frequencies in Fig. 9 are used for each temperature. The broken curves show the theoretical curves in the case of no frequency change. As is seen in Fig. 10, the solid curves explain well the experimental data.

Deviation of the experimental data near T_c from the solid curve may be caused by the effect of the phase transition. However, the relaxation rate just around the phase transition point cannot be obtained in the present experiment, and the relation between the temperature-dependent relaxation rate and the structural phase transition is not clear.

5. Coherent phonon spectroscopy in Ca-doped $SrTiO_3$

5.1. Angular dependence of the coherent phonon signal

The angular dependence of the coherent phonon signal in Ca-doped $SrTiO_3$ observed at 50 K is shown in Fig. 12(a), where the angle between the [100] axis of the crystal and the polarization plane of the pump pulse is 0°, 25°, and 45°. The angle between the polarization planes of the pump and probe pulses is fixed to 45°. Vertical axis is the ellipticity in electric field amplitude of the transmitted probe. At zero delay, a large signal due to the optical Kerr effect, whose width is determined by the laser-pulse width, appears. After that, damped oscillations of coherent phonons are observed. For the 0° pumping a 0.5 ps period signal appears while it disappears for the 45° pumping and a 2 ps period signal appears. TheFourier transform of the coherent phonon signals in Fig. 12(a) is shown in Fig. 12(b). The oscillation frequency of the signalfor the 0° pumping is 1.9 THz and that for the 45° pumping is 0.5 THz. For other pumping angles both frequency components coexist in the coherent phonon signal. These phonon modes are the soft modes related to the structural phase transition at T_{c1}=180 K. From the oscillation frequencies the 1.9 THz component corresponds to the A_{1g} mode, and the 0.5 THz components to the E_g mode, which are assigned from the modes of pure $SrTiO_3$ [7] and from that of $Sr_{1-x}Ca_xTiO_3$ (x=0.007) [8].

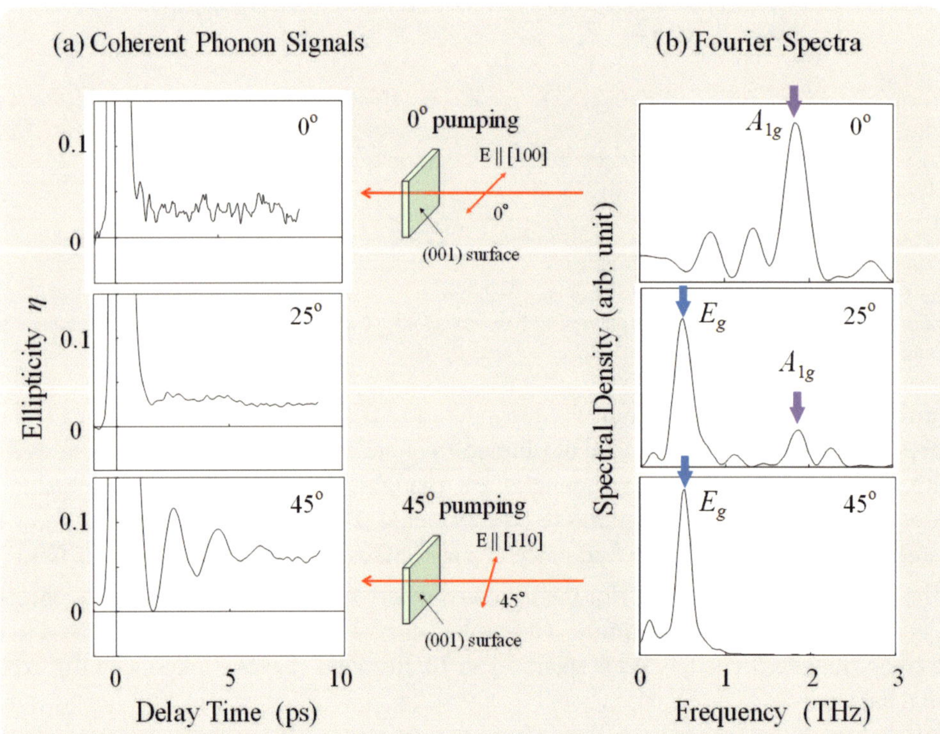

Figure 12. (a) Coherent phonon signals in Ca-doped $SrTiO_3$ observed at 50 K for the 0°, 25°, and 45° pumping, where the angle between the [100] axis of the crystal and the polarization direction of the pump pulse is changed. (b) Fourier transform of the coherent phonon signals in (a).

http://dx.doi.org/10.5772/45744

5.2. Temperature dependence of the coherent phonon signal

The temperature dependence of the coherent phonon signal in Ca-doped $SrTiO_3$ observed for the 45° pumping, which corresponds to the E_g mode, is shown in Fig. 13. At 6 K some oscillation components, which have different frequencies, are superposed. As the temperature is increased, the number of the oscillation components is decreased, the oscillation period becomes longer and the relaxation time becomes shorter. At T_{c1}=180 K, which is the structural phase-transition temperature obtained from the birefringence measurement, no signal of coherent phonons is observed.

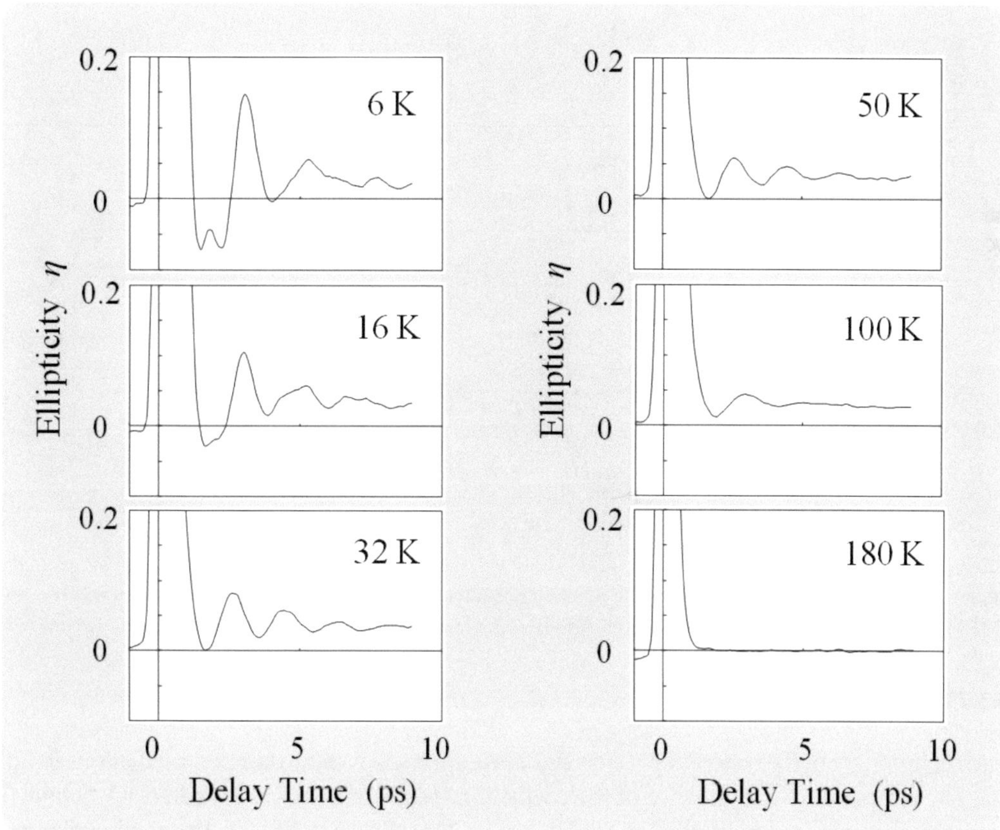

Figure 13. Temperature dependence of the coherent phonon signal in Ca-doped SrTiO3 observed for the 45° pumping.

The temperature dependence of the coherent phonon signal in Ca-doped $SrTiO_3$ observed for the 25° pumping are shown in Figs. 14 and 15. Figure 14(a) shows the coherent phonon signals below the ferroelectric phase-transition temperature T_{c2} and Fig. 15(a) shows those above T_{c2}. Figure 14(b) shows the Fourier transform of the coherent phonon signals in Fig. 14(a), where the peaks with arrows 1, 2, and 3 correspond to the ferroelectric phonon modes [8]. Figure 15(b) shows the Fourier transform of the coherent phonon signals in Fig. 15(a), where modes 1, 2, and 3 disappear but the A_{1g} and E_g modes related to the structural phase-transition remain.

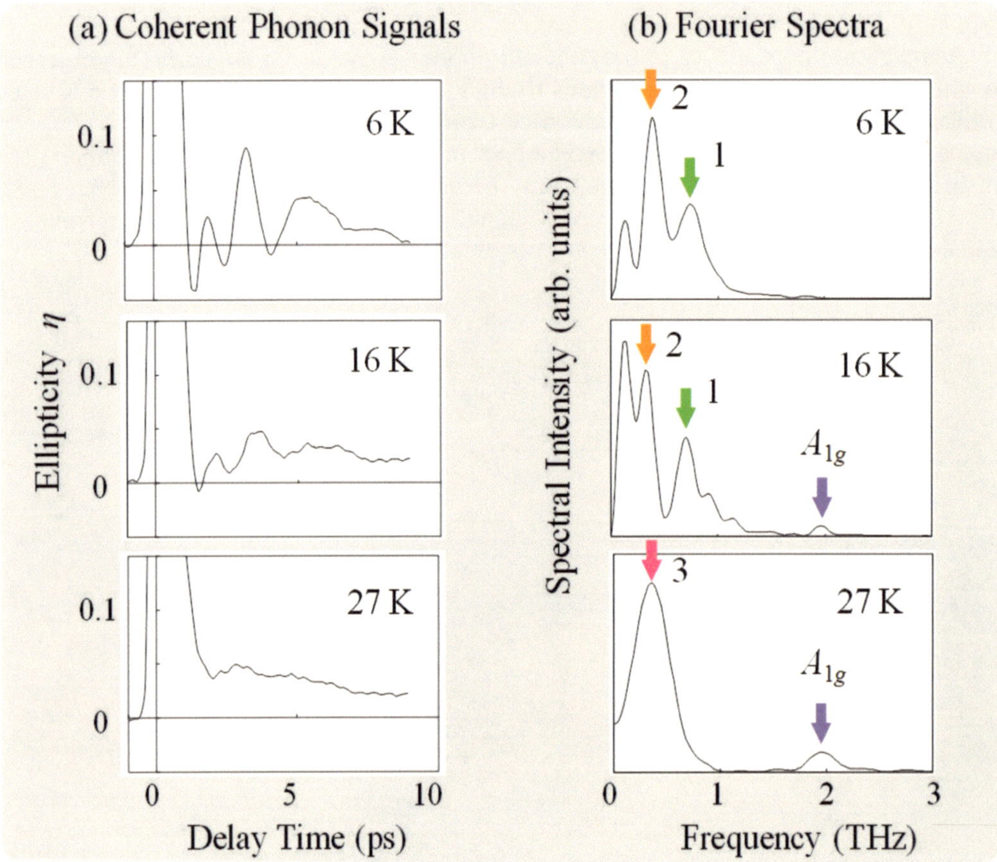

Figure 14. (a) Temperature dependence of the coherent phonon signal in Ca-doped $SrTiO_3$ observed for the 25° pumping below the ferroelectric phase-transition temperature T_{c2}. (b) Fourier transform of the coherent phonon signals in (a).

5.3. Phonon frequencies

Each component of the coherent phonon signal is expressed well by damped oscillations in Eq. (1). The temperature dependence of the oscillation frequencies in Ca-doped $SrTiO_3$ obtained from the observed coherent phonon signals below T_{c1} is shown in Fig. 16. The solid circles are the oscillation frequencies for the A_{1g} and E_g modes. The triangles are that for modes 1, 2, and 3 which are related to the ferroelectric phase transition. As for the A_{1g} and E_g modes, as the temperature is increased from 6 K, the oscillation frequencies decrease and approach to zero at the structural phase-transition temperature T_{c1} for both modes. This result is consistent with the temperature dependence of phonon frequency observed by Raman scattering [7] and coherent phonons in section 4.3 for pure $SrTiO_3$ except for the phase-transition temperature. The broken curves describe a temperature dependence of the form $\omega \propto (T_c - T)^n$. The experimental results for the temperature region between T_{c1} and T_{c2} are explained well by n=0.5 for both modes. Below the ferroelectric phase-transition temperature T_{c2}, another mode appears at 0.9 THz. It is considered that the doubly degenerate E_g mode is split into two components under the tetragonal-to-rhombohedrallattice distortion.

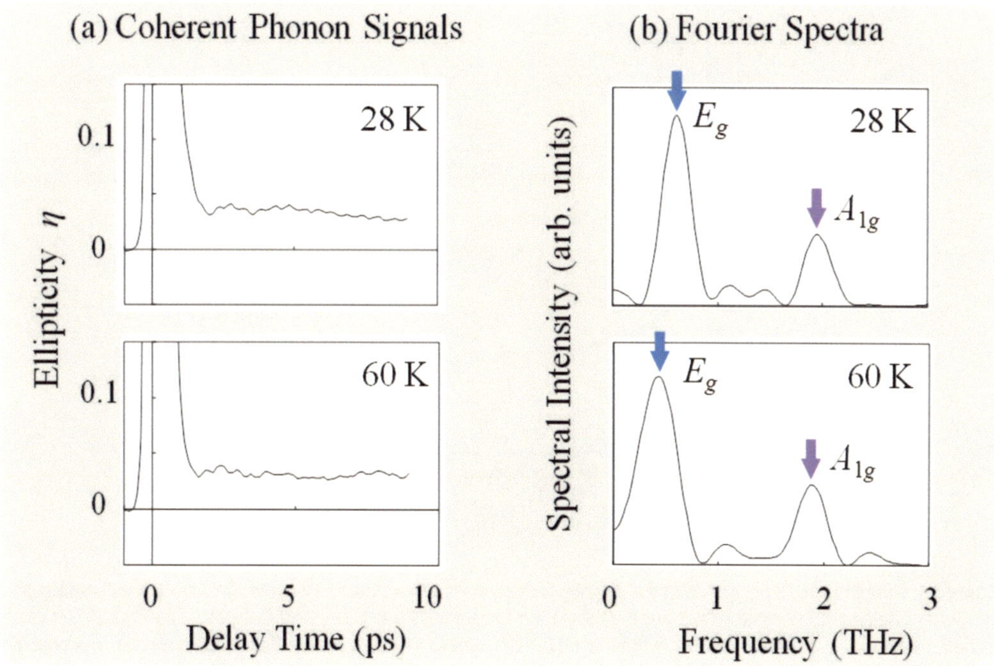

Figure 15. (a) Temperature dependence of the coherent phonon signal in Ca-doped $SrTiO_3$ observed for the 25° pumping above the ferroelectric phase-transition temperature T_{c2}. (b) Fourier transform of the coherent phonon signals in (a).

The temperature dependence of the phonon frequencies, which are related to the ferroelectric phase transition, obtained from the observed coherent phonon signals below T_{c2} is shown in Fig. 17, where only the frequencies of the reproducible peaks in the Fourier spectrum are plotted. While the lowest peaks at 6 and 16 K in Fig. 14, for example, may be the third mode observed in the Raman experiment [8], the frequencies of the peaks with poor reproducibility are not plotted in Figs. 16 and 17. Under dark illumination, two phonon modes 1 and 2 are softened toward about 25 K and degenerate into mode 3, which seems to be softened toward the ferroelectric phase-transition temperature T_{c2}=28 K. The lift of degeneracy for mode 3 below T_{c2} suggests that another structural deformation occurs at 25 K. In the Raman-scattering experiment [8], three modes deriving from the TO_1 mode has been observed. The three modes do not all split off at the same temperature; the highest-energy component splits off at the ferroelectric transition temperature while the other two split off at the temperature about 3 K lower. These two temperatures may correspond to the two temperatures, 28 and 25 K, observed in our experiment, and the temperature differences are nearly equal to each other. The existence of the second structural deformation at 25 K is consistent with the Raman-scattering data.

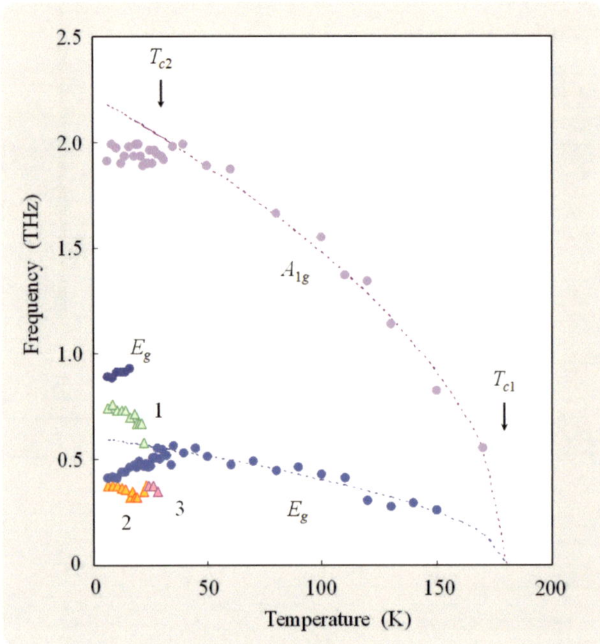

Figure 16. Temperature dependence of the phonon frequencies in Ca-doped SrTiO3 obtained from the coherent phonon signals below T_{c1}. The solid circles are the oscillation frequencies for the A_{1g} and E_g modes. The broken curves describe a temperature dependence of the form $\omega \propto (T_c - T)^n$, where T_c=180 K and n=0.5 for both modes. The triangles are that for modes 1, 2, and 3 related to the ferroelectric phase transition.

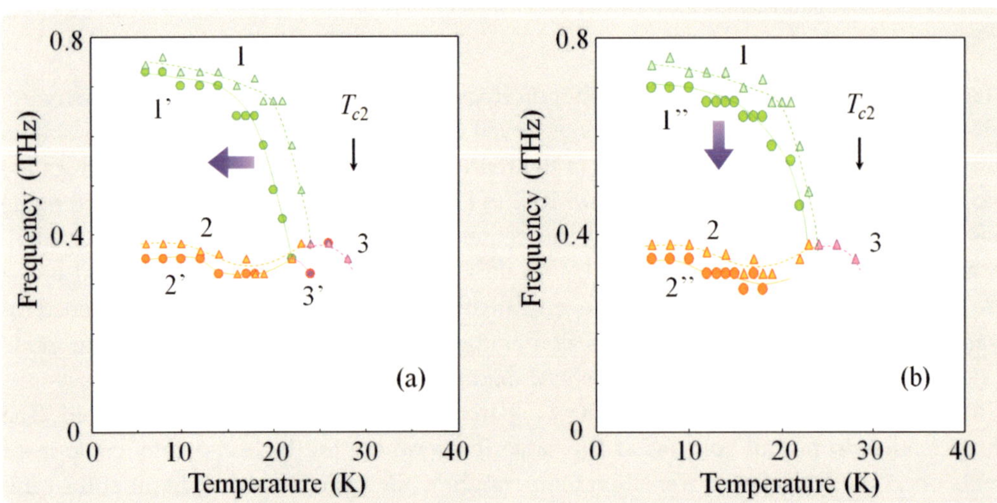

Figure 17. Temperature dependence of the phonon frequencies (a) under dark (1, 2, 3) and under UV (1',2',3') illumination below T_{c2}, which correspond the phonon modes related to the ferroelectric phase transition, and (b)under dark (1, 2, 3) and after UV(1'',2'',3'') illumination, where the intensity of the UV illumination is 7 mW/mm^2.

5.4. UV-illumination effect

In order to examine the UV-illumination effect, two types of measurements, under the UV illumination and after the UV illumination, are carried out. The temperature dependence of the phonon frequencies under the UV illumination is shown in Fig. 17(a), where the intensity of the UV illumination is 7 mW/mm^2. As is seen in Fig. 17(a), the temperature, toward which the two modes 1 and 2 are softened and degenerate into mode 3, shifts to the lower temperature side. The temperature shift due to the UV-illumination effect is ~3 K.

Doped Ca ions behave as permanent dipoles and ferroelectric clusters are formed around Ca dipoles with high polarizability of the host crystal. The ferroelectric transition is caused by the ordering of randomly distributed Ca dipoles. The UV-illumination-induced T_{c2} reduction is related to the screening of the internal macroscopic field by UV-excited carriers. Under the UV-illumination non equilibrium carriers appear, which are captured by traps and screen the polarization field. The ordering is prevented by the photo excited carriers. Thus, the screening effect due to the UV-excited carriers weakens the Coulomb interaction between dipoles, and the transition temperature is decreased.

The theoretical T_{c2} reduction under the UV illumination ΔT_{c2} is given by [11]

$$\frac{\Delta T_{c2}}{T_{c2}(0)} = 1 - \left(1 + \lambda a + \frac{\gamma}{4\pi}\lambda^2 a^2\right)\exp(-\lambda a), \tag{3}$$

where $T_{c2}(0)$ is the transition temperature before the UV illumination, λ is the inverse of the screening length, a is the mean separation between the dipoles, and γ is the number of the nearest-neighbor dipoles. The expression of λ is presented by $\lambda = \sqrt{ne^2 / \varepsilon\varepsilon_0 k_B T}$, where n is the carrier concentration which depends on the UV-illumination intensity and ε is the relative dielectric constant. In the present case, the domain is large enough and the dipole-dipole interaction is expressed as a simple Coulomb interaction. The UV-illumination-intensity dependence of the T_{c2} reduction for $Sr_{1-x}Ca_xTiO_3$ (x=0. 011) was analyzed by using Eq. (3). Assuming that the carrier concentration is proportional to the UV-illumination intensity, the fitting result reproduced well the experimental result obtained in the measurement of dielectric constants [11]. Here we estimate the inverse λ of the screening length at the UV-illumination intensity of 7 mW/mm^2. As a result of the measurement of dielectric constant [11], the carrier concentration is n=2. 5× 10^{17} cm^{-3} at 3 mW/mm^2. From this value we can estimate the carrier concentration to be n=5. 8× 10^{17} cm^{-3} at 7 mW/mm^2. The value of the inverse of the screening length is obtained as

$$\lambda = \sqrt{\frac{ne^2}{\varepsilon\varepsilon_0 k_B T}} = \sqrt{\frac{8\pi n a_B E_{1s}}{\varepsilon k_B T}} = 1.1 \times 10^7 \text{ cm}^{-1}, \tag{4}$$

where a_B and E_{1s} are Bohr radius and the energy of the hydrogen atom in the 1s ground state, respectively, and we use the dielectric constant ε=4 in the visible region. Substituting the value of λ into Eq. (3), the shift of the transition temperature can be estimated to be ΔT_{c2}~8 K. This estimation is not inconsistent with the observed values of the temperature shift in the experiment of coherent phonons, in which the value of the observed transition temperature shift is ~3 K.

The temperature dependence of the phonon frequencies after the UV illumination is shown in Fig. 17(b), where theUV illumination of 7 mW/mm^2 is on before the coherent phonon measurement but is off during the measurement. In this case, on the other hand, the shift of the softening temperature for modes 1 and 2 is not clear. The phonon frequencies for modes 1 and 2 are decreased while the coherent phonon signal for mode 3 is not observed. The relaxation time of the UV-illumination induced carriers is on the order of milliseconds below 30 K for $SrTiO_3$ [21]. As is seen in Fig. 17(b), however, it is suggested that the UV-illumination effect, frequency decrease for modes 1 and 2 and disappearance of mode-3 signal, remains for at least several minutes, even if the UV illumination is switched off, although the T_{c2} shifting effect disappears immediately.

6. Summary

We observed coherent phonons in pure and Ca-doped $SrTiO_3$ using ultrafast polarization spectroscopy to study the ultrafast dynamics of soft phonon modes and their UV-illumination effect. Coherent phonons are generated by linearly polarized pump pulses. The time-dependent linear birefringence induced by the generated coherent phonons is detected asa change of the polarization of the probe pulses. A high detection sensitivity of ~10^{-5} in polarization change, which corresponds to the change Δn=2× 10^{-9} of the refractive index for a 1 mm sample, has been achieved in our detection system.

In $SrTiO_3$, damped oscillations of coherent phonons for the A_{1g} and E_g modes, which contribute to the structural phase transition at 105 K, were observed. The temperature dependences of the frequency and the relaxation rate of the observed coherent phonons were measured. Softening of the phonon frequencies was observed. The phonon relaxation is explained well by a decay model of a frequency-changing phonon, in which the optical phonon decays into two acoustic phonons due to the anharmonic phonon-phonon coupling.

We observed the temperature dependences of the birefringence and the coherent phonon signal to investigate the doping-induced ferroelectric phase transition in Ca-doped$SrTiO_3$ with the Ca concentration of x=0.011. In the birefringence measurement, it was confirmed that the structural phase-transition temperature is T_{c1}=180 K. Coherent phonons were observed by using ultrafast polarization spectroscopy. The damped oscillations of coherent phonons for the A_{1g} and E_g modes, which contribute to the structural phase transition at T_{c1}=180 K, and for the modes 1, 2, and3, which contribute to the ferroelectric phase transition at T_{c2}=28 K, were observed. The phonon frequencies were obtained from the observed signals of coherent phonons, and their softening toward each phase-transition temperature was observed. Another structural deformation at 25 K was found in addition to the ferroelectric phase transition at T_{c2}. It was found that the UV illumination causes the shift of the ferroelectric phase-transition temperature toward the lower temperature side, and the temperature

http://dx.doi.org/10.5772/45744

shift is ~3 K. The decrease in the phonon frequencies after the UV illumination suggests the UV-illumination effect remains even if the UV illumination is switched off.

It was shown that the coherent phonon spectroscopy in the time domain is a very useful approach to study the soft phonon modes and their UV-illumination effect in dielectric materials.

Acknowledgement

We would like to thank Dr. Y. Koyama for experimental help and Dr. Y. Yamada and Prof. K. Tanaka for providing us the samples of Ca-doped $SrTiO_3$.

Author details

Toshiro Kohmoto

Graduate School of Science, Kobe University, Japan

References

[1] K. A. Müller and H. Burkard, Phys. Rev. B 19, 3593 (1979).

[2] H. Uwe and T. Sakudo, Phys. Rev. B 13, 271 (1976).

[3] M. Itoh, R. Wang, Y. Inaguma, T. Yamaguchi, Y.-J. Shan, and T.Nakamura, Phys. Rev. Lett. 82, 3540 (1999).

[4] B. E. Vugmeister and M. P. Glinchuk, Rev. Mod. Phys. 62, 993 (1990).

[5] J. G. Bednorz and K. A. Müller, Phys. Rev. Lett. 52, 2289 (1984).

[6] J. F. Scott, Rev. Mod. Phys. 46, 83 (1974), and references therein.

[7] P. A. Fleury, J. F. Scott, and J. M. Worlock, Phys. Rev. Lett. 21,16 (1968).

[8] U. Bianchi, W. Kleemann, and J. G. Bednorz, J. Phys.: Condens. Matter 6, 1229 (1994).

[9] U. Bianchi, J. Dec, W. Kleemann, and J. G. Bednorz, Phys. Rev. B 51, 8737 (1995).

[10] Bürgel, W. Kleemann, and U. Bianchi, Phys. Rev. B 53, 5222 (1995).

[11] Y. Yamada and K. Tanaka, J. Phys. Soc. Jpn. 77, 5 (2008).

[12] G. Geneste and J.-M. Kiat, Phys. Rev. B 77, 174101 (2008).

[13] M. Takesada, T. Yagi, M. Itoh, and S. Koshihara, J. Phys. Soc. Jpn. 72, 37 (2003).

[14] T. Hasegawa, S. Mouri, Y. Yamada, and K. Tanaka, J. Phys. Soc. Jpn. 72, 41 (2003).

[15] G. Godefroy, P. Jullien, and L. Cai, Ferroelectrics 13, 309 (1976).

[16] S. Ueda, I. Tatsuzaki, and Y. Shindo, Phys. Rev. Lett. 18, 453 (1967).

[17] Y. Yamada and K. Tanaka, J. Lumin. 112, 259 (2005).

[18] T. P. Dougherty, G. P. Wiederrecht, K. A. Nelson, M. H. Garrett,H. P. Jensen, and C. Warde, Science 258, 770 (1992).

[19] H. J. Bakker, S. Hunsche, and H. Kurz, Rev. Mod. Phys. 70, 523 (1998).

[20] W. G. Nilsen and J. G. Skinner, J. Chem. Phys. 48, 2240 (1968).

[21] T. Hasegawa, M. Shirai, and K. Tanaka, J. Lumin. 87-89, 1217 (2000).

[22] Y.-X. Yan, E. B. Gamble, Jr., and K. A. Nelson, J. Chem. Phys.83, 5391 (1985).

[23] G. A. Garrett, T. F. Albrecht, J. F. Whitaker, and R. Merlin, Phys. Rev. Lett. 77, 3661 (1996).

[24] T. Kohmoto, K. Tada, T. Moriyasu, and Y. Fukuda,Phys. Rev. B 74, 064303 (2006).

[25] Y. Koyama, T. Moriyasu, H. Okamura, Y. Yamada, K. Tanaka, and T. Kohmoto, Phys. Rev. B 81, 024104 (2010).

[26] T. Kohmoto, Y. Fukuda, M. Kunitomo, and K. Isoda, Phys. Rev. B 62, 579 (2000).

[27] R. V. Jones, Proc. R. Soc. London, Ser. A 349, 423 (1976).

[28] E. Courtens, Phys. Rev. Lett. 29, 1380 (1972).

[29] K. Watanabe, N. Takagi, and Y. Matsumoto, Phys. Rev. Lett. 92,057401 (2004).

[30] F. Vallée, Phys. Rev. B 49, 2460 (1994).

[31] M. Hase, K. Mizoguchi, H. Harima, S. I. Nakashima, and K. Sakai, Phys. Rev. B 58, 5448 (1998).

[32] M. Hase, K. Ishioka, J. Demsar, K. Ushida, and M. Kitajima, Phys. Rev. B 71, 184301 (2005).

Chapter 13

Raman Scattering Study on the Phase Transition Dynamics of Ferroelectric Oxides

Hiroki Taniguchi, Hiroki Moriwake,
Toshirou Yagi and Mitsuru Itoh

Additional information is available at the end of the chapter

http://dx.doi.org/10.5772/52059

1. Introduction

Ferroelectric phase transitions are conventionally divided into two types: an order-disorder and a displacive-type.[1] In the former one, which is frequently seen in hydrogen-bond-type ferroelectrics such as KH_2PO_4 (KDP), local dipole moments μs are randomly distributed between opposite directions in the paraelectric phase, leading to zero macroscopic net polarization P (= $\Sigma\mu$). A spontaneous polarization in the ferroelectric phase is then driven by their ordering through the ferroelectric phase transition. In the later type, in contrast, there are no dipole moments in the paraelectric phase. The spontaneous polarization in the ferroelectric phase stems from relative polar displacement between cationic and anionic sublattices. It is induced by freezing of a so-called ferroelectric "soft mode", which is known as a strongly anharmonic optical phonon mode at Γ-point in the Brillouin zone (BZ). The displacive-type transition is often found in the ferroelectric oxides as represented by perovskite-type compounds such as $PbTiO_3$. Since the ferroelectric oxides have been widely applied for the electronic devices, such as actuators, sensors, and memories, due to their chemical stability, the large spontaneous polarization, and the relatively high transition temperature, a better understanding of the soft mode behavior would be important from viewpoints of both application and fundamental sciences. Spectroscopic techniques employing light scattering, infrared absorption, and neutron scattering have been generally utilized for investigating the dynamics of the soft mode. Among them, Raman scattering has an advantage especially in a low-frequency region, which is significant to resolve the critical dynamics of the soft mode near the transition temperature. In the present review, we will discuss the soft mode behavior at the ferroelectric phase transition by referring the Raman scattering studies on $CdTiO_3$ and ^{18}O-substituted $SrTiO_3$. Furthermore, a microscopic origin of the soft mode is figured out from a viewpoint of local chemical bonds with the help of first-principles calculations.

2. Ferroelectric soft mode in the classical scheme

Since a concept of the soft mode was proposed, many theoretical and experimental studies have been conducted to clarify its dynamics.[2-4] In this section, we discuss the soft mode behavior in the classical scheme.

2.1. Theoretical background of the ferroelectric soft mode in the classical scheme

The potential of the soft mode is approximately described by accounting only a biquadratic phonon-phonon interaction as follows:

$$U_0 = \frac{1}{2} M(s,0)\omega_{\text{bare}}^2(s,0) Q^2(s,0) + \sum_{\lambda}\sum_{k} \frac{1}{2} J(s,0;\lambda,k) Q^2(s,0) Q^2(\lambda,k) \tag{1}$$

where $M(s,0)$, $\omega_{\text{bare}}(s,0)$, $Q(s,0)$ in the first term are effective mass, bare frequency, and displacement of the soft mode at the Γ-point. The $J(s,0;\lambda,k)$ in the second term denotes an interaction between the soft mode at the Γ-point and the optical phonon mode λ at the wave vector k, where the summation runs over all blanches and BZ. The equation can be transformed by a pseudo-harmonic approximation as

$$\begin{aligned} U_0 &\approx \frac{1}{2} M(s,0)\omega_{\text{bare}}^2(s,0) Q^2(s,0) + \sum_{\lambda}\sum_{k} \frac{1}{2} J(s,0;\lambda,k) Q^2(s,0) \left\langle Q^2(\lambda,k)\right\rangle \\ &= \frac{1}{2} M(s,0)\left[\omega_{\text{bare}}^2(s,0) + \sum_{\lambda}\sum_{k} \frac{J(s,0;\lambda,k)}{M(s,0)} \left\langle Q^2(\lambda,k)\right\rangle\right] Q^2(s,0) \\ &\equiv \frac{1}{2} M(s,0)\omega_s^2 Q^2(s,0) \end{aligned} \tag{2}$$

where the inside of the brackets is redefined by the soft mode frequency ω_s. In the classical regime, the mean square displacement $<Q^2(\lambda,k)>$ is proportional to $k_B T$, where k_B is a Boltzmann constant. The soft mode frequency is thus expressed as

$$\omega_s^2 \equiv \omega^2(s,0) = \omega_{\text{bare}}^2(s,0) + \sum_{\lambda}\sum_{k} \frac{J(s,0;\lambda,k)}{M(s,0)} \frac{k_B T}{M(\lambda,k)\omega^2(\lambda,k)} \tag{3}$$

Since ω_s is ideally zero at T_c of the second-order phase transition,

$$\omega_s^2 = \omega_{\text{bare}}^2(s,0) + \sum_{\lambda}\sum_{k} \frac{J(s,0;\lambda,k)}{M(s,0)} \frac{k_B T_c}{M(\lambda,k)\omega^2(\lambda,k)} = 0 \tag{4}$$

leading to

http://dx.doi.org/10.5772/52059

$$\omega_{\text{bare}}^2(s,0) = -\sum_{\lambda}\sum_{k}\frac{J(s,0;\lambda,k)}{M(s,0)}\frac{k_B T_c}{M(\lambda,k)\omega^2(\lambda,k)} \tag{5}$$

We thus obtain

$$\omega_s^2 = \sum_{\lambda}\sum_{k}\frac{J(s,0;\lambda,k)}{M(s,0)}\frac{k_B T}{M(\lambda,k)\omega^2(\lambda,k)} - \sum_{\lambda}\sum_{k}\frac{J(s,0;\lambda,k)}{M(s,0)}\frac{k_B T_c}{M(\lambda,k)\omega^2(\lambda,k)} \tag{6}$$

Defining C as

$$C \equiv \left[\sum_{\lambda}\sum_{k}\frac{J(s,0;\lambda,k)}{M(s,0)}\frac{k_B}{M(\lambda,k)\omega^2(\lambda,k)}\right]^{1/2} \tag{7}$$

it finally comes to Cochran's law

$$\omega_s = C(T - T_c)^{1/2} \tag{8}$$

where

$$T_c \equiv -\omega_{\text{bare}}^2(s,0)\left[\sum_{\lambda}\sum_{k}\frac{J(s,0;\lambda,k)}{M(s,0)}\frac{k_B}{M(\lambda,k)\omega^2(\lambda,k)}\right]^{-1} \tag{9}$$

As seen in the above equation, the soft mode decreases its frequency with approaching T_c, and finally freezes to induce the ferroelectric phase transition. In the second-order phase transition, in particular, the displacement of the soft mode $Q(s,0)$ corresponds to the polar displacement in the ferroelectric phase. It should be noted that the bare frequency $\omega_{\text{bare}}(s,0)$ of the soft mode is assumed to be imaginary at zero-Kelvin. The softening of the soft mode connects to the divergent increase of the dielectric constant through Lyddane-Sachs-Teller (LST) relation,

$$\varepsilon' \propto \frac{1}{\omega_s^2} \tag{10}$$

2.2. Experimental observation of the ferroelectric soft mode in the classical scheme

The experimental observation of the typical soft mode behavior in the perovskite-type ferroelectric oxide, $CdTiO_3$, is presented in this section.

$CdTiO_3$ possesses an orthorhombic *Pnma* structure at room temperature due to ($a^+b^-b^-$)-type octahedral rotations in the Glazer's notation from the prototypical Cubic *Pm*-3*m* structure. The $CdTiO_3$ undergoes ferroelectric phase transition into the orthorhombic $Pna2_1$ phase around 85 K.[5-7]

The confocal micro-Raman scattering is the one of useful techniques to investigate the soft mode dynamics in the displacive-type ferroelectric phase transition due to the following reason; In general, the complicated domain structure forms across the ferroelectric phase transition, where the principle axis of the crystal orients to various directions, which are allowed by the symmetry relation between the paraelectric and the ferroelectric phases. The size of the individual domain generally ranges over several nanometers to microns. When the domains are smaller than the radiated area of the incident laser, the observed Raman spectrum is composed of signals from differently oriented domains, leading to difficulty in the precise spectral analyses. With application of the confocal micro-Raman scattering, on the other hand, we can selectively observe the spectrum from the single domain region, because its spatial resolution reaches sub-microns not only for lateral but also depth directions.[8] Note that the Raman scattering can observe the soft mode only in the non-centrosymmetric phase due to the selection rule, therefore the critical dynamics of the soft mode is in principle investigated in the ferroelectric phase below T_c.

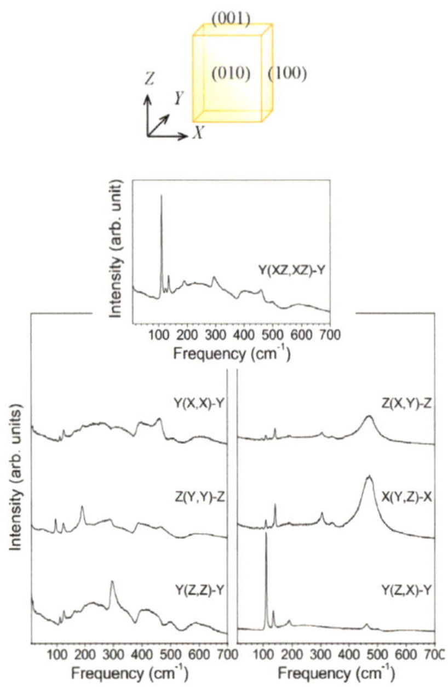

Figure 1. Raman spectra of $CdTiO_3$ at room temperature observed with several scattering geometries. See text for configuration notations, for instance Y(X,X)-Y.

Flux-grown colorless single crystals of $CdTiO_3$ with a rectangular solid shape were used for this study. The two samples have dimensions of approximately 0.3×0.2×0.1 mm^3 and

0.2×0.1×0.1 mm³. Since the present samples were twinned, we carefully determined the directions of the axes in the observed area by checking the angular dependence of the Raman spectra with the confocal micro-Raman system, whose spatial resolution is around 1 μm (Fig. 1). Here we use laboratory coordinates X, Y, and Z, corresponding respectively to the crystallographic axes of [100], [010], and [001] in the paraelectric phase, to denote the Raman scattering geometries. For example, XZ(Y,Y)-Z-X means that the incident laser polarized parallel to the [010] direction penetrates the sample along the [101] direction, and the [010]-polarized scattered light is collected with the backscattering geometry. The scattered light was analyzed by a Jovin-Yvon triple monochromator T64000 with the subtractive dispersion mode, and was finally observed by a liquid-N_2-cooled CCD camera. The frequency resolution of the present experiment was better than 1 cm^{-1} The temperature of the sample was controlled by an Oxford microstat with a temperature stability of ±0.01 K.

Figure 2 shows the temperature dependence of the Raman spectrum of $CdTiO_3$, which is observed in Y(ZX,ZX)-Y scattering geometry with several temperatures from 85.0 K to 40.0 K. In the vicinity of T_c, in particular, spectra were observed with the temperature intervals of 0.5 K. As shown in the figures, the soft mode is seen at around 60 cm^{-1} at 40 K. It gradually softens toward zero-frxuency as approaching T_c ~ 85 K. The spectral profile of the soft mode is analyzed by a damped-harmonic-oscillator (DHO) model;

$$I(\omega) = \frac{B\omega_0^2\Gamma\omega}{(\omega_0^2 - \omega^2)^2 + \Gamma^2\omega^2} \tag{11}$$

where B, ω_0, and Γ denote amplitude, harmonic frequency, and a damping constant of the soft mode, respectively. The soft mode frequency ω_s is defined here as

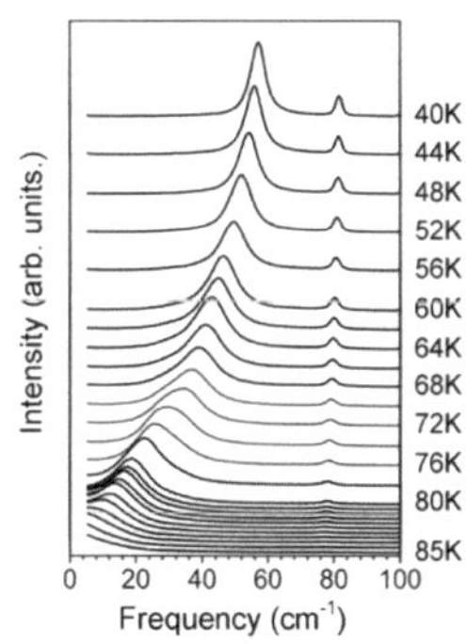

Figure 2. Temperature Dependence of the Raman spectrum of $CdTiO_3$ observed in the Y(ZX,ZX)-Y scattering geometry, and from 40K to 85 K on heating.

$$\omega_s^2 = \sqrt{\omega_0^2 - \Gamma^2} \tag{12}$$

The temperature dependence of the soft mode frequency and the damping constant, which are determined by the analyses, are plotted by solid circles and open squares in Fig. 3. Synchronizing with the softening of the soft mode, the damping constant is increased as approaching T_c.

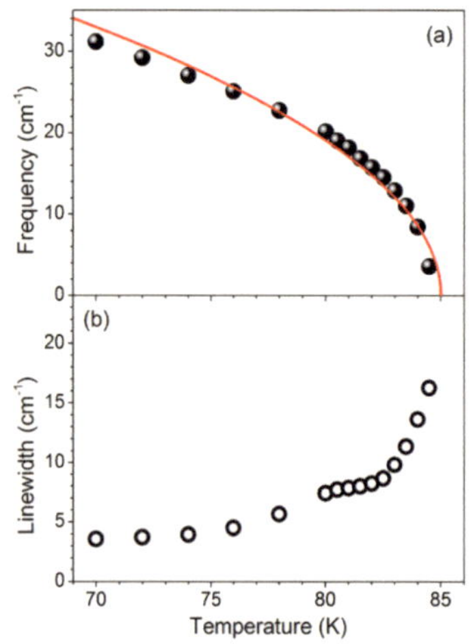

Figure 3. Temperature Dependencies of (a) the soft mode frequency and (b) the damping constant of $CdTiO_3$. The solid line in (a) is calculated by Cochran's law (see text).

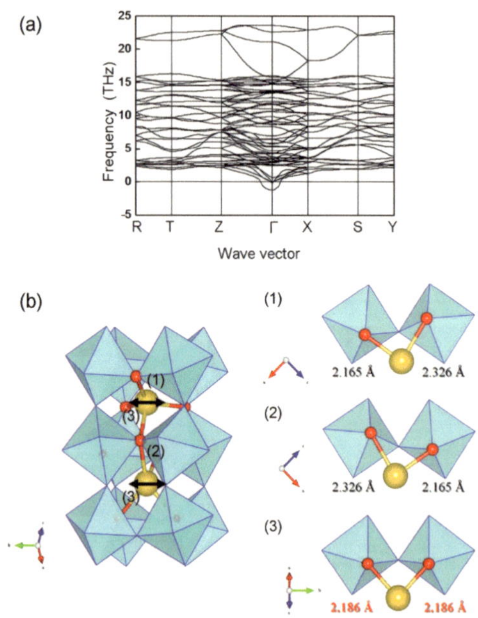

Figure 4. a) Phonon dispersion in $CdTiO_3$ obtained by the first-principles calculations. (b) The displacement pattern of the soft mode, and configurations of three O-Cd-O bonds along (1) [-101], (2) [101], and (3) [010] directions.

http://dx.doi.org/10.5772/52059

The temperature dependence of the soft mode frequency obeys the Cochran's low (Eq. 1.8) with $C = 8.5$ and $T_c = 85$ K as expressed by the solid curve in the figure, indicating the typical displacive-type ferroelectric phase transition on $CdTiO_3$.

The displacement pattern of the soft mode is obtained by first principle calculations. The calculations were conducted with a pseudopotential method based on a density functional perturbation theory with norm-conserving pseudopotentials, which was implemented in the CASTEP code.[9] Figure 4a presents phonon dispersion curves of $CdTiO_3$ in the paraelectric *Pnma* phase, which was obtained by the first-principle calculations. As seen in the figure, the soft mode, which is indicated by an imaginary frequency, is observed at Γ-point in Brillouin zone. According to the calculation, a displacement pattern of the Γ-point soft mode is composed of relative displacement of Ti with respective to the octahedra and an asymmetric stretching of O-Cd-O bond. It has been previously proposed that the hybridization between O-2p and empty Ti-3d orbitals triggers the non-centrosymmetric displacement of Ti thorough a second-order Jahn-Teller (SOJT) effect to induce the ferroelectricity as exemplified by $BaTiO_3$.[10] The SOJT effect, however, does not provide clear understanding on the role of *A*-site ion. In the case of $CdTiO_3$, in particular, the ferroelectricity disappears when the Cd is replaced by Ca, though the $CaTiO_3$ is an isomorph of $CdTiO_3$. Therefore, the ferroelectricity in $CdTiO_3$ can not be explained only by the *B*-O orbital hybridization. Here, we focus on the asymmetric stretching of O-Cd-O to elucidate the mechanism of ferroelectricity in $CdTiO_3$. The asymmetric stretching in the Γ-point soft mode is schematically illustrated by arrows in the right panel in Fig. 4b, where large and small spheres denote Cd and O ions, respectively. Ti and O ions that do not participate in the O-Cd-O bonds are omitted for simplicity. In a prototypical *Pm*-3*m* structure of the perovskite-type oxide, an *A*-site ion is surrounded by neighboring twelve O ions. In the $CdTiO_3$ of paraelectric *Pnma* structure, however, the number of closest oxygen ions for Cd ion is reduced to four due to the octahedral rotations. (Note that the octahedral rotations in the *Pnma* structure are different from each other according to the direction; the rotation is in-phase for a [010] direction whereas those for [101] and [-101] are anti-phase.) As a result, $CdTiO_3$ has three different O-Cd-O bonds along [-101], [101], and [010] directions, whose equilibrium structures are respectively denoted by (1), (2), and (3) in the Fig. 2(b). The result of first-principles calculations indicates that the soft mode displacement is inherently dominated by the asymmetric stretching of the O-Cd-O bond along the [010] direction denoted by (3). By comparing three equilibrium structures for O-Cd-O bonds presented in the right-hand side of Fig. 4b, it is found that the O-Cd-O bond along [010] direction is symmetric with the Cd-O bond lengths of 2.186 Å in the paraelectric *Pnma* phase, in contrast the other two are composed of non-equivalent bonds having 2.326 and 2.165 Å. Therefore, there is a remaining degree of freedom for further O-Cd-O off-center ordering due to the covalent bonding along the [010] direction. This remaining degree of freedom would be the origin of the ferroelectric soft mode.

3. A role of covalency in the ferroelectric soft mode

As indicated above, the covalency of the *A*-site ion plays an important role in the ferroelectricity in addition to that of *B*-site ion. The *A*-site substitution with an isovalent ion is instructive for the better understanding on the mechanism of the ferroelectricity in the perovskite-type oxides. This section is thus devoted to the effect of the isovalent Ca-substitution on the ferroelectricity of $CdTiO_3$. [11, 12]

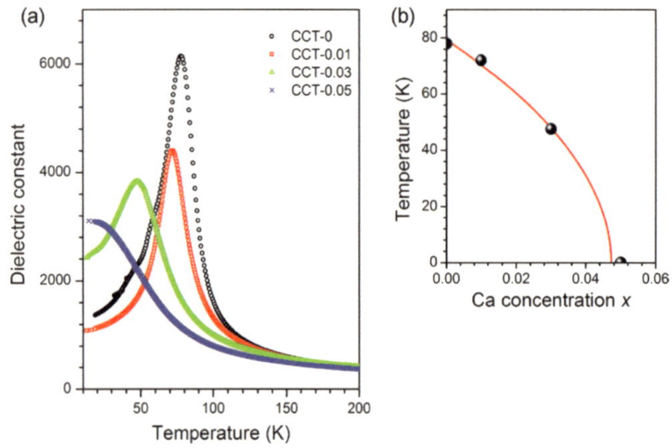

Figure 5. a) Dielectric constants of CCT-*x*. (b) Ca concentration dependence of the ferroelectric phase transition temperature T_c. Solid curve denotes Curie–Weiss fit of data points for CCT-0, 0.01, and 0.03.

Ceramics form of Ca-doped $CdTiO_3$, $Cd_{1-x}Ca_xTiO_3$ (CCT-*x*) with $x = 0, 0.01, 0.03, 0.05$, and 0.07, were fabricated by conventional solid state reactions from stoichiometric mixtures of CdO (3N), $CaCO_3$ (4N), and TiO_2 (3N). The mixtures were calcined during 24 h in air at 975 ~ 1175ºC. Then the reground powders were pressed at 4 ton/cm^2, and sintered at 1200 ~ 1230ºC. Dielectric measurements were performed by a LCZ meter.

Temperature dependencies of the dielectric constants in CCT-*x* are presented in Fig. 5a. As seen in the figure, the dielectric constant of CCT-0 divergently increases and culminates at the phase transition temperature T_c ~ 85 K. The T_c decreases with Ca doping, and the peak transforms into a low-temperature plateau at $x = 0.05$, indicating a quantum paraelectric state as discussed later. The Ca concentration dependence of T_c, which is estimated from the dielectric peaks, is shown in Fig. 5b. The fit curve with conventional Curie-Weiss law (a red line in the figure) indicates that the ferroelectricity in the CCT-*x* system is suppressed with a value of *x* larger than $x_c = 0.047$.

The temperature dependence of the soft mode frequency is presented in Fig. 6 as a function of *x*. The soft modes in CCT-*x* increase in frequency and intensity with distance from T_c on cooling, exhibiting typical behavior of the soft-mode-driven phase transition. According to the Lyddane–Sachs–Teller relationship (Eq. 1.10), the squared soft mode frequency ω_s^2 and the dielectric constant ε' are inversely proportional as mentioned in the preceding section. Therefore, if the dielectric properties of the CCT-*x* system are governed by the lattice dynamics, the soft mode frequency at the lowest temperature should decrease with increasing x because the dielectric constant near 0 K increases with x (see Fig. 5a). The Raman spectra at 4 K presented in the bottom panels in Fig. 6 show that the soft mode softens as x increases, indicating a displacive-type phase transition of the CCT-x system. This result suggests that the phase transition of CCT-x can be discussed in terms of lattice dynamics. Note that, the soft mode become observable at low temperature in CCT-0.05 and -0.07, although it does not undergo the ferroelectric phase transition within a finite temperature range. This is a typical characteristic of precursory softening of the soft mode in the quantum paraelectric state.[13, 14]

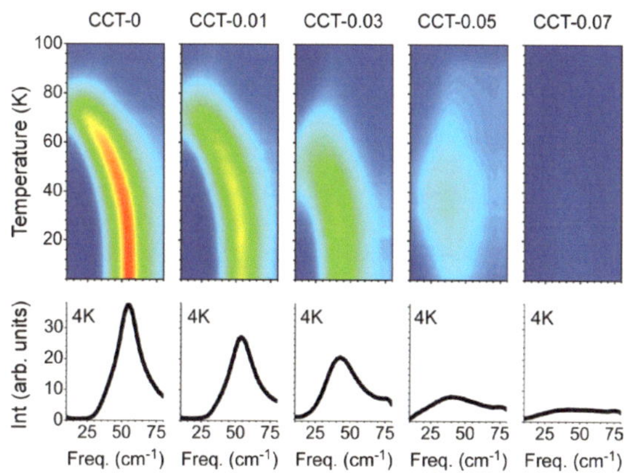

Figure 6. Contour plots for the temperature dependence of soft mode spectra in CCT-x (upper panels). Bottom panels show the low-frequency spectra observed at ~ 4 K.

Figure 7 presents the partial electronic density of states (p-DOS) for $CdTiO_3$ (left) and $CaTiO_3$ (right) obtained by the first-principles calculations, where the p-orbital of oxygen, the s-orbital of Cd and Ca, and the total DOS are denoted by oppositely hatched and blank areas, respectively. As shown in the figure, the s-orbital of Cd ion has strong hybridization with the p-orbital of oxygen. In marked contrast, the s-orbital of Ca has little hybridization, indicating larger A-O covalency in $CdTiO_3$ than $CaTiO_3$. This characteristic substantially agrees with the difference in electronegativity of Cd and Ca ions, where those in Cd and Ca are 1.7 and 1.0, respectively. A calculated charge density distributions around the O-Cd(Ca)-O bonds along the [010] direction are indicated in Fig. 8. As shown in the panels, larger charge density is observed between Cd and O ions in $CdTiO_3$, confirming its strong covalency. In contrast, Ca-O bond is nearly ionic. If the Cd(Ca) bonding is ionic, Cd(Ca) ion in the O-Cd(Ca)-O bond tends to locate a centric position. Therefore, Ca-substitution suppresses the freezing of the asymmetric stretching vibration with the off-centering of Cd(Ca) by decreasing the covalency. Since the asymmetric stretching of the O-Cd-O bond along [010] direction is suggested to be the origin of the soft mode, Ca-substitution results in the suppression of the softening of the soft mode. The present result confirms that the ferroelectric instability of CCT-x stems from A-site covalency.

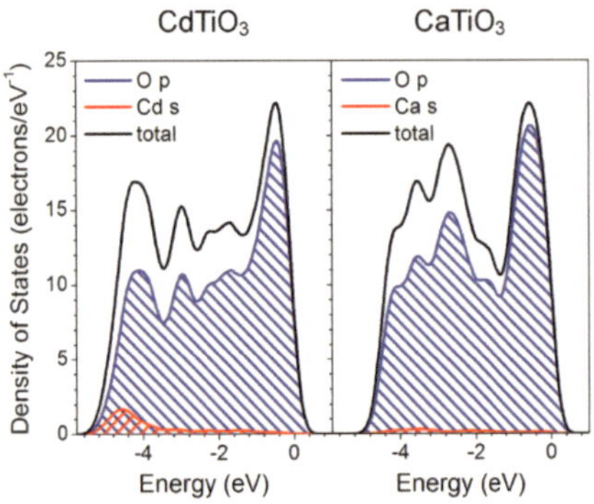

Figure 7. Partial electron density of states (p-DOS) for $CdTiO_3$ (left) and $CaTiO_3$ (right). Blue, red, and black curves represent *p* orbital of oxygen ions, *s* orbital of Cd and Ca ions, and total DOS.

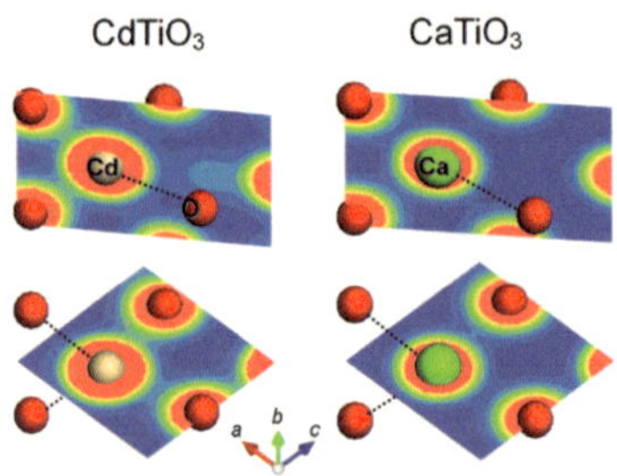

Figure 8. Cross sections of charge density around 4-coordinated Cd (left) and Ca (right) ions obtained by first principle calculations. Note that the upper and lower cross sections include O–Cd(Ca)–O bonds, which are involved in O–Cd(Ca)–O chains along the *b* and *a* direction, respectively.

4. Ferroelectric Soft Mode in the Quantum Scheme

It has been known that the quantum fluctuation plays a non-negligible role in the phase transition dynamics when the T_c goes down near 0 K, where the transition is suppressed though the dielectric permittivity reaches to several tens thousand. This effect is known as "quantum paraelectricity", which was first proposed as an origin of the giant dielectric plateau of $SrTiO_3$ at the low-temperature.[15] Twenty years later, it has been discovered that an isotope substitution with ^{18}O induces the ferroelectricity of $SrTiO_3$, attracting considerable attention of researchers.[16] In this section, the dynamics of quantum para/ferroelectric is discussed from a view point of the soft mode.

4.1. Theoretical background of the ferroelectric soft mode in the quantum scheme

In the classical case as mentioned before, the soft mode frequency can be described as Eq. 1.3. At the low-temperature, where the influence of the quantum fluctuation can not be ignored, the approximation $<Q^2(\lambda, k)> \sim k_B T$ is not appropriate. Therefore, Eq. 1.3 should be modified with quantum statistic treatment as

$$\omega_s^2 = \omega_{\text{bare}}^2(s,0) + \sum_\lambda \sum_k \frac{J(s,0;\lambda,k)}{M(s,0)} \frac{\hbar}{2M(\lambda,k)\omega(\lambda,k)} \coth \frac{\hbar\omega(\lambda,k)}{2k_B T} \tag{13}$$

At the sufficiently low-temperature, where only the Γ-point soft mode that is the optical phonon with lowest energy can be thermally excited, any other optical phonons are not effective in the second term except for $\lambda = s$ at $k = 0$. The soft mode frequency can thus be expressed as

$$\omega_s^2 = \omega_{\text{bare}}^2(s,0) + \frac{J(s,0;s,0)}{2M^2(s,0)} \frac{\hbar}{\omega_s} \coth \frac{\hbar\omega_s}{2k_B T} \tag{14}$$

As in the classical case,

$$\omega_s^2 = \omega_{\text{bare}}^2(s,0) + \frac{J(s,0;s,0)}{2M^2(s,0)} \frac{\hbar}{\omega_s} \coth \frac{\hbar\omega_s}{2k_B T_c} = 0 \tag{15}$$

gives

$$\omega_{\text{bare}}^2(s,0) = -\frac{J(s,0;s,0)}{2M^2(s,0)} \frac{\hbar}{\omega_s} \coth \frac{\hbar\omega_s}{2k_B T_c} \tag{16}$$

We thus obtain the soft mode frequency at the low-temperature,

$$\omega_s = C' \left(\frac{T_1}{2} \coth \frac{T_1}{2T} - T_c' \right)^{1/2} \tag{17}$$

with

$$C' \equiv \left[\frac{J(s,0;s,0)}{M^2(s,0)} \frac{k_B}{\omega_s^2} \right]^{1/2}, \quad T_1 \equiv \frac{\hbar\omega_s}{k_B}, \quad T_c' \equiv \frac{T_1}{2} \coth \frac{T_1}{2T_c} \tag{18}$$

In this treatment, the temperature dependence of the squared soft mode frequency, which is inversely proportional to the dielectric permittivity, in no longer linear with respect to temperature, but saturated with the constant value near 0 K. The qualitative behavior of the soft mode in the quantum para/ferroelectrics is schematically illustrated as a function of T_1 with the fixed T'_c in Fig. 9.

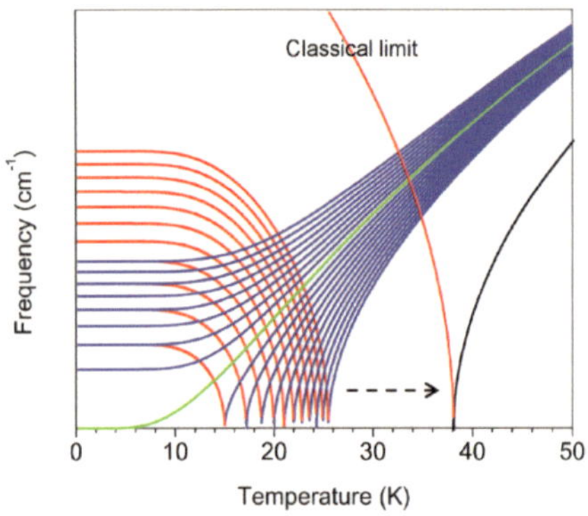

Figure 9. The schematic illustration describing the variation of the temperature dependence of the soft mode frequency as a function of T_1. See text for the detail.

The classical limit is also presented for comparison. Note that the value of T_1 is varied with constant intervals. As presented in the figure, the phase transition is completely suppressed when T_1 is sufficiently large, whereas it recovers as decreasing T_1. Interestingly, the T_1 dependence of the transition temperature becomes extremely sensitive when the transition temperature is close to 0K, suggesting that the quantum paraelectric-ferroelectric transition can be induced by subtle perturbation such as the isotope substitutions. [17]

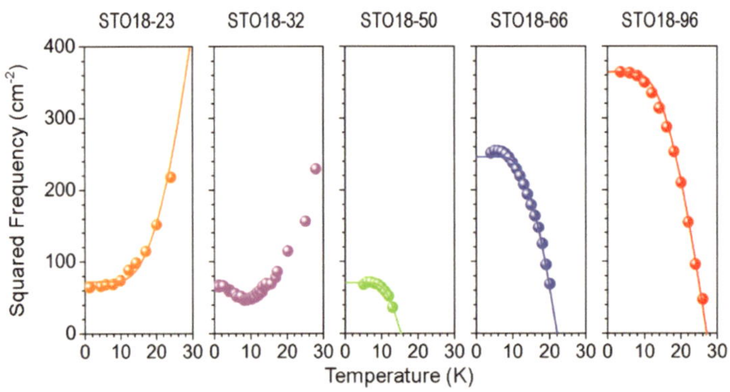

Figure 10. Temperature dependencies of the squared frequencies of the soft mode in STO18-*x* with various *x*. Solid lines in the figures denote temperature dependencies calculated by Eq. 3.7 with optimized control parameters (see text).

4.2. Experimental observation of the ferroelectric soft mode in the quantum scheme

The isotopically induced phase transition of the quantum paraelectric $SrTiO_3$ is a good example for the quantum paraelectric-ferroelectric phase transition driven by the soft mode.

Here we show the Raman scattering study on the soft mode in $SrTi(^{16}O_{1-x}{}^{18}O_x)_3$ (STO18-100x) as functions of temperature and the isotope substitution rate x. The experiments were performed with the Raman scattering geometries of X(YY)-X, where X, Y, and Z are denoted by the cubic coordination of $SrTiO_3$.[13,14,18]

Figure 10 presents the temperature dependence of the squared soft mode frequency in STO18-100x for 0.23, 0.32, 0.50 0.66, and 0.96, observed in the X(YY)-X scattering geometry, respectively. In x = 0.50, 0.66, and 0.96, the soft mode frequency is shown only for the ferroelectric phase. As indicated in the figure, the softening of the soft mode in STO18-23 saturates at the low-temperature region near 0 K, showing excellent agreement with the theory. Since the square of the soft mode frequency is inversely proportional to the dielectric permittivity as indicated by the LST-relation as mentioned before, it is clear that the dielectric plateau in $SrTiO_3$ stems from the soft mode dynamics. Note that the soft mode is nominally Raman inactive in the centrosymmetric structure as for the paraelectric phase of $SrTiO_3$ due to the selection rule. In the present case, however, the centrosymmetry is locally broken in the low-temperature region of STO18-100x ($x < x_c \sim 0.32$), leading to the observation of the soft mode even in the macroscopic centrosymmetry. The local non-centrosymmetric regions grow with ^{18}O-substitution to activate the soft mode spectrum even in the paraelectric phase as indicated in Fig. 11. A mechanism of such defect-induced Raman process and an expected spectral profile are discussed in detail in Ref. [13] and [14]. The date points for STO18-x ($x < x_c$) in Fig. 10 are obtained by the spectral fitting with the defect-induced Raman scattering model.

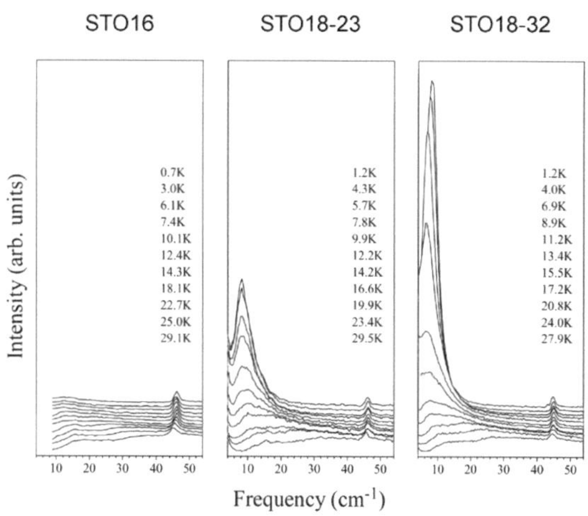

Figure 11. Temperature dependencies of the Raman spectra in STO18-x (x = 0 [STO16: the pure $SrTiO_3$], 0.23, and 0.32).

With the isotope substitution, softening of the soft mode is enhanced and the phase transition takes place above the critical concentration x_c as manifested by the hardening of the soft mode on cooling in the low temperature region. The transition temperature elevates as increasing x. The temperature dependencies of the soft modes in all samples except for that in

STO18-32 were examined by the generalized quantum Curie-Weiss law, which was proposed in Ref. 19, where the power of Eq. 3.5 is modified by $\gamma/2$;

$$\omega_s = C'\left(\frac{T_1}{2}\coth\frac{T_1}{2T} - T_c'\right)^{\gamma/2} \tag{19}$$

The γ is the one of critical exponents, which characterizes the critical behavior of the susceptibility. As seen in the figure, the plots are well reproduced by the fitting with the systematic variations of T_1 and γ as indicated in Fig. 12.

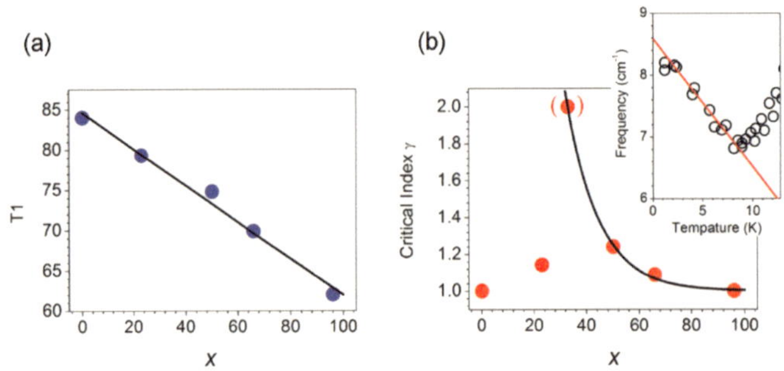

Figure 12. a) The x dependence of T_1obtained by the fitting. The dashed line is a guide for the eye. (b) The x dependence of critical exponent γ. In STO18-23, -50, -66, -96, γ is determined by the fitting of the observed soft mode frequency with Eq. 3.7. In STO18-32, we directly determined γ = 2 from the linear temperature dependence of the soft mode frequency with T, as seen in the inset. The solid lines are eye guides.

In STO18-32, the strongly rounded soft mode behavior keeps us from fitting with Eq. 3.7. However, we can determine $\gamma = 2$ for STO18-32 from the obvious linear temperature dependence of the soft mode frequency, as seen in the inset of the panel (b). It should be noted here that $\gamma = 2$ corresponds to the theoretical value for the quantum ferroelectric phase transition [20–22], and is caused by the quantum mechanical noncommutativity between kinetic and potential energy. These results show experimentally that the quantum phase transition of STO18-x is an ideal soft mode-type transition driven by direct control of the quantum fluctuation with the ^{18}O-substitution. An origin of the strongly rounded softening was suggested in our previous study to be the nanoscopic phase coexistence between ferroelectric and paraelectric region. (See Ref. 18 for the detail)

5. Summary

In the present review, we overviewed the soft mode behavior in both classical and quantum schemes through Raman scattering experiment in the $CdTiO_3$ and the ^{18}O-substituted $SrTiO_3$. The analyses show excellent agreement qualitatively between the fundamental theory and experimental observations. The systematic studies with Raman scattering experi-

http://dx.doi.org/10.5772/52059

ments and the first-principles calculations on Ca-substituted $CdTiO_3$ clarified that the covalency between constituent cation and oxygen plays an essential role in the origin of the soft mode. We hope the present review serves better understanding on the mechanism of displacive-type ferroelectric phase transition.

Author details

Hiroki Taniguchi[1*], Hiroki Moriwake[2], Toshirou Yagi[3] and Mitsuru Itoh[1]

*Address all correspondence to:

1 Materials and Science Laboratory, Tokyo Institute of Technology, Yokohama, Japan

2 Nanostructures Research Laboratory, Japan Fine Ceramics Center, Nagoya, Japan

3 Research Institute for Electronic Science, Hokkaido University, Sapporo, Japan

References

[1] Line, M. E., & Glass, A. M. (1997). Principles and Applications of Ferroelectrics and Related Materials,. Claredon, Oxford,.

[2] Cochran, W. (1960). Crystal stability and the theory of ferroelectricity. *Adv. Phys.*, 9, 387.

[3] Cochran, W. (1961). Crystal stability and the theory of ferroelectricity part II. Piezoelectric crystals. *Adv. Phys.*, 10, 401 EOF-420 EOF.

[4] Scott, J. F. (1974). Soft-mode spectroscopy: Experimental studies of structural phase transitions. Rev. Mod. Phys. ., 46, 83 EOF-128 EOF.

[5] Shan, Y. J., Mori, H., Imoto, H., & Itoh, M. (2002). Ferroelectric Phase Transition in Perovskite Oxide $CdTiO_3$. *Ferroelectrics*, 381 EOF-386 EOF.

[6] Shan, Y. J., Mori, H., Tezuka, K., Imoto, H., & Itoh, M. (2003). Ferroelectric Phase Transition in $CdTiO_3$ Single Crystal. *Ferroelectrics*, 107 EOF-112 EOF.

[7] Moriwake, H., Kuwabara, A., Fisher, C. A. J., Taniguchi, H., Itoh, M., & Tanaka, I. (2011). First-principles calculations of lattice dynamics in $CdTiO_3$ and $CaTiO_3$: Phase stability and ferroelectricity. *Phys. Rev. B*, 84, 104.

[8] Taniguchi, H., Shan, Y. J., Mori, H., & Itoh, M. (2007). Critical soft-mode dynamics and unusual anticrossing in $CdTiO_3$ studied by Raman scattering. *Phys. Rev. B*, 76, 212103.

[9] Clark, S. J., Segall, M. D., Pickard, C. J., Hasnip, P. J., Probert, M. I. J., Refson, K., & Payne, M. C. (2005). First principles methods using CASTEP. *Z. Kristallogr*, 220, 567.

[10] Bersuker, I. B. (1995). Recent development of the vibronic theory of ferroelectricity:. *Ferroelectrics*, 164, 75.

[11] Taniguchi, H., Soon, H. P., Shimizu, T., Moriwake, H., Shan, Y. J., & Itoh, M. (2011). Mechanism for suppression of ferroelectricity in $Cd_{1-x}Ca_xTiO_3$. *Phys. Rev. B*, 84, 174106.

[12] Taniguchi, H., Soon, H. P., Moriwake, H., Shan, Y. J., & Itoh, M. (2012). Effect of Ca-Substitution on $CdTiO_3$ Studied by Raman Scattering and First Principles Calculations. *Ferroelectrics*, 268 EOF-273 EOF.

[13] Taniguchi, H., Takesada, M., Itoh, M., & Yagi, T. (2004). Effect of oxygen isotope exchange on ferroelectric microregion in $SrTiO_3$ studied by Raman scattering. *J. Phys. Soc. Jpn.*, 73, 3262 EOF-3265 EOF.

[14] Taniguchi, H., Takesada, M., Itoh, M., & Yagi, T. (2005). Isotope effect on the soft-mode dynamics of $SrTiO_3$ studied by Raman scattering. *Phys. Rev. B*, 72, 064111.

[15] Müller, K. A., Burkard, H., & Sr Ti, O. (1979). An intrinsic quantum paraelectric below 4 K. *Phys. Rev. B*, 19, 3593.

[16] Itoh, M., Wang, R., Inaguma, Y., Yamaguchi, Y., Shan, Y. J., & Nakamura, T. (1999). Ferroelectricity Induced by Oxygen Isotope Exchange in Strontium Titanate Perovskite. *Phys. Rev. Lett.*, 82, 3540 EOF-3543 EOF.

[17] Tokunaga, M., & Aikawa, Y. (2010). Isotope-Induced Ferroelectric Phase Transition in Strontium Titanate Only by a Decrease in Zero-Point Vibration Frequency. *J. Phys. Soc. Jpn.*, 79, 024707.

[18] Taniguchi, H., Itoh, M., & Yagi, T. (2007). Ideal Soft Mode-Type Quantum Phase Transition and Phase Coexistence at Quantum Critical Point in ^{18}O-Exchanged $SrTiO_3$. *Phys. Rev. Lett.*, 99, 017602 EOF.

[19] Dec, J., & Kleemann, W. (1998). From barrett to generalized quantum curie-weiss law. *Solid State Commun.*, 106, 695 EOF.

[20] Vojta, M. (2003). Quantum phase transitions. *Rep. Prog. Phys.*, 66, 2069.

[21] Sondhi, S. L., Girvin, S. M., Carini, J. P., & Shahar, D. (1997). Continuous quantum phase transitions. *Rev. Mod. Phys.*, 69, 315 EOF-333 EOF.

[22] Schneider, T., & Stoll, E. F. (1976). Quantum effects in an n-component vector model for structural phase transitions. *Phys. Rev. B*, 13, 1123 EOF-1130 EOF.

Chapter 14

Electromechanical Coupling Multiaxial Experimental and Micro-Constitutive Model Study of $Pb(Mg_{1/3}Nb_{2/3})O_3$- $0.32PbTiO_3$ Ferroelectric Single Crystal

Wan Qiang, Chen Changqing and Shen Yapeng

Additional information is available at the end of the chapter

http://dx.doi.org/10.5772/52540

1. Introduction

Compared to polycrystalline ferroelectric ceramics such as $Pb(Zr_{1/2}Ti_{1/2})$-O_3 (PZT), domain engineered relaxor ferroelectric single crystals $Pb(Zn_{1/3}Nb_{2/3})O_3$-$x$$PbTiO_3$ (PZN-xPT) and $Pb(Mg_{1/3}Nb_{2/3})O_3$-$x$$PbTiO_3$ (PMN-xPT) show greatly enhanced electromechanical properties: the piezoelectric coefficient d_{33} and electrically induced strain of <001> oriented single crystals direction can respectively reach 2500pC/N and 1.7%, or even greater [1],[2]. PZN-xPT and PMN-xPT with outstanding properties are usually near the morphotropic phase boundary (MPB), separating rhombohedra (R) and tetragonal (T) phases [1], [4], [5]. For example, the MPB is located within a range of x=0.275-0.33 for PMN-xPT [6], [7]. Recent investigations have also shown that the MPB of PMN-xPT and PZN-xPT is actually in a multi-phase state at room temperature [8], [9]. e.g., the coexistence of rhombohedra (R) and monoclinic (M) phases at x=0.33 or rhombohedra and tetragonal phases at x=0.32. Other phases can also exist near MPB under stress and/or electric field loading. Experiment studies [10], [11], [12] and first-principles calculations [4], [13]reveal that two different homogeneous polarization rotation pathways are present between rhombohedra and tetragonal phases under an electric field. It is found that three kinds of intermediate monoclinic phases M_A, M_B and M_C are associated with the two pathways [9], [10]. Orthorhombic phase has also been observed between the rhombohedra and tetragonal phases when PMN-PT is loaded with strong electric field along the <110> direction [9],[14]. Although the mechanism underlying the high performance of these crystals is not completely clear, the existence of the intermediate phases and the associated phase transitions are believed to be one of the main reasons [1], [3], [4], [8], [10].

Note that PZN-PT and PMN-PT single crystals usually experience electric and/or mechanical loading during their in-service life. It has been shown that externally applied loading has significant effect on the properties of these crystals [1], [5]-[28]. A number of studies have focused on the loading induced behavior of <001> and <110> oriented anisotropic PZN-*x*PT and PMN-*x*PT crystals [1], [5], [9], [15]-[19]. It is has been shown loading in the form of electric field [1]-[21], and stress[21]-[26], can lead to polarization switching and phase transitions, which changes the crystal phase and domain structure of these single crystals, and hence dramatically alter their electromechanical properties. A mature level understanding of their responses to electrical, mechanical and temperature loading condition is thus essential to fulfill the applications of these crystals.

Rhombohedral phase of PZN-*x*PT($0<x<0.1$) and PMN-*x*PT($0<x<0.35$) exhibit excellent electromechanical properties along {001} and {110} orientations compared to the {111}(cubic cell reference) spontaneous polarization direction. A number of studies have focused on the loading induced behavior of {001}, {011} and {111} oriented PZN-*x*PT and PMN-*x*PT single crystals. It has been shown that loadings in the form of electric field and stress can lead to polarization rotation and phase transition, which change the crystal phase and domain structure of these single crystals, and hence dramatically alter their electromechanical properties. When an electric field is applied along the {001} direction of PZN-PT, polarization rotation occurs from {111} towards {001} via either M_A or M_B, depending the composition (e.g., the rotation is R-M_A-T for PMN-4.5PT). For {110}-oriented PMN-PT, polarization rotates from {111} to {110} via M_B when electric field is along {110} and ends up with an orthorhombic phase when the electric field exceeds a critical value. These single crystals can be polarized into a single domain state with {111} oriented electric field. Most available studies on the effect of bias stress on the crystal behavior focus only on uni-polar electric field loading. However, it has been shown that the uni-polar and bi-polar responses of these crystals can be very different.

A number of studies have focused on experimental, but the constitutive model of ferroelectric single is absent. Huber [29] and Bhattacharya[30], [31] et.al established a model based on the micromechanical method, but the simulation is not well compare to experimental result.

At present, a mature model to explain the stress-strain behavior of ferroelectric single is absent all along. In this study, electric field induced "butterfly" curves and polarization loops for a set of compressive bias stress of {001}, {011} and {111} poled PMN-0.32PT single crystals will be explored by systematical experiment study. The effects of the compressive bias stress on the material properties along these three crystallographic directions of PMN-0.32PT single crystals will be quantified. The underlying mechanisms for the observed feature will be explained in terms of phase transformation or domain switching,depending on the crystallographic direction.The stress-strain curves along <001> crystallographic direction of ferroelectric single crystals $BaTiO_3$ will be calculated in the first principle method to validate polarization rotation model. Finally, Based on the experimental phase transformation mechanism of ferroelectric single crystal, a constitutive model of ferroelectric single crystal is proposed based on micromechanical method. This constitutive model is facility and high computational efficiency.

http://dx.doi.org/10.5772/52540

2. Experimental methodology

At the test room temperature, PMN-0.32PT single crystals used in this study are of morphotropic composition, and in the rhombohedral phase, very close to MPB. The pseudo-cubic {001}, {011} and {111} directions of these crystals are determined by x-ray diffraction (XRD). Pellet-like specimens of dimensions $5\times5\times3mm^3$ are then cut from these crystals, with the normal of the $5\times5mm^2$ major specimen surfaces along the pseudo-cubic {001}, {011} or {111} direction. All specimens are electroded with silver on the 5×5mm major surfaces and poled along the {001}, {011} and {111} orientations (i.e., the specimen thickness direction) under a field of 1.5kV/mm. Note that there are eight possible dipole orientations along the body diagonal directions of unpoled PMN-0.32PT single crystals (i.e., the <111> direction). When an electric poling field is applied to the crystals along the {001} direction, a multi-domain structure can be produced, comprising four degenerate states. For {011} direction poled single crystals, the number of degenerate states is two. The single crystals can be poled into a single domain state when they are poled along the {111} direction. The states of {001}, {011} and {111} poled PMN-0.32PT single crystals are sketched in Fig. 1, where solid arrows refer to the domain states induced by poling (also labeled as type 1 in Figs. 1b and 1c), dotted arrows represent possible domains switched from type 1 domains upon loading or unloading, and vice versa (also labeled as type 2,).

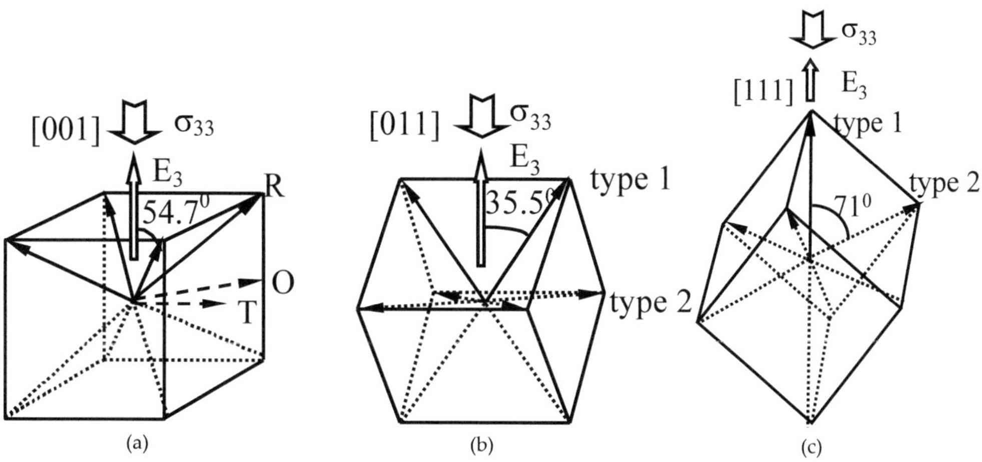

Figure 1. (a) Multidomain rhombohedral crystal obtained by poling along the {001} orientation. The solid line arrows show possible directions of the polarization vector in a fully {001} poled rhombohedral crystal, the dashed line arrows show possible directions of the polarization vector in orthorhombic and tetragonal phases crystal. (b) Two domain rhombohedral crystal obtained by {011} poling. The solid line arrows indicate type 1 domains in a fully {011} poled crystal and dashed line arrows indicate possible type 2 domains under compressive stress. (c) Monodomain rhombohedral crystal poled along the {111} orientation. The solid and dashed line arrows have similar means to (b). The hollow arrows in Fig. a ~ Fig. c show the electric field and stress loading. The small rhombohedral distortion is neglected and all numbers and notations refer to a quasi-cubic unit cell.

Since the focus of this study is to explore the effect of bias stress on the electromechanical properties of PMN-0.32PT single crystals along different crystallographic directions, experimental setup is adapted from Ref. [31] (Fig.2) to allow simultaneously imposing uniaxial stress and electric field to the specimen along the thickness direction. Mechanical load is applied by a servo-hydraulic materials test system (MTS) and electric field is applied to the specimen using a high voltage power amplifier. Once the specimen is placed in the fixture, a compressive bias stress with magnitude of at least 0.4MPa is maintained throughout the test to ensure electrical contact. Stress controlled loading instead of displacement controlled loading is adopted during the test, so that the specimen is not clamped but is free to move longitudinally when electric field E_3 and mechanical stress σ_{33} loading are applied, where subscript 3 refers to the thickness direction of the samples, corresponding to the {001}, {011} and {111} direction of {001}, {011} and {111} oriented PMN-0.32PT single crystals, respectively. During test, polarization P_3 (or electric displacement) is measured using a modified Sawyer-Tower bridge, and the deformation (i.e., the strain) is monitored by two pairs of strain gauges (in total four strain gauges used) mounted on the four $5\times3mm^2$ surfaces: one pair placed on two opposite $5\times3mm^2$ surfaces is applied to measure the longitudinal normal strain ε_{33} along the {001}, {011} and {111} direction and another pair to measure the transverse normal strain ε_{11} in the direction perpendicular to the {001}, {011} and {111} direction respectively. Output of the strain gauges during deformation is recorded by a computer through a multiple-channel analog-to-digital (AD) converter.

The first set of tests is performed for electric field loading of triangular wave form of magnitude 0.5kV/mm and frequency 0.02Hz, free of stress loading. The low frequency is chosen to mimic quasistatic electric loading, which is of particular interest in this study [32]. Unless stated otherwise, this loading frequency for the electric field is used throughout the following test. The second set of tests consists of mechanical loading upon short circuited samples. The samples are compressed to –40MPa and unloaded to –0.4MPa at loading and unloading rate of 5MPa/min, followed by an electric field which is sufficiently large to remove the residual stress and strain to re-polarize the samples. In the third set of tests, triangular wave form electric field is applied to the samples which are simultaneously subject to co-axial constant compressive stress preload. The magnitude of the preload is varied from test to test and is in the range between 0 and -40MPa. Note that there is a time-dependent effect of the depolarization and strain responses under constant compressive stress. To minimize this effect, each electric field loading starts after a holding time of 150 seconds for a new stress preloading. It is found that three cycles of electric field loading and unloading are sufficient to produce stabilized response for each constant prestress, and the results for the last cycle are reported in the following.

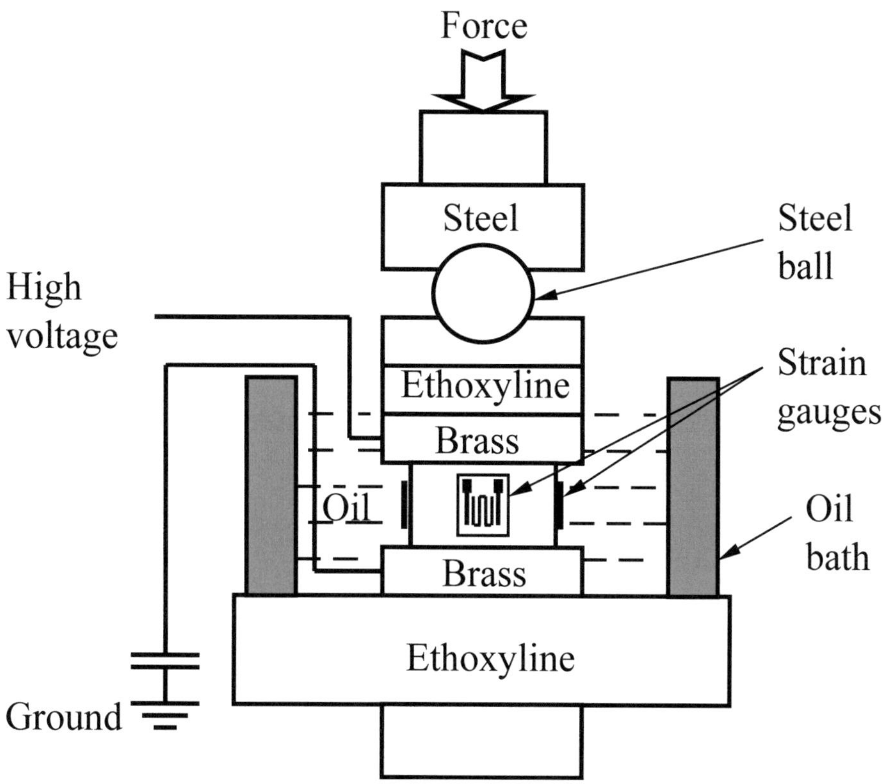

Figure 2. The sketch of experimental set

3. Experimental results and discussion

3.1. Crystallographic dependence of electric behavior and piezoelectric properties

The measured electric field induced polarization hysteresis loops and butterfly curves for {001}, {011} and {111} oriented poled PMN-0.32PT single crystals without stress loading are shown in Figs. 3a and 3b. The remnant polarizations P_r (defined as the polarization value at zero electric field), the coercive electric fields E_c (defined as the electric field value at zero polarization) and the piezoelectric coefficients d_{33} (defined as $d_{33}=\Delta\varepsilon_{33}/\Delta E_3$ where ΔE_3 is limited between –0.05 and +0.05 kV/mm, namely the slops of the $\varepsilon_{33}-E_3$curves as the electric field passes through zero) depend strongly on the crystallographic orientation. From Figs. 3a and 3b, one can calculate that $P_{r_{[001]}}$, $P_{r_{[011]}}$ and $P_{r_{[111]}}$ are 0.247, 0.324 and 0.395C/m^2, $E_{c_{[001]}}$, $E_{c_{[011]}}$ and $E_{c_{[111]}}$ are 0.255, 0.298 and 0.216kV/mm, and $d_{33_{[001]}}$, $d_{33_{[011]}}$ and $d_{33_{[111]}}$ are 1828, 1049 and 200pC/N, respectively. It is noticed that there are eight possible polarization orientations along the pseudo-cubic {111} for un-poled rhombohedral PMN-0.32PT single crystals. Upon poling, the dipoles switch as close as possible to the applied electric field direction: For {001} poled crystals, there are four equivalent polar vectors along the {111} orientation, with an inclined angle of -54.7º from the poling field (Fig. 1a); For {011} poled crystals, there are two equivalent polar vectors along the {111} direction (labeled as type 1 in Fig. 1b); For {111} poled crystals, there is one polar vector along {111} (type 1 in Fig. 1c). According to the domain configurations in Fig. 1, the remnant polarizations $P_{r_{[001]}}$ and $P_{r_{[011]}}$ are approximately related to $P_{r_{[111]}}$ by $P_{r_{[001]}} = P_{r_{[111]}} / \sqrt{3}$ and $P_{r_{[011]}} = \sqrt{2} P_{r_{[111]}} / \sqrt{3}$, respectively. By taking the measured value for $P_{r_{[111]}}$ (0.395C/m^2), $P_{r_{[001]}}$ and $P_{r_{[011]}}$ are predicted to be 0.228 and 0.322C/m^2, respectively, which are very close to the measured ones ($P_{r_{[001]}}$ =0.227C/m^2 and $P_{r_{[011]}}$=0.324C/m^2). Therefore, the measured results are consistent with the domain configurations shown in Fig. 1.

It is also seen from Figs. 3a and 3b that the coercive field is lowest for the {111} oriented crystals, and becomes successively higher for {001} and {110} orientations. This is same to Ref.25 except for {110} orientation. This trend of coercive field is due to two reasons: One is due to reorientation driving force being proportional to the component of electric field aligned with the rhombohedral direction; The other one is due to the domain switching process. In {111}-oriented PMN-0.32PT single crystals, there are two types of domains (shown as type 1 and type 2 in Fig.1c). When the electric field is decreased from 0.5kV/mm to –0.05kV/mm, the strain first decreases linearly (see Fig.3b). When the electric field is decreased further, the type 1 domain switches to type 2 domains, leading to abrupt displacement change. When the electric field exceeds the coercive field –0.216kV/mm, the type 2 domains switch back to type 1 domain, recovering the deformation. In type 2 domain state, three equivalent polar vectors with an angle of 71º from the {111} direction can coexist and are separated by domain walls across which the normal components of electric displacement and displacement jump are zero. Ideally, this type of domain walls has no associated local stress or electric field. So the existing of type 2 domain state and the largest component of electric field along polarization direction induces the lowest coercive field in {111} orientation poled crystal. In the {011} orientation crystals, there are also two types of domains (i.e., type 1 and type 2 domain in Fig. 1b), the domain switching process is similar to {111}-poled crystals. In the type 2 domain state, however, the four possible polar vectors are perpendicular to the applied electric field (Fig. 1b). It is thus difficult to switch type 2 domain to type 1 domain only by applying electric field. Both the type 2 domain state and the smallest component of E contribute to the largest coercive field in {011} oriented crystals. In {001} oriented crystals, there will be no associated local stress or electric field, similar to type 2 domain state of {111} poled crystals. This feature again renders domain switching easy. On the other hand, the domain structure of {001} poled crystals is stable [4] and the component of electric field is smaller, giving the coercive field higher than that of {111} poled crystals but lower than that of {011} oriented crystals (see, Figs. 3a and 3b).

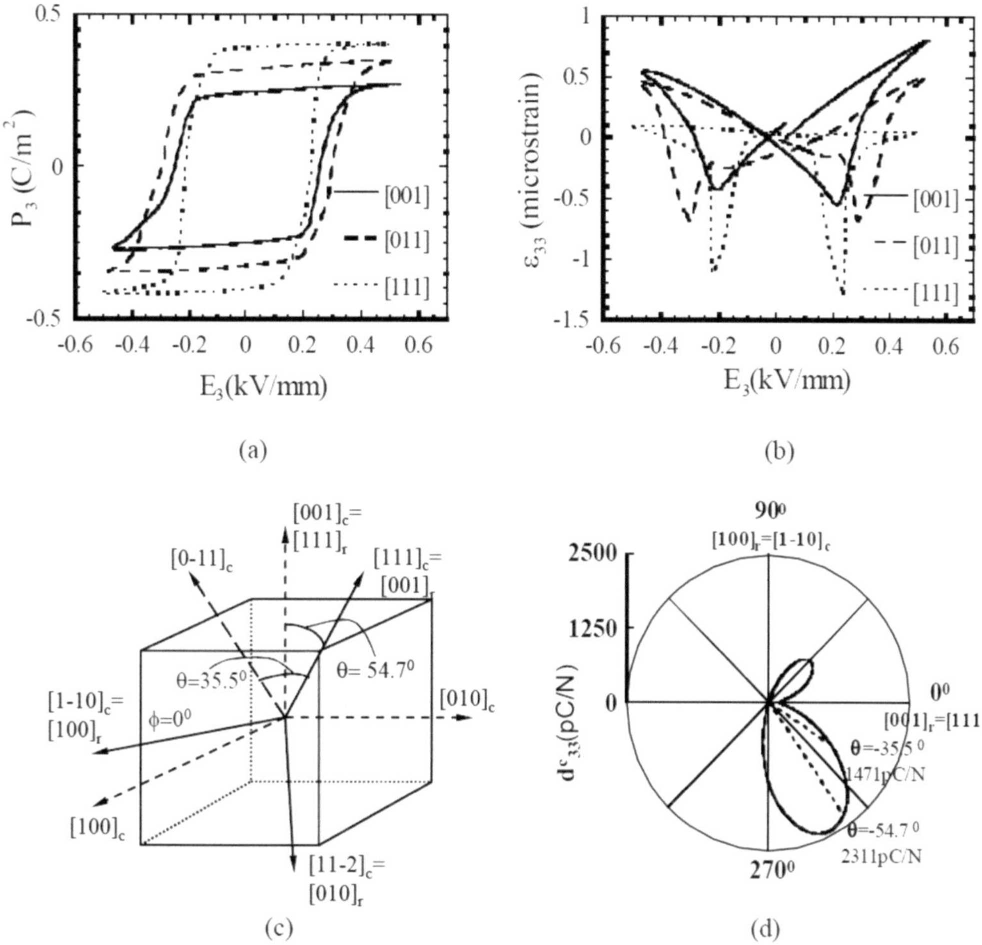

Figure 3. Electric field induced polarization and strain responses for {001}, {011} and {111}-oriented crystals of PMN-0.32PT: (a) P_3-E_3 curves and (b) $\varepsilon_{33}-E_3$ curves. (c) Relationship between rhombohedral (solid lines) and cubic (dashed lines) coordinate systems. Subscripts *r* and *c* denote directions with respect to the rhombohedral and cubic coordinate systems, respectively. (d) Orientational dependence of piezoelectric coefficient d^*_{33} of PMN-0.32PT in the polar plane.

Some researchers attribute the high piezoelectric coefficients along {001} and {011} oriented ferroelectric single crystals to the engineered domain state [4, 11]. It has also reported that, however, the piezoelectric coefficient along the {001} direction of single crystals with mono-domain structure is comparable to that of crystals with multi-domain structure [17], implying the origins of the high piezoelectric constants of PMN-0.32PT single crystals may not be due to the engineered domain state. Instead, it could be due to the effect of crystal lattice properties. To further explore this issue, we follow Ref. [17] to calculate the piezoelectric coefficients d^*_{ij} along an arbitrary direction in a mono-domain crystal. For a direction defined by the Euler angles (φ, θ, ψ) (see Fig. 3c), d^*_{ij} is related to d_{ij} (measured along the principal crystallographic axes) by,

$$d^*_{33}(\phi,\theta) = d_{33}\cos^3\theta + (d_{15} + d_{31})\cos\theta\sin^2\theta - d_{22}\cos\phi\sin^3\theta(\cos^2\phi - 3\sin^2\phi) \tag{1}$$

With d_{33}=200pC/N taken from our measured value and d_{15}=4100, d_{31}=–90 and d_{22}=1340pC/N, $d^*_{33[111]}(0, \theta)$ for PMN-0.32PT single crystals can be calculated and is shown in Fig. 3d. Note that in the case of ϕ=0 θ=-35.5º and -54.7º correspond to {011} and {001} direction, respectively. $d^*_{33[011]}$=1471pC/N and $d^*_{33[001]}$=2311pC/N can be inferred from Figs. 3d, which are not far away from our measured values ($d_{33[011]}$=1049 pC/N and $d_{33[001]}$=1828 pC/N). The small discrepancy between the predictions and measurements is believed to be due to the fact that the predictions are based on ideal mono-domain single crystals while the material parameters used in Eq. (1) are actually from less ideal mono-domain crystals (in fact, they are more or less multi-domain crystals). Nevertheless, we can conclude that the dominant contribution to the large {001} and {011} piezoelectric response should be the crystal anisotropy other than engineered domain state.

3.2. Crystallographic dependence of stress induced strain and polarization responses

Figure 4 shows the measured $\sigma_{33}-\varepsilon_{33}$ and $\sigma_{33}-P_3$ curves for {001}, {011} and {111} oriented short circuited samples. The results indicate obvious crystallographic anisotropy in stress inducing responses. From domain switching viewpoint, there should be no significant deformation in the thickness dimension of {001} poled crystal samples under stress loading, except for the elastic strain. However the longitudinal strain of {001} is contractive with a maximum magnitude of 0.3% under –40MPa. The contractive strain of {001} oriented crystals is about twice as much as those of {011} (about –0.18%) and {111} (about –0.17%) oriented crystals under a loading of –40MPa (Fig. 4a). This abnormal behavior of {001}-oriented crystals lies in that the mechanism underlying the stress induced response of {001}-oriented crystals is the R to O and T phase transition (see, Fig. 1a) rather than domain switching for {011} and {111} oriented crystals: the lattice distortion due to phase transition induces large deformation in the thickness dimension of {001} poled crystals. In {011} and {111}- poled crystals, on the contrary, the type 2 multi-domain state induced by compressive stress is the stable and preferred state (Figs. 1b and 1c), and can form more easily by domain switching than phase transformation. Therefore, the stress induced strain and polarization curves in {011} and {111} orientation crystals are similar to those of ferroelectric polycrystals of which domain switching is also the dominant deformation mechanism (Figs. 4a and 4b). During the unloading of {001} oriented crystals (corresponding to the returning to R phase of the unstable O phase), the $\sigma_{33}-\varepsilon_{33}$ and $\sigma_{33}-P_3$ curves show obvious nonlinear behavior. Note that the remnant strain and polarization at the end of unloading are attributed to the stable T phase which does not switches back to the R phase. For {011} and {111}-oriented crystals, however, there is only domain switching (i.e., switching between type 1 and type 2 domains) and no phase transformation occurs. During unloading, the $\sigma_{33}-\varepsilon_{33}$ curves and $\sigma_{33}-P_3$ curves of {011} and {111} show linear response since almost no domain switches back (Figs. 4a and 4b).

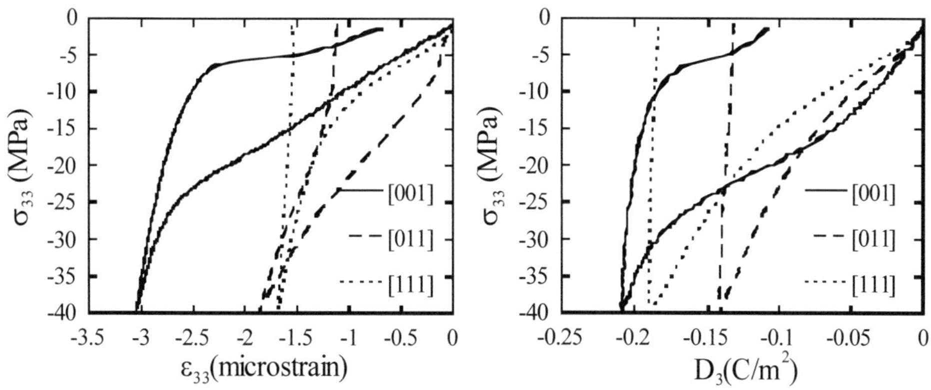

Figure 4. Compressive loading and unloading stress cycles induced polarization and strain responses for {001}, {011} and {111}-oriented crystals of PMN-0.32PT: (a) $\sigma_{33}-\varepsilon_{33}$ curves and (b) $\sigma_{33}-P_3$ curves.

The stress cycles of {001} poled crystals can be explained by the polarization vector rotation mechanism sketched in Fig. 5 as follows. In general two most possible mechanisms responsible for the observed features of {001} poled crystals in Figs.4 are domain switching and polarization rotation associated phase transformation. Recall that the PMN-0.32PT single crystals considered here are in the R phase close to MPB (although in reality there could also be M or T phase in these crystals near MPB [8], [9] the R phase is nevertheless the dominant phase). It is noted that that, upon application of a field along the <001> poling axis of R phase domain engineered PMN-0.32PT single crystals, only four of the eight polarization orientations are possible, i.e., <111>, <$\bar{1}$11>, <1$\bar{1}$1>, and <$\bar{1}\bar{1}$1>. Since the <001> components of these four polar vectors are completely equivalent, each domain wall cannot move under an external electric field along the <001> direction owing to the equivalent domain wall energies [32]. In other words, no ferroelectric domain switching is possible. Equally, no ferroelastic domain switching is expected when a compressive stress is applied along the <001> axis. (For PMN-PT systems with orthorhombic 4O <001> domain engineered structure [33], 60º switching from <011> to <110> would be driven by an applied stress. But, this is not what we are considering). Such facts partly rationalize our hypothesis that the underlying mechanism associated with the observed behavior of the R-phased PMN-0.32PT single crystals considered in this paper is phase transformation. Nevertheless, it should be emphasized that since no in situ diffraction observation is conducted to properly identify the stress induced phase transformation in PMN-0.32PT single crystals, discussions in this paper on phase transition are solely inferred from the measured stress induced strain and polarization curves.

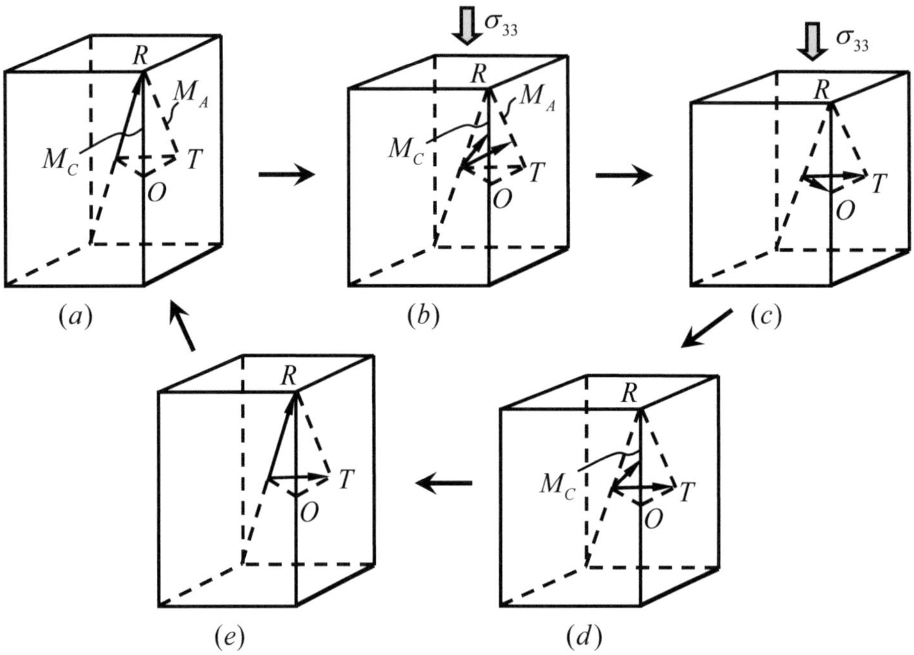

Figure 5. Schematic drawing of the polarization rotation from R to T and O phases due to <001> direction compression: (a) the initial state, (b) polarization starts to rotate from R to O and T via two pathways M_A and M_C, (c) R phase completely transformed, (d) O phase starts to switch back to R phase upon unloading, (e) O phase completely switched back to R phase. The states (a) to (e) correspond to the stress levels marked in Figs. 2(a) and 2(b). At the initial poled state (a), PMN-0.32PT possesses four equivalent <111>-oriented polarizations vector close to <001> direction though only one polarization rotation processes is shown here.

Fig. 5 illustrates the polarization rotation mechanism of PMN-0.32PT single crystals under stress loading and unloading cycle. The polarization vector states shown in Figs. 5(a)-5(e) correspond to the stress levels marked on the curves in Figs. 4 for the -40MPa stress cycle. In the initial poled state, PMN-0.32PT single crystals possess four equivalent <111> polarizations, with only one shown in Fig. 5a for the sake of clarity. Upon loading, polarization vector starts to rotate from R to O and T phases through intermediate phases M_A and M_C, when the compressive stress exceeds about 15MPa in magnitude (Fig. 5(b)). This gives a mixture of R and T phases and remarkable augment in the polarization and strain change around σ_{33} =-20MPa (Figs. 4). When the presence of an electric field along <110> direction, polarization rotation only occurs from R to O under compression along <001> direction. In the absence of electric field along <110>, however, almost all the polarization vectors may switch to T and O phases when the compression exceed 30MPa in magnitude (Fig. 5(c)). When the magnitude of the compressive loading is further increased, there is no more polarization rotation and the PMN-0.32PT single crystal shows linear response in Figs. 4. It is noted that O phase is usually unstable (its free energy balance between R and T phases depends on the electric and mechanical loading history [9], [14]). Upon unloading, the O phase switches back to R phase when the compression lower than -10MPa (Figs. 4, and 5(d)). On the other hand T phase is a stable phase. It will not switch back to R phase upon unloading, leading to remnant strain of about -0.07% for and remnant polarization of about -0.13C/m^2 at the end of the -40MPa stress cycle (Figs. 4(a) and 4(b), Fig. 5(e)). After the stress cycle, a unipolar electric field of 0.5kV/mm in magnitude is applied to re-pole the single crystal to its initial state, i.e., the polarization vector completely switches back. As a result, the remnant strain and polarization diminish to zero at the end of the re-polarization process (Figs. 5(a) and 5(e)).

3.3. Crystallographic dependence of electric field induced behavior at constant bias compressive stress

The $\varepsilon_{33}-E_3$ "butterfly" curves and P_3-E_3 hysteresis loops of {001}, {011} and {111} poled crystals under different constant compressive bias stresses are shown in Fig. 6. Dependence of the electric coercive E_c, remnant polarization P_r, dielectric permittivity χ_{33}, piezoelectric coefficient d_{33} and aggregate strain $\Delta\varepsilon$ for {001}, {011} and {111} oriented crystals on the compressive bias stress is summarized in Fig. 7. The aggregate strain $\Delta\varepsilon$ is defined as the difference between the maximum and minimum strain for a complete responsive butterfly curves. Similar to d_{33}, χ_{33} are calculated by $\Delta P_3/\Delta E_3$ where ΔP_3 is the polarization difference between –0.05 and +0.05kV/mm. The calculated χ_{33} within such a small field range is almost equal to the slope of P_3-E_3 hysteresis loops when the electric field passes through zero. The calculated χ_{33} includes both the reversible (intrinsic dielectric property) and irreversible (extrinsic domain switching and phase transformation related property) contributions of the material, which is generally higher than the permittivity measured by a dynamic method [32].

The influence of the preloaded compressive stress on the aggregate strain $\Delta\varepsilon$, remnant polarization P_r and piezoelectric coefficient d_{33} seems to be similar for {001}, {011} and {111} oriented crystals: The remnant polarization P_r decreases with increasing the magnitude of the compressive prestress; The aggregate strain $\Delta\varepsilon$ and piezoelectric coefficient d_{33} first increase and then decrease with the magnitude of the prestress increasing. As suggested earlier, however, the underlying mechanisms for the electromechanical behavior of {001} {011}, and {111} oriented crystals are different. The change of the dielectric permittivity χ_{33} {001} oriented crystals under compressive stress is similar to that of χ_{33} induced by temperature: near the phase transformation temperature there is a peak in the χ_{33} curves, As is shown in Fig. 7c, there is a small peak near –6MPa and a big peak near –20MPa in χ_{33} curves for {001} oriented crystals. This may be due to R-M (near –6MPa) and M-R and M-O phase transformation (near –20MPa). On the other hand, there is no obvious peak in χ_{33} curves for {011} and {111}-oriented crystals (Fig.7c). The change of χ_{33} curves induced by compressive stress and the aforementioned stress induced strain and polarization responses are consistent with the hypothesis that there is phase transformation for {001}-orientated crystals under compressive stress.

As suggested by Fu et al. [7], for {001} oriented crystal origins of large aggregate strain and high piezoelectric coefficient at a moderate compressive bias stress (i.e., around -20MPa for the single crystals considered here) may be attributed to the stress induced intermediate states between rhombohedral and tetragonal phases. Under a bias stress of about -20MPa, PMN-0.32PT single crystals, after a phase transformation, are in a state of monoclinic phase which has a larger c/a (c and a are lattice parameters) and a smaller polarization component along the field direction than those of rhombohedral phase [20], implying electric field induced greater aggregate strain and piezoelectric coefficient (Figs. 7d and 7e). With the magnitude of the compressive bias stress increased further (i.e., σ_{33}=-30 and -40MPa), the {001} oriented single crystals are in the state of a mixture of orthorhombic and tetragonal phases. Although O and T phases also have large c to a ratio, the presence of a large compression has an opposite effect (i.e., preventing larger deformation induced by electric field) and results in small aggregate strain (Fig.7e).

As is noted, the polarization rotation introduced by compression gives rise to θ in $d_{33}^{*}(\phi, \theta)$ in the range between -54.7º and -90º (the angle between loading and polarization direction of T or O phase) under -20MPa, and results in larger d_{33}^{*} in accordance with Eq. (1) (see in Fig. 7d). When the magnitude of the compressive bias stress increases further, θ approaches 90º and d_{33}^{*} becomes smaller. Meanwhile, the remnant polarization P_r and coercive field E_c decrease monotonically with the applied compressive bias stress because of the decreased component polarization along the {001} direction under compression (Fig.7b).

Under zero stress, the initial state of {011}-oriented crystals is of multi-domain with two equivalent polarization directions (Fig.1b). For {111}-oriented crystals, the initial state is of mono-domain with polarization direction in the {111} direction (Fig.1c). When the applied electric field decreases from 0.5kV/mm to $-E_c$, the strain and polarization decrease linearly. Upon approaching $-E_c$, domain state switches to four polar domain state for {011} oriented crystals (type 2 in Fig. 1b) and three polar domain state for {111} oriented crystals (type 2 in Fig. 1c), giving the jumps in strain and polarization responses. When electric field exceeds $-E_c$, domain complete second switching, from four polar domain state to two polar domain state for {011} oriented crystals and from three polar domain state to mono-domain state for {111} oriented crystals. When the electric field reaches -0.5kV/mm, the strain and polarization change linearly again (Fig. 7).

Note that compressive stress can induce domain switching in {011} and {111}-orientated crystals. When compressive stress is superimposed on the samples, the shapes of $\varepsilon_{33}-E_3$ and P_3-E_3 curves are different from those under the zero stress state (Fig.6c-f). Due to compression inducing depolarization, the remnant polarization decreases in {011} and {111} oriented crystals (Fig. 7b). A low compressive stress (e.g., -5MPa for {011}, -7MPa for {111}) can lead to type 1 to type 2 domain switching. As a result, more domains take part in switching during the electric field loading cycle and larger aggregate strain$\Delta\varepsilon$ than under zero compressive stress is observed. This partly explains the observed features in Fig. 7e.

http://dx.doi.org/10.5772/52540

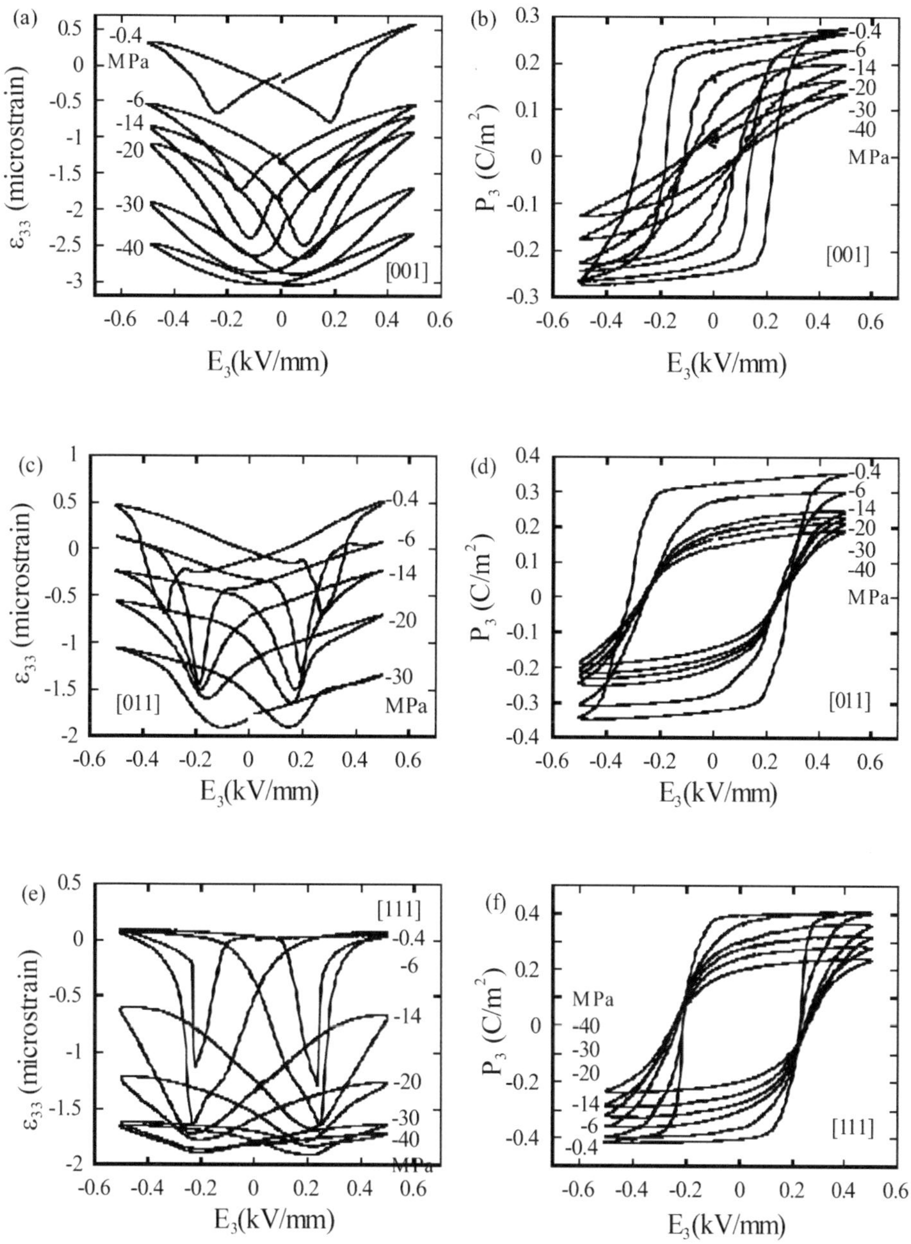

Figure 6. Electric field induced $\varepsilon_{33}-E_3$ and P_3-E_3curves at different compressive bias stresses: (a) and (b) for {001}-oriented; (c) and (d) for {011}-oriented, (e) and (f) for {111}-oriented.

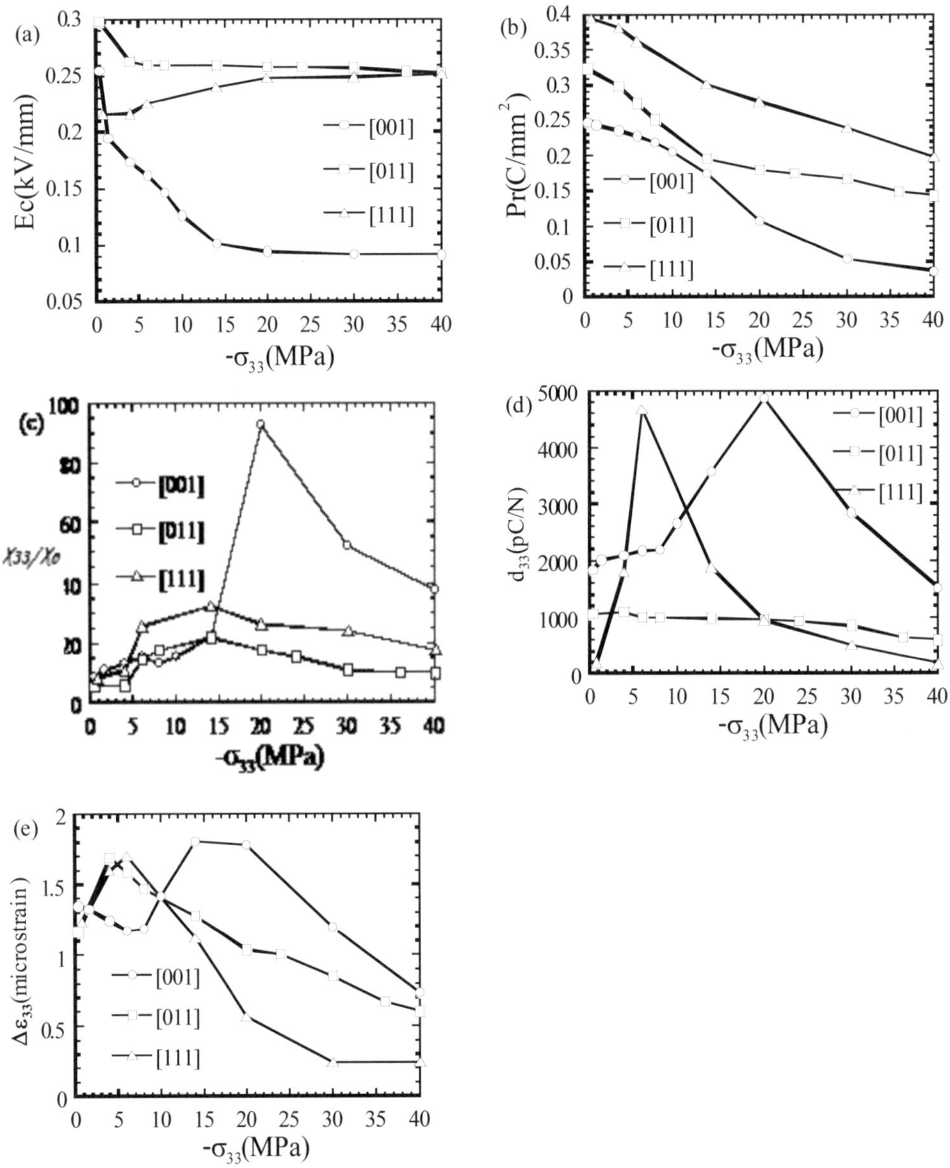

Figure 7. Effect of compressive bias stress on (a) the coercive field E_c, (b) remnant polarization P_r, (c) piezoelectric constant d_{33}, (d) relative dielectric constant X_{33}/X_0, and (e) and aggregate strain$\Delta\varepsilon_{33}$.

http://dx.doi.org/10.5772/52540

4. The first principle calculation of stress-strain

4.1. Calculation methodology

PMN-PT and $BaTiO_3$ have the similar ABO_3 structure, so they have the similar ferroelectric properties. There is only Ti^{4+} particle in B site of $BaTiO_3$, however, there is not only Ti^{4+} but also minim Mn^{4+} and Ni^{2+} particle in B site of PMN-PT. So the single cell of $BaTiO_3$ is convenient in calculation and it keeps the similar ferroelectric to PMN-PT.

In this paper, we calculate a single cell of BaTiO3, the single cell should be the smallest periodic reduplicate cell, it includes one Ti^{4+} particle, three O^{2-} particles and one Ba^{2+} particle (Fig.8). It is suggested that the initialized state of $BaTiO_3$ ferroelectric single crystal is R phase after {001} oriented polarization. Refer to literature [34], the crystal lattice constant of R phase $BaTiO_3$ ferroelectric single crystal is 4.001A, the coordinate of particle in single cell is shown in table 1.The loading along {001} direction is carried out through application increasing strain by degrees. Stress and other parameters under each strain level is calculated by VASP. Calculation under each strain level include two steps. Firstly, the particle coordinate of single cell under each strain level is calcluated by first principle molecular dynamic method. Secondly, Stress and other parameters are obtained through relaxation that is based on the first result. Each increment of strain in this paper is 0.5%, untill 4%.

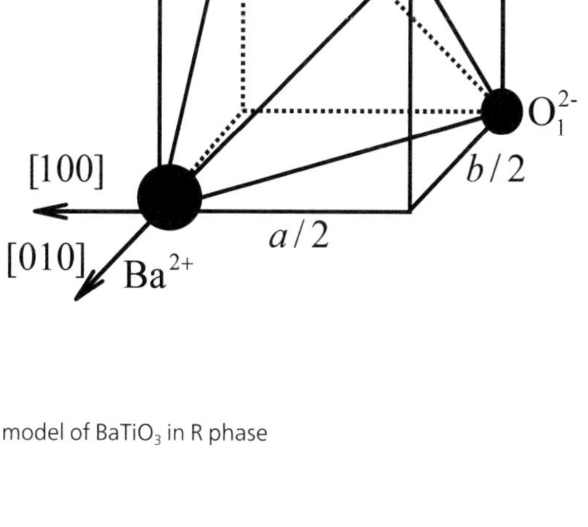

Figure 8. The cell calculation model of $BaTiO_3$ in R phase

4.2. Discussion

In order to validate that the method a mentioned in this paper is correct, we calculate the elasticity constant C_{33} of T phase $BaTiO_3$ using the method mentioned before. Fig.9 shows that the C_{33} is 180GPa, which is approach to the experimental value 189± 8Gpa reported in refer[34]. This means that the method used in this paper is trusty.

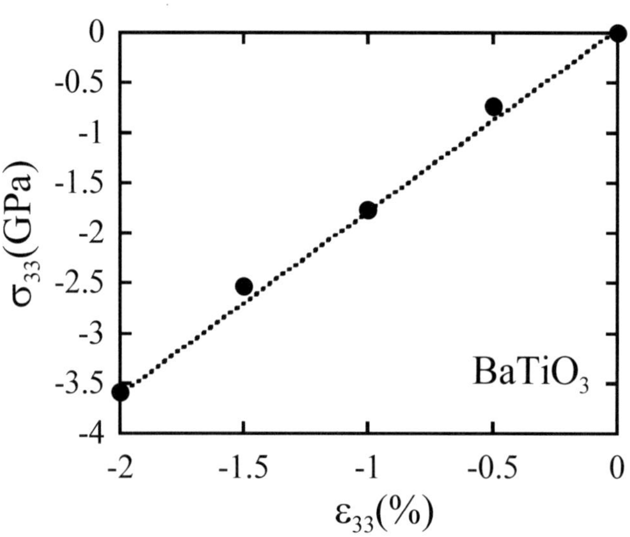

Figure 9. The stress-train curve of $BaTiO_3$ ferrorlctric single crystal obtained by the method amentioned in this paper, the {001} oriented $BaTiO_3$ ferroelectric single crystal is in T phase.

Stress-strain curve of {001} orientated R phase $BaTiO_3$ calculated in first principle method is shown in Fig.10(a). Fig.10(a) shows that the result has simlar nonlinear behavior to the experimental result of PMN-0.32PT during the loading that sketched in Fig.4(a), namely, there is obvious "a,b,c" steps during loading. We know that the nonlinear behavior of PMN-0.32PT shown in Fig.4(a)should be polarization rotation (R→M→O and R→M→T). The R→M→O is corresponding to the processing of polarization vector P_R switching to P_O, which is shown in Fig.10(b). table 2 is calculated coordinate of particle in $BaTiO_3$, the coordinate is correspond to point A in fig.10(a). Table 3 is coordinate of particle in O phase $BaTiO_3$ ferroelectric single crystal from refer [35]. Data in table 2 are equal to those in table 3, which indicates that "A" point in fig.10(a) should be O phase. With increasing strain, the coordinates of particle are unchangeable after "A", this indicates $BaTiO_3$ ferroelectric single crystal is stable in O phase after "A". From the first principle calculation, we testify that the PMN-0.32PT ferroelectric single crystal undergoes polarization rotation(R→M→O), this prove the polarization rotation model is reasonable.

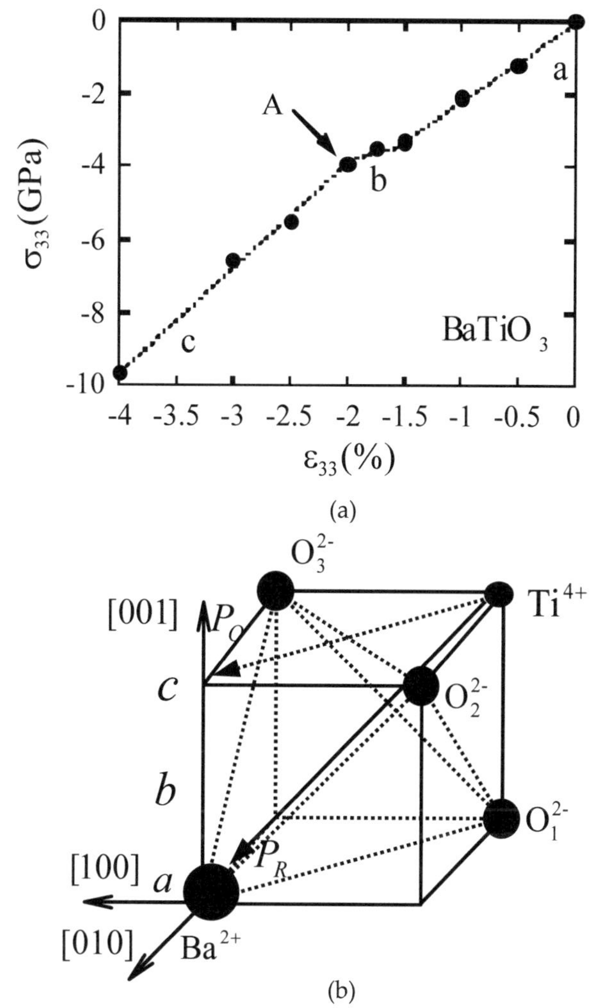

Figure 10. (a)The stress-strain curve of {001} oriented $BaTiO_3$ ferroelectric single crystal calculated in first principle calculation, the scatter point is calculation value and the broken line fit the scatter point.(b) The sketch map of polarization rotation during the calculation. The "a,b,c" letter in (a) is correspond to the "a,b,c" letter in (b), the letter indicates different polarization states during loading.

5. The viscoplastic model of ferroelection single

5.1. The viscoplastic model

From the experimental analysis, it is known that the <001>-oriented PMN-0.32PT single crystal undergo the R→M→T and R→M→O phase transformation under the compression. In order to establish a compact model and keep the essence of experiment, R→O phase transformation of PMN-0.32PT is considered in this study. Polarization rotation cause bulk deformation and shear deformation associated with the slip planes. It is assumed that the corresponding deformation associated with the slip planes is shear dominated, a feature similar to that of the multi-slip system of the crystal plasticity. This similarity renders possible to use crystal plasticity models to describe the transformation deformation of PMN-0.32PT single crystal.

The poled ferroelectric single is four domain state, the polarization vector is along <111>, <$\bar{1}$ 11>, <1$\bar{1}$1> and <$\bar{1}\bar{1}$1> respectively. In this study, it is assumed that the polarization vector of R phase can switch to the polarization vector of O phase, such as <111> vector switches to <001>, the vector also can switch back from O phase to R phase. The polarization rotation has an analogy to crystal plasticity slip, so we suggest there is eight slip systems in PMN-0.32PT ferroelectric single crystal. According to the crystalgraphic theory, we suggest ferroelectric single crystal has possible 8 variants; transformation of variants can be characterized by the habit planes illustrated in Fig.11. where n denotes the unit normal to the habit plane and s refers to the direction of transformation.

In general, a criterion (i.e., the phase transformation criterion) exists for the phase transformation of ferroelectric single crystal, and the material is assumed to undergo phase transformation when at least one of the 8 variants satisfies the phase transformation criterion. This is detailed as follows. Upon loading, the condition to produce R→O polarization rotation on a specified habit plane is that the driving force G of that plane reaches the critical value G_{0O}. The driving force is composed of the chemical driving force G_{chem} and mechanical driving force G_{mech}[36]

$$G = G_{chem} + G_{mech} = G_{0O} \tag{2}$$

A similar condition holds for reverse transformation from O→R polarization rotation with a critical value G_{0A}

$$G = G_{chem} + G_{mech} = G_{0A} \tag{3}$$

In Eqs. (2) and (3), the mechanical driving force can be expressed by

$$G^{r}_{mech} = \tau^{a}\gamma^{*} + E^{a}P^{*} \tag{4}$$

Where τ^{a} and E^{a} are the resolved stress on the "a" transformation system, and γ^{*} and P^{*} denote the associated transformation strain and transformation polarization. Following the crystal theory of plasticity, the resolved stress τ^{a} of the variant "a" is related to the stress tensor σ_{ij} and the Schmid factor a^{a}_{ij} by

$$\tau^a = \sum_{i,j=1}^{3} \alpha_{ij}^a \sigma_{ij} \tag{5}$$

where the Schmid factor is defined as follow

$$\alpha_{ij}^a = \frac{1}{2}(m_i^a n_j^a + m_j^a n_i^a) \tag{6}$$

with m_i and n_i being the unit normal to the habit plane and shear direction to the variant "a", respectively. The resolved stress E^a of the variant "a" is related to the electrical field tensor E_i

$$E^a = \sum_{i=1}^{3} E_i s_i \tag{7}$$

The chemical driving force in Eqs. (2) and (3) is assumed to be a linear function of the temperature

$$G_{chem} = \beta(T - T_0) \tag{8}$$

where T and T_0 denote the temperature in the single crystal and the equilibrium temperature respectively, and β is the stress-temperature coefficient. Note that the equilibrium temperature T_0 is defined as the average of the starting temperature of O phase transformation and that of R phase transformation, that is

$$T_0 = \frac{1}{2}(O_s + T_s) \tag{9}$$

When applying the rate independent crystal theory based model, Eqs. (2) and (3), to simulate the behavior of ferroelectric single crystals, one of the most computationally consuming tasks is to determine the set of instantaneously active transformation systems among the 8 possible variants at crystal level. This determination is usually achieved by an iterative procedure and must be carried out at each loading step, requiring extensive computation. Note that in the crystal theory of plasticity, a similar problem exists whilst in a rate dependent viscoplastic version of crystal theory of plasticity, determination of the set of active transformation systems is not necessary. As a result, computation effort can be reduced significantly. Following this idea, a viscoplastic version of Eqs. (21) and (32) are proposed and employed in this study. In the viscoplastic crystal model for PMN-0.32PT ferroelectric single crystals all transformation systems are assumed to be instantaneously active of varying extent, which is governed by a rate dependent viscoplastic law. In this paper, the phase transformation of variant "a" is assumed to comply with the following power law of viscoplasticity,

$$\dot{f}^a = \dot{f}_0 \left|\frac{G^a}{G_0}\right|^{\frac{1}{m}-1} \frac{G^a}{G_0} \left|\frac{c}{c_0}\right|^{1/k} \tag{10}$$

where $\dot{f}^a$ is phase fraction transformation rate of variant "a", $\dot{f}_0$ is reference phase fraction transformation rate, G^a is the driving force of variant "a", G_0 is the critical driving force, exponents k and m are material parameters dictating the rate effect, c depends upon the phase fraction, c_0 the reference value at initial state.

The phase fraction of variant "a" is calculated as,

$$\xi^a = f^a / f_0 \tag{11}$$

Summation over all possible variants provides the phase fraction for the whole ferroelectric single crystal, that is

$$\xi = \sum_{a=1}^{8} \xi^a \tag{12}$$

The incremental form of Eq. (12) can be written as

$$d\xi^a = df^a / f_0 \tag{13}$$

The transformation strain tensor ε_{ij}^{tr}, which is associated with $d f^a$, can be obtained as

$$d\varepsilon_{ij}^{tr} = \sum_{r=1}^{8} \alpha_{ij}^a \gamma^* df^a \tag{14}$$

where α_{ij} is Schmid factor. $d\varepsilon_{ij}^{tr}$ is increment of phase train. γ^* is the maximal strain transformation during loading. $d f^a$ is increment of phase fraction.

Elastic strain and electrical field induced strain during loading can expressed as

$$d\varepsilon_{ij}^{e} = C_{ijkl} d\sigma_{ij} \qquad d\varepsilon_{ij}^{E} = d_{kij} dE_k \tag{15}$$

Increment of the total strain $d\varepsilon_{ij}$ can then expressed as the sum of an elastic component $d\varepsilon_{ij}^{e}$, a electrical field induced component $d\varepsilon_{ij}^{E}$ and a transformation induced component $d\varepsilon_{ij}^{tr}$,

$$d\varepsilon_{ij} = d\varepsilon_{ij}^{e} + d\varepsilon_{ij}^{E} + d\varepsilon_{ij}^{tr} \tag{16}$$

5.2. Numerical results

To validate the model, the corresponding calculation result is compared to experimental stress-strain curve. The material parameters used in this study are as follow [37, 38]. Material parameters of R phase are C_{11}^{R}=9.2GPa, C_{12}^{R}=10.3GPa, C_{44}^{R}=6.9GPa. γ^{*}=0.0033, G_{0R}=0.5073MJ/m^3, β=0.004MPa/K, $1/m$=11, $1/k$=1.0. T=403, T_0=298K. Material parameters of O phase are C_{11}^{O}=38GPa, C_{12}^{O}=40GPaC_{44}^{O}=28GPa. G_{0O}=0.276MJ/m^3 β=0.039MPa/K, $1/m$=11.5, $1/k$=1.1. Fig.12 shows the model predicted and experimental measured response. Result in fig.8 shows that stress-strain response of PMN-0.32PT can be predicted by the developed constitutive model, with quantitative agreement.

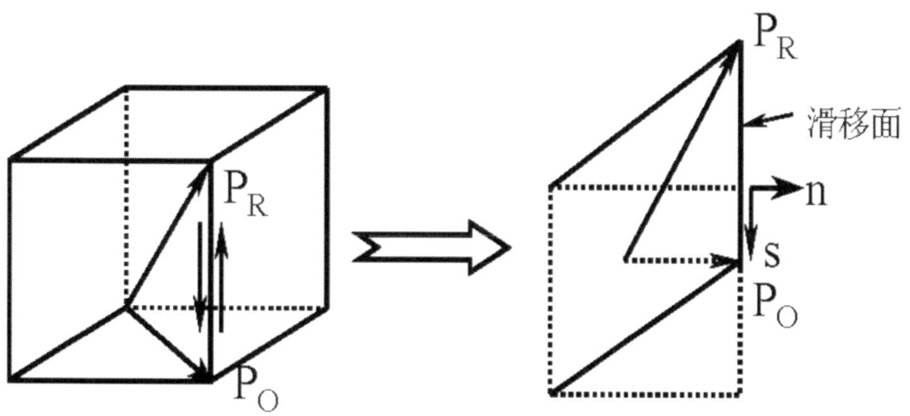

Figure 11. The comparison between ferroelectric phase transformation and plastic slip of single crystal, *n* is direction normal to slip surface, *s* is direction along phase transformation.

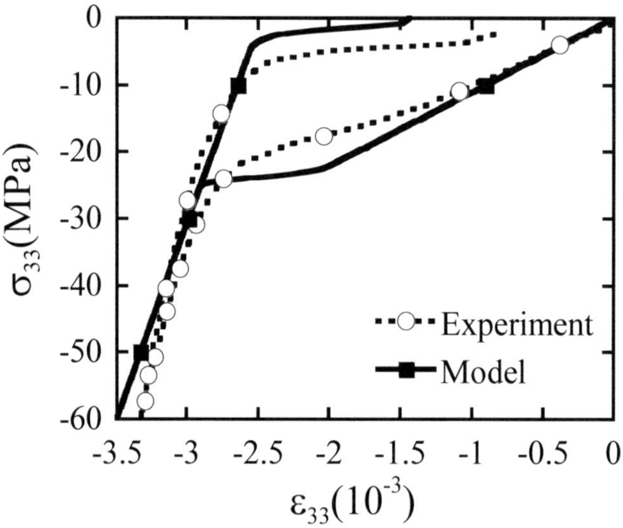

Figure 12. The comparison between experimental and simulation stress-strain curve along <001> direction of ferroelectric single crystal.

	100	010	001
Ba^{2+}	0	0	0
$O_1{}^{2-}$	0.5133	0.5133	0.0192
$O_2{}^{2-}$	0.5133	0.0192	0.5133
$O_3{}^{2-}$	0.0192	0.5133	0.5133
Ti^{4+}	0.489	0.489	0.489

Table 1. The coordinate of particle in R phase $BaTiO_3$ ferroelectric single crystal cell [34]

	100	010	001
Ba^{2+}	0	0	0
$O_1{}^{2-}$	0.514	0.514	0.01
$O_2{}^{2-}$	0.514	0.016	0.521
$O_3{}^{2-}$	0.016	0.521	0.499
Ti^{4+}	0.4846	0.4846	0.499

Table 2. The calculated coordinate that is correspond to point A in fig.6(a)

	100	010	001
Ba^{2+}	0	0	0
O_1^{2-}	0.5144	0.5144	0
O_2^{2-}	0.5	0.0162	0.523
O_3^{2-}	0.0162	0.523	0.5
Ti^{4+}	0.4857	0.4857	0.5

Table 3. The coordinate of particle in O phase $BaTiO_3$ ferroelectric single crystal cell from refer.

6. Conclusion and discussion

In this paper, Stress induced strain and polarization, and electric field induced "butterfly" curves and polarization loops for a set of compressive bias stress for {001}, {011} and {111} poled PMN-0.32PT single crystals are experimentally explored. Obtained results indicate that high piezoelectric responses of PMN-0.32PT single crystals are controlled by the anisotropy of the crystals and the multi-domain structure (i.e., engineered domain structure) has a relatively minor effect. Analysis shows that in all three directions the electric field induced aggregate strain $\Delta\varepsilon$ and piezoelectric constant d_{33} increase with increasing the magnitude of the compressive bias stress. However, when the magnitude of the compressive bias stress is further increased the electric field induce $\Delta\varepsilon$ and d_{33} decrease. As a result, an optimized compressive bias stress exists for the purpose of enhancing the electromechanical properties of {001} and {111} oriented PMN-0.32PT single crystals. These results have apparent importance in the design of actuators and sensors using PMN-0.32PT single crystals. It is found that the observed stress induced strain and polarization in {001}-oriented PMN-0.32PT can be described by a polarization rotation mechanism, i.e., polarization rotates from rhombohedral (R) to orthohombic (O) and tetragonal (T) phases through the intermediate Monoclinic (M) phase during loading, and O to R transition during unloading. However, domain switching is believed to be the main mechanism dictating the electromechanical behavior of {011} and {111} oriented PMN-0.32PT single crystals. polarization rotation model is developed to explain the observed behaviors of PMN-0.32PT. The stress-strain curve along <001> crystallographic direction of ferroelectric single crystal $BaTiO_3$ is calculated with the first principle method. Obtained results show that the R→M→O polarization rotation (phase transformation) takes place in rhombohedral $BaTiO_3$ ferroelectric single crystals under compression, which is consistent with the polarization rotation model. Based on the polarization rotation model, a constitutive model of PMN-0.32PT is proposed based on micromechanical model. It is shown that the developed model can faithfully capture the key characteristic of the observed constitutive behavior of <001> oriented PMN-0.32PT.

Acknowledgments

The authors are grateful for the financial supported by the Natural Science Foundation of China (No. 10425210,10802081) and the Ministry of Education of China.

Author details

Wan Qiang[1], Chen Changqing[2] and Shen Yapeng[2]

1 Institute of Structural Mechanics, China Academy of Engineering Physics, Mianyang, Sichuan, China

2 Key Laboratory, School of Astronautics and Aeronautics, Xi'an Jiaotong University, Xi'an, China

References

[1] Park S. Shrout TR. Ultrahigh strain and piezoelectric behavior in relaxor based ferroelectric single crystals. Journal of Applied Physics 1997:82,1804-1812.

[2] Service RF. Shape-changing Crystals Get Shifter. Science 1997:275,1878-1880.

[3] Liu SF. Park S. Shrout TR. Cross LE. Electric field dependence of piezoelectric properties for rhombohedral $0.955Pb(Zn_{1/3}Nb_{2/3})O_3$– $0.045PbTiO_3$ single crystals Journal of Applied Physics 1999:85,2810-2815.

[4] Fu H. Cohen RE. Polarization Rotation Mechanism for Ultrahigh Electromechanical Response in Single-crystal Pizeoelectrics. Nature(London) 2000:403,281-283.

[5] Choi SW. Shrout TR. Jang SJ. Bhalla AS. Morphotropic phase boundary in Pb(Mg1/3Nb2/3)O3-PbTiO3 system. Mater. Lett. 1989:8(6-7),253–255

[6] Luo HS. Xu GS. et. Compositional Homogeneity and Electrical Properties of Lead Magnesium Niobate Titanate Single Crystals Grown by a Modified Bridgman Technique. Jpn. J. Appl. Phys. 2000:39,5581-5585.

[7] Sreemoolanadhan H. Dielectric ceramics in the BaO-Ln2O3-5TiO2 composition. Ferroelectrics. 1996:189,43-46.

[8] Xu GS. Luo HS. Xu HQ. Third ferroelectric phase in PMNT single crystals near the morphotropic phase boundary composition. Phys. Rev. B, 2001:64(2),020102-020105.

[9] Viehland D. Li JF. Amin A. Electromechanical and elastic isotropy in the (011) plane of $0.7Pb(Mg_{1/3}Nb_{2/3})O_3$-$0.3PbTiO_3$ crystals: Inhomogeneous shearing of polarization. Journal of Applied Physics 2002:92,3985-3990.

[10] Noheda B. Zhong Z. Cox DE. Shirane G. Park SE. Electric-field-induced phase transitions in rhombohedral $Pb(Zn_{1/3}Nb_{2/3})_{1-x}Ti_xO_3$. Physics Review B. 2002:65,224101-224104.

[11] Ye ZG. Noheda B. Dong M. et al. Monoclinic phase in the relaxor-based piezoelectric/ferroelectric $Pb(Mg_{1/3}Nb_{2/3})O_3$-$PbTiO_3$ system. Physics Review B 2001:64,184114-184117.

[12] Ge WW. Luo CT. Zhang QH. Devreugd CP. et al. Ultrahigh electromechanical response in (1−x)(Na0.5Bi0.5)TiO3-xBaTiO3 single-crystals via polarization extension. Journal of Applied Physics 2012:111, 093508-093511.

[13] Xing ZP. Xu K. Dai G. Li JF. Viehland D. Giant magnetoelectric torque effect and multicoupling in two phases ferromagnetic/piezoelectric system. Journal of Applied Physics 2011:110,104510-104514

[14] Tu CS. Tsai CL. Schmidt VH. Luo HS. Dielectric, hypersonic, and domain anomalies of $Pb(Mg_{1/3}Nb_{2/3})O_3$-x$PbTiO_3$ single crystals. Journal of Applied Physics 2001:89(12), 7908-7916.

[15] Farokhipoor S. Noheda B. Conduction through 71° Domain Walls in $BiFeO_3$ Thin Films. Physics Review Letter 2011:107,127601-127605.

[16] Topolov VY. Cao H. Viehland D. Correlation between non-180° domain structures in (1−x)PbA1/3Nb2/3O3−xPbTiO3 single crystals (A = Mg or Zn) under an applied (001) electric field. Journal of Applied Physics 2007:102,024103-024106

[17] Christelle J. Li JF. Viehland D. Investigation of polarization switching in (001)c, (110)c, and (111)c oriented $Pb(Zn_{1/3}Nb_{2/3})O_3$–4.5%$PbTiO_3$ crystals. Journal of Applied Physics 2004:95,5671-5675.

[18] Cao H. Li JF. Viehland D. Electric-field-induced orthorhombic to monoclinic MB phase transition in [111] electric field cooled $Pb(Mg_{1/3}Nb_{2/3}O_3)$–30%$PbTiO_3$ crystals. Journal of Applied Physics 2006:100,084102-084106

[19] Chen KP. Zhang XW. Luo HS. Electric-field-induced phase transition in <001>-oriented $Pb(Mg_{1/3}Nb_{2/3})O_3$-$PbTiO_3$ single crystals. Journal of Physics: Condensed Matter 2002:14 (29),571-576.

[20] Tu CS. Tsai CL. Chen LF. Luo HS. Dielectric properties of relaxor ferroelectric $Pb(Mg_{1/3}Nb_{2/3})O_3$-x$PbTiO_3$ single crystals. Ferroelectrics 2001:261(1-4),831-836.

[21] Catalan G. Janssens A. Rispens G. Polar Domains in Lead Titanate Films under Tensile Strain. Physics Review Letter 2006:96,127602-127606.

[22] Viehland D. Powers J. Ewart L. Ferroelastic switching and elastic nonlinearity in <001>-oriented $Pb(Mg_{1/3}Nb_{2/3})O_3$–$PbTiO_3$ and $Pb(Zn_{1/3}Nb_{2/3})O_3$–$PbTiO_3$ crystals. Journal of Applied Physics. 2000:88,4907-4912

[23] Viehland D. Li JF. Investigations of electrostrictive $Pb(Zn_{1/3}Nb_{2/3})O_3$–$PbTiO_3$ ceramics under high-power drive conditions: Importance of compositional fluctuations on residual hysteresis. Journal of Applied Physics 2001:89,1826-1829.

[24] Viehland D. Powers J. Effect of uniaxial stress on the electromechanical properties of $0.7Pb(Mg_{1/3}Nb_{2/3})O_3$–$0.3PbTiO_3$ crystals and ceramics. Journal of Applied Physics 2001:89,1820-1824.

[25] Viehland D. Ewart L. Powers J. et al. Stress dependence of the electromechanical properties of <001> -oriented $Pb(Mg_{1/3}Nb_{2/3})O_3$–$PbTiO_3$ crystals: Performance advantages and limitations. Journal of Applied Physics 2001:90,2479-2484

[26] Viehland D. Li JF. Anhysteretic field-induced rhombhohedral to orthorhombic transformation in <110> -oriented $0.7Pb(Mg_{1/3}Nb_{2/3})O_3$–$0.3PbTiO_3$ crystals. Journal of Applied Physics 2002:92,7690-7696

[27] Noheda B. Cox DE. Shirane G. Park SE. Polarization Rotation via a Monoclinic Phase in the Piezoelectric 92% $PbZn_{1/3}Nb_{2/3}O_3$-8% $PbTiO_3$. Physics Review Letter 2001:86,3891-3895.

[28] Shang JK. Tan X. Indentation-induced domain switching in PMN-PT piezoelectric crystal. Acta Materialia 2001:49,2993-2999.

[29] Sani A. Noheda B. Kornev IA. High-pressure phases in highly piezoelectric $PbZr_{0.52}Ti_{0.48}O_3$. Physics Review B 2004:69,020105-020108.

[30] Burcsu E. Ravichandran G. Bhattacharya K. Large strain electrostrictive actuation in barium titanate. Applied Physics Letter 2000:77,1698-1670.

[31] Shu YC. Bhattacharya K. Domain patterns and macroscopic behaviour of ferroelectric materials. Philosophical Magazine Part B 2001:81(12),2021-2054.

[32] Yu HF. Zeng HR. Wang HX. Li GR. et al. Domain structures in tetragonal $Pb(Mg_{1/3}Nb_{2/3})O_3$-$PbTiO_3$ single crystals studied by piezoresponse force microscopy. Solid State Communications 2005:133,311-314.

[33] Ye ZG. Bing Y. Gao J. Bokov AA. Development of ferroelectric order in relaxor (1-x)$Pb(Mg_{1/3}Nb_{2/3})O_3$–x$PbTiO_3$(0<~x<~0.15). Physics Review B. 2003:67,104104-104107.

[34] Ghosez P. Gonze X. Michenaud JP. First-principles characterization of the four phases of barium titanate. Ferroelectrics 1999:220,1-15.

[35] Shuo D. Yong Z. Long LY. Elastic and Piezoelectric Properties of Ce:$BaTiO_3$ Single Crystals.Chinese Physics Letter 2005:22(7),1790-1792.

[36] Feng ZY. Tan OK. Zhu WG. et al. Aging-induced giant recoverable electrostrain in Fe-doped $0.62Pb(Mg_{1/3}Nb_{2/3})O_3$-$0.38PbTiO_3$ single crystals. Applied Physics Letters 2008:92, 142910-142914.

[37] Weaver PM. Cain MG. Stewart M. Room temperature synthesis and one-dimensional self-assembly of interlaced Ni nanodiscs under magnetic field. J. Phys. D: Appl. Phys. 2010:43,275002-275006.

[38] Desheng F. Takahiro A. Hiroki T. Ferroelectricity and electromechanical coupling in (1− x)$AgNbO_3$-x$NaNbO_3$ solid solutions. Applied Physics Letter 2011:99,012904-012907.

Chapter 15

'Universal' Synthesis of PZT (1-X)/X Submicrometric Structures Using Highly Stable Colloidal Dispersions: A Bottom-Up Approach

A. Suárez-Gómez, J.M. Saniger-Blesa and
F. Calderón-Piñar

Additional information is available at the end of the chapter

http://dx.doi.org/10.5772/51996

1. Introduction

The synthesis of nanostructured ferroelectric materials has been a subject of increasing interest for more than a decade due to their possible applications as nonvolatile/dynamic random access memories (NVRAMs/DRAMs), tunnel effect capacitors for high frequency microwave applications, infrared detectors, micro-electromechanical systems (MEMs) and electro-optical modulators among others. These materials, when fully integrated into appropriately designed micro-systems, are promising candidates for robotics sensing and for future medical procedures requiring an in situ, real time and nondestructive monitoring with very high sensitivity.

As a logical consequence, the quest for nanostructured ferroelectric materials has also led to modify the synthesis routes usually considered for obtaining conventional bulk materials and thin films. One of those routes, the sol–gel method, overcomes the inherent limitations of the conventional powders-based synthesis routes when dealing with molecular homogeneity and, therefore, it has been extensively studied and applied in different scenarios in order to obtain not only thin films, nanotubes or nanorods, but also submicron grains due to the distinctive dielectric and ferro/piezoelectric features that are associated with size related effects when average grain size is well below 1 μm [1]-[5]and because of the potential use of these materials in a plethora of nanodevices [6].

One of the backbones of the ferroelectrics industry is the Lead zirconate titanate [PZT: $Pb(Zr_{1-x}Ti_x)O_3$] ceramic system, a well-known ABO_3 perovskite with a wide range of industrial applications since the early 1950s of the 20th century. Its dielectric and electromechanical properties have made possible to develop and implement a myriad of devices such as underwater sonar systems, sensors, actuators, accelerometers, ultrasonic equipment, imaging devices, microphones and multiple active and passive damping systems for the car industry.

The sol–gel synthesis of PZT ceramics has evolved a lot since it was first reported in the mid-1980s of the last century [7]-[9]; thermal treatments, precursors, additives, diluents and stabilizers have been incorporated and/or modified in order to achieve less hydrolysable, more stable, compounds to suit a specific need. In the case of the synthesis of thin films, nanotubes or nanorods, sol–gel routes have been used in many different ways: from the well-known spin-coating method to the electrophoretic deposition on a given substrate [10] or the insertion on nanoporous templates [11]. Submicron and nanosized PZT particles are also important in the fabrication of highly dense bulk ceramics which, even nowadays, comprise almost the entirety of the electroceramics market. In this particular case, several works have been published in which PZT powders are synthesized first via sol–gel and then put to sinter for densification [12]-[15]. In this procedure, sintering temperatures tend to lower and densification attains notably high values mostly due to a higher Gibb's free energy per unit surface area while some material properties strongly dependent on grain size are also enhanced [16].

Accordingly, in this chapter we will be devoted to analyze the feasibility of a 'customized' synthesis of $PbZr_{1-x}Ti_xO_3$ [PZT (1-x)/x] nano/submicrometric structures by using a sol-gel based colloidal dispersion as a precursor solution. This study will be done on the basis of a 'bottom-up' approach as it will take into account (i) Synthesis route, (ii) Properties of colloidal dispersions and (iii) Final crystallization. We will try to thoroughly illustrate every synthesis step while paying special attention to the physicochemical depiction of some phenomena that not always are sufficiently described or explained. It has to be pointed out that most of the main procedures, discussions and results contained herein could be easily extrapolated to a wide range of materials, not exclusively PZT-based ones.

2. Sol-gel based synthesis route

The complexity of the intermediate reactions, one of the few handicaps of the sol–gel method, makes almost mandatory a step by step study of the synthesis method. The chemical reactivity of precursors is a well-known key feature determining the nature of the intermediate organic ligands and the control of the hydrolysis rate in the final sol–gel. Several studies have been made for particular synthesis routes both theoretical and experimental and many possible reaction pathways have been proposed and dissected. Particularly, the -chemistry of metal alkoxides has been intensively studied by Sanchez et al. [17] and Sayer and coworker [18]. It is our purpose here to analyze the different reaction steps and intermediates involved in the PZT (1-x)/x sol–gel synthesis using propoxides as starting metal alkoxides.

2.1. Synthesis

The followed sol-gel route was the acetic acid, acetylacetone and 2-methoxyethanol propoxy-based sol-gel method as illustrated in Figure 1. Starting reagents for the sol-gel PZT (1-

x)/x solutions were: (1) lead(II) acetate trihydrate ($Pb(OAc)_2{\bullet}3H_2O$, Mallinckrodt Baker, Inc., 99.8% pure), (2) glacial acetic acid (HOAc, Mallinckrodt Baker, Inc., 99.7% pure), (3) acetylacetone (AcacH, Sigma-Aldrich Co., 99% pure), (4) 2-methoxyethanol (2-MOE, Mallinckrodt Baker, Inc., 100% pure), (5) zirconium(IV) propoxide ($Zr(OPr)_4$, Sigma-Aldrich Co., 70 wt.% in 1-propanol), and (6) titanium(IV) propoxide ($Ti(O^iPr)_4$, Sigma-Aldrich Co., 97% pure).

First, lead acetate was dissolved in acetic acid with a 1:3 molar ratio while stirred and refluxed at 115 °C during 3 h for dehydration and homogeneity purposes. After this step, a thick transparent solution was obtained which will be referred hereafter as solution A. On a separate process, stoichiometric amounts of zirconium and titanium propoxides were mixed with acetylacetone on a 1:2 molar ratio in order to avoid fast hydrolysis of reactants. This mixture was stirred and refluxed at 90 °C during 4 h forming a clear yellow solution referred hereafter as solution B. When this solution cooled down, the precipitation of several needle shaped crystals was verified. In order to keep stoichiometry unaffected, we then repeated the solution B procedure to isolate and characterize some of those crystals by single crystal XRD.

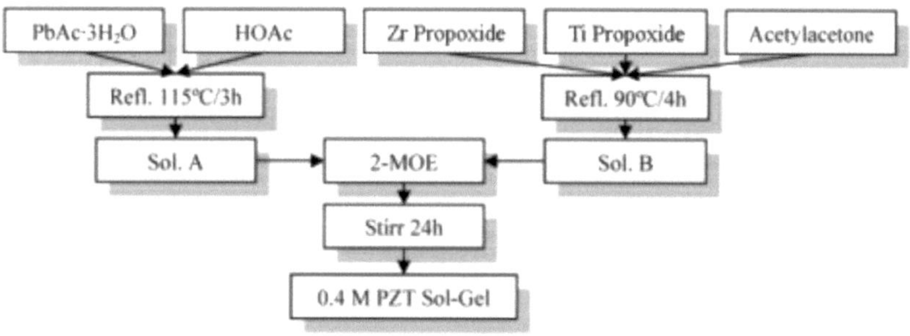

Figure 1. Flow chart depicting the basic experimental procedure followed in order to obtain a 0.4M PZT (1-x)/x sol-gel based precursor.

Solutions A and B were then mixed together as appropriate amounts of solvent (2-MOE) were slowly added for complete dilution of the precipitated crystals and for controlling the PZT concentration on the final solution. Afterwards, a light yellow solution was obtained after stirring for 24 h at room temperature.

It must be stated here that we devoted our work to several Zr:Ti molar ratios that are commonly used in practical applications: *(i)* PZT 25/75, *(ii)* PZT 53/47, *(iii)* PZT 60/40, *(iv)* PZT 80/20 and *(v)* PZT 95/05. Besides, we also tried to cover a relatively wide concentration range for every PZT (1-x)/x sol-gel precursor under study: from 0.05 M to 0.4 M. Due to these facts, we will only focus on the discussion of representative samples instead of discussing irrelevant data of samples that showed no significant discrepancies between each other. As depicted in Figure 1, this section will analyze the synthesis of a 0.4 M PZT 53/47 precursor solution.

2.2. Experimental techniques

Every reactant and intermediate product was analyzed using FT-IR and Raman spectroscopies. As was described earlier, single crystal XRD was also carried out for the solution B acicular precipitates.

2.2.1. Raman spectroscopy

Raman characterization was made on an Almega XR Dispersive Raman spectrometer equipped with an Olympus microscope (BX51). An Olympus 10x objective (N.A. = 0.25) was used both for focusing the laser on the sample, with a spot size ~ 5 μm, and collecting the scattered light in a 180° backscattering configuration. The scattered light was detected by a CCD detector, thermoelectrically cooled to -50 °C. The spectrometer used a grating (675 lines/mm) to resolve the scattered radiation and a notch filter to block the Rayleigh light. The pinhole of the monochromator was set at 25μm. The Raman spectra were accumulated over 25 s with a resolution of ~ 4 cm^{-1} in the 100-2000 cm^{-1} interval. The excitation source was a 532 nm radiation from a Nd:YVO_4 laser (frequency-doubled) and the incident power at the sample was of ~ 8 mW.

2.2.2. FT-IR spectroscopy

FT-IR analysis was carried out on a Thermo Nicolet Nexus 670 FT-IR in transmission mode with a resolution of ~ 4 cm^{-1} in the 400-2000 cm^{-1} interval. The excitation source was a Helium-Neon laser light incident on a KBr compact target containing ~ 0.5 % in weight of the sample of interest.

2.2.3. Single crystal XRD

Single crystal XRD experiments were carried out for selected specimens. A Bruker SMART APEX CCD-based X-ray three circle diffractometer was employed for crystal screening, unit cell determination and data collection. A Van Guard 40x microscope was used to identify suitable samples and the goniometer was controlled using the SMART software suite. The X-ray radiation employed was generated from a Mo sealed X-ray tube (Kα= 0.70173 Å with a potential of 50 kV and a current of 30 mA) and filtered with a graphite monochromator in the parallel mode (175 mm collimator with 0.5mm pinholes).

2.3. Solution A

Figure 2 shows, from bottom to top, the IR and Raman spectra of starting acetic acid and lead acetate as well as the final product, solution A, after stirring and refluxing. There are several features in common between these spectra due to the organic nature of ligands. Basically, bands corresponding to the CH_2 and CH_3 groups in the 1300-1400 cm^{-1} interval as well as in the low frequency range [19].

The glacial acetic acid spectra reveal several representative peaks found at 619 cm^{-1} (RA: Raman active), 889-893 cm^{-1} (RA, IRA: Infrared active), 1294 cm^{-1} (IRA), 1410 cm^{-1} (IRA), 1668

cm^{-1} (RA), 1716 cm^{-1} (IRA) and 1757 cm^{-1} (IRA) for this molecule. Those peaks correspond, respectively, to the τ(C-C=O), ν(C-C), $\delta_S(CH_3)$, $\delta_A(CH_3)$, ν(C=O), $\nu(C{=}O)_{dimer}$ and $\nu(C{=}O)_{monomer}$ vibrations [20],[21]. The clear splitting detected for the C=O dimeric and monomeric stretch vibrations somewhat evidences the presence of some small amounts of water in the acidic medium that, for the purpose of our study, will not be taken into consideration.

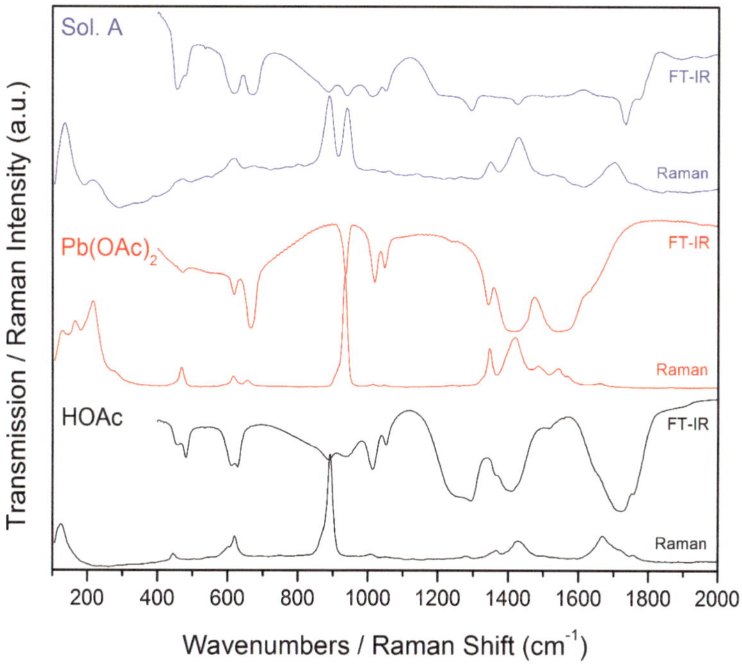

Figure 2. Infrared and Raman spectra of the reactants involved in solution A formation.

On the other hand, the lead(II) acetate trihydrate spectra also feature some representative peaks located at 216 cm^{-1} (RA), 615-617 cm^{-1} (RA, IRA), 665 cm^{-1} (IRA), 934-935 cm^{-1} (RA, IRA), 1342-1346 cm^{-1} (RA, IRA), 1417-1420 cm^{-1} (RA, IRA) and 1541-1543 cm^{-1} (RA, IRA) related to the ν(Pb-O), ϱ(COO), $\delta_S(COO)$, ν_S(C-C), $\delta_S(CH_3)$, ν_S(C-O) and ν_A(C=O) vibrational modes, respectively [20],[22]. It is well known that the acetate ligands can complex a metal ion in three different ways (monodentate, bidentate chelating and bridged) and that, unfortunately, none of these can be uniquely identified by symmetry considerations. Traditionally, the various types of bonding have been identified by the magnitude of the difference between symmetric ν_S(C-O) and asymmetric ν_A(C=O) vibrations. In our case, this difference is 122 cm^{-1} indicating a bidentate chelating coordination for the acetate-metal complex that has been widely accepted for lead(II) acetate even though the Δν criterion has led sometimes to incorrect conclusions [20],[22].

Solution A vibrational spectra shown in Figure 2 evidences the expected dilution of lead(II) acetate in acetic acid: there are no new vibrational modes and a strong band overlapping is seen. There is some band shift due to the overlapping and is worth to notice the weakening of the lead acetate Raman active Pb-O band at 216 cm^{-1} when in solution. Unfortunately, a noticeable fluorescence in the Raman spectrum could not be avoided and this fact made very difficult to carry on an appropriate analysis with useful data.

Nevertheless, and according to our results, the formation of Solution A could be fairly described by the following equation:

$$\begin{aligned} Pb(OAc)_2 \cdot 3H_2O_{(s)} + 3HOAc_{(Aq)} \xrightarrow[t-3hrs.]{T-115^{\circ}C} Pb^{2+}_{(Aq)} + 3\alpha H_3O^{+}_{(Aq)} + (2+3\alpha)OAc^{-}_{(Aq)} + \\ +3(1-\alpha)HOAc_{(Aq)} \end{aligned} \qquad (1)$$

Where α is the HOAc dissociation degree under our experimental conditions and where we have also assumed, on the simplest approach, neglectable losses due to evaporation of H_2O and HOAc during the whole process.

2.4. Solution B

FT-IR spectra for Solution B precursors, Figure 3, showed some features like the ones discussed above. A Raman spectra based analysis for these compounds could not be completed due to the strong fluorescence exhibited by the zirconium and titanium alkoxides at our fixed operating laser wavelength.

The titanium propoxide IR spectrum exhibit sharp bands at 1377-1464 cm^{-1} corresponding, respectively, to the stretching and bending vibrations of the aliphatic CH_3 groups. A distinctive single peak at 1011 cm^{-1} corresponds to the propoxy- Ti-O-C vibration and a similar behavior is found for zirconium propoxide with Zr-O-C vibrations located at 1142 cm^{-1}[18].

The acetylacetone vibrational spectrum shows the main features attributed to the most probable staggered conformation of this compound. This conformation has some typical weak IR active vibrations in the low frequency range, as it is shown. Modes at 509, 554 and 640 cm^{-1} correspond to the in plane ring deformation (Δ ring), out of plane ring deformation (Γ ring) and Δ ring + $\varrho(CH_3)$ modes, respectively [23].

As described above, when reaction took place, some crystals precipitated short after solution B reached room temperature; IR spectra of these single crystals are also shown in Figure 3. Unlike solution A, a reaction is now verified by the shifting and/or reinforcement of bands associated to reactants vibrations. A discussion of several mechanisms for this kind of reaction has been reviewed by several authors taking into account, primarily, the mixing conditions, the reactivity of metal alkoxides and the Acac/Alkoxides molar ratio [17],[18],[24].

2.4.1. *Single crystal XRD characterization*

At this stage, a suitable colorless parallelepiped 0.356 mm × 0.162 mm × 0.066 mm was chosen from a representative sample of crystals of the same habit. After the determination of a suitable cell, it was refined by nonlinear least squares and Brava is lattice procedures. The unit cell was then verified by examination of the (h k l) overlays on several frames of data, including zone photographs, and no super cell or erroneous reflections were observed. A search performed on the Cambridge Structural Database and updates using the program Conquest afforded 77 coincidences within 1% of the longest length of a monoclinic C-centered cell whose lattice parameters and reported structure are shown in Figure 4(a). Two entries, those with ACACZR and ZZZADD CCD reference code, in addition, revealed chemical coincidences (both in composition and stoichiometry) for the compound tetrakis(acetylacetonate-O,O')-zirconium(IV) or $Zr(Acac)_4$.

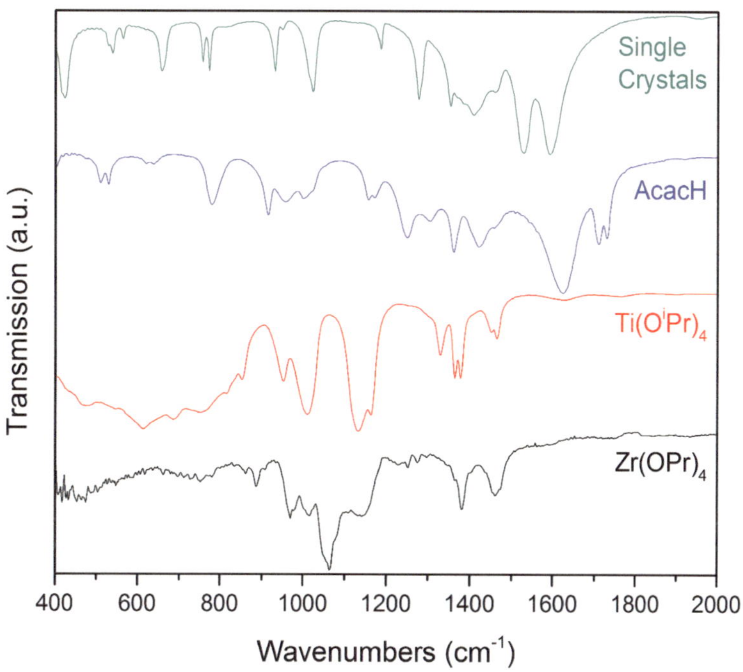

Figure 3. Infrared spectra of the reactants involved in solution B formation. The IR spectra for the resultant precipitated single crystals are also shown.

According to these data, the $Zr(Acac)_4$ structure was simulated using the Accelrys Materials Studio 3D visualizer environment [25]as shown in Figure 4(b). In this case, the acetylacetone reaction with the Zr propoxide results in the formation of an oxo cluster where the metallic atom changes its coordination number from 4 to 8 which is the highest possible value for zirconium. Metallic cations are now bonded to the acetylacetonate chelating ligand giving rise to a less hydrolysable organic complex.

In the case of the titanium propoxide reaction, and taking into account our experimental conditions, it is not a bad assumption to consider the formation of a fully chelated organometallic complex, just as it was described for zirconium. Given the maximum coordination number of 6 for titanium, a $Ti(Acac)_2(OPr)_2$ compound will be the most likely to expect as is also confirmed by earlier reports that consider the multiple chelation routes for titanium propoxide [18],[20]. Figure 4(c) shows a very simple Materials Studio 3D modeling for the most probable configuration of $Ti(Acac)_2(OPr)_2$ as determined by the Forcite package [25] on a single step relaxation and energy minimization routine.

Accordingly, bands located at 1280, 1370-1440 and 1530-1595 cm^{-1} in the crystals spectrum, shown in Figure 3, can be assigned to the ν(C-CH_3:Acac), δ(CH_3:Acac) and ν(C-C) + ν(C-O:Acac) vibrational modes, respectively [26].

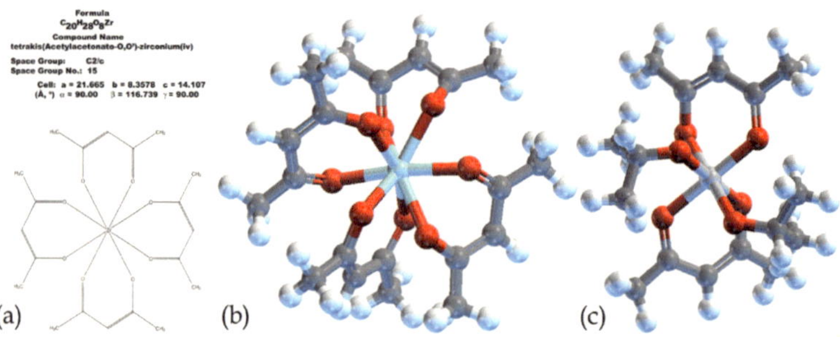

Figure 4. Fully chelated metal-acetylacetonate complexes. (a) $Zr(Acac)_4$ as reported by single crystal XRD charateriza-tion, (b) $Zr(Acac)_4$ structure simulated by Materials Studio (MS) according to its crystallographic data and (c) $Ti(Acac)_2(OPr)_2$ simulated by MS according to its most probable configuration.

2.4.2. Chemical reaction

As seen before, solution B can be thought as the resulting chelated metal complexes mixed with residual isopropanol. In this case, an appropriate chemical equation for the reaction could be:

$$\begin{aligned} &0.53Zr\left(O^i\,\mathrm{Pr}\right)_{4(Aq)} + 0.47Ti\left(O^i\,\mathrm{Pr}\right)_{4(Aq)} + 2AcacH_{(Aq)} \xrightarrow[t\sim4hrs.]{T\sim90^\circ C} 2HO^i\,\mathrm{Pr}_{(Aq)} + yZr\left(Acac\right)_{4(S)} + \\ &+\left(1-2y\right)Ti\left(Acac\right)_2\left(O^i\,\mathrm{Pr}\right)_{2(Aq)} + \left(0.53-y\right)Zr\left(O^i\,\mathrm{Pr}\right)_{4(Aq)} + \left(2y-0.53\right)Ti\left(O^i\,\mathrm{Pr}\right)_{4(Aq)} \end{aligned} \tag{2}$$

where we have assumed that all the acetylacetone reacts with the alkoxides thus forming the chelated organometallic complexes. As we see, there is still a fraction of nonchelated reac-tants due to the insufficient amount of acetylacetone needed for that purpose. Let us find now the exact amount of acetylacetone required for obtaining fully chelated organometallic complexes without any further byproducts. We could rewrite equation [2] as:

$$\begin{aligned} &0.53Zr\left(O^i\,\mathrm{Pr}\right)_{4(Aq)} + 0.47Ti\left(O^i\,\mathrm{Pr}\right)_{4(Aq)} + (2+\Delta)AcacH_{(Aq)} \xrightarrow[t\sim4hrs.]{T\sim90^\circ C} (2+\Delta)HO^i\,\mathrm{Pr}_{(Aq)} + \\ &+0.53Zr\left(Acac\right)_{4(S)} + 0.47Ti\left(Acac\right)_2\left(O^i\,\mathrm{Pr}\right)_{2(Aq)} \end{aligned} \tag{3}$$

Now, it is obvious that $\Delta = 1.06$ implying an alkoxides: acacH molar ratio of 1:3.06 for full chelation of metallic centers. Equation [3] can now be written for any given Zr:Ti ratio (1-x)/x:

$$(1-x)Zr\left(O^i\,\mathrm{Pr}\right)_{4(Aq)} + xTi\left(O^i\,\mathrm{Pr}\right)_{4(Aq)} + 2(2-x)AcacH_{(Aq)} \xrightarrow[t\sim 4hrs.]{T\sim 90^\circ C} 2(2-x)HO^i\,\mathrm{Pr}_{(Aq)} + \\ +(1-x)Zr\left(Acac\right)_{4(S)} + xTi\left(Acac\right)_2\left(O^i\,\mathrm{Pr}\right)_{2(Aq)} \tag{4}$$

This equation represents an extension for the initially proposed sol-gel based route in order to obtain fully chelated and no hydrolysable precursor solutions for PZT based ferroelectric materials. It allows us to synthesize PZT at any Zr:Ti ratio while maximizing solution stability and offering a noticeable repetitiveness for both small and large scale manufacturing and processing.

On the following section we will analyze some features of different sol-gel based PZT (1-x)/x precursors when synthesized using the universal 1:2(2-x) alkoxides: acacH molar ratio already discussed above.

3. Some properties of the PZT precursors

In the final step of the synthesis, when mixing Solutions A and B with 2-methoxyethanol, the dissolved lead acetate complex can react with 2-MOE forming a very stable acetate–methoxyethoxy lead complex [27] that, along with the chelated metal complexes already formed, may lead to turn this final solution into a hydrophobic sol, poorly hydrolysable and, therefore, very stable.

Stability issues tend to be crucial when using sol-gel based precursors in research, small scale applications and industry. Therefore, it is very important to keep control over some parameters that affects the solution stability and, most of all, the average particles size. Particularly, particles size can be an indicator of some undesirable processes that could be taking place in the sol: aggregation, flocculation and sedimentation; aggregation, even though is a reversible process, is a good indicator of instability since the other two processes, which are not reversible, generally follows after some time.

In this respect, the aging time dependences of some physical parameters directly related to the stability of colloidal dispersions must be explored and discussed. Two of the most important parameters that should be taken into account are pH and the average particles size. The first one is determinant for fixing the thickness of the ionic layers surrounding any given charged colloidal particle (Stern layers) while affecting, at the same time, the Zeta potential, the electrophoretic mobility and the aggregation mechanisms as well as the apparent hydrodynamic particle size. As a consequence, the average particles size is the final result of the conjugate action of all the physicochemical variables hardly mentioned before. It is also the definitive experimental variable on which any post processing technique should be based on as it explicitly defines the size range of the so called "building blocks" for bottom-up studies or applications.

Currently, there are no known extensive reports in the literature directly concerned to the study of the time dependence of the average particles size or pH in a PZT precursor sol. It must be noted that, in the case of magnetic nanoparticles, some studies have been carried out [28]-[30]and all of them stress the strong correlation between the size of the colloidal particle and the final properties of the conceived structure, whether it will be nano- or not.

3.1. pH

For this study, we synthesized PZT (1-x)/x precursor solutions by using the same route described in 2.1 with the only difference being the propoxides: acacH molar ratio. In this case, and for the rest of our text, that ratio will be 1:2(2-x) for full chelation of the metal alkoxides.

As it was stated earlier, we will avoid again showing and discussing irrelevant data of samples that showed no significant discrepancies between each other and, because of this, this section will analyze the pH vs. Concentration vs. aging time behaviors of several PZT 53/47 precursor solutions with concentrations ranging from 0.05 M to 0.37 M in a time interval of up to 4 months of stocking.

As can be seen in Figure 5, the acidity/basicity of the solutions clearly shows a tendency with both sols concentration and aging time. Figure 5(a) shows the measured pH vs concentration for the studied sols at several aging steps and, as expected, the more concentrated sample implies the more acidic medium which is consistent with the chemical reaction proposed for sols synthesis.

Figure 5(b), on the other hand, shows the measured pH vs. time behavior. As a general tendency, sols pH dependency with aging time can be divided in three stages which are highlighted in the graph: (i) pH increases notably in the first days after which (ii) it decreases to values somehow close to the initial ones. In this moment, (iii) pH starts to rise again but with a slower time gradient than on stage (i). In our opinion, for the understanding of the pH vs. aging time behavior, we have to take into account the coexistence of different competitive processes right after sols were prepared. In this way, a qualitative description could be done as follows [31]: (i) Remnant unreacted acetic acid evaporates during the first days implying a decrement on the acidity of sols. At this point, (ii) particles are less positively charged and are able to aggregate via condensation reactions followed by the formation of small portions of alcohol (1-propanol) and thus implying a more acidic environment as seen in Figure 5. The number of polyanions per aggregate chain must be limited, however, by the high chemical stability of the chelated metal complexes and that is why the aggregation process does not imply polymerization and/or gelation as it has been reported for more hydrophilic sols. Shortly after condensation rate vanishes, (iii) it is possible for the residual alcohols to evaporate as solutions age thus allowing pH to rise slowly, as seen in Figure 5 for the more aged solutions.

At this scenario, however, hydrolysis is not expected due to (a) the complete chelation of the metal complexes, (b) to the short lengths of the already formed polymeric chains and, therefore, (c) to the small amount of alcohol that could be formed afterwards. After 4 months of aging, no noticeable changes in pH values were detected. Moreover, the stability of these solutions could be eye inspected by verifying neither the absence of sedimentation nor precipitation of single particles or aggregates after almost 1 year of stocking.

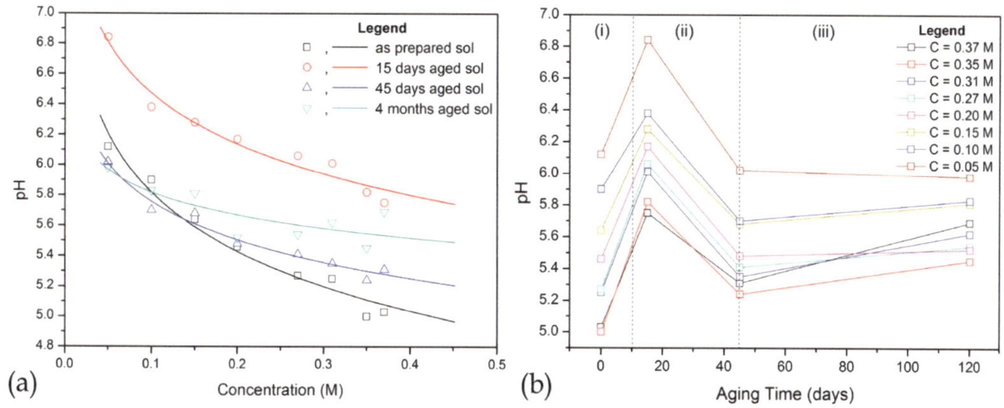

Figure 5. a) pH vs. concentration dependencies for the PZT 53/47 samples understudy measured at different time intervals. Semi-log fittings to guide the eye are also drawn. (b) Aging effects on pH for the same samples. Three distinctive stages were detected and are highlighted in the graph.

3.2. Mean particle size

3.2.1. Experimental technique

Particles size measurements were carried out by means of the dynamic light scattering (DLS) technique in a Zetasizer Nano ZS90 manufactured by Malvern Instruments Ltd. equipped with a HeNe laser source. Approximately 1.5 ml of as synthesized sols were stored in polystyrene cells (DTS0012 cells) provided by the same manufacturer and then size distributions curves were recorded from the very first day until 125 days after synthesis. For this purpose, several solvent parameters were needed for further processing of dispersed light intensity, namely viscosity ($\eta_{2\text{-MOE}} \sim 1.5410$ cP), dielectric constant ($\varepsilon_{2\text{-MOE}} \sim 16.9$) and refraction index ($n_{2\text{-MOE}} \sim 1.33$) [32]. After each measurement, a nonlinear least squares fitting (NLLSF) was made to the experimental data according to a log-normal distribution function:

$$f_{\ln}(x,\beta_i) = \beta_0 + \frac{\beta_1}{\sqrt{\pi}\beta_2}\frac{1}{x}\exp\left[-\frac{1}{2\beta_2^2}\ln\left(\frac{x}{\beta_3}\right)^2\right] \tag{5}$$

Where β_I are fitting parameters. It must be stressed that this fitting was carried out only for size intervals where unimodal distribution curves were measured.

3.2.2. Concentration dependence

Figure 6 shows the results obtained for particles size distribution for two concentrations under study (a PZT 53/47 precursor) after different time lapses that are also shown in the figure.

According to what is shown, an appropriate discussion relaying on the acidity/particles size dependence cannot be easily established: oscillations in the pH values, see Figure 5, during aging are not followed by oscillations in the mean particles size values. Even though this feature is not fully understood, it may be due to the short lengths of the oligomeric chains present in the solution and to the weak sensitivity of the completely chelated metal complexes to the ionic strength of the solvent medium. Another possible explanation could be given in terms of the irreversibility associated to the formation of these chains when the solvent medium is not acidic enough to break the corresponding bonds and/or coordinations.

As can be seen, as prepared sols featured multimodal distributions with some small percent of particles that could even have sizes higher than 1μm. As solutions are aged, these distributions tend to be unimodal featuring a mean particles size well below 10 nm. We consider that this is a remarkable result of this work as it shows that there is no rule of thumb closely related to the need of using fresh, as prepared solutions for synthesizing several kinds of nanodimensional systems by means of electrophoretic deposition(EPD) or simple template immersion, [33]-[36]. Moreover, in this case we could have chosen solutions aged for 20 days or more for bulk, thin films or nanostructures synthesis due to the better homogeneity regarding particles size distribution in both cases under study.

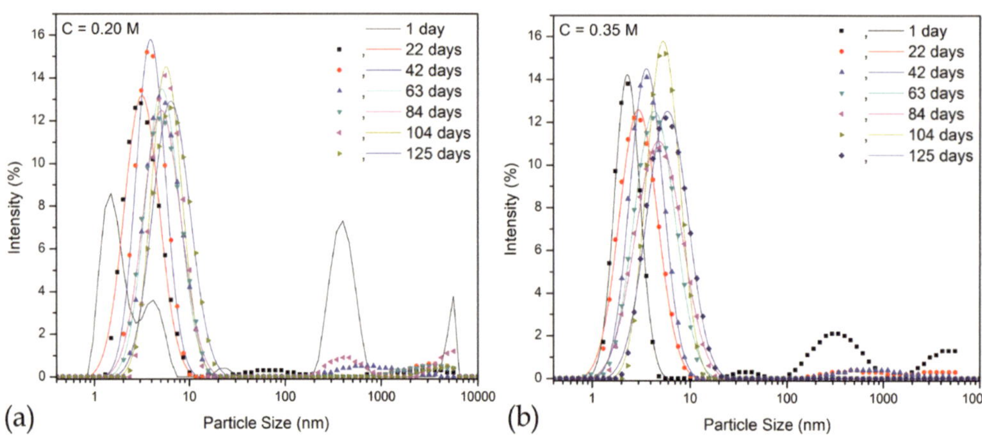

Figure 6. Particles size distribution functions for the analysed PZT 53/47 precursor solutions and for several aging times in the whole measurement range. Log-normal fittings for the more populated size intervals are also depicted.

This behavior could be explained, in principle, on the basis of the chemical reaction which was described for obtaining what we called solution B, see Section 2.4. As we stated, by the end of this step we observed the crystallization of the compound $Zr(Acac)_4$. Such crystallization does not occur after mixing solutions A and B in the presence of 2-MOE which can be attributed (I) to the higher acidity of the final A+B+2-MOE solution, (II) to the solubility of

$Zr(Acac)_4$ in 2-MOE, (III) to a dilution process while mixing A+ B, or (IV) to a combined effect of all of the above. Must be noted that the less acidic solution (C = 0.20M) possesses, at the same time, the less unimodal distribution function which is an expected result in concordance to the relationship between ionic strength and particles size.

Thus, it seems plausible to assume solution B as a dispersion containing the $Ti(Acac)_2(OPr)_2$ compound at molecular level and submicrometric aggregates of $Zr(Acac)_4$ that do not precipitate; these are redissolved afterwards when mixed with the acetate rich solution A and 2-MOE. After stirring the resultant sol for one day, the size of some percentage of the $Zr(Acac)_4$ aggregates still ranges between 100 and 1000 nm; after 20 days, however, size is well below 10 nm. From these results, and somehow confirming our previous discussion, we can also see that the less concentrated solution features a considerable amount of particles with sizes well above 50nm even after 100 days of stocking. For this reason, we will focus our attention from now on in precursor solutions with 0.35 M aged during the first 30-35 days, a time period that, according to our results, seems to be critical in the colloidal stability of as synthesized sols.

3.2.3. *Aging time dependence*

As it was said before, all samples under study featured a noticeable stabilization when aged for about 1 month. Figure 7 shows this aging behavior for two PZT (1-x)/x precursor solutions concentrated at 0.35 M. In all cases, it could be observed a similar time evolution as the one described earlier for Figure 6.

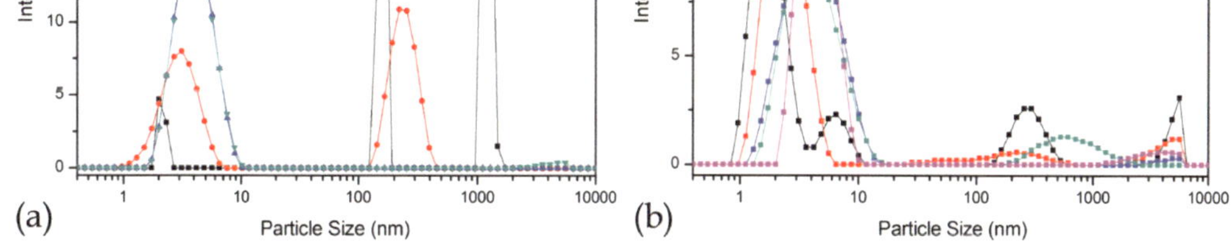

Figure 7. Aging behavior, as depicted by the time evolution of particles size distribution curves, for two PZT (1-x)/x precursor solutions concentrated at 0.35 M: (a) PZT 25/75 and (b) PZT 60/40.

On the other hand, Figure 8 shows the whole dataset of the measured mean particles size for the synthesized solutions. Fitting curves shown here were determined by considering a time dependence given by $\langle d \rangle = C_1 t^{1/(1-\lambda)} + C_2$ that, according to the classical Smoluchowski theory in the van Dongen–Ernst approximation, describes appropriately a nongelling system of dispersed clusters. Moreover, in the same approximation, the small values of $\lambda(\lambda<<1)$ are usually associated with a Brownian diffusion limited aggregation process vastly dominated by collisions between larger clusters with smaller ones [31].

Another view of the aging process could be illustrated by means of the dimensionless normalized distribution curves, not shown here. In that representation, the broadening of the distribution curves with aging is an indicator of very likely aggregation processes and somehow will help to complete the kinetic analysis that we have been through in this section.

In a way of summarizing our results, it could be said that, just after synthesis, several populations of particles were detected until homogenized a few days later; then particles tend to grow very slowly with time according to an almost linear law (corresponding to that describing a nongelling system) and aggregation is expected. Moreover, these cases fit very well in the Brownian diffusion limited aggregation type where smaller clusters stick to bigger ones when they collide. The slow but evident increase in the mean cluster size, as well as the noticeable broadening in the distribution peaks, must warn us about an undesirable and irreversible precipitation process for, hypothetically, $t \rightarrow \infty$.

The high stability shown by our samples even after 4months of stocking and the small values for mean particles size (well below 10 nm) are good indicators that the complete chelation of the organometallic compounds plays a key role in keeping short, and poorly reactive, oligomeric chains. An aggregation process dominated exclusively by Brownian motion is highly desirable when solution stability needs to be maximized.

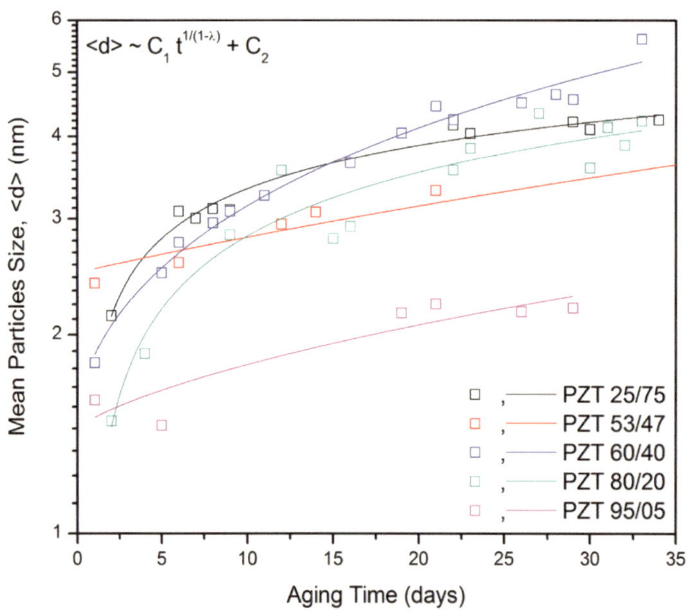

Figure 8. Mean particles size vs. aging time dependence for the systems under study. Fitting curves obeying a $t^{1/(1-\lambda)}$scaling law are depicted.

http://dx.doi.org/10.5772/51996

Our samples were also characterized by the High Resolution Transmission Electron Microscopy (HR-TEM) technique. A few drops of the synthesized sols were deposited on copper grids, evaporated afterwards and put under a JEOL JEM2200 microscope with Omega filter and a spherical aberration corrector; some of the obtained images are shown in Figure 9 for a PZT 95/05 precursor solution. As expected, particles are likely to coalesce as the solvent evaporated prior to characterization, see Figure 9(a) and (b), and some of those nanoparticles are shown in higher magnification images in Figure 9(c) and (d). Most of the features regarding particle size that were discussed earlier were corroborated by means of this technique for all the precursor solutions.

At this point, and considering what has been discussed until now, it must be highlighted that a rigorous control of the chelation rate, pH and aging time could give us a chance to "tune" the average nanoparticles size and/or the colloidal stability as desired. By looking back at Figure 5 through Figure 8 one may feel free to choose to explore several "working points" according to our research or technological needs for every (1-x)/x Zr:Ti ratio in the PZT system.

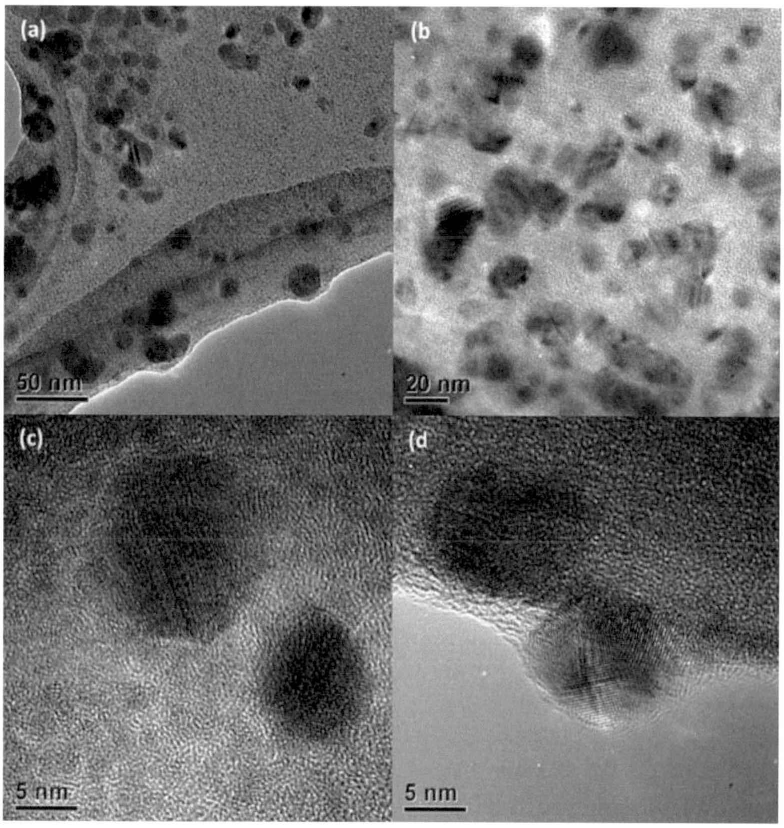

Figure 9. HR-TEMimages at different magnifications for a PZT 95/05 precursor solution, with C = 0.35 M and aged for 2 months.

This, in fact, is a powerful tool provided by the physicochemical phenomena and mechanisms previously discussed and, generally speaking, could be extended to other material systems of technological interest at a moderate cost.

4. Crystallization route

The final A + B + 2-MOE solutions were dried at 100°C for several days and, after that, they were thermally treated in order to analyze the phase evolution from the amorphous PZT sol-gel network to the expected final Perovskite structure. Due to fundamental similarities among the different samples we will be discussing only the case where Zr:Ti ratio is 53/47, concentrated at 0.35 M and aged for 2 months.

4.1. Experimental techniques

Crystallization was monitored by means of FT-IR and Raman vibrational spectroscopies, see Sections 2.2.1 and 2.2.2, and by X-Ray Diffraction (XRD) and Scanning Electron Microscopy (SEM) techniques after treating the samples for 12 hours at certain temperatures that were previously chosen. For the sake of clarity, we decided to divide the crystallization study in two temperature intervals: (i) $100 \leq T \leq 510$°C and (ii) $550 \leq T \leq 900$°C. At this point, it is important to say that, for powders heated at 850°C and 900°C, the treatment was carried out for just 2 hours due to the high volatility of lead for 850°C and beyond.

Scanning Electron Microscopy (SEM) was carried out on fine ground powders in a Leica Cambridge Stereoscan 440 microscope and the X-Ray powder diffraction patterns were recorded over a 20-60° 2θ range on a Bruker D8 Advance diffractometer with filtered CuKα radiation. In this technique, the identified phases were indexed by comparing the resulting diffraction patterns with those of similar compositions reported in the International Union of Crystallography (IUCr) JCPDS-ICDD database.

4.2. $100 \leq T \leq 510$ °C

In this temperature range the material goes through very noticeable phase transformations; the initial amorphous powder starts to show some crystallinity for about 500°C, right after the combustion of the remaining organic species. Figure 10 shows these first crystallization stages as recorded by the Raman (a) and FT-IR (b) vibrational spectroscopies. The vibrational modes associated to the remnant organic compounds coexisting after heat treating at 100 °C are also highlighted in this graph, Figure 10(b), and it must be noticed that they correspond, basically, to the lead acetate and to the completely chelated organometallic complexes. More explicitly, numbered modes in Figure 2(b) have been assigned to: (1)$Pb(OAc)_2$:ϱ(COO), (2)$Pb(OAc)_2$:δ_S(COO), (3)$Pb(OAc)_2$:ν_S(C-C), (4)$Ti(OPr)_2(Acac)_2$:Ti-Acac, (5)$Zr(Acac)_4$:Zr-Acac,(6)$Pb(OAc)_2$:$\delta_S(CH_3)$, (7)$Ti(OPr)_2(Acac)_2/Zr(Acac)_4$:$\delta(CH_3$:Acac) + $Pb(OAc)_2$:ν_S(CO)and (8)$Ti(OPr)_2(Acac)_2/Zr(Acac)_4$:[ν(C-C) + ν(C-O:Acac)] + $Pb(OAc)_2$:ν_A(C=O)[26],[37].

http://dx.doi.org/10.5772/51996

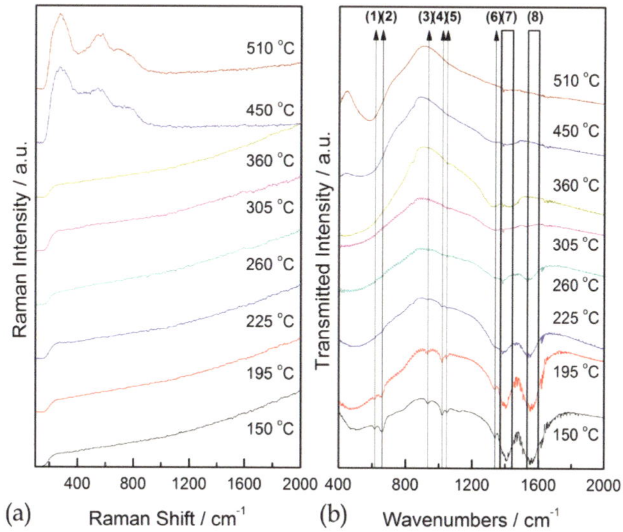

Figure 10. Phase transformations for 100 °C≤ T≤ 510 °C as registered by Raman (a) and FT-IR (b) vibrational spectroscopies; in this case, remnant vibrations due to the presence of acetate and chelated metal complexes are highlighted.

Until 225°C, IR spectra showed no variation; the intensity ratios between the detected modes are almost the same and, on the other hand, Raman spectra featured no signal given the strong presence of π-bonded organic species. Afterwards, IR spectra revealed the almost complete decomposition of acetates and a decrement of Metal:Acac complexes while any noticeable Raman signal is still absent. Just after T = 450°C the active Raman modes normally assigned to a PZT 53/47 started to show up. After treating at this temperature, the FT-IR spectra reveals the formation and definition of the A1(3TO) vibrational mode (~600 cm^{-1}) which is also characteristic for this compound [38]-[41].

Another view of the whole process in this temperature range could be given, as seen in Figure 11, by means of XRD characterization. XRD patterns clearly illustrate the phase transformation exhibited by an amorphous material turning into crystalline; as for the Raman spectra, crystallinity starts to get noticed at T = 450°C. However, these patterns revealed the existence of well defined peaks near 28° that disappear almost entirely at 510°C.

These maxima are usually associated to an intermediate metastable phase belonging to the Fluorite crystal system (F) which is mainly characterized by the spatial disarrangement of the oxygen atoms, vacancies and metal cations while coexisting with some carbonates and/or oxides that still remain in the material. Afterwards, when carbonates and oxides react, a new metastable phase is formed but this new arrangement tends to be more ordered than the previous fluorite. This intermediate phase is considered as belonging to the Pyrochlore crystal system (P or Pyr) and consisting of a 2 x 2 x 2 ordered fluorite cell with oxygen and metallic vacancies or, from another point of view, consisting of a 2 x 2 x 2 non stoichiometric Perovskite (Per) cell. This rigorous differentiation between Fluorite and Pyrochlore is not always considered in literature and it can be subtle sometimes, especially when working with a well known compound as PZT is. Anyway, in Figure 11 we have denoted by F (Fluorite) and P (Pyrochlore) the diffraction maxima associated with each of these phases and according to the indexed cubic structures reported in reference [42] (a = 5.25 Å) and in reference [43] (a = 10.48Å), respectively.

Up to this point, we have seen the almost complete elimination of organic ligands with heating as well as the amorphous/crystalline phase transformation going through two intermediate metastable phases. In the next subsection we will carry on a similar study for higher temperatures while exploring the formation of pure Perovskite phase with morphological quality.

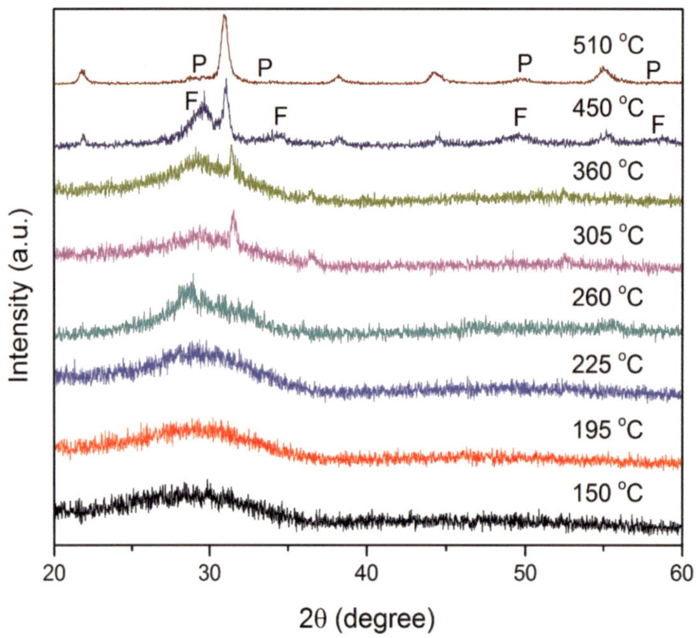

Figure 11. Thermal evolution of PZT 53/47 precursor powders when100 °C≤ T≤ 510 °C as recorded by XRD patterns.

4.3. 550≤ T ≤ 900 °C

Figure 12 shows Raman (b) and FT-IR (b) spectra for heat treated powders in the 550 – 900 °C range. Unlike the previously studied temperature range, vibrational spectra did not show drastic or very noticeable changes. The FT-IR spectra features the only IR active band for aPZT-R3m in the 400-1500 cm^{-1} range (A1(3TO)); this band, associated to the extensional vibrations of the Perovskite BO_6 octahedra, shifts slightly to higher frequencies while gets narrower as temperature increases. This shifting suggests a more compact octahedral structure while a narrower band could be the evidence of a better 'environment' around the octahedron or, in other words, a better local stoichiometry [44] that seems to be minimum for T = 800°C, an indicative of an optimum crystallization. Raman spectra featured vibrational modes that tend to define better as temperature increases.

http://dx.doi.org/10.5772/51996

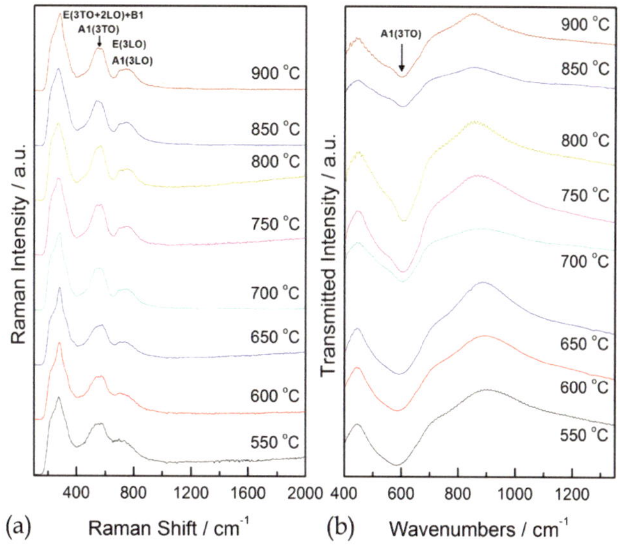

Figure 12. Phase transformations for 550 °C≤ T≤ 900 °C as registered by Raman (a) and FT-IR (b) vibrational spectroscopies.

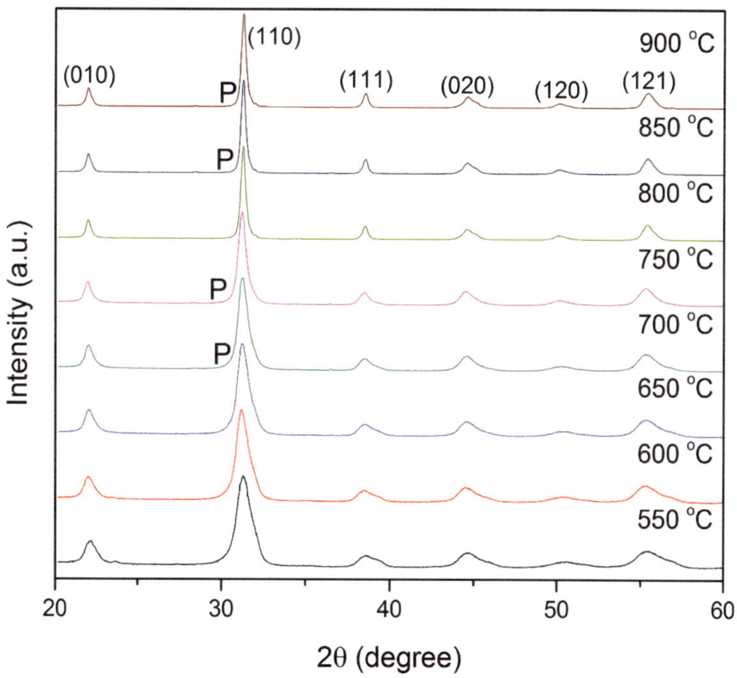

Figure 13. Thermal evolution of the powders under study when 550 °C ≤ T ≤ 900 °C as evidenced by XRD patterns in the 20 - 60 deg. 2θ interval. Regions where the most intense Pyr peak showed up are denoted by P.

XRD patterns depicted in Figure 13 allow us to notice the temperature evolution of the distinctive Perovskite diffraction peaks. Peaks indexing has been done according to a PZT 53/47, $F_{R(HT)}$; sym: R3m. Besides, regions where the most intense Pyr peak showed up (~28.5°) are denoted by P; the Pyr phase is present in a small percentage (%Pyr ~ 1 %) except for powders treated at T = 800°C which must be chosen as the appropriate crystallization temperature for this material. The final Perovskite phase homogenizes and tends to be predominant as temperature increases until lead losses become noticeable.

The rest of the PZT compositions that were chosen for this study, PZT (1-x)/x with C = 0.35 M and 2 months aged, were also treated at 800 °C in order to obtain the appropiate Perovskite phase; Figure 14 shows the corresponding XRD patterns for each one of them. As seen in the graph, all samples attained a perfect crystallization in the corresponding phase and the proper pattern indexing as well as the phase indentification are also shown.

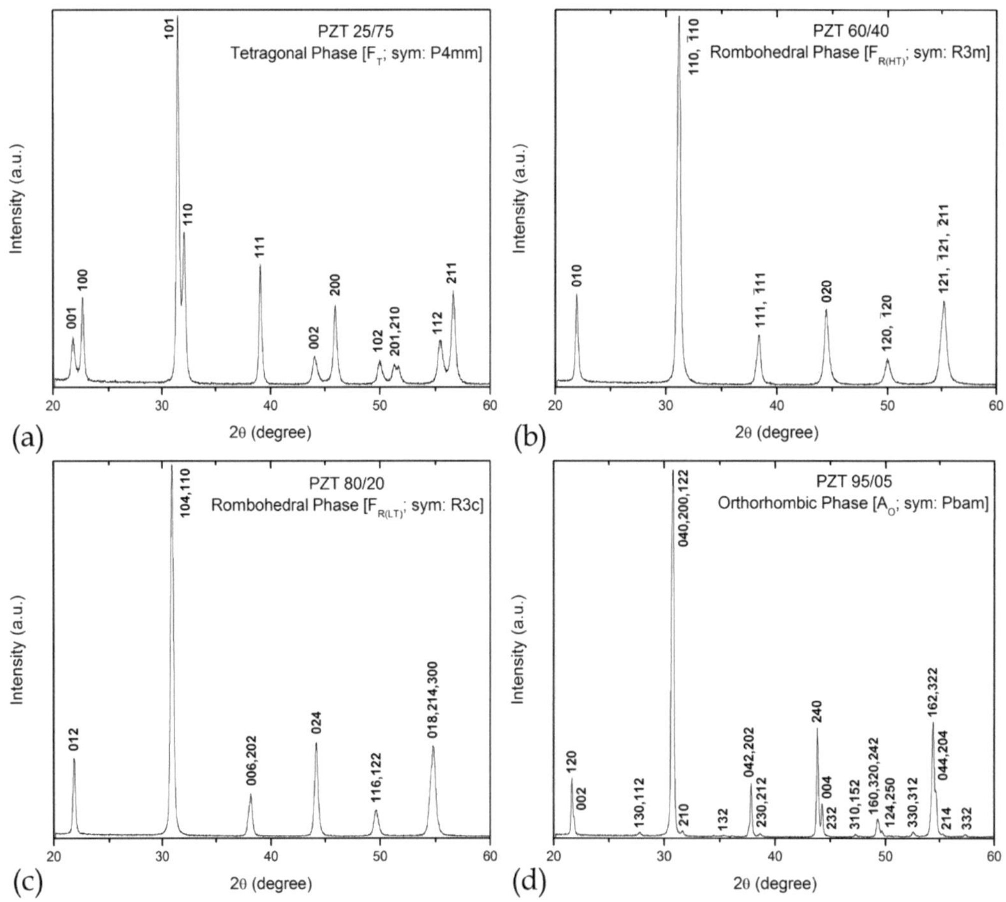

Figure 14. XRD patterns for PZT (1-x)/x under study: (a) PZT 25/75, (b) PZT 60/40, (c) PZT 80/20 and (d) PZT 95/05.

http://dx.doi.org/10.5772/51996

On the other hand, Figure 15 shows SEM micrographs of crystallized PZT (1-x)/x powdered samples. As illustrated, a granular structure of fair morphological uniformity is found in all cases along with a remarkable submicrometric average grain size. Another interesting feature is the low porosity found in all compositions, somehow resembling the granular structure of sintered bulk samples, which certainly favors the formation of a high density ferroelectric material.

As it was said at the beginning of this chapter, this kind of sub-micron granular structures, or nanoceramics, allows the technological exploitation of dielectric and ferro/piezoelectric size related features. It is not shown here, but that very same samples have been resynthesized a couple of times with starting reactants being bought to different companies and final results showed almost the same granular and morphological quality. Repetitiveness is also a bonus when working with chemical routes of synthesis that tend to be less straightforward, even though cheaper, than physical ones.

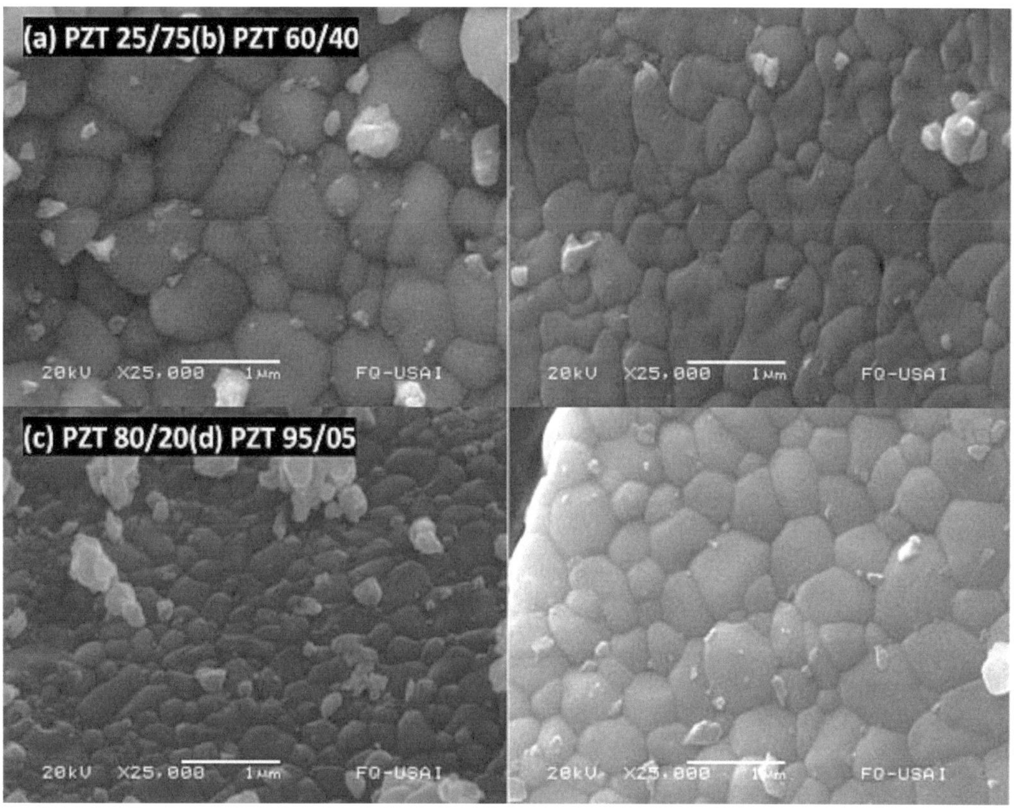

Figure 15. SEM micrographs showing the morphological quality of the PZT (1-x)/x powders when crystallized at 800 °C: (a) PZT 25/75, (b) PZT 60/40, (c) PZT 80/20 and (d) PZT 95/05.

5. Chapter remarks

Through this chapter, we tried to expose some relevant results directly concerned with a feasible 'universal' controlled synthesis of nano/submicrometric grain-sized PZT (1-x)/x piezoelectric structures that, technically speaking, potentially enhance the performance of current commercially bulk based devices by exploiting the size effects related phenomena that arise, as is commonly accepted, for grain sizes below 1 μm. Under the light of this study, the tentative bottom-up "design" of any desired nano/submicrometric PZT (1-x)/x structure seems to be a plausible and successful task that, methodologically at least, could be expanded to more complex materials systems.

Further reading regarding the work that has been shown here can be found in references [45]-[48]. On the other hand, references [49]-[60] will also provide the reader with very recent research papers in this field that undoubtedly extend the applicability and versatility of what has been discussed.

Author details

A. Suárez-Gómez[1], J.M. Saniger-Blesa[2] and F. Calderón-Piñar[3]

1 UdG-CUVALLES, Carr. Guadalajara-Ameca, Ameca, Jalisco, México

2 CCADET-UNAM, Cd. Universitaria, Coyoacán, México D.F., México

3 Fac. de Física/IMRE, San Lázaro y L, Univ. de la Habana, Habana , Cuba

References

[1] McCauley D, Newnham RE, Randall CA. Intrinsic size effects in a barium titanate glass-ceramic. Journal of the American Ceramic Society1998;81(4)979-987.

[2] Frey MH, Xu Z, Han P, Payne DA. The role of interfaces on an apparent grain size effect on the dielectric properties for ferroelectric barium titanate ceramics. Ferroelectrics1998;206(1)337-353.

[3] Zhang L, Zhong LW, Wang CL, Zhang PL, Wang YG. Finite-size effects in ferroelectric solid solution $Ba_xSr_{1-x}TiO_3$. Journal of Physics D: Applied Physics 1999;32 546-551.

[4] Park Y, Knowles KM, Cho K. Particle-size effect on the ferroelectric phase transition in $PbSc_{1/2}Ta_{1/2}O_3$ ceramics. Journal of Applied Physics 1998;83(11)5702-5708.

[5] Desu S.B., Ramesh R., Tuttle B.A., Jones R.E., Yoo I.K., editors. Ferroelectric Thin Films V (MRS Symposium Proceedings Series Vol. 433), San Francisco, CA, 1996.

[6] Wang ZL. The new field of nanopiezotronics. Materials Today 2007;10(5) 20-28.

[7] Budd K.D., Dey S.K., Payne DA. Sol-Gel Processing of $PbTiO_3$-$PbZrO_3$ and PLZT Thin Films. Proceedings of the British Ceramic Society 1985;36107–121.

[8] Schneller T, Waser R. Chemical Solution Deposition of Ferroelectric Thin Films: State of the Art and Recent Trends. Ferroelectrics 2002;267(1) 293–301.

[9] Majumder SB, Bhaskar S, Katiyar RS. Critical issues in sol-gel derived ferroelectric thin films: A review. Integrated Ferroelectrics 2002:42(1)245–292.

[10] Besra L, Liu M. A review on fundamentals and applications of electrophoretic deposition (EPD). Progress in Materials Science 2007;52(1) 1-61.

[11] Shantha Shankar K, Raychaudhuri AK. Fabrication of nanowires of multicomponent oxides: Review of recent advances. Materials Science and Engineering: C 2005;25(5–8) 738-751.

[12] Ogihara T, Kaneko H, Mizutani N, Kato M. Preparation of monodispersed lead zirconate-titanate fine particles. Journal of Materials Science Letters 1988;7(8) 867-869.

[13] Lakeman CDE, Payne DA. Processing Effects in the Sol–Gel Preparation of PZT Dried Gels, Powders, and Ferroelectric Thin Layers. Journal of the American Ceramic Society 1992;75(11) 3091-3096.

[14] Wang FP, Yu YJ, Jiang ZH, Zhao LC. Synthesis of $Pb_{1-x}Eu_x(Zr_{0.52}Ti_{0.48})O_3$ nanopowders by a modified sol–gel process using zirconium oxynitrate source. Materials Chemistry and Physics 2003;77(1)10-13.

[15] Brunckova H, Medvecky L, Briancin J, Saksi K. Influence of hydrolysis conditions of the acetate sol–gel process on the stoichiometry of PZT powders. Ceramics International 2004;30(3) 453-460.

[16] Praveenkumar B, Sreenivasalu G, Kumar HH, Kharat DK, Balasubramanian M,Murty BS. Size effect studies on nanocrystalline $Pb(Zr_{0.53}Ti_{0.47})O_3$ synthesized by mechanical activation route. Materials Chemistry and Physics 2009;117(2–3) 338-342.

[17] Livage J, Henry M, Sanchez C. Sol-gel chemistry of transition metal oxides. Progress in Solid State Chemistry 1988;18(4)259–342.

[18] Sedlar M, Sayer M. Reactivity of titanium isopropoxide, zirconium propoxide and niobium ethoxide in the system of 2-Methoxyethanol, 2,4-Pentanedione and water. Journal of Sol–Gel Science and Technology 1995;5(1) 27–40.

[19] Nakamoto K. Infrared and Raman Spectra of Inorganic and CoordinationCompounds. John Wiley & Sons; 1997.

[20] Doeuff S, Henry M, Sanchez C, Livage J. Hydrolisis of titanium alkoxides: Modification of the molecular precursor by acetic acid. Journal of Non-Crystalline Solids 1987;89(1-2) 206–216.

[21] Lewandowski H, Koglin E, Meier RJ. Computational study of the infrared spectrum of acetic acid, its cyclic dimer, and its methyl ester. Vibrational Spectroscopy 2005;39(1) 15–22.

[22] Yang MM, Crerar DA, Irish DE. A Raman spectroscopic study of lead and zinc acetate complexes in hydrothermal solutions. Geochimica et Cosmochimica Acta 1989;53(2) 319–326.

[23] Tayyari SF, Milani-nejad F. Vibrational assignment of acetylacetone. Spectrochimica Acta A: Molecular and Biomolecular Spectroscopy2000;56(14) 2679–2691.

[24] Schubert U, Husing N, Lorenz A. Hybrid Inorganic-Organic Materials by Sol-Gel Processing of Organofunctional Metal Alkoxides. Chemistry of Materials 7(11) 2010–2027.

[25] Accelrys Inc., San Diego, CA USA. http://accelrys.com/products/materials-studio/

[26] Schwartz RW, Voigt JA, Boyle TJ, Christenson TA, Buchheit CD,"Control of Thin Film Processing Behavior Through Precursor Structural Modifications. Ceramic Engineering and Science Proceedings 1995;16(5)1045–1056.

[27] Kolb U, Gutwerk D, Beudert R, Bertagnolli H. An IR- and EXAFS-study of the precursor system lead(II) acetate trihydrate, dissolved in methanol and 2-methoxyethanol. Journal of Non-Crystalline Solids 1997;217(2-3) 162-166.

[28] Zeisgerber M, Dutz S, Lehnert J, Müller R. Measurement of the distribution parameters of size and magnetic properties of magnetic nanoparticles for medical applications. Journal of Physics: Conference Series 2009;149(1) 012115.

[29] Ortega D, Garitaondia JS, Ramírez-del-Solar M, Barrera-Solano C, Domínguez M. Implications of nanoparticle concentration and size distribution in the superparamagnetic behaviour of aging-improved maghemite xerogels. The European Physics Journal D: Atomic, Molecular, Optical and Plasma Physics 2009;52(1-3) 19-22.

[30] Maity D, Choo SG, Yi J, Ding J, Xue JM. Synthesis of magnetite nanoparticles via a solvent-free thermal decomposition route. Journal of Magnetism and Magnetic Materials 2009;321(9) 1256-1259.

[31] Brinker CJ, Scherer GW. Sol–gel Science: The Physics and Chemistry of Sol–gel Processing. Academic Press Inc., San Diego; 1990.

[32] Das B, Hazra DK. Conductance of selected Alkali Metal Salts in Aqueous Binary Mixtures of 2-Methoxyethanol at 25 °C. Journal of Solution Chemistry 1998;27(11) 1021-1031.

[33] Valdés-Solís T, Marbán G, Fuertes AB. Preparation of Nanosized Perovskites and Spinels through a Silica Xerogel Template Route. Chemistry of Materials 2005;17(8) 1919-1922.

[34] Steinhart M, Jia Z, Schaper AK, Wehrspohn RB, Gösele U, Wendorff JH. Palladium Nanotubes with Tailored Wall Morphologies. Advanced Materials 2003;15(9) 706-709.

[35] Limmer SJ, Hubler TL, Cao G. Nanorods of Various Oxides and Hierarchically Structured Mesoporous Silica by Sol-Gel Electrophoresis. Journal of Sol–Gel Science and Technology 2003;26(1-3) 577-581.

[36] Urban JJ, Yun WS, Gu Q, Park H. Synthesis of Single-Crystalline Perovskite Nanorods Composed of Barium Titanate and Strontium Titanate. Journal of the American Chemical Society 2002;124(7) 1186-1187.

[37] Hardy A, Van Werde K, Vanhoyland G, Van Bael MK, Mullens J, Van Poucke LC. Study of the decomposition of an aqueous metal–chelate gel precursor for $(Bi,La)_4Ti_3O_{12}$ by means of TGA–FTIR, TGA–MS and HT-DRIFT. Thermochimica Acta 2003;397(1-2)143-153.

[38] Petzelt J, Ostapchuk T. Infrared and Raman spectroscopy of some ferroelectric perovskite films and ceramics. Journal of Optoelectronics and Advanced Materials 2003;5(3) 725-733.

[39] Meng JF, Katiyar RS, Zou GT, Wang XH. Raman Phonon Modes and Ferroelectric Phase Transitions in Nanocrystalline Lead Zirconate Titanate. Physica Status Solidi A: Applications and Materials Science 1997;164(2)851-862.

[40] Burns G, Scott BA. Raman Spectra of Polycrystalline Solids; Application to the $PbTi_{1-x}Zr_xO_3$ System. Physical Review Letters 1970;25(17) 1191-1194.

[41] Camargo ER, Leite ER, Longo E. Synthesis and characterization of lead zirconate titanate powders obtained by the oxidant peroxo method. Journal of Alloys and Compounds 2009;469(1-2)523-528.

[42] Wilkinson AP, Speck JS, Cheetham AK, Natarajan S, Thomas JM. In Situ X-ray Diffraction Study of Crystallization Kinetics in $PbZr_{1-x}Ti_xO_3$ (PZT, x = 0.0, 0.55, 1.0). Chemistry of Materials 1994;6(6)750-754.

[43] Kwok CK, Desu SB. Pyrochlore to perovskite phase transformation in sol-gel derived lead-zirconate-titanate thin films. Applied Physics Letters 1992;60(12)1430-1432.

[44] Lakeman CDE, Xu Z, Payne DA. Rapid Thermal Processing of Sol-Gel Derived PZT 53/47 Thin Layers. Proceedings of the Ninth IEEE International Symposium onApplications of Ferroelectrics, 1994. ISAF '94. 1995404-407.

[45] Suárez-Gómez A, Sato-Berrú R, Toscano RA, Saniger-Blesa JM, Calderón-Piñar F. On the synthesis and crystallization process of nanocrystalline PZT powders obtained by a hybrid sol–gel alkoxides route. Journal of Alloys and Compounds 2008;450(1-2) 380-386.

[46] Suárez-Gómez A, Saniger-Blesa JM, Calderón-Piñar F. The effects of aging and concentration on some interesting Sol-gel parameters: A feasibility study for PZT nano-

particles insertion on in-house prepared PAA matrices via electrophoresis. Journal of Electroceramics 2009;22(1-3) 136-144.

[47] Suárez-Gómez A, Saniger-Blesa JM, Calderón-Piñar F. A study on the stability of a PZT precursor solution based on the time evolutionof mean particles size and pH. Materials Chemistry and Physics 2010;123(1) 304-308.

[48] Suárez-Gómez A, Saniger-Blesa JM, Calderón-Piñar F. A Crystallization Study of Nanocrystalline PZT 53/47 Granular Arrays Using a Sol-Gel Based Precursor. Journal of Materials Science & Technology 2011;27(6) 489-496.

[49] Meng Q, Zhu K, Pang X, Qiu J, Shao B, Ji H. Sol-hydrothermal synthesis and characterization of lead zirconate titanate fine particles. Advanced Powder Technology 2012;In Press DOI:10.1016/j.apt.2012.06.004.

[50] Chiolerio A, Quaglio M, Lamberti A, Celegato F, Balma D, Allia P. Magnetoelastic coupling in multilayered ferroelectric/ferromagnetic thin films: A quantitative evaluation. Applied Surface Science 2012;258(20) 8072–8077.

[51] Bastani Y, Bassiri-Gharb N. Enhanced dielectric and piezoelectric response in PZT superlattice-like films by leveraging spontaneous Zr/Ti gradient formation. Acta Materialia 2012;60(3) 1346–1352.

[52] Stawski TM, Besselink R, Veldhuis SA, Castricum HL, Blank DHA, ten Elshof JE. Time-resolved small angle X-ray scattering study of sol–gel precursor solutions of lead zirconate titanate and zirconia. Journal of Colloid and Interface Science 2012;369(1) 184–192.

[53] Goel P, Vijayan N, Biradar AM. Complex impedance studies of low temperature synthesized fine grain PZT/CeO_2 nanocomposites. Ceramics International 2012;38(4) 3047–3055.

[54] Yin Y, Ye H, Zhan W, Hong L, Ma H, Xu J. Preparation and characterization of unimorph actuators based on piezoelectric $Pb(Zr_{0.52}Ti_{0.48})O_3$ materials. Sensors and Actuators A: Physical 2011;171(2) 332–339.

[55] Sachdeva A, Tandon RP. Effect of sol composition on dielectric and ferroelectric properties of PZT composite films. Ceramics International 2012;38(2) 1331–1339.

[56] Bochenek D, Skulski R, Wawrzała P, Brzezińska D. Dielectric and ferroelectric properties and electric conductivity of sol–gel derived PBZT ceramics. Journal of Alloys and Compounds 2011;509(17) 5356–5363.

[57] Chen W, Zhu W, Chen XF, Sun LL. Preparation of $(Ni_{0.5}Zn_{0.5})Fe_2O_4/Pb(Zr_{0.53}Ti_{0.47})O_3$ thick films and their magnetic and ferroelectric properties. Materials Chemistry and Physics 2011;127(1–2) 70–73.

[58] Zhong C, Wang X, Li L. Fabrication and characterization of high curie temperature $BiSc_{1/2}Fe_{1/2}O_3$–$PbTiO_3$ piezoelectric films by a sol–gel process. Ceramics International 2012;38(S1) S237–S240.

[59] Fuentes-Fernandez E, Baldenegro-Perez L, Quevedo-Lopez M, Gnade B, Hande A, Shah P, Alshareef HN. Optimization of $Pb(Zr_{0.53},Ti_{0.47})O_3$ films for micropower generation using integrated cantilevers. Solid-State Electronics 2011;63(1) 89–93.

[60] Kim SH, Koo CY, Lee J, Jiang W, Kingon AI. Enhanced dielectric and piezoelectric properties of low-temperature processed $Pb(Zr,Ti)O_3$ thick films prepared by hybrid deposition technique with chemical solution infiltration process. Materials Letters 2011;65(19–20) 3041–3043.

Chapter 16

Phase Transitions, Dielectric and Ferroelectric Properties of Lead-free NBT-BT Thin Films

N. D. Scarisoreanu, R. Birjega, A. Andrei, M. Dinescu,
F. Craciun and C. Galassi

Additional information is available at the end of the chapter

http://dx.doi.org/10.5772/ 52395

1. Introduction

Ferroelectric perovskites based on $Na_{0.5}Bi_{0.5}TiO_3$ (NBT) are considered among the most promising lead-free candidate materials to substitute $Pb(Zr_{1-x}Ti_x)O_3$ (PZT) in devices designed to respect standards and environmental laws. Taking into account the toxicity of lead-based systems, there are numerous lead-free piezoelectric materials under investigation in worldwide spread laboratories for replacing PZT in future devices. Constant efforts are made to find viable replacements for all these materials containing harmful elements.

Solid-solution systems based on lead-free perovskites like $Na_{0.5}K_{0.5}NbO_3$ (NKN), $BaTiO_3$ (BT), $Na_{0.5}Bi_{0.5}TiO_3$ (NBT) or bismuth layered-structured $SrBi_2Ta_2O_9$ (SBT), and $SrBi_2Nb_2O_9$ (SBN) are considered as viable alternatives for replacing lead-based materials. For example, $(K,Na)NbO_3$–$LiTaO_3$–$LiSbO_3$ alkaline niobate ceramics exhibit a d_{33} piezoelectric coefficient up to 416 pC/N together with Curie temperature T_c around 526 K, as reported by Saito *et al* [1]. Sodium/bismuth titanate (NBT) belongs to the bismuth-based perovskites in which the A-site atom is replaced. The crystalline structure, phase transitions and physical properties have been intensively studied since the discovery of the material in 1960 by Smolensky *et al* [2]. NBT has a relatively high depolarization temperature, T_d = 470 K, high remanent polarization, 38 $\mu C/cm^2$ and piezoelectric coefficient d_{33} = 125 pC/N [3]. However, owing to the high value of the coercive field and high electrical conductivity, NBT cannot be easily polarized, therefore different A-site substitutions have been attempted to avoid this drawback.

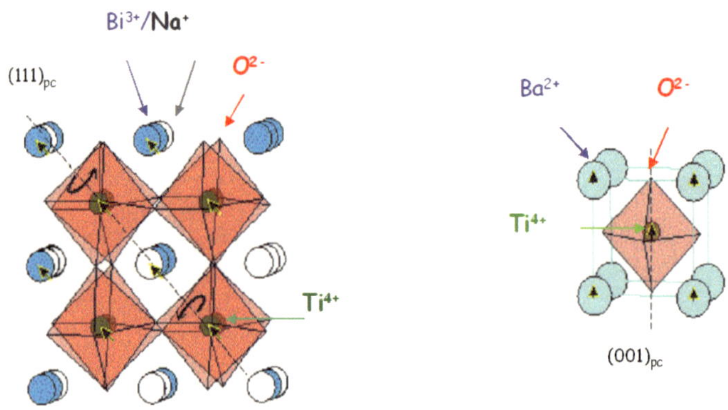

Figure 1. The end-members of perovskite NBT-BT: rhombohedral NBT and tetragonal BT. Cations Na^{+}/Bi^{3+} and Ba^{2+} occupy the A-sites while Ti^{4+} occupies B-sites (oxygen octahedra centers).

The solid solution with $BaTiO_3$, (1-x) NBT-x BT shows a morphotropic phase boundary (MPB) between the rhombohedral and the tetragonal phase, at x between 0.06 and 0.07 for which the material properties are considerably improved. Indeed d_{33} values up to 450 pC/N, and huge electric field-induced strain have been reported [4, 5]. Figure 1 shows the crystalline structures of the NBT and BT end members at room temperature. Perovskite structure deformations include oxygen octahedral rotations around different axis and cation shifts, therefore giving rise to a complex succession of ferroelastic and ferroelectric phase transformations with temperature variation.

Due to this polymorphic structure, NBT and NBT-BT have been also intensively studied in order to clarify their complicated phase transitions, which still pose questions [6].Structural and polar transformations in NBT-BT are more complicated than in other perovskite solid solutions, also due to the strong disorder of the A-sites occupied by Na^{+}, Bi^{3+} or Ba^{2+} ions, with different valence, mass and ionic radius. NBT transforms successively, from the high temperature cubic paraelectric into tetragonal antiferroelectric (or ferrielectric) and further into a rhombohedral ferroelectric phase [6]. In solid solution with BT, the ground ferroelectric phase changes from rhombohedral *R3c* to tetragonal ferroelectric *P4mm*, at the so-called morphotropic phase boundary (MPB) ($x \approx 0.06$-0.07) [5, 7, 8]. The phase diagram of NBT-BT bulk material, mainly based on dielectric measurements, was completed by Cordero *et al* by performing direct anelastic measurements, the border between tetragonal and cubic phases being evidenced [9, 10, 11].

For NBT-BT thin films growth many techniques have been used. Guo *et al*. have investigated NBT-BT-based tri-layered films prepared by chemical solution deposition as a possible solu-

http://dx.doi.org/10.5772/ 52395

tion to the problem of avoiding leakage currents under high electric fields [12, 13]. Using pulsed laser deposition (PLD), Duclère *et al.* have reported the heteroepitaxial growth of NBT thin films on epitaxial platinum electrodes supported on a sapphire substrate [14]. More recently, M.Bousquet *et al* have described the electrical properties of (110)- oriented NBT thin films deposited by laser ablation on (110)Pt/(110) $SrTiO_3$ substrates [15]. They reported the coexistence of two kinds of grains with different shapes in the films, flat and elongated grains corresponding to (100) and (110) oriented NBT crystallites. The effects of Bi- excess in target on the dielectric and ferroelectric properties of the films have been also presented; the reported values for relative permittivity and remnant polarization were $\varepsilon_r \approx 225\text{-}410$ and 14 $\mu C/cm^2$, respectively. Furthermore, very recently, the electrical properties of (100)-oriented $Na_{0.5}Bi_{0.5}TiO_3\text{-}BiFeO_3$ thin films deposited by sol-gel have been reported by Qin et al, aiming to important applications such as photovoltaic devices [16].

However, despite the fact that ferroelectric materials with MPB have enhanced ferroelectric and piezoelectric properties, it is difficult to transpose them in thin films since MPB is limited to a small composition range. Almost all the physical parameters involved in thin films deposition like the substrate type, the microstructure and stress have strong impact on their physical properties [17]. In some previous papers we have investigated the role of different deposition parameters on NBT-BT film growth and properties [18, 19]. In this chapter, we discuss the role of certain experimental conditions like deposition temperature and substrate type, as well as of the amount of BT present in the target on crystalline structure, microstructure, dielectric properties, phase transition temperatures and stability limits of ferroelectric phases in NBT-BT thin films produced by PLD.

2. Experimental method

Pulsed laser deposition (PLD) was used for the film growth. The targets with composition $(NBT)_{1-x}(BT)_x$ (x = 0.06-0.08), further called NBT-BT6 and NBT-BT8, have been prepared following the mixed oxide route and sintered at 1150 °C for 2 h. The sintering was performed in crucibles with the sample surrounded with NBT pack, in order to avoid the loss of Na and Bi, which occurs at temperatures over 1000 °C; more details can be found in Ref.11. X-ray diffraction analysis evidenced the obtaining of pure perovskite phase. The microstructure of the sintered targets was investigated on polished and etched surfaces by scanning electron microscopy. The observed grain sizes were 2-10 μm.

For the film deposition, a Surelite II Nd:YAG pulsed laser with wavelength of 265 nm, pulse duration of 5 ns and frequency 10 Hz, has been employed. The laser fluence was set at 1.6 J/cm^2. The films were grown on Nb:STO and $Pt/TiO_2/SiO_2/Si$ substrates, placed at a distance of about 4.3 cm from the target. Different sets of films have been grown at different substrate temperatures, ranging between 650-730 °C. Deposition and after-deposition cooling were performed in flowing oxygen atmosphere (0.3-0.6 mbar) to favour the formation of perovskite phase without oxygen vacancies. Chemical composition was checked via SIMS technique using a Hiden SIMS/SNMS system. The thickness of the thin films, evaluated by spectroellipsometry, was between 300-500 nm.

For the investigation of the crystalline structure of the targets and films, a PANalytical X'pert MRD diffractometer in Bragg-Brentano geometry was used. The measurements were performed with a step size of 0.02^0 and with a scanning time on step of 25 s or 250 s, depending on the angular range.

The film surface morphology was examined by AFM (model XE100, Park Systems). Piezoelectric force microscopy measurements were performed with a PFM system which includes a lock-in amplifier SR-830 and a dc- high voltage amplifier WMA-280. Conductive all-metal Pt tips were employed for these measurements were the switching characteristcs of the films have been tested.

Several Au electrode dots with an area of about 0.22 mm^2 have been evaporated through a mask on the films for electrical characterization. Polarization hysteresis was measured by using a Radiant Technology RT66A ferroelectric test system, in the virtual ground mode. The dielectric measurements were carried out in a frequency range between 200 Hz and 1 MHz using an HP 4194A impedance analyzer and an HP 4284A LCR meter with a four wire probe. The measurements were performed at 1.5 K/min between 300 and 570 K in a Delta Design climatic chamber model 9023 A (on targets) and in a Linkam variable-temperature stage (model HFS 600E) on films.

3. Results and discussion.

3.1. Growth mode of NBT-BT thin films.

The microstructure of ceramic thin films is one of the most important factors that influence their physical properties. Since the growth mode of thin films is strongly dependent on the substrate type, we investigated the deposition of NBT-BT thin films on two different types of substrates:

1. single crystal $SrTiO_3$: Nb (Nb:STO) and
2. $Pt/TiO_2/SiO_2/Si$.

The AFM pictures obtained on the two sets of films show important microstructural differences, mainly due to different growth mechanisms. In Figure 2 we show AFM images taken on a NBT-BT6 film deposited on Nb: STO monocrystalline substrate at 650 ^{0}C. It can be observed that a first stage of growth resulting into a continuous layer stops when the critical thickness for misfit dislocations (probably a few tens of nm) is reached. After that, the growth continues in platelet-like form (see details in Fig. 2 b). If the deposition temperature is not sufficiently high to favor material exchange between platelets via surface migration, successive layers will grow on the top of the first islands and the growth will result into a discontinuous layer. This explains the platelet-like aspect of the film shown in Figure 2.

http://dx.doi.org/10.5772/ 52395

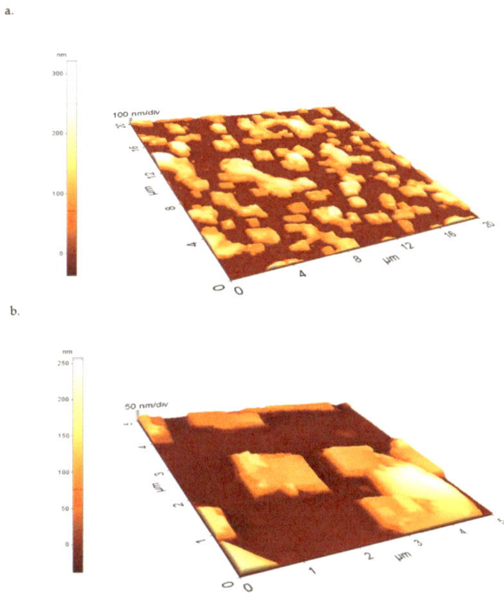

Figure 2. AFM images of NBT-BT6% films deposited on Nb:STO substrates at temperature of 650 °C. The displayed surfaces are 20x20 μm^2 (a) and 5x5 μm^2 (b).

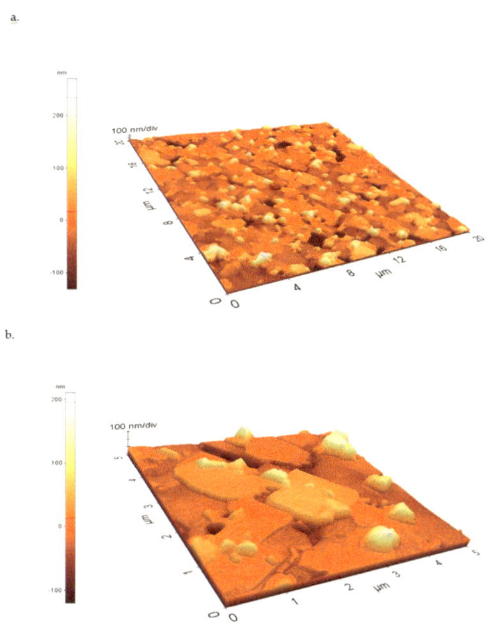

Figure 3. AFM images of NBT-BT6% films deposited on Nb:STO substrates at 700°C. The displayed surfaces are 20x20 μm^2 (a) and 5x5 μm^2 (b).

However, raising the substrate temperature to 700 ^{0}C during the deposition of a second set of films while keeping constant all the other parameters, including the number of laser pulses, produces a uniform layer of continuous platelets, on top of which new islands nucleate (Figure. 3).

A rather different morphology is displayed by NBT-BT films grown on $Pt/TiO_2/SiO_2/Si$ (Fig. 4 and 5). In this case, the growth progresses from the beginning in island-like form since the polycrystalline Pt layer provide the nucleation sites for their formation. Moreover, these NBT-BT islands grow on the Pt layer without preserving a unique orientation, due to the same reason. Instead films grown on Nb:STO monocrystalline structures are uniaxially (001)-oriented, as it will be shown in the next section.

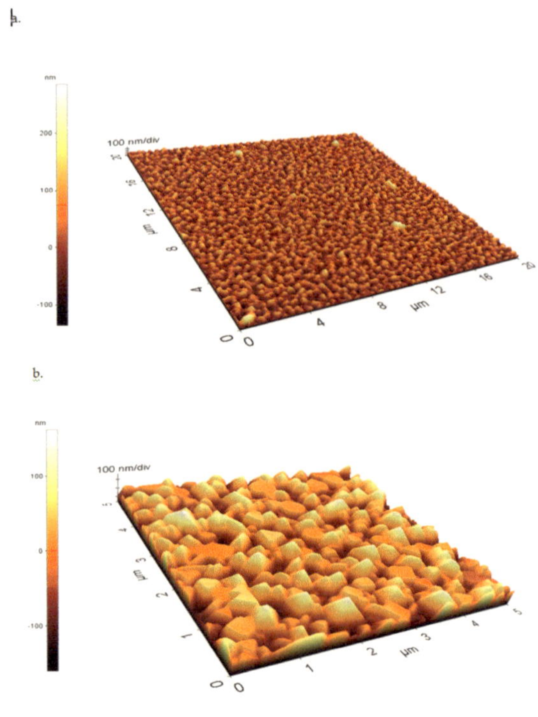

Figure 4. AFM images of NBT-BT6% films deposited on $Pt/TiO_2/SiO_2/Si$ substrate at a temperature of 700 °C. The displayed surfaces are 20x20 μm^2 (a) and 5x5 μm^2 (b).

A fine microstructure with grain size ranging from a few tens of nm up to a few hundred of nm is displayed by NBT-BT6 films (Fig. 4). We note the striking difference with bulk samples microstructures (not shown here), which consists of crystallites of 1-10 µm size.

A similar fine microstructure is displayed by NBT-BT8% films grown on $Pt/TiO_2/SiO_2/Si$ (Fig. 5 a). However, the enlarged AFM image displayed in Fig 5 b) reveales a somewhat different aspect with triangular nanograins lying in plane.

http://dx.doi.org/10.5772/ 52395

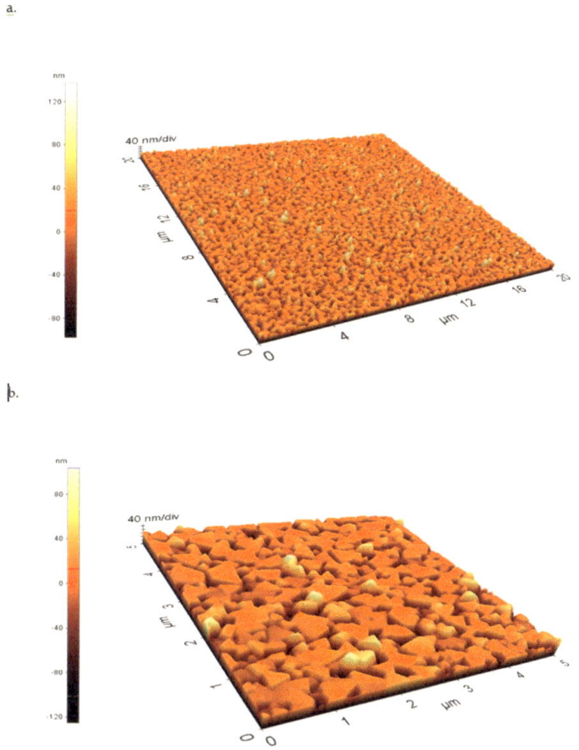

Figure 5. AFM images of NBT-BT8% films deposited on $Pt/TiO_2/SiO_2/Si$ substrate at a temperature of 700 ° C. The displayed surfaces are 20x20 μm^2 (a) and 5x5 μm^2 (b).

3.2. Crystalline structure

The XRD spectrum of NBT-BT6 target corresponds to a mixture of rhombohedral R3c and tetragonal P4mm phases, as shown by the splitting of (111) and (200), (012) and (024) rhombohedral peaks in the bottom pattern in Fig. 6 [20]. The main Miller index of the rhombohedral phase are depicted horizontally on the bottom of the figures while those of the tetragonal phase vertically above. On the same graph, the pattern corresponding to the NBT-BT6 film grown on $Pt/TiO_2/SiO_2/Si$ at 700 ^{0}C is given. The curve corresponding to NBT-BT6/$Pt/TiO_2/SiO_2/Si$ film deposited in the same conditions but at a substrate temperature of 650^0 exhibits similar features as we had reported and is not presented here [18].The as deposited thin films exhibit pure perovskite phase with symmetry congruent with that of the target. The reflection peaks indicate a randomly oriented structure, consistent with the polycrystalline nature of the films.

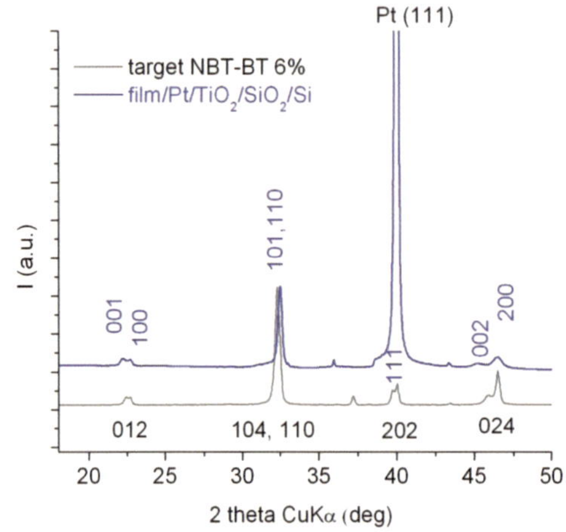

Figure 6. XRD spectra of NBT-BT6% deposited on Pt/TiO_2/SiO_2/Si substrate. The bottom pattern corresponds to the target

Figure 7 displays the XRD patterns of NBT-BT6 films deposited at two temperatures, 650 ^{0}C and 700 ^{0}C, around the (100)/(001) and (200)/(002) reflections of the Nb:STO substrate. The spectra indicate the epitaxial growth of NBT-BT6% films on the Nb:STO substrate at the two temperatures. This feature is congruent with the microstructure shown in the previous section (Fig. 2 and Fig. 3), consisting of large platelet-like crystallites which preserve the same axis of orientation with the monocrystalline substrate.

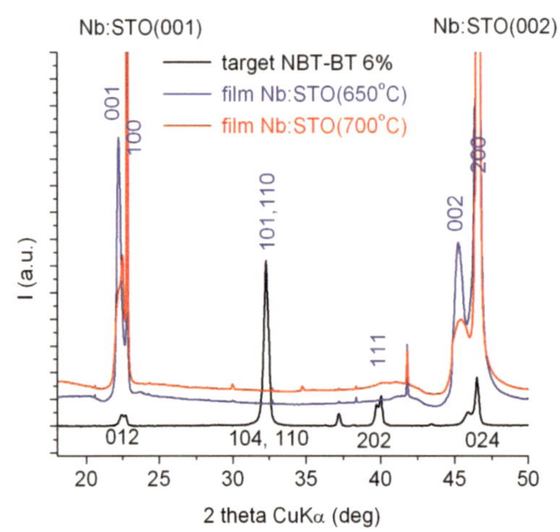

Figure 7. XRD spectra of NBT-BT6% /Nb:STONb:STO films deposited at 650 °C and 700 °C. The grey pattern represents the Nb:STONb:STO target reflection peaks.

Figure 8 shows the XRD patterns of NBT-BT8 films grown on $Pt/TiO_2/SiO_2/Si$ and Nb:STO substrates at 700 ^0C. The grey pattern represents the NBT-BT8 target spectrum, which corresponds to the tetragonal P4mm symmetry. It can be observed that, similar to the previous composition, the growth on single crystal Nb:STO substrate produces an epitaxial film, while the growth on $Pt/TiO_2/SiO_2/Si$ substrate results into a polycrystalline randomly oriented film.

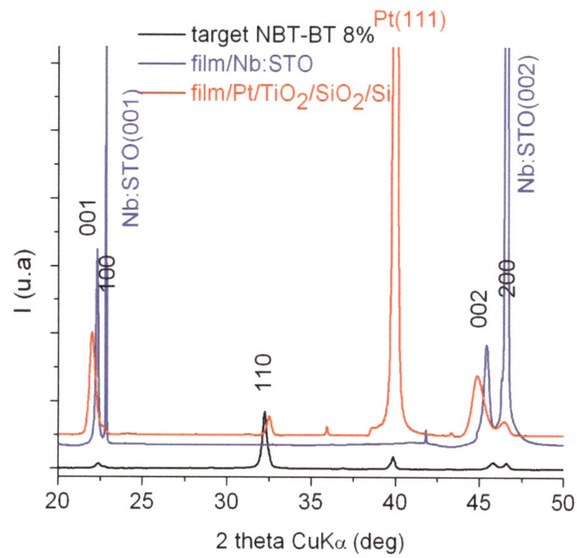

Figure 8. XRD spectra of NBT-BT8% deposited on $Pt/TiO_2/SiO_2/Si$ and on Nb:STONb:STO substrates

3.3. Dielectric and ferroelectric properties

The dielectric and ferroelectric properties of NBT-BT thin films have been evaluated on capacitors formed by evaporating through a mask an array of gold electrode dots with an area of about 0.22 mm^2 on the surface of films grown on $Pt/TiO_2/SiO_2/Si$ and Nb:STO substrates. The bottom electrode was formed by the Pt layer in the first case or by the Nb:STO substrate itself in the second case.

The piezoresponce force microscopy results are presented in Figure 9. The full-Pt tips were brought in contact with the surface of the sample and then a *dc* bias and test *ac* bias were applied between the tip and the bottom electrode of the samples. The *dc* bias was generated by a high voltage amplifier and the *ac* bias was generated by a lock-in amplifier. The same lock-in amplifier was used to analyse the vertical deflection signal from the PSPD, in order to extract the amplitude and the phase of the cantilever oscillations induced by the local deformation of the sample due to the applied *dc* bias. The NBT-BT6/Pt/Si thin films show good switching behavior, the piezoelectric hysteresis and pronounced imprint (not showed here) confirming the piezoelectric and ferroelectric characteristics. The dependence of effective piezoelectric coefficient $d_{33}{}^{eff}$ on the applied electric field is given in Figure 9. The locally measured values with the highest being around $d_{33}{}^{eff} \approx 83$ pm/V, are even higher then for previouslly reported values for pure NBT or lead-based thin films, such as $Pb(ZrTi)O_3$ or $PbTiO_3$ [21, 22]. However, these $d_{33}{}^{eff}$ are a bit smaller then NBT-BT6 ceramics which are reported to be more than 100 pm/V [21]. The reasons for these smaller values are related with the film's porosity, but also with the clamping effect which occurs because the PFM tip- applied electric field will piezoelectrically deform only a small fraction of the film. The rest of the sample will restrict the relative deformation of this small fraction, resulting a lower value for $d_{33}{}^{eff}$ [5, 22].

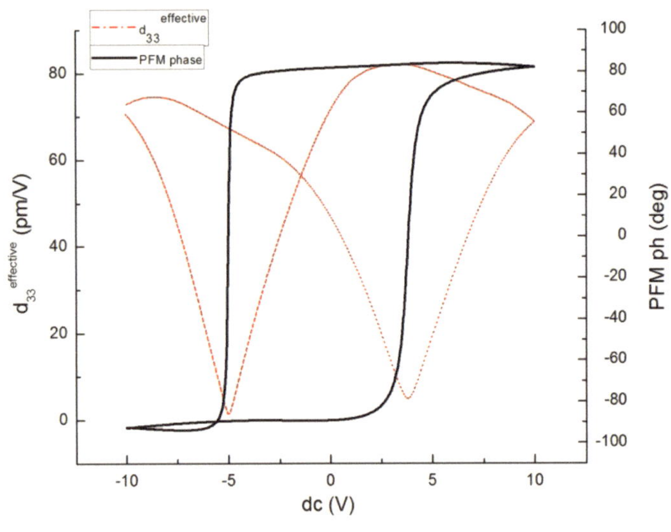

Figure 9. The piezoresponse measurements performed on NBT-BT6 thin films.

In Fig. 10 the room temperature dielectric properties of NBT-BT6 films deposited on $Pt/TiO_2/SiO_2/Si$ at different substrate temperatures, 650 ^{0}C and 730 ^{0}C, have been compared in the frequency range 100 Hz-1 MHz. Films grown at 650 ^{0}C show a higher dielectric constant ($\varepsilon' \sim 1000$), in the order of magnitude of the bulk values ($\varepsilon'_{bulk} \sim 1900$), while films grown at 730 ^{0}C show lower values ($\varepsilon' \sim 700$). The dielectric loss values are instead comparable in the two samples, and similar to bulk values.

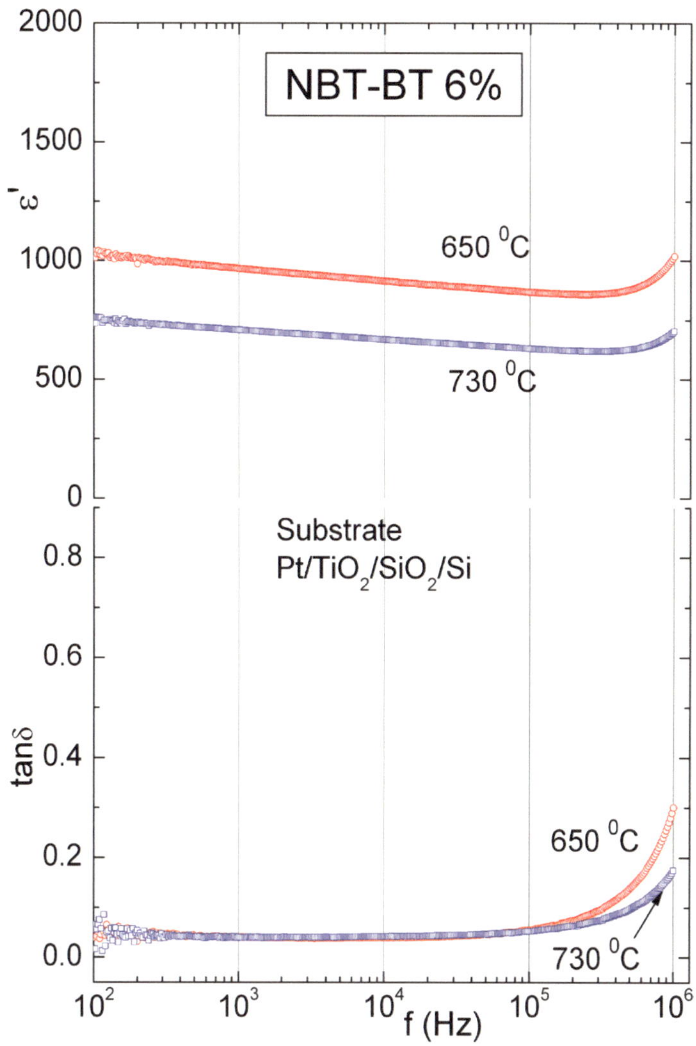

Figure 10. Room temperature dielectric constant ε' and loss tanδ variation with frequency for NBT-BT6% films deposited at different temperatures on $Pt/TiO_2/SiO_2/Si$.

Figure 11 displays the room temperature dielectric constant and dielectric loss in the frequency range 100 Hz-1 MHz for NBT-BT8 films grown on $Pt/TiO_2/SiO_2/Si$ at different temperatures: 650, 700 and 730 ^{0}C. Unlike the previous composition, in this case growth at higher substrate temperatures was beneficial for the improvement of dielectric properties, at least in the frequency domain up to a few hundred kHz. Above this frequency there is a strong increase of dielectric loss. Since an increase is registered also in the dielectric constant, this could be caused by a relaxation mechanism which is active at room temperature at these frequencies, like e.g. free charge relaxation.

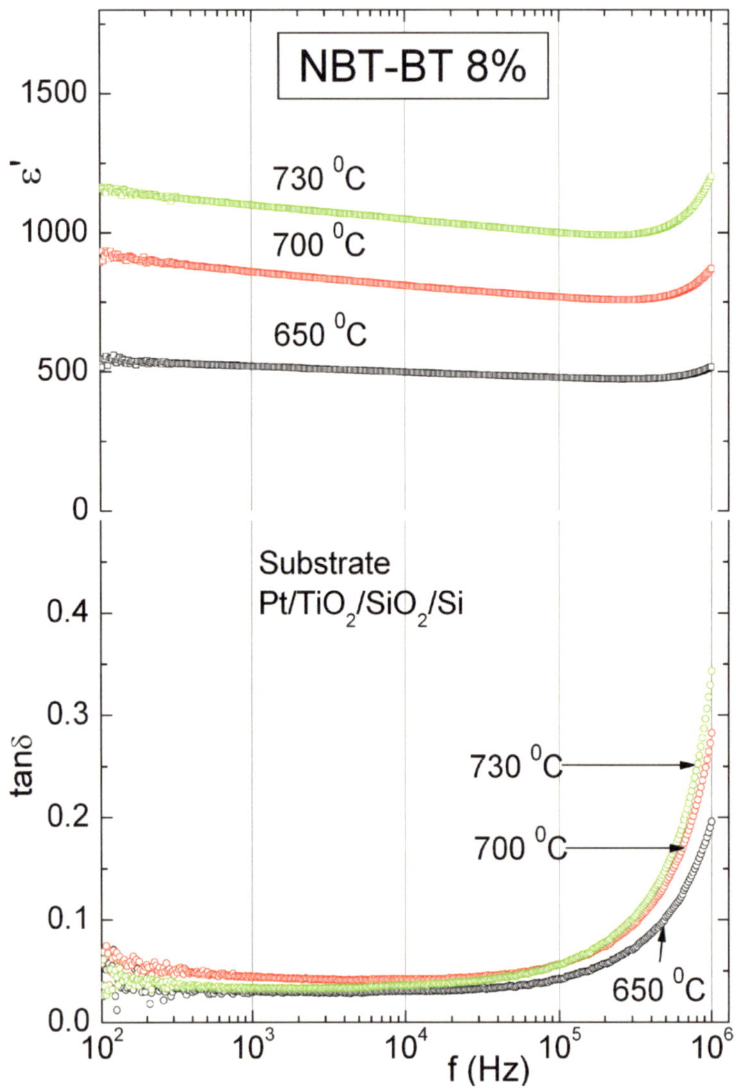

Figure 11. Room temperature dielectric constant and loss variation with frequency for NBT-BT8% films deposited at different temperatures on $Pt/TiO_2/SiO_2/Si$

Polarization hysteresis measurements on NBT-BT6 films grown on $Pt/TiO_2/SiO_2/Si$ are shown in Fig. 12. Spontaneous polarization was about 30 $\mu C/cm^2$ and the remnant polarization was about 10 $\mu C/cm^2$. The rather high value of coercive field (100 kV/cm) could be explained by the presence of intrinsic strain and pinning defects.

Dielectric and ferroelectric properties measurements on films deposited on Nb:STO substrates have been less reliable, probably due to the presence of a non-ohmic contact at the NBT-BT film – semiconductor Nb:STO interface. However PFM measurements (not shown

http://dx.doi.org/10.5772/ 52395

here) evidenced good piezoelectric response, which indicates good intrinsic dielectric and ferroelectric properties, although quantitative values are difficult to extract.

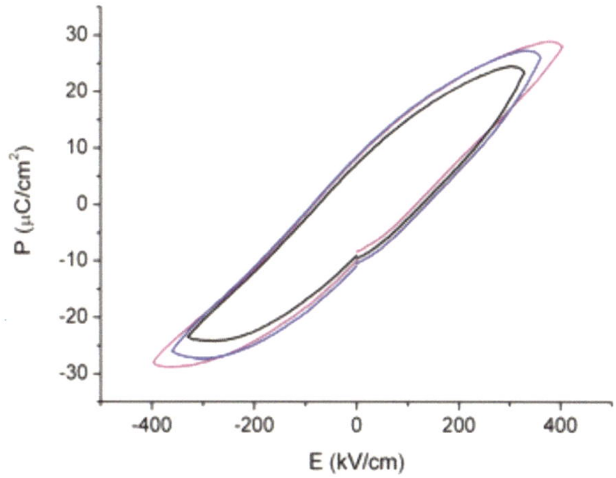

Figure 12. Polarization-electric field hysteresis loop measured on a NBT-BT6% film deposited on $Pt/TiO_2/SiO_2/Si$

3.4. Phase transitions

Phase transitions in ferroelectric materials are accompanied by anomalies of complex dielectric permittivity variation with temperature, generally narrow peaks or steps, depending on the type of phase transformation. However NBT-BT compositions near the morphotropic phase boundary behave as relaxors, due to the cation disorder. This is evidenced in Fig. 13 for NBT-BT8 bulk material. The main characteristic of a relaxor ferroelectric is a broad dielectric peak at a temperature T_m which is not related to a structural transformation. This is due to a wide distribution of relaxation times which characterizes the dielectric response of polar nanoregions. This peak shifts with the increasing of the measurement frequency toward higher temperatures. Thus the dielectric maximum of NBT-BT8 shifts from about 497 K at 200 Hz to about 512 K at 100 kHz. A similar dependence is obeyed also by the dielectric loss.

In Fig. 14 the variation of dielectric constant and loss with temperature for a NBT-BT8 film deposited on $Pt/TiO_2/SiO_2/Si$ is shown. The maximum of the dielectric constant occurs at about 485 K, not far from bulk T_m. However the anomaly is characteristic of a well-behaved phase transition, since the peak temperature T_m does not shift with frequency. A similar qualitative behavior was observed also on the dielectric permittivity variation with temperature for NBT-BT6 films (not shown here).

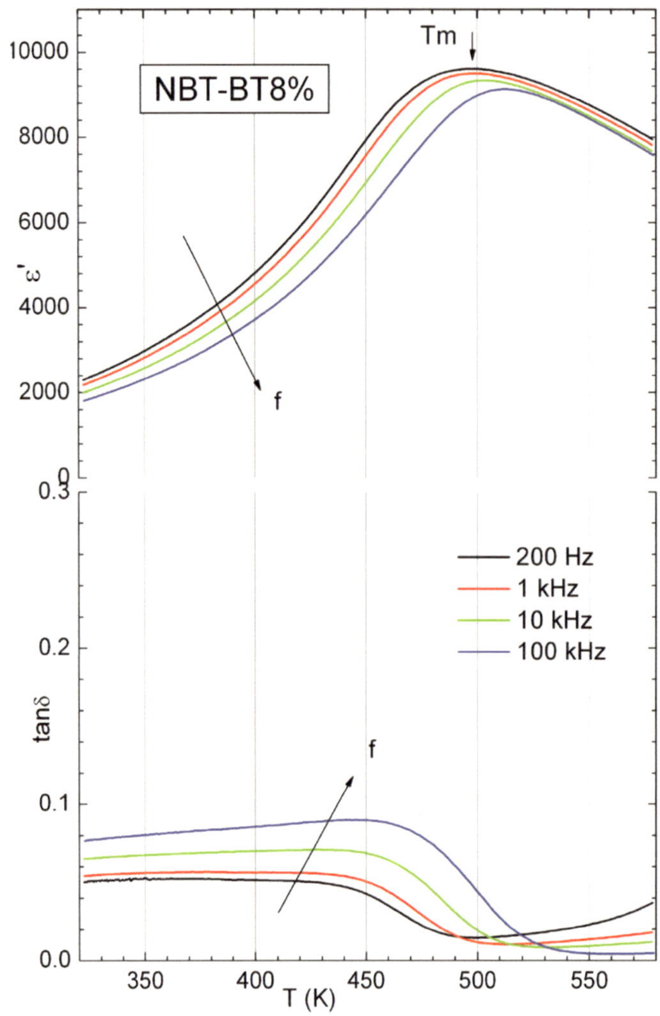

Figure 13. The dielectric permittivity and loss variation with temperature measured on a NBT-BT8% bulk sample at different frequencies. The long arrows mark the increasing of frequency.

While the dielectric permittivity peak position Tm ~ 485 K does not shift with measuring frequency, peak height is strongly dependent on it, decreasing for higher frequencies. This can be attributed to the possible presence of a non-polar dielectric layer, which does not influence the general behavior at phase transition, but can modify the value of the dielectric constant [23]. Generally these interface layers can have strong frequency-dependent dielectric properties which influence the overall properties of the heterostructures. However we stress again that the dependence on temperature and phase transition temperatures can be influenced only in the limits of a monotonous contribution, since the dielectric behaviour of non-polar layers is free of temperature anomalies. Indeed the stronger variation with temperature of the peak intensity at T_m could be attributed to the non-polar layer contribution at higher temperatures due to conductivity variation.

The temperature Td where a strong increase of dielectric loss and dielectric constant occurs marks the ferroelectric – antiferroelectric phase transition which, in bulk samples with the same composition is visible only in the poled state. It is called also depolarization temperature.

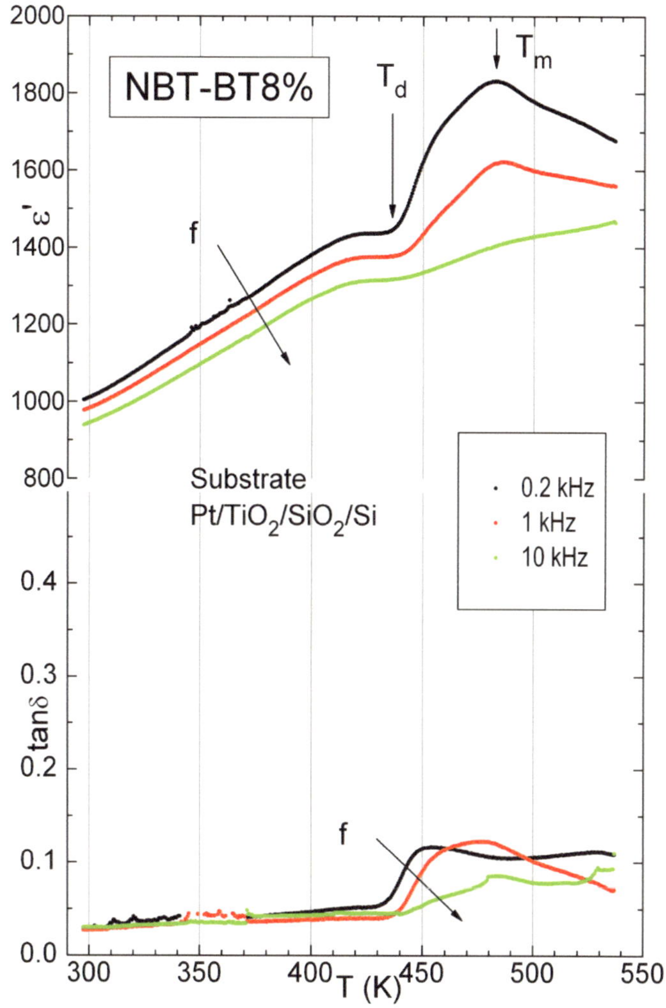

Figure 14. The dielectric permittivity and loss variation with temperature measured on a NBT-BT8% thin film at different frequencies. The long arrows mark the increasing of frequency.

An apparent frequency dependence of the step increase in tanδ which marks T_d is visible on the lower curves in Fig. 14. This could be attributed to a partial relaxor behavior due to mixed nanodomains-normal ferroelectric domains, which can be find also, but in proportion displaced to the nanodomain limit, in bulk samples. We must remark that the films were not poled, since no bias electric field was applied. Therefore the occurrence of a ferroelectric ground state in the NBT-BT films, in striking contrast with relaxor bulk samples with the same composition, must be generated by some intrinsic differences between ceramic bulk samples and ceramic thin films. The first and most obvious reason could be related to the constraining stress of the substrate on the thin films. This is strong in epitaxially grown thin films with a thickness generally below 100 nm, but it should be almost absent in polycrystalline films, randomly oriented and with a thickness of several hundreds of nm. This last one is the case for NBT-BT thin films deposited on $Pt/TiO_2/SiO_2/Si$ substrates. The second reason could be related to the strong differences in the microstructures of ceramic thin films and bulk materials with the same composition. Therefore the occurrence of a ferroelectric ground state instead of a relaxor state in NBT-BT films, as well as the occurrence of a true ferroelectric phase transition could be due to the constrainig imposed by the nanograin boundaries on the ensemble of polar nanoregions.

4. Conclusions

In summary, we have investigated the role of deposition temperature and substrate type as well as the amount of BT present in the target on crystalline structure, microstructure, dielectric properties, phase transition temperatures and stability limits of ferroelectric phases in NBT-BT thin films grown by pulsed laser deposition. We have successfully deposited pure perovskite epitaxial films on single-crystal Nb:STO substrates. Successful growth of NBT-BT films on platinized silicon substrates has been achieved. Good dielectric and ferroelectric properties, comparable with bulk values, have been obtained. The NBT-BT6/Pt/Si thin films show a classic switching behavior, the piezoelectric hysteresis and pronounced imprint confirming the piezoelectric and ferroelectric characteristics. The locally measured value of effective piezoeletric coefficient $d_{33}^{\ eff}$ was around 83 pm/V, higher to the previouslly reported values for pure NBT or lead-based thin films. An enhanced stability of ferroelectric phase in thin films with respect to bulk has been observed and explained by their peculiar nanocrystalline microstructure.

Author details

N. D. Scarisoreanu[1*], R. Birjega[1], A. Andrei[1], M. Dinescu[1], F. Craciun[2] and C. Galassi[3]

*Address all correspondence to: snae@nipne.ro

1 NILPRP, National Institute for Laser, Plasma & Radiation Physics, Bucharest, Romania

2 CNR-ISC, Istituto dei Sistemi Complessi, Area della Ricerca Roma-Tor Vergata, Roma, Italy

3 CNR-ISTEC, Istituto di Scienza e Tecnologia dei Materiali Ceramici, Faenza, Italy

References

[1] Saito, Y., Takao, H., Tani, T., Nonoyama, T., Takatori, K., Homma, T., Nagaya, T., & Nakamura, M. (2004). Lead-free piezoceramics. *Nature*, 432, 84.

[2] Smolenski, G. A., Isupov, V. A., Agranovskaya, A. I., & Krainik, N. N. (1961). New ferroelectrics of complex composition, Sov. *Phys. Solid State*, 2, 2651-196.

[3] Chiang, Y. M., Farrey, G. W., & Soukhojak, A. N. (1998). Lead-free high-strain single-crystal piezoelectrics in the alkaline-bismuth-titanate perovskite family. *Appl. Phys. Lett*, 73, 3683.

[4] Takenaka, T., & Nagata, H. (1999). Present Status of Non-Lead-Based Piezoelectric-Ceramics. *Key Engineering Materials*, 157/158, 57.

[5] Takenaka, T, Maruyama, K.-I., & Sakata, K. (1991). (Bi1/2Na1/2)TiO3-BaTiO3 System for Lead-Free Piezoelectric Ceramics. *Jpn. J. Appl. Phys*, 30, 2236-2239.

[6] Jones, G. O., & Thomas, P. A. (2002). Investigation of the structure and phase transitions in the novel A-site substituted distorted perovskite compound Na0.5Bi0.5TiO3, Acta Crystallogr. *Sect. B: Struct. Sci*, 58, 168 -178 .

[7] Hiruma, Y., Watanabe, Y., Nagata, H., & Takenaka, T. (2007). Phase Transition Temperatures of Divalent and Trivalent Ions Substituted (Bi1/2Na1/2)TiO3 Ceramics, Key Eng. *Mater*, 350, 93.

[8] Glazer, A. M. (1972). The classification of tilted octahedra in perovskites, Acta Crystallogr. *Sect. B: Struct. Sci*, 28, 3384.

[9] W.-van, B., Eerd, D., Damjanovic, N., Klein, N., Setter, , & Trodhal, J. (2010). Structural complexity of (Na0.5Bi0.5)TiOP3BaTiO3 as revealed by Raman spectroscopy. *Phys. Rev. B*, 82, 104112.

[10] Ma, C., & Tan, X. (2010). Phase diagram of unpoled lead-free (1-x)(Bi/sub 1/2/Na/sub 1/2/)TiO/sub 3-x/BaTiO/sub 3/ ceramics. *Solid State Commun*, 150, 1497-1500.

[11] Cordero, F., Craciun, F., Trequattrini, F., Mercadelli, E., & Galassi, C. (2010). Phase transitions and phase diagram of the ferroelectric perovskite (Na0.5Bi0.5)1−xBaxTiO3 by anelastic and dielectric measurements. *Phys. Rev. B*, 81, 144124.

[12] Guo, Y. P., Akai, D. S., Sawada, K., & Ishida, M. (2008). Dielectric and ferroelectric properties of highly (100)-oriented (Na0.5Bi0.5)0.94Ba0.06TiO3 thin films grown on LaNiO3/γ-Al2O3/Si substrates by chemical solution deposition. *Solid State Sciences*, 10, 929.

[13] Guo, Y. P., Akai, D. S., Sawada, K., Ishida, M., & Gu, M. Y. (2009). Structure and electrical properties of trilayered BaTiO3/(Na0.5Bi0.5)TiO3BaTiO3/BaTiO3 thin films deposited on Si substrate. *Solid State Communications*, 149, 14.

[14] Duclère, J.R., Cibert, C., Boulle, A., Dorcet, V., Marchet, P., Champeaux, C., Catherinot, A., Députier, S., & Guilloux-Viry, M. (2008). Lead-free Na0.5Bi0.5TiO3 ferroelec-

tric thin films grown by Pulsed Laser Deposition on epitaxial platinum bottom electrodes. *Thin Solid Films*, 517, 592.

[15] Bousquet, M., , J., Ducle`re, R., Gautier, B., Boulle, A., Wu, A., De´, S., putier, D., Fasquelle, F., Re´mondie`, re. D., Albertini, C., Champeaux, P., Marchet, M., Guilloux- , Viry, & Vilarinho, P. (2012). Electrical properties of (110) epitaxial lead-free ferroelectric Na0.5Bi0.5TiO3 thin films grown by pulsed laser deposition: Macroscopic and nanoscale data. *J. Appl. Phys*, 111, 104 -106.

[16] Qin, W., Guo, Y., Guo, B., & Gu, M. (2012). Dielectric and optical properties of BiFeO3-(Na0.5Bi0.5)TiO3 thin films deposited on Si substrate using LaNiO3 as buffer layer for photovoltaic devices. *J. Alloys Compd*, 513, 154-158.

[17] Shaw, T. M., Trolier Mc Kinstry, S., & Mc Intyre, P. C. (2000). The properties of ferroelectric thin films at small dimensions. *Annu. Rev. Mater. Sci*, 30, 263.

[18] Scarisoreanu, N., Craciun, F., Chis, A., Birjega, R., Moldovan, A., Galassi, C., & Dinescu, M. (2010). Lead-free ferroelectric thin films obtained by pulsed laser deposition. *Appl. Phys A*, 101, 747 -751 .

[19] Scarisoreanu, N., Craciun, F., Ion, V., Birjega, S., & Dinescu, M. (2007). Structural and electrical characterization of lead-free ferroelectric Na/sub 1/2/Bi/sub 1/2/TiO/sub 3/-BaTiO/sub 3/ thin films obtained by PLD and RF-PLD. *Appl. Surf. Sci*, 254, 1292.

[20] Picht, G., Töpfer, J., & Henning, E. (2010). Structural properties of (Bi0.5Na0.5)1−xBaxTiO3 lead-free piezoelectric ceramics. *J. Eur. Ceram. Soc*, 30, 3445.

[21] Takenaka, T. (1999). Piezoelectric Properties of Some Lead-Free Ferroelectric Ceramics. *Ferroelectrics*, 230, 87-98.

[22] Morelli, A., Sriram, Venkatesan, Palasantzas, , Palasantzas, Kooi, G, & De Hosson, J. Th. M. (2009). Piezoelectric properties of PbTiO3 thin films characterized with piezoresponse force and high resolution transmission electron microscopy. *J. Appl. Phys*, 105, 064106.

[23] Craciun, F., & Dinescu, M. (2007). Piezoelectrics", in "Pulsed laser deposition of thin films: applications-led growth of functional materials". *edited by R. Eason (Wiley-Interscience, John Wiley & Sons, Hoboken, New Jersey, U.S.A.,)*, 487-532.

Chapter 17

Thin-Film Process Technology for Ferroelectric Application

Koukou Suu

Additional information is available at the end of the chapter

http://dx.doi.org/10.5772/54360

1. Introduction

Recently thin-film ferroelectrics such as Pb(Zr, Ti)O_3 (PZT) and (Ba, Sr)TiO_3 (BST) have been utilized to form advanced semiconductor and electronic devices including Ferroelectric Random Access memory(FeRAM), actuators composing gyro meters, portable camera modules, and tunable devices for smart phone applications and so on. Processing technology of ferroelectric materials is one of the most important technologies to enable the above- mentioned advanced devices and their productions.

In this paper, we will report our development results of ferroelectric thin film processing technologies including sputtering, MOCVD and plasma etching as well as manufacturing processes for FeRAM, MEMS(actuators, tunable devices) and ultra-high density probe memory.

2. FeRAM technology

Ferroelectric Random Access Memory (FeRAM) is a main candidate of next generation nonvolatile random access memory. As its density has been continuously increasing, its applications have spread from RFID card to Nonvolatile latch circuit with low power consumption. FeRAM has a ferroelectric thin films sandwiched by rare metal electrodes and we have established mass production capability on integrated FeRAM solution, consisting of electrode film deposition, PZT thin-film deposition, etching/ashing, anneal and passivation deposition.

2.1. Mass productive sputtering technology for perovskite oxide thin-film

We have been developing mass production technology for perovskite oxide thin-film such as PZT for a long period [1-5], since we consider these materials as the most promising candidate for ferroelectric material used in FeRAMs (Ferroelectric Random Access Memories) as a non-volatile memory device, piezoelectric MEMS (Micro Electro Mechanical System) devices due to its longer period of research, existence of actual production, manufacturing capability within the tolerable temperature range of general Si LSI technology.

The sputtering was selected for mass production technology of perovskite oxide thin-films owing to the following factors: (1) Good compatibility with conventional Si LSI processes. (2) Superb controllability of film quality (e.g. film composition), which enabled relatively easy thin film deposition. (3) Better possibilities of obtaining uniform surfaces in large diameter substrates (e.g. 6-8 inch). (4) Sputtering was plasma processing which was promising for deposition and heat treatment at low temperature. (5) Feasibility of high-speed deposition. (6) Same deposition method as electrodes (Pt, Ir, Ru, etc.), which will facilitate in-situ integration. (7) Present difficulties and lack of future potentialities in other technologies.

With emphasis on both mass production capability and advanced process capacity of perovskite oxide thin-film sputtering, we consider the following factors as important for development:

1. Throughput

Compared to the other processes (e.g. electrode deposition), the deposition speed of perovskite oxide material sputtering was considerably slower and thereby limited the throughput. Though two ways for improving the throughput, high-speed deposition and thinner film deposition, are considered, the former is more promising for the improvement of throughput, while the latter is apt to cause deterioration of film characteristics.

2. Control of film composition

Film composition determines other film qualities (crystal structure, electric characteristics). Since volatile elements were included in the perovskite oxide materials, they were sensitive and easily fluctuated according to temperature or plasma status. Film composition control is the fundamental factor in this process.

3. Uniformity over large diameter substrates

Large diameter substrates up to 8 inch were expected to be used for mass production of perovskite oxide thin-film. Film thickness and uniformity of film quality were the keys to mass production.

4. Process stability / reliability

Under the circumstances where there were many unknown factors such as new materials, ceramic targets, insulator sputtering, etc., process stability / reliability was the more pertinent factor. In fact, there was a problem in film composition that changed over time.

5. Prevention of particles

While the characteristics, such as ceramic targets or insulating thin film, increased the mechanical factors (e.g. adherence, thermal expansion) for particle occurrence, they were also

http://dx.doi.org/10.5772/54360

producing electric factors (e.g. dielectric breakdown due to charge-up) and made the measures difficult to obtain.

2.2. Optimization of sputtering processes for perovskite oxide thin-film

We adapted the RF magnetron sputtering method for perovskite oxide thin-film sputtering. As for the PZT thin-films, sputtering methods included high-temperature deposition, where film deposition was made at substrate temperatures above 500°C [6, 7], and low temperature deposition, where the films were deposited at room temperature and then crystallized by post-annealing process. The improvement in high-speed deposition, film composition control, and stability of sputtering processes is described in the following.

If the ceramic target with inferior thermal conductivity is used, application of high power leads to the destruction of the target. By adopting the backing plate with high cooling efficiency and high-density target, higher deposition rate by an increase in sputtering power is achieved.

The film composition control can be translated into the volatile element (e.g. Lead) control within film. Various factors that influence the volatile element within film are thought to exist (Fig. 1) and we have investigated the influence of sputtering conditions, strength of magnetic field and electric potential of substrate.

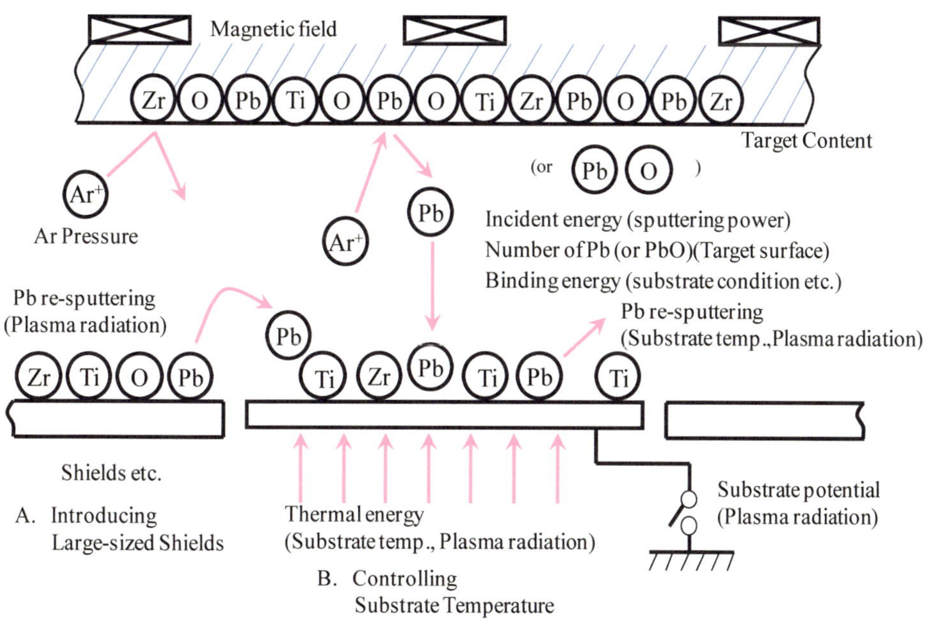

Figure 1. Various factors that influence content with PZT Film.

In addition to the known characteristics in perovskite oxide thin-film sputtering that volatile elements easily fluctuate, it was confirmed that volatile elements within film were unstable even when the deposition was performed under identical conditions. The conceivable causes for that phenomenon were the instability of the target, fluctuation of temperature in the sputtering chamber, and variation of plasma status over time.

Main problem in perovskite oxide thin-film is the change in the volatile element content with the passage of sputtering time. As for the PZT, continuous sputtering of 2.0 kWh showed approximately a 30% decline in lead content compared to that at the beginning. It was determined that the reason for this problem was the change in plasma status. As shown in Fig. 2, when insulating PZT film adhered to the shields of the ground potential, charge-up occurred, the impedance of the system changed, plasma was pushed to the center of the chamber, exposure to plasma was enhanced, and as a consequence, Pb content within film is reduced. In order to stabilize the status of plasma, we installed a stable anode, that is, an anode that avoided charge-up due to the adhesion of insulating PZT film and maintained the role as an anode. Consequently, as can be seen in Fig. 3, stability of Pb content within film in continuous sputtering has been confirmed.

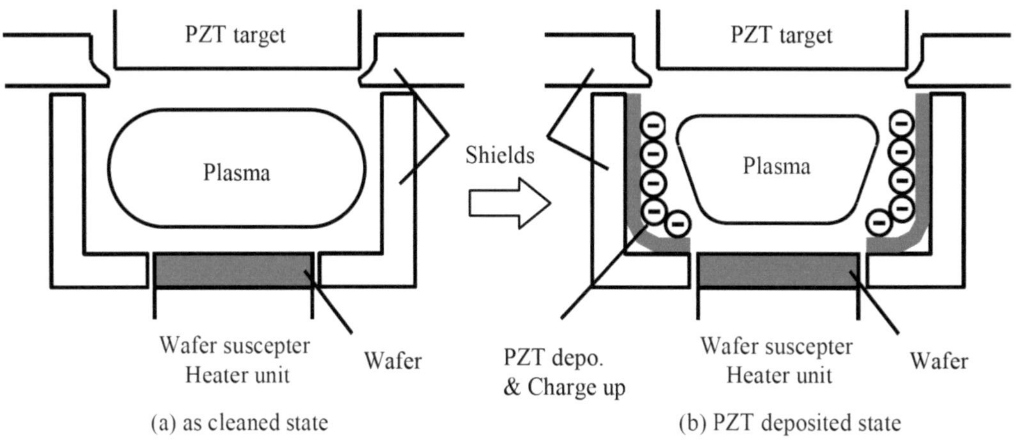

Figure 2. Change in plasma status which is responsible for Pb content variation.

Fig. 4 shows an example of thickness and Pb content uniformity in a PZT film on 8″ substrate. Both thickness and Pb content uniformity varied according to the deposition condition and was minimized around the sputtering pressure of 1.0 Pa. Thickness and Pb content uniformity represent good result as low as ±1.9% and ±1.1%, respectively. These uniformities were also confirmed as stable, thereby satisfactorily meeting the requirements of mass production.

http://dx.doi.org/10.5772/54360

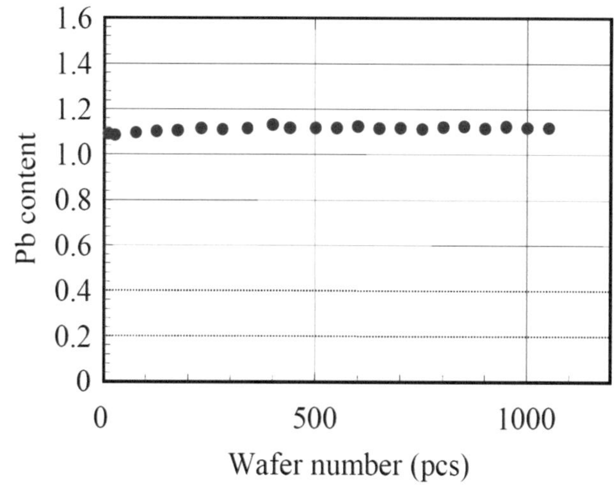

Figure 3. Stable transition of Pb content within film in continuous sputtering.

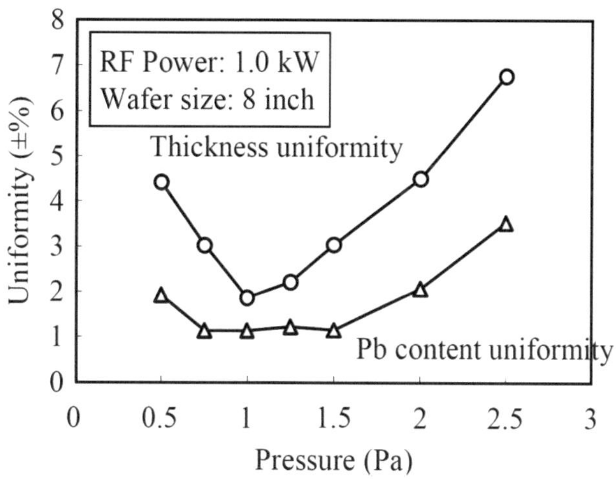

Figure 4. Uniformity in 8 inch area.

2.3. Sputtered PZT thin-films for non-volatile memory application

In this section, we introduce one of its achievements, the ferroelectric characteristics of PLZT capacitors for FeRAM. We used the multi chamber type mass production sputtering system equipped with an exclusive sputtering module for ferroelectric materials, CERAUS ZX-1000 from ULVAC. This system has the following features in addition to features such as easy maintenance, short exhausting time, short down time, etc.

1. Can mount 300 mm (12 inch) in diameter targets and process large diameter substrates of 200 mm. At present, deposition of PZT thin-films on 6 and 8 inch substrates are performed using 12 inch single ceramic target.

2. Including the heat chamber, this system has five process chambers, thereby achieving high flexibility. The system is presently executing the following in-situ processes as standard: pre-heating of substrate → substrate sputtering (e.g. Ti, TiN, Pt) → ferroelectric material sputtering.

3. As a substrate heating mechanism, this system was capable of precise and rapid heating in a wide range from low to high temperatures, with the aid of an electrostatic chuck type hot plate, in addition to lamp heating.

4. This system used RF sputtering for ferroelectric deposition and counters RF noises.

Fig. 5 shows the transition of switching charge (Q_{SW}) and saturation characteristics of a Pt/PZT(200 nm)/Pt capacitor measured at 5 V. Q_{SW} with 5 V applied was approximately 34 μC/cm^2 and saturation voltage of 90% ($V_{90\%}$) was 3.1-3.2 V. The composition of PLZT film, which contained added Ca and Sr for the improvement of retention and imprint characteristics, was excellent.

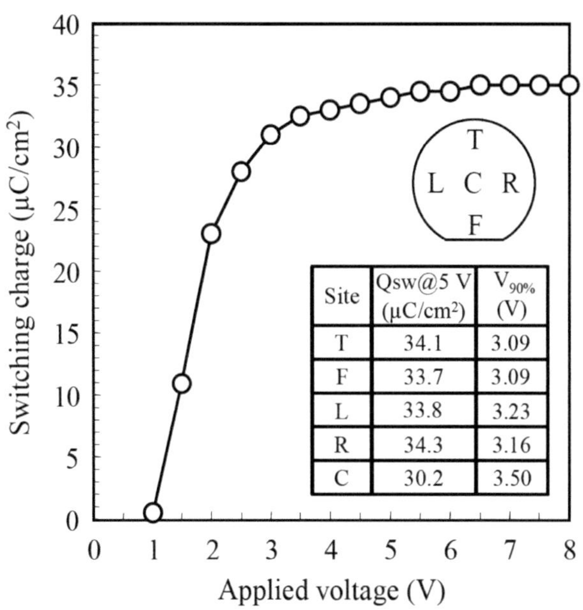

Site	Qsw@5 V (μC/cm²)	V90% (V)
T	34.1	3.09
F	33.7	3.09
L	33.8	3.23
R	34.3	3.16
C	30.2	3.50

Figure 5. Transition of switching charge of Pt/PZT(200 nm)/Pt capacitor.

Fig. 6 shows the stability of Q_{SW} in continuous sputtering of 1000 substrates. The result showed high stability sufficient for mass production. Also, reference data was the data before chamber cleaning. The fact that it was equivalent to the data after chamber cleaning demonstrates the reliability of the system process.

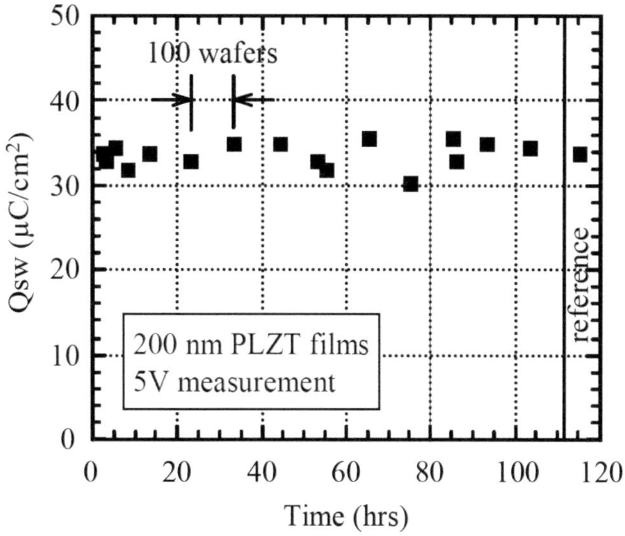

Figure 6. Stability of Qsw in continuous sputtering of 1000 substrates.

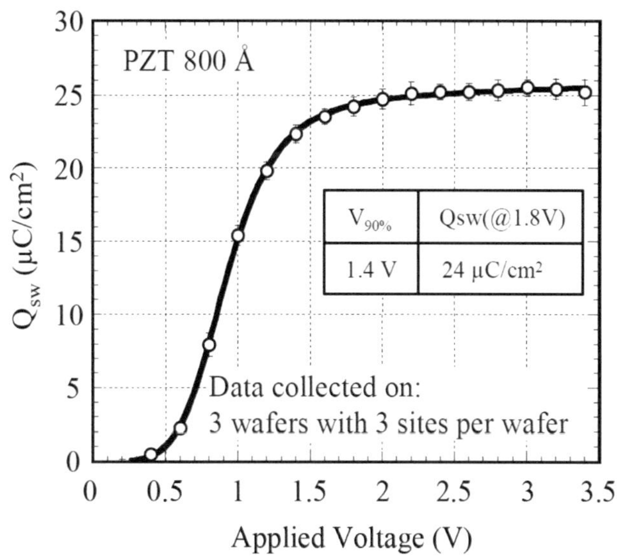

Figure 7. Transition of switching charge of IrOx/PZT(80 nm)/Pt capacitor.

Further improvement of ferroelectric performancs is needed because scaled thinner capacitor (sub 100 nm) is demanded for next generation. Pb deficient surface layer is confirmed to be responsible for degradation of ferroelectric performance. Bottom electrode and PZT deposition process were modified to further improve the ferroelectric performance and achieve thinner capacitor with good performances. [8]. Fig. 7 shows the Q_{SW} transition curve of a IrOx/PZT(80 nm)/Pt capacitor measured at 3 V. Q_{SW} with 1.8 V applied was approximately 24 $\mu C/cm^2$ and $V_{90\%}$ was 1.4 V. Fig. 8 also shows the fatigue characteristics of voltage application at 2 V. Q_{SW} was not decreased even after switching in 10^9 cycles.

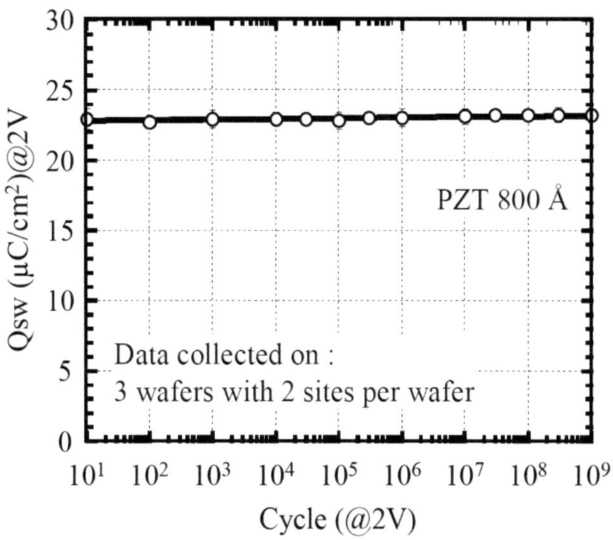

Figure 8. Fatigue characteristics of IrOx/PZT(80 nm)/Pt capacitor.

From above results, sputter-derived PZT capacitor was proved to be suitable for 0.18 μm technology node.

2.4. MOCVD technology for non-volatile memory application

For next-generation FeRAMs beyond 0.18 μm, because it is necessary to achieve further larger packing densities and integration with logic devices, thinner films of ferroelectric associated with shrinking of the thickness of ferroelectric capacitors and the three-dimensional structures associated with shrinking of the capacitor areas are demanded. In addition, it is necessary to achieve thinner films of ferroelectric in parallel with its higher quality in order to meet the demand of still lower voltage drives for device performance. It is said that MOCVD (Metal organic chemical vapor deposition) technology can meet these demands. While depositions are physically made by plasma collision in sputtering processes, depositions are made on heated substrates by chemical reaction among source gases in MOCVD. Therefore, denser crystalline films are easily obtainable, and it is possible to achieve thinner films and higher quality. In addition, uniform deposition can be obtained also on three-dimensional structure, and it is considered that good step coverage can be obtained. Because of these features, MOCVD is the prime candidate of deposition technologies for next-generation FeRAMs in mass production, and its development is making progress.

http://dx.doi.org/10.5772/54360

First, the features of MOCVD are briefly explained. In the method of MOCVD, raw material that was changed into organic metal with high vapor pressure is led to a substrate, and is thermally reacted with a reactant gas (such as oxygen) on the substrate to form a deposition. Methods for the gasification of organic metals are classified into two groups. One of them is the sublimation of solid raw materials (sublimation method). After a certain vapor pressure is obtained by heating a solid raw material using a heater, it is transported with carrier gas. In the second method, a liquid raw material or solid raw material is melted in an organic solvent. Then, after the solution is vaporized, it is transported with carrier gas (vaporization method). In the case of the former, the deposition rate is small. In the case of the latter, there is a problem of the instability of vaporization. For the target of mass-production system development, we adopted the latter vaporization method from the beginning of the development. The problems that need to be resolved are as follows (see Fig. 9):

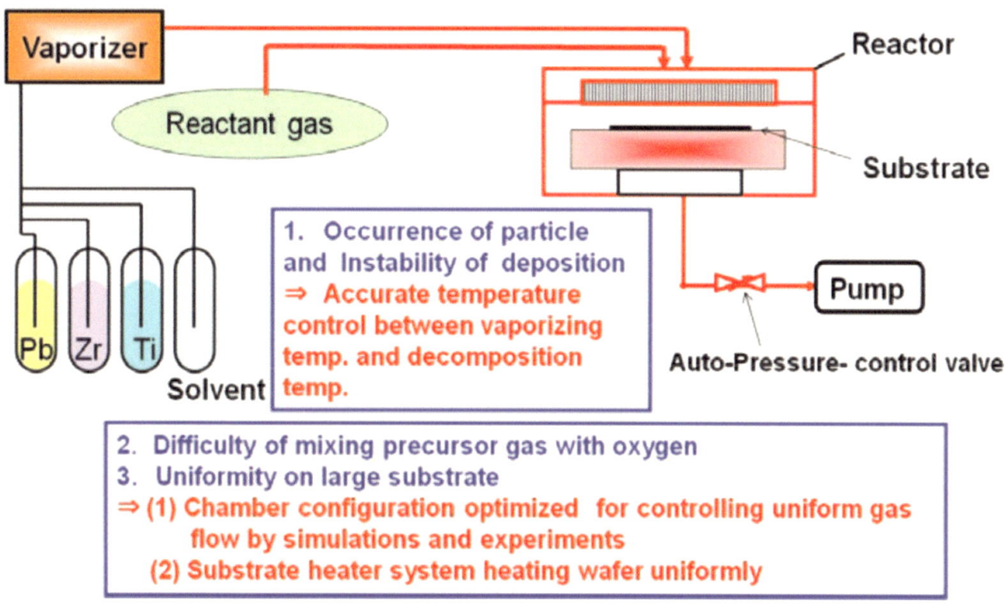

Figure 9. Features of MOCVD Module

1. Vaporization and transport of multi-element raw materials In MOCVD using the vaporization of a ferroelectric solution, because a multi-element organic metal material is vaporized and transported, stable vaporization is difficult, and there is a possibility that precipitation or decomposition may occur in a pipe during transport.

2. Mixture of gases whose molecular weights are mutually different It is difficult to uniformly mix a reactant gas such as oxygen and an organic raw material whose molecular weights are vastly different.

3. Uniformity of large-diameter substrates For the film thickness and the composition of a multi-element oxide thin film PZT, in-plane uniformity in 8 inches is required.

4. PTZ film properties Because of high-temperature processes and in-situ crystallization in the processes, the correlation between the kind of a raw material and a process is sensitive, and the control of a film composition and crystalline orientation is difficult. Therefore, necessary performance of films is difficult to obtain.

In order to establish MOCVD mass-production technology, it is absolutely necessary to resolve the above problems in parallel, and advanced vaporization technology, mixing technology, and reaction control technology are simultaneously required. We combined ferroelectric deposition technology for FeRAMs, module design technology, and CVD equipment technology, which were provided to users until now, and completed full-fledged MOCVD equipment for mass production. The major features include the following four points:

1. The reproducibility in the continuous operation of vaporization and the gas transport to a substrate that controls condensation and decomposition were achieved by accurate temperature control for each part of equipment and the optimization of vaporization conditions. Consequently, film thicknesses within ±2% and the reproducibility of the PZT composition were kept with no mechanical maintenance, and a running test for 1,000 substrates was successful as shown in Fig. 10. [9]

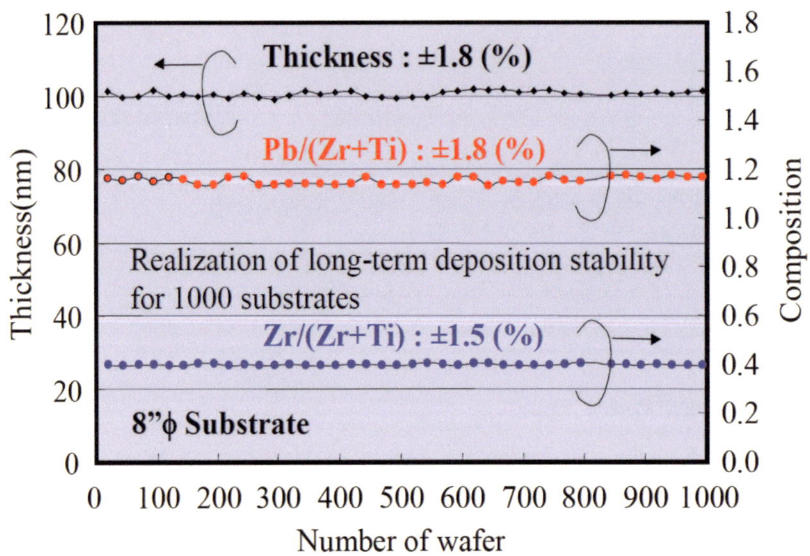

Figure 10. Reproducibility in MOCVD system (film thickness and PZT composition)

2. Because the design of the equipment was conducted in consideration of the flow of gases and the mixture between a raw material gas and a reactant gas on the basis of simulations, film thickness distribution within ±3% on 8 inch substrates was achieved as shown in Fig. 11.

http://dx.doi.org/10.5772/54360

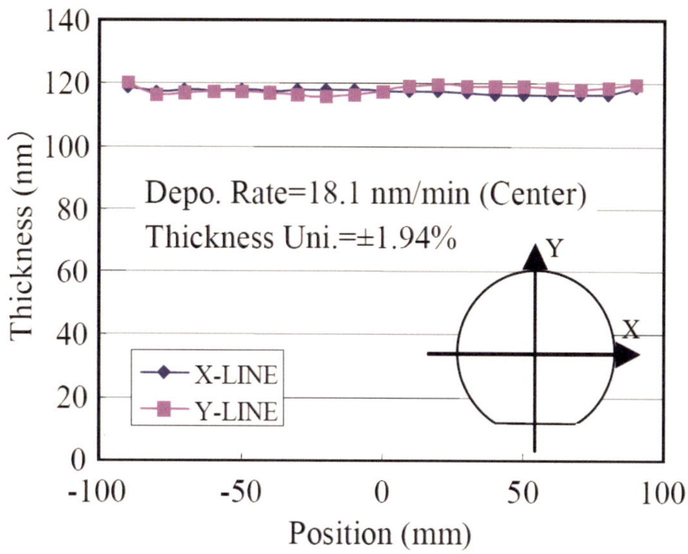

Figure 11. In-plane uniformity of PZT film thickness

3. Because of the development of a new heater, including the optimization of the shape of the heater, the temperature distribution of 8 inch substrates can be continually controlled within ±3ºC. Consequently, in-plane uniformity within ±3% was achieved for both film thicknesses and the composition as shown in Fig. 12. [10]

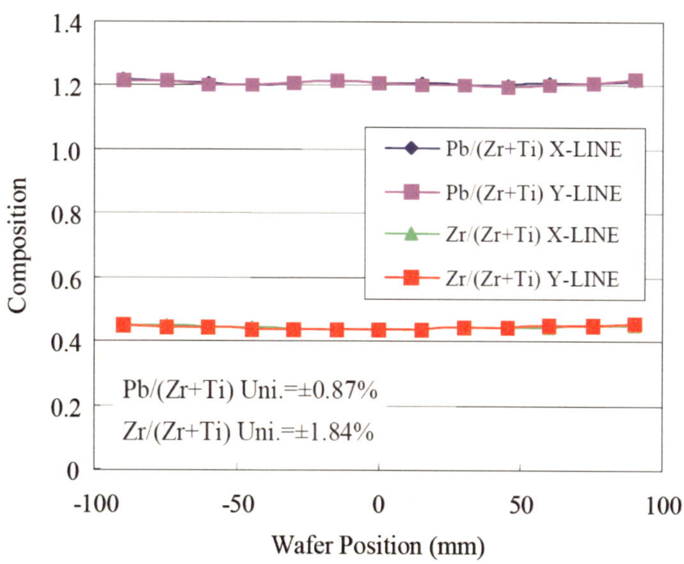

Figure 12. In-plane uniformity of PZT composition

4. By taking advantage of the progress of the equipment hardware as described above, the optimization of processes was conducted, and PZT thin films can be controlled to ensure preferential orientation in the <111> direction as shown in Fig. 13. Consequently, the formation of PZT films within 100 nm that have capacitor properties with a 1.5 V low voltage drive is achieved as shown in Fig. 14. [11,12] In addition excellent endurance properties which are over 10^{10}cycles were obtained for 73nm-PZT in Fig 15.

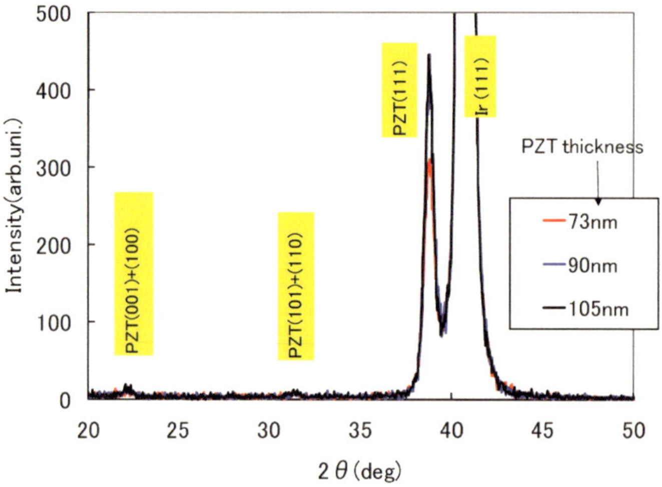

Figure 13. XRD spectrum for various thickness PZT

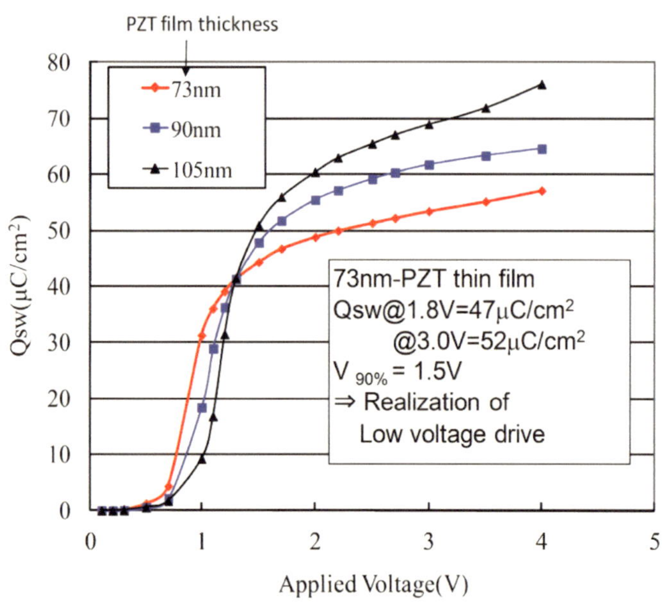

Figure 14. Characteristics of MOCVD-PZT film (applied voltage dependence of switching charge)

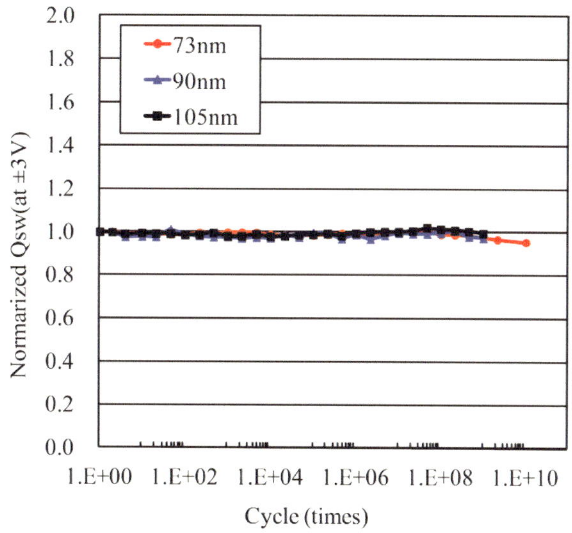

Figure 15. Endurance properties of MOCVD-PZT films

Recently mass production tool for 300mm Si wafer was developed and excellent in-plane uniformity less than 1.5% for thickness and PZT composition were obtained.

2.5. Etching technology

2.5.1. Issues of ferroelectric etching technology

Conventionally, the piezoelectric elements have been fabricated by chemical wet etching [13] or argon ion milling. With the miniaturization of MEMS, there have been increasing demands for dry etching with the excellent shape controllability as semiconductor technology. Recently dry etching technique for MEMS using PZT was reported. [14,15] The Pt, Ir and other rare metal electrodes and the PZT ferroelectric thin films that compose piezoelectric elements react poorly with halogen gases and their halides have low vapor pressures. For these reasons, these materials are called hard-to-etch materials. The following technical issues are important for dry etching of the PZT ferroelectric thin films for not only FeRAM but also MEMS productions:

1. Etching selectivity to resist mask and the bottom rare metal electrode

A piezoelectric element film consists of PZT with a thickness of several micrometers and the rare metal electrodes with a thickness of about 100 nm. Generally, the bottom electrode is left after the PZT etching. Therefore, a low etching rate for the bottom electrode, the so-called high etching selectivity, is important as a PZT etching condition.

2. Adhesion of conductive deposit to the pattern sidewalls and damage to PZT

The materials are hard to etch, and their etching products easily adhere to the pattern sidewalls, and result in leaks between the top and bottom electrodes. What is worse, the pattern sidewalls are exposed to reactive gas plasma during etching, and tend to suffer lead and oxygen coming out and other damages.

3. Plasma stability during continuous processing

Adhesion of etching products to chamber walls, especially the RF introduction window that generates plasma, causes instabilities of plasma and deteriorates the etching rate and the shape reproducibility. Avoiding of adhesion of etching products to chamber walls is important for mass production.

4. Uniformity of etching rate within wafer

As in the case of (1), to stop the thick PZT at the thin bottom electrode after etching, the uniformity of etching rate within wafer is important.

2.5.2. Ferroelectric etching systems and process for mass production

As the piezoelectric PZT etching systems, this section explains about Apios NE series made by ULVAC, Inc. The etching module is equipped with the ISM (Inductively Super Magnetron) plasma source that can generate low-pressure and high-density plasma. Fig. 16 shows a drawing of the etching module. Table I shows the comparison between the normal ICP type plasma source and the ISM plasma source. The RF antenna is mounted in the upper part of the etching chamber, so that RF is introduced through the quartz window into the etching chamber to generate plasma. The uniformity of the etching rate within wafer can be easily optimized by positioning permanent magnets under the antenna. Fig. 17 shows the uniformity of PZT etching rate within 6 inch wafer. A high uniformity (<+/-5%) was realized by means of optimization to permanent magnet layout. A STAR electrode is provided between the antenna and the quartz window to control adhesion of etching products to the quartz window by applying RF to the STAR electrode. The substrate is held on the electrostatic chuck. The substrate temperature is controlled by introducing Helium to the backside of the substrate. The ion energy is controlled by applying RF power to the substrate. The materials to be etched are non-volatile, and the etching products adhere to the shield located in the chamber. The temperature of the shield is kept constant by heater, so process is high stability.

	ISM	ICP
Plasma density (cm^{-3})	$1\times10^{10} \sim 1\times10^{11}$	$5\times10^{9} \sim 5\times10^{10}$
Operating pressure (Pa)	0.07<P<7	0.5<P<50
Uniformity	Optimized magnetic layout	Determined by chamber structure
Damage	Plasma density and substrate bias can be controlled independently.	
Repeatability, stability	Less re-deposition results in Low-pressure etching -> better repeatability	High pressure process causes re-deposition.
Maintenance	Chamber structure is simple for easy maintenance.	

Table 1. Inductively super magnetron (ISM) plasma performances.

http://dx.doi.org/10.5772/54360

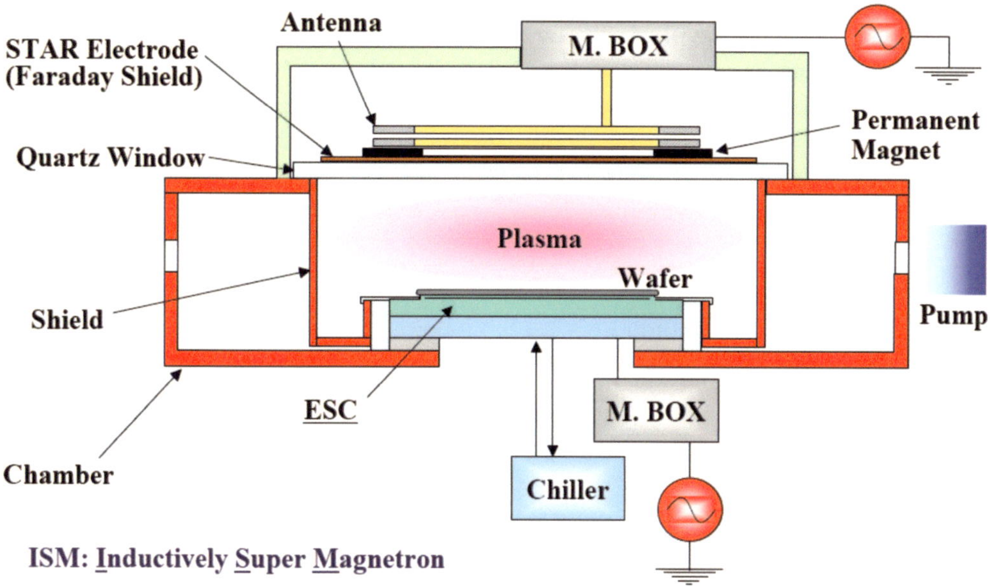

Figure 16. Etching chamber with inductively super magnetron (ISM) plasma performances

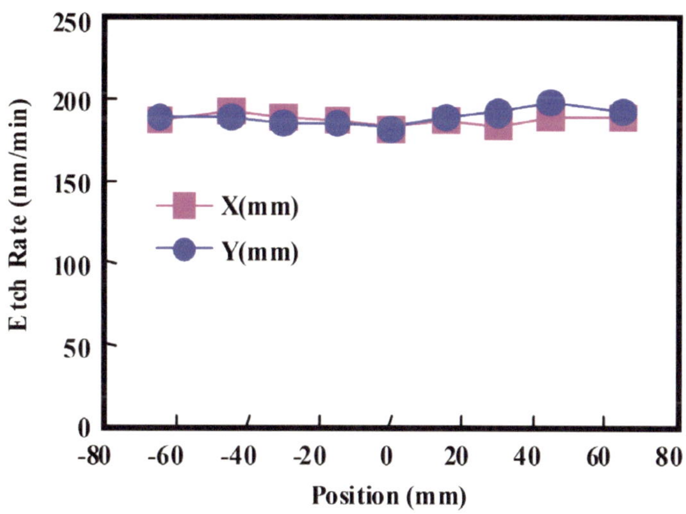

Figure 17. PZT etching uniformity

2.5.3. Mass productive high temperature etching technology for high dense FeRAM

Dry etching techniques are used for the patterning of FeRAM device. It is difficult to etch the material of FeRAM such as the noble metal and PZT, because these have low reactivity with halogen gas plasma and these halides have low vapor pressure. The photo-resist is used as etching mask for the patterning of FeRAM memory cell. But the etching selectivity to photo-resist is low. Therefore the etching profile becomes low taper angle. Since FeRAM device shrinks down recently(0.35-μm-design rule or lower), high temperature etching was developed for high density FeRAM device. Furthermore, FeRAM device changed to the stack structure from planer structure. At that time, the top electrode, PZT and the bottom electrode are etched in a series by using of hard-mask (Fig. 18). [16]

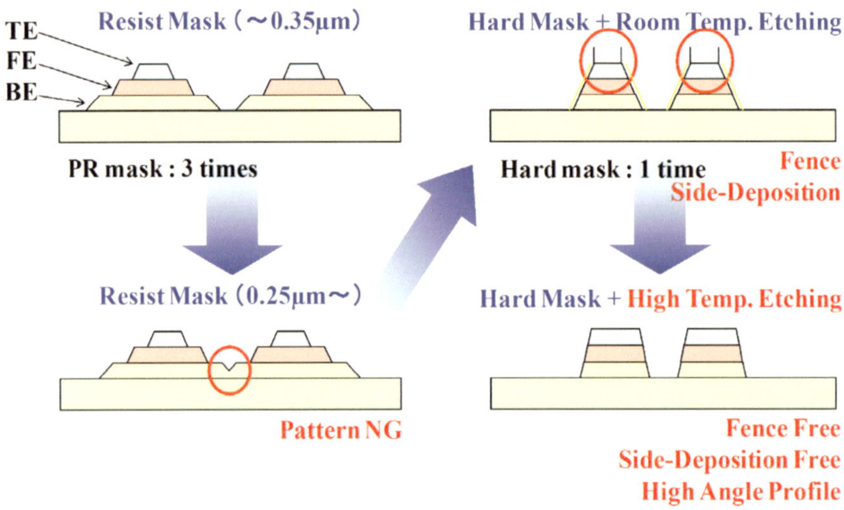

Figure 18. Necessity of high-temperature etch

This section explains high temperature etching system ULHITE series made by ULVAC, Inc. This etching module is equipped with ISM plasma source. Most important feature is the novel electro-static-chuck (ESC) type hot plate stage at a temperature up to 450°C, and this stage can be supplied high bias power. The process chamber, variable conductance valve, pumps and gas exhaust are heated up. The deposition shields equipped in process chamber are also heated up to 200°C. The effect is reducing the deposition during the etch process, and high process stability is achieved.

Fig. 19 shows repeatability of Ir etching time for 300-wafers running test. The plasma cleaning were carried out in every 25-wafers. Excellent stability was confirmed. For next generation FeRAM devices, the continuous etching among the top electrode, ferroelectrics and bottom electrode by use of one mask is need. And the high taper angle and no sidewall deposition are needed for etching process.

Fig. 20 shows the results of FeRAM capacitor etching of stack structure dependence of the stage temperature. The halogen gases were used in etching process. Thus the higher profile can be obtained by higher stage temperature. Fig. 21 shows repeatability of etching profile for 300-wafers running test. Excellent repeatability of etching profile was confirmed.

http://dx.doi.org/10.5772/54360

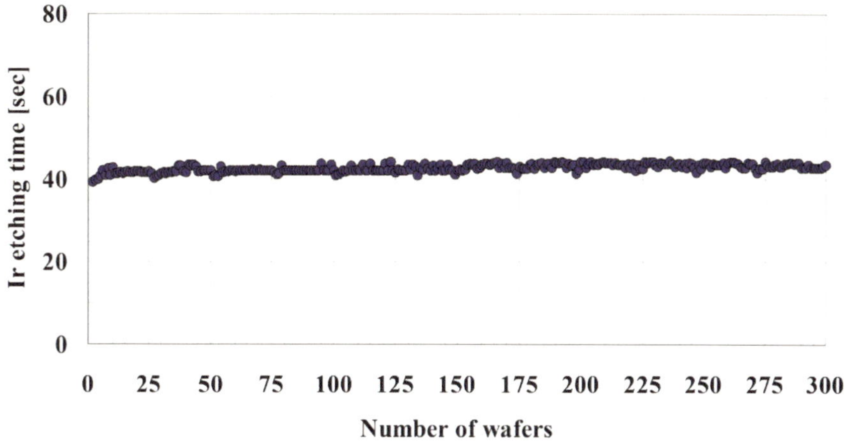

Figure 19. Repeatability of Ir etching time for 300-wafers running test

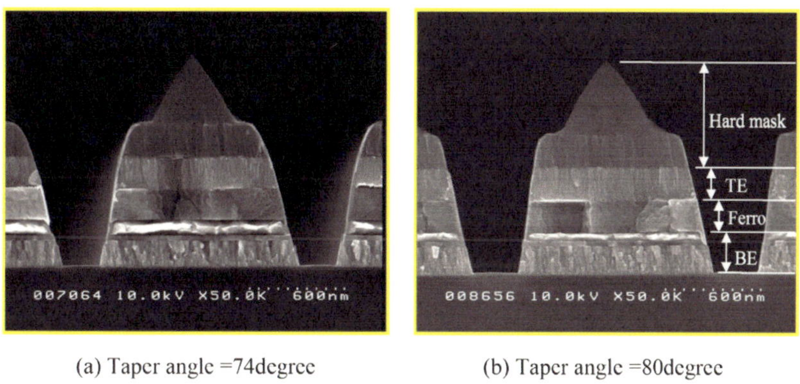

Figure 20. Stage temperature dependency of taper angle ((a) 300°C, (b) 400°C)

Usually, high density device is required for non-volatile memory. Therefore the high temperature etching technique contributes to realize the next generation FeRAM devices production.

3. Technology for MEMS application

3.1. Sputtered PZT piezoelectric films for MEMS application

A multi-chamber type mass production sputtering systems for electronic devices SME-200 equipped with an exclusive sputtering module described above. PZT films have been deposited on 6 inch diameter silicon substrates.

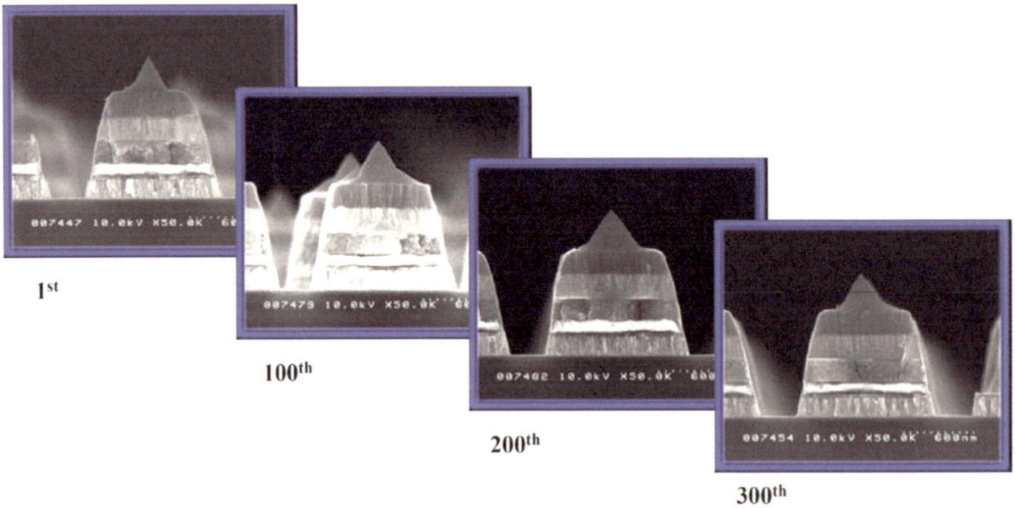

Figure 21. Repeatability of etching profile for 300-wafers running test

The platinum bottom electrodes, whose orientation is (111), were deposited on the substrate. The PZT films were deposited under Ar/O_2 mixed gas atmosphere of 0.5 Pa. Substrate temperature was heated up to around 550°C. After the deposition, PZT films were conducted with no thermal treatments such as post-annealing. PZT films were deposited with relatively high growth rate about 2.1 μm/h and these thicknesses were from 0.5 to 3.0 μm in consideration of piezoelectric MEMS applications. The ceramic target with Zr/Ti ratio of 52/48, in which 30 mol % excess PbO was added for the compensation of the lead re-evaporation from the films, was used in order to obtain PZT films near the stoichiometric composition. After the PZT deposition, top electrode 100-nm-thick Pt was deposited by the dc sputtering method.

For the measurement of the piezoelectric properties of the PZT films, Rectangular beams (cantilevers) with the size of about 30 mm (3 cm) × 3 mm were prepared. Polarization and displacement in these films were simultaneously observed using the laser doppler vibrometer (Graphtec AT-3600) and the laser interferometer (Graphtec AT-0023) which were attached to a ferroelectric test system.

Fig. 22 shows relationship between Pb composition, deposition rate and repeatability. As a result, stable transition of Pb content within film in continuous sputtering has been confirmed. As can be seen from the figure, the change in the deposition rate was 2.1 μm±1.4%, and changing Pb composition was 1.0±0.1% in the short running for 35 pieces (total thickness; 105 μm).

Fig. 23 shows XRD patterns of as-sputtered PZT film. No pyrochlore phase can be confirmed and the film appears to be almost perovskite phase with the preferred orientation to (001) or (100). It confirmed that the uniformity of the crystalline property were good in 6 inch area. In addition, the dielectric constant of these samples was measured as shown in Fig.24. As a result, excellent properties with the dielectric constant over 1000 and its uniformity of ±4.7% were confirmed.

http://dx.doi.org/10.5772/54360

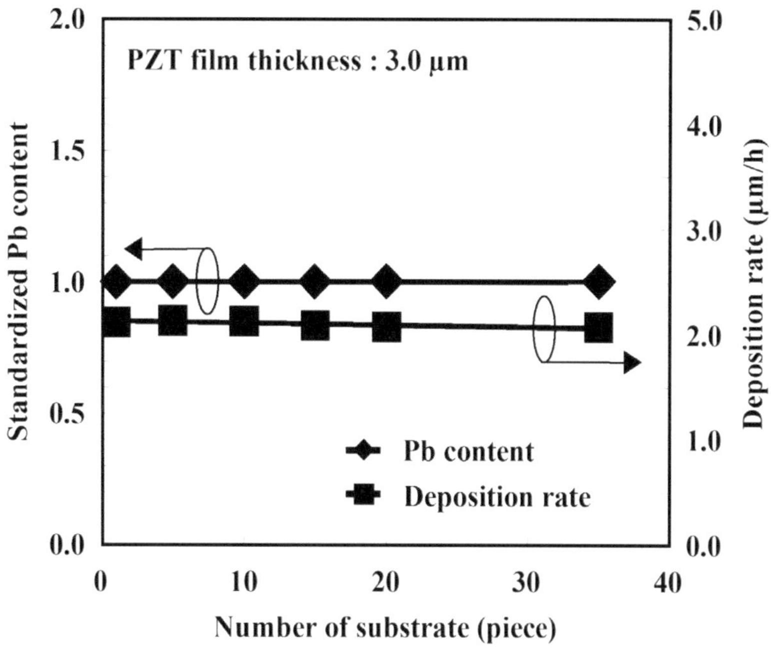

Figure 22. Relationship between Pb composition, deposition rate and repeatability

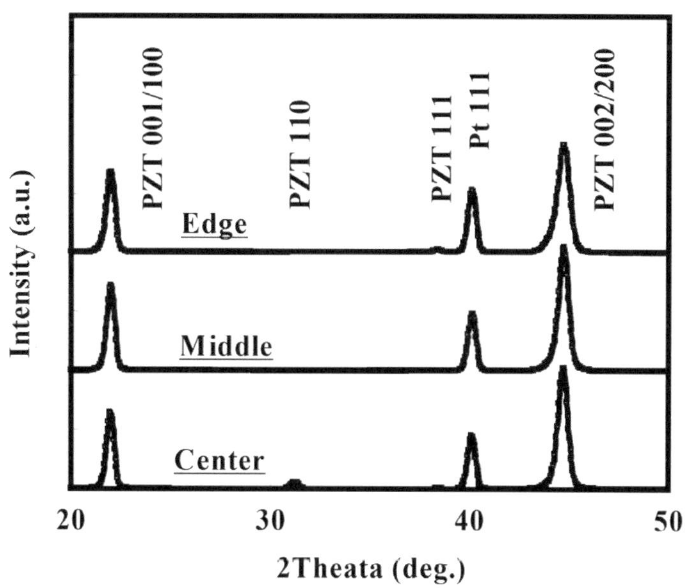

Figure 23. Crystallization uniformity of PZT films Deposited on 6-inches substrate

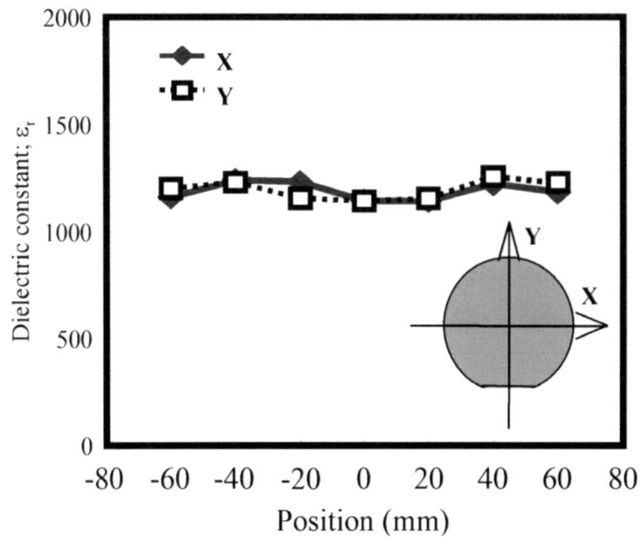

Figure 24. Dielectric constant of PZT films deposited on 6 inch substrate

Piezoelectric properties were finally confirmed for PZT films by checking the cantilever as shown in Fig. 25. Large piezoelectric coefficient from -60 to -120 pm/V was observed in our PZT films.

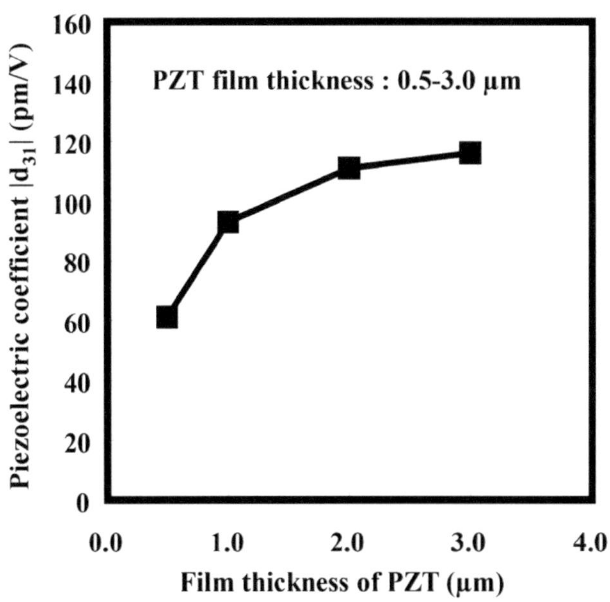

Figure 25. Relationship between piezoelectric coefficient and PZT film thickness

3.2. Etching technology of PZT piezoelectric Films for MEMS application

Additionally etching technology for mass production is developed. Since high etching rate is required in MEMS PZT process because of large PZT thickness, etching selectivity to the bottom electrode(Pt) and uniformity are important. In Fig. 26 we show etching rate profile of both electrode(Pt) and PZT film. It has been found that etching rate of Pt is smaller than PZT over whole wafers, implying that Pt film is the role of an etch stop layer.

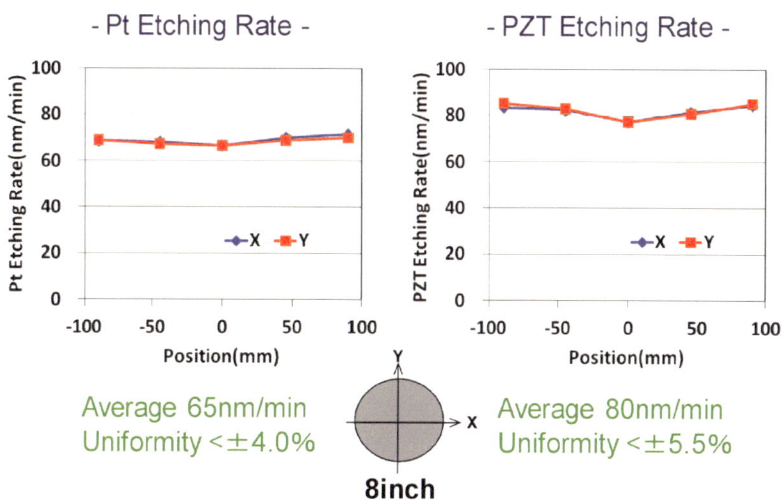

Figure 26. Etching rate and in-plane uniformity of Pt & PZT

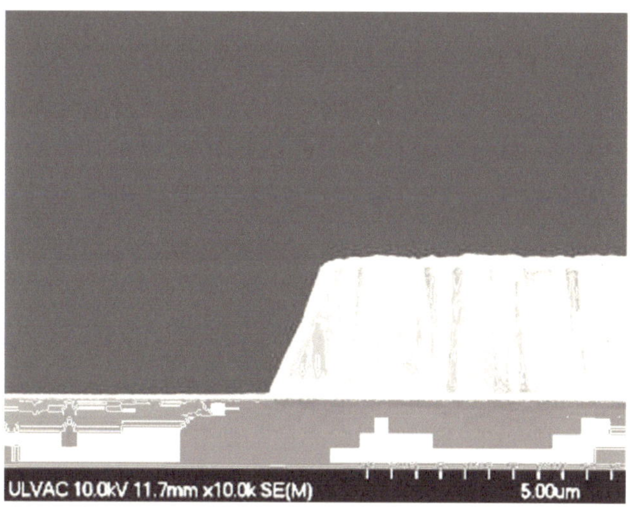

Figure 27. SEM image of Pt/PZT etching profile

Fig. 27 shows the SEM image after the piezoelectric element was etched and the resist mask was removed. The film composition is Pt/PZT/Pt=100 nm/3 μm/100 nm, and the top Pt electrode and PZT were continuously etched by using a 5-μm-thick photo resist as a mask. Chlorine and fluorine mixed gases are used for PZT etching. The etching shape (taper angle) is about 65º, and nothing adhered to the pattern sidewalls. Despite 20% of over etching, the bottom Pt electrode was hardly etched. This indicates that a high etching selectivity to Pt was achieved. Fig. 28 shows the dependency of the etching rate and the taper angle on the bias power. As the bias power increases, the etching rate increases linearly. When 400 W was applied, the PZT etching rate of 190 nm/min. was achieved. The taper angle also increased gradually as the bias power increased.

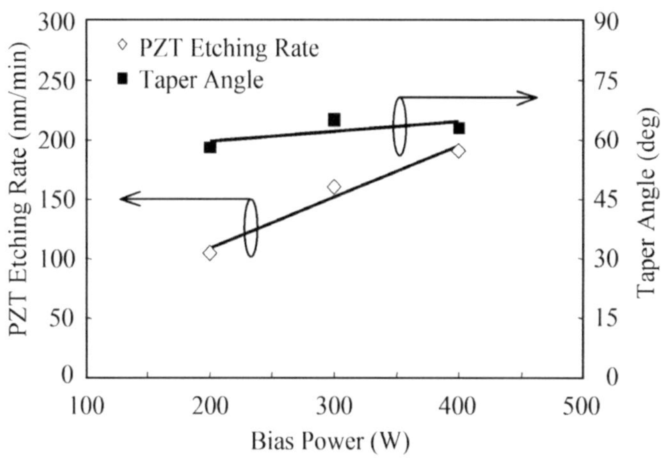

Figure 28. Dependence of PZT etching rate and taper angle on the substrate bias power

Figure 29. Pt/PZT etching profile of ϕ50mm piezoelectric element array

Fig. 29 shows a 50-μm-diameter actuator element array fabricated by dry etching. Fig. 30 shows that the remanent polarization (Pr value) of the piezoelectric thin-film actuator with a 3-μm-thick PZT film was 40.5-42.8 μC/cm² and the coercive electric field (E_c value) was 44.5-46.0 kV/cm at an applied voltage of 30 V, and the characteristics without the dependency on element size (30-300 μm diameter) were obtained. The damage to the PZT piezoelectric thin-film actuator caused by the dry etching is considered to be negligible. The displacement of the PZT thin-film actuator was measured by contact-AFM. Fig.31 shows that a displacement of about 4 nm was obtained at 3-μm-thick PZT film, 30-μm-diameter element size, and an applied electric field of 100 kV/cm. It was clarified that the processing of PZT piezoelectric thin-film actuators by dry etching is very effective.

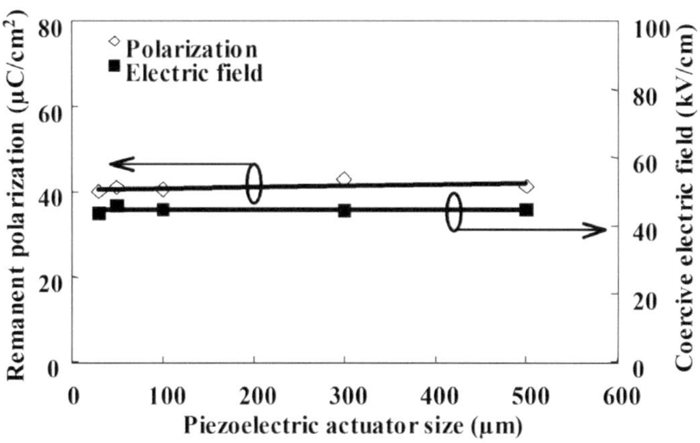

Figure 30. Dependence of ferroelectric properties on piezoelectric actuator size

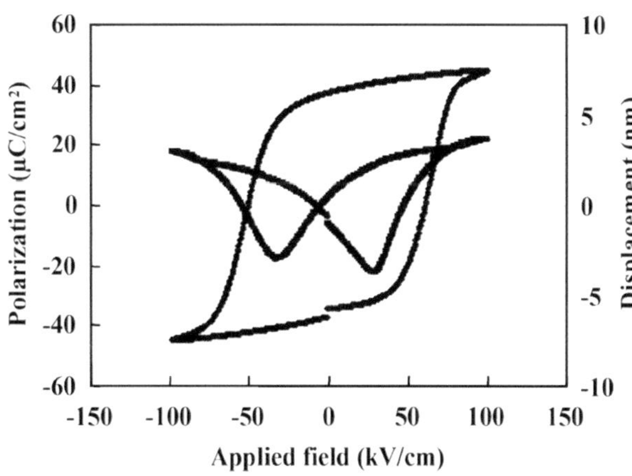

Figure 31. Ferroelectric and displacement properties of Pt/PZT/Pt element with 30mm diameter etching rate and taper angle on the substrate bias power

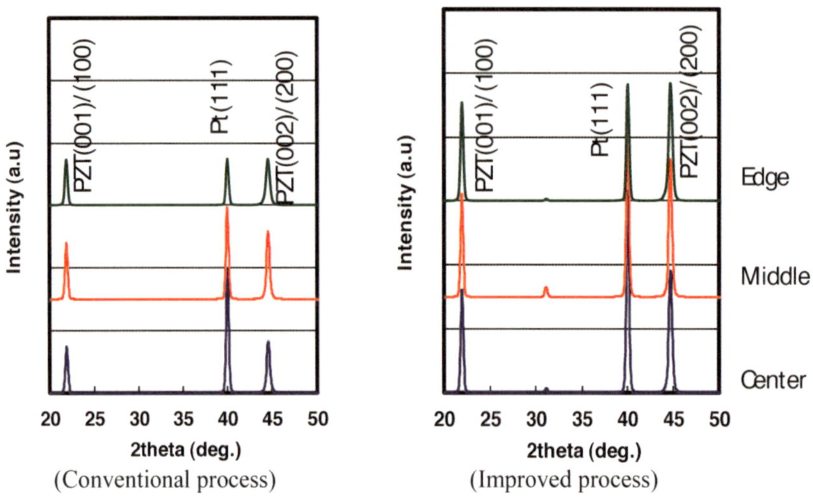

Figure 32. XRD Patterns of PZT film before and after improvement on 8inch Pt/Ti/SiO_2/Si Substrate

3.3. Further development for MEMS application

We also note that post in-situ treatment [17] has been recently established, aiming to improve crystalline property and piezoelectric coefficient in our PZT films. Fig. 32 shows the results before and after improvement of crystalline properties. Peak intensity of a/c axis of PZT film deposited by improved process is approximately two times higher of peak intensity of PZT film deposited by conventional process uniformity over 8 inch wafers. As a consequence good piezoelectric coefficient (12.9C/m^2) and large breakdown voltage (68V) were obtained for PZT film deposited by improved sputtering process in Fig 33.

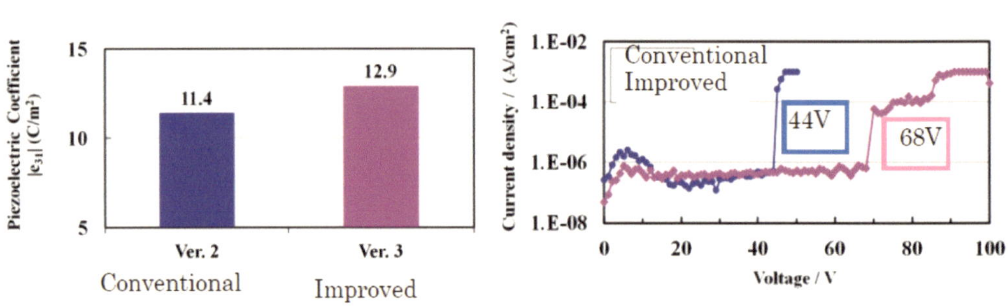

Figure 33. Improvement of piezoelectric coefficient(Left) and break down voltage(Right)

Good piezoelectric performance which is based on mass production technology including excellent in-plane uniformity and process rate of sputtering and etching was obtained in above evaluation.

3.4. Sputtered BST thin-films for capacitor and RF tunable devices

$(Ba,Sr)TiO_3$ (BST) is expected to use as the thin film capacitor, RF tunable component [18, 19] on-chip capacitor [20] for its excellent dielectric behavior. In these applications, decoupling, high permittivity and high tunability are required as dielectric characteristics. High permittivity is thought to be primarily important character as dielectric characteristics. BST thin films have been deposited by some deposition techniques. Among these techniques, RF magnetron sputtering is thought to be suitable for mass-production because of its stability and reproducibility as well as good film performances. In this section, BST films deposition by RF magnetron sputtering and an approach to deposit BST films with higher permittivity were described.

BST films were also deposited by an RF magnetron sputtering method using sintered BST ceramic targets. Basic sputtering conditions such as RF power, deposition temperature and so on were varied. As a top electrode, Pt dot with 0.5 mm in diameter was deposited by DC magnetron sputtering. After top electrode deposition, BST capacitors were post-annealed for 1 hour in oxygen atmosphere under 1 atm using resistive heat furnace.

Fundamental properties of BST thin films such as dielectric constant, crystalline quality (XRD), surface morphology (SEM) were investigated. Agilent 4284A LCR meter was used for dielectric properties measurement and measuring cycle and volts alternating current were 1 kHz and 1 Vrms, respectively.

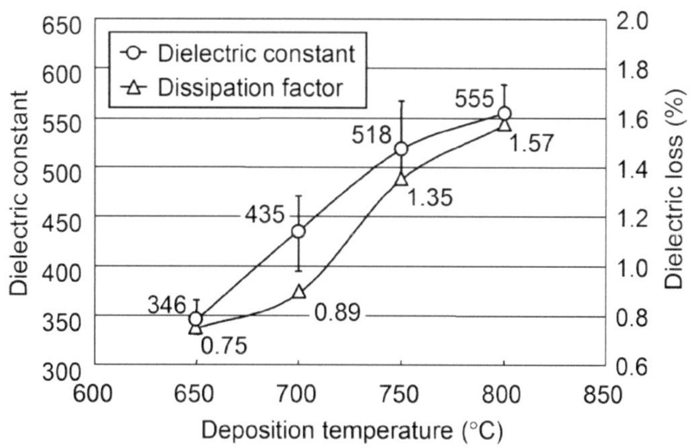

Figure 34. Relationship between dielectric constant and deposition temperature

The relationship between dielectric constant and deposition temperature was investigated as shown in Fig. 34. Ba/Sr ratio of BST target was 50/50. RF power was 1500 W. Ar/O_2 flow ratio was 3. Sputtering pressure was 2.0 Pa. BST film thickness was 100 nm. As a result, deposition rate was 4.8 nm/min and almost constant irrespective of deposition temperature. As can be seen in this figure, dielectric constant is strongly dependent on deposition temperature and increasing with deposition temperature increasing. XRD patterns of these films are also shown in Fig. 35. We can see in this figure, BST grains are randomly orientated and XRD peak intensities from BST films are increasing with deposition temperature increasing. So, it is thought that this represents the relationship between dielectric constant and deposition temperature.

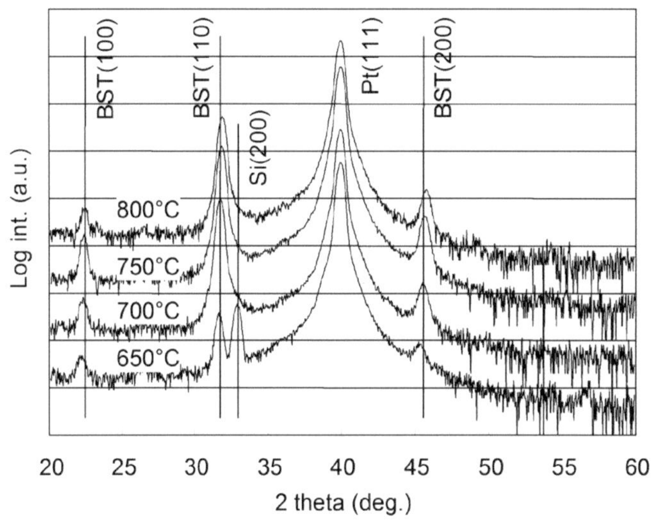

Figure 35. Relationship between XRD pattern and deposition temperature

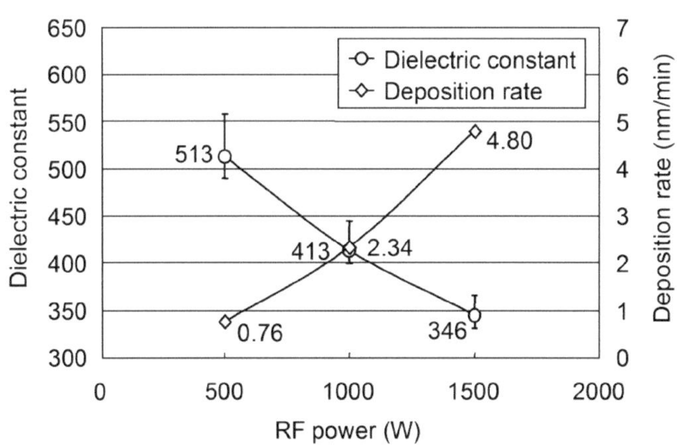

Figure 36. Relationship between dielectric constant, deposition rate and RF power temperature

The relationship between dielectric constant, deposition rate and RF power is shown in Fig. 36. Deposition temperature was 650°C in this relationship, while other conditions were not changed. We can see in this figure, there is trade-off relationship between dielectric constant and deposition rate and higher dielectric constant can be obtained at low sputtering power of 500 W in this experiment. BST morphology (SEM photographs) dependence on RF power is also shown in Fig. 37. As can be seen in this figure, BST film deposited at 500 W (b) has dense and uniform columnar structure. Possibly it represents some relationship between dielectric constant and RF power. It is speculated that there is a long time for atomic migration and the growth of BST crystal is encouraged under such low deposition rate around 1 nm/min.

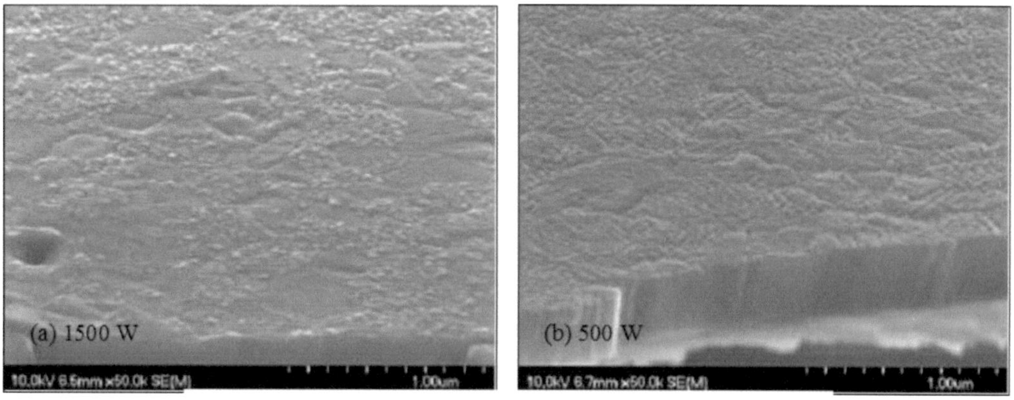

Figure 37. BST morphology dependence on RF power

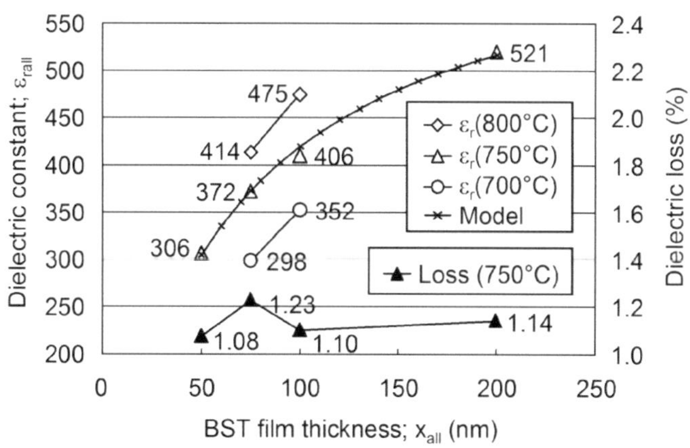

Figure 38. Relationship between dielectric constant and BST film thickness

Furthermore, the relationship between dielectric constant and BST film thickness is shown in Fig. 38. Ba/Sr ratio of BST target was 70/30 in this experiment. Deposition temperature was varied from 700°C to 800°C as a parameter, while other conditions were not changed.

We can see in this figure, dielectric constant is also strongly and non-linearly dependent on BST film thickness.

If BST capacitor is assumed to be a simple series connected capacitor between transition layer near BST/Pt bottom electrode interface and main layer which represents BST film excluding transition layer, it is speculated that there have been the transition layer in this BST film. [21]

As a simple experiment, effects of gas flow sequence for BST deposition was investigated. Gas flow sequence, that is ON/OFF step was changed, while Ar/O_2 flow ratio was not changed. Dielectric constant was improved by introducing of oxygen gas before BST deposition. It is speculated in this experiment that BST/Pt interface was improved because the oxygen vacancies of BST in this region were reduced. Therefore, dielectric properties are noticeably influenced by gas flow sequencing variation.

4. Technology for ferroelectric probe memory

Technology for ferroelectric probe memory was developed with the deposition technology for FeRAM manufacturing. Hard-disk using the magnetic recording is the one of the major storage device, but the recording density will reach the limit in the near future by exteriorization of the magnetization disappearance by the heat disturbance. Ferroelectric probe memory is ultrahigh-density memory applied Scanning Probe Microscope (SPM) and ferroelectric property.

Ferroelectric perovskite Pb(Zr, Ti)O_3 (PZT) is polarized by electric field. For scanning probe memory device, ferroelectric layer required atomically-smooth surface and low leakage current to achieve large recording density by forming polarized domains as unit cell size. Therefore it is necessary to grow the epitaxial ferroelectric thin films oriented to *c*-axis direction.

In this time we prepared epitaxial growth PZT/ $SrRuO_3$ (SRO) thin films on single crystal $SrTiO_3$ (STO) substrate for probe memory device, and evaluated the film properties and the recording density with Piezoresponse force microscopy (PFM). The SRO films were deposited on STO (100) single crystal substrate with DC sputtering. After that, the PZT film was deposited by MOCVD process.

Fig. 39 shows the AFM image of PZT / SRO / STO deposited at the optimal Pb flux. The surface morphology of PZT has step structure and smooth terraces similar to SRO/STO substrate before PZT deposition. This surface profile indicates layer-by-layer growth of PZT on SRO / STO substrate [22] [23] [24]. Fig. 40 shows XRD spectrum of PZT deposited at Pb flux of 0.160ml/min. There are only PZT, SRO and STO peaks. The fringe pattern appears around PZT (001) and (002) peaks because of the smooth interface between PZT and SRO. Full-width at half-maximum (FWHM) of PZT (002) with X-ray rocking curve is 0.129degree for optimal Pb flux of 0.160ml/min. Rms of surface roughness and PZT(002) FWHM of X-ray rocking curve have a similar trend relative to Pb flux.

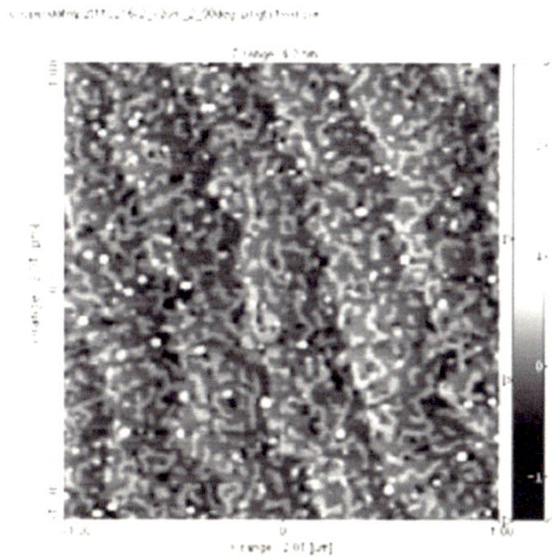

Figure 39. AFM image of PZT at Pb flux of 0.160ml/min (Scanning range is 2um square)

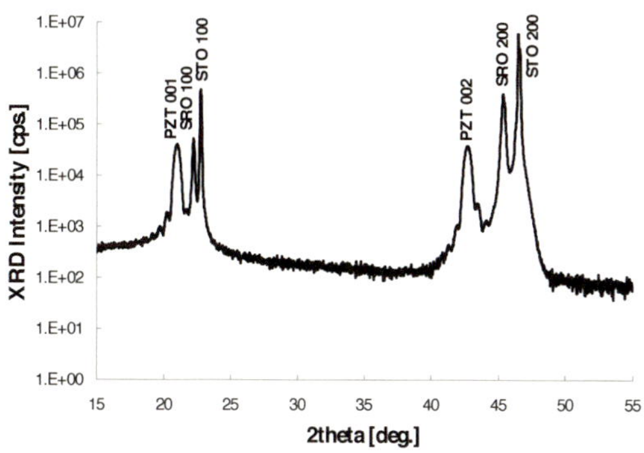

Figure 40. XRD spectrum of PZT / SRO /STO

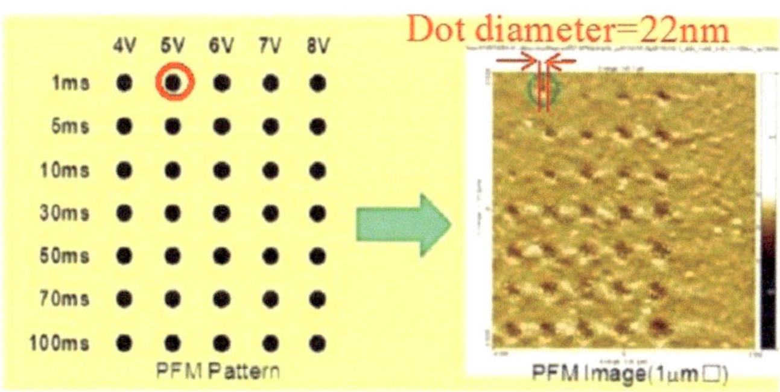

Figure 41. Writing Voltage & Writing Time Dependency of Dot diameter – [Small diameter (22nm) of recording-dot was obtained for 5V & 1ms writing condition. (PZT/SRO/c-STO) ・ Estimated Recording density: 1.4Tbit/inch^2.]

PFM measurement is carried out with various biases between 4V and 8V and pulse widths between 1msec and 100msec with $PtIr_5$ coatedprobe (tip diameter is 25nm). Fig. 41 shows PFM write-condition and PFM image of the result. Polarization-inverted dot diameter has a minimum of 20nm at the bias of 5V and pulse width of 1msec. Recording density is estimated on the assumption that this tiny dot is arrayed with pitch of 20nm. From the calculation result, device sample deposited under the optimal growth condition achieves large bit density of 1.4Tb/inch^2.

5. Conclusion

We have been developing thin film process technologies for ferroelectric application of advanced semiconductor and electronics usage for 20 years or more, and completed the ferroelectric thin film solutions (sputtering, MOCVD, and etching) that became a de facto standard. These technologies will support a wide variety of convenient energy-saving devices such as FeRAM, MEMS production (actuators composing gyro meters, portable camera modules for smart phone applications, tunable devices and so on), and ultra-high density probe memory.

Author details

Koukou Suu

Institute of Semiconductor and Electronics Technologies, ULVAC, Inc., Shizuoka, Japan

References

[1] K. Suu, A. Osawa, N. Tani, M. Ishikawa, K. Nakamura, T. Ozawa, K. Sameshima, A. Kamisawa and H. Takasu: Jpn. J. Appl. Phys. 35 (1996) 4967.

[2] K. Suu, A. Osawa, N. Tani, M. Ishikawa, K. Nakamura, T. Ozawa, K. Sameshima, A. Kamisawa and H. Takasu: Integr. Ferroelectr. 14 (1997) 59.

[3] K. Suu, A. Osawa, Y. Nishioka and N. Tani: Jpn. J. Appl. Phys. 36 (1997) 5789.

[4] K. Suu, Y. Nishioka, A. Osawa and N. Tani: Oyo Buturi 65 (1996) 1248. (in Japanese)

[5] K. Suu: Proc. Semicon Korea Tech. Symp., 1998 p. 255.

[6] N. Inoue, Y. Maejima and Y. Hayashi: Int. Electron Device Meet. Tech. Dig., 1997 p. 605.

[7] N. Inoue, T. Takeuchi and Y. Hayashi: Int. Electron Device Meet. Tech. Dig., 1998 p. 819.

[8] F. Chu, G. Fox, T. Davenport, Y. Miyaguchi and K. Suu: Integr. Ferroelectr. 48 (2002) 161.

[9] T. Yamada, T. Masuda, M. Kajinuma, H. Uchida, M. Uematsu, K. Suu and M. Ishikawa : IFFF2002 abstract, 4 (2002) 37.

[10] T. Masuda, M. Kajinuma, T. Yamada, H. Uchida, M.Uematsu, K. Suu and M. Ishikawa : Integr. Ferroelectr., 46 (2002) 66

[11] Y. Nishioka, T. Jinbo, T. Yamada, T. Masuda, M. Kajinuma, M. Uematsu, K. Suu and M. Ishikawa: Integr. Ferroelectr., 59 (2003) 1445

[12] Y. Nishioka, T. Masuda, M. Kajinuma, T. Yamada, M. Uematsu and K. Suu: MRS Fall Meeting Proceedings, 784-C7 (2003) 6

[13] K. Zheng, J. Lu, and J. Chu: Jpn. J. Appl. Phys. 43 (2004) 3934.

[14] Y. Kokaze, M. Endo, M. Ueda, and K. Suu: 17th Int. Symp. Integrated Ferroelectrics, 2005, 5-26-P.

[15] Y. Kokaze, I Kimura, M. Endo, M. Ueda, S. Kikuchi, Y. Nishioka, and K. Suu: Jpn. J. Appl. Phys. 46 (2007) 282.

[16] M. Endo, M. Ueda, Y. Kokaze, M. Ozawa, T. Nakamura, K. Suu: SEMI Technology Symposium 2002; December 5, 2002

[17] Suu et.al in preparation

[18] I. P. Koutsaroff, T. Bernacki, M. Zelner, A. Cervin-Layry, A. Kassam, P. Woo, L. Woodward, A. Patel, Mat. Res. Soc. Symp. Proc. Vol. 762, C5.8.1 (2003).

[19] X. H. Zhu, J. M. Zhu, S. H. Zhou, Z. G. Liu, N. B. Ming, S. G. Lu, H. L. W. Chen, C. L. Choy, Journal of Electronic Materials, 32 (2003) 1125.

[20] M. C. Werner, I. Banerjee, P. C. McIntyre, N. Tani, M. Tanimura, Appl. Phys. Lett., 77 (2000) 1209.

[21] T. Jimbo, I. Kimura, Y. Nishioka and K. Suu, Mat. Res. Soc. Symp. Proc. Vol. 784, C7.8.1 (2003).

[22] P.-E. Janolin, B. Fraisse, F. Le Marrec, and B. Dkhil, Appl. Phys. Lett. 90, 212904 (2007)

[23] Keisuke Saito, Toshiyuki Kurosawa, Takao Akai, Shintaro Yokoyama, Hitoshi Morioka, Takahiro Oikawa, and Hiroshi Funakubo, Mater. Res. Soc. Symp. Proc. 748, U13.4.1-U13.4.6 (2003).

[24] H. Hu, C. J. Peng and S. B. Krupanidhi, Thin Solid Films, 223 (1993) 327 333I. G. Baek, et al,: Tech. Dig. Int. Electron Devices Meet., San Francisco, 2004, 23, 6, p58

Chapter 18

Ferroelectrics at the Nanoscale: A First Principle Approach

Matías Núñez

Additional information is available at the end of the chapter

http://dx.doi.org/10.5772/52268

1. Introduction

The field of ferroelectric materials is driven by its possible use in various micro-electronic devices that take advantage of their multifunctional properties. The existence of a switchable spontaneous polarization (see Figure 1) is at the basis of the design of non volatile ferroelectric random access memories (FERAMs), where one bit of information can be stored by assigning one value of the Boolean algebra ("1" or "0") to each of the polarization states. Also, the high dielectric permitivity of ferroelectrics makes them possible candidates to replace silica as the gate dielectric in metal-oxide-semiconductor field effect transistors (MOSFETs). Their piezoelectric behavior enables them to convert mechanical energy in electrical energy and vice versa. Their pyroelectric properties are the basis for highly sensitive infrared room temperature detectors.

The current miniaturization of microelectronic technology, imposed by the semiconductor industry, raises the question of possible size effects on the properties of the components. Except for the case of the gate dielectric problem in MOSFET's, the thickness of the films used in contemporary applications is still far away from the thickness range where size effects become a concern; therefore the question at the moment is merely academic. Nevertheless, it is very possible that the fundamental limits of materials might be reached in the future.

Despite the efforts and advancement in the field, both theoretically and experimentally (see a recent review in Ref.[1]), many questions still remain open. The main reason for the poor understanding of some of the size effects on ferroelectricity is the vast amount of different effects that compete and might modify the delicate balance between long range dipole-dipole electrostatic interactions and the short range forces, whose subtle equilibrium is known to be at the origin of the ferroelectric instability.

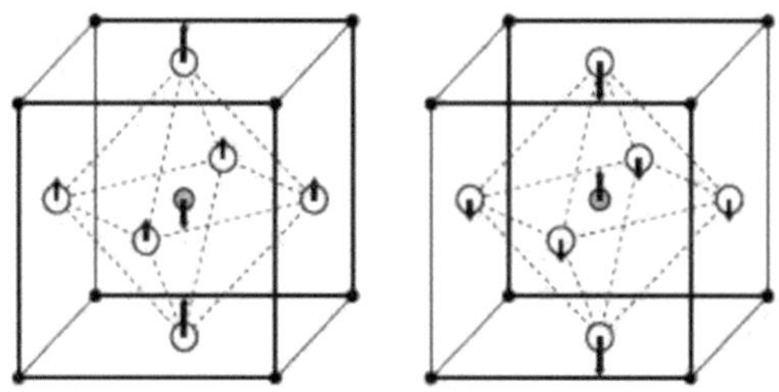

Figure 1. Tetragonal ferroelectric structure of $BaTiO_3$. Solid, shaded and empty circles represent Ba, Ti, O atoms, respectively. The arrows indicate the atomic displacements, exaggerated for clarity. The origin has been kept at the Ba site. Two structures, with polarization along [001], are shown. Application of a sufficiently large electric field causes the system to switch between the two states, reversing the polarization. in principle, we can assign one value of the Boolean algebra ("1" or "0") to each of the polarization states.

The next section will give an overview of the main concepts involved in the modern theory of polarization. A definition of a local polarization will be given in terms of the centers of the wannier functions associated with the band structure of the system. Next, some basic electrostatic notions related with ferroelectric films will be given, in particular the concept of depolarization field and screening by metal contacts. Finally, the results of this research work: applying the layer polarization concept, where we show the hidden structure of the polarization at the nanoscale.

2. Modern theory of polarization

Classically, the macroscopic polarization in dielectric media is defined to be an intensive quantity that quantifies the electric dipole moment per unit volume [2–4]. Definitions along these lines work well for finite systems but have important conceptual problems when applied to periodic crystalline systems because there is no unique choice of cell boundaries [5].

In the early 1990s a new viewpoint emerged and lead to the development of a microscopic theory [6–8]. It starts by recognizing that the bulk macroscopic polarization cannot be determined, not even in principle, from a knowledge of the periodic charge distribution of the polarized crystalline material. This establishes a fundamental difference between finite systems (e.g., molecules, clusters, etc.) and infinite periodic ones. For the first case, the dipole moment can be easily expressed in terms of the charge distribution. While for periodic systems one focuses on differences in polarization between two states of the crystal that can be connected by an adiabatic switching process [6]. The polarization difference is then equal to the integrated transient macroscopic current that flows through the sample during the switching process.

Therefore, the macroscopic polarization of an extended system is, according to the modern viewpoint, a dynamical property of the current in the adiabatic limit. The charge density is a property connected with the square modulus of the wavefunction, while the current also has a dependence on the phase. Indeed, it turns out that in the modern theory of polarization [7–9], the polarization difference is related to the Berry phase [10, 11], defined over the manifold of Bloch orbitals. The theory not only defines what polarization really is,

http://dx.doi.org/10.5772/52268

but also proposes a powerful algorithm for computing macroscopic polarizations from first principles.

The modern theory can be equivalently reformulated using localized Wannier functions [12–15] instead of extended Bloch orbitals. The electronic contribution to the macroscopic polarization **P** is then expressed in terms of the dipole of the Wannier charge distribution associated with one unit cell. In this way, **P** is reformulated as a property of a localized charge distribution which goes back to the classic definition of polarization. However, one has to bear in mind that the phases of the Bloch orbitals are essential for building the Wannier functions. They are needed to specify how the periodic charge distribution should be decomposed into localized ones. The knowledge of the periodic charge distribution of the polarized dielectric is not enough to determine the wannier functions.

The main concepts

Here I will review the central topics uncovered in the early 1990s, often known as the modern theory of polarization. Understanding this background is necessary in order follow this chapter. The general idea is to consider the change in polarization of a crystal as it undergoes a slow change and relate this to the current that flows during this adiabatic evolution of the system. These ideas will lead to an expression for the polarization that does not take the form of an expectation value of an operator, but takes the form of a Berry phase, which is a geometrical phase property of a closed manifold (the Brillouin zone) on which a set of vectors, the occupied Bloch states, are defined.

As a classical example of a Berry phase, lett's think of the paralel transport of a vector along a loop on a shpere (for instance a compass needle carried in a car traveling on the surface of the Earth). After completing a closed path, or loop, the vector will be back at the original starting point but it will be rotated with respect to the direction it was pointing at when it started the trip. The reason for this rotation is purely geometrical topological and intrinsicaly connected to the curvature of the sphere and would not exist if the the vector would be parallel transported along a flat manyfold, like a plane, or cylinder. This rotation angle is related to the integral of the curvature on the surface bounded by the loop.

Now lets see the quantum counterpart and its connection to the definition of Polarization in a solid. Let's assume that the crystal Hamiltonian H_λ depends smoothly on parameter λ and has Bloch eigenvectors obeying $H_\lambda|\psi_{\lambda,n\mathbf{k}}\rangle = E_{\lambda,n\mathbf{k}}|\psi_{\lambda,n\mathbf{k}}\rangle$ and that λ changes slowly with time, so that it is correct to use the adiabatic approximation.

Since the spatially averaged current density is just $\mathbf{j} = \frac{d\mathbf{P}}{dt} = \left(\frac{d\mathbf{P}}{d\lambda}\right)\left(\frac{d\lambda}{dt}\right)$ we can write the change in polarization during some time interval as

$$\triangle\mathbf{P} = \int \mathbf{j}(t)dt \qquad (1)$$

and is phrased in terms of the current density that is physically flowing through the crystal as the systems traverses some adiabatic path. It can also be written in terms of the parameter λ as:

$$\triangle \mathbf{P} = \int \frac{d\mathbf{P}}{d\lambda} d\lambda. \tag{2}$$

Then, from Ref.[7]

$$\frac{d\mathbf{P}}{d\lambda} = \frac{ie}{(2\pi)^3} \sum_n \int_{BZ} d\mathbf{k} \langle \nabla_{\mathbf{k}} u_{n\mathbf{k}} | \frac{du_{n\mathbf{k}}}{d\lambda} \rangle + c.c. \tag{3}$$

and the rate of change of polarization with λ is a property of the occupied bands only. This expression can be integrated with respect to λ to obtain

$$\mathbf{P}(\lambda) = \frac{ie}{(2\pi)^3} \sum_n \int_{BZ} d\mathbf{k} \langle u_{\lambda,n\mathbf{k}} | \nabla_{\mathbf{k}} | u_{\lambda,n\mathbf{k}} \rangle. \tag{4}$$

It can be verified by taking the λ derivative of both sides of Eq.4 and comparing with Eq.3. The result is independent of the particular path of $\lambda(t)$ in time, and depends only on the final value of λ as long the change is adiabatically slow. We can associate the physical polarization of state λ with $\mathbf{P}(\lambda)$ and drop the λ label. Eq. 4 can be written then as

$$\mathbf{P} = \frac{e}{(2\pi)^3} Im \sum_n \int_{BZ} d\mathbf{k} \langle u_{n\mathbf{k}} | \nabla_{\mathbf{k}} | u_{n\mathbf{k}} \rangle, \tag{5}$$

and this is the electronic contribution to the polarization. To obtain the total polarization it must be added the nuclear (or ionic) contribution

$$\mathbf{P}_{ion} = \frac{1}{\Omega} \sum_s Z_s \mathbf{r}_s,$$

where the sum is over atoms s having core charge Z_s and spatial position $\mathbf{r}_s$ in the unit cell of volume Ω.

The equation 5 is the main result of the modern theory of polarization It states that the electronic contribution to the polarization of crystalline insulator may be expressed as a Brillouin zone integral of an operator $i\nabla_{\mathbf{k}}$. However, it is not a common quantum mechanical operator, the result of its action on the wavefunctions $(i\nabla_{\mathbf{k}}|u_{n\mathbf{k}}\rangle)$ depends on the relative phases of the Bloch functions at different $\mathbf{k}$.

To understand the nature of the integrand in Eq.5 I want to recall the fundamental paper by Zak [16] in which he postulates the existence of a Berry phase associated with each one of the bands of a one dimensional crystalline solid. He shows how the variation of k over the entire Brillouin zone produces the appearance of a Berry phase. Moreover, he associates this phase, that is a characteristic feature of the whole band, with the band center operator [17].

http://dx.doi.org/10.5772/52268

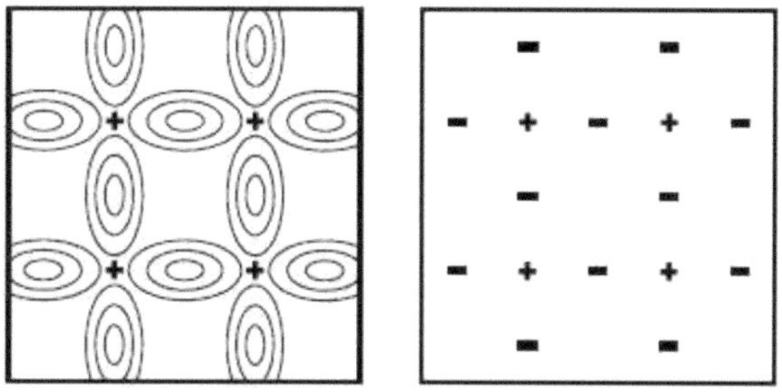

Figure 2. From Ref.[18] Mapping of the distributed charge density onto the centers of charge of the Wannier functions.

Its eigenvalues turn out to give the average positions of an electron in different bands which coincide with the symmetry centers of the space group of the solid [17] These previous ideas set the stage for the crucial meaning of the expression in Eq.5.

In three dimensions, the Brillouin zone can be regarded as a closed 3-torus obtained by the identifying points $u_{n\mathbf{k}} = u_{n,\mathbf{k}+\mathbf{G}_j}$ where $\mathbf{G}_j$ are the three primitive reciprocal lattice vectors. Then, the electronic polarization contribution of each band can be written as

$$\mathbf{P}_n = \frac{-e}{2\pi\Omega}\sum_j \Phi_{nj}\mathbf{R}_j \tag{6}$$

where $\mathbf{R}_j$ is the real space primitive translational corresponding to $\mathbf{G}_j$, and the Berry phase for band n in direction j is

$$\Phi_{nj} = -\frac{\Omega}{(2\pi)^3} Im \int_{BZ} d^3k \langle u_{n\mathbf{k}}|\mathbf{G}_j.\nabla_{\mathbf{k}}|u_{n\mathbf{k}}\rangle. \tag{7}$$

More details and discussion, including the equivalent of Eq.7 for the case of connected multiple bands, may be found in Refs.[7] and [8].

Reformulation in terms of Wannier Functions.

We can rewrite Eq.5 in terms of the Wannier functions (WF's), which brings an alternative and more intuitive way of thinking about it. The WF's and Bloch functions can be regarded as two different orthonormal representations of the same occupied Hilbert space. The WF's are localized functions $w_{n\mathbf{R}}(\mathbf{r})$ that are labeled by band n and unit cell $\mathbf{R}$ and constructed by carrying out a unitary transformation of the Bloch states $\psi_{n\mathbf{k}}$. They are constructed via a Fourier transform of the form

$$|w_{n\mathbf{R}}\rangle = \frac{\Omega}{(2\pi)^3}\int_{BZ} d\mathbf{k} e^{i\mathbf{k}.\mathbf{R}}|\psi_{n\mathbf{k}}\rangle \tag{8}$$

where the Bloch states are normalized in one unit cell. There is some freedom in the choice of the these WF's, as the transformation is not unique. In particular, a set of Bloch functions

$$|\psi_{n\mathbf{k}'}\rangle = e^{-i\beta_n(\mathbf{k})}|\psi_{n\mathbf{k}}\rangle \tag{9}$$

results in WF's $w_{n\mathbf{R}'}$ which are different from the $w_{n\mathbf{R}}$. Usually, this "gauge" is set by some criterion that keeps the WF's well localized in real space, such as the minimum quadratic spread criterion introduced by Marzari and Vanderbilt [19]. However, it is expected that any physical quantity, such as the electronic polarization arising from band n, should be invariant with respect to the phase change $\beta_n(\mathbf{k})$.

Once we have obtained the WF's, we can locate the "wannier centers" $\mathbf{r}_{n\mathbf{R}} = \langle w_{n\mathbf{R}}|\mathbf{r}|w_{n\mathbf{R}}\rangle$. It can be shown that

$$\mathbf{r}_{n\mathbf{R}} = \mathbf{R} + \sum_j \frac{\phi_{nj}}{2\pi}\mathbf{R}_j, \tag{10}$$

where ϕ_{nj} is given by Eq.7. In simple words, the location of the n'th Wannier center in the unit cell is just given by the three Berry phases ϕ_{nj} of band n in the primitive lattice vector directions $\mathbf{R}_j$. The key result, is that the polarization is just related to the Wannier centers by

$$\mathbf{P} = -\frac{e}{\Omega}\sum_n \mathbf{r}_{nO}. \tag{11}$$

The Berry Phase theory can then be regarded as providing a mapping of the distributed quantum mechanical electronic charge density onto a lattice of negative point charges $-e$ (see Fig.2).

3. Layer polarization definition

One issue that has received much attention theoretically is how to quantify the concept of *local polarization*. This can be very useful in understanding the enhancement or suppression of spontaneous polarization; or even to find new spatial patterns of polarization that would remain hidden otherwise, as is the case that I will show in a following section. It can also be essential for characterizing and understanding interface contributions to such properties.

But first I will give a review of the concept introduced in 2006 by Wu.*et al.* [20] which is the one used here. As explained in the previous section, the modern theory of polarization establishes that the polarization of a crystal is expressed in the Wannier representation as the contribution of the ionic and electronic charge:

$$\mathbf{P} = \frac{1}{\Omega}\sum_s Q_s\mathbf{R}_s - \frac{2e}{\Omega}\sum_m \mathbf{r}_m, \tag{12}$$

http://dx.doi.org/10.5772/52268

where s and m run over ion cores (of charge Q_s located at $\mathbf{R}_s$) and Wannier centers (of charge -2e located at $\mathbf{r}_m$), respectively, in the unit cell of volume Ω.

They defined the Layer polarization (LP) along z for superlattices built from II-VI ABO_3 perovskites such as $BaTiO_3$, $SrTiO_3$, and $PbTiO_3$. For these cases, they were able to decompose the system into neutral layers (that is AO and BO_2 subunits) and define a layer polarization

$$p_j = \frac{1}{S} \sum_{s \in j} Q_s \mathbf{R}_{sz} - \frac{2e}{\Omega} \sum_{m \in j} z_m, \tag{13}$$

in which the sums are restricted to entities belonging to layer j. S is the basal cell area and we are now focusing only on z components. The LP p_j thus defined has units of dipole moment per unit area. The total polarization, with units of dipole moment per volume, is exactly related to the sum of LP's via

$$P_z = \frac{1}{c} \sum_j p_j \tag{14}$$

where $c = \frac{\Omega}{S}$ is the supercell lattice constant along z.

They propose two conditions to be able to associate a physical meaning to the LP definition: (i) resolve the arbitrariness associated with the positions of the Wannier centers, and (ii) associate without ambiguity the right number of Wannier centers to each layer.

As it will be shown in the next section, we applied this concept to thin ferroelectric films sandwiched between metal contacts. These kind of systems, in which metals and oxides coexist, bring new challenges. Condition (i) was satisfied using maximally localized wannier functions plus a disentanglement procedure (Ref.[21]) and (ii) was satisfied, even close to the metal oxide interfaces.

4. Depolarization field

When dealing with ferroelectric thin films, a finite polarization normal to the surface will give rise to a depolarizing field and a huge amount of electrostatic energy, enough to be able to suppress the ferroelectric instability. In order to preserve ferroelectricity and minimize the total energy, the depolarization field must be screened, either by free charges coming from metallic electrodes or by breaking into a domain structure. I will go into more details about this, but first a basic electrostatic background will be given.

Influence of the electrical boundary conditions

In the case of a free standing slab of a ferroelectric material, a uniform polarization P with an out of plane direction will appear and originate a surface polarization charge density

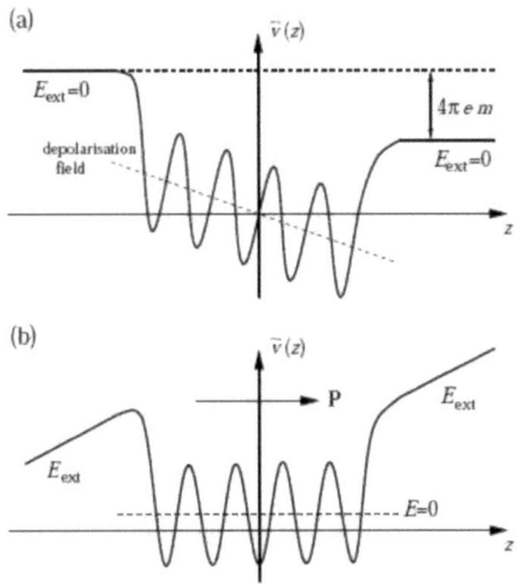

Figure 3. Schematic representation of the planar averaged electrostatic potential (fill line) of a slab with polarization perpendicular to the surface under (a)$D = 0$ boundary conditions (vanishing external field) and b) a vanishing internal electric field $E = 0$. Plannar averages are taken on the (x, y) planes parallel to the surface. Dashed lines represent an average over unit cell of the planar averages. An estimate of the macroscopic internal field inside the slab can be obtained from their slope. m stands for the dipole moment parallel to the surface normal and e for the electron charge. From Ref.[23]

$$\sigma_{pol} = \mathbf{P}.\hat{n}, \tag{15}$$

where $\hat{n}$ is a unit vector normal to the surface pointing outward. The charge density is proportional to the magnitude of the normal component of the polarization inside the material [22] and is positive one one side of the slab and negative on the other one. In order the determine the electric fields generated by the surface charge density, we need to know the electrical boundary conditions of the problem and solve the Poisson equation. For the general case, let's suppose that an external electric field E_{ext}, perpendicular to the surface, is applied in the vacuum region. There, the electric displacement $\mathbf{D} = \mathbf{E} + 4\pi\mathbf{P}$ equals the applied electrical field $D = E_{ext}$ as the polarization is zero. The normal component of D is conserved across the ferroelectric/vacuum interface,

$$E + 4\pi P = E_{ext} \tag{16}$$

Now we can see two extreme cases. One of them is that of zero external electric field, $E_{ext} = 0$, in which case the displacement vector is null while the internal field inside the slab is $E = -4\pi P$ (see Fig.3a). The absence of an external electric field and the presence of the surface polarization charges produce an internal field that points in the direction opposite to that of the polarization. Therefore, it tends to restore the paraelectric configuration and that is the reason of its name, *Depolarization field*. The coupling between the polarization and this internal field is so big, that any atomic relaxation under these circumstances will end up with the atoms back in the centrosymmetric non polar configuration.

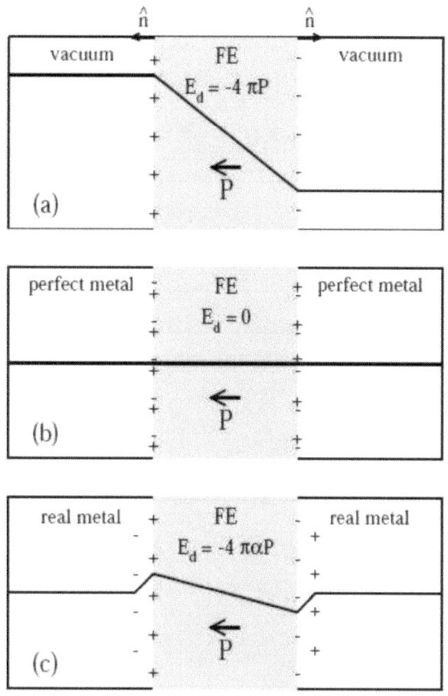

Figure 4. Depolarization field for an isolated free standing slab in the absence of an external electric field (a), and for a ferroelectric capacitor with perfect screening (ideal metal) (b) and a real metallic electrode (c) under short circuit boundary conditions. From Ref.[1]

Another case, a vanishing internal electric field $E = 0$ corresponds to a case where no field opposes the polarization and a spontaneous polarization might arise (Fig.3b). In this case, it is the continuity of the normal component electric displacement vector that establishes the value of it,

$$E_{ext} = 4\pi P_s. \tag{17}$$

These two cases illustrate how a polarization perpendicular to the surfaces can exist or not depending on the electrical boundary conditions. For simulating a real system, the mechanical boundary conditions should be included via atomic relaxations. These cases have been extensively analyzed [23–26] and have opened the ground for more realistic systems..

Metal contacts

The next level of approximation to a real case,is the case of having the ferroelectric thin film sandwiched between metal contacts (short circuit boundary conditions). In the case of a *perfect* metal, the surface charges σ_{pol} will be exactly compensated at the interfaces by the compensation charges σ_{com} and we will have a null depolarization field (see Fig.4b). In this case, both $\sigma's$ will lie in a sheet right at the interface between the metal and the ferroelectric electrode. However, in the case of *real electrodes*, the compensation of the polarization charges will be incomplete. In this case,the compensation charge σ_{com} will be spread over a finite distance inside the electrode and will create an interface dipole density at each of the ferroelectric/metal interfaces (see Fig.4c). Due to the short circuit boundary conditions, an amount of charge must flow from one interface to the other, creating a residual depolarization field inside the thin film. There are various models for expressing this field [24, 27–29], shown below is one described in Ref.[30]:

$$E_{dep} = -\frac{8\pi P\lambda}{d} \tag{18}$$

where d is the film length, and λ is the effective screening length, proportional to the spatial spread of the interface dipole.

Electrostatic energy of the thin film

In the presence of a residual depolarizing field the energy of the thin film U can be approximated by

$$E(P_0) = U(P_0) + E_{elec}(P_0), \tag{19}$$

where U is the internal energy under zero field. If we assume that the interface effects are hidden in the screening length parameter, then they only appear through the depolarization field. This is a rough approximation but we will introduce later a model that will relax this condition, but let's keep it for now. Following this, then

$$U(P_0) = mU_{FE}(P_0) \tag{20}$$

where m stands for the number of unit cells of the film. U_{FE} can be calculated from bulk DFT calculations. Its dependence on P_0 has typically the shape of a double well.

The electrostatic energy $E_{elec}(P_0)$ can be approximated by [1, 31]

$$E_{elec}(P_0) = m\Omega\left(-E_{dep}.P_0\right) + O\left(E_{dep}\right)^2 \tag{21}$$

The depolarization field can be calculated from first principles by taking the electrostatic average [32, 33] or can be expressed using Eq.18. In this case, Eq.19 can be rewritten as

$$E(P_0) = U(P_0) + \frac{8\pi\lambda P_0^2}{d} + O\left(P_0^4\right). \tag{22}$$

The electrostatic energy, the second term, is positive, meaning that the effect of the depolarization field is to suppress the ferroelectric instability by rescaling the quadratic term of the double well. This simple model has been implemented [34, 35] in order to extend the

http://dx.doi.org/10.5772/52268

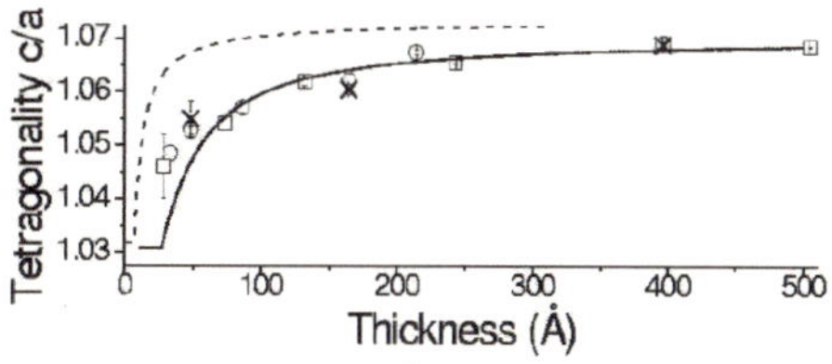

Figure 5. Evolution of the c/a ratio with the film thickness for monodomain $PbTiO_3$ films grown epitaxially on top of Nb-doped $SrTiO_3$. With circles and squares, the experimental results. The dashed line is the phenomenological theory prediction. Solid lines correspond to the first principles model Hamiltonian results. From Ref. [35].

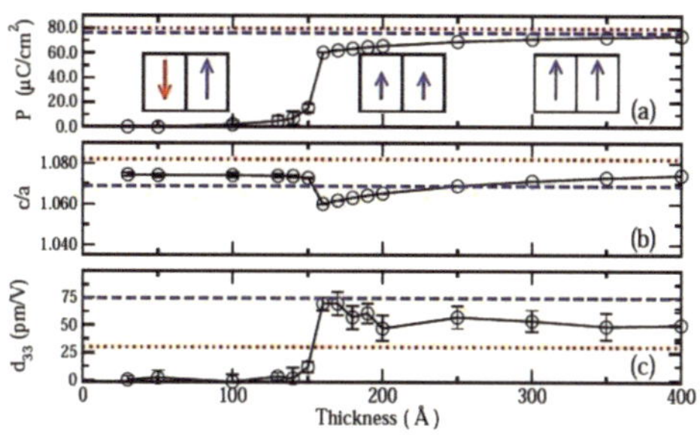

Figure 6. Theoretical predictions of the thickness dependence of the normal average polarization P, the tetragonality c/a, and the out of plane piezoelectric constant d_{33} at room temperature for $PbTiO_3$ thin films grown on a $SrRuO_3/SrTiO_3$ substrate. Values of the quantities at the bulk level are represented by dashed lines for the unstrained configuration, and with dotted lines for a geometry under the strain imposed by the substrate. The evolution of the system, from a monodomain configuration at large thickness, (where the depolarization field E_d is small), to a 180o stripe domain structure in order to minimize the energy associated with E_d, is represented in the inset. From [37]

first principle model Hamiltonian to thin films. It also has been used with the result of first principle calculations in Ref.[30]. In this work, we will use it as a starting point for a toy model that would mimic the result of our first principle calculations.

4.1. Effects of the depolarization field

There are two ways to minimize the energy due to the depolarization field , i) a reduction of the polarization, ii) a reduction of the depolarization field itself. This can be seen by inspecting the first term in Eq.21.

As an example of case i) it is shown in Fig.5 the experimental results for the tetragonality in function of the film thickness for monodomain $PbTiO_3$ epitaxially grown on Nb-doped $SrTiO_3$ [35]. The polarization is strongly coupled to strain in perovskites oxides [36], therefore a reduction of the polarization must be accompanied by a reduction of the tetragonality of the system. The incomplete screening of the depolarization field is the driving force for the reduction of the polarization. The structure remains in a monodomain at all thickness.

Case ii) might arise when there are no compensation charges provided by the electrodes or when the compensation charges do not provide an efficient enough screening of the polarization charges. In this case the system might break up into 180o stripe domains to reduce the magnitude of the surface dipole density (see Fig.6).

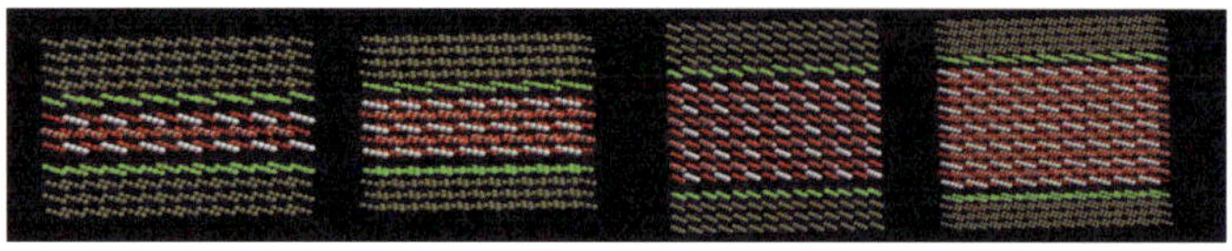

Figure 7. We simulated ferroelectric films at different thicknesses, sandwiched between metal contacts.

The reason why some systems remain in a monodomain configuration while other similar heterostructure break up into domains remains an open question that requires further clarification.

In a seminal theoretical paper, Junquera *et al.*[30] has suggested that the appearance of ferroelectricity in thin films of $BaTiO_3$, with $SrRuO_3$ contacts, should be limited to a thicknesses of more than six unit cells. Below that limit, ferroelectricity would effectively be canceled by the strength of the depolarization field (DF) inside the oxide produced by unscreened charges at the interfaces, which, in that work, were calculated for the first time in a first principles framework. More recently this picture has been broadened, and a variety of studies have shown both experimentally and with more realistic ab initio simulations that thin oxide films can indeed maintain some ferroelectricity at thicknesses even smaller than six unit cells.[37–40]

5. The hidden nature of ferroelectricity at the nanoscale

The size limit of ferroelectricity in ultrathin films can be understood using a simple electrostatics argument: as the thickness of the oxide film is reduced, the intensity of the DF is made stronger,[30] thus increasing the electrostatic energy of the system. In a ferroelectric crystal, this energy contribution can be minimized in two ways: either by breaking the polarization pattern in 180o stripe domains [41–44] to reduce the magnitude of the surface dipole density, or by a reduction of the ionic polarization while the system remains in a monodomain state.[35, 45] The final polarization pattern depends on the individual material and/or structure, and understanding which system modification will occur under what conditions is still a matter of debate.

At small dimensions, the spatial confinement and the interface details become extremely relevant since they determine the spatial localization of the electrons and thus influence directly the ferroelectric characteristics and, more in general, all the electronic properties of the system.[46] Therefore, to clarify the behavior of ferroelectric thin films in the nanometer regime it is crucial to obtain a precise description not only of the geometry and electronic properties of the film but first and foremost of the polarization profile inside the oxide.

In order to obtain a complete description of ferroelectricity at the nanoscale we have exploited the notion of layer polarization (LP). This idea, based on modern theory of polarization and the concept of maximally localized Wannier functions (see Sec.2), was recently introduced to describe the layer-by-layer modulation of polarization in ferroelectric superlattices (Sec.3) .

We have applied this method to a model metal-ferroelectric system ($BaTiO_3$/Pt) (see Fig.7) and obtained a detailed spatial profile of the polarization in the oxide region in the direction of growth that accounts not only for the ionic displacements but also for the spatial rearrangement of the electron density. It is important to note that using more standard methods to evaluate macroscopic polarization, this information would have remained hidden

http://dx.doi.org/10.5772/52268

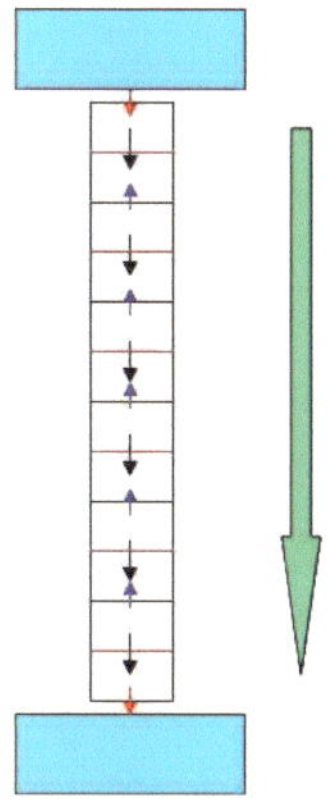

Figure 8. The thin film oxide develops a ferrielectric [47] pattern of polarization that exhibits antiferroelectric properties along the growth axis and with a net total polarization. This pattern would remain hidden by using more standard methods for calculating the local polarization.

by the averaging procedure, or would have produced misleading results. In fact, by considering exclusively the ionic displacements (rumpling) as a measure of polarization, we do not account for the full electronic contributions to the local polarization and, depending on the system size, we might obtain results that would not describe accurately the physical behavior of the thin film.

Here, we show that the ferroelectric structures associated with small length-scales are more complex than previously thought, mainly due to the redistribution of the electrons under the constraints imposed by the interfaces. In particular, the oxide develops a ferrielectric [47] pattern of polarization that exhibit antiferroelectric properties along the growth axis and with a net total polarization (see Fig.8).

This unbalanced dipole structure is particularly evident in Fig.10A, where we show how positive rumpling on all the atomic layers of a thin $BaTiO_3$ film correspond actually to layer dipoles of opposite signs. This is a consequence of the interplay between the orientation and magnitude of the dipoles with the DF, their mutual interaction and the nature of the interfaces. A simple analytic model of a ferroelectric thin film, where these effects are explicitly taken into account, can capture all these features and it will be discussed in detail at the end.

We have simulated thin films of $BaTiO_3$ between Pt metal contacts in a (001) stack, where the oxide is terminated with a BaO plane at both interfaces. In this configuration, the metal atoms are directly bonded to the O atoms of the oxide plane.[30, 48] The different supercell constructed in this way can be labeled as Pt/(BaO-TiO2)m-BaO/Pt with m=1,2,4,6. Nine atomic planes of Pt have been found to be sufficient to simulate the contacts under short-circuit boundary conditions. The in plane lattice constant of the supercell is set equal to that of $BaTiO_3$ at 0 K (3.991)[38], and kept it fixed in all calculations while all other geometrical parameters (atomic positions and intra-layer distances) were fully relaxed. All simulations have been performed using Density Functional Theory (DFT) within the Local Density Approximation, with ultrasoft pseudopotentials and a plane waves basis set.

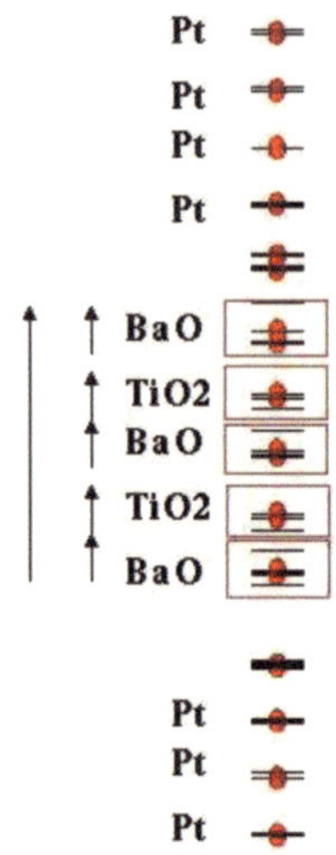

Figure 9. The set of MLWFs so obtained cluster around each atomic layer in sets containing the correct number of Wannier function needed to keep the layer charge neutral. Here we can see the z position (horizontal lines lines) of the centers of the Wannier functions and the position of the ions (indicated with circles). To calculate the LP of each layer, we use Eq.12.

We used the expression of Eq.12 to calculate the LP for each plane of the thin film. A is the in plane area of the cell, and we are only interested on the z components (parallel to the growth direction). It is important to stress that in the definition of the LP we have explicitly included the full ionic and electronic information necessary to define unambiguously the local value of the polarization. The essential step for the evaluation of the layer polarizations p_j is the determination of the Wannier centers z_m. We used the maximally localized Wannier functions (MLWFs) algorithm originally proposed by Marzari and Vanderbilt[19] as implemented in the WanT code[1]. Since the valence bands of the oxide and the bands of the metal are mixed in the full supercell calculation, a disentanglement procedure (Ref.[21]) was applied before starting the localization algorithm[19, 49].

The set of MLWFs so obtained cluster around each atomic layer in sets containing the correct number of Wannier function needed to keep the layer charge neutral (Fig.9). This is a crucial condition for the validity of the definition of LP's in Eq.12 and might not be attainable in systems with more covalent nature.[20] In fact, the meaningful use of this concept is based on the condition of being able to decompose the system into neutral layers (TiO and BaO2 units in the case of perovskite $BaTiO_3$) and resolve the arbitrariness associated with the positions of the Wannier centers. This allows us to associate each sets of Wannier centers to the corresponding crystal plane without ambiguity. By doing so we go back to the classic idea of defining dipoles in spatial neutral units, along the lines of the Clausius-Mossotti limit.v Moreover, we can associate each LP value p_j with its correspondent (dimensionally correct) dipole p_jA, which gives precise information about the orientation and magnitude of the atomic layer dipoles inside the crystal.

[1] code by A. Ferretti, B. Bonferroni, A. Calzolari, and M. Buongiorno Nardelli, (http://www.wannier-transport.org). In the non-self-consistent calculation for obtaining the MLWFs we used MP k meshes 2x2xN with N=16,14,12,10 for m=1,2,4,6 respectively. I. Souza, N. Marzari, and D. Vanderbilt, Phys. Rev. B 65, 035109 (2001). ote

The LP profile calculated for the metal-oxide systems provides a detailed local description of the polarization, which is superior and even contradictory to the information obtained from empirical evaluations based solely on ionic displacements. As an illustration, in Fig.10A we display the rumpling profile (cation-oxygen normal distance within each atomic layer.) for a thin film comprised of six $BaTiO_3$ unit cells (m6). From the ionic displacement profile one would be tempted to conclude that the film is in a ferroelectric domain configuration, since the rumpling values (and hence the *classical* local dipoles) associated to each plane are positive. Instead, if the correspondent LP profile is analyzed, a completely different picture emerges, where the main feature is an alternation of positive and negative values of the LP in the individual atomic layers of the oxide film. This is clearly shown in Fig.10B, where we plot the LPs values computed for m6. Here the central BaO planes (the BaO planes at the interface behave differently and will be discussed later) have positive dipoles while the TiO2 planes have negative ones. The first are aligned with the direction of the DF, while latter oppose it. A schematic illustration of the associated dipoles is drawn above the plot. This result implies that what could have been interpreted as a ferroelectric domain using standard geometrical information, in fact displays a much more complex spatial pattern of polarization. We believe that this effect is comparable to the onset of the formation of two-dimensional domain islands commonly observed both theoretically and experimentally in ferroelectric films.[37] In that case, the domains are composed of adjacent 180o domains in order to minimize the electrostatic energy associated with the DF. Here, our results show that the system can reduce its electrostatic energy in an alternative way: the energetic interplay between the dipoles, the DF and the interface bonding drives the individual atomic layer dipoles to arrange themselves in an uncompensated dipole pattern along the growth direction (see inset of Fig.10A), or, in other words, the system undergoes a ferroelectric-to-ferrielectric phase transition. Similarly to the formation of two-dimensional islands, we also predict the existence of a critical thickness (CT) above which the system exists in a true ferroelectric domain with all the dipoles pointing in the same direction (see Fig8B, where the LP's values for bulk $BaTiO_3$ are shown with triangles).

As the thickness is reduced, the magnitude of all the dipoles is diminished by the increase of the DF that opposes them. Due to the different physical properties of consecutive layers (BaO and TiO2), their associated dipoles reduce their magnitude at different rates. At the CT, the BaO layers will have null dipoles and eventually they will flip layer polarization direction, aligning with the DF in order to minimize the associated internal energy and creating a ferrielectric dipole pattern (FDP).

The formation and characteristics of the FDP are determined by the bonding properties of the interfaces that pin the LP values of the outer layers of the oxide film. At small film thicknesses the interface effects become dominant and induce the overall spatial variation of the LP's that determines the orientation and magnitude of the layer dipoles along the structure. We can see how the FDP observed in the m6 geometry is increasingly influenced by the locality of the interfaces as the thickness is reduced to m4, m2 and m1 (see Fig.10C,D,E respectively). The middle crystal planes become more and more influenced by the interface environment, while the FDP becomes irregular, and finally disappear when the film is comprised by just 3 layers (i.e. one unit cell, m1), leaving instead a centro-symmetric paraelectric structure (Fig.10E). It is important to note that exclusively electronic effects drive this series of regime transitions. In fact, the ionic polarization, directly proportional to the ionic rumpling, remains almost

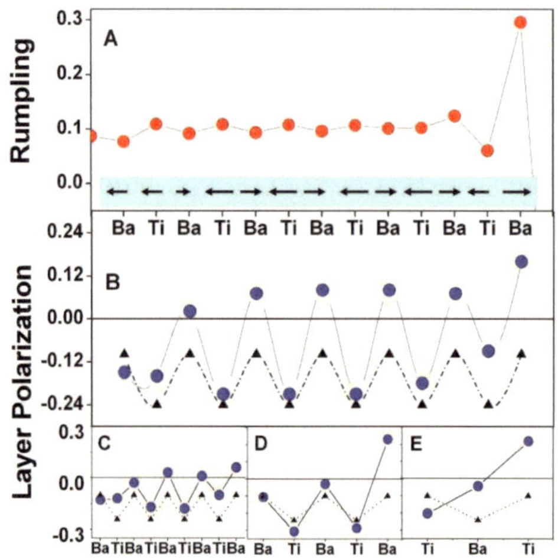

Figure 10. A. Rumpling (cation-oxygen perpendicular distance (in Å) within each atomic layer) profile for six unit cells (m6) of $BaTiO_3$ sandwiched between Pt contacts. (on the horizontal axis Ti and Ba indicate the position of the individual TiO2 and BaO planes). B. Solid circles (blue on line): layer polarization (LP) profile for the same system m6 in units of $10^-10C/m$. The values for the equivalent planes in bulk $BaTiO_3$ are shown with black triangles. The orientation of the individual layer dipoles is displayed by arrows in the bottom of the panel. C.D.E. LP profiles for m4, m2 and m1. As the thickness is reduced the FDP is modified by the increasingly dominant interface effects.

constant for all the thicknesses considered, and it is only the electronic polarization that changes its values in the different systems.

These results show that the ferroelectric response of a thin film is indeed critically influenced by the interface properties of the system and that the analysis of the local variation of layer polarization captures completely the physical characteristics of the ferroelectric. As the size of the film is increased, the DF decreases and the dipoles that are oriented along the DF direction get smaller, while the interface effects lose relevance. At the critical size, the dipoles flip direction and a ferroelectric domain structure is established again. Eventually, the dipoles will relax to the bulk structure, showed with triangles in Fig.10

Model

We further illustrate these physical principles by developing a simple classical model of a ferroelectric thin film that reproduces the appearance of a FDP. We extended the model introduced in Ref. ([30]) by discretizing the thin film along the growth direction in each of its component unit cells in order to introduce a local spatial description of the polarization with the introduction of the *local polarization*, P_i. Each individual crystal unit cell is comprised of two consecutive layers of BaO and TiO2, therefore P_i could be obtained by adding each individual LP and dividing by the lattice constant in the stacking direction, c. The full polarization profile of the film is completely specified by the full set of LPs $(p_1, p_2, ..., p_{2M-1}, p_{2M})$ associated with each crystal plane of the oxide film (see Fig.11). We can write the total energy of the thin film composed by M unit cells as:

http://dx.doi.org/10.5772/52268

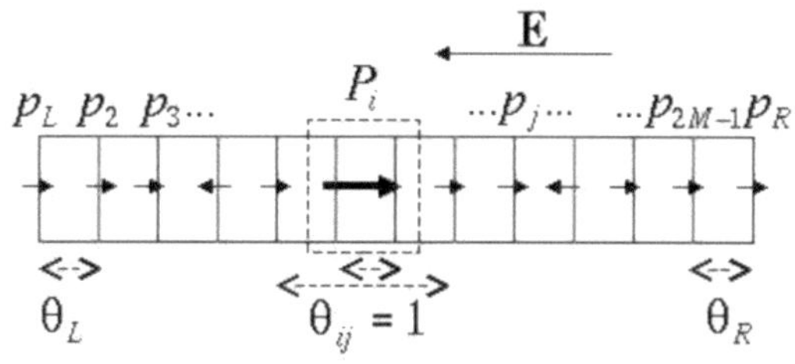

Figure 11. Each individual crystal unit cell is comprised of two consecutive layers of BaO and TiO2. A *local polarization* P_i can be associated with it. It is calculated by adding both individual LP's and dividing by the lattice constant in the stacking direction, c. We can associate a dipole $p_i A$ to each plane i, separated from its next neighbors by distance d_{ij}. The full polarization profile of the film is completely specified by the full set of dipoles $(p_L, p_2, ..., p_{2M-1}, p_R)$. The LP at the interfaces (p_L and p_R), are kept fixed. Two distinct interaction parameters θ_L and θ_R, establish the locality of the interface environment.[50] The interaction parameter θ_{ij} for the internal dipoles is arbitrarily chosen to be unity.

$$E = \sum_{i=1}^{M}(-aP_i^2 + bP_i^4) + \sum_{i=1}^{M}(-\Omega \mathbf{P}_i.\mathbf{E}) + U_{int}(p_L, p_2, ..., p_{2M-1}, p_R), \tag{23}$$

where each term of the first sum accounts for the internal energy of the individual unit cells under zero electric field. The parameters a and b are obtained from an ab initio calculation of the total energy of a $BaTiO_3$ bulk for different ionic displacements along the soft mode.[51] The contribution from the depolarization field E is included in the second sum [22, 30] where Ω is the volume of the unit cell. Each one of its terms favors energetically the local dipoles that follow the same direction than the DF. The third sum represents the classical electrostatic energy for a series of dipoles separated by a distance d_{ij},

$$U_{int} = -\frac{A^2}{8\pi\epsilon}\sum_{i,j=1}^{2M}\frac{\theta_{ij}p_j p_i}{d_{ij}^3} \tag{24}$$

The above sum can be separated in three contributions: two that account for the interaction between the interface dipoles with the internal ones, and another one that accounts for the mutual electrostatic interaction between the internal dipoles. We have chosen to assign a fixed magnitude and direction to the interface layer dipoles (p_L and p_R), with two distinct interaction parameters θ_L and θ_R, thus establishing the locality of the interface environment.[50] The interaction parameter θ_{ij} for the internal dipoles is arbitrarily chosen to be unity.

With the constraint imposed by the simplicity of the model, we look for the energetically more favorable spatial polarization profile as a function of the film thickness. The spatial profile of polarization that characterize the film can be expressed as the vector $(p_L, p_{TiO_2}, p_{BaO}, p_{TiO_2}, ..., p_{BaO}, p_{TiO_2}, p_R)$ where a ferroelectric domain and a FDP will have all dipoles parallel or antiparallel respectively. Note that the total energy per unit cell $(E' = \frac{E}{M})$ an be expressed as function of only two single variables p_{TiO_2} and p_{BaO} in the form:

$$E'\left(p_{TiO_2}, p_{BaO}\right) = -a\left(\frac{p_{TiO_2} + p_{BaO}}{c}\right)^2 + b\left(\frac{p_{TiO_2} + p_{BaO}}{c}\right)^4 + \Omega E\left(\frac{p_{TiO_2} + p_{BaO}}{c}\right)$$
$$-\frac{A^2}{4\pi\epsilon}\left(\frac{\alpha p_{BaO}}{Mc^3} + \frac{p_R p_L}{M\left(2M-1\right)c^3} + \frac{1}{M}\sum_{i=3}^{2M-1}\frac{p_L p_i}{d_{Lj}^3} + \frac{1}{M}\sum_{i=3}^{2M-1}\frac{p_i p_R}{d_{iR}^3} + \frac{1}{2M}\sum_{i,j=4}^{2M-2}\frac{p_i p_j}{d_{ij}^3}\right) \quad (25)$$

where we used Eqs.2324 and we define an overall interface parameter $\alpha = \left(\theta_R p_R + \theta_L p_L\right)$. Note that the thickness of the film is given by the number of unit cells M and directly affects the magnitude of the depolarization field E. The DF is calculated from an ab initio calculation for a supercell with the same size M and using a well-established averaging technique for the electrostatic potential inside the film.[50]

As $M \to \infty$, the terms in Eq.25 that describe the mutual interaction between the dipoles converge to constant values, the DF vanishes and only the first two terms remain in E?, recovering the bulk properties. Only for small M these terms contribute appreciably to the total energy of the system. This is true in particular for the interface term that contains the parameter (fourth term in Eq. 25) that carries the information that defines the interfaces. We can estimate the interface parameters θ_R, θ_L, p_L, p_R from our ab initio simulations[2] and we assume that their numerical values will not depend upon the film thickness (as observed in the actual calculations, at least to first order).

As we vary the thickness of the ferroelectric film, we observe a change of the total energy landscape consequence of the interplay between the different energetic contributions. This is clearly seen in Fig. 12, where we show contour plots of $E'\left(p_{TiO_2}, p_{BaO}\right)$ for different values of M. On the left, a diagram with the correspondent spatial distribution of dipoles is indicated. For thick films, the spatial distribution of dipoles that minimizes the total energy forms a domain (both LP's in the BaO and TiO2 planes have the same sign), as can be seen in the position of the minimum in the contour plot in Fig.12A. As the number of unit cells is lowered, the minimum shifts to lower values of p_{BaO} until at a critical thickness (Fig. 12B) it becomes zero. Further reduction of the number of layers M flips only the dipoles belonging to BaO planes thus establishing a FDP (see Fig. 12D). In conclusion, this simple toy model captures the general physics of the thin film, and illustrates the intimate relation between the interface characteristics, the thickness of the film, and the existence of an FDP.

Thus, by exploiting the concept of layer polarization in the description of ferroelectric thin films between metal contacts, we have been able to obtain detailed information on the modulation of polarization at the nanoscale and to understand the constraining effects of the interfaces in the determination of the ferroelectric response of the system. Our results shows that when film thicknesses reach a critical value, the ferroelectric system responds via a transition from the bulk ferroelectric structure to a ferrielectric antidipole pattern, where individual atomic layers acquire uncompensated opposing dipoles. This state arises as

[2] The DF does not change much for the films thicknesses considered in this work. Therefore the value used as input for the model was calculated from the m6 averaged electrostatic potential ($E = 2.1410^14V/m$). The values for both interface layer dipoles used are $(p_R + p_L)/c = 1.0C/m^2$). In the model we also chose $c/a = 0.98$, respecting the tetragonal structure of the internal cells of the oxide film.

http://dx.doi.org/10.5772/52268

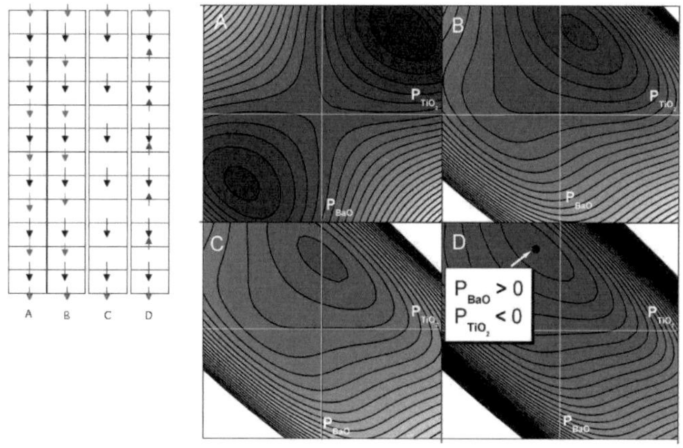

Figure 12. Energy landscape $E'(p_{TiO_2}, p_{BaO})$ for different values of θ/M (film thickness). The interaction parameters for both interfaces were considered equal, $theta_R = \theta_L = \theta$ and were kept fixed as the film size M is varied. The remaining parameters of the model have been derived from our ab initio calculations. A. $\theta/M = 0$, bulk limit where the usual double well potential produces two equivalent minima with all the layer dipoles parallel. B.$\theta/M = 3$, as the thickness is reduced the interface effects gain relevance and start to modify the energy landscape. The layer dipoles associated with the BaO planes reduce their magnitude in order to minimize the total energy. C. At $\theta/M = 5$ the system reaches a critical thickness where these dipoles become zero and eventually flip orientation. D. For $\theta/M = 10$ the system display a ferrielectric dipole pattern.

consequence of the complex energetic competition between the interface effects, the DF, and orientation and mutual interaction of the layer dipoles. The appearance of this particular ferrielectric state can be understood using a simple phenomenological model where the interface effects are explicitly taken into account.

These results suggest the possibility that such FDPs could be the normal state of a ferroelectric thin film at the nanoscale, even combined with the formation of two-dimensional island domains. It would be tempting to link the formation of such FDPs to the appearance or not of two-dimensional island below the critical thickness, and to understand the ultimate dependence upon the detailed interface structure and the nature of the metal contact. The next section will explore in detail the former point.

6. Tuning of polarization in metal-ferroelectric junctions

In this section we study the paradigmatic case of a $BaTiO_3$ film between Pt contacts, already introduced in the previous section, where we modify the ferroelectric properties by a selective control of the chemical species present at the interface. In particular, by inserting a single layer of a different metal (Au,Cu) we demonstrate that we are able to tune the polarization of the ferroelectric film via the modification of the screening properties of the composite metal contacts and through the change in the geometrical constraints at the interface.

6.1. Methods and discussion

We have simulated thin films of $BaTiO_3$ between Pt metal contacts in a (001) supercell. The oxide is terminated with a BaO plane at both interfaces, with the metal atoms directly bonded to the O atoms of the oxide planei,ii. The supercells constructed in this way can be labeled as $Pt/(M)/(BaO - TiO2)_m - BaO/(M)/Pt$ with m=1,2,4,6 and M=(Pt, Au or Cu). Nine

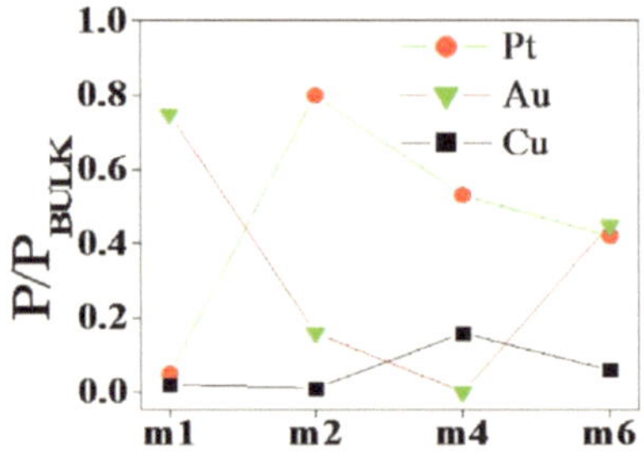

Figure 13. Normalized polarization for different number of oxide layers and interface phases. The local polarization is calculated for each oxide layer, added up and divided by the $BaTiO_3$ bulk polarization calculated in the same way with equivalent numbers of layers.

atomic metal planes (including the intralayer) have been found to be sufficient to simulate the contacts under short-circuit boundary conditions. We used the experimental in-plane lattice parameter of $BaTiO_3$ (3.99Å), and kept it fixed in all calculations. All simulations have been performed using Density Functional Theory (DFT) within the Local Density Approximation, using ultrasoft pseudopotentials and a plane waves basis set.[52]

A detailed analysis of the polarization at the nanoscale is critical for a complete description of the physical properties of the ferroelectric thin film. We have used Modern Theory of Polarization [6, 8, 18, 53, 54] and in particular the concept of layer polarization (LP)[20] to evaluate the polarization of the different structures and extract the information on the local profile of polarization at the nanoscale.

Turning to the results, in Fig.13 we show the total polarization (normalized to the bulk value) in $BaTiO_3$ films of different thickness between composite metal contacts. These results clearly elucidate our claim: the polarization of the film is critically affected by the contact geometry at all the thicknesses we have considered and a residual polarization can be observed in films as thin as one unit cell.

In fact, a one unit cell thick $BaTiO_3$ film between plain Pt contacts is still in a ferroelectric state with a polarization 10% of the ideal bulk, and the introduction of a Au intralayer enhances this value up to 70% of that! In contrast, thicker (6 unit cells) oxide films display similar ferroelectric characteristics with 50% of the bulk polarization, a clear indication of the rapid decay of the interface effects with the film thickness. It is worth to note that at variance with the previous cases, a Cu intralayer induces a paraelectric (or almost paraelectric) behavior for all thicknesses. We will come back to this point later in the discussion.

In order to understand better the behavior of ferroelectricity in such ultrathin films, we computed the layer-resolved spatial profile of the polarization, which is quantified by values of the layer polarization along [3] the structure. This is shown in Fig.14 for m4 films. When only Pt is present at the contact, the oxide develops a pattern of polarization composed of

[3] The local polarization is defined as $P_j = \frac{p_j}{c_j}$ for each layer j where c_j is half the distance between neighboring cations and p_i is the calculated layer polarization[20]. For the outermost layers, we had to make a somewhat arbitrary choice for c_j and used the distance between the outermost cation and the opposite metal plus half the distance between the cation and the one belonging to the second oxide layer.

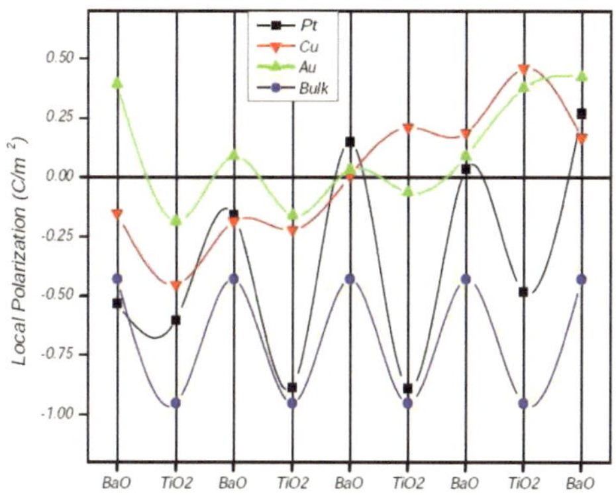

Figure 14. Local Polarization for a unit cell of size m4. Only one layer of Cu at the interface betwen the $BaTiO_3$ and Pt is enough to drive the system from a ferroelectric structure (squared dots) into a paraelectric one (red triangular dots).

consecutive alternated signs along the direction perpendicular to the interface (see Section 5) From the behavior of the local polarization it is clear that while the pristine Pt (curve with black squares) interface display a distinct ferroelectric behavior, a single layer of Cu (red triangular dots) at the interface between $BaTiO_3$ and Pt is sufficient to stabilize the system in a paraelectric (non polar) structure (the character of the structure correlates with the symmetry of the polarization profile: an asymmetric pattern correspond to a polar geometry and hence to a ferroelectric behavior; a centro-symmetric pattern on the contrary gives rise to a paraelectric behavior). If we exchange Cu for Au (curve with green triangles), the polarization is restored, although smaller that in the Pt case, and the whole polarization profile is substantially changed.

These effects vary as the size of the film changes. In thinner films the local interface details have strong effects, while they are smoothed out in thicker systems. This can be seen in Fig.15 A-D where we have plotted the LP profile for different sizes of the oxide and different metal intralayers at the interfaces. In general we notice that the Pt contact maintain the system in a ferroelectric state for all the sizes although the polarization profiles vary greatly with the system size. The same effect is observed with the addition of an Au intralayer. These changes are more noticeable in the thinner systems (m1 and m2) while for the thicker samples both systems share a similar polarization profile with a higher overall polarization in the Au case. This behavior can be directly associated with the screening of the depolarization field [4] at the metallic contacts. We have estimated the depolarization field from the macroscopic average of the electrostatic potential.[30] Indeed we observed a decrease of the DF with the introduction of the Au layer. This clearly demonstrate that Au has better screening properties compared to Pt and in consequence a Au intralayer enhances the overall polarization of the system. A similar reasoning is more difficult to do in the smaller systems due to the strong asymmetry

[4] We used the maximally localized Wannier functions (N. Marzari, and D. Vanderbilt, Phys. Rev. B 56, 12847 (1997)) as implemented in the WanT code (A. Ferretti, B. Bonferroni, A. Calzolari, and M. Buongiorno Nardelli, http://www.wannier-transport.org) for the determination of the Wannier centers. The disentanglement procedure (I. Souza, N. Marzari, and D. Vanderbilt, Phys. Rev. B 65, 035109 (2001)) was applied before starting the localization algorithm, since the bands of the metal and the valence bands of the oxide are mixed in the full supercell calculation.

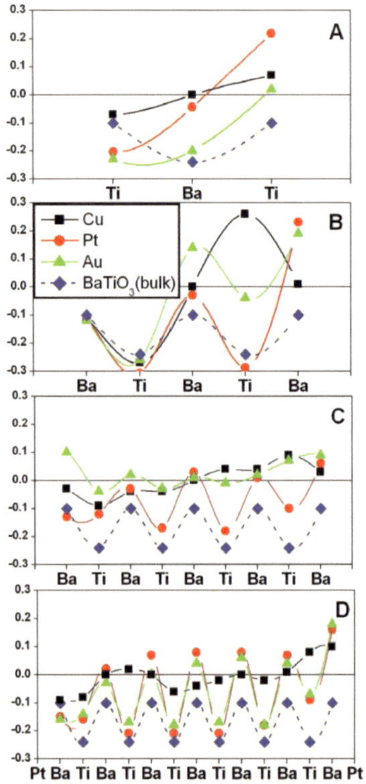

Figure 15. Layer polarization for different metal interfaces and films thicknesses A)m1, B)m2, C)m4, D)m6. As reference, the LPs for bulk $BaTiO_3$ are indicated with blue circles and dotted lines for each case. Metals intralayers used are Cu, Au, Pt. Case of m6 with Cu layer: this result is in fact an artificial effect due to the known LDA underestimation of the band gap. The Fermi level of the system overlaps the conduction band and starts filling states that should be empty. We expect the same issue for $m > m6$, or with the use of metals with a higher work function. The calculations for $m < m6$ are be correct under the DFT?LDA framework, as the Fermi level is safely below the conduction band.

introduced by the proximity of the interfaces and the uncertainty in the evaluation of the macroscopic averages. However, the qualitative behavior remains the same.

Contrary to Pt and Au, Cu intralayer do not stabilize any ferroelectric distortion at any thickness . This is a strong indication that Cu screening is weaker and not sufficient to reduce the depolarization field inside the oxide.

The above behavior is obviously correlated with the redistribution of charge across the interfaces in the different cases. The latter is quantified by the modification of the band alignment induced by the metal intralayer, and its influence on the screening properties of the metal contact. In fact, the knowledge of the band alignment, or Schottky barrier Height (SBH), allows us to define the properties of the interface phase between the metal and oxide. In a previous work,[51] we have demonstrated the correlation between the local structure of the interface composed by a crystalline oxide (BaO) and a *d* metal, and how we can tune the SBH by controlling the relative overlap of the local density of states of the different atoms close to the interface. Indeed we have found an almost complete similarity between that BaO/metal interface and the $BaTiO_3$/metal interface of this work. In both cases we considered a BaO terminated oxide slabs, that show very similar characteristics (as also indicated by the fact that the interfaces between BaO and $BaTiO_3$ have almost zero valence band offset[52]).

Indeed, we find similar behaviors for the band offsets of the interfaces BaO/M/Pd (see Ref. [52]) and $BaTiO_3$/M/Pt when Au, Cu and Pt are used as interlayer M. Furthermore, the SBH for the $BaTiO_3$/metal interfaces follow the same ascending order for each metal intralayer (Au (0.8 eV) < Pt (1.1 eV) < Cu (1.4 eV)) in both systems (values given correspond to the case of m4)[5] This trend suggests a relation between the band offsets of the interfaces and the screening properties of the metals, that is to be expected given the strong correlation between the charge transfer at the interface and the SBH.[51] In particular, if a strong dipole is established at the interface, due to a high SBH, as in the case of Cu, fewer charges will be available for screening the depolarization field and the system will not support a ferroelectric distortion. Following Ref. [51] we can state the following phenomenological rule: metal intralayers that reduce the SBH at the interfaces will enhance the screening properties of the contact and stabilize the ferroelectric distortion.

We have used Density Functinal Theory and the layer polarization concept to analyze the effect of the interface structure on thin ferroelectric films between metal contacts. As the size of the film is reduced, the interface effects become strong and influence dramatically the spatial polarization profile of the system. We monitored these changes as different metal layers were introduced at the interface, modifying the local structure of the interface, the band alignment and in turn affecting the screening nature of the metal contact. In this way we are able to tune the ferroelectric state by carefully choosing the metal interlayer.

Author details

Matías Núñez

Consejo Nacional de Investigaciones Científicas y Técnicas (CONICET), Argentina
Centro Atomico Bariloche, U-A Tecnologia de Materiales y Dispositivos, Division Materiales Nuclkheares, Argentina
Instituto de Ciencias Basicas, Universidad Nacional de Cuyo, Argentina
Previously at North Carolina State University, Raeligh, North Carolina, USA

References

[1] P. Ghosez and J. Junquera. First-Principles Modeling of Ferroelectric Oxide Nanostructures. *ArXiv Condensed Matter e-prints*, May 2006.

[2] L. D. Landau and E. M. Lifshitz. *Electrodynamics of Continuous Media.* Pergamon Press, Oxford, 1984.

[3] C. Kittel. *Introduction to Solid State Physics.* Wiley, New York, 7 edition, 1996.

[5] We obtained the SBH values of the $BaTiO_3$/metal interfaces by calculating the energy difference between the Fermi level and the upper edge of the valence band, by direct inspection of the LDOS of the systems projected on of each crystal plane.

[4] N. W. Ashcroft and N. D. Mermin. *Solid State Physics.* Saunders, Philadelphia, 1976.

[5] Richard M. Martin. Comment on calculations of electric polarization in crystals. *Phys. Rev. B*, 9(4):1998–1999, Feb 1974.

[6] R. Resta. Theory of the electric polarization in crystals. *Ferroelectrics*, 136, 1992.

[7] R.D. King-Smith and David Vanderbilt. Theory of polarization of crystalline solids. *Phys. Rev. B*, 47(3):1651–1654, Jan 1993.

[8] Raffaele Resta. Macroscopic polarization in crystalline dielectrics: the geometric phase approach. *Rev. Mod. Phys.*, 66(3):899–915, Jul 1994.

[9] Raffaele Resta. Quantum-mechanical position operator in extended systems. *Phys. Rev. Lett.*, 80(9):1800–1803, Mar 1998.

[10] A. Shapere and F. Wilczek. *Geometric Phases in Physics.* World Scientific, Singapore, 1989.

[11] Raffaele Resta. Manifestations of berry's phase in molecules and condensed matter. *Journal of Physics: Condensed Matter*, 12(9):R107–R143, 2000.

[12] Gregory H. Wannier. The structure of electronic excitation levels in insulating crystals. *Phys. Rev.*, 52(3):191–197, Aug 1937.

[13] W. Kohn. Analytic properties of bloch waves and wannier functions. *Phys. Rev.*, 115(4):809–821, Aug 1959.

[14] E.I. Blount. Formalism of band theory. *Solid State Physics*, 13(305), 1962.

[15] Nicola Marzari and David Vanderbilt. Maximally localized generalized wannier functions for composite energy bands. *Phys. Rev. B*, 56(20):12847–12865, Nov 1997.

[16] J. Zak. Berry's phase for energy bands in solids. *Phys. Rev. Lett.*, 62(23):2747–2750, Jun 1989.

[17] J. Zak. Band center—a conserved quantity in solids. *Phys. Rev. Lett.*, 48(5):359–362, Feb 1982.

[18] D. Vanderbilt and R. Resta. *Conceptual foundations of materials properties: A standard model for calculation of ground- and excited-state properties.*, chapter Quantum electrostatics of insulators: Polarization, Wannier functions, and electric fields. Elsevier, The Netherlands, 2006.

[19] N. Marzari and D. Vanderbilt. Maximally localized generalized wannier functions for composite energy bands. *Phys. Rev. B*, 56(20):12847–12865, 1997.

[20] X. Wu, O. Dieguez, K. M. Rabe, and D. Vanderbilt. . *Phys. Rev. Lett.*, 97(107602), 2006.

[21] I. Souza, N. Marzari, and D. Vanderbilt. Maximally localized wannier functions for entangled energy bands. *Phys. Rev. B*, 65(035109), 2001.

[22] J.D. Jackson. *Classical electrodynamics*. Wiley, 3 edition, 1998.

[23] B. Meyer and David Vanderbilt. Ab initio study of *batio*3 and *pbtio*3 surfaces in external electric fields. *Phys. Rev. B*, 63(20):205426, May 2001.

[24] Ph. Ghosez and K. M. Rabe. Microscopic model of ferroelectricity in stress-free pbtio[sub 3] ultrathin films. *Applied Physics Letters*, 76(19):2767–2769, 2000.

[25] T. Maruyama, M. Saitoh, I. Sakai, T. Hidaka, Y. Yano, and T. Noguchi. Growth and characterization of 10-nm-thick c-axis oriented epitaxial pbzr[sub 0.25]ti[sub 0.75]o[sub 3] thin films on (100)si substrate. *Applied Physics Letters*, 73(24):3524–3526, 1998.

[26] Lennart Bengtsson. Dipole correction for surface supercell calculations. *Phys. Rev. B*, 59(19):12301–12304, May 1999.

[27] Batra I. P. and Silverman B. D. *Sol. State Comm.*, 11(291), 1972.

[28] P.B.Littlewood M.Dawber1, P.Chandra and J.F. Scott. Depolarization corrections to the coercive field in thin-film ferroelectric. *J. Phys.: Condens. Matter*, 15:L393–L398, 2003.

[29] A. M. Bratkovsky and A. P. Levanyuk. Very large dielectric response of thin ferroelectric films with the dead layers. *Phys. Rev. B*, 63(13):132103, Mar 2001.

[30] J. Junquera and P. Ghosez. . *Nature*, 422(506), 2003.

[31] Na Sai, Karin M. Rabe, and David Vanderbilt. Theory of structural response to macroscopic electric fields in ferroelectric systems. *Phys. Rev. B*, 66(10):104108, Sep 2002.

[32] Alfonso Balderschi, Stefano Baroni, and Raffaele Resta. Band offsets in lattice-matched heterojunctions: A model and first-principles calculations for gaas/alas. *Phys. Rev. Lett.*, 61(6):734–737, Aug 1988.

[33] L. Colombo, R. Resta, and S. Baroni. Valence-band offsets at strained si/ge interfaces. *Phys. Rev. B*, 44(11):5572–5579, Sep 1991.

[34] Igor Kornev, Huaxiang Fu, and L. Bellaiche. Ultrathin films of ferroelectric solid solutions under a residual depolarizing field. *Phys. Rev. Lett.*, 93(19):196104, Nov 2004.

[35] Celine Lichtensteiger, Jean-Marc Triscone, Javier Junquera, and Philippe Ghosez. Ferroelectricity and tetragonality in ultrathin pbtio[sub 3] films. *Physical Review Letters*, 94(4):047603, 2005.

[36] R.E. Cohen. *Nature*, 358(136), 1992.

[37] V. Nagarajan, J. Junquera, J. Q. He, C. L. Jia, R. Waser, K. Lee, Y. K. Kim, S. Baik, T. Zhao, R. Ramesh, Ph. Ghosez, and K. M. Rabe. Scaling of structure and electrical properties in ultrathin epitaxial ferroelectric heterostructures. *Journal of Applied Physics*, 100(5):051609, 2006.

[38] C.-G. Duan, R.F. Sabirianov, W.-N. Mei, S.S. Jaswal, and E.Y. Tsymbal. Interface effect on ferroelectricity at the nanoscale. *Nano Letters*, 6(3):483–487, 2006.

[39] G. Gerra, A. K. Tagantsev, N. Setter, and K. Parlinski Parlinski. Ionic Polarizability of Conductive Metal Oxides and Critical Thickness for Ferroelectricity in BaTiO3. *Physical Review Letters*, 96(10):107603–+, March 2006.

[40] N. Sai, A. M. Kolpak, and A. M. Rappe. . *Phys. Rev.B*, 72(020101(R)), 2005.

[41] Toshito Mitsui and Jiro Furuichi. Domain structure of rochelle salt and k*h*2p*o*4. *Phys. Rev.*, 90(2):193–202, Apr 1953.

[42] Zhongqing Wu, Ningdong Huang, Zhirong Liu, Jian Wu, Wenhui Duan, Bing-Lin Gu, and Xiao-Wen Zhang. Ferroelectricity in pb (zr0.5 ti0.5) o3 thin films: Critical thickness and 180o stripe domains. *Phys. Rev. B*, 70(10):104108, Sep 2004.

[43] M. Sepliarsky, S. R. Phillpot, D. Wolf, M. G. Stachiotti, and R. L. Migoni. Atomic-level simulation of ferroelectricity in perovskite solid solutions. *Applied Physics Letters*, 76(26):3986–3988, 2000.

[44] S. K. Streiffer, J. A. Eastman, D. D. Fong, Carol Thompson, A. Munkholm, M. V. Ramana Murty, O. Auciello, G. R. Bai, and G. B. Stephenson. Observation of nanoscale 180° stripe domains in ferroelectric $pbtio3$ thin films. *Phys. Rev. Lett.*, 89(6):067601, Jul 2002.

[45] Y. S. Kim, D. H. Kim, J. D. Kim, Y. J. Chang, T. W. Noh, J. H. Kong, K. Char, Y. D. Park, S. D. Bu, J.-G. Yoon, and J.-S. Chung. Critical thickness of ultrathin ferroelectric batio[sub 3] films. *Applied Physics Letters*, 86(10):102907, 2005.

[46] R. A. McKee, F. J. Walker, M. Buongiorno Nardelli, W. A. Shelton, and G. M. Stocks. The Interface Phase and the Schottky Barrier for a Crystalline Dielectric on Silicon. *Science*, 300(5626):1726–1730, 2003.

[47] C. F. Pulvari. *Phys. Rev.*, 120(1670), 1960.

[48] Nicola A. Spaldin. MATERIALS SCIENCE: Fundamental Size Limits in Ferroelectricity. *Science*, 304(5677):1606–1607, 2004.

[49] K.S. Thygensen, L. B. Hansen, and K. W. Jacobsen. Partly Occupied Wannier Functions. *Phys. Rev. Lett.*, 94(026405), 2005.

[50] S. Baroni, R. Resta, A. Baldereschi, and M. Peressi. *Spectroscopy of Semiconductor Microstructures*, chapter "Can we tune the band offset at semiconductor heterojunctions?", pages 251–271. Plenum, London, 1989.

[51] Matias Nunez and Marco Buongiorno Nardelli. First-principles theory of metal-alkaline earth oxide interfaces. *Physical Review B (Condensed Matter and Materials Physics)*, 73(23):235422, 2006.

[52] J. Junquera, M. Zimmerman, P. Orejon, and P. Ghosez. First-principles calculation of the band offset at BaO/BaTiO3 and SrO/SrTiO3 interfaces. *Phys. Rev. B*, 67(153327), 2003.

[53] R. Raffaele. Macroscopic polarization in crystalline dielectrics: the geometric phase approach. *Rev. Mod. Phys.*, 66(3):899–915, Jul 1994.

[54] R.D. King-Smith and D. Vanderbilt. Theory of polarization of crystalline solids. *Phys. Rev. B*, 47(3):1651–1654, Jan 1993.

Chapter 19

The Influence of Vanadium Doping on the Physical and Electrical Properties of Non-Volatile Random Access Memory Using the BTV, BLTV, and BNTV Oxide Thin Films

Kai-Huang Chen, Chien-Min Cheng, Sean Wu,
Chin-Hsiung Liao and Jen-Hwan Tsai

Additional information is available at the end of the chapter

http://dx.doi.org/10.5772/52203

1. Introduction

Recently, the various functional thin films were widely focused on the applications in non-volatile random access memory (NvRAM), such as smart cards and portable electrical devices utilizing excellent memory characteristics, high storage capacity, long retention cycles, low electric consumption, non-volatility, and high speed readout. Additionally, the various non-volatile random access memory devices such as, ferroelectric random access memory (FeRAM), magnetron memory (MRAM), resistance random access memory (RRAM), and flash memory were widely discussed and investigated [1-9]. However, the high volatile pollution elements and high fabrication cost of the complex composition material were serious difficult problems for applications in integrated circuit semiconductor processing. For this reason, the simple binary metal oxide materials such as ZnO, Al_2O_3, TiO_2, and Ta_2O_5 were widely considered and investigated for the various functional electronic product applications in resistance random access memory devices [10-12].

The (ABO_3) pervoskite and bismuth layer structured ferroelectrics (BLSFs) were excellent candidate materials for ferroelectric random access memories (FeRAMs) such as in smart cards and portable electric devices utilizing their low electric consumption, nonvolatility, high speed readout. The ABO_3 structure materials for ferroelectric oxide exhibit high remnant polarization and low coercive filed. Such as $Pb(Zr,Ti)O_3$ (PZT), $Sr_2Bi_2Ta_2O_9$ (SBT), $SrTiO_3$ (ST), $Ba(Zr,Ti)O_3$ (BZ1T9), and $(Ba,Sr)TiO_3$ (BST) were widely studied and discussed for large storage capacity FeRAM devices. The (Ba,Sr)TiO_3 and Ba(Ti,Zr)O_3 ferroelectric materials were also expected to substitute the PZT or SBT memory materials and improve the environmental pollution because of their low pollution problem [9-15]. In addition, the high dielectric constant and low leakage current density of zirconium and strontium-doped $BaTiO_3$ thin films were applied for the further application in the high density dynamic random access memory (DRAM) [16-20].

Bismuth titanate system based materials were an important role for FeRAMs applications. The bismuth titanate system were given in a general formula of bismuth layer structure ferroelectric, $(Bi_2O_2)^{2+}(A_{n-1}B_nO_{3n+1})^{2-}$ (A=Bi, B=Ti). The high leakage current, high dielectric loss and domain pinning of bismuth titanate system based materials were caused by defects, bismuth vacancies and oxygen vacancies. These defects and oxygen vacancies were attributed from the volatilization of Bi_2O_3 of bismuth contents at elevated temperature [21-23].

1.1. ABO_3 pervoskite structure material system

For ABO_3 pervoskite structure such as, $BaTiO_3$ and BZ1T9, the excellent electrical and ferroelectric properties were obtained and found. For SOP concept, the ferroelectric BZ1T9 thin film on ITO substrate were investigated and discussed. For crystallization and grain grow of ferroelectric thin films, the crystal orientation and preferred phase of different substrates were important factors for ferroelectric thin films of MIM structures.

The XRD patterns of BZ1T9 thin films with 40% oxygen concentration on Pt/Ti/SiO_2/Si substrates from our previous study were shown in Fig. 1 [24-25]. The (111) and (011) peaks of the BZ1T9 thin films on Pt/Ti/SiO_2/Si substrates were compared with those on ITO substrates. The strongest and sharpest peak was observed along the Pt(111) crystal plane. This suggests that the BZ1T9 films grew epitaxially with the Pt(111) bottom electrode. However, the (111) peaks of BZ1T9 thin films were not observed for (400) and (440) ITO substrates. Therefore, we determined that the crystallinity and deposition rate of BZ1T9 thin films on ITO substrates differed from those in these study [24-27].

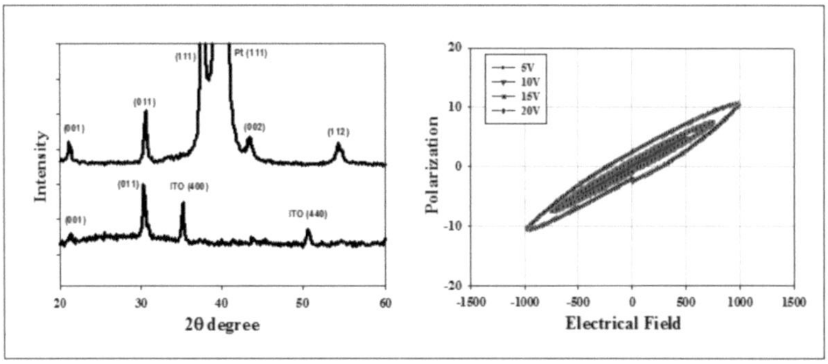

Figure 1. a) XRD patterns of as-deposited thin films on the ITO/glass and Pt substrates, and (b) *p-E* curves of thin films.

The polarization versus applied electrical field (*p-E*) curves of as-deposited BZ1T9 thin films were shown in Fig. 1(a). As the applied voltage increases, the remanent polarization of thin films increases from 0.5 to 2.5 μC/cm². In addition, the $2P_r$ and coercive field calculated and were about 5 μC/cm² and 250 kV/cm, respectively. According to our previous study, the BZ1T9 thin film deposited at high temperature exhibited high dielectric constant and high leakage current density because of its polycrystalline structure [24].

1.2. Bismuth Layer Ferroelectric Structure material system

The XRD patterns of as-deposited $Bi_4Ti_3O_{12}$ thin films and ferroelectric thin films under 500~700 °C rapid thermal annealing (RTA) process were compared in Fig. 2(a). From the results obtained, the (002) and (117) peaks of as-deposited $Bi_4Ti_3O_{12}$ thin film under the optimal sputtering parameters were found. The strong intensity of XRD peaks of $Bi_4Ti_3O_{12}$ thin film under the 700 °C RTA post-treatment were is found. They were (008), (006), (020) and (117) peaks, respectively. Compared the XRD patterns shown in Fig. 2, the crystalline intensity of (111) plane has no apparent increase as the as-deposited process is used and has apparent increase as the RTA-treated process was used. And a smaller full width at half maximum value (FWHM) is revealed in the RTA-treated $Bi_4Ti_3O_{12}$ thin films under the 700 °C post-treatment. This result suggests that crystal structure of $Bi_4Ti_3O_{12}$ thin films were improved in RTA-treated process.

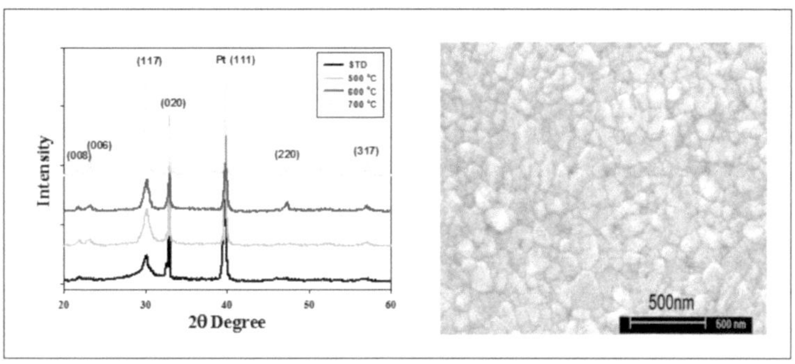

Figure 2. a) XRD patterns of as-deposited $Bi_4Ti_3O_{12}$ thin films, and (b)the SEM morphology of as-deposited $Bi_4Ti_3O_{12}$ films.

The surface morphology observations of as-deposited $Bi_4Ti_3O_{12}$ thin films under the 700 °C RTA processes were shown in Fig. 2(b). For the as-deposited $Bi_4Ti_3O_{12}$ thin films, the morphology reveals a smooth surface and the grain growth were not observed. The grain size and boundary of $Bi_4Ti_3O_{12}$ thin films increased while the annealing temperature increased to 700 °C. In RTA annealed $Bi_4Ti_3O_{12}$ thin films, the maximum grain size were about 200 nm and the average grain size is 100 nm. The thickness of annealed $Bi_4Ti_3O_{12}$ thin films were calculated and found from the SEM cross-section images. The thickness of the deposited $Bi_4Ti_3O_{12}$ thin films is about 800 nm and the deposited rate of $Bi_4Ti_3O_{12}$ thin films is about 14 nm/mim.

2. Experimental Detail

S. Y. Wu firstly reported that an MFS transistor fabricated by using bismuth titanate in 1974 [28-29]. The first ferroelectric memory device was fabricated by replacing the gate oxide of a conventional metal-oxide-semiconductor (MOS) transistor with a ferroelectric material. However, the interface and interaction problem between the silicon substrate and ferroelectric films were very important factors during the high temperature processes in 1TC structure. To overcome the interface and interaction problem, the silicon dioxide and silicon nitride films were used as the buffer layer. The low remnant polarization and high operation voltage of 1TC were also be induced by gate oxide structure with double-layer ferroelectric silicon dioxide thin films. Sugibuchi et al. provided a 50 nm silicon dioxide thin film between the $Bi_4Ti_3O_{12}$ layer and the silicon substrate [30].

The ferroelectric ceramic target prepared, the raw materials were mixed and fabricated by solid state reaction method. After mixing and ball-milling, the mixture was dried, grounded, and calcined for some time. Then, the pressed ferroelectric ceramic target with a diameter of two inches was sintered in ambient air. The base pressure of the deposited chamber was brought down 1×10^{-7} mTorr prior to deposition. The target was placed away from the Pt/Ti/SiO_2/Si and SiO_2/Si substrate. For metal-ferroelectric-metal (MFM) capacitor structure, the Pt and the Ti were deposited by dc sputtering using pure argon plasma as bottom electrodes. The SiO_2 thin films were prepared by dry oxidation technology. The metal-ferroelectric-insulator-semiconductor (MFIS) and metal-ferroelectric-metal (MFM) structures were shown in Fig. 3.

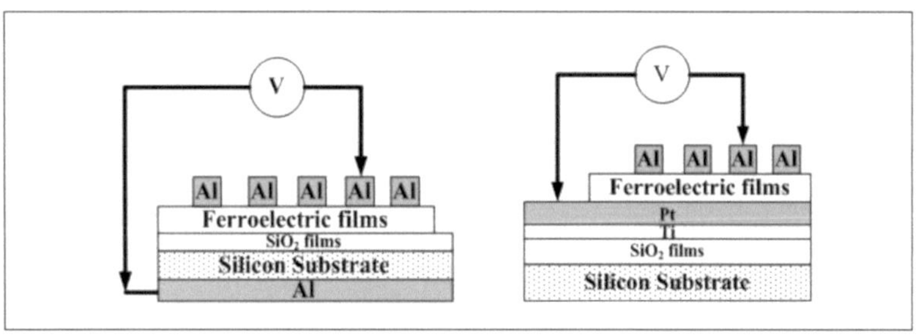

Figure 3. a) Metal-ferroelectric-insulator-semiconductor (MFIS) structure, and (b) Metal ferroelectric-metal (MFM) structure.

For the physical properties of ferroelectric thin films obtained, the thickness and surface morphology of ferroelectric thin films were observed by field effect scanning electron microscopy (FeSEM). The crystal structure of ferroelectric thin films were characterized by an X-ray diffraction (XRD) measurement using a Ni-filtered CuKα radiation. The capacitance-voltage (C-V) properties were measured as a function of applied voltage by using a Hewlett-Packard (HP 4284A) impedance gain phase analyzer. The current curves versus the applied voltage (I-V characteristics) of the ferroelectric thin films were measured by a Hewlett-Packard (HP 4156) semiconductor parameter analyzer.

3. Results and Discussion

3.1. Large memory window in the vanadium doped $Bi_4Ti_3O_{12}$ (BTV) thin films

The XRD pattern was used to identify the crystalline structures of as-deposited BTV thin films, labeled "vanadium doped at 550 °C," with various depositing parameters. From the XRD pattern, we found that the optimal deposition parameters of as-deposited BTV thin films were RF power of 130 W, chamber pressures of 10 mtorr and oxygen concentrations of 25%. The crystalline orientations of (117), (008) and (200) planes were apparently observed in the films. It was found that all of the films consisted of a single phase of a bismuth layered structure showing the preferred (008) and (117) orientation. Both films were well c-axis oriented, but BTV thin film was more c-axis oriented than BIT, labeled "undoped at 550 °C". For the polycrystalline BTV thin films, the (117) peak was the strongest peak and the intensity of the (008) peak was 10% of (117) peak intensity. An obvious change in the orientation due to the substitution was observed except for the degree of the (117) orientation for BTV films. In addition, the XRD patterns of the as-deposited BTV thin films deposited using optimal parameters at room and 550 °C substrate temperatures were observed in Fig. 2. This result indicated that the crystalline characteristics of BTV thin films deposited at 550 °C were better than those of BTV thin films at room temperature. The crystalline and dielectric characteristics of as-deposited BTV thin films were influenced by substrate temperatures. The electrical characteristics of as-deposited BTV thin films at substrate temperatures of 550 °C under optimal parameters will be further developed.

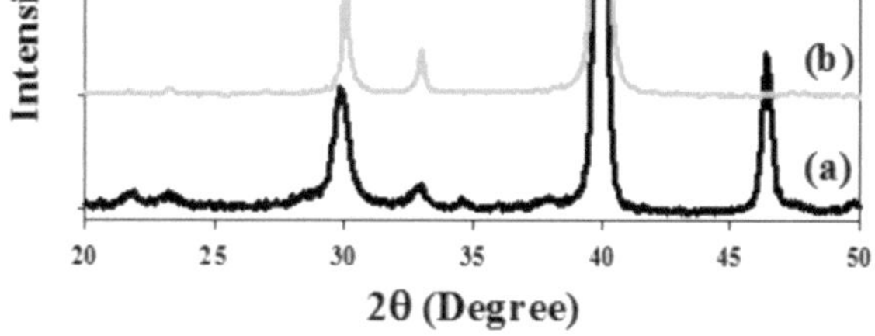

Figure 4. The XRD patterns of (a) undoped at 550 °C (b) vanadium doped at R.T. and (c) vanadium doped at 550 °C thin films deposited using optimal parameters.

In Fig. 5, circular-like grains with 150 nm width were observed with scanning electron microscopy (SEM) for as-deposited BTV thin films. From the cross-sectional SEM image, film thicknesses were measured to be 742 nm. As the depositing time increases from 30 and 60, to 120 min, the thickness of as-deposited BTV thin films increases linearly from 197 and 386, to 742 nm, respectively, as the depositing rate decreases from 6.57 and 6.43, to 6.18 nm.

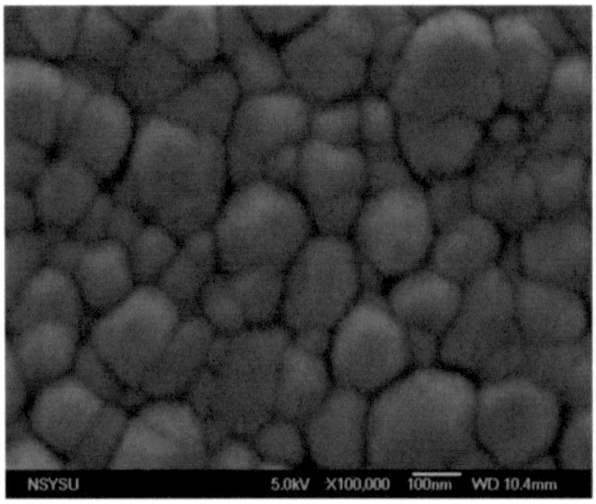

Figure 5. The surface micro structure morphology of the BTV thin films deposited using optimal parameters.

Figure 6(a) compares the change in the capacitance versus the applied voltage (*C-V*) for the un-doped and vanadium doped thin films. Based on Fig. 4, the capacitances of the BIT thin films appear to increase due to the vanadium dopant. We found that the capacitances of BTV thin films increased from 1.3 to 4.5 nF. As suggested by Fig. 4, the improvement in the dielectric constants of the BTV thin films can be attributed to the compensation of the oxygen vacancy and the improvement in the B-site substitution of the ABO_3 phase in the BTV thin films [31-34].

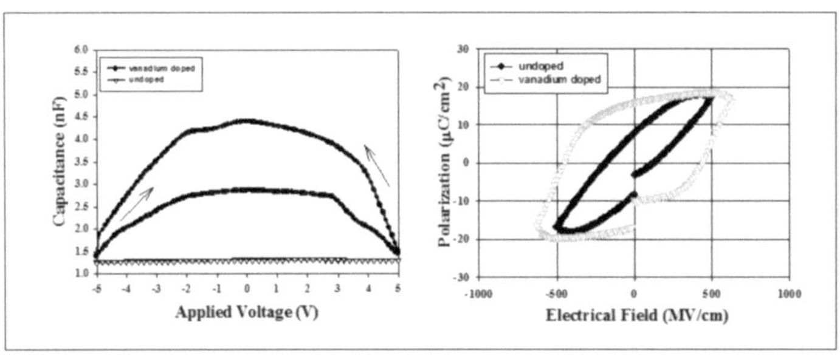

Figure 6. a) The P-E characteristics of vanadium doped and undoped thin films, and (b) The normalization C-V curves of vanadium doped and undoped thin films.

Figure 6 shows ferroelectric hysteresis loops of BIT and as-deposited BTV thin film capacitors measured with a ferroelectric tester (Radiant Technologies RT66A). The as-deposited BTV thin films, labeled "vanadium doped," clearly show ferroelectricity. The remanent polarization and coercive field were 23 μC/cm^2 and 450 kV/cm. Comparing the vanadium doped and undoped BIT thin films, the remanent polarization (2Pr) would be increased form 16μC/cm^2 for undoped BIT thin films to 23 μC/cm^2 for vanadium doped. However, the coercive field of as-deposited BTV thin films would be increased to 450 kV/cm. These results indicated that the substitution of vanadium was effective for the appearance of ferroelectricity at 550 °C. The 2Pr value and the Ec value were larger than those reported in Refs. [35-36], and the 2Pr value was smaller and the Ec value was larger than those reported in [37]. Based on above results, it was found that the simultaneous substitutions for B-site are effective to derive enough ferroelectricity by accelerating the domain nucleation and pinning relaxation caused by B-site substitution [31-37].

The leakage current density versus applied voltage curves of as-deposited BTV thin films for different depositing time on the MFIS structure were be found. We found that the leakage current density of undoped BIT thin films, labeled "undoped," were larger than those of vanadium-doped BIT thin films. This result indicated that the substitution of B-site in ABO_3 perovskite structure for BTV thin films was effective in lowering leakage current density. Besides, the thickness of BTV thin films has an apparent influence on the leakage current density of BTV thin films, and that will have an apparent influence on the other electrical characteristics of BTV thin films. At an electric field of 0.5 MV/cm, the leakage current density critically decreases from the 3.0 ×10^{-7} A/cm2 for 30 min-deposited BTV thin films to around 3×10^{-8}A/cm^2 and 2×10^{-8} A/cm^2 for 60 and 120 min-deposited BTV thin films.

Figure 7(a) show the capacitance versus applied voltage (*C-V*) curves of as-deposited vanadium doped BTV and un-doped BIT thin films. The applied voltages, which are first changed from -20 to 20 V and then returned to -20 V, are used to measure the capacitance voltage characteristics (*C-V*) of the MFIS structures. For the vanadium doped thin films, the memory window of MFIS structure increased from 5 to 15 V, and the threshold voltage decreased from 7 to 3 V. This result demonstrated that the lower threshold voltage and decreased oxygen vacancy in undoped BIT thin films had been improved from the *C-V* curves measured.

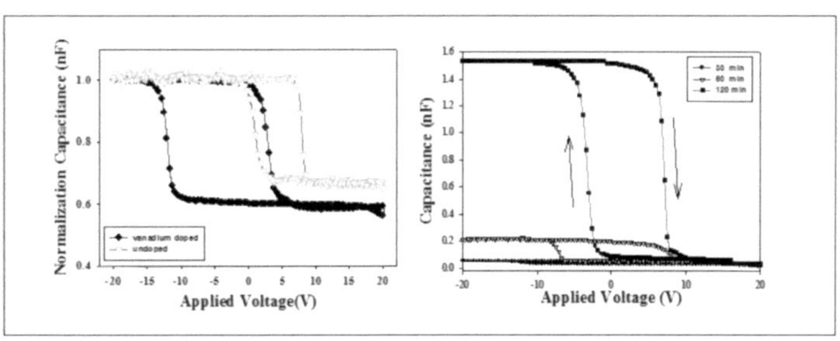

Figure 7. a) The P-E characteristics of vanadium doped and undoped thin films, and (b) The normalization *C-V* curves of vanadium doped and undoped thin films.

Figure 7(b) shows the *C-V* curves of 30 min-deposited BTV thin films, and Figure 7(b) compares the *C-V* curves of BTV thin films deposited at different depositing time. Figure 7(b) also shows that the capacitance at the applied voltage value of 0 V critically increases as the depositing time increases. The memory window would be decreased from 15.1 and 13.4, to 10.6 V as the depositing time increased. Two reasons may be the cause that the increases capacitance of as-deposited BTV thin films in the MFIS structure for the different depositing time. First, the decrease leakage current density was attributed to the increase thickness of as-deposited BTV thin films in Fig. 4. Second, the different thickness and high dielectric constant of as-deposited BTV thin films are also an important factor. In Fig. 7, the capacitance of as-deposited BTV thin films for MFIS structure could be calculated from Eq. 1 and Eq. 2:

$$C = \varepsilon_0 \varepsilon_r \frac{A}{d} \quad (1)$$

$$\varepsilon_r = \frac{(\varepsilon_1 d_2 + \varepsilon_2 d_1)}{(d_1 + d_2)} \quad (2)$$

The ε1 and d1 are the effective dielectric constant and the total thickness of silicon and SiO_2 layer. The ε2 and d2 are the effective dielectric constant and the thickness of as-deposited BTV thin film layer. The relative dielectric constants of Si and SiO_2 are 11.7 and 3.9. The εr value (ε1) of silicon and SiO_2 layer is much smaller than that (ε2) of as-deposited BTV thin films. The ε1×d2 is the unchanged value and the ε2×d1 value increases with the increase of depositing time. In Eq. (2), the ε1×d2 increases more quickly as d2 increases. In Eq. (1), the C value will increase as the εr value increases.

For memory window characteristics at applied voltage of 0 volts, the upper and lower capacitance values of as-deposited BTV thin films for 30 min depositing time were 0.056 and 0.033 nF, respectively. For 60 min and 120 min depositing time, they were 0.215~0.048 nF and 1.515 ~0.105 nF, respectively. The change ratios at zero voltage were defined in Eq.(3) from these experimental results:

$$ratio = \frac{(C_u - C_l)}{C_u} \quad (3)$$

where Cu and Cl are the upper and lower capacitance values.

The capacitance change ratios of as-deposited BTV thin films for different depositing time were 41, 73 and 93%, respectively. From above statements, the good switching characteristics of ferroelectric polarization could be attributed to memory windows ratio and the thinner thickness of as-deposited BTV thin film for the depositing time of 30 min. These results indicted the upper and lower capacitance of memory window would be decreased by lowering the thickness of SiO_2 layer.

3.2. The Influence of Lanthanum Doping on the Physical and Electrical Properties of BTV (BLTV) Ferroelectric Thin Films

For MFM structures, the crystal orientation and preferred phase of ferroelectric thin films on Pt/Ti/SiO_2/Si substrates was important factor. The x-ray diffraction (XRD) patterns of BLTV and BTV thin films prepared by rf magnetron sputtering were be found. From the XRD pattern, the BLTV and BTV thin film were polycrystalline structure. The (004), (006), (008), and (117) peaks were observed in the XRD pattern. All of thin films consisted of a single phase of a bismuth layered structure showing the preferred (117) orientation. All of thin films were exhibited well c axis orientation. The change in the orientation of BLTV thin films due to the substitution was not observed. The degree of the (117) orientation relative to the (001) orientation of BLTV thin films dominant was shown.

Figure 8. The surface micro structure of as-deposited (a) BTV and (b) BLTV thin films.

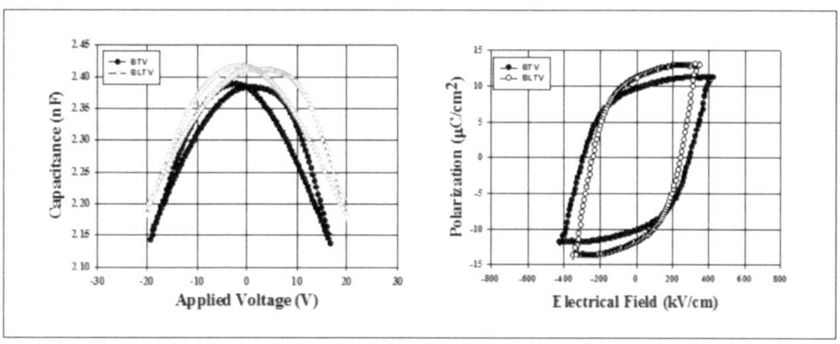

Figure 9. a)The *C-V* characteristics of as-deposited BTV and BLTV thin films. (b) The polarization versus electrical field characteristics of as-deposited BTV and BLTV thin films.

In Fig. 8a, rod-like and circular-board grains with 250 nm length and 150 nm width were observed with scanning electron microscopy (SEM) for as-deposited BTV films. The small grain was gold element in preparation for the SEM sample. However, the BLTV thin films exhibited a great quantity of 400 nm length and 100 nm width rod-like grain structure in Fig 8 b. The rod-like grain size of BLTV thin films was larger than those of BTV. We induced that the bismuth vacancies of BTV thin films compensate for lanthanum addition and microstructure were improved in BLTV thin films. From the cross-sectional SEM image, average thin film thicknesses for MFIS structure were about 610 nm. The average thickness of thin films for MFM structure was about 672 nm.

Figure 9(a) shows the change in the capacitance versus the applied voltage (*C-V*) of the BTV and BLTV thin films in MFM structure measured at 100 kHz. The applied voltages, which were first changed from -20 to 20 V and then returned to -20 V, were used to measure the capacitance voltage characteristics (CV). The BLTV thin films exhibited high capacitance than those of BTV thin films. We found that the capacitances of the lanthanum-doped BTV thin films were increased from 2.38 to 2.42 nF.

Figure 9(b) shows the *p-E* curves of the different ferroelectric thin films under applied voltage of 18V from the Sawyer–Tower circuits. The remanent polarization of non-doped, vanadium-doped, and lanthanum-doped ferroelectric thin films linearly was increased from 5, 10 to 11 μC/cm^2, respectively. The coercive filed of non-doped, vanadium-doped, and lanthanum-doped ferroelectric thin films were about 300, 300, and 250 kV/cm, respectively. The ferroelectric properties of lanthanum-doped and vanadium-doped BIT thin films were improved and found.

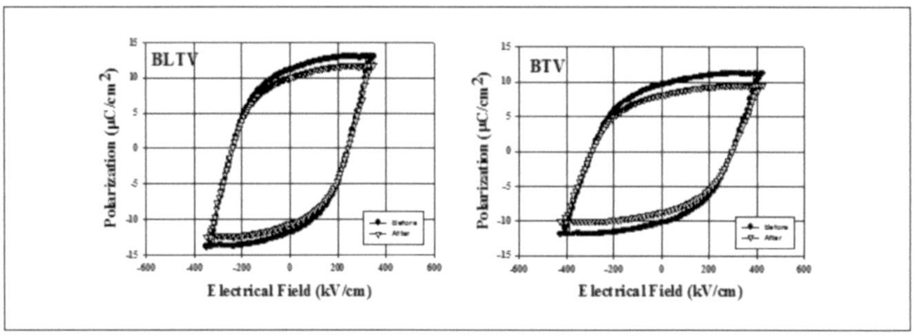

Figure 10. a)The *C-V* characteristics of as-deposited BTV and BLTV thin films. (b) The polarization versus electrical field characteristics of as-deposited BTV and BLTV thin films.

The fatigue characteristics for ferroelectric thin films were the time dependent change of the polarization state. After a long time, the polarization loss of ferroelectric thin films was affected by oxygen vacancies, defect, and space charge in the memory device. Figure 10 shows the polarization versus electrical field (*p-E*) properties of the different ferroelectric thin films before and after the switching of 10^9 cycles. The remnant polarization loss of BLTV and BTV thin films were about 9% and 15% of initial polarization value, respectively. The remnant polarization of BLTV thin films were little changed after the switching cycles. The fatigue behavior and domain pinning were improved by lanthanum and vanadium addition in BIT thin films. To reduce bismuth defect and oxygen vacancy, the high-valence cation substituted for the A-site of BLTV thin films were observed.

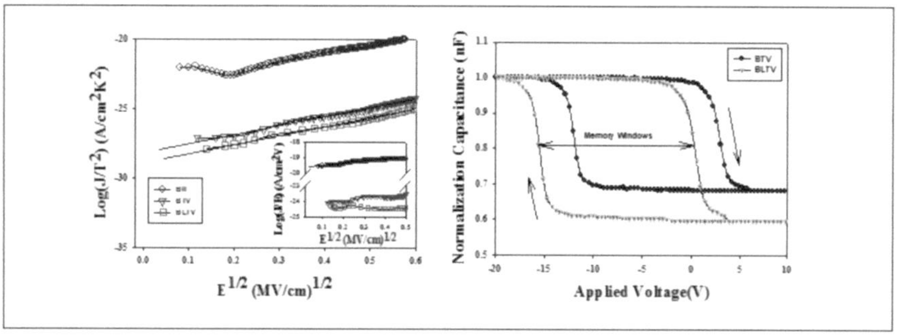

Figure 11. a) The (J/T^2) and (J/E) versus $E^{1/2}$ curves of as-deposited BIT, BTV, and BLTV thin films. (b) The normalization capacitance versus applied voltage curves of as-deposited BTV and BLTV thin films.

From the *J-E* curves, the conduction mechanism of as-deposited thin films was also proved the bismuth defect and oxygen vacancy results. For MFIS structure, the interface and interaction problem between of the silicon and ferroelectric films was serious important. The effect of the bound polarization charge from the carrier in the silicon substrate was observed. To overcome this problem, the insulator films of buffer layer were used for MFIS structure [28-30, 37-38].

Figure 11(a) shows the leakage current density versus electrical field (*J-E*) characteristics of as-deposited BTV and BLTV thin films for MFIS structure. The leakage current density of the as-deposited BLTV thin films were about one order of magnitude lower than those of the non-doped lanthanum thin films. However, the lanthanum and vanadium doped BIT thin films were lower than those of BIT thin films. We suggested that low leakage current density attributed to substituting a bismuth ion with a lanthanum ion at A-site for lanthanum doped BTV thin films. To discuss bismuth ion substituting by lanthanum ion effect, the leakage current versus electrical field curves of BLTV thin films were fitted to Schottky emission and Poole–Frankel transport models. The inset of Fig. 11(a) shows the JE characteristics for BIT, BTV, and BLTV thin films in terms of J/E as vertical axis and $E^{1/2}$ as horizontal axis. The fitting curves were straight in this figure and a JE curve of thin films was the Poole–Frankel emission model. The high leakage current was attributed to the bismuth vacancies and oxygen vacancies of as-deposited BIT thin films [39–42]. Park et al. also suggested that the enhanced stability of TiO_6 octahedra against oxygen vacancies for fatigue resistance of lanthanum doped BIT thin films [31].

In a previous study, the low threshold voltage of ferroelectric thin films was attributed by bismuth and oxygen vacancy [9]. The threshold voltage for the lanthanum-doped BTV thin films of MFIS structure was improved from 5 to 3 V. The memory functional effect and depletion delay of the MFIS structure was caused by remanent polarization of ferroelectric thin films in CV curves. In this study, the memory window was increased from 15 to 18 V. The large memory window of lanthanum-doped BTV thin films was also proved by *p-E* curves in Fig. 11(b). As a result, the improvement in the capacitance of the BLTV thin films was attributed to the compensation of the oxygen vacancy of the BLSF structure. Additionally, the memory window of ferroelectric thin films was changed by the sweeping speed. The ferroelectric capacitance and threshold voltage of MFIS structure slightly decreases as the sweeping speed increases. That was influenced by the mobile ions and charge injection between as-deposited ferroelectric thin films and metal electrode as the sweeping speed increased [9].

3.3. The Influence of Neodymium Doping on the Physical and Electrical Properties of BTV (BNTV) Ferroelectric Thin Films

Figure 12(a) shows x-ray diffraction patterns of the as-deposited ferroelectric thin films for different oxygen concentration on ITO substrate. From the XRD patterns, we found that the ferroelectric thin films exhibited polycrystalline structure. In addition, the (117), (008), and (220) peaks were observed in the XRD pattern. The intensity of the (117) peak of the ferroelectric thin films increases linearly as the oxygen concentration increases from 0 to 40%. The intensity of the (117) peak of the as-deposited ferroelectric thin films decreases at oxygen concentration from 40 to 60%. As shown in Fig. 3, the (117) preferred phase and smallest full-width-half-magnitude (FWHM) value were exhibited by the as-deposited ferroelectric thin film with the 40% oxygen concentration. The polycrystalline structure of the as-deposited $(Bi_{3.25}Nd_{0.75})(Ti_{2.9}V_{0.1})O_{12}$ ferroelectric thin film was optimal at 40% oxygen concentration.

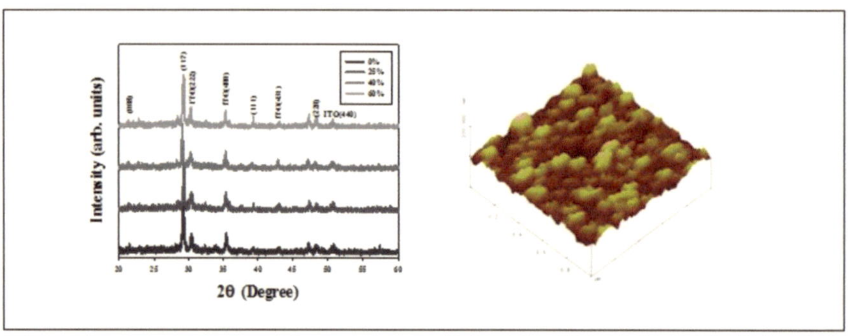

Figure 12. a) The x-ray diffraction patterns of the as-deposited $(Bi_{3.25}Nd_{0.75})(Ti_{2.9}V_{0.1})O_{12}$ ferroelectric thin film for the different oxygen concentration. (b) The surface morphology of the as-deposited $(Bi_{3.25}Nd_{0.75})(Ti_{2.9}V_{0.1})O_{12}$ ferroelectric thin film for 40% oxygen concentration.

Besides, the interface between an electrode and the as-deposited $(Bi_{3.25}Nd_{0.75})(Ti_{2.9}V_{0.1})O_{12}$ thin films was an important factor that seriously influences the physical and electrical properties of the MIM capacitor structure. Therefore, the surface roughness of the as-deposited $(Bi_{3.25}Nd_{0.75})(Ti_{2.9}V_{0.1})O_{12}$ thin films for 40% oxygen concentration was absolutely determined and calculated. Figure 12(b) shows the surface roughness of the as-deposited $(Bi_{3.25}Nd_{0.75})(Ti_{2.9}V_{0.1})O_{12}$ ferroelectric thin film from the AFM images. The roughness of the ferroelectric thin film were 4.278nm. The surface roughness of the as-deposited ferroelectric thin films increases with the oxygen concentration. Therefore, we assume that the surface roughness of the as-deposited ferroelectric thin films increases due to an increase in the crystallinity with oxygen concentration.

http://dx.doi.org/10.5772/52203

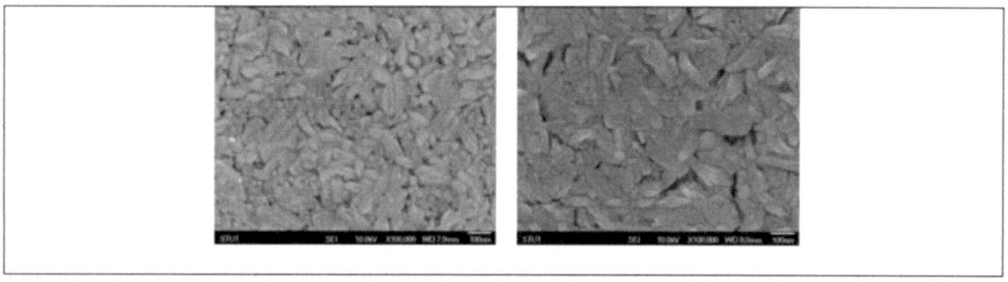

Figure 13. The surface morphology of the as-deposited $(Bi_{3.25}Nd_{0.75})(Ti_{2.9}V_{0.1})O_{12}$ ferroelectric thin film for 40% oxygen concentration. The surface micro-structure of the as-deposited $(Bi_{3.25}Nd_{0.75})(Ti_{2.9}V_{0.1})O_{12}$ ferroelectric thin film for (a) 25% and (b) 40% oxygen concentration.

From the SEM images in Fig. 13, the surface morphology and grain size of the as-deposited $(Bi_{3.25}Nd_{0.75})(Ti_{2.9}V_{0.1})O_{12}$ thin films for 25 and 40% oxygen concentration were observed. The grain size of the as-deposited $(Bi_{3.25}Nd_{0.75})(Ti_{2.9}V_{0.1})O_{12}$ thin films were about 110 nm and 50 nm, respectively. We deduced that grain size changed caused by the different oxygen xoncentration.

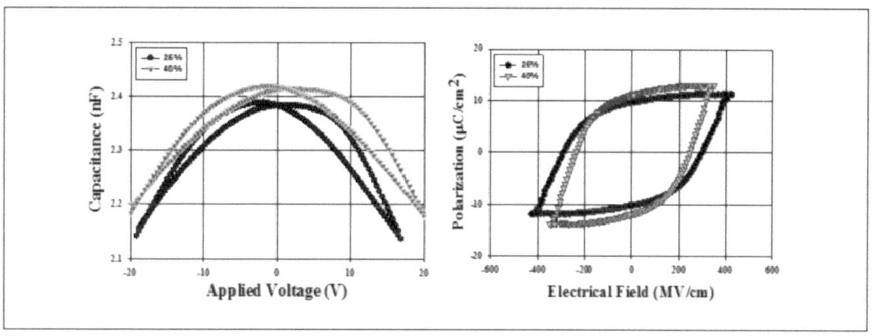

Figure 14. a) The capacitance versus applied voltage (C-V) properties of the as-deposited $(Bi_{3.25}Nd_{0.75})(Ti_{2.9}V_{0.1})O_{12}$ ferroelectric thin film. (b) The polarization versus applied electrical field (p-E) properties of the as-deposited $(Bi_{3.25}Nd_{0.75})(Ti_{2.9}V_{0.1})O_{12}$ ferroelectric thin film.

Figure 14(a) shows the *C-V* characteristics measured with the MFM capacitor structure for the as-deposited $(Bi_{3.25}Nd_{0.75})(Ti_{2.9}V_{0.1})O_{12}$ thin films deposited under various oxygen concentrations. The applied bias voltage is adjusted from -20 to 20V. The capacitance of ferroelectric thin films first increases with the increase of oxygen concentration and reaches the maximum value in the 40 % oxygen atmosphere. Then the capacitance apparently decreases in the further increase of oxygen to 60 %. This variation of capacitance has the similar results with the XRD patterns and the AFM images. The polarization versus applied electrical field (*p-E*) curves of the as-deposited $(Bi_{3.25}Nd_{0.75})(Ti_{2.9}V_{0.1})O_{12}$ ferroelectric thin film at the frequency of 1kHz were shown in fig. 14(b). From the *p-E* curves results, the remnant polarization of ferroelectric thin films were 11 μC/cm², as the coercive filed of 220 kV/cm. In addition, the remnant polarization and coercive filed of ferroelectric thin films for 40% oxygen concentration were about 10 μC/cm² and 300 kV/cm. From the experimental measurement, this result was attributed to the suitable oxygen concentration sample as compared to that of the as-deposited ferroelectric thin film. As the oxygen/argon mixtures were used as the depositing atmosphere, the defects and oxygen vacancies in ferroelectric thin films were filled and compensated by oxygen gas, and the leakage current density were decreased. The low leakage current density will reveal in the 40%-oxygen-deposited ferroelectric thin films. For that the capacitance will be increased and the leakage current density will be decreased. As the applied voltage of 15V was used, the leakage current density of ferroelectric thin films deposited at 40% oxygen concentration is about $1{\times}10^{-9}$A/cm^2.

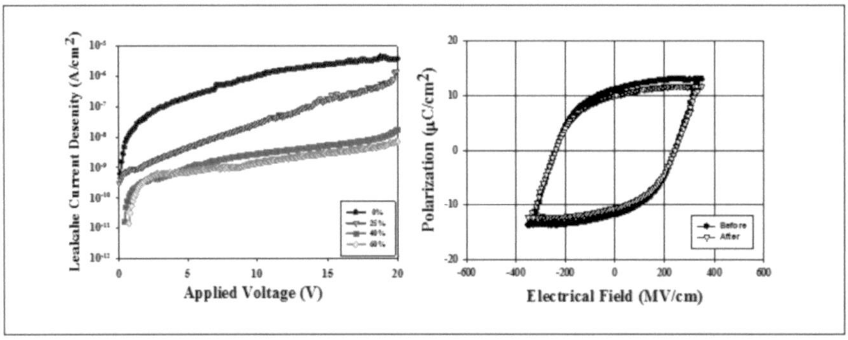

Figure 15. a) The leakage current density versus applied voltage (J-E) properties of the as-deposited $(Bi_{3.25}Nd_{0.75})(Ti_{2.9}V_{0.1})O_{12}$ ferroelectric thin film.(b) The retention and fatigue properties of the as-deposited ferroelectric thin film corresponding hysteresis loop before and after fatigue test.

The retention and fatigue properties for the as-deposited $(Bi_{3.25}Nd_{0.75})(Ti_{2.9}V_{0.1})O_{12}$ ferroelectric thin films were the time dependent change of the polarization state. After long time testing, the polarization loss of the as-deposited ferroelectric thin films was affected by oxygen vacancies, defect, and space charges in the memory device test. Figure 15 shows the polarization versus electrical field (*p-E*) properties of the as-deposited $(Bi_{3.25}Nd_{0.75})(Ti_{2.9}V_{0.1})O_{12}$ ferroelectric thin films before and after the switching of 10^9 cycles. The remnant polarization loss of ferroelectric thin films was about 9% of initial polarization value, respectively. The remnant polarization of ferroelectric thin films was little changed after the test cycles. The fatigue behavior and domain pinning were improved by neodymium and vanadium addition in BIT thin films. To improve bismuth and oxygen vacancy, the high-valence cation substituted for the A-site of $(Bi_{3.25}Nd_{0.75})(Ti_{2.9}V_{0.1})O_{12}$ thin films were observed.

3.4. Bipolar Resistive Switching Properties of Transparent Vanadium Oxide (V_2O_5) Resistive Random Access Memory

Figure 16(b) shows x-ray diffraction patterns of the as-deposited vanadium oxide thin films for 60% oxygen concentration on ITO substrate prepared by different sintering temperature. From the XRD patterns, we found that the vanadium oxide thin films exhibited polycrystalline structure. In addition, the (110), (222), and (400) peaks were observed in the XRD pattern. The intensity of the (110) peak of the thin films increases linearly as the sintering temperature increases from 400 to 550 °C. The intensity of the (110) peak of the as-deposited thin films decreases at sintering temperature from 550 to 600 °C. As shown in Fig. 16, the (110) preferred phase and smallest full-width-half-magnitude (FWHM) value were exhibited by the as-deposited vanadium oxide thin film with the sintering temperature of 550 °C. The polycrystalline structure of the as-deposited vanadium oxide thin film was optimal at 550 °C sintering temperature.

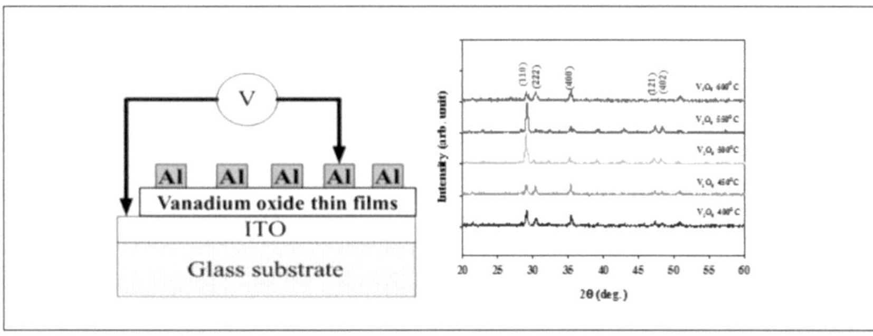

Figure 16. a) Metal-insulator-metal (MIM) structure using as-deposited vanadium oxide thin films. (b) The x-ray diffraction patterns of the as-deposited vanadium oxide thin films on ITO substrate for different sintering temperature.

The thickness of as-depoisted vandium oxide thin films for different sintering temperature was determined by SEM morphology. As the oxygen concentration increases from 0 to 60%, the thickness of as-deposited vandium oxide thin films linearly decreases. In addition, the deposition rate of as-deposited vanadium oxide thin films with 60% oxygen concentration was 2.62 nm/min. The decreases in the deposition rate and thickness of as-depoisted vanadium oxide thin films might be affected by the decrease in Ar/O_2 ratio. The Ar/O_2 ratio was adjusted using argon gas to generate the plasma on the surface of the as-depoisted vanadium oxide ceramic target during sputtering. Figure 17 shows the surface morphology for the as-deposited and 500 °C sintered vanadium oxide thin films. We found that the grain size of 500 °C sintered vanadium oxide thin films were larger than others. The better resistance properties might be caused by this reason.

Figure 18 shows the current vsersus applied voltage (*I-V*) properties of vanadium oxide thin films for the different sintering temperature. After the staring forming process, the device reached a low resistance state (LRS) and high resistance state (HRS). By sweeping the bias to negative over the reset voltage, a gradual decrease of current was presented to switch the cells from LRS to HRS (reset process). Additionally, the cell turns back to LRS while applying a larger positive bias than the set voltage (set process). All of the vanadium oxide thin film were exhibit the bipolar behavior. The *I-V* properties of as-deposited vanadium oxide thin films of 60% oxygen concentration was about 1×10^{-4}A/cm^2 when an applied electrical voltage of 0.1V. During the rf sputtering deposition process, oxygen vacancies appear in the as-deposited vanadium oxide thin films. The defects and oxygen vacancies of as-deposited vanadium oxide thin films were filled and compensated for to different extents at different oxygen concentrations. In addition, the smallest leakage current density of as-deposited vanadium oxide thin films was obtained at an oxygen concentration of 40%. As shown in Fig. 18, the high leakage current density and thin films of as-deposited vanadium oxide thin films for 60% oxygen concentration were attributed to low argon sputtering gas concentration.

Figure 17. The surface morphology for (a) as-deposited (b) 500°C sintered vanadium oxide thin films.

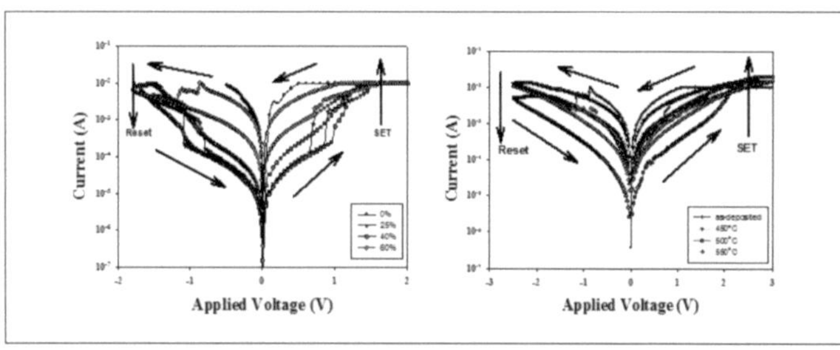

Figure 18. a) Typical *I-V* characteristics of vanadium oxide thin films for the different oxygen concentration. (b) Typical *I-V* characteristics of the as-deposited vanadium oxide thin films for different sintering temperature.

In addition, The transport current of the vanadium oxide thin films decreases linearly as the sintering temperature increases from 450 to 500 °C. The transport current of the as-deposited vanadium oxide thin films increases at sintering temperature from 500 to 550 °C. We found the as-deposited vanadium oxide thin films prepared by 500 °C sintering temperature were exhibited the large the on/off ratio resistance properties. In addition, the switching cycling was measured another type of reliability and retention characteristics were observed. There was a slight flucation of resistance in the HRS and LRS states, and the stable bipolar witching property was observed during 20 cycles. The results show remarkable reliability performance of the resistance random access memory devices for nonvolatile memory applications.

4. Conclusion

In conclusion, BIT, BTV, BNTV, and BLTV thin films were prepared by rf magnetron sputtering. We confirmed that all thin films on $Pt/Ti/SiO_2/Si$ substrate well crystallized by XRD

http://dx.doi.org/10.5772/52203

analysis. The BNTV, BLTV, and BTV shows clear ferroelectricity form the *p-E* curves. The remnant polarization properties of BLTV thin film decreased by 9%, while that of the BTV decreased by 15% after the fatigue test with 10^9 switching cycles. Fatigue behavior in ferroelectric capacitors was attributed to oxygen and bismuth vacancies. The leakage current density of as-deposited BLTV and BTV thin films were lower than those of BIT, which were were attributed to the decrease of oxygen and ismuth vacancies after vanadium and lanthanum addition. The conduction mechanism of as-deposited BTV and BLTV thin films were also proved these results in *J-E* curves. We indicated that small ions substitution for A and B site of BLSF structure was effective decreased the oxygen and bismuth vacancies. Finally, the low threshold voltage and memory window of BLTV thin films were improved from the *C-V* curves measured. In addition, the metal oxide thin films such as, TiO_2, Ta_2O_5, Al_2O_3, and CuO were widely investigated and discussed for applications in nonvolatile resistive random access memory (RRAM) devices. The nonvolatile resistive random access memory (RRAM) devices were well developed and studied because of their structural simplicity, high density and low power, read / write speed (about 10^1 ~ 10^3ns), high operating cycles (> 10^{13}) and other non-volatile advantages. Therefore, the electrical switching properties of the vanadium oxide thin films for nonvolatile resistive random access memory (RRAM) device were observed.

Acknowledgements

The authors will acknowledge to Prof. Ting-Chang Chang and Prof. Cheng-Fu Yang. Additionally, this work will acknowledge the financial support of the National Science Council of the Republic of China (NSC 99-2221-E-272-003) and (NSC 100-2221-E-272-002).

Author details

Kai-Huang Chen[1], Chien-Min Cheng[2], Sean Wu[3], Chin-Hsiung Liao[4] and Jen-Hwan Tsai[4]

1 Department of Electronics Engineering/Tung-Fang Design University/Taiwan,R.O.C., Taiwan

2 Department of Electronic Engineering/Southern Taiwan University of Science and Technology/Taiwan, R.O.C., Taiwan

3 Department of Electronics Engineering/Tung-Fang Design University/Taiwan,R.O.C., Taiwan

4 Department of Mathematics and Physics/Chinese Air Force Academy/Taiwan, R.O.C., Taiwan

References

[1] Shannigrahi, R., & Jang, H. M. (2001). *Appl. Phys. Lett.*, 79, 1051.

[2] Hong, S. K., Suh, C. W., Lee, C. G., Lee, S. W., Hang, E. Y., & Kang, N. S. (2000). *Appl. Phys. Lett.*, 77-76.

[3] Xiong, S. B., & Sakai, S. (1999). *Appl. Phys. Lett.*, 75.

[4] Kim, J. S., & Yoon, S. G. (2000). *Vac. Soc. Technol.*, B, 18, 216.

[5] Wu, T. B., Wu, C. M., & Chen, M. L. (1996). *Appl. Phys. Lett.*, 69, 2659.

[6] Chen, K. H., Chen, Y. C., Yang, C. F., & Chang, T. C. (2008). *J. Phys. Chem. Solids*, 69, 461.

[7] Yang, C. F., Chen, K. H., Chen, Y. C., & Chang, T. C. (2007). *IEEE Trans. Ultrason. Ferroelectr. Freq. Control*, 54.

[8] Yang, C. F., Chen, K. H., Chen, Y. C., & Chang, T. C. (2008). *Appl. Phys. A-Mater. Sci. Process*, 90.

[9] Chen, K. H., Chen, Y. C., Chen, Z. S., Yang, C. F., & Chang, T. C. (2007). *Appl. Phys. A-Mater. Sci. Process*, 89, 533.

[10] Wu, S., Lin, Z. X., Lee, M. S., & Ro, R. (2007). *Appl. Phys. Lett.*, 102, 084908.

[11] Lee, S., Song, E. B., Kim, S., Seo, D. H., Seo, S., Won, K. T., & Wang, K. L. (2012). *Appl. Phys. Lett.*, 100, 023109.

[12] Yang, Y., Jin, L., & Yang, X. D. (2012). *Appl. Phys. Lett.*, 100, 031103.

[13] Scotta, J. F., Paz, C. A., & de Araujoa, B. M. (1994). *J. Alloys, Compounds*, 211 451.

[14] Araujo, C. A., Cuchiaro, J. D., Mc Millian, L. D., Scott, M. C., & Scott, J. F. (1995). *Nature (London)*, 374, 627.

[15] Park, B. H., Kang, B. S., Bu, S. D., Noh, T. W., Lee, J., & Jo, W. (1999). *Nature (London)*, 401, 682.

[16] Leu, C. C., Yao, L. R., Hsu, C. P., & Hu, C. T. (2010). *J. Electrochem. Soc.*, 157, 3, G85.

[17] Chen, K. H., Chen, Y. C., Yang, C. F., & Chang, T. C. (2008). *J. Phys. Chem. Solids*, 69.

[18] Yang, C. F., Chen, K. H., Chen, Y. C., & Chang, T. C. (2007). *IEEE Trans. Ultrason. Ferroelectr. Freq. Control*, 54 1726.

[19] Yang, C. F., Chen, K. H., Chen, Y. C., & Chang, T. C. (2008). *Appl. Phys. A*, 90 329.

[20] Chen, K. H., Chen, Y. C., Chen, Z. S., Yang, C. F., & Chang, T. C. (2007). *Appl. Phys. A*, 89 533.

[21] Watanabe, T., Funakubo, H., Osada, M., Noguchi, Y., & Miyayama, M. (2002). *Appl. Phys. Lett.*, 80 1.

[22] Kim, S. S., Song, T. K., Kim, J. K., & Kim, J. (2002). *J. Appl. Phys.*, 92 4.

[23] Noguchi, Y., & Miyayama, M. (2001). *Appl. Phys. Lett.*, 78 13.

[24] Velu, G., Legrand, C., Tharaud, O., Chapoton, A., Remiens, D., & Horowitz, G. (2001). *Appl. Phys. Lett.*, 79 659.

[25] Chen, K. H., Yang, C. F., Chang, C. H., & Lin, Y. J. (2009). *Jpn. J. Appl. Phys.*, 48 091401.

[26] Miao, J., Yuan, J., Wu, H., Yang, S. B., Xu, B., Cao, L. X., & Zhao, B. R. (2001). *Appl. Phys. Lett.*, 90 022903.

[27] Chen, K. H., Chen, Y. C., Chia, W. K., Chen, Z. S., Yang, C. F., & Chung, H. H. (2008). *Key Eng. Mater.*, 368-372.

[28] Wu, S. Y. (1974). *IEEE Trans. Electron Devices*, 21 499.

[29] Wu, S. Y. (1976). *Ferroelectrics*, 11 379.

[30] Sugibuchi, K., Kurogi, Y., & Endo, N. (1975). J. Appl. Phys. 46 2877. .

[31] Park, B. H., Kang, B. S., Bu, S. D., Noh, T. W., Lee, L., & Joe, W. (1999). *Nature (London)*, 401 682.

[32] Noguchi, Y., Miwa, I., Goshima, Y., & Miyayama, M. (2000). *Jpn. J. Appl. Phys*, 39 L1259.

[33] Noguchi, Y., & Miyayama, M. (2001). *Appl. Phys. Lett.*, 78 1903.

[34] Friessnegg, T., Aggarwal, S., Ramesh, R., Nielsen, B., Poindexter, E. H., & Keeble, D. J. (2000). *Appl. Phys. Lett.*, 77 127.

[35] Watanabe, T., Funakubo, H., Osada, M., Noguchi, Y., & Miyayama, M. (2002). *Appl. Phys. Lett.*, 80 1.

[36] Kim, S. S., Song, T. K., Kim, J. K., & Kim, J. (2002). *J. Appl. Phys.*, 92, 4.

[37] Noguchi, Y., & Miyayama, M. (2001). *Appl. Phys. Lett.*, 78 13.

[38] Rost, T. A., Lin, H., & Rabson, T. A. (1991). *Appl. Phys. Lett.*, 59 3654.

[39] Rost, T. A., Lin, H., Rabson, T. A., Baumann, R. C., & Callahan, D. C. (1991). IEEE Trans. Ultrason. Ferroelectr. Freq. Control, , 38

[40] Fleischer, S., Lai, P. T., & Cheng, Y. C. (1994). *J. Appl. Phys.*, 73 8353.

[41] Mihara, T., & Watanabe, H. (1995). *Part I, Jpn. J. Appl. Phys.*, 34 5664.

[42] Lin, Y. B., & Lee, J. Y. (2000). *J. Appl. Phys.*, 87 1841.

Chapter 20

Nanoscale Ferroelectric Films, Strips and Boxes

Jeffrey F. Webb

Additional information is available at the end of the chapter

http://dx.doi.org/10.5772/52040

1. Introduction

A major characteristic of ferroelectric materials is that they posses a non-zero polarization in the absence of any applied electric field provided that the temperature is below a certain critical temperature[1]; in the literature this is often called a spontaneous polarization, or an equilibrium polarization. Another important defining characteristic is that this polarization can be reversed by applying an electric field. This is the basis of a useful application of ferroelectrics in which binary information can be stored according to the polarization direction which can be switched between two states by an electric field. For example, a thin film of ferroelectric can be switched to different states at different regions by suitably patterned electrodes, thus creating a ferroelectric random access memory[2].

With the increasing miniaturization of devices it becomes important to investigate size-effects in ferroelectrics. Thin film geometries are of interest in which one spatial dimension is confined, as well as strip geometries in which two-dimensions are confined, and, what here we will call box geometry in which all three dimensions are confined. By confinement we mean that surfaces of the ferroelectric that intersect on at least one line going through the central region of the ferroelectric are sufficiently close to this region to make their presence cause a non-negligible effect on the ferroelectric as a whole so that the behaviour is different from what would be expected in a bulk region far from any surface. The geometries chosen are typical of the sort that can be fabricated in micro or nanoelectronics and fit conveniently into a Cartesian coordinate system.

The aim of this chapter is to show how the equilibrium polarization can be calculated in general for a confined volume of ferroelectric and this is applied to the three aforementioned geometries. In fact such calculations for thin films are well established[3–7]; here they are included with the other geometries, which are not so well studied, for completeness, and because they fit logical into the theoretical framework to be presented. A general problem of a similar nature to the one of interest here, but applied to spheres and cylinders has been considered by Morison's et al.[8]. This work employed approximate analytical solutions to the problem of calculating the polarization and avoided a full three-dimensional application of the general theory.

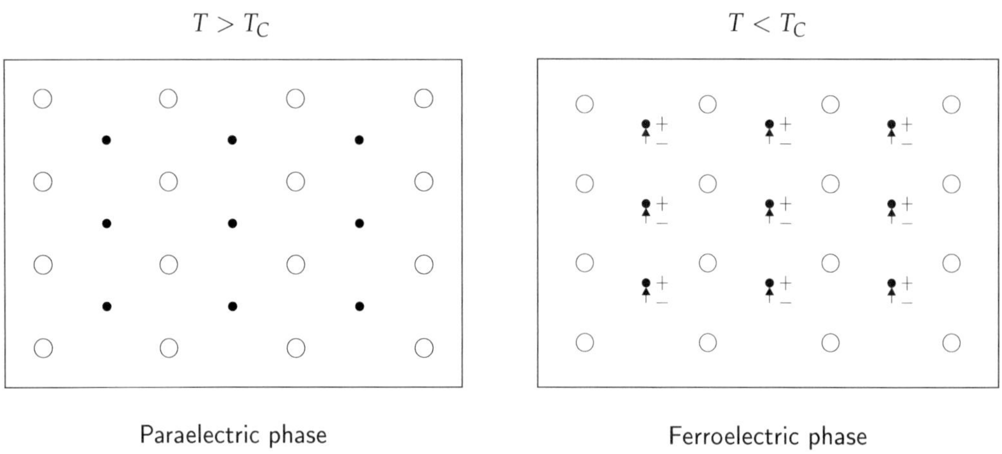

Figure 1. A two-dimensional representation of how a crystal structure changes symmetry when passing between the paraelectric ($T > T_C$) and ferroelectric ($T < T_C$) phases

2. Foundations of the Landau-Devonshire theory of ferroelectrics

2.1. Bulk ferroelectrics

Since Landau-Devonshire theory is the theoretical basis for the work in this chapter it will be introduced here. Further details can be found elsewhere such as in Refs. [1, 9, 10]. In this section we consider the case of a bulk ferroelectric; the effect of surfaces will be introduced after that.

It is common to state that the starting point for Landau-Devonshire theory is the free energy per unit volume of a bulk ferroelectric expressed as an expansion in powers of the polarization P. The observed equilibrium polarization P_0 is then given by the value of P that minimizes the free energy. Here, however, following Strukov and Lenanyuk[10], we outline how this expansion comes from a statistical thermodynamic treatment involving an incomplete Gibbs potential, since this gives more insight into the theory. First a few more details about the nature of ferroelectrics.

When a ferroelectric crystal is above the critical temperature T_C there is no spontaneous polarization; this is the paraelectric or high symmetry phase in which the crystal structure is of higher symmetry than the ferroelectric phase below T_C since the appearance of the spontaneous polarization lowers the symmetry, as is evident from the two-dimensional representation in Fig. 1. Thus we see that during the transition from the paraelectric to ferroelectric state, a structural phase transition occurs. A consequence of this is that as the temperature approaches T_C the structure becomes easy to distort giving rise to anomalous peaks in some of the properties such as the dielectric function ϵ, as illustrated in Fig. 2. In fact, in the ferroelectric phase, a crystal may have different domains such that the polarization in adjacent domains is not in the same direction[9], however it is possible to produce single-domain crystals—by cooling through T_C in the presence of an electric field in one of the possible polarization directions for instance—and that is what we consider here.

http://dx.doi.org/10.5772/52040

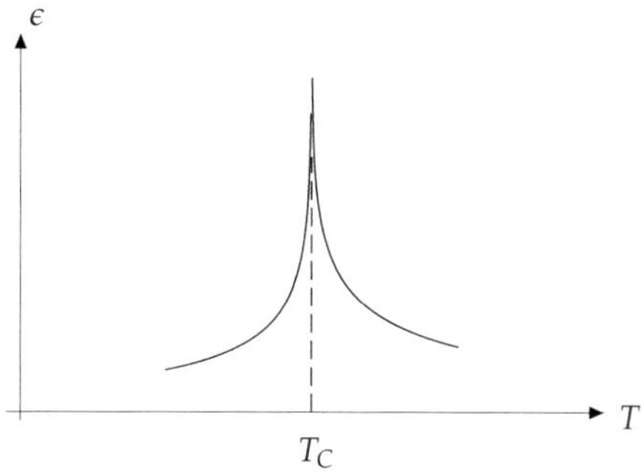

Figure 2. A peak in the dielectric function ϵ occurs as $T \to T_C$.

In the simplest case, then, a ferroelectric phase transition can be described by a single parameter that in Fig. 1 would be describable by the displacement of the sublattice from its central position. Since the paraelectric phase is of higher symmetry than the ferroelectric phase we can consider it to be more orderly and so the parameter is sometimes referred to as an order parameter. Actually the atoms are ions with charge and the corner-atom lattice would have a charge opposite to that of the interior sub-lattice. From the macroscopic point of view the corner atoms can be considered to produce a focus of charge at the center of each rectangle of opposite sign to the atoms that are displaced during the phase transition (see Fig. 1 where the centers of positive and negative charge are marked for the ferroelectric phase). For this reason the order parameter is often taken to be the polarization, which is the dipole moment per unit volume. For the time being, however, we will stay with the displacement representation of the order parameter as this is more directly related to the structural transition. If the displacement is denoted by η then it is clear that in the ferroelectric phase $\eta \neq 0$ and in the paraelectric phase $\eta = 0$.

The description above implies that each center ion is displaced in the same direction and this is the case for what are termed displacive ferroelectrics. There is another possibility, however, in which the displacement is not the same for every unit cell. For example, in some cells the displacement could be up while in others it could be down. Macroscopically ferroelectric and paraelectric phases can still result but now the ferroelectric phase occurs when statistically there are more displaced in one direction than in the other; and the paraelectric phase corresponds to the number of atoms displaced upwards is statistically equal to the number displaced downwards. In this case the ferroelectric is called an order-disorder ferroelectric. In terms of the macroscopic crystal symmetry however displacive and order-disorder ferroelectrics are equivalent and both can be described in general by Landau-Devonshire theory. The crystal structure of order-disorder ferroelectrics tends to be more complicated than for the displacive type[9, 10].

The basis of the Landau-Devonshire theory is that in the phase transition the ferroelectric phase may be represented by a distorted symmetrical phase. As outlined above this is a macroscopic view and the theory is phenomenological, as will be brought out further below. It turns out that the symmetry elements lost by the crystal at the transition temperature is sufficient information for a description of the anomalies of practically all of the thermodynamic properties of the crystal[10].

To make further progress consider the crystal as a system of N interacting particles with potential energy $U(\mathbf{r}_1, \ldots, \mathbf{r}_N)$. According to Gibbs[11, Chapter 8], the equilibrium thermodynamic potential at pressure p and temperature T associated with the potential energy of interaction of the particles is given by

$$\Phi(p, T) = -k_B T \ln Z, \tag{1}$$

where k_B is Boltzmann's constant and

$$Z = \int_{-\infty}^{\infty} \exp\left[-U(\mathbf{r}_1, \ldots, \mathbf{r}_N)/(k_B T)\right] \prod_i d\mathbf{r}_i. \tag{2}$$

The expression for Z in Eq. (2) comes from the probability that the value of the radius vector of the first particle lies between $\mathbf{r}_1$ and $\mathbf{r}_1 + d\mathbf{r}_1$ and similarly for the other vectors, being given by

$$dw = C \exp\left[-U(\mathbf{r}_1, \ldots, \mathbf{r}_N)/(k_B T)\right] \prod_i d\mathbf{r}_i. \tag{3}$$

Since integration of all of the variables must yield unity, $C = 1/Z$. The probability distribution can therefore be written as

$$dw = \exp\{[\Phi(p, T) - U(\mathbf{r}_1, \ldots, \mathbf{r}_N)] / (k_B T)\} \prod_i d\mathbf{r}_i. \tag{4}$$

A Gibbs thermodynamic potential for nonequilibrium states can also be formulated by using a larger number of variables than only p and T. The new variables are introduced via a linear transformation:

$$\mathbf{r}_1, \ldots, \mathbf{r}_N \rightarrow \xi_1, \ldots, \xi_{3N}. \tag{5}$$

We choose one of these variables to be η, which describes the nonequilibrium state of interest as explained above, by setting $\xi_1 = \eta$. Now

$$dw = \exp\{[\Phi(p, T) - U(\eta, \xi_2 \ldots, \xi_{3N})] / (k_B T)\} d\eta d\xi_2 \cdots \xi_{3N}, \tag{6}$$

the probability that the variables will lie in the ranges η to $\eta + d\eta$, ξ_2 to $\xi_2 + d\xi_2$, and so on. Focusing on the variable of interest η the probability of finding the system in a state in which this variable is in the range $\eta + d\eta$ is

http://dx.doi.org/10.5772/52040

$$dw(\eta) = d\eta \int_{-\infty}^{\infty} \exp\{[\Phi(p,T) - U(\eta, \xi_2, \ldots, \xi_{3N})] / (k_B T)\} d\xi_2 \cdots \xi_{3N}. \tag{7}$$

The thermodynamic potential is now also a function of η, and we write

$$\Phi(p,T,\eta) = -k_B T \ln \int_{-\infty}^{\infty} \exp\{[-U(\eta, \xi_2, \ldots, \xi_{3N})] / (k_B T)\} d\xi_2 \cdots \xi_{3N}, \tag{8}$$

and

$$dw(\eta) = \exp\{[\Phi(p,T) - \Phi(p,T,\eta)] / (k_B T)\} d\eta. \tag{9}$$

Here it can be seen that the value of η that minimizes $\Phi(p,T,\eta)$ will maximize the distribution function and so this value corresponds to the equilibrium value of η. If $\Phi(p,T,\eta)$ is known, the usual equilibrium function $\Phi(p,T)$ can be found, using Eq. (9), from

$$\Phi(p,T) = -k_B T \ln \int_{-\infty}^{\infty} \exp\{[-\Phi(\eta)] / (k_B T)\} d\eta. \tag{10}$$

Now, let the equilibrium value of η be η_0—which corresponds to the minimum value of $\Phi(p,T,\eta)$—and expand $\Phi(p,T,\eta)$ in a series about the point $\eta = \eta_0$:

$$\Phi(p,T,\eta) = \Phi(p,T,\eta_0) + \Delta\Phi(p,T,\eta-\eta_0) = \Phi(p,T,\eta_0) + \frac{1}{2}A(p,T)(\eta-\eta_0)^2 + \cdots. \tag{11}$$

Using Eq. (10), $\Phi(p,T)$ can be written as

$$\Phi(p,T) = \Phi(p,T,\eta_0) - k_B T \ln \int_{-\infty}^{\infty} \exp\{[-\Delta\Phi(p,T,\eta-\eta_0)] / (k_B T)\} d\eta. \tag{12}$$

In Eq. (12) the first term is the minimum of the potential as a function of p, T and η, and we are interested in the minimizing value $\eta = \eta_0$ for given values of p and T. The second term is the contribution to the thermodynamic potential of fluctuations in η with order of magnitude given by the thermal energy per degree of freedom $K_B T/2$. However since we are dealing with a large number of particles in a macroscopic sample ($\sim 10^{23}$) the fluctuations are very small and the second term can be neglected to give

$$\Phi(p,T) = \Phi(p,T,\eta_0). \tag{13}$$

Note that if the total number of degrees of freedom (left unintegrated as the single degree η was) required to describe the system is comparable to N, then the neglect of the second term is no longer valid and Eq. (10) must be used in place of Eq. (13). We will not be concerned with such situations here, but very close to the transition temperature the Landau-Devonshire theory breaks down and the fluctuations must be considered, which can be done with the aid of Eq. (10), as discussed in Ref. [10].

In general thermodynamic functions which contain extra variables that remain unintegrated as η was above, are nonequilibrium thermodynamic functions known as incomplete functions. The general form is

$$\Phi(p, T, \eta_1, \eta_2, \ldots, , \eta_n) = \Phi_0(p, T) + \Phi_1(p, T, \eta_1, \eta_2, \ldots, , \eta_n). \tag{14}$$

The equilibrium values of the η_i are found by minimizing $\Phi(p, T, \eta_1, \eta_2, \ldots, , \eta_n)$ according to

$$\frac{\partial \Phi}{\partial \eta_1} = 0, \ldots, \frac{\partial \Phi}{\partial \eta_n} = 0, \tag{15}$$

which determine a set of minimizing values $\eta_{oi}(p, T)$. Substituting these values into Eq. (14) determines the equilibrium thermodynamic function

$$\Phi(p, T) = \Phi_0(p, T) + \Phi_1(p, T, \eta_{01}, \eta_{02}, \ldots, , \eta_{0n}). \tag{16}$$

This does not take fluctuations of the η_{0i} into account but the corresponding error will be small provided that $n \ll 3N$.

Although we are only considering a ferroelectric phase transition of the simplest type describable by introducing a single extra variable η into the thermodynamic potential the general form shows how to treat more complicated cases for which the polarization may occur in more ways than along one line, as we will see later. Polarization in three-dimensionss can be treated with three parameters, η_1 to η_3.

Returning to a one-component case we now show how the form of the thermodynamic potential can be worked out. As explained, in a ferroelectric phase transition we are dealing with displacements of certain atoms or groups of atoms; the structure of the nonsymmetric phase can be obtained from the structure of the symmetric phase by small displacements. Although the same lowering of symmetry in a phase transition can occur with different types of ordering implying that the choice of the order parameter is ambiguous, it turns out that the character of the anomalies of the physical properties in the transitions can be elucidated for any particular relationship between the order parameter and the displacements as long at the appearance of the order parameter leads to a symmetry change corresponding to that of the crystal's[10, 12].

In view of this we can take the order parameter as the extra variable of the incomplete thermodynamic potential. The phase transition can then be described at a given pressure if $\Phi(p, T, \eta)$ has a minimum at $\eta = 0$ in the symmetric phase ($T > T_C$), and at least two

http://dx.doi.org/10.5772/52040

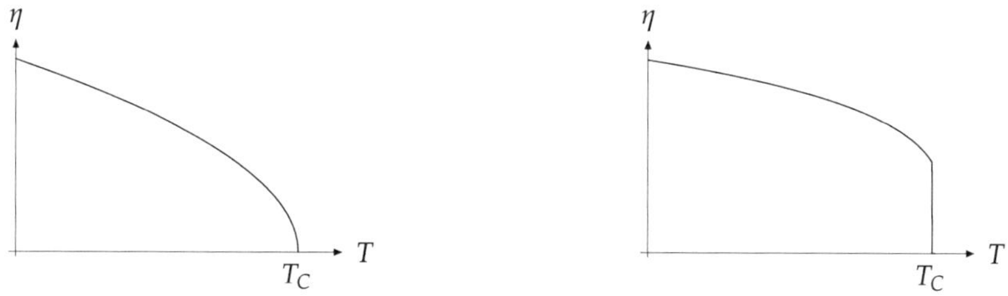

(a) (b)

Figure 3. (a) Second order phase transition in which the order parameter decreases continuously to zero. (b) First order phase transition in which the order parameter displays a discontinuity. There is a continuous decrease in η before, which means that it is a first-order transition close to a second order transition. at $T = T_C$.

minima[1] at $\eta \neq 0$ in the nonsymmetric phase ($T < T_C$). The function $\Phi(p, T, \eta)$ must be consistent with this as T changes through the transition temperature so that it is determined for both the symmetric and nonsymmetric phases. These requirements place restrictions on the dependence of the potential Φ on η: Φ is a scalar function characterizing the physical properties of ferroelectric crystals, and as such must be invariant under any symmetry operations on the symmetric phase consistent with the crystal symmetry of this phase.

Before exploring further the form of Φ that describes the phase transition the difference between first and second order phase changes will be discussed. In a second order transition η decreases continuously to zero as the temperature is lowered through T_C; in a first order transition, on the other hand, there is a discontinuous jump of η to zero at $T = T_C$. Both situations are illustrated in Fig. 3. The first order transition shown in this figure is in fact close to a second order transition and so there is a continuous decrease in η before the jump and the jump is not very large. Crystals can exhibit first or second order transitions depending on the crystal structure[9, 10]. Either case can be treated by Landau-Devonshire theory, as long as the first order case is close to a second order transition. Turning back now to the development of an expression for Φ, we see that in the vicinity of a second order transition it is permissible to deal only with small lattice distortions, that is, small η. Hence the thermodynamic potential can be expanded into a series in η, with p and T treated as parameters, and we obtain

$$\Phi(p, T, \eta) = \Phi(p, T, 0) + \Phi'(p, T, 0)\eta + \frac{1}{2}\Phi'(p, T, 0)\eta^2 + \cdots . \tag{17}$$

To represent a given crystal this expression must be invariant under symmetry operations that correspond to the symmetry elements of the crystal. A ferroelectric phase transition resulting in a spontaneous polarization occurs as long as the order parameter transforms as a vector component. Only crystal symmetries for which this is true are ferroelectric and then the order parameter can be considered to be equivalent to a component of the spontaneous polarization (this point is discussed in more detail in Refs.[9, 10, 12]).

[1] Part of the definition of a ferroelectric is that it has a reversible polarization in the nonsymmetric phase so that at least two minimum are required; otherwise the reverse direction would not be an equilibrium state.

$2/m$	1	2	m	$\bar{1}$
$\eta = P$	P	P	$-P$	$-P$

Table 1. Transformation of the order parameter

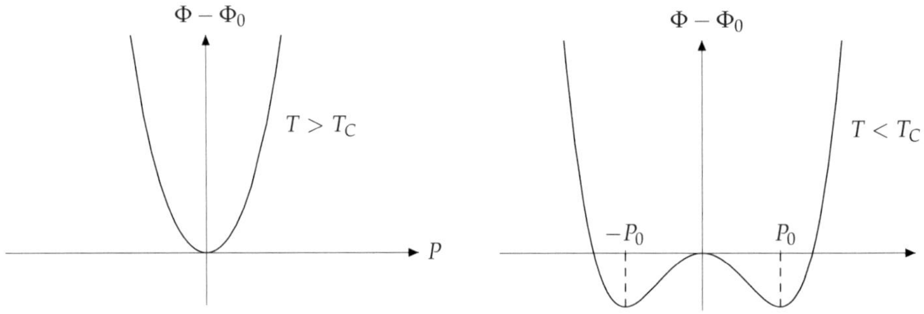

Figure 4. The form of $\Phi - \Phi_0$ that satisfies the basic requirements for a ferroelectric phase transition. When $T > T_C$ there is single minimum at $P = 0$ and there is no spontaneous polarization; for $T < T_C$ a spontaneous polarization can exist at one of the minima at $P = \pm P_0$.

To make further progress we consider a definite example of a symmetry which allows the order parameter to transform as a vector and so corresponds to a polarization having a single component; to reflect this correspondence we also make the notation change $\eta \rightarrow P$ where P is the polarization component.[2] Consider a second-order phase transition such that the change into the ferroelectric state at the critical temperature is accompanied by a symmetry change from the group $2/m$ (symmetry elements: 1, $\bar{1}$, m, 2) to 2 (symmetry elements: $1, 2$). The ferroelectric triglycine sulphate shows this symmetry change during its phase transition[13, 14]. During the transition to the nonsymmetric phase the crystal loses the inversion center and the plane of symmetry, $\bar{1}$ and m. These disappearing symmetry elements correspond to the symmetry operations that reverse the sign of the order parameter $\eta = P$, as is shown in Table 1. Since Φ must be invariant under any transformations of the symmetrical phase it is clear from this that it cannot include terms linear in P or odd powers of P, since such terms would be changed by the operations m and $\bar{1}$.

Therefore the first $\eta = P$ dependent term in Eq. (17) will be of the form $A(p,T)P^2$. For simplicity we now assume that the pressure is at some fixed value and discuss the temperature dependence. The expression for Φ must reflect the fact that at $T > T_C$ there is a minimum at $P = 0$ so that the equilibrium state (represented by the minimum) is one in which $P = 0$, and when $T < T_C$ there will be two minima for which $P \neq 0$ so that the possible equilibrium states occur at nonzero values of P. This is satisfied if $A(T)$ passes continuously from $A(T < T_C) < 0$ to $A(T > T_C) > 0$ when $T > T_C$, with $A(T = T_C) = 0$, provided that the next allowable term $B(T)P^4$ (remembering that odd terms have the wrong symmetry) is included with $B(T) > 0$, as can be seen from Fig. 4. This results in a thermodynamic potential of the form

$$\Phi = \Phi_0 + A(T)P^2 + B(T)P^4. \tag{18}$$

[2] The general case in which the polarization has three components would be covered by three order parameters and the correspondence would be $(\eta_1 \rightarrow P_1, \eta_2 \rightarrow P_2, \eta_3 \rightarrow P_3)$, with $\mathbf{P} = (P_1, P_2, P_3)$.

http://dx.doi.org/10.5772/52040

Generally the exact form of the temperature dependence may be difficult to find. However near the phase transition temperature a series expansion in powers of $T - T_C$ (at a given pressure) can be used to give

$$A(T) = A(T_C) + A_1(T - T_C)A_2(T - T_C)^2 + \cdots \tag{19}$$

$$B(T) = B(T_C) + B_1(T - T_C)B_2(T - T_C)^2 + \cdots . \tag{20}$$

Taking into account the above mentioned properties: $A(T_C) = 0$ and $B(T_C) > 0$, the simplest forms retaining the essential non-zero first terms gives

$$A(T) = A_1(T - T_C) = \frac{1}{2}a(T - T_C) \tag{21}$$

$$B(T) = B(T_C) = \frac{1}{4}b. \tag{22}$$

The numerical factors are introduced for convenience, as will be clear shortly. We thus obtain, for a given pressure,

$$\Phi(T, P) = \Phi_0(T) + \frac{1}{2}a(T - T_C)P^2 + \frac{1}{4}bP^4. \tag{23}$$

Usually it is sufficient to consider the energy due to the presence of the spontaneous polarization per unit volume (we are dealing with a bulk sample assumed uniform of the crystal volume) so that it is convenient to write, for a volume v of ferroelectric,

$$F = \frac{\Phi(T, P) - \Phi_0(T)}{v} = \frac{1}{2}\alpha(T - T_C)P^2 + \frac{1}{4}\beta P^4, \tag{24}$$

where $\alpha = a/v > 0$, $\beta = b/v > 0$, and we refer to F as the Gibbs free energy density.

The equilibrium polarization P_0 is now found by minimizing F from the conditions

$$\frac{\partial F}{\partial P} = 0, \quad \text{and} \quad \frac{\partial^2 F}{\partial P^2} > 0.$$

From this we find

$$P_0^2 = \begin{cases} 0 & \text{for } T > T_C, \\ -\frac{\alpha(T - T_C)}{\beta} & \text{for } T < T_C. \end{cases} \tag{25}$$

Here we see that it is necessary to differentiate to find P_0 which explains why the numerical coefficients $1/2$ and $1/4$ were introduced into the expansion of the thermodynamic potential.

The form of F in Eq. (24) is sufficient to describe a second order phase transition since $P_0 \to 0$ continuously as $T \to T_C$. For second order transitions in which there is a discontinuous jump, Eq. (24) is insufficient. However a first-order transitions near to a second-order transition the

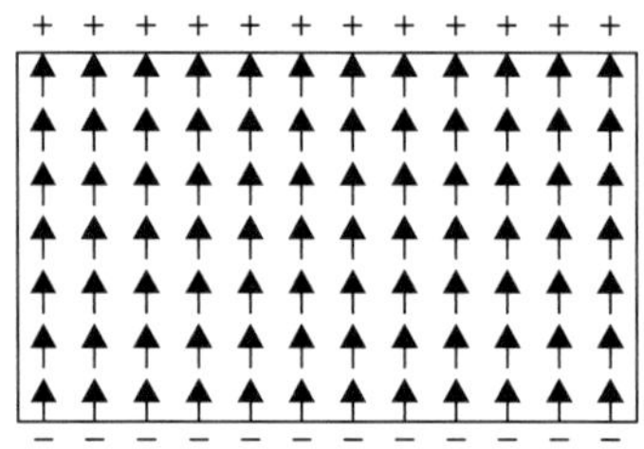

Figure 5. A component of polarization that is directed along a normal to the plane of the film when $T < T_C$. The resulting surface charges with opposite sides having signs of opposite charge, creates an electric field which in the opposite direction to the polarization.

phase change can be described by adding a further term in P^6 (a term in P^5, since it is an odd power, does not fulfill the symmetry requirements) which results in an energy density given by

$$F = \frac{1}{2}\alpha(T - T_C)P^2 + \frac{1}{4}\beta P^4 + \frac{1}{6}\gamma P^6. \tag{26}$$

But, whereas $\beta > 0$ for a second order transition, for F to exhibit a discontinuous jump it is necessary to choose $\beta < 0$ with $\gamma > 0$. More on this can be found in Lines and Glass[9].

We have thus shown in some detail how it is that a relatively simple free energy expression can account for the main characteristics of a ferroelectric, with the observed equilibrium polarization found by minimizing the free energy. Other terms can be added to the free energy to account for various external influences. For example an external electric field with component E along the direction of polarization can be accounted by first substituting the equilibrium polarization P_0 into F and adding a term EP_0. The corresponding susceptibility is then given by $\chi = \frac{\partial^2 F}{\partial E^2}$.

Note that Landau-Devonshire theory is a phenomenological theory of the macroscopic properties of the ferroelectric: the coefficients, α, β, etc., are not derivable from it; instead they must be found from experiment, or, in some cases, can be found from first-principle calculations[1].

2.2. Extension to thin films

Here a free-standing thin film is considered. If the equilibrium polarization below T_C has a component that is aligned with a normal to the plane of the film, then depolarization effects due to the space charge that appears at the surfaces, as illustrated in Fig. 5, must be taken into account. At first we avoid this complication by assuming a polarization that is in-plane. Also to remain with a one dimensional treatment for now we assume that this this is along the x direction of a Cartesian coordinate system and that the plane surfaces of the film are normal to the z axis at $z = 0$ and $z = L$, where L is the thickness of the film. The effect of the surfaces is such that the polarization near them may differ from the bulk value. For the case just described this implies that $P = P(z)$. It can be shown[1, 3, 15] that this can be accounted for in the free energy by adding surface terms involving P^2 integrated over the surfaces, and a gradient term $|dp/dz|$. The free energy is now given by

$$G = \left(\iint dxdy \right) \int_0^L f(P, dP/dz)\, dz + \frac{1}{2}\frac{D}{d} \left(\iint dx\, dy \right) \left[P^2(0) + P^2(L) \right], \tag{27}$$

in which

$$f(P, dP/dz) = \frac{1}{2}AP^2 + \frac{1}{4}BP^4 + \frac{1}{6}CP^6 + \frac{1}{2}D\left(\frac{dP}{dz}\right)^2, \tag{28}$$

the temperature dependence is in A through $A = a(T - T_C)$, and the integrals over x and y are factored out of the three dimensional integral over a volume of the film because P does not vary in these directions. Thus any surface area can be chosen to integrate over so that, if this areas is S, then $\iint dxdy = S$, and we can write a free energy per unit area as

$$F = \frac{G}{S} = \int_0^L dz \left[\frac{1}{2}AP^2 + \frac{1}{4}BP^4 + \frac{1}{6}CP^6 + \frac{1}{2}D\left(\frac{dP}{dz}\right)^2 \right] + \frac{1}{2}D\left[\frac{P^2(0)}{d} + \frac{P^2(L)}{d} \right]. \tag{29}$$

Here since the film is free standing with identical surface properties for both surfaces the surface terms at $z = 0$ and $z = L$ involve the same factor, $1/d$, which simplifies the problem. If the film was not free standing, with for example, one surface an interface with a substrate, then this could be modeled by introducing different factors[6] so that the last term in square brackets in Eq. (29) would be replaced by $P(0)/d_1 + P(L)/d_2$.

The equilibrium polarization is still the P that minimizes F, but now the problem is to find the function $P(z) = P_0(z)$ that minimizes F rather than a finite set of values. So we are dealing with the minimization of a functional and the classical methods from the calculus of variations[16, Chapter 4] may be used, which cast the problem into the form of a differential equation known as the Euler-Lagrange equation and the function that minimizes F is the solution to this equation for a set of boundary conditions that also need to be specified. In this case, as will be shown below the boundary conditions used are implicit in Eq. (29) and are equivalent to allowing the polarization to be free at the film surfaces rather than fixed. More details on the method for the general three dimensional case are given in Section 4. Here we simply state that the classical methods lead to the following Euler-Lagrange equation and boundary conditions for the equilibrium polarization in the film:

$$D\frac{d^2P_0}{dz^2} - AP_0 - BP_0^3 - CP_0^5 = 0. \tag{30}$$

With boundary conditions

$$\frac{dP_0}{dz} - \frac{1}{d}P_0 = 0, \quad \text{at } z = 0, \tag{31}$$

and

$$\frac{dP_0}{dz} + \frac{1}{d}P_0 = 0, \quad \text{at } z = L. \tag{32}$$

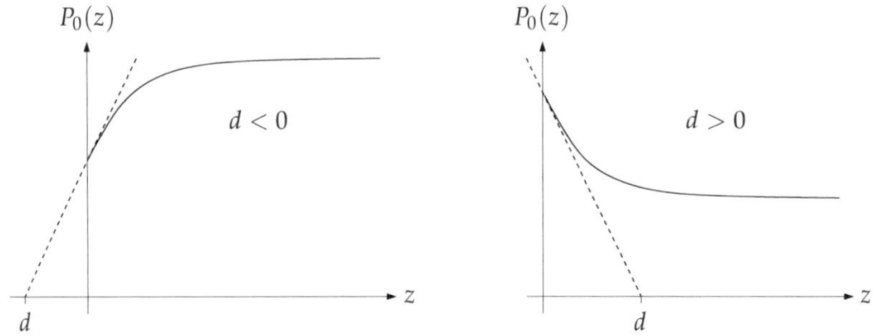

Figure 6. The extrapolation length d for one of the film surfaces at $z = 0$. For $d < 0$ the polarization turns down as the surface is approached; for $d > 0$ it turns up; this is also true for the surface at $z = L$ (not shown). Here the extrapolated gradient line crosses the z-axis at $z = 0 + d$; for the surface at $z = L$ it would cross at $z = L + d$.

It can be seen from the boundary conditions that d can be interpreted as an extrapolation length, as illustrated in Fig. 6. For $d < 0$ the polarization turns upwards as it approaches a surface; for $d > 0$ it turns down as a surface is approached.

Now the task is to solve the differential equation—Eq. (30) which is nonlinear—subject to the boundary conditions. For first order transitions in which $C \neq 0$ the equation must be solved numerically[17, 18]. However, for the simpler case of second-order transitions ($C = 0$) even though the equation is still nonlinear, an analytical solution can be found in terms of elliptic functions[3, 19, 20] and is given by

$$P_0(z) = P_1 \,\mathrm{sn}\left[K(\lambda) - \frac{z - L/2}{\zeta}, \lambda\right], \quad \text{for } d < 0, \tag{33}$$

where sn is the Jacobian elliptic function with modulus $\lambda = P_1/P_2$, $K(\lambda)$ is the complete elliptic integral of the first kind (see Ref. [21] for more on elliptic functions and integrals),

$$P_1^2 = -\frac{A}{B} - \sqrt{\frac{A^2}{B^2} - \frac{4G}{B}}, \tag{34}$$

$$P_2^2 = -\frac{A}{B} + \sqrt{\frac{A^2}{B^2} - \frac{4G}{B}}, \tag{35}$$

$$\zeta = \frac{1}{P_2}\sqrt{\frac{2D}{B}}, \tag{36}$$

and G is found by substituting it into the boundary conditions and solving the resulting transcendental equations numerically.

A plot of the polarization profile according to Eq. (33) is given in Fig. 7 for parameter values given in the figure caption. The polarization throughout the film decreases continuously to zero as the temperature increases and approaches T_C. So there are two states: a ferroelectric state below T_C for which there is a polarization everywhere in the film and a paraelectric state above T_C for which the polarization is zero everywhere.

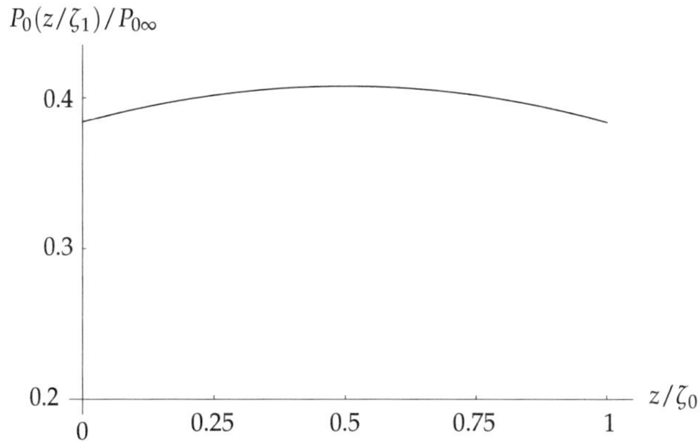

Figure 7. Polarization in a thin film plotted from Eq. (33) and the boundary conditions in Eqs. (31) and (32). Dimensionless variables and parameters used are: $P_{0\infty} = (aT_C/B)^{1/2}$ (the spontaneous polarization for a bulk ferroelectric), $\zeta_0 = [2D/(aT_C)]^{1/2}$, $\zeta_1 = P_{0\infty}/P_2$ (P_2 is given by Eq. (35)), $\Delta T' = (T - T_C)/T_C = -0.4$, $L' = L/\zeta_0 = 1$, $d' = 4L'$, $G' = 4GB/(a/T_C)^2 = 0.105$ (found from the boundary conditions).

The case $d > 0$ is complicated by the appearance of a surface term between T_C and a lower temperature T_{C0} for which the regions near the film surfaces have a spontaneous polarization but the polarization in the interior is still zero. Only below T_{C0} is there a nonzero polarization in the interior of the film. The form of the solution, although still expressible in terms of an elliptic function is somewhat different for the surface state, as is discussed elsewhere[3, 19, 20]

The above was for in-plane polarization which, as has been explained, avoids the need to consider a depolarization field induced by the polarization. In general, however, there may be an out-of-plane component that would give rise to a depolarization field. Theoretical descriptions of the depolarization field have been considered by several authors[5, 7, 22, 23]. For the case in which there are no free charges at the surfaces it can be shown[5, 22] from Maxwell's equations that a term

$$\frac{1}{2\epsilon_0\epsilon_\infty}\left[P\left(P - \langle P\rangle\right)\right] \tag{37}$$

added to the energy density in Eq. (28) accounts for the effect of the depolarization field provided that P is now taken to be directed along a normal to the plane of the the film (as it is in Fig. 5). In Eq. (37) ϵ_0 is the permittivity of free space and ϵ_∞ is the background dielectric constant of the film[5, 24], and

$$\langle P\rangle = \frac{1}{L}\int_0^L P\,dz, \tag{38}$$

which is the average value of $P(z)$ in the film.

3. The free energy density appropriate for a three-dimensional treatment of a ferroelectric crystal

The general form of the free energy density for the interior of a ferroelectric crystal in three-dimensions can be written as series expansion in components of the polarization, together with terms taking into account the influence of surfaces and depolarization fields. The region occupied by the interior of the ferroelectric (the surface energy will appear later as a surface energy density integrated over the surface), is taken to be the interior of a rectangular box defined by

$$V = \{\mathbf{x} \mid x_i \in (l_i^-, l_i^+), i = 1,2,3\}, \tag{39}$$

so that the sides of the box are given by $l_i = l_i^+ - l_i^-$, $i = 1,2,3$. Extending the formalism in the previous sections to three dimensions, the energy density can be written

$$f(\mathbf{x}, \mathbf{P}, \mathbf{P}_{x_1}, \mathbf{P}_{x_2}, \mathbf{P}_{x_3}, \langle \mathbf{P} \rangle) = f_{\text{series}}(\mathbf{x}, \mathbf{P}) + f_{\text{grad}}(\mathbf{x}, \mathbf{P}_{x_1}, \mathbf{P}_{x_2}, \mathbf{P}_{x_3}) + f_{\text{dep}}(\mathbf{x}, \mathbf{P}, \langle \mathbf{P} \rangle) \tag{40}$$

The notation used for the arguments, which will be useful when applying calculus of variations methods in the next section, is as follows: $\mathbf{x} = (x_1, x_2, x_3)$, for a position in the crystal in Cartesian coordinates; $\mathbf{P} = (P_1, P_2, P_3)$, the polarization vector; $P_i = P_i(x_1, x_2, x_3)$, $i = 1,2,3$; $\mathbf{P}_{x_i} = \partial \mathbf{P}/\partial x_i$, $i = 1,2,3$, and the averages vector is $\langle \mathbf{P} \rangle = (\langle P_1 \rangle, \langle P_2 \rangle, \langle P_3 \rangle)$ in which

$$\langle P_i \rangle = \frac{1}{l_i} \int_{l_i^-}^{l_i^+} P_i \, dx_i. \tag{41}$$

The terms on the right of Eq. (40) are given next. The first of these terms f_{series} is the series expansion part of the free energy (due to the polarization, without a constant first term) in terms of the components of the polarization. A general series expansion in these terms can be written as

$$f_{\text{series}}(\mathbf{x}, \mathbf{P}) = \sum_{i=1}^{\infty} \frac{1}{i} \left(\sum_{\alpha_1 \geqslant \cdots \geqslant \alpha_i} A_{\alpha_1 \cdots \alpha_i} P_{\alpha_1} \cdots P_{\alpha_i} \right). \tag{42}$$

Here each α_i runs from x_1 to x_3 and the notation $\alpha_1 \cdots \alpha_i$ indicates that all permutations of $P_{\alpha_1} \cdots P_{\alpha_i}$ of the ith term, which would otherwise appear as separate, have already been summed: the inequalities insure that only one permutation is present for the ith term with coefficient $A_{\alpha_1 \cdots \alpha_i}$. However in practice, depending on the crystal symmetry, terms which are not separately invariant under the symmetry transformations of the symmetry group of the crystal, will not appear.

The term f_{grad}, a gradient term, is given by

$$f_{\text{grad}}(\mathbf{x}, \mathbf{P}_{x_1}, \mathbf{P}_{x_2}, \mathbf{P}_{x_3}) = \frac{1}{2}\delta \left[|\nabla P_1|^2 + |\nabla P_2|^2 + |\nabla P_3|^2 \right] = \frac{1}{2}\delta \left[\left| \frac{\partial \mathbf{P}}{\partial x_1} \right|^2 + \left| \frac{\partial \mathbf{P}}{\partial x_2} \right|^2 + \left| \frac{\partial \mathbf{P}}{\partial x_3} \right|^2 \right]. \tag{43}$$

It is easy to show that $\frac{1}{2}\delta \left[|\nabla P_1|^2 + |\nabla P_2|^2 + |\nabla P_3|^2 \right]$ is equal to the right hand side of Eq. (43). Both forms are shown so that it can be seen that f_{grad} involves gradient terms. However the other form is sometimes convenient to work with. Note that in general there are other symmetry allowed terms such as $\left[\left| \frac{\partial \mathbf{P}}{\partial x_1} \right|^2 \left| \frac{\partial \mathbf{P}}{\partial x_2} \right|^2 \right]$. However usually such terms are ignored

– a practice which will be followed here. Nonetheless, in the future it may be interesting to study such terms to extended the formalism given here.

Finally f_{dep} is the term due to the depolarization field for the general case in which the polarization has components along x_1, x_2 and x_3. It is given by

$$f_{\text{dep}}(\mathbf{x}, \mathbf{P}, \langle \mathbf{P} \rangle) = \frac{1}{2\epsilon_0\epsilon_\infty}\Big[P_1(P_1 - \langle P_1\rangle) + P_2(P_2 - \langle P_2\rangle) + P_3(P_3 - \langle P_3\rangle)\Big]. \tag{44}$$

Having dealt with the energy density for the interior of the ferroelectric the remaining energy density is the surface energy density at any point on the surface of the ferroelectric. This will be proportional to P^2, where $P = |\mathbf{P}|$ and, as will be evident below, it is convenient to write it as

$$f_{\text{surf}}(\mathbf{x}, \mathbf{P}) = \frac{\delta}{2d(\mathbf{x})} P^2(\mathbf{x}) \quad \text{for} \quad \mathbf{x} \in S, \tag{45}$$

where S is the entire surface of the ferroelectric box given by

$$S = S_{x_1=l_1^-} \cup S_{x_{1=l_1^+}} \cup S_{x_2=l_2^-} \cup S_{x_2=l_2^+} \cup S_{x_3=l_3^-} \cup S_{x_3=l_3^+} \tag{46}$$

where the sides of the box at $x_i = l_i^\mp$ are given by

$$S_{x_j=l_j^\mp} = \{\mathbf{x} \,|\, x_i = l_i^\mp, x_{\sigma^j(i)} \in [l_i^-, l_i^+], j = 1, 2\}, \tag{47}$$

where we have introduced the cyclic operator σ which performs the operation $x_1 \mapsto x_2 \mapsto x_3 \mapsto x_1$, so that

$$\sigma(x_i) = x_{j+1} \pmod 3; \tag{48}$$

also σ applied n times where $n \geqslant 0$ is denoted by σ^n, and the definition of the operator is extended to when the x_i are arguments of a function h such that

$$\sigma(h(x_1, \ldots, x_3)) = h(\sigma(x_1), \ldots, \sigma(x_3)). \tag{49}$$

It is easy to see that many of the terms in the free energy can be written in a shortened form with this notation because successive terms can often be generated by cyclic permutations of a starting term; later this operator notation will be useful when the free energy for the box is inserted into the Euler-Lagrange equations in Section 5.

Showing the form of the free energy densities for the ferroelectric box helps pave a way for dealing with the general problem of finding the minimum of the free energy density for an arbitrary volume of ferroelectric, which will be discussed next.

4. The calculus of variations applied to the polarization of a ferroelectric of arbitrary shape in three dimensions

The next step towards working out the equilibrium polarization for ferroelectric boxes and strips—an extension of the theory for thin films—is to work out in general the free energy for an arbitrary volume of ferroelectric. The minimization of this, as for the thin film, involves the minimization of a functional, which involves the calculus of variations and will be dealt with using classical methods.

It can be seen from the previous section that the general form of the free energy for a ferroelectric in the region V in $\mathbb{R}^3$ bounded by the closed surface S is given by

$$G = \int_V f\left(\mathbf{x}, \mathbf{P}, \mathbf{P}_{x_1}, \mathbf{P}_{x_2}, \mathbf{P}_{x_3}, \langle \mathbf{P} \rangle\right) dV + \int_S f_{\text{surf}}\left(\mathbf{x}, \mathbf{P}\right) dS. \tag{50}$$

Here G depends on whatever function $\mathbf{P}$ is, so the domain of G is a function space and the equilibrium polarization distribution is that function $\mathbf{P}_0$ in the space which minimizes G. At this point certain boundary conditions could be imposed on $\mathbf{P}$ according to the physical situation. For a ferroelectric crystal no such conditions are imposed which implies that $\mathbf{P}$ is free at the boundaries. However natural boundary conditions (that do not fix $\mathbf{P}$) will emerge from the minimization as we will see.

The minimization of the functional G in Eq. (50) will be carried out using a classical technique due to Euler[16] which involves considering a variation around $\mathbf{P}_0 = (P_{01}, P_{02}, P_{03})$ in function space. To this end we write

$$\mathbf{P}(\mathbf{x}) = \mathbf{P}_0(\mathbf{x}) + \epsilon \boldsymbol{\eta}(\mathbf{x}), \tag{51}$$

in which ϵ is a variable parameter and $\boldsymbol{\eta} = (\eta_1, \eta_2, \eta_3)$ is an arbitrary vector function. G can now be expressed as

$$G(\epsilon) = \int_V f\left(\mathbf{x}, \mathbf{P}_0 + \epsilon\boldsymbol{\eta}, \mathbf{P}_{0x_1} + \epsilon\boldsymbol{\eta}_{x_1}, \mathbf{P}_{0x_2} + \epsilon\boldsymbol{\eta}_{x_2}, \mathbf{P}_{0x_3} + \epsilon\boldsymbol{\eta}_{x_3}, \langle \mathbf{P}_0 + \epsilon\boldsymbol{\eta} \rangle\right), \tag{52}$$

where

$$\mathbf{P}_{0x_i} = \frac{\partial \mathbf{P}_0}{\partial x_i}. \tag{53}$$

The key idea for finding the minimizing function $\mathbf{P}_0$ is that a necessary condition for its existence is

$$\left.\frac{dG}{d\epsilon}\right|_{\epsilon=0} = 0.$$

http://dx.doi.org/10.5772/52040

Applying this condition to Eq. (52), and using the chain rule when differentiating the integrands, we find

$$\left.\frac{dG}{d\epsilon}\right|_{\epsilon=0} = \int_V \{(\nabla_{\mathbf{P}_0} f)\cdot\boldsymbol{\eta} + \left(\nabla_{\mathbf{P}_{0x_1}} f\right)\cdot\boldsymbol{\eta}_{x_1} + \left(\nabla_{\mathbf{P}_{0x_2}} f\right)\cdot\boldsymbol{\eta}_{x_2} + \left(\nabla_{\mathbf{P}_{0x_3}} f\right)\cdot\boldsymbol{\eta}_{x_3} + \left(\nabla_{\langle\mathbf{P}_0\rangle} f\right)\cdot\langle\boldsymbol{\eta}\rangle\}\, dV + \int_S (\nabla_{\mathbf{P}_0} f_{\text{surf}})\cdot\boldsymbol{\eta}\, dS = 0, \quad (54)$$

in which the notation

$$(\nabla_{\mathbf{A}} f)\cdot\mathbf{B} = \frac{\partial f}{\partial A_1}B_1 + \frac{\partial f}{\partial A_2}B_2 + \frac{\partial f}{\partial A_3}B_3 \quad (55)$$

has been introduced.

If all terms in the integrands in Eq. (54) could be written as a dot product with $\boldsymbol{\eta}$, as is the case for the first and last terms on the right of Eq. (54), then $\boldsymbol{\eta}$ would appear as a single factor in the volume and surface integrands. Eq. (54) could then be written in the form $\int_V \boldsymbol{\eta}\cdot\boldsymbol{\Lambda}_1 + \int_S \boldsymbol{\eta}\cdot\boldsymbol{\Lambda}_2$, where $\boldsymbol{\Lambda}_1$ and $\boldsymbol{\Lambda}_2$ do not involve $\boldsymbol{\eta}$. As shown below, the Euler-Lagrange equations and boundary conditions will then follow by setting $\boldsymbol{\Lambda}_1 = \boldsymbol{\Lambda}_2 = 0$, which will satisfy Eq. (54). Therefore we now proceed to express Eq. (54) in the $\boldsymbol{\eta}$ factored form.

To facilitate this we start by writing Eq. (54) as

$$\left.\frac{dG}{d\epsilon}\right|_{\epsilon=0} = \int_V \{(\nabla_{\mathbf{P}_0} f)\cdot\boldsymbol{\eta} + \mathcal{I}_1 + \mathcal{I}_2\}\, dV + \int_S (\nabla_{\mathbf{P}_0} f_{\text{surf}})\cdot\boldsymbol{\eta}\, dS = 0, \quad (56)$$

where

$$\mathcal{I}_1 = \left(\nabla_{\mathbf{P}_{0x_1}} f\right)\cdot\boldsymbol{\eta}_{x_1} + \left(\nabla_{\mathbf{P}_{0x_2}} f\right)\cdot\boldsymbol{\eta}_{x_2} + \left(\nabla_{\mathbf{P}_{0x_3}} f\right)\cdot\boldsymbol{\eta}_{x_3}, \quad (57)$$

$$\mathcal{I}_2 = \left(\nabla_{\langle\mathbf{P}_0\rangle} f\right)\cdot\langle\boldsymbol{\eta}\rangle. \quad (58)$$

Now we want to find $\boldsymbol{\Lambda}_1$ and $\boldsymbol{\Lambda}_2$ such that $\mathcal{I}_1 = \boldsymbol{\eta}\cdot\boldsymbol{\Lambda}_1$ and $\mathcal{I}_2 = \boldsymbol{\eta}\cdot\boldsymbol{\Lambda}_2$.

Starting with $\mathcal{I}_1$, by expanding the dot products and re-grouping the terms, it is easy to show that it can be rewritten as

$$\mathcal{I}_1 = (\nabla_{P_{o1}\mathbf{x}} f)\cdot(\nabla\eta_1) + (\nabla_{P_{o2}\mathbf{x}} f)\cdot(\nabla\eta_2) + (\nabla_{P_{o3}\mathbf{x}} f)\cdot(\nabla\eta_3), \quad (59)$$

in which the following notation has been used:

$$\nabla\mathbf{A} = \left(\frac{\partial A_1}{\partial x_1}, \frac{\partial A_2}{\partial x_2}, \frac{\partial A_3}{\partial x_3}\right) = \frac{\partial A_1}{\partial x_1}\hat{\mathbf{x}}_1 + \frac{\partial A_2}{\partial x_2}\hat{\mathbf{x}}_2 + \frac{\partial A_3}{\partial x_3}\hat{\mathbf{x}}_3, \quad (60)$$

the normal gradient operation, where $\hat{\mathbf{x}}_i$, $i = 1,2,3$, are unit vectors along the corresponding x_i axes of a Cartesian coordinate system. Also,

$$\boldsymbol{\nabla}_{A_{ix}} = \left(\frac{\partial}{\partial A_{ix_1}}, \frac{\partial}{\partial A_{ix_2}}, \frac{\partial}{\partial A_{ix_3}} \right), \tag{61}$$

with $A_{ix_j} = \partial A_i / \partial x_j$.

Now, we utilize the vector identity

$$(\boldsymbol{\nabla}\phi) \cdot \mathbf{A} \equiv \boldsymbol{\nabla} \cdot (\phi \mathbf{A}) - \phi \left(\boldsymbol{\nabla} \cdot \mathbf{A} \right), \tag{62}$$

together with the divergence theorem

$$\int_V \boldsymbol{\nabla} \cdot \mathbf{A} \, dV = \oint_S \mathbf{A} \cdot (\hat{\mathbf{n}} \, dS), \tag{63}$$

where at each point on the surface S, $\hat{\mathbf{n}} = \hat{\mathbf{n}}(\mathbf{x}|_S)$ is a unit normal vector directed outwards (from the enclosed volume). Using Eqs. (62) and (63), it follows that

$$\int_V \mathcal{I}_1 \, dV = -\int_V \boldsymbol{\eta} \cdot \left[\sum_{i=1}^{3} \boldsymbol{\nabla} \cdot \left(\boldsymbol{\nabla}_{P_{0i}\mathbf{x}} f \right) \hat{\mathbf{x}}_i \right] dV + \int_S \boldsymbol{\eta} \cdot \left[\sum_{i=1}^{3} \left\{ \left(\boldsymbol{\nabla}_{P_{0i}\mathbf{x}} f \right) \cdot \hat{\mathbf{n}} \right\} \hat{\mathbf{x}}_i \right] dS. \tag{64}$$

Next the $\mathcal{I}_2$ term is dealt with. But here is is better to look at it together with the integral from the outset, by considering

$$\int_V \mathcal{I}_2 \, dV = \int_V \left(\boldsymbol{\nabla}_{\langle \mathbf{P}_0 \rangle} f \right) \cdot \langle \boldsymbol{\eta} \rangle \, dV = \int_V \left(\sum_{i=1}^{3} \frac{\partial f}{\partial \langle P_i \rangle} \langle \eta_1 \rangle \right) dV. \tag{65}$$

The reason for considering $\mathcal{I}_2$ inside the integral is that, remembering that $\langle \eta_i \rangle$ is proportional to $\int_{l_i^-}^{l_i^+} \eta_i \, dx_i$, Eq. (65) can be written as

$$\int_V \mathcal{I}_2 \, dV = \int_V \left(\sum_{i=1}^{3} \eta_i \left\langle \frac{\partial f}{\partial P_{0i}} \right\rangle \right) dV = \int_V \boldsymbol{\eta} \cdot \left\langle \boldsymbol{\nabla}_{\langle \mathbf{P}_0 \rangle} f \right\rangle dV, \tag{66}$$

by changing the order of integration. In doing this we have used the fact that $dV \equiv dx_1 dx_2 dx_3$ and $\int_V \equiv \int_{x_1} \int_{x_2} \int_{x_3}$.

The goal of factoring out $\boldsymbol{\eta}$ involving a dot product has now been achieved and Eq. (54) can be written, using Eqs. (64) and (66), as

http://dx.doi.org/10.5772/52040

$$\int_V \boldsymbol{\eta} \cdot \left[\nabla_{\mathbf{P}_0} f - \sum_{i=1}^{3} \nabla \cdot \left(\nabla_{P_{0i}\mathbf{x}} f \right) \hat{\mathbf{x}}_i + \left\langle \nabla_{\langle \mathbf{P}_0 \rangle} f \right\rangle \right] dV$$
$$+ \int_S \boldsymbol{\eta} \cdot \left[\sum_{i=1}^{3} \left\{ \left(\nabla_{P_{0i}\mathbf{x}} f \right) \cdot \hat{\mathbf{n}} \right\} \hat{\mathbf{x}}_i + \nabla_{\mathbf{P}_0} f_{\text{surf}} \right] dS = 0. \quad (67)$$

However, at first we consider a restricted function space such that $\boldsymbol{\eta}$ is zero on S; then the surface integral in Eq. (67) vanishes resulting in

$$\int_V \boldsymbol{\eta} \cdot \left[\nabla_{\mathbf{P}_0} f - \sum_{i=1}^{3} \nabla \cdot \left(\nabla_{P_{0i}\mathbf{x}} f \right) \hat{\mathbf{x}}_i + \left\langle \nabla_{\langle \mathbf{P}_0 \rangle} f \right\rangle \right] dV = 0. \quad (68)$$

For this to be zero for any $\boldsymbol{\eta}$ in the restricted space, it follows from the lemma of du Bois-Reymond[25], that

$$\nabla_{\mathbf{P}_0} f - \sum_{i=1}^{3} \nabla \cdot \left(\nabla_{P_{0i}\mathbf{x}} f \right) \hat{\mathbf{x}}_i + \left\langle \nabla_{\langle \mathbf{P}_0 \rangle} f \right\rangle = 0, \quad (69)$$

which can be rewritten as

$$\frac{\partial f}{\partial P_{0i}} - \nabla \cdot \left(\nabla_{P_{0i\mathbf{x}}} f \right) + \left\langle \frac{\partial f}{\partial \langle P_{0i} \rangle} \right\rangle = 0, \quad i = 1, 2, 3. \quad (70)$$

These form the Euler-Lagrange equations that we seek. It is clear that the lifting of the restriction on $\boldsymbol{\eta}$ so that it is not necessarily zero on S, means that the Euler-Lagrange equations still result from Eq. (67), provided the lemma of du Bois-Reymond is applied to the surface integral term. This implies the boundary conditions

$$\left(\nabla_{P_{i\mathbf{x}}} f \right) \cdot \hat{\mathbf{n}} + \frac{\partial f_{\text{surf}}}{\partial P_i} \quad \text{on } S, \quad i = 1, 2, 3. \quad (71)$$

For any i the first term is a natural boundary condition arising from the minimization (not a prescribed condition), and the second term is due to the surface energy.

Thus we have shown that the equilibrium polarization for a ferroelectric of arbitrary shape in the region V is found by solving the Euler-Lagrange equations in Eq. (70), subject to the boundary conditions in Eq. (71). This formulation can be applied to finding the polarization in ferroelectrics with the geometries, box, strip and film, as will be shown in the next section for the case of a ferroelectric exhibiting a symmetric phase of cubic symmetry.

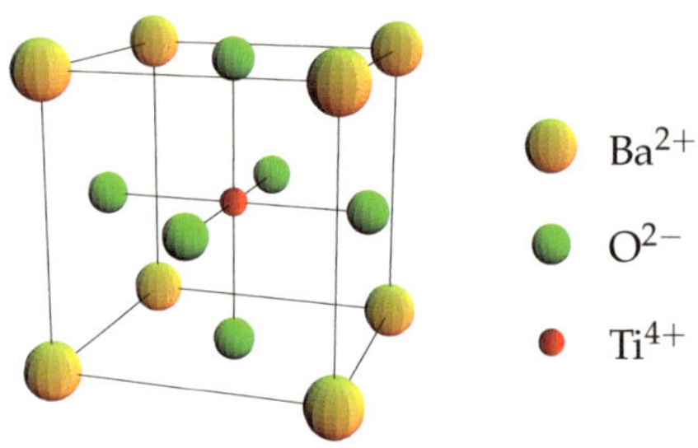

Figure 8. A structural unit for barium titanate, $BaTiO_3$.

5. The equilibrium Polarization in ferroelectric boxes, strips and films for cubic symmetry

Cubic symmetry is chosen because it is the symmetry of the popular perovskite ferroelectrics such as barium titanate ($BaTiO_3$) and lead titanate ($PbTiO_3$). A structural unit for barium titanate is represented in Fig. 8

The interest is in nanoscale boxes, strips and films. The formalism itself could easily be applied to thicker films, however for confinement in the nanoscale (in all three dimensions for boxes, in two dimensions in strips and in one for films) it is expected that the influence of the change in the polarization as it approaches a surface will have a more significant effect that at lager scales where the bulk properties would be more dominant, and so is sometimes referred to as a size effect.

5.1. Polarization for a ferroelectric nano-box

For a cubic crystal only those terms in f given by Eq. (40) that are invariant under the symmetry 16 operations of the cubic symmetry group[26] are allowed.[3] It can be shown[10, 12] that these are given by

$$f_{\text{cubic}}(\mathbf{x}, \mathbf{P}, \mathbf{P}_{x_1}, \mathbf{P}_{x_2}, \mathbf{P}_{x_3}, \langle \mathbf{P} \rangle) = \alpha_1(T - T_C)P^2 + \sum_{i=2}^{3} \frac{1}{2i}\alpha_i (P^2)^i$$
$$+ \frac{1}{2}\beta_2 \sum_{i=0}^{2} \sigma^i \left(P_1^2 P_2^2\right) + \frac{1}{2}\beta_3 \sum_{i=0}^{2} \sigma^i \left(P_1^4 (P_2^2 + P_3^2)\right) + \frac{1}{2}\gamma_3 P_1^2 P_2^2 P_3^2$$
$$+ \frac{1}{2}\delta \sum_{i=1}^{3} \left| \frac{\partial \mathbf{P}}{\partial x_i} \right|^2 + \frac{1}{2\epsilon_0 \epsilon_\infty} \sum_{i=1}^{3} P_i (P_i - \langle P_i \rangle), \quad (72)$$

where the cyclic operator defined by Eqs. (48) and (49) is being used and, to lighten the notation a little we make the replacement $\mathbf{P}_0 \to \mathbf{P} \Rightarrow P_{oi} \to P_i$, although the subscript that distinguishes the equilibrium polarization from the general function space remains implied.

[3] The operations are: $1, \bar{1}, 4, 4^{-1}, \bar{4}, \bar{4}^{-1}, 2', 2'', 2_x, 2_y, 2_z, m_1, m_2, m_3, m_4, m_5$.

http://dx.doi.org/10.5772/52040

The Euler-Lagrange equations that need to be solved to find the equilibrium polarization in the nano-box are then found from Eq. (70) with $f = f_{\text{cubic}}$. The steps will not be shown here but the cyclic operator can make the working easier and it is straightforward to show that the required Euler-Lagrange equations are given by

$$\alpha_1(T - T_C)P_i + P_i \sum_{j=2}^{3} \alpha_j (P^2)^{j-1} + \beta_2 P_i \sum_{j=1}^{2} \sigma^j (P_j^2) \\ + \beta_3 P_i \left[2P_i^2 \sum_{j=1}^{2} \sigma^j (P_i^2) + \sum_{j=2}^{2} \sigma^j (P_j^4) \right] + \delta \nabla^2 P_i + \frac{1}{\epsilon_0 \epsilon_\infty} (P_i - \langle P_i \rangle), \quad i = 1,2,3. \quad (73)$$

The surface term is given by Eq. (45), but here we assume that $d(\mathbf{x}|_S)$ is constant over each surface of the box. The boundary conditions in Eq. (71) can then be shown to reduce to

$$\mp \frac{\partial P_i}{\partial x_i} + \frac{1}{d_i^{\mp}} = 0 \ \forall \mathbf{x} \in S_{x_i = l_j^{\mp}} \quad i = 1,2,3. \quad (74)$$

For a free standing box it is reasonable to have $d_i^{\mp} = d \ \forall i = 1,2,3$.

The solution of Eqs. (73) and (74) has not yet been obtained but progress is being made on it and will be reported later. Eq. (73) is a nonlinear partial differential equation, and as such it is unlikely that an analytical solution can be found. Instead, a numerical approach or an approximate solution using trial functions with adjustable parameters could be employed. The box geometry is likely to be amenable to a finite difference numerical solution. Dealing with a simpler symmetry such as that discussed in the introduction would make the problem less computationally intensive but still not amenable to an analytical solution, as will now be discussed.

If Eq. (72) were reduced to a simpler form by putting $\alpha_3 = \beta_2 = \beta_3 = \gamma_3 = 0$, neglecting the depolarization terms, and arranging for the polarization to be aligned along only one coordinate axis; the problem would then be something like a three-dimensional form of the thin-film case discussed in Section 2.2. It might then be tempting to try a separation of variables approach to solving the Euler-Lagrange equations with the hope that the solution would be of the form $P(x_1)P(x_2)P(x_3)$ in which each factor is a solution of the form of the thin film solution given in Section 2.2. However any attempt to do this would fail to produce an exact solution to the three-dimensional problem due to the nonlinearity of a term cubic in the polarization. However it might be useful to treat $P(x_1)P(x_2)P(x_3)$ as an approximate function and introduce into it some variable parameters that could be optimized to find the best approximation. This idea is currently being explored but as yet there are no final conclusions.

5.2. Polarization for a ferroelectric nano-strip

The case of a strip in which confinement along two of the axes can be thought of as a special case of the box in which one side, l_1 say, approaches infinity such that $\mathbf{P}(\mathbf{x}) = \mathbf{P}(x_1, x_2, x_3) \rightarrow$

$\mathbf{P}(x_2, x_3)$, and is constant with respect to x_3 since the boundaries in this direction have no influence, which also implies that the depolarization terms only involve $\langle \mathbf{P} \rangle = (\langle P_2 \rangle, \langle P_3 \rangle)$.

Now the relevant quantity is a free energy density per unit length, since for a section of the strip between $x_1 = 0$ and $x_1 = l_1'$ the free energy is

$$G = \int_0^{l_1'} dx_1 \int_A f(\mathbf{x}, \mathbf{P}, \mathbf{P}_{x_2}, \mathbf{P}_{x_3}, \langle \mathbf{P} \rangle)\, dA + \int_0^{l_1'} dx_1 \int_\Gamma f_{\text{surf}}(\mathbf{x}, \mathbf{P}) d\Gamma, \tag{75}$$

leading to an energy density

$$F = \frac{G}{l_1'} = \int_A f(\mathbf{x}, \mathbf{P}, \mathbf{P}_{x_2}, \mathbf{P}_{x_3}, \langle \mathbf{P} \rangle)\, dA + \int_\Gamma f_{\text{surf}}(\mathbf{x}, \mathbf{P}) d\Gamma, \tag{76}$$

where $\int_\Gamma$ is a line integral around the path Γ that traces a rectangular cross section of the strip normal to x_1, $\int_A \equiv \int_{l_2^-}^{l_2^+} \int_{l_3^-}^{l_3^+}$, and $dA = dx_2 dx_3$.

For $f = f_{\text{cubic}}$ (Eq. (72)), the Euler-Lagrange equations are similar to Eq. (73) but P_1 is constant, and so $\nabla^2 = \frac{\partial^2}{\partial x_2^2} + \frac{\partial^2}{\partial x_3^2}$, and the depolarization term for $i = 1$ reduces to zero. Also the boundary conditions are given by Eq. (74), but with $i = 2, 3$. The solution is again best approached by numerical or other approximate methods. Work on this is in progress.

The comments made in at the end of the previous section for the ferroelectric nano-box with regard to simplifying it to a three dimensional form of the problem in Section 2.2 also apply here: for the same reason it is not possible to express an exact solution of the form $P(x_2)P(x_3)$, where P has the form of the solution given in Section 2.2. Again it may be possible to utilize such a solution as an approximate function. This needs to be investigated further.

5.3. Polarization for a ferroelectric nano-film

For a film confined in the x_3 direction, a similar argument to the one given for a nano-strip yields $\mathbf{P} = \mathbf{P}(x_3)$, $\langle \mathbf{P} \rangle = \langle P_3 \rangle \hat{\mathbf{x}}_3$, and a free energy density

$$F = \frac{G}{l_1' l_2'} = \int_{l_3^-}^{l_3^+} f(x_3, \mathbf{P}, \mathbf{P}_{x_3}, \langle P_3 \rangle)\, dx_3. \tag{77}$$

With $f = f_{\text{cubic}}$ given by Eq. (72), the Euler-Lagrange equations again take the form of Eq. (73); now with P_1 and P_2 constant so that the depolarization terms reduce to zero for $i = 1, 2$, and $\nabla^2 \to \frac{\partial^2}{\partial x_3^2}$. The boundary conditions are given by Eq. (74), with $i = 3$ and $\frac{\partial}{\partial x_3} \to \frac{d}{dx_3}$. An analytical solution, even for this one-dimensional case is still not likely to exist unless f is reduced to the simpler higher symmetry form for second order transitions and in-plane polarization discussed in Section 2.2. However it is amenable to numerical solution or other approximate methods. A similar thin film problem, but involving strain in the ferroelectric as well, which we do no consider, has been solved numerically using a finite difference method by Wang and Zhang[7].

http://dx.doi.org/10.5772/52040

6. Conclusion

The foundations of Landau-Devonshire theory have been introduced and it has been used to develop a general formulation for the calculation of the spontaneous polarization in a ferroelectric nano-box which is confined to the nanoscale in all three dimensions, and for which the influence of the surfaces is expected to be more pronounced than for large scales. From the formalism for the box calculations it has been shown how to deal with a nano-strip which is confined to the nanoscale in two dimensions and to a nano-film with such confinement in only one dimension. The thin film case has been fairly well studied, but much less work has been done on the other geometries, box and strip. Such work is timely given the increasing use of nanoscale structures in electronic devices which include ferroelectric materials, for example ferroelectric random access memories.

A particular example of how the formalism can applied to a ferroelectric crystal of cubic symmetry has been given for all three geometries. Many ferroelectric materials have this symmetry. However the general formalism presented is not restricted to this symmetry, any symmetry can be handled through knowledge of which terms in the free energy needed to be dropped, the criterion being that only terms invariant under the symmetry operations of the symmetry group are allowed.

Future work will involve investigating the numerical or approximate function solutions to the Euler-Lagrange equations which need to be solved in order to be able to plot the spontaneous polarization. Such work is in progress and will be reported in due course.

Acknowledgements

The author would like to thank Manas Kumar Halder for useful and enlightening discussions on several of the ideas presented in this chapter.

Author details

Jeffrey F. Webb

Faculty of Engineering, Computing and Science, Swinburne University of Technology, Sarawak Campus, Kuching, Sarawak, Malaysia

References

[1] P. Chandra and P. B. Littlewood. A Landau primer for ferroelectrics. In K. Rabe, Ch. H. Ahn, and J. M. Triscone, editors, *Physics of Ferroelectrics*, volume 105 of *Topics in Applied Physics*, page 69. Springer, Heidelberg, 2007.

[2] J. F. Scott. The physics of ferroelectric ceramic thin films for memory applications. *Ferroelectr. Rev.*, 1:1, 1998.

[3] D. R. Tilley and B. Zeks. Landau theory of phase transtions in thick films. *Solid State Commun.*, 49:823, 1984.

[4] D. R. Tilley. Phase transitions in thin films. In N. Setter and E. L. Colla, editors, *Ferroelectric Ceramics*, page 163. Birkhäuser Verlag, Berlin, 1993.

[5] D. R. Tilley. Finite-size effects on phase transitions in ferroelectrics. In Carlos Paz de Araujo, J. F. Scott, and G. W. Taylor, editors, *Ferroelectric Thin Films: Synthesis and Basic Properties*, Integrated Ferroelectric Devices and Technologies, page 11. Gordon and Breach, Amsterdam, 1996.

[6] J. F. Webb. Theory of size effects in ferroelectric ceramic thin films on metal substrates. *J. Electroceram.*, 16:463, 2006.

[7] J. Wang and T. Y. Zhang. Influence of depolarization field on polarization states in epitaxial ferroelectric thin films with nonequally biaxial misfit strains. *Physical Review B*, 77:014104–1 to 014104–7, 2008.

[8] Anna N. Morozovska, Maya D. Glinchuk, and Eugene A. Eliseev. Phase transitions induced by confinement of ferroic nanoparticles. *Physical Review B*, 76:014102–1 to 014102–13, 2007.

[9] M. E. Lines and A. M. Glass. *Principles and Applications of Ferroelectrics and Related Materials*. Clarendon, Oxford, UK, 1977.

[10] B. A. Strukov and A. P. Lenanyuk. *Ferroelectric Phenomena in Crystals*. Springer, Berlin, 1998.

[11] R. C. Tolman. *The Principles of Statistical Thermodynamics*. Clarendon Press, Oxford, 1938.

[12] J. C. Toledano and P. Toledano. *The Landau Theory of Phase Transitions*. World Scientific, Singapore, 1987.

[13] R. Blinc, S. Detoni, and M. Pintar. Nature of the ferroelectric transition in triglycine sulfate. *Physical Review Letters*, 124:1036–1038, 1961.

[14] K. Itoh and T. Mitsui. Studies of the crystal structure of triglycine sulfate in connection with its ferroelectric phase transition. *Ferroelectrics*, 5:235–251, 1973.

[15] M. G. Cottam, D. R. Tilley, and B. Zeks. Theory of surface modes in ferroelectrics. *J. Phys C*, 17:1793–1823, 1984.

[16] R. Courant and D. Hilbert. *Methods of Mathematical Physics*, volume 1. Interscience, New York, 1953.

[17] X. Gerbaux and A. Hadni. *Static and dynamic properties of ferroelectric thin film memories*. PhD thesis, University of Colorado, 1990.

[18] E. K. Tan, J. Osman, and D. R. Tilley. First-order phase transitions in ferroelectric films. *Solid State Communications*, 116:61–65, 2000.

[19] K. H. Chew, L. H. Ong, J. Osman, and D. R. Tilley. Theory of far-infrared reflection and transmission by ferroelectric thin films. *J. Opt. Soc. Am B*, 18:1512, 2001.

[20] J. F. Webb. Harmonic generation in nanoscale ferroelectric films. In Mickaël Lallart, editor, *Ferroelectrics - Characterization and Modeling*, chapter 26. Intech, Rijeka, Croatia, 2011.

[21] M. Abramowitz and I. A. Stegun, editors. *Handbook of Mathematical Functions*. Dover, New York, 1972.

[22] R. Kretschmer and K. Binder. Surface effects on phase transitions in ferroelectrics and dipolar magnets. *Phys. Rev. B*, 20:1065–1076, 1979.

[23] M.D. Glinchuk, E.A. Eliseev, and V.A. Stephanovich. Ferroelectric thin film properties—depolarization field and renormalization of a "bulk" free energy coefficients. *J. Appl. Phys.*, 93:1150–1159, 2003.

[24] D. L. Mills. *Nonlinear Optics*. Springer, Berlin, second edition, 1998.

[25] E. W. Honson. On the fundamental lemma of the calculus of variations and on some related theorems. In *Proceedings of the London Mathematical Society*, volume s2-11, pages 17–28, 1913.

[26] J. Nye. *Physical Properties of Crystals. Their Representation by Tensors and Matrices*. Oxford University Press, Oxford, 1964.

Chapter 21

Emerging Applications of Ferroelectric Nanoparticles in Materials Technologies, Biology and Medicine

Yuriy Garbovskiy, Olena Zribi and
Anatoliy Glushchenko

Additional information is available at the end of the chapter

http://dx.doi.org/10.5772/52516

1. Introduction

Consider an insulating system with non-zero spontaneous polarization P_s (dielectric dipole moment per unit volume). If an applied external electric field E that is greater than the so-called coercive field E_c can reverse P_s then our system is a ferroelectric system. Ferroelectricity has a long and exciting history described in [1,2]. In the beginning of its historical development (the Rochelle salt period) ferroelectricity was considered an academic curiosity with no practical applications. There was little theoretical interest due to the quality of the ferroelectric materials (very fragile and water-soluble) existing at that time. The discovery of ferroelectricity in robust ceramic materials (barium titanate) during World War II launched a new era of rapid progress in the field. The structural simplicity of barium titanate stimulated numerous theoretical works, while its physical properties were utilized in many devices. Since that time, ferroelectric response has been found in a wide range of materials, including inorganic, organic, and biological species. According to [3] there are 72 families of ferroelectrics presented in Landolt–Börnstein-Vol.III/36 (LB III/36). Forty-nine of these families are inorganic crystals (19 families of oxides + 30 families of crystals other than oxides), and 23 families are organic crystals, liquid crystals, and polymers.

The enormously broad range of materials exhibiting ferroelectricity and the variety of their physical properties result in numerous applications of bulk ferroelectrics [4]. Table 1 shows the connections between different physical effects exhibited by bulk ferroelectrics and their applications.

Recent advances in nanotechnologies, especially in nanoinstrumentation (for example, scanning probe microscopy [5]) and materials nanofabrication [6], allowed the direct probing of ferroelectricity at the nanoscale. The new and unexplored world of nanoscale ferroelectrics (nanoparticles of different shapes and sizes, nanofilms, nanopatterned structures, etc.) raised fundamental questions and stimulated very active research in both academic and industrial sectors [7]. As a result, a new era of nanoscale ferroelectrics was launched. Novel effects, associated with reduced dimensions and found in nanoscaleferroelectrics, highlighted exciting possibilities for new applications reviewed recently in [8]. Almost all of the attention for the mentioned review [8] was devoted to the *thin film* nanoscale device structures (which can be easily integrated with a Si chip) with focus on ultrafast switching, electrocaloric coolers for computers, phase-array radar, three-dimensional trenched capacitors for dynamic random access memories, room temperature magnetic field detectors, and miniature X-ray and neutron sources. So far, we have not found a coherent review summarizing the actual and possible applications of ferroelectric *nanoparticles*. Our book chapter is an attempt to describe and analyze the state of the field of applications of ferroelectric nanoparticles with focus on materials technologies, medicine, and biology.

Physical effect/property	Applications
Ferroelectric hysteresis	Nonvolatile computer information storage
High relative permittivities (several thousands)	Capacitors
Direct piezoelectric effect	Sensors (microphones, accelerometers, hydrophones, etc.)
Converse piezoelectric effect	Actuators, ultrasonic generators, resonators, filters
Pyroelectric effect	Uncooled infra-red detectors
Electro-optic effects	Laser Q-switches, optical shutters and integrated optical (photonic) devices
Nonlinear optical effects	Laser frequency doubling, optical mixing, including four-wave mixing and holographic information storage
Coupling between stress and birefringence	Radar signal processing
Positive temperature coefficient of resistance (PTCR)	Electric-motor overload-protection devices and self-stabilizing ceramic heating elements

Table 1. Applications of bulk ferroelectrics

2. Methods of ferroelectric nanoparticles production: Nano-powders and nano-colloids

The existing methods that produce ferroelectric nanoparticles are numerous and can be classified as physical, chemical, and biological (Fig. 1). The primary goal of each method is to control size, shape, morphology, and crystallinity of nanoparticles to produce a desirable effect. This task represents a real challenge, and, as a result, there are no well-defined boundaries between physical, chemical, or biological methods. Moreover, in many cases, a combination of at least two methods (for example, physical and chemical methods, i.e. a physical-chemical approach) is required in order to fabricate good quality particles (small sizes, narrow size distribution, ferroelectric phase).

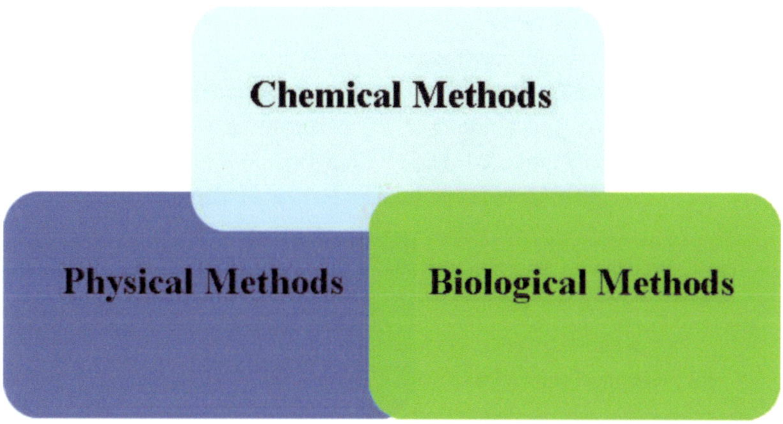

Figure 1. Methods used to fabricate ferroelectric nanoparticles

2.1. Chemical methods

The most widely used chemical methods for the synthesis of ferroelectric nanoparticles are: 1) solid-state reaction; 2) sol-gel technique; 3) solvothermal method; 4) hydrothermal method; and 5) molten salt method.

Table 2 shows selected examples of these methods applied to synthesis of ferroelectric nanoparticles of $BaTiO_3$. As can be seen from Table 2, the ultra-fine ferroelectric nanoparticles (<10 nm) in almost all cases except [16] are synthesized in a cubic phase which is not ferroelectric. The tetragonal phase (with ferroelectric response) is possible for relatively large particles (~50-70 nm). This fact can be critical for certain types of applications, and will be discussed later.

Shape of nanoparticles; Ref.	Synthesis method; Raw materials	Relevant parameters (size, morphology, crystallinity, spontaneous polarization etc.); Measurement and characterization methods	Possible applications & comments
Nanotubes; [10]	Wet chemical route at low temperature (50°C); H_2TiO_3 nanotubes; ethanol/water mixture with 25% ethanol by volume; $Ba(OH)_2{\cdot}8H_2O$	Uniform $BaTiO_3$ nanotubes, a *cubic phase*, average diameter = 10 nm and wall thickness = 3 nm at room temperature; Powder X-ray diffraction (XRD), field-emission scanning electron microscopy (FE-SEM), transmission electron microscopy (TEM), Raman spectroscopy, and X-ray photoelectron spectroscopy	Promising microwave-absorbing materials
Nanocrystals; [11]	Sol-gel technique; Barium titanium ethyl hexano-isopropoxide, $BaTi(O_2CC_7H_{15})(OCH(CH_3)_2)_5$; a mixture of diphenyl ether, $(C_5H_5)_2O$; stabilizing agent oleic acid, $CH_3(CH_2)_7CHdCH(CH_2)_7$-$CO_2H$, at 140 °C, under argon or nitrogen	Monodisperse nanoparticles with diameters ranging from 6 to 12 nm, *cubic phase;* XRD,TEM	Multilayer ceramic capacitors
Dense polycrystalline aggregates (~80 nm) of nanocrystals (~30 nm); [12]	Solid-state reaction as a function of temperature (400–8000C), time (1–24 hr); Nanocrystalline TiO_2, ultrafine $BaCO_3$ and submicrometer $BaCO_3$ were intensively mixed in an aqueous suspension for 24 hr using polyethylene jars and zirconia media. The polymer (ammonium polyacrylate) was required for the formation of a monolayer on the particle surface	Nanoparticle size is 70 nm. Specific surface area up to ~$15 m^2/g$, *tetragonal phase*; TG and DTA analysis, XRD, SEM	Multilayer ceramic capacitors
Polyhedral with hexagonal outline in shape; [13]	Molten salt method; Hydroxide octahydrate ($Ba(OH)_2{\cdot}8H_2O$), titanium dioxide (TiO_2), and the eutectic salts (NaCl–KCl), 600-900 °C	50 nm, *cubic phase*; XRD, Fourier transform infrared spectrometry, UV–Vis diffuse reflectance spectra, and field emission SEM	Nanoparticles are well dispersed

Shape of nanoparticles; Ref.	Synthesis method; Raw materials	Relevant parameters (size, morphology, crystallinity, spontaneous polarization etc.); Measurement and characterization methods	Possible applications & comments
Single crystalline nanoparticles; [14]	Solvothermal method; $Ba(CH_3COO)_2$ and $Ti(OC_4H_9)_4$; the autoclave was maintained at 200 °C for 12 hr	5-20 nm, *cubic phase*; XRD, TEM, SAED, FTIR	Dense bulk nanocomposites
Nanoparticles; [15]	Hydrothermal method; $(Ba,Sr)(OiPr)_2$, Ti(OiPr), EtOH, 330–400 °C, 16-30 MPa	15-50 nm,*cubic or tetragonal phase*; XRD, TEM	Dense bulk nanocomposites
Nanopowders; [9]	Combined wet-chemical and rapid calcination process; $BaCO_3$–TiO_2 precursors	125 nm, *tetragonal phase*, remnant polarization $P_r = 1.64$ μC/cm^2, the coercive field, $E_c = 4.91$ kV/cm; (cigar-like loop);surface area 7.96 m^2/g XRD, TEM, TGA, SMPS, SAED,dielectric spectroscopy	Multilayer ceramic capacitor
Nanocrystals; [16]	One-step solvothermal route; Tetra-n-butyl titanate, Barium hydroxide octahydrate, solvent (Diethylene glycol), surfactant (Polyvinyl Pyrrolidone)	5 nm, *tetragonal phase*; XRD, TEM, HRTEM, Raman spectroscopy	Multilayer ceramic capacitor

Table 2. Chemical methods

2.2. Physical methods

Dry and wet mechanical grindings are methods of choice for inexpensive nanoparticle preparation. For technological applications, wet grinding is preferable because it allows more options to control the size of the nanoparticles. During the last few years, substantial progress was made in the field of ferroelectric nanoparticle preparation by means of wet mechanical grinding [17]. Generally, three components are needed: raw material to grind (micron-sized powders of $BaTiO_3$); surfactant (which covers the particle surface and prevents their aggregation and overheating during grinding; oleic acid is a good choice for $BaTiO_3$); and fluid carrier (both raw material and surfactant are mixed with fluid carrier; heptane is widely used to grind $BaTiO_3$). An extensive list of references, as well as recent achievements in the field, are discussed in the review [18]. Table 3 shows how particle sizes depend upon grinding time [19].

As is seen from Table 3, wet mechanical grinding can produce 9 nm nanoparticles, and such small particles are still in ferroelectric phase. It is important to remember that $BaTiO_3$ nanoparticles of the same sizes synthesized by the majority of chemical methods are not ferroelectric – the method used to fabricate nanoparticles really does matter. The ability to produce ferroelectric nanoparticles of very small sizes is a defining characteristic of this physical method (see [18] for more references).

Grinding time, hrs	3	7	10	16	20	24
Average particle diameter, nm	26	14.9	11.6	9.5	9.2	9

Table 3. $BaTiO_3$ particle sizes as a function of grinding time (liquid carrier – heptane; two-station PM200 planetary ball mill from Retsch)

The net dipole moment of ferroelectric nanoparticles allows them to be harvested by using an inhomogeneous electric field. The harvesting concept proposed in [20] is based on the fact that dipoles experience a translational force only when exposed to a field gradient. For a given linear field gradient and assuming a single ferroelectric domain, the net translational force on a dipole scales proportionally with the particle size. The Brownian motion effects also become progressively more pronounced at smaller particle sizes and so the required field strength for successful separation scales nonlinearly as the particle size is reduced.

Both gas-phase and liquid-phase harvesting methods were proposed and tested successfully, and single ferroelectric monodomain nanoparticles as small as 9 nm from mechanically ground nanoparticles were selectively harvested [20]. In contrast to many reports on the lack of ferroelectricity for nanoparticles below 10 nm (see Table 2), the harvested nanoparticles do maintain ferroelectricity. The ferroelectric response of such tiny nanoparticles was attributed to the existence of an induced surface strain as a result of the grinding process [20]. The lack of a mechanically induced strain in similarly sized but chemically produced nanoparticles accounts for the absence of ferroelectricity in these materials. The concept of stress-induced ferroelectricity was verified in further experiments, which are reviewed in [18]. It was found [21] that the spontaneous polarization of the nanoparticles is four to five times larger than the spontaneous polarization of the bulk raw materials. To obtain this result the following two conditions must be satisfied: (1) a nonpolar solvent for the nanoparticle suspension and (2) nonaggregated nanoparticles. Under these conditions, for 9 nm nanoparticles, the values of 100–120 $\mu C/cm^2$ and 8.9 10^{-23}C cm have been measured for the spontaneous polarization and dipole moment, respectively. The aggregation of ferroelectric nanoparticles masks the ferroelectric response due to the partial compensation of the dipole moments of the individual particles. Finally, we can conclude that recent advances in the production of uniform, monodomain, highly ferroelectric nanoparticles indicate that this field has reached its maturity. The particles can now be reliably prepared to certain specifications and characteristics [18].

2.3. Physical-chemical methods

In order to prepare ferroelectric nanoparticles with controllable sizes and shapes a combination of chemical methods with external physical factors (for example, a chemical reaction in the presence of electromagnetic fields or mechanical milling) is needed. Table 4 shows several examples of these physical-chemical methods for the case of $BaTiO_3$.

As can be seen from Table 4, chemical reactions under the presence of an external driver (microwave, ultrasonic, milling, heat, pressure) are able to produce very fine particles (~5-10 nm) with tetragonal structures.

Shape of nanoparticles; Ref.	**Synthesis method; Raw materials**	**Relevant parameters (size, morphology, crystallinity, spontaneous polarization etc.); Measurement and characterization methods**	**Possible applications & comments**
Truncated nanocubes; [22]	Microwave (2.45 GHz, output power ~ 800W) hydrothermal; High-purity reagents barium nitrate (Sigma Aldrich, purity ≥99.99%), titanium isoproxide (Sigma Aldrich, purity ≥97%), nitric acid (ACS reagent 70%), ammonium hydroxide (Sigma Aldrich), and glycine (Sigma-Aldrich, purity ≥99%)	Cross section 70 ± 9 nm in [100] projection; *tetragonal structure*; remnant polarization is 15.5 $\mu C/cm^2$; saturation polarization is 19.3 $\mu C/cm^2$; XRD, high resolution TEM, impedance spectroscopy	Charge storage devices
Nearly spherical nanoparticles; [23]	Sol–gel-hydrothermal method under an oxygen (partial pressure is ~60 bar); $TiCl_4$; HCl; $BaCl_2{\cdot}2H_2O$; deionized water, stirring and N_2 bubbling, NaOH	Sizes range from 50 to 75nm; transition of $BaTiO_3$ from *cubic* to a *pseudotetragonal phase* with the increase ofthe reaction temperature from 80 to 220 °C; remnant polarization P_r = 2.2 C/cm^2; the coercive field, E_c = 3.2 kV/cm (cigar-like loop); XRD, TEM, Raman and dielectric spectroscopy; energy-dispersive X-ray spectroscopy; AFM	Oxygen atmosphere decreases sizes of the nanoparticles (for example , at 200 °C sizes decrease from 72.54 nm to 49.54 nm)
Nanocrystals; [24, 25]	Direct synthesis from solution (DSS) – mechanochemical synthesis; Anhydrous $Ba(OH)_2$ and tetrabutyl titanate [Ti $(OC_4H_9)_4$] Mechanical milling at the rate of 200 rpm, room temperature	7 nm, *tetragonal structure*; XRD, TEM, Raman spectroscopy	Chemical reaction during milling
Large aggregates of 5–10 nm small nanocrystals; [26]	Sonochemical synthesis; $BaCl_2$; $TiCl_4$; NaOH	100 nm aggregates, *cubic phase*; XRD, TEM, DLS, SAED	Multilayer ceramic capacitors

Shape of nanoparticles; Ref.	Synthesis method; Raw materials	Relevant parameters (size, morphology, crystallinity, spontaneous polarization etc.); Measurement and characterization methods	Possible applications & comments
Single nanocrystals; [27]	RF-Plasma chemical vapor deposition (CVD); iso-propoxide ($Ti(OiPr)_4$, bis-dipivaloylmethanate ($Ba(DPM)_2$	12.2-15.4 nm; XRD, TEM, SAED, EDX	Multilayer ceramic capacitors
Nanofibers; [28, 29, 30]	Electrospinning, sol-gel; barium acetate, PVP, titanium isopropoxide, ethanol, acetic acid	80-190 nm in diameter, 0.1 mm in length; XRD, TEM, SAED	Piezo, microwave material, sensors, capacitors
Nanopowders; [31]	Flame spray pyrolysis; barium carbonate and titanium tetra-*iso*-propoxide, citric acid	70 nm, *tetragonal structure*; XRD, TEM, SEM	Multilayer ceramic capacitors

Table 4. Physical-chemical methods

Another interesting subject is the possible application of ceramic nanofibers produced by the method of electrospinning. These possible applications include: nanofiber-based supports for catalysts, nanofiber-based photocatalysts, nanofiber-based membrane for filtration, nanofiber-based sensors, nanofiber-based photoelectrodes for photovoltaic cells, nanofiber-based electrodes for lithium batteries, nanofiber-based electrode-supports for fuel cells, and materials for implants [28, 29, 30].

2.4. Biological methods

The proper combination of chemical and physical methods (Table 4) allows the production of very small ferroelectric nanoparticles (5-10 nm). However, toxic chemicals and/or high temperatures/pressures are needed in most cases. These requirements substantially limit the possible biomedical applications of ferroelectric nanoparticles. Biological methods were proposed as eco-friendly "green" alternatives to existing chemical and physical methods. The biosynthesis of different types of nanoparticles was reviewed recently in a few papers [32, 33, 34, 35], but biological methods that produce specifically ferroelectric nanomaterials are not very numerous. For example, the first review [32] lists nearly a hundred variations for the synthesis of nanoparticles by microorganisms, and only a few of them produce ferroelectric nanoparticles. The biological methods applied to synthesize the nanoparticles of the most widely known ferroelectric $BaTiO_3$ are summarized in Table 5 (papers published before 2008 were discussed in [36]). The references listed in this table show that most of the biological methods mentioned employ some chemical or physical steps as well.

http://dx.doi.org/10.5772/52516

Shape of nanoparticles; Ref.	Synthesis method; Raw materials	Relevant parameters (size, morphology, crystallinity, spontaneous polarization etc.); Measurement and characterization methods	Comments
Aggregates (~500 nm) of nanocrystals(~ 50 nm); [37]	The peptides BT1 and BT2induce the room-temperature *precipitation* of $BaTiO_3$; Aqueous precursor solution composed of barium acetate ($Ba(OOCCH_3)_2$) potassium bis(oxalato) oxotitanate(IV) ($K_2[TiO(C_2O_4)_2]2H_2O$); pH 6.8	50-100 nm; *tetragonal phase* remnant polarization P_r ~ 2-3 C/cm^2; the coercive field, E_c = 5-6 kV/cm (cigar-like loop); SEM, TEM, SAED and XRD analysis, dielectric measurements	Rapid room-temperature synthesis of ferroelectric (tetragonal) $BaTiO_3$ within 2 hr
Quasi-spherical nanoparticles; [38]	*Lactobacillus* - assisted biosynthesis; $BaCO_3$ and TiO_2 solid state synthesis $BaTiO_3$; Slush of micron-sized $BaTiO_3$ particles+ Lactic acid Bacillus spore tablets	20-80 nm, *single-phase tetragonal structure*; XRD, TEM	Extracellular synthesis
Spherical nanoparticles; [39]	(*Fusarium oxysporum*) fungus-assisted biosynthesis of $BaTiO_3$ nanoparticles; $(CH_3COO)_2Ba$ and K_2TiF_6 + fungal micelles of *Fusarium oxysporum*	4-5 nm, *tetragonal structure*; XRD, DCS, TEM, SAED, XPS, SPM	Extracellular synthesis
Spherical nanoparticles; [40]	Peptide nanorings used as templates; $BaTi(O_2CC_7\text{-}H_{15})[OCH(CH_3)_2]_5$	6-12 nm, *tetragonal structure*; AFM, TEM, XRD, EFM, Raman	4 days
Spherical nanoparticles; [41]	(*Saccharomyces cerevisiae*) baker's yeast-assisted biosynthesis of $BaTiO_3$ nanoparticles; $BaCO_3$ and TiO_2 solid state synthesis $BaTiO_3$, Slurry of large $BaTiO_3$ + yeast culture	8-21 nm, *single-phase hexagonal structure*; XRD, TEM	Extracellular synthesis
Spherical nanoparticles; [42]	Kinetically controlled vapor diffusion Single source, bimetallic alkoxide with the vapor diffusion of a hydrolytic catalyst (H_2O)	6-8 nm, *cubic phase*; XRD, TEM	~250 g.

Table 5. Biological methods

One can note from Table 5 that these biological methods can be roughly grouped into two categories: the synthesis of ferroelectric nanoparticles using a number of seed chemicals and some type of biological or bio-inspired system, and the creation of nanoparticles using large particles of ready ferroelectric material and some type of biological system. Most of the methods that are currently available fall into the first category; the methods in the second category have appeared only very recently.

The earliest discoveries of nanoparticle synthesis by microorganisms (fungus, *Lactobacillus* and yeast) involve metal, alloy, and metal oxide nanoparticles. With a lag of one to two years, this research was followed by attempts at ferroelectric nanoparticle synthesis by the same living systems [32]. Fungus (*Fusarium oxysporum,* commonly found in soil) has been shown to synthesize extracellularly ferroelectric nanoparticles of barium titanate (4-5 nm of average size) [39, 43] at room temperature, producing small ferroelectric nanoparticles on a scale that has been previously inaccessible. This method falls into the first category, since fungus produces barium titanate from more than one seed chemical. *Lactobacillus* (bacteria commonly used to curdle milk and produce buttermilk) has been shown to synthesize ferroelectric nanoparticles (with sizes ranging from 20 to 80 nm) from the slush of large barium titanate particles [38]. Baker's yeast (*Saccharomyces cerevisiae*) has also been attempted as a biosynthesis agent; the barium titanate nanoparticles received were on average ~10nm [41]. Both the yeast- and lactobacillus-mediated production of barium titanate nanoparticles fall into the second category: they start with large particles of ready barium titanate in water and produce small (under 100 nm) nanoparticles extracellularly.

Peptide nanorings provide another interesting biomimetic route to the template-mediated synthesis of $BaTiO_3$ and $SrTiO_3$ nanoparticles [40]. Developed in 2006, it was the very first method of obtaining ferroelectric barium titanate nanoparticles at room temperature (many previous room-temperature methods had trouble growing barium titanate not in the cubic phase).

In this category of biological methods, one has to mention bioinspired methods of dispersing nanoparticles and producing stable colloids. Take for examplethe production of stable nanoparticle suspensions using microbial-derived surfactants [44] where usual garden variety surfactants are replaced with bio-derived materials that are significantly smaller in size and that help prevent aggregation of multiple nanoparticles during sol-gel synthesis. This process relies on some physical/chemical steps, just like the majority of other biological methods of ferroelectric nanoparticle production that have been developed to date.

These bio-inspired methods show great promise because they produce relatively small ferroelectric nanoparticles at room (or low) temperature compared to conventional physical and chemical methods, many of which require lengthy processes, use of high temperatures, harsh chemicals, etc.

http://dx.doi.org/10.5772/52516

3. Applications of ferroelectric nanoparticles in materials technologies

3.1. Multilayered capacitors and nanocomposites

As was previously mentioned in the introduction, the era of applied ferroelectricity was launched after the first reliable ferroelectric material – ceramic $BaTiO_3$ – was discovered [1]. The very high dielectric constants (~1000) of bulk ferroelectric materials were applied in the manufacturing of discrete and multilayered ceramic capacitors. Advances in the production of ferroelectric nanoparticles broadened this application to a large extent. Currently, nanosized $BaTiO_3$ powders are successfully used for manufacturing miniaturized and highly volume-efficient multilayer ceramic capacitors (MLCCs, for more details see review [45]).

Typical MLCCs utilize pellets of nanopowders sintered at high temperatures, producing structures with high rigidity that can impose certain limitations. Nanocomposites made of flexible polymers and ferroelectric nanoparticles overcome the restrictions caused by rigidity of sintered structures. Polymeric materials doped with ferroelectric nanoparticles are of considerable interest as a solution for processable high permittivity materials for various electronic applications: volume efficient multilayer capacitors, high-energy-density capacitors, embedded capacitors, and gate insulators in organic field effect transistors [46].

Both components (polymer and ferroelectric nanoparticle) of such nanocomposites are essential. The polymeric material brings flexibility and the ability to cover a large surface area, while nanoparticles share high values of dielectric permittivity with the matrix-enhancing effective dielectric permittivity of the composite. It's not easy to achieve homogeneous dispersion of ferroelectric nanoparticles in a polymeric matrix because of the high surface energy of nanoparticles, which usually leads to aggregation and phase separation, resulting in poor processability of the films and a high defect density. In order to make high-quality nanocomposites (good solution processability, low leakage current, high permittivity, and high dielectric strength), surface modification of nanoparticles is needed. The surface modification of nanoparticles decreases attractive forces between them, thus preventing aggregation. In addition, the proper design of a surface agent can increase interactions between the nanoparticles and the polymeric matrix. As a result, a quite homogeneous distribution of the nanoparticles in the polymeric matrix can be achieved. However, there are several obstacles that prevent nanocomposites "polymer/ferroelectric nanoparticle" from mass commercialization. The most common are poor temperature stability of dielectric constants and large dielectric losses. The proper design of both the nanoparticles (material size, shape, surface modification) and the polymer matrix can overcome these problems. For example, Table 6 shows short descriptions of the relatively successful (in terms of material performance) nanocomposites "polymer/$BaTiO_3$". All of them are optically transparent, have high dielectric permittivity (25-30), and moderate losses (loss tangent is in the range 0.01-0.05 for different materials).

Composition (polymer/ nanoparticle)	Particle surface functionalization	Dielectric permitti-vity of nanocom-posite	Particle size &breakdown electric field	Ref.
Barium titanate / poly(vinylidene fluoride-*co*-hexafluoro propylene), volume concentration < 80%	Phosphonic acid surface-modified $BaTiO_3$ nanoparticles	35	30-50 nm >164 V/μm	[46]
Barium titanate/polyimide ($BaTiO_3$/PI); volume concentration < 40%	Core-shell structure; polyamic acid (PAA) was used to cover particle	20	70-100 nm 67 MV/m	[47]
Poly(vinylidene fluoride)/ Barium titanate (PVDF/ $BaTiO_3$); volume concentration 10%- 30%	Surface hydroxylated $BaTiO_3$	18-25	85-100 nm	[48]
Barium titanate/polyimide ($BaTiO_3$/PI); volume concentration < 50%	$BaTiO_3$ nanoparticles were prepared by the alkoxide route	30	30-50 nm 50-80 nm	[49]
Barium titanate/poly(methyl methacrylate) ($BaTiO_3$/PMMA); volume concentration 53%	$BaTiO_3$ nanoparticles were modified with a silane coupling agent (3-methacryloxypropyltrimethoxysilane)	36	16-21 nm	[50]
Barium titanate/polyimide ($BaTiO_3$/PI); volume concentration 59%	$BaTiO_3$ nanoparticles are surface modified by phthalimide with the aid of a silane coupling agent as a scaffold	37	20 nm	[51]

Table 6. Nanocomposites of polymer/ferroelectric nanoparticles

3.2. Ferroelectric nanoparticles and liquid crystals: Display and non-display applications

It is an established fact that modern technologies require novel and highly advanced materials. During the last decade, nanochemistry has developed a dozen conceptually different pathways of synthesis to satisfy the constantly growing demand for new materials [52]. However, chemical methods are elaborate, time consuming, and expensive. Although they are generally accepted as universal, nanochemistry methods do not work perfectly in certain cases. For example, it's very difficult (but not impossible) to produce 5-10 nm ferroelectric nanoparticles with tetragonal structures through chemical methods (see Table 2). In many cases, the chemical methods used to produce novel materials can be efficiently supplemented or even replaced with non-synthetic ones. The addition of ferroelectric nanoparticles to different materials (polymers; see Table 6, liquids, liquid crystals etc.) is a good example of a non-synthetic method used to develop materials with improved properties. This subsection will discuss how ferroelectric nanoparticles can modify properties of liquid crystals – materials that found numerous applications in the display industry, photonics, optical processing, and the biotech industry.

The concept of ferroelectric colloids in liquid crystals was created by Yu. Reznikov and co-authors [53] – liquid crystal material was doped with stabilized ferroelectric nanoparticles of low concentration (~0.3 %). Ferroelectric nanoparticles share their intrinsic properties with the liquid crystals matrix due to the alignment within the liquid crystal and interactions between mesogenic molecules. The low concentration of nanoparticles and their stabilization with surfactant (oleic acid) provided the stability of such systems. This classic paper reported on the main features of such colloids: (1) a lower threshold voltage by a factor of 1.7; (2) an enhanced dielectric anisotropy by a factor of 2; (3) a linear electro-optical response (the sensitivity to the sign of an applied electric field). It should be pointed out that pure nematics are not sensitive to the sign of an electric field – this property is intrinsic to ferroelectric liquid crystals rather than to nematics. The first experimental results stimulated a very active global interest in the field. More than 100 papers were published during the last decade, and, in the last few years, several review papers summarized the most important results [18, 54, 55]. The review paper [54] published this year by the founder of this field is of special interest since it comprehensively discusses the past, the present, and the future of the liquid crystal colloids.

Ferroelectric nanoparticles embedded into a liquid crystal host strongly interact with the surrounding mesogenic molecules, due to the strong permanent electric field of this particle. These interactions can affect the basic physical parameters of liquid crystals: birefringence, dielectric permittivity, elastic constants, viscosity, electrical and thermal conductivity, temperatures of phase transitions, etc. There are two basic mechanisms responsible for the observed effects: (1) the increase of the orientation coupling between mesogenic molecules; and (2) the direct contribution of the permanent polarization of the particle [54]. Experimental data suggests that the first scenario (the increase of the orientation coupling between mesogenic molecules) is more likely to happen in the case of single component liquid crystals [54]. The second mechanism (associated with the direct contribution of the permanent polarization of the ferroelectric nanoparticle to the physical properties of liquid crystals) is the primary factor in the case of multi-component liquid crystal mixtures [54].

So far most of the reported experimental data describes the properties of multi-component nematics: liquid crystals doped with ferroelectric nanoparticles. The electric field generated by ferroelectric nanoparticles can cause micro- and/or nano-separation of such mixturesandaffect the macroscopic properties of the liquid crystal colloid. As a result, in some cases, the data found for single component liquid crystals (5CB) is different from the data obtained for multi-component mixtures. Nevertheless, an analysis of the existing literature [18] shows that most of the published experimental data report that ferroelectric nanoparticles embedded in liquid crystals at low concentrations: (1) enhance dielectric permittivity, dielectric anisotropy, and optical birefringence; (2) lower the switching voltage U_{th} for the Freedericksz transition; (3) increase the orientational order parameter S and the isotropic–nematic transition temperature T_{NI}; and (4) reduce the switching times needed to reorient liquid crystals by an external electric field. It should be pointed out that in all of these cases, wet grinding was used to prepare the ferroelectric nanoparticles. Few papers claim results opposite to the ones (1)-(4) listed above; however, nanoparticles used to make a liquid crystal colloid in this case were not ferroelectric [18]. The important conclusion to take from these findings is that a strong ferroelectric response of ferroelectric nanoparticles is a key factor leading to all "positive" effects (1)-(4).

The stability of liquid crystal colloids is the most challenging problem that prevents the transition of liquid crystal colloids of ferroelectric nanoparticles from the academic sector to the industrial domain. During the last few years, the general focus of the research has shifted toward this problem, and the proper surface functionalization of nanoparticles is now being considered as one of the most important factors affecting the stability of liquid crystal colloids. The shape and size of nanoparticles are also of the utmost importance – it was found that 10-20 nm ferroelectric nanoparticles affect the properties of a liquid crystal host much morethan the same but larger nanoparticles(~100 nm).

In summary, ferroelectric nanoparticles can modify the intrinsic properties of liquid crystal materials without time-consuming and expensive chemical synthesis. Experimental data on the enhancement of electro-optical, optical, and nonlinear-optical responses of such materials shows the strong potential of ferroelectric nanoparticles for improving the "practical" properties of liquid crystals, especially for those materials where the method of chemical synthesis has reached its limit. Such modified materials are very attractive and suitable for use in switchable lenses, displays, and beam steering, as well as other light-controlling devices (spatial light modulators, tunable filters etc.) [18].

4. Ferroelectric nanoparticles in biology

In the last decade, ferroelectric nanoparticle areas of application veered toward biology and medicine. The first challenge to address when introducing nanoparticles to biological and medical systems is how to make them stable in media that is bio-compatible, namely in aqueous solutions. If the ferroelectric nanoparticles are not coated with a stabilizing agent, the intrinsic properties of such nanoparticles very often lead to unwanted effects such as aggregation and precipitation, as well as the leaching of some ions (leading to change in particle properties) in aqueous solutions. There has been a tremendous amount of effort in the last twenty years to create many species of ferroelectric nanoparticles with different surface coatings [56and references within]. For example, such stabilizing agents as polyacrylic acid, polymethacrylic acid, polyaspartic acid, (aminomethyl) phosphonic acid and poly-l-lysine were tried on barium titanate nanoparticles. A large variety of possible coatings yields different surface properties of said nanoparticles – different thickness and charge of the surface layer, different strength of stabilizing agent adhesion, different resulting particle size – all of it leading to very different interactions with biological objects such as cells and cellular components. Such water-soluble ferroelectric nanoparticles can be further functionalized with fluorescent markers or antibodies [57], and they have been recently observed within a number of mammalian cells [58, 59, 60] in vitro. The majority of ferroelectric nanoparticles that have been successfully functionalized are in the middle of nano size range (~100 nm or more), and the surface functionalization very frequently adds a thick surface layer, resulting in particles that are a few hundred nanometers in size. This poses a challenge which a few researchers are attempting to overcome through bio-inspired surfactants [44]. The surface functionalization of nanoparticles can be crucial for a desired effect in a biosystem – the response is often cell-specific, and biocompatibilty has to be experimentally tested for each new combination of nanoparticle and cell [61]. For example, a recent study [62] indicates that larger (over 200 nm) barium titanate nanoparticles have been successfully used with and without surface modification and they cause no toxic response in endothelial cell cultures.

Current efforts in this area concentrate on carrying these findings through to the systems of cells and tissue in vivo [63, 64]. Such biocompatible nanoparticles can be used in a variety of ways: as imaging agents [58, 59, 60, 64, 65], biocompatible nanoprobes [62, 66, 67, 68], cell proliferation agents [65], etc. Ferroelectric nanoparticles are currently being widely explored for applications in the area of medical imaging, e.g. for enhancing contrast in second harmonic generating methods for imaging deep tissue in vitro and in vivo [64, 69] for such important applications as screening for genetic diseases or cancer[70].

5. Ferroelectric nanoparticles applications in medicine and medical engineering

As mentioned above, ferroelectric nanoparticles have become widely used in material science and engineering applications, but biological and medical applications of these fascinating nanoparticles has only begun to be explored in the last decade. It has been recently discovered that they can be used for cell imaging or the detection of malignancies in lung cancer studies, or they can be functionalized to induce cell proliferation. Ferroelectric materials have a non-centrosymmetric crystalline structure, and are thus capable of generating a second harmonic of light [67]. This distinctive feature of ferroelectrics is the basis for a growing number of applications of ferroelectric nanoparticles as imaging/diagnostic agents and nanoprobes in optical imaging [64].

Optical imaging and detection methods are the most widespread among biological and medical communities. For example, second harmonic generation imaging has been successfully used for detection of *osteogenesis imperfecta* in biopsies of human skin [71] and lung cancer [72]. To improve contrast, many of the imaging methods rely on imaging probes, such as fluorescent markers or quantum dots [73]. This is the case for the SHG imaging technique: the SHG signal from biological cell components is often weak, so novel nanoprobes have been introduced to enhance contrast. SHG probes are photo-stable, and do not bleach or blink unlike conventional fluorescent probes. By definition, second harmonic generating nanoprobes (such as ferroelectric nanoparticles) are capable of converting two photons of light into one photon of half the incident wavelength [74]. This second-harmonic light can be detected using methods of nonlinear optical microscopy. Nonlinear optical properties of ferroelectric nanomaterials can be used for optical phase conjugation [75] and nonlinear microscopy [62, 76] – these properties have allowed them to spread to the area of medical sensors.

This also gives the ferroelectric SHG particle an edge in cell and tissue imaging in vivo. For example, recent advances in this area include imaging a tail of a living mouse with the aid of barium titanate nanoparticles [69].

The intrinsic large values of the dielectric permittivities of ferroelectric nanoparticles suggest their use to enhance the dielectric contrast of materials, such as polymers [77 and references within] and biological tissue [70]. These unique properties of ferroelectric nanoparticles lead to their novel use as contrast-enhancing agents for microwave tomography, which is a method of non-invasive assessment and diagnostics of soft tissues (such as detecting malignancies)[70].Recently, ferroelectric electro spun nanofibers also emerged in various biomedical areas including medical prostheses, tissue engineering, wound dressing, and drug delivery [28,29,30].

In conclusion, ferroelectric materials found a wide variety of biomedical applications in the last decade – and the list is constantly growing [78]. The ferroelectric material (e.g. barium titanate) used in medical implants has been known to accelerate osteogenesis [79], and the same material in nanoparticle form works both as an SHG probe to detect Osteogenesis Imperfecta [71] and, through microwave tomography, to detect lung cancer [70]. We expect more applications will become possible if other effects, such as piezoelectricity, ferroelectric hysteresis or stress-birefringence coupling, are exploited with biology and medicine in mind.

6. Conclusions

A review of recently published research papers allows us to conclude that the unique properties of ferroelectric nanoparticles offer an enormous range of applications, especially in material technologies, biology, and medicine. However, since such a conclusion has become an academic cliché, we would like to make just a few realistic and optimistic comments on the subject.

Discussing any application, we have to be more specific and distinguish between potential applications and actual (i.e. mass commercialized) ones. So far, the "applied" field of ferroelectric nanoparticles is at the stage of high potential for commercialization rather than real mass-commercialization. However, after the very first proof-of-concept research studies were completed just recently, a serious shift toward real applications was initiated. The research community realized that the controllable preparation of nanoparticles (size, shape), their proper surface functionalization, and the homogeneous dispersion into host material (liquid crystals, polymers, biological species) are among the most critical steps on the way to mass-commercialization. Many applications (especially bio-medical) require a large-scale preparation of mono-dispersed, very fine (5-10 nm) ferroelectric nanoparticles in the tetragonal phase – and only recently has substantial progress been made in this direction (see section 2). However, such ultra-fine nanoparticles were mostly tested in material technologies (ferroelectric liquid crystal colloids), while bio-medical methods utilized relatively large ferroelectric nanoparticles (~50-100 nm) (sections 4-5). The bioconjugation of 5-10 nm ferroelectric nanoparticles can extend the area of biomedical application even more, since these nanoparticles can be used as both passive and active nanoprobes.

The applied science of ferroelectric nanoparticles is in the beginning of its development. Many fundamental and applied problems still need to be solved before the potential applications will be converted into actual ones. Nevertheless, despite the fact that the emerging problems in applying nanoparticles in material technologies and biomedicine are still numerous [80], we can conclude that a good balance between purely academic and purely applied foci of research is the key toward mass commercialization of ferroelectric nanoparticles in all applied fields mentioned above.

Acknowledgements

The authors are very grateful to all coauthors of the papers related to the topic of ferroelectric nanoparticles. We acknowledge also the support of the UCCS Center for the Biofrontiers Institute, University of Colorado at Colorado Springs. A great portion of the work described in this review has been supported by the grants from the NSF #1102332 "Liquid Crystal Signal Processing Devices for Microwave and Millimeter Wave Operation" and # 1010508 "STTR Phase 1: Design, Fabrication, and Characterization of Ferroelectric Nanoparticle Doped Liquid Crystal / Polymer Composites."

List of abbreviations

AFM atomic force microscopy

DCS differential scanning calorimetry

DLS dynamic light scattering

DTA differential thermal analysis

EDX energy dispersive X-rayspectroscopy

EFM electrostatic force microscopy

FE-SEM field-emission scanning electron microscopy

FTIR Fourier transform infrared spectroscopy

HRTEM high resolution transmission electron microscopy

MLCC multilayer ceramic capacitors

PAA polyamic acid

PTCR positive temperature coefficient of resistance

PMMA poly(methyl methacrylate)

PVDF poly(vinylidene fluoride)

SAED selected area electron diffraction

SEM scanning electron microscopy

SHG probe second harmonic generating probe

SMPS scanning mobility particle size

SPM scanning probe microscopy

TEM transmission electron microscopy

TG or TGA thermo gravimetric analysis

XPSX-ray photoelectron spectroscopy

XRDX-ray diffraction

Author details

Yuriy Garbovskiy, Olena Zribi and Anatoliy Glushchenko

UCCS Center for the Biofrontiers Institute, Department of Physics, University of Colorado at Colorado Springs, Colorado Springs, Colorado, USA

References

[1] Haertling, G.H. Ferroelectric ceramics: history and technology. *Journal of the American Ceramic Society*82, 797-818 (1999).

[2] Cross, L.E. & Newnham, R.E. in *High-Technology Ceramics—Past, Present,and Future* 289–305 (American Ceramic Society, Westerville, OH, 1987).

[3] Mitsui, T. in Springer Handbook of Condensed Matter and Materials Data (ed. Warlimont, W.M.a.H.) 903 – 938 (Springer, Berlin, 2006).

[4] Whatmore, R. in Springer Handbook of Electronic and Photonic Materials (ed. Capper, S.K.a.P.) 597 – 623 (Springer, Berlin, 2006).

[5] Kalinin, S.V. et al. Nanoscale Electromechanics of Ferroelectric and Biological Systems: A New Dimension in Scanning Probe Microscopy. *Annual Review of Materials Research*37, 189-238 (2007).

[6] Gruverman, A. & Kholkin, A. Nanoscale ferroelectrics: processing, characterization and future trends. *Reports on Progress in Physics*69, 2443-74 (2006).

[7] Ahn, C.H., Rabe, K.M. & Triscone, J.M. Ferroelectricity at the nanoscale: local polarization in oxide thin films and heterostructures. *Science*303, 488-91 (2004).

[8] Scott, J.F. Applications of modern ferroelectrics. *Science*315, 954-9 (2007).

[9] Sarkar, D. Synthesis and Properties of $BaTiO_3$ Nanopowders. *Journal of the American Ceramic Society*94, 106-110 (2011).

[10] Zhu, Y.-F., Zhang, L., Natsuki, T., Fu, Y.-Q. & Ni, Q.-Q. Facile Synthesis of $BaTiO_3$ Nanotubes and Their Microwave Absorption Properties. *ACS Applied Materials & Interfaces*4, 2101-2106 (2012).

[11] O'Brien, S., Brus, L. & Murray, C.B. Synthesis of Monodisperse Nanoparticles of Barium Titanate: Toward a Generalized Strategy of Oxide Nanoparticle Synthesis. *Journal of the American Chemical Society*123, 12085-12086 (2001).

[12] Buscaglia, M.T., Bassoli, M., Buscaglia, V. & Vormberg, R. Solid-State Synthesis of Nanocrystalline $BaTiO_3$: Reaction Kinetics and Powder Properties. *Journal of the American Ceramic Society*91, 2862-2869 (2008).

[13] Zhang, Y., Wang, L. & Xue, D. Molten salt route of well dispersive barium titanate nanoparticles. *Powder Technology*217, 629-633 (2012).

[14] Wei, X. et al. Synthesis of Highly Dispersed Barium Titanate Nanoparticles by a Novel Solvothermal Method. *Journal of the American Ceramic Society*91, 315-318 (2008).

[15] Hayashi, H. & Hakuta, Y. Hydrothermal Synthesis of Metal Oxide Nanoparticles in Supercritical Water. *Materials*3, 3794-3817 (2010).

[16] Zhang, H. et al. Fabrication of Monodispersed 5-nm $BaTiO_3$ Nanocrystals with Narrow Size Distribution via One-Step Solvothermal Route. *Journal of the American Ceramic Society*94, 3220-3222 (2011).

[17] Atkuri, H. et al. Preparation of ferroelectric nanoparticles for their use in liquid crystalline colloids. *Journal of Optics A: Pure and Applied Optics*11, 024006 (2009).

[18] Garbovskiy, Y. & Glushchenko, A. in Solid State Physics (ed. Camley, R.E.) 1 – 74 (2010).

[19] Cook, G. et al. Nanoparticle doped organic-inorganic hybridphotorefractives. *Optics Express*16, 4015-4022 (2008).

[20] Cook, G. et al. Harvesting single ferroelectric domain stressed nanoparticles for optical and ferroic applications. *Journal of Applied Physics*108, 064309-4 (2010).

[21] Basun, S.A., Cook, G., Reshetnyak, V.Y., Glushchenko, A.V. & Evans, D.R. Dipole moment and spontaneous polarization of ferroelectric nanoparticles in a nonpolar fluid suspension. *Physical Review B*84, 024105 (2011).

[22] Swaminathan, V. et al. Microwave synthesis of noncentrosymmetric $BaTiO_3$ truncated nanocubes for charge storage applications. *ACS Appl Mater Interfaces*2, 3037-42 (2010).

[23] Fuentes, S. et al. Synthesis and characterization of $BaTiO_3$ nanoparticles in oxygenatmosphere. 505, 568–572 (2010).

[24] Qi, J.Q. et al. Direct synthesis of ultrafine tetragonal $BaTiO_3$ nanoparticles at room temperature. *Nanoscale Research Letters*6, 466 (2011).

[25] Kong, L.B., Zhang, T.S., Ma, J. & Boey, F. Progress in synthesis of ferroelectricceramicmaterialsviahigh-energymechanochemicaltechnique. 53, 207–322 (2008).

[26] Dang, F. et al. Aneweffect of ultrasonication on the formation of $BaTiO_3$ nanoparticles. 17, 310–314 (2010).

[27] Suzuki, K. & Kijima, K. Well-crystallized barium titanate nanoparticles prepared by plasma chemical vapor deposition. 58, 1650–1654 (2004).

[28] Ramaseshan, R., Sundarrajan, S., Jose, R. & Ramakrishna, S. Nanostructured ceramics by electrospinning. *Journal of Applied Physics*102, 111101-17 (2007).

[29] He, Y. et al. Humidity sensing properties of $BaTiO_3$ nanofiber prepared via electrospinning. *Sensors and Actuators B: Chemical*146, 98-102 (2010).

[30] Dai, Y., Liu, W., Formo, E., Sun, Y. & Xia, Y. Ceramic nanofibers fabricated by electrospinning and their applications in catalysis, environmental science, and energy technology. *Polymers for Advanced Technologies*22, 326-338 (2011).

[31] Jung, D.S., Hong, S.K., Cho, J.S. & Kang, Y.C. Nano-sizedbariumtitanatepowders with tetragonalcrystalstructureprepared by flamespraypyrolysis. 28, 109–115 (2008).

[32] Li, X., Xu, H., Chen, Z.-S. & Chen, G. Biosynthesis of Nanoparticles by Microorganisms and Their Applications. *Journal of Nanomaterials*2011 (2011).

[33] Chen, C.-L. & Rosi, N.L. Peptide-Based Methods for the Preparation of Nanostructured Inorganic Materials. *Angewandte Chemie International Edition*49, 1924-1942 (2010).

[34] Briggs, B.D. & Knecht, M.R. Nanotechnology Meets Biology: Peptide-based Methods for the Fabrication of Functional Materials. *The Journal of Physical Chemistry Letters*3, 405-418 (2012).

[35] Dhillon, G.S., Brar, S.K., Kaur, S. & Verma, M. Green approach for nanoparticle biosynthesis by fungi: current trends and applications. *Critical Reviews in Biotechnology*32, 49-73 (2011).

[36] Beier, C.W., Cuevas, M.A. & Brutchey, R.L. Room-Temperature Synthetic Pathways to Barium Titanate Nanocrystals. *Small*4, 2102-2106 (2008).

[37] Ahmad, G. et al. Rapid Bioenabled Formation of Ferroelectric $BaTiO_3$ at Room Temperature from an Aqueous Salt Solution at Near Neutral pH. *Journal of the American Chemical Society*130, 4-5 (2007).

[38] Jha, A.K. & Prasad, K. Ferroelectric $BaTiO_3$ nanoparticles: Biosynthesis and characterization. *Colloids and Surfaces B: Biointerfaces*75, 330-334 (2010).

[39] Bansal, V., Poddar, P., Ahmad, A. & Sastry, M. Room-Temperature Biosynthesis of Ferroelectric Barium Titanate Nanoparticles. *Journal of the American Chemical Society*128, 11958-11963 (2006).

[40] Nuraje, N. et al. Room Temperature Synthesis of Ferroelectric Barium Titanate Nanoparticles Using Peptide Nanorings as Templates. *Advanced Materials*18, 807-811 (2006).

[41] Jha, A.K. & Prasad, K. Synthesis of $BaTiO_3$ nanoparticles: A new sustainable green approach. *Integrated Ferroelectrics*117, 49-54 (2010).

[42] Ould-Ely, T. et al. Large-scale engineered synthesis of $BaTiO_3$ nanoparticles using low-temperature bioinspired principles. *Nature Protocols*6, 97-104 (2011).

[43] Bansal, V., Bharde, A., Ramanathan, R. & Bhargava, S.K. Inorganic materials using 'unusual' microorganisms. *Advances in Colloid and Interface Science*.

[44] Kamiya, H. et al. Preparation of Highly Dispersed Ultrafine Barium Titanate Powder by Using Microbial-Derived Surfactant. *Journal of the American Ceramic Society*86, 2011-2018 (2012).

[45] Pithan, C., Hennings, D. & Waser, R. Progress in the Synthesis of Nanocrystalline $BaTiO_3$ Powders for MLCC. *International Journal of Applied Ceramic Technology*2, 1-14 (2005).

[46] Kim, P. et al. High Energy Density Nanocomposites Based on Surface-Modified $BaTiO_3$ and a Ferroelectric Polymer. *ACS Nano*3, 2581-2592 (2009).

[47] Dang, Z.-M. et al. Fabrication and Dielectric Characterization of Advanced $BaTiO_3$/Polyimide Nanocomposite Films with High Thermal Stability. *Advanced Functional Materials*18, 1509-1517 (2008).

[48] Zhou, T. et al. Improving Dielectric Properties of $BaTiO_3$/Ferroelectric Polymer Composites by Employing Surface Hydroxylated $BaTiO_3$ Nanoparticles. *ACS Applied Materials & Interfaces*3, 2184-2188 (2011).

[49] Fan, B.-H., Zha, J.-W., Wang, D.-R., Zhao, J. & Dang, Z.-M. Experimental study and theoretical prediction of dielectric permittivity in $BaTiO_3$/polyimide nanocomposite films. *Applied Physics Letters*100, 092903-4 (2012).

[50] Nagao, D., Kinoshita, T., Watanabe, A. & Konno, M. Fabrication of highly refractive, transparent $BaTiO_3$/poly(methyl methacrylate) composite films with high permittivities. *Polymer International*60, 1180-1184 (2011).

[51] Abe, K., Nagao, D., Watanabe, A. & Konno, M. Fabrication of highly refractive barium-titanate-incorporated polyimide nanocomposite films with high permittivity and thermal stability. *Polymer International*, n/a-n/a (2012).

[52] Balzani, V. Nanochemistry: A Chemical Approach to Nanomaterials. Geoffrey A. Ozin and André C. Arsenault. *Small*2, 678-679 (2006).

[53] Reznikov, Y. et al. Ferroelectric nematic suspension. *Applied Physics Letters*82, 1917-1919 (2003).

[54] Reznikov, Y. in Liquid Crystals Beyond Displays: Chemistry, Physics, and Applications (ed. Li, Q.) 403-426 (John Wiley & Sons, Inc., 2012).

[55] Liang, H.-H. & Lee, J.-Y. Enhanced Electro-Optical Properties of Liquid Crystals Devices by Doping with Ferroelectric Nanoparticles. *Ferroelectrics - Material Aspects*, 193-210 (2011).

[56] Ciofani, G. et al. Preparation of stable dispersion of barium titanate nanoparticles: Potential applications in biomedicine. *Colloids and Surfaces B: Biointerfaces*76, 535–543 (2010).

[57] Hsieh, C.-L., Grange, R., Pu, Y. & Psaltis, D. Bioconjugation of barium titanate nanocrystals with immunoglobulin G antibody for second harmonic radiation imaging probes. *Biomaterials*31, 2272–2277 (2010).

[58] Hsieh, C.-L., Grange, R., Pu, Y. & Psaltis, D. (eds. Campagnola, P.J., Stelzer, E.H.K. & von Bally, G.) 73670D-6 (SPIE, Munich, Germany, 2009).

[59] Hsieh, C.-L., Grange, R., Pu, Y. & Psaltis, D. Three-dimensional harmonic holographic microcopy using nanoparticles as probes for cell imaging. *Optics Express*17, 2880-2891 (2009).

[60] Grange, R., Hsieh, C.L., Pu, Y. & Psaltis, D. in European Conference on Lasers and Electro-Optics 2009 and the European Quantum Electronics Conference. CLEO Europe - EQEC 2009. 1-1 (2009).

[61] Ciofani, G. et al. in Piezoelectric Nanomaterials for Biomedical Applications 213-238 (Springer Berlin Heidelberg, 2012).

[62] Yust, B.G., Razavi, N., Pedraza, F. & Sardar, D.K. (eds. Cartwright, A.N. & Nicolau, D.V.) 82310H-6 (SPIE, San Francisco, California, USA, 2012).

[63] Zipfel, W.R. et al. Live tissue intrinsic emission microscopy using multiphoton-excited native fluorescence and second harmonic generation. *Proceedings of the National Academy of Sciences*100, 7075-7080 (2003).

[64] Pantazis, P., Maloney, J., Wu, D. & Fraser, S.E. Second harmonic generating (SHG) nanoprobes for in vivo imaging. *Proceedings of the National Academy of Sciences*107, 14535-14540 (2010).

[65] Ciofani, G., Ricotti, L. & Mattoli, V. Preparation, characterization and in vitro testing of poly(lactic-co-glycolic) acid/barium titanate nanoparticle composites for enhanced cellular proliferation. *Biomedical Microdevices*13, 255-266 (2011).

[66] Pantazis, P., Pu, Y., Psaltis, D. & Fraser, S. (eds. Periasamy, A. & So, P.T.C.) 71831P-5 (SPIE, San Jose, CA, USA, 2009).

[67] Horiuchi, N. Imaging: Second-harmonic nanoprobes. *Nature Photonics*5, 7-7 (2011).

[68] Hsieh, C.-L., Grange, R., Pu, Y. & Psaltis, D. (eds. Mohseni, H. & Razeghi, M.) 77590T-6 (SPIE, San Diego, California, USA, 2010).

[69] Hsieh, C.-L., Lanvin, T., Grange, R., Pu, Y. & Psaltis, D. in Conference on Lasers and Electro-Optics (CLEO), 2011 1-2 (2011).

[70] Semenov, S., Pham, N. & Egot-Lemaire, S. in World Congress on Medical Physics and Biomedical Engineering (eds. Dössel, O. & Schlegel, W.C.) 311-313 (Springer Berlin Heidelberg, Munich, Germany, 2009).

[71] Adur, J. et al. (eds. Periasamy, A., Konig, K. & So, P.T.C.) 82263P-7 (SPIE, San Francisco, California, USA, 2012).

[72] Wang, C.-C. et al. Differentiation of normal and cancerous lung tissues by multiphoton imaging. *Journal of Biomedical Optics*14, 044034-4 (2009).

[73] Chan, W.C.W. in Advances in Experimental Medicine and Biology 208 (Springer Science+Business Media, LLC, 2007).

[74] Dempsey, W.P., Fraser, S.E. & Pantazis, P. SHG nanoprobes: Advancing harmonic imaging in biology. *BioEssays*34, 351-360 (2012).

[75] Yust, B.G., Sardar, D.K. & Tsin, A. (eds. Cartwright, A.N. & Nicolau, D.V.) 79080G-7 (SPIE, San Francisco, California, USA, 2011).

[76] Ganeev, R.A., Suzuki, M., Baba, M., Ichihara, M. & Kuroda, H. Low- and high-order nonlinear optical properties of $BaTiO_3$ and $SrTiO_3$ nanoparticles. *Journal of the Optical Society of America B*25, 325-333 (2008).

[77] Lai, K.T., Nair, B.G. & Semenov, S. Optical and microwave studies of ferroelectric nanoparticles for application in biomedical imaging. *Microwave and Optical Technology Letters*54, 11-13 (2012).

[78] Ciofani, G. et al. 987-990 (KINTEX, Korea, 2010).

[79] Furuya, K., Morita, Y., Tanaka, K., Katayama, T. & Nakamachi, E. (eds. Martin-Palma, R.J. & Lakhtakia, A.) 79750U-6 (SPIE, San Diego, California, USA, 2011).

[80] Florence, A.T. "Targeting" nanoparticles: The constraints of physical laws and physical barriers. (2012).

Chapter 22

Photorefractive Effect in Ferroelectric Liquid Crystals

Takeo Sasaki

Additional information is available at the end of the chapter

http://dx.doi.org/10.5772/52921

1. Introduction

Ferroelectric liquid crystals have been attracting great interest for their application in photorefractive devices. The photorefractive effect is a phenomenon by which a change in refractive index is induced by the interference of laser beams [1, 2]. Dynamic holograms are easily realized by the photorefractive effect. Holograms generate three-dimensional images of objects. They are produced by recording interference fringes generated by light reflected from an object and a reference light (Figure 1). A hologram diffracts incident light to produce a three-dimensional images of an object. 3D displays are expected to be widely used as next-generation displays. However, current 3D displays are essentially stereograms. However, holographic displays that can realize natural 3D images are anticipated.

The photorefractive effect has the potential to realize dynamic holograms by recording holograms as a change in the refractive index of the medium [1, 2]. The photorefractive effect induces a change in the refractive index by a mechanism involving both photovoltaic and electro-optic effects (Figure 2). When two laser beams interfere in an organic photorefractive material, a charge generation occurs at the bright positions of the interference fringes. The generated charges diffuse or drift within the material. Since the mobilities of positive and negative charges are different in most organic materials, a charge separated state is formed. The charge with the higher mobility diffuses over a longer distance than the charge with lower mobility, such that while the low mobility charge stays in the bright areas, the high mobility charge moves to the dark areas. The bright and dark positions of the interference fringes are thus charged with opposite polarities and an internal electric field (space-charge field) is generated in the area between the bright and dark positions. The refractive index of this area between the bright and dark positions is changed through the electro-optic effect. Thus, a refractive index grating (or hologram) is formed. One material class that exhibits high photorefractivity is that of glassy photoconductive polymers doped with high concentrations of D-π-A chromophores, in which donor and acceptor groups are attached to a πconjugate system [3-5]. In order to obtain photorefractivity in polymer materials, a high electric field of 10–50 V/μm is usually applied to a polymer film [6-8]. This electric field is necessary to increase charge generation efficiency. Significantly, the photorefractive effect permits two-beam coupling. It can be used to coherently amplify signal beams, and so has the potential to be used in a wide range of optical technologies as a transistor in electric circuits. With the aim of developing a 3D display, a multiplex hologram has been demonstrated using a photorefractive polymer film [9,10]. Clear 3D images were recorded in the film. However, the slow response (~100 ms) and the high electric field (30–50 V/μm) required to activate the photorefractive effect in the polymer materials both need to be improved.

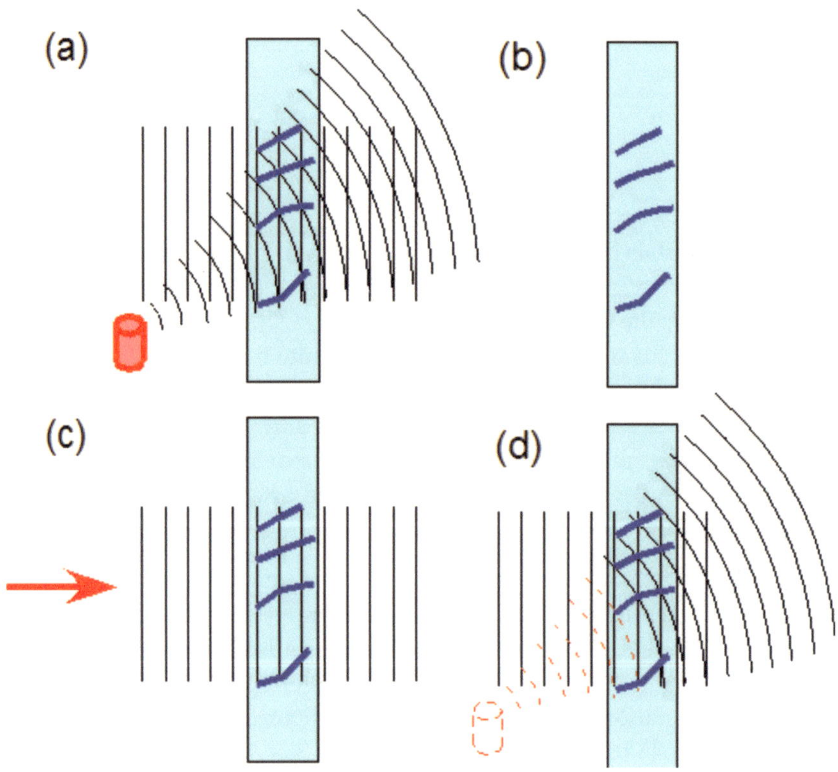

Figure 1. Hologram recording and image reconstruction. (a) A light reflected from an object is interfered with by a reference light in a photo-sensitive material, such as a photopolymer. (b) The interference fringe is recorded in a photosensitive material (hologram). (c) A readout beam is irradiated by the hologram. (d) The readout beam is diffracted by the hologram and the object image is reconstructed.

Ferroelectric liquid crystals are expected to be used as high performance photorefractive materials [11-13]. A photorefractive ferroelectric liquid crystal with a fast response of 5 ms has been reported [14]. The photorefractive effect has been reported in surface-stabilized ferroelectric liquid crystals (SS-FLCs) doped with a photoconductive compound. Liquid crystals are classified into several groups. The most well known are nematic liquid crystals and smectic liquid crystals.

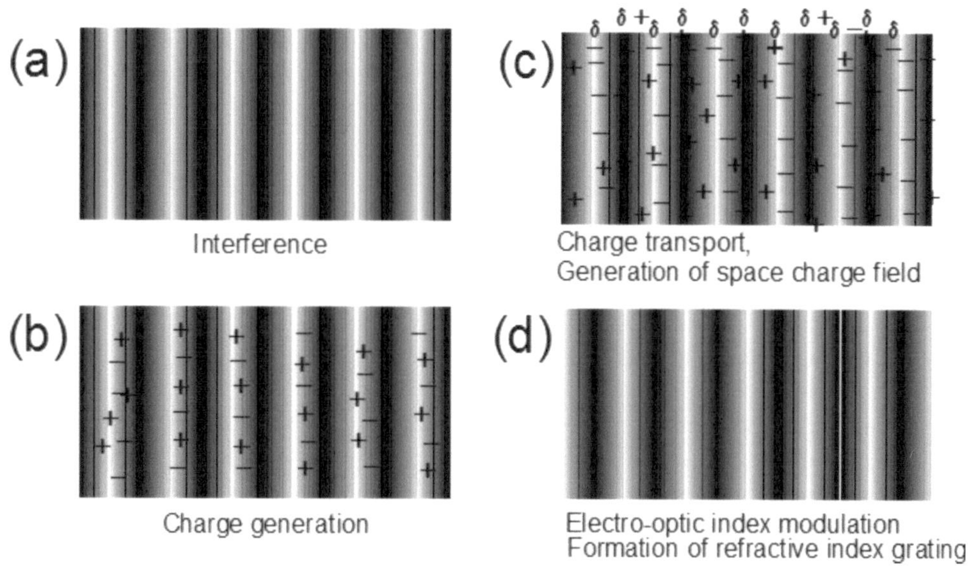

Figure 2. Schematic illustration of the mechanism of the photorefractive effect. (a) Two laser beams interfere in the photorefractive material; (b) charge generation occurs at the light areas of the interference fringes; (c) electrons are trapped at the trap sites in the light areas and holes migrate by diffusion or drift in the presence of an external electric field and generate an internal electric field between the light and dark positions; (d) the refractive index of the corresponding area is altered by the internal electric field generated.

Nematic liquid crystals are used in LC displays. On the other hand, smectic liquid crystals are very viscous and, hence, are not utilized in any practical applications. Ferroelectric liquid crystals (FLCs) belong to the class of smectic liquid crystals that have a layered structure (Figure 3). The molecular structure of a typical FLC contains a chiral unit, a carbonyl group, a central core - which is a rigid rod-like structure, such as biphenyl, phenylpyrimidine or phenylbenzoate - and a flexible alkyl chain [15]. Thus, the dipole moment of an FLC molecule is perpendicular to the molecular long axis. FLCs exhibit a chiral smectic C phase (SmC*) that possesses a helical structure. Compared to nematic LCs, FLCs are more crystal than liquid and the preparation of fine FLC films requires several sophisticated techniques. Obtaining a uniformly aligned, defect-free, surface-stabilized FLC (SS-FLC, Figure 3) using a single FLC compound is very difficult, and mixtures of several LC compounds are usually used to obtain fine SS-FLC films.

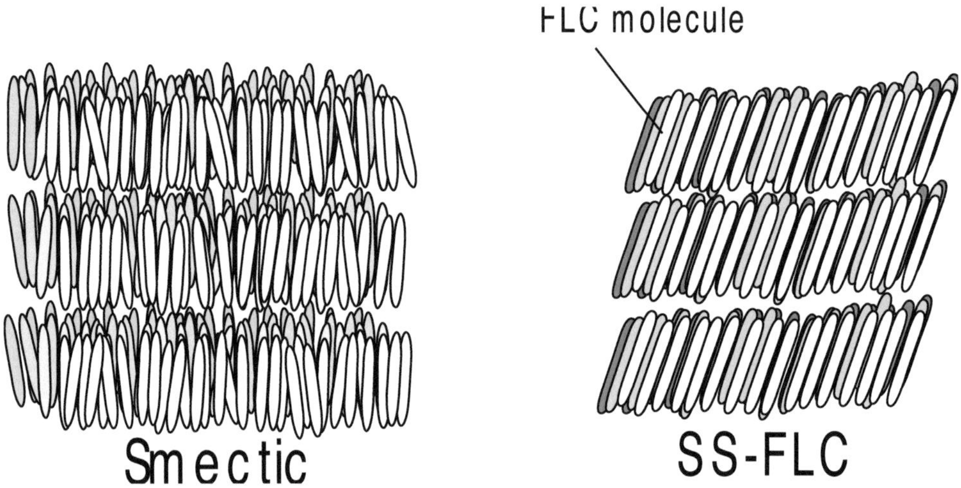

Figure 3. Structures of the smectic phase and the surface stabilized smectic-C phase (SS-FLC).

The FLC mixtures are composed of the base LC - which is also a mixture of several LC compounds - and a chiral compound. The chiral compound introduces a helical structure into the LC phase through supramolecular interactions. It should be mentioned here that in order to observe ferroelectricity in these materials, the ferroelectric liquid crystals must be formed into thin films. The thickness of the film must be within a few micrometers. When an FLC is sandwiched between glass plates to form a film a few micrometers thick, the helical structure of the smectic C phase uncoils and a surface-stabilized state (SS-state) is formed in which spontaneous polarization (Ps) appears. For display applications, the thickness of the film is usually 2 μm. In such thin films, FLC molecules can align in only two directions. This state is called a surface-stabilized state (SS-state). The alignment direction of the FLC molecules changes according to the direction of the spontaneous polarization (Figure 4). The direction of the spontaneous polarization is governed by the photoinduced internal electric field, giving rise to a refractive index grating with properties dependent on the direction of polarization.

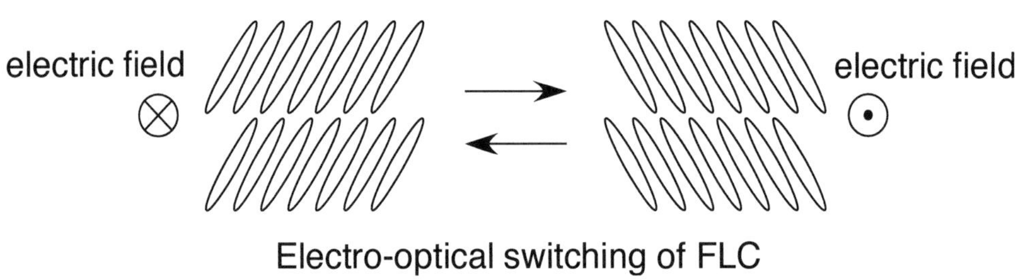

Figure 4. Electro-optical switching in the surface-stabilized state of FLCs.

Figure 5 shows a schematic illustration of the mechanism of the photorefractive effect in FLCs. When laser beams interfere in a mixture of an FLC and a photoconductive compound, charge separation occurs between bright and dark positions and an internal electric field is produced. The internal electric field alters the direction of spontaneous polarization in the area between the bright and dark positions of the interference fringes, which induces a periodic change in the orientation of the FLC molecules. This is different from the processes that occur in other photorefractive materials in that the molecular dipole rather than the bulk polarization responds to the internal electric field. Since the switching of FLC molecules is due to the response of bulk polarization, the switching is extremely fast.

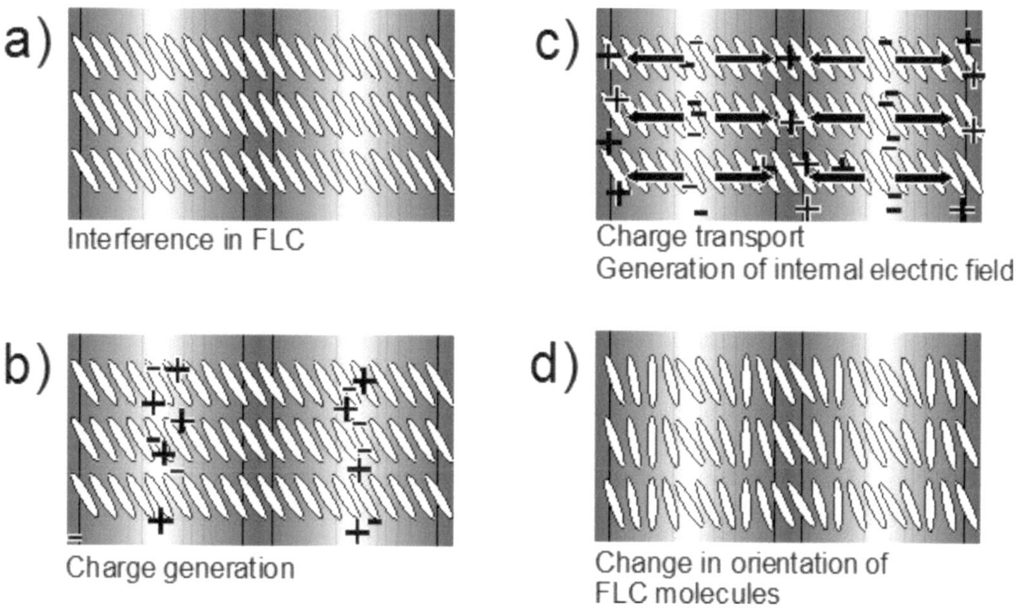

Figure 5. Schematic illustration of the mechanism of the photorefractive effect in FLCs. (a) Two laser beams interfere in the surface-stabilized state of the FLC/photoconductive compound mixture; (b) charge generation occurs at the bright areas of the interference fringes; (c) electrons are trapped at the trap sites in the bright areas, holes migrate by diffusion or drift in the presence of an external electric field to generate an internal electric field between the bright and dark positions; (d) the orientation of the spontaneous polarization vector (i.e., the orientation of mesogens in the FLCs) is altered by the internal electric field.

2. Characteristics of the photorefractive effect

Since a change in the refractive index via the photorefractive effect occurs in the areas between the bright and dark positions of the interference fringe, the phase of the resulting index grating is shifted from the interference fringe. This is characteristic of the photorefractive effect that the phase of the refractive index grating is $\pi/2$-shifted from the interference fringe. When the material is photochemically active and is not photorefractive, a photochemical reaction takes place at the bright areas and a refractive index grating with the same phase as that of the interference fringe is formed (Figure 6(a)).

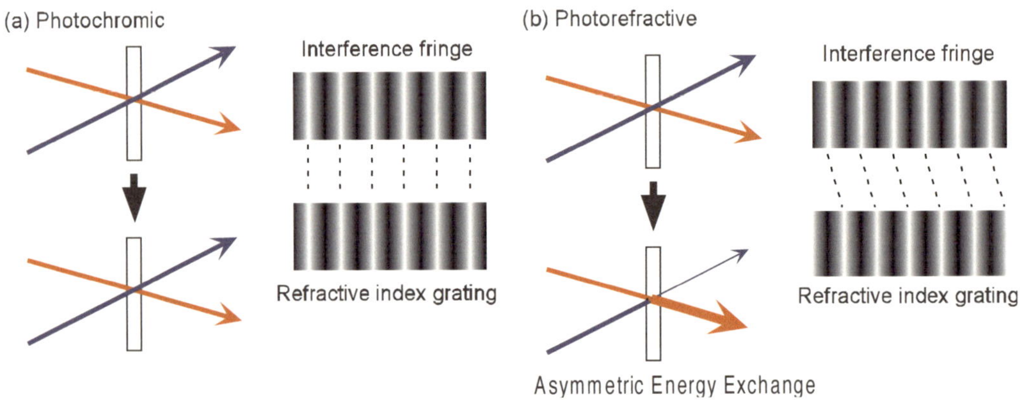

Figure 6. (a) Photochromic grating, and (b) photorefractive grating.

The interfering laser beams are diffracted by this grating; however, the apparent transmitted intensities of the laser beams do not change because the diffraction is symmetric. Beam 1 is diffracted in the direction of beam 2 and beam 2 is diffracted in the direction of beam 1. However, if the material is photorefractive, the phase of the refractive index grating is shifted from that of the interference fringes, and this affects the propagation of the two beams. Beam 1 is energetically coupled with beam 2 for the two laser beams. Consequently, the apparent transmitted intensity of beam 1 increases and that of beam 2 decreases (Figure 6(b)). This phenomenon is termed 'asymmetric energy exchange' in the two-beam coupling experiment. The photorefractivity of a material is confirmed by the occurrence of this asymmetric energy exchange.

3. Measurement of photorefractivity

The photorefractive effect is evaluated by a two-beam coupling method and by a four-wave mixing experiment. Figure 7(a) shows a schematic illustration of the experimental setup used for the two-beam coupling method. A p-polarized beam from a laser is divided into two beams by a beam splitter and the beams are interfered within the sample film. An electric field is applied to the sample using a high voltage supply unit. This external electric field is applied in order to increase the efficiency of charge generation in the film. The change in the transmitted beam intensity is monitored. If a material is photorefractive, an asymmetric energy exchange is observed. The magnitude of photorefractivity is evaluated using a parameter called the gain coefficient, which is calculated from the change in the transmitted intensity of the laser beams induced through the two-beam coupling [1]. In order to calculate the two-beam coupling gain coefficient, it must be determined whether the

http://dx.doi.org/10.5772/52921

diffraction condition is within the Bragg regime or within the Raman-Nath regime. These diffraction conditions are distinguished by a dimensionless parameter Q.

$$Q = 2\pi lL/n\Lambda^2 \tag{1}$$

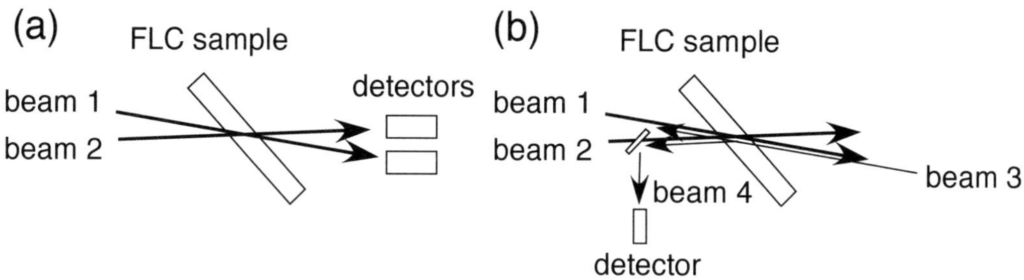

Figure 7. Schematic illustrations of the experimental set-up for the (a) two-beam coupling, and (b) four-wave-mixing techniques.

$Q > 1$ is defined as the Bragg regime of optical diffraction. In this regime, multiple scattering is not permitted, and only one order of diffraction is produced. Conversely, $Q < 1$ is defined as the Raman-Nath regime of optical diffraction. In this regime, many orders of diffraction can be observed. Usually, $Q > 10$ is required to guarantee that the diffraction is entirely within the Bragg regime. When the diffraction is in the Bragg diffraction regime, the two-beam coupling gain coefficient Γ (cm^{-1}) is calculated according to the following equation:

$$\Gamma = \frac{1}{D}\ln\left(\frac{gm}{1+m-g}\right) \tag{2}$$

where $D = L/\cos(\theta)$ is the interaction path for the signal beam (L=sample thickness, θ =propagation angle of the signal beam in the sample), g is the ratio of the intensities of the signal beam behind the sample with and without a pump beam, and m is the ratio of the beam intensities (pump/signal) in front of the sample.

A schematic illustration of the experimental setup used for the four-wave mixing experiment is shown in Figure 7(b). S-polarized writing beams are interfered in the sample film and the diffraction of a p-polarized probe beam, counter-propagating to one of the writing beams, is measured. The diffracted beam intensity is typically measured as a function of time, applied (external) electric field and writing beam intensities, etc. The diffraction efficiency is defined as the ratio of the intensity of the diffracted beam and the intensity of the probe beam that is transmitted when no grating is present in the sample due to the writing beams. In probing the grating, it is important that beam 3 does not affect the grating or interact with the writing beams. This can be ensured by making the probe beam much weaker than the writing beams and by having the probe beam polarized orthogonal to the writing beams.

4. Photorefractive effect of FLCs

4.1. Two-beam coupling experiments on FLCs

The photorefractive effect in an FLC was first reported by Wasielewsky et al. in 2000 [11]. Since then, details of photorefractivity in FLC materials have been further investigated by Sasaki et al. and Golemme et al. [12-14]. The photorefractive effect in a mixture of an FLC and a photoconductive compound was measured in a two-beam coupling experiment using a 488 nm Ar^+ laser. The structures of the photoconductive compounds used are shown in Figure 8. A commercially available FLC, SCE8 (Clariant), was used. CDH was used as a photoconductive compound and TNF was used as a sensitizer. The concentrations of CDH and TNF were 2 wt% and 0.1 wt% respectively. The samples were injected into a 10-μm-gap glass cell equipped with 1 cm^2 ITO electrodes and a polyimide alignment layer (Figure 9).

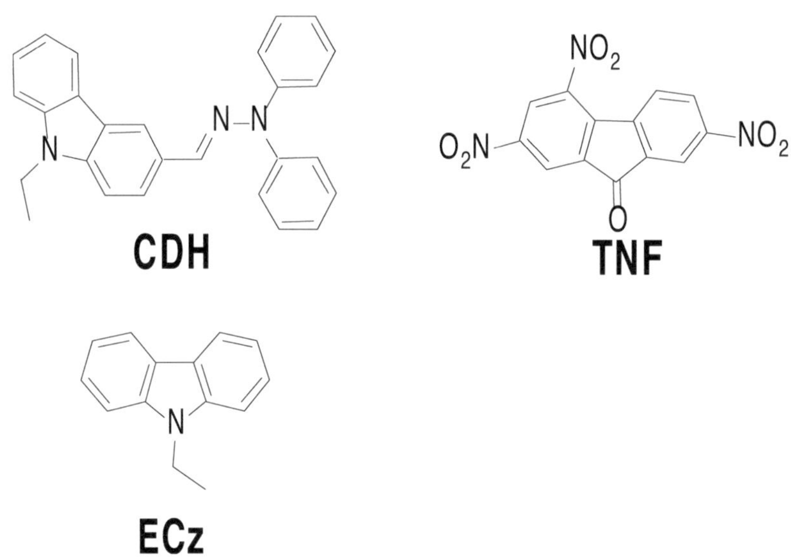

Figure 8. Structures of the photoconductive compound CDH, ECz and the sensitizer TNF

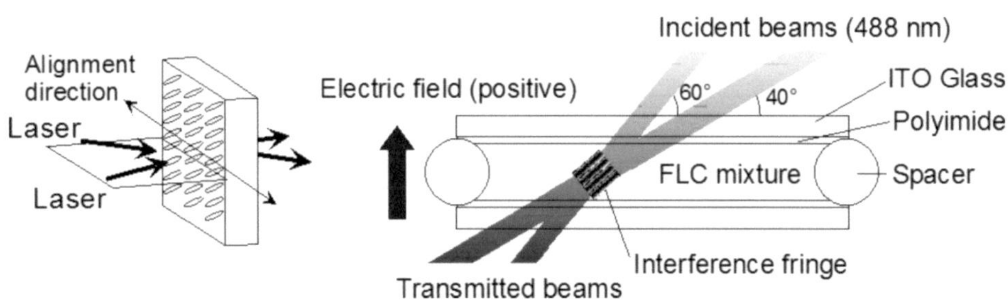

Figure 9. Laser beam incidence condition and the structure of the LC cell.

Figure 10 shows a typical example of asymmetric energy exchange observed in the FLC(SCE8)/CDH/TNF sample under an applied DC electric field of 0.1 V/μm [12]. The interference of the divided beams in the sample resulted in the increased transmittance of one beam and the decreased transmittance of the other. The change in the transmitted intensities of the two beams is completely symmetric, as can be seen in Figure 10. This indicates that the phase of the refractive index grating is shifted from that of the interference fringes. The grating formation was within the Bragg diffraction regime and no higher-order diffraction was observed under the conditions used.

The temperature dependence of the gain coefficient of SCE8 doped with 2 wt% CDH and 0.1 wt% TNF is shown in Figure 11(a). Asymmetric energy exchange was observed only at temperatures below 46°C. The spontaneous polarization of the identical sample is plotted as a function of temperature in Figure 11(b).

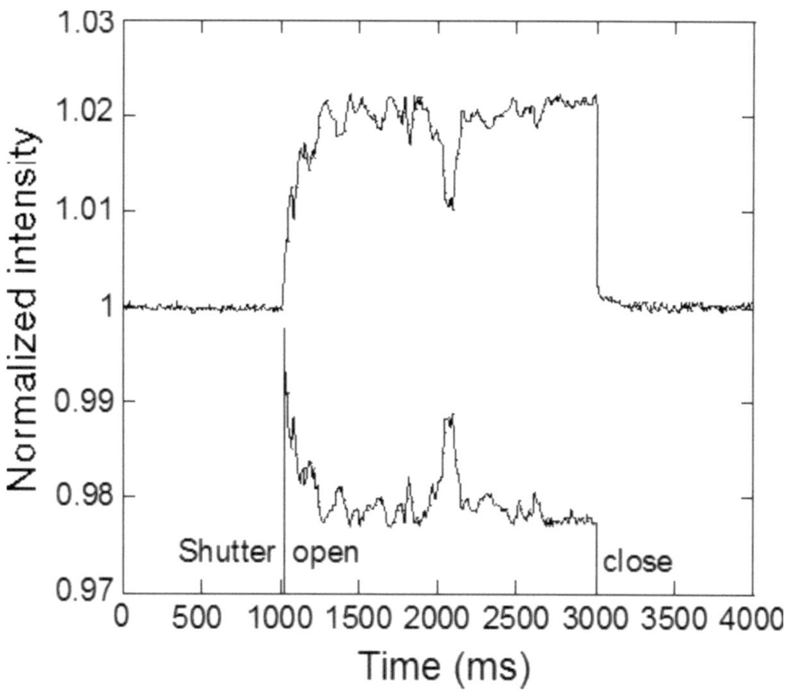

Figure 10. A typical example of asymmetric energy exchange observed in an FLC (SCE8) mixed with 2 wt% CDH and 0.1 wt% TNF. An electric field of +0.3 V/μm was applied to the sample.

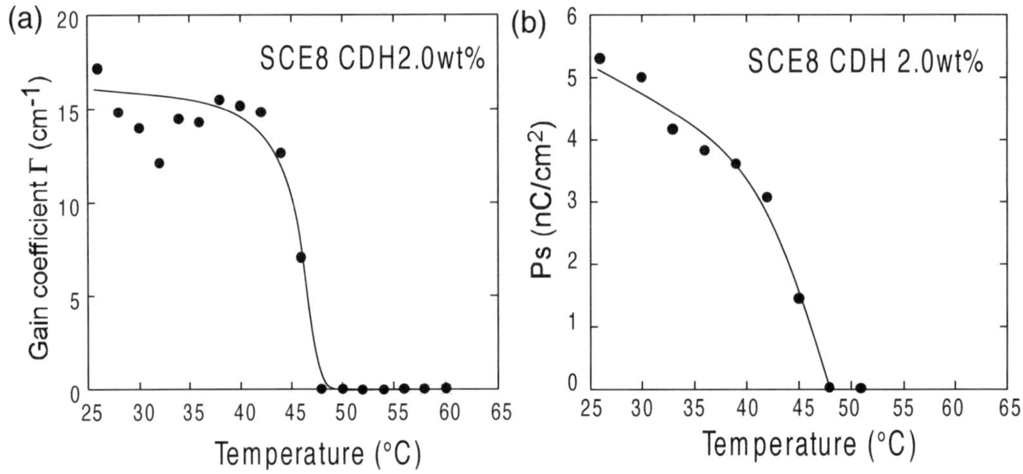

Figure 11. Temperature dependence of (a) the gain coefficient, and (b) the spontaneous polarization of an FLC (SCE8) mixed with 2 wt% CDH and 0.1 wt% TNF. For two-beam coupling experiments, an electric field of 0.1 V/μm was applied to the sample.

Similarly, the spontaneous polarization vanished when the temperature was raised above 46°C. Thus, asymmetric energy exchange was observed only in the temperature range in which the sample exhibited ferroelectric properties; in other words, the SmC* phase. Since the molecular dipole moment of the FLCs is small and the dipole moment is aligned perpendicular to the molecular axis, large changes in the orientation of the molecular axis cannot be induced by an internal electric field in the SmA or N* phase of the FLCs. However, in the SmC* phase, reorientation associated with spontaneous polarization occurs due to the internal electric field. The spontaneous polarization also causes the orientation of FLC molecules in the corresponding area to change accordingly. A maximum resolution of 0.8 μm was obtained in this sample.

4.2. Effect of the magnitude of the applied electric field

In polymeric photorefractive materials, the strength of the externally applied electric field is a very important factor. The external electric field is necessary to increase the charge separation efficiency sufficiently to induce a photorefractive effect. In other words, the photorefractivity of the polymer is obtained only with the application of a few V/μm electric fields. The thickness of the polymeric photorefractive material commonly reported is about 100 μm, so the voltage necessary to induce the photorefractive effect is a few kV. On the other hand, the photorefractive effect in FLCs can be induced by applying a very weak external electric field. The maximum gain coefficient for the FLC (SCE8) sample was obtained using an electric field strength of only 0.2–0.4 V/μm. The thickness of the FLC sample is typically 10 μm, so that the voltage necessary to induce the photorefractive effect is only a few V. The dependence of the gain coefficient of a mixture of FLC (SCE8)/CDH/TNF on the strength of the electric field is shown in Figure 12. The gain coefficient of SCE8 doped with 0.5–1 wt% CDH increased with the strength of the external electric field. However, the gain coefficient of

SCE8 doped with 2 wt% CDH decreased when the external electric field exceeded 0.4 V/μm. The same tendency was observed for M4851/050 as well. The formation of an orientational grating is enhanced when the external electric field is increased from 0 to 0.2 V/μm as a result of induced charge separation under a higher external electric field. However, when the external electric field exceeded 0.2 V/μm, a number of zigzag defects appeared in the surface-stabilized state. These defects cause light scattering and result in a decrease in the gain coefficient.

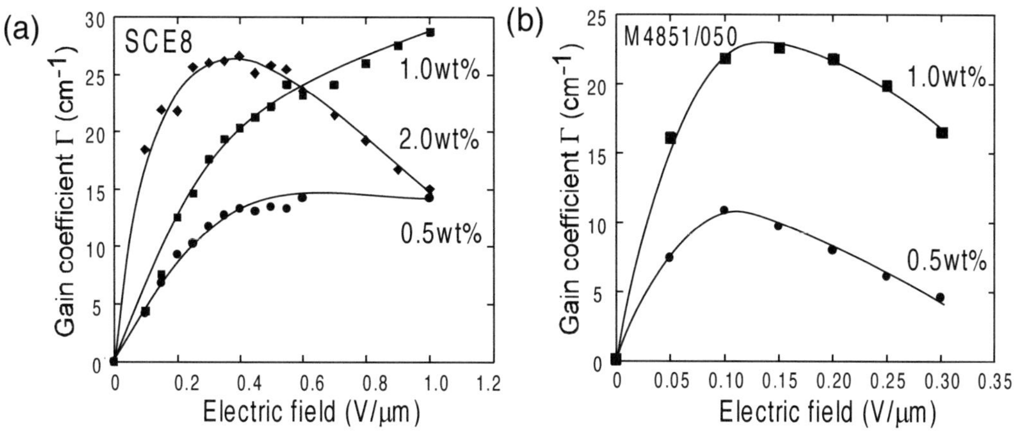

Figure 12. Electric field dependence of the gain coefficient of SCE8 and M4851/050 mixed with several concentrations of CDH and 0.1 wt% TNF in a 10 μm-gap cell measured at 30 °C.

4.3. Refractive index grating formation time

The formation of a refractive index grating involves charge separation and reorientation. The index grating formation time is affected by these two processes and both may be rate-determining steps. The refractive index grating formation times in SCE8 and M4851/050 were determined based on the simplest single-carrier model of photorefractivity [1,2], wherein the gain transient is exponential. The rising signal of the diffracted beam was fitted using a single exponential function, shown in equation (3).

$$\gamma(t) - 1 = (\gamma - 1)[1 - \exp(-t/\tau)]^2 \tag{3}$$

Here, $\gamma(t)$ represents the transmitted beam intensity at time t divided by the initial intensity ($\gamma(t) = I(t)/I0$) and τ is the formation time. The grating formation time in SCE8/CDH/TNF is plotted as a function of the strength of the external electric field in Figure 13(a). The grating formation time decreased with increasing electric field strength due to the increased efficiency of charge generation. The formation time was shorter at higher temperatures, corresponding to a decrease in the viscosity of the FLC with increasing temperature. The formation time for SCE8 was found to be 20 ms at 30°C. As shown in Figure 13(b), the formation time for M4851/050 was found to be independent of the magnitude of the external electric field, with a time of 80-90 ms for M4851/050 doped with 1 wt% CDH and 0.1 wt% TNF. This is slower than for SCE8, although the spontaneous polarization of M4851/050 (-14 nC/cm^2) is larger than that of SCE8 (-4.5 nC/cm^2) and the response time of the electro-optical switching (the flipping of spontaneous polarization) to an electric field (±10 V in a 2 μm cell) is shorter for M4851/050. The slower formation of the refractive index grating in M4851/050 is likely due to the poor homogeneity of the SS-state and charge mobility.

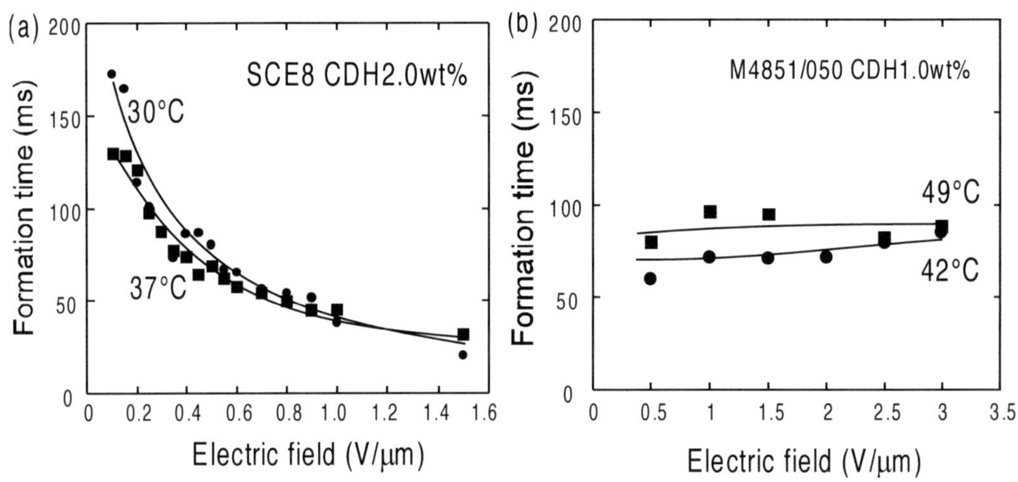

Figure 13. Electric field dependence of the index grating formation time. (a) SCE8 mixed with 2 wt% CDH and 0.1 wt % TNF in the two-beam coupling experiment. ▪, measured at 30 °C (T/T_{SmC^*-SmA}=0.95); ▪, measured at 36 °C (T/T_{SmC^*-SmA}=0.97). (b) M4851/050 mixed with 1 wt% CDH and 0.1 wt% TNF in a two-beam coupling experiment. ▪, measured at 42 °C (T/T_{SmC^*-SmA}=0.95); ▪, measured at 49 °C (T/T_{SmC^*-SmA}=0.97).

4.4. Formation mechanism of the internal electric field in FLCs

Since the photorefractive effect is induced by the photoinduced internal electric field, the mechanism of the formation of the space-charge field in the FLC medium is important. The two-beam coupling gain coefficients of mixtures of FLC (SCE8) and photoconductive compounds under a DC field were investigated as a function of the concentration of TNF (electron acceptor). The photoconductive compounds - CDH, ECz and TNF (Figure 8) - were used in this examination. When an electron donor with a large molecular size (CDH) relative to the TNF was used as the photoconductive compound, the gain coefficient was strongly affected by the concentration of TNF (Figure 14(a)). However, when ethylcarbazole (ECz) - the molecular size of which is almost the same as that of TNF - was used, the gain coefficient was less affected by the TNF concentration (Figure 14(b)).

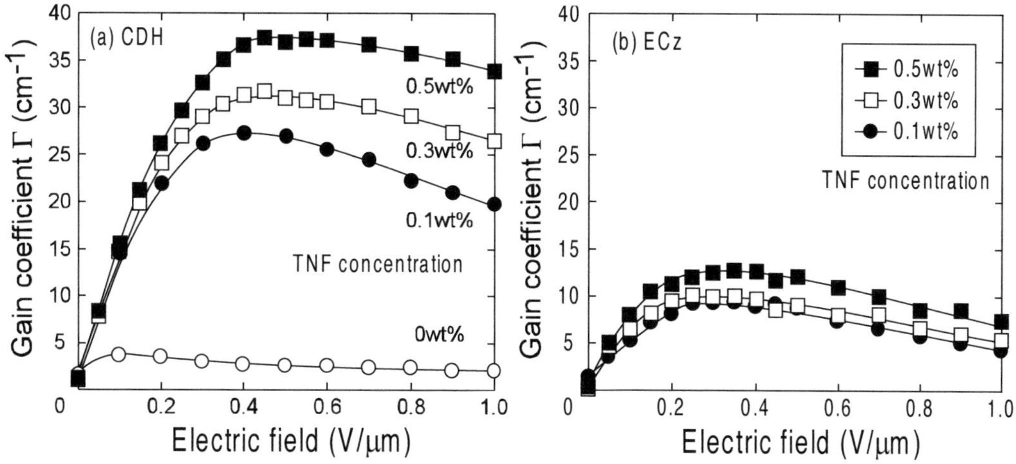

Figure 14. Dependence of the TNF concentration on the gain coefficients of an FLC doped with photoconductive dopants. (a) SCE8 doped with 2 wt% CDH, and (b) SCE8 doped with 2 wt% ECz. An electric field of ± 0.5 V/μm, 100 Hz was applied.

These findings suggest that ionic conduction plays a major role in the formation of the space-charge field. The mobility of the CDH cation is smaller than that of the TNF anion, and this difference in mobility is thought to be the origin of the charge separation. In this case, the magnitude of the internal electric field is dominated by the concentration of the ionic species. On the other hand, the difference in the mobilities of ECz and TNF is small and, thus, less effective charge separation is induced, indicating that the internal electric field is independent of the concentration of ionic species.

4.5. Photorefractive effect in FLC mixtures containing photoconductive chiral compounds

4.5.1. Photoconductive chiral dopants

The photorefractive effect of FLC mixtures containing photoconductive chiral dopants has been investigated [15]. The structures of the LC compounds, the electron acceptor trinitrofluorenone (TNF) and the photoconductive chiral compounds are shown in Figure 15. The mixing ratio of 8PP8 and 8PP10 was set to 1:1 because the 1:1 mixture exhibits the SmC phase over the widest temperature range. Hereafter, the 1:1 mixture of 8PP8 and 8PP10 is referred to as the base LC. The concentration of TNF was 0.1 wt%. Four photoconductive chiral compounds with the terthiophene chromophore (3T-2MB, 3T-2OC, 3T-OXO, and 3T-CF3) were synthesized. The base LC, TNF and a photoconductive chiral compound were dissolved in dichloroethane and the solvent was evaporated. The mixture was then dried in a vacuum at room temperature for one week. The samples were subsequently injected into a 10-μm-gap glass cell equipped with 1-cm^2 ITO electrodes and a polyimide alignment layer for the measurements. The base LC, which is a 1:1 mixture of 8PP8 and 8PP10, was mixed with the photoconductive chiral dopants and the electron acceptor (TNF). The concentration of TNF was set to 0.1 wt%. The terthiophene chiral dopants showed high miscibility with the phenyl pyrimidine-type smectic LC. The chiral smectic C (SmC*) phase appeared in all of the mixtures of the base LC and the chiral dopants. With the increase of the concentration of the chiral dopants, the temperature range of the SmC* phase and the chiral nematic (N*) phase were reduced, whereas that of the smectic A (SmA) phase was enhanced. The miscibility of the 3T-CF3 with the base LC was the lowest among the four chiral dopants. It was considered that the dipole moment of the trifluoromethyl substituted group is large, so that the 3T-CF3 molecules tend to aggregate. All the samples exhibited absorption maxima at 394 nm. Absorption at 488 nm (wavelength of the laser used) was small. The absorption spectra were not changed when TNF (1.0 × 10-5 mol/L) was added to the solution. The small absorption at the laser wavelength is advantageous for minimizing the optical loss. The photocurrents in the mixtures of the base LC, photoconductive chiral dopants and TNF were measured. As shown in Figure 16, the samples were good insulators in the dark. When a 488 nm laser irradiated the samples, photocurrents were clearly observed. The magnitudes of the photocurrents were slightly different in the four samples. The only difference in the molecular structures of these compounds is the chiral substituent. Thus, the difference in the photocurrent cannot be attributed to the difference in the molecular structure. It was considered that the miscibility of the photoconductive chiral compounds to the LC and the homogeneity of the LC phase affected the magnitude of the photocurrent.

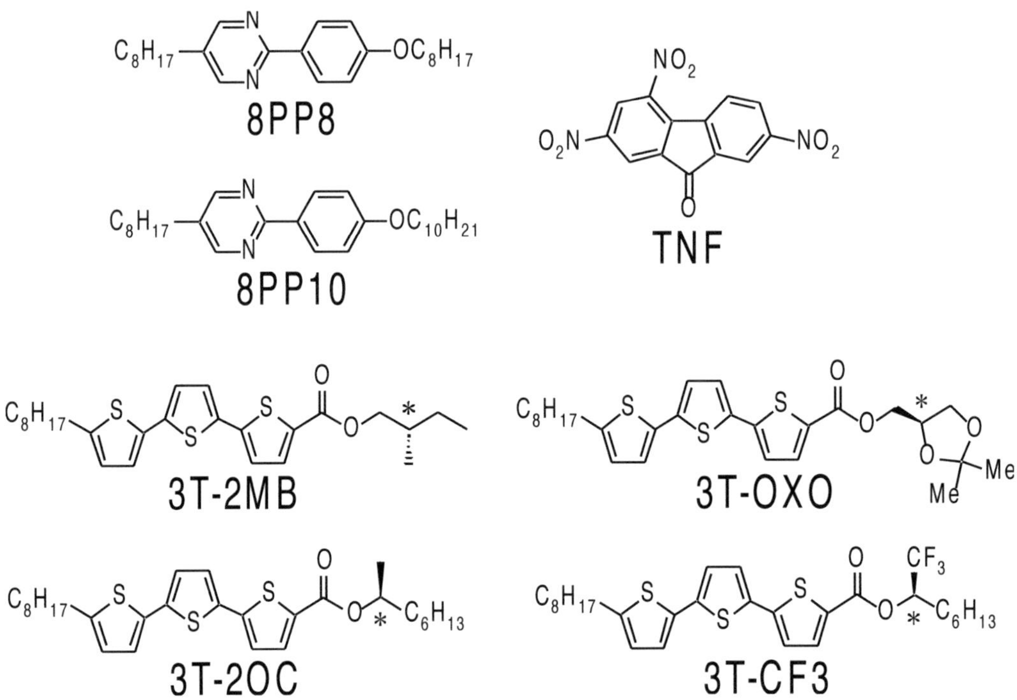

Figure 15. Structures of the smectic LCs (8PP8 and 8PP10), photoconductive chiral dopants (3T-2MB, 3T-2OC, 3T-OXO and 3T-CF3) and the sensitizer TNF used in this work.

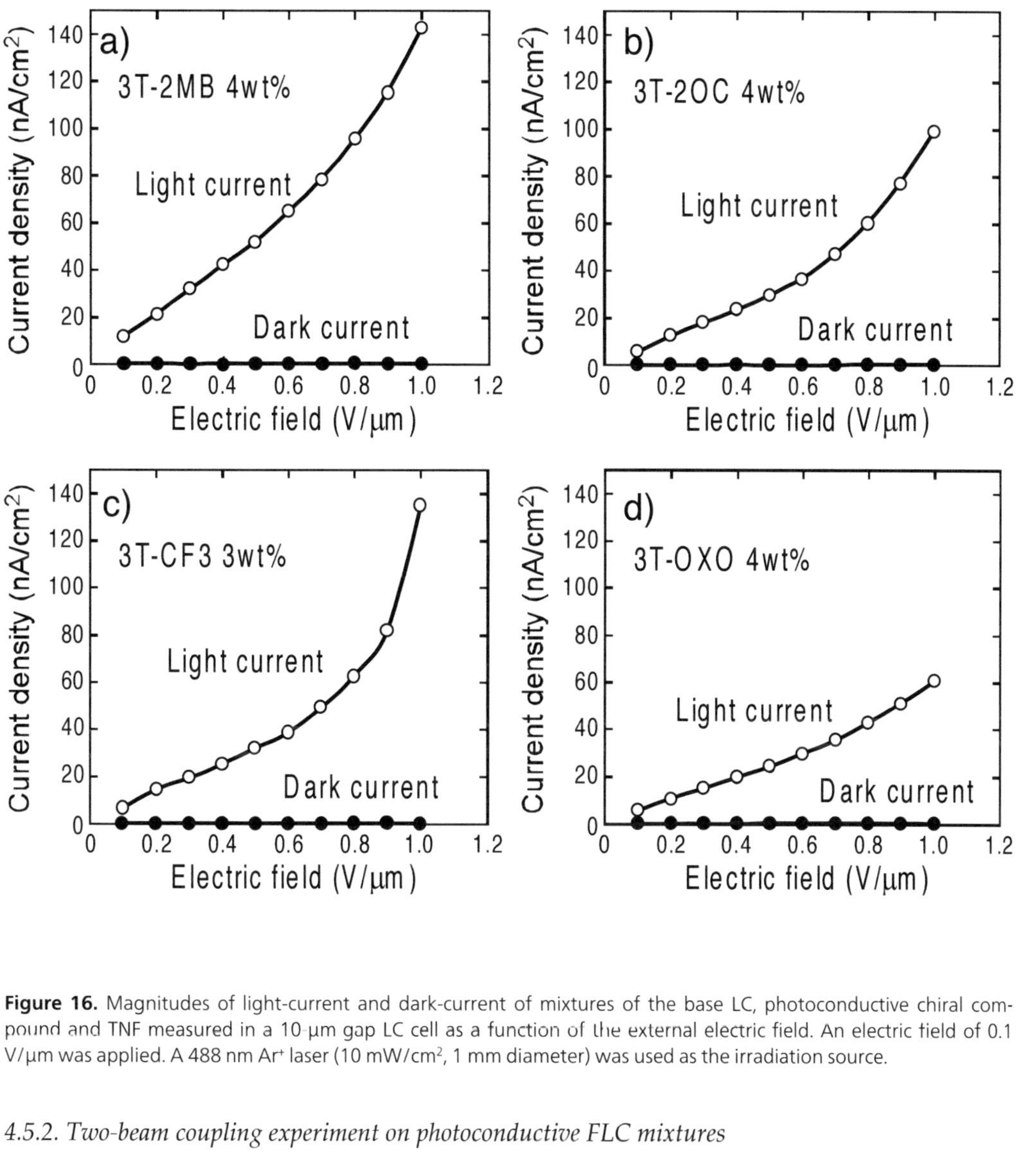

Figure 16. Magnitudes of light-current and dark-current of mixtures of the base LC, photoconductive chiral compound and TNF measured in a 10 μm gap LC cell as a function of the external electric field. An electric field of 0.1 V/μm was applied. A 488 nm Ar^+ laser (10 mW/cm^2, 1 mm diameter) was used as the irradiation source.

4.5.2. Two-beam coupling experiment on photoconductive FLC mixtures

The photorefractive effect was measured in a two-beam coupling experiment. A linearly polarized beam from an Ar^+ laser (488 nm, continuous wave) was divided into two by a beam splitter, each of which was then interfered in the sample film. A p-polarized beam was used. The laser intensity was 2.5 mW for each beam (1 mm diameter). The incident beam angles to the glass plane were 30° and 50°. Each interval of the interference fringe was 1.87 μm. Figure 17 shows typical examples of the asymmetric energy exchange observed in a mixture of the base LC, 3T-2MB and TNF at 30 ºC with the application of an electric field of 0.2–0.4 V/μm.

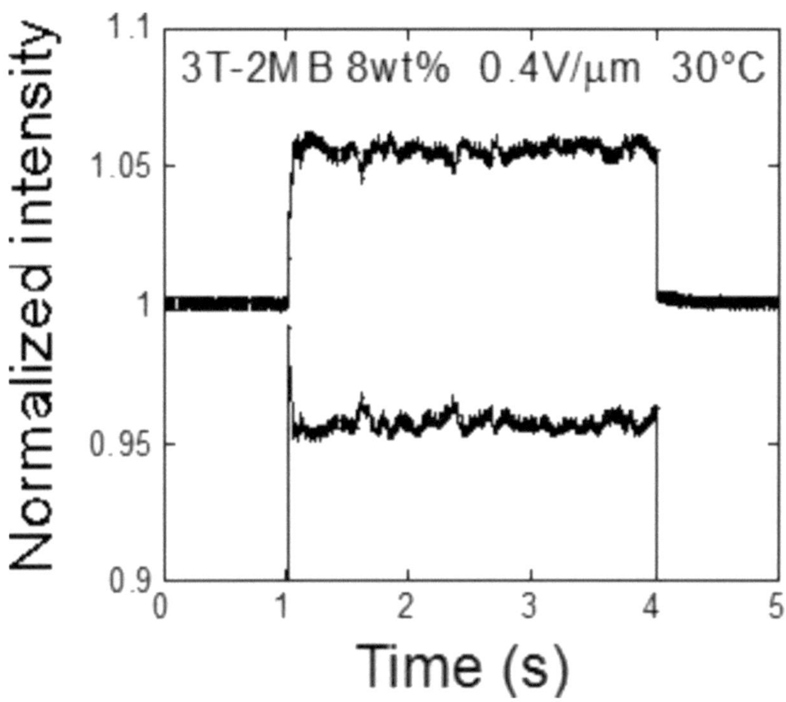

Figure 17. Examples of the results of two-beam coupling experiments for mixtures of the base LC, 3T-2MB and TNF measured at 30 °C.

The interference of the divided beams in the sample resulted in the increased transmittance of one of the beams and the decreased transmittance of the other beam. These transmittance characteristics were reversed when the polarity of the applied electric field was reversed. Asymmetric energy exchange was only observed when an electric field was applied, indicating that beam coupling was not caused by a thermal grating. With an increased concentration of 3T-2MB, the magnitude of the gain coefficient also increased. In order to calculate the two-beam coupling gain coefficient, the diffraction condition needs to be correctly identified.

http://dx.doi.org/10.5772/52921

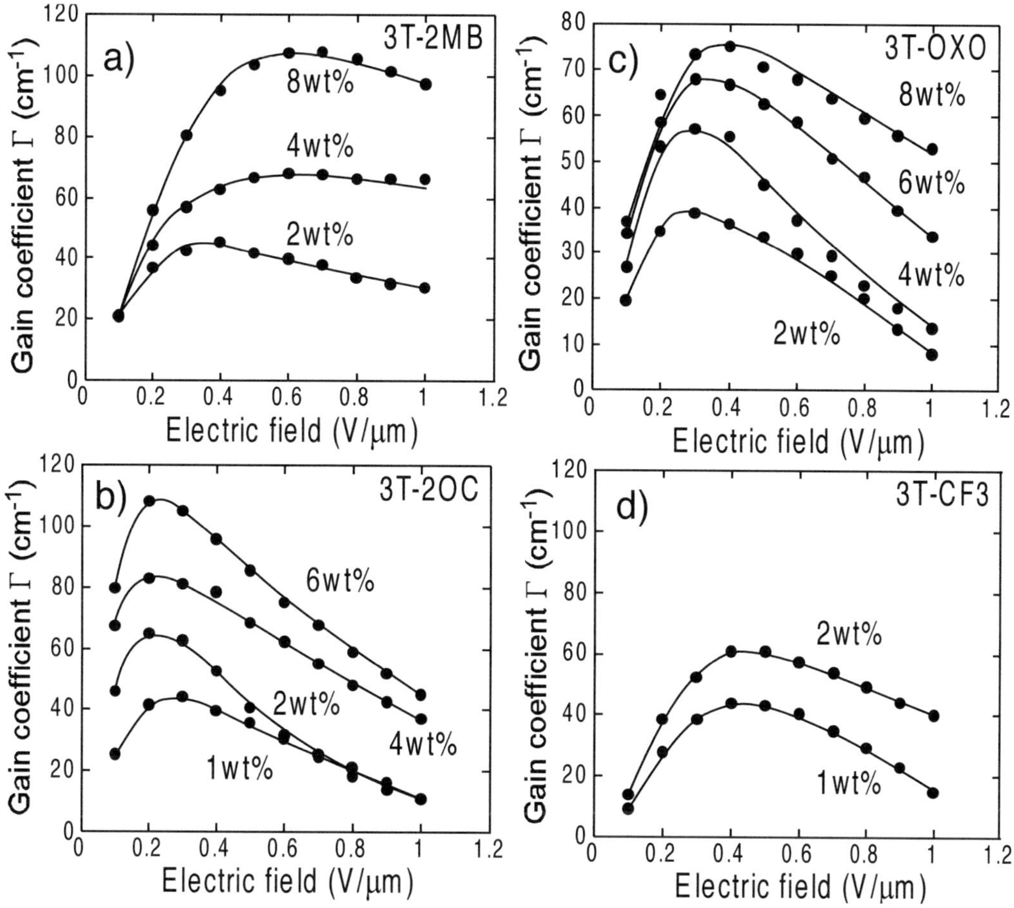

Figure 18. Electric field dependence of the gain coefficients of the mixtures of the base LC, photoconductive chiral compounds and TNF. (a) 3T-2MB; (b) 3T-2OC; (c) 3T-OXO; (d) 3T-CF3.

The difference in the gain coefficients in mixtures of the base LC, photoconductive chiral dopants (3T-2MB, 3T-2OC, 3T-OXO, and 3T-CF3) and TNF was investigated. All the samples formed a finely aligned SS-state in 10-μm-gap cells with a LX-1400 polyimide alignment layer and exhibited clear photorefractivity in the ferroelectric phase. The asymmetric energy exchange was only observed in the temperature range in which the sample exhibits ferroelectric properties (SmC* phase). The gain coefficients of the samples are plotted as a function of the magnitude of the external electric field in Figure 18. The gain coefficient increased with the strength of the external electric field up to 0.2-0.6 V/μm and then decreased with the strength of the external electric field. As the concentration of the photoconductive chiral dopants increased, so did the gain coefficient. This may be due to the increased density of charge carriers in the FLC medium and an increase in the magnitude of Ps. All the samples exhibited relatively large photorefractivity. A gain coefficient higher than 100 cm^{-1} was obtained in the 3T-2OC (6 wt%) sample with the application of only 0.2 V/μm. This means that a voltage of only 2 V is needed to obtain the gain coefficient of 100 cm^{-1} in a 10 μm FLC sample. In the FLCs reported previously, gain coefficients of only 50-60 cm^{-1} were obtained with the application of a 1 V/μm electric field. A gain coefficient higher than 100 cm^{-1} was also obtained in the 3T-2MB sample with an applied electric field of 0.5 V/μm. In order to obtain photorefractivity in polymer materials, the application of a high electric field of 10-50 V/μm to the polymer film is typically required. The small electric field necessary for the activation of the photorefractive effect in FLCs is thus a great advantage for their use in photorefractive devices. The miscibility of 3T-CF3 with the base LC was low and could be mixed with the base LC at concentrations lower than 2 wt%. The grating formation time in the mixtures of the base LC, photoconductive chiral dopants and TNF is plotted as a function of the strength of the external electric field in Figure 19. The grating formation time decreased with an increased electric field strength due to the increased efficiency of charge generation. The shortest formation time was obtained as 5–8 ms with a 1 V/μm external electric field in all chiral compounds. The 3T-CF3 sample exhibited the fastest response. This was because of the larger polarity of the 3T-CF3.

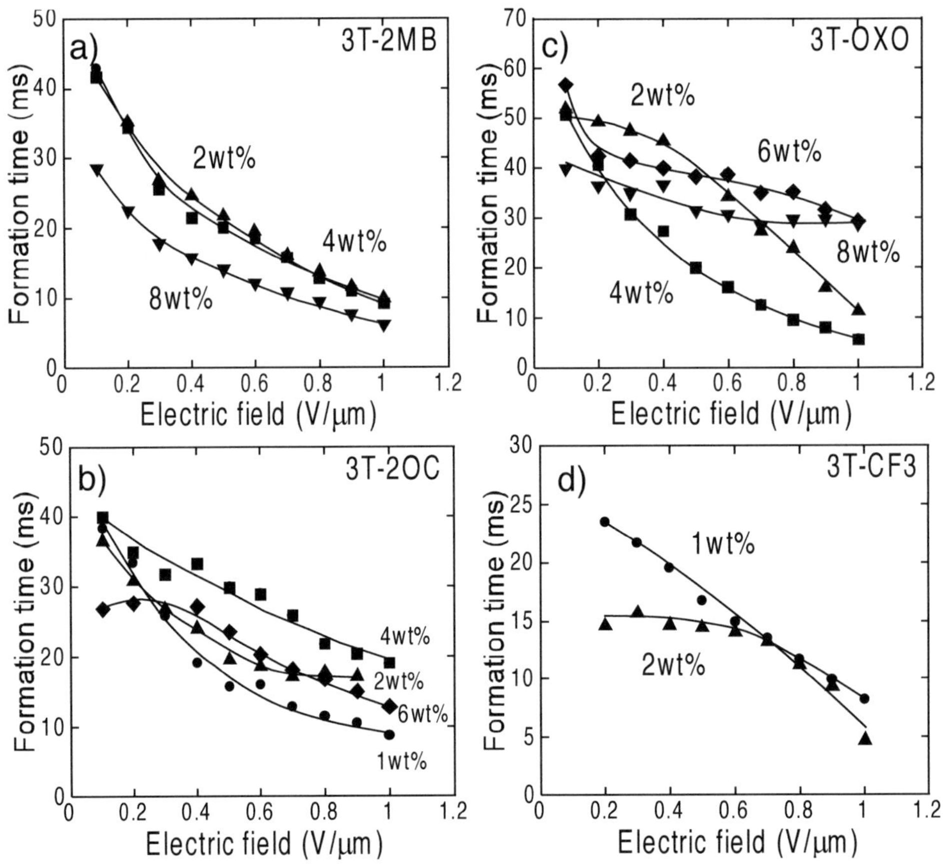

Figure 19. Electric field dependence of the index grating formation times of mixtures of the base LC, photoconductive chiral compounds and TNF measured at 30 °C. (a) 3T-2MB; (b) 3T-2OC; (c) 3T-OXO; (d) 3T-CF3.

http://dx.doi.org/10.5772/52921

4.5.3. Photorefractive effect in a ternary mixture of a SmC liquid crystal doped with the photoconductive chiral compound

The photorefractive effect of the mixture shown in Figure 20 was investigated. The gain coefficients of the samples were measured as a function of the applied electric field strength (Figure 21(a)). The gain coefficient was calculated to be higher than 800 cm^{-1} in the 10 wt% sample with the application of only 1 V/μm. This gain coefficient is eight times higher than that of the FLCs described in the previous section. It was considered that a higher transparency of the ternary LC mixture contributed to the high gain coefficient. The small electric field required to activate the photorefractive effect in FLCs is a great advantage for photorefractive devices. The response time decreased with an increased electric field strength due to the increased charge separation efficiency (Figure 21(b)). The shortest formation time obtained was 8 ms for an external electric field of 1.5 V/μm. The large gain and fast response are advantageous for realizing optical devices, such as real-time image amplifiers and accurate measurement devices.

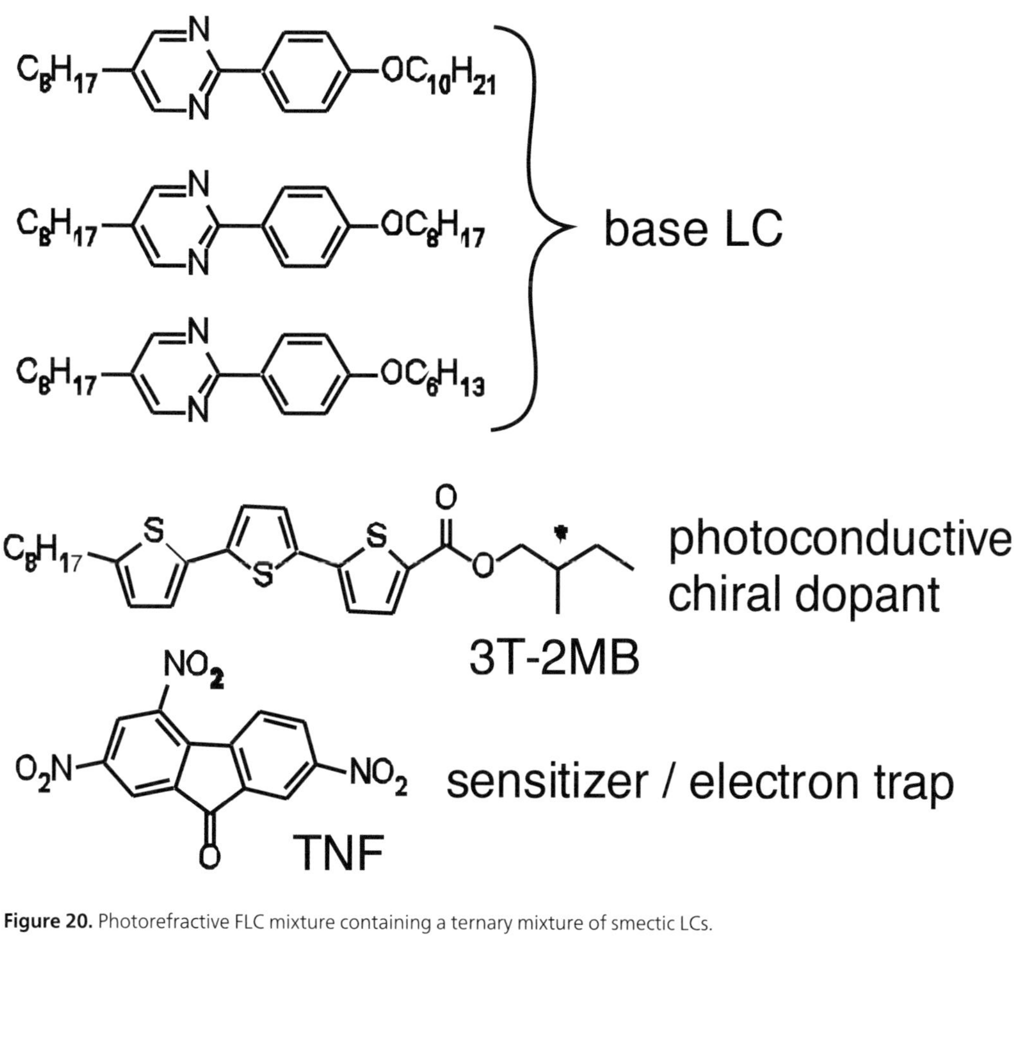

Figure 20. Photorefractive FLC mixture containing a ternary mixture of smectic LCs.

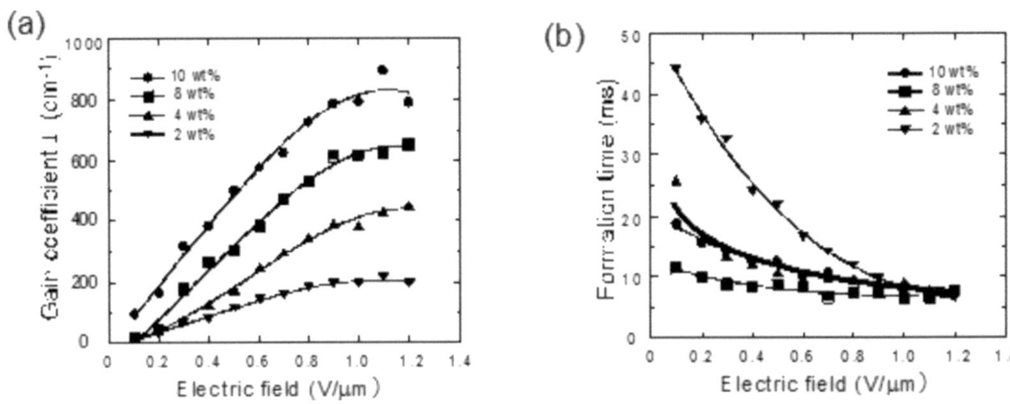

Figure 21. (a) Electric field dependence of gain coefficients of mixtures of the base LC, 3T-2MB and TNF (0.1 wt%) measured at 30 °C. The 3T-2MB concentration was within the range 2–10 wt%. (b) Refractive index grating formation time (response time) of mixtures of the base LC, 3T-2MB and TNF (0.1 wt%) measured at 30 °C. The 3T-2MB concentration was within the range 2–10 wt%.

4.5.4. Formation of dynamic holograms in FLC mixtures

The formation of a dynamic hologram was demonstrated. A computer-generated animation was displayed on a spatial light modulator (SLM). A 488 nm beam from a DPSS laser was irradiated on the SLM and the reflected beam was incident on the FLC sample. A reference beam interfered with the beam from the SLM in the FLC sample. The refractive index grating formed was within the Raman–Nath regime, in which multiple scattering is allowed. A 633 nm beam from a He–Ne laser was irradiated on the FLC sample and the diffraction was observed. A moving image of the animation on the computer monitor was observed in the diffracted beam (Figure 22). No image retention was observed, which means that the hologram image (refractive index grating) formed in the FLC was rewritten sufficiently rapidly to project the reproduction of a smooth holographic movie.

http://dx.doi.org/10.5772/52921

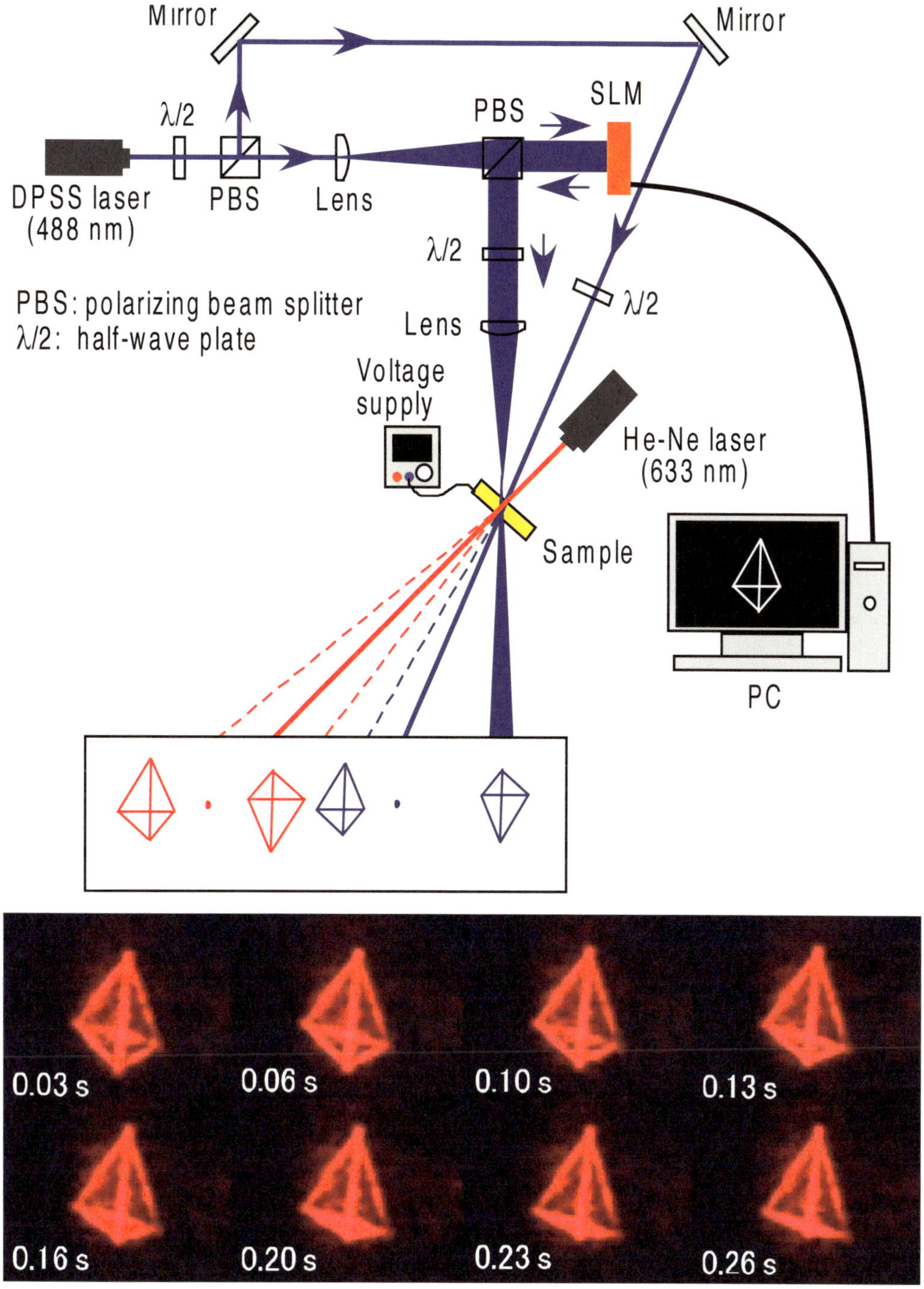

Figure 22. Dynamic hologram formation experiment on an FLC sample. A computer-generated animation was displayed on the SLM. The SLM modulated the object beam (488 nm), which was irradiated on the FLC sample and interfered with the reference beam. The readout beam (633 nm) was irradiated on the FLC and diffraction was observed.

5. Photorefractive effect in FLCs with the application of an AC field

5.1. Formation of dynamic holograms based on the spatial modulation of the molecular motion of FLCs

The formation of dynamic holograms based on the spatial modulation of the molecular motion of ferroelectric liquid crystals (FLCs) was demonstrated [16, 17]. The switching movement of an FLC molecule is essentially a rotational motion along a conical surface. When an alternating triangular-waveform voltage is applied, the FLC molecules uniformly go through a consecutive rotational switching motion along a conical surface (Figure 23(a)). If a photoconductive FLC material is used, this rotational motion can be modulated by illuminating the material with interfering laser beams, as shown in Figure 23. The internal electric field vector is directed along the interference fringe wave vector and, in many cases, it differs from the direction of the applied alternating electric field. Thus, the total electric field at each moment on the FLC molecules is altered by the presence of the internal electric field. The consecutive rotational motion of the FLC molecules in the areas between the light and dark positions of the interference fringes is biased by the internal electric field. Consequently, a grating based on the spatial difference in the rotational motion (or switching motion) of the FLC molecules is created. This grating is different from those currently used for holograms, wherein changes in the static properties of a medium - such as absorbance, transparency, film thickness and molecular orientation - are induced by photochemical reactions.

The formation of a motion-mode grating was examined using a SCE8/CDH/TNF mixture in a 10 μm-gap-cell. The formation of a holographic grating and the occurrence of a phase shift between the formed grating and the interference fringes were examined in this experiment. An alternating triangular-waveform electric field (0 to ±1 V/μm, and 1 kHz to 3 MHz) was applied to the sample. Under the effect of an alternating triangular-waveform electric field of ±0.5 V/μm - 100 kHz in the two-beam coupling experiment - the FLC molecules exhibited a consecutive switching motion. This switching was confirmed using a polarizing microscope equipped with a photodetector Figure 24. Figure 25(a) shows the transmitted intensities of the laser beams through the FLC/CDH/TNF mixture upon the application of an alternating electric field as a function of time. The interference of the divided beams in the sample resulted in the increased transmittance of one of the beams and the decreased transmittance of the other. The incident beam conditions were the same as those used in the DC experiment.

Although the transmitted intensity of the laser beam oscillates due to the switching motion of the FLC molecules, the average intensities of the beams were symmetrically changed, as shown in Figure 25. This indicates that a diffraction grating based on the spatial difference in the rotational switching motion of FLC molecules was formed. The symmetric change in the transmittance of the two beams proves that the phase of the motion-mode grating shifted from that of the interference fringe. No higher-order diffraction was observed under this condition. The spacing of the grating was calculated to be 0.9 μm (1,100 lines/mm). Figure 26 shows the temperature dependence of the gain coefficient obtained under an AC electric field. Asymmetric energy exchange was observed only at those temperatures at which the FLC/CDH/TNF mixture exhibits a ferroelectric phase. This suggests that ferroelectricity or the switching movement of SmC* is necessary for beam coupling under an AC electric field.

http://dx.doi.org/10.5772/52921

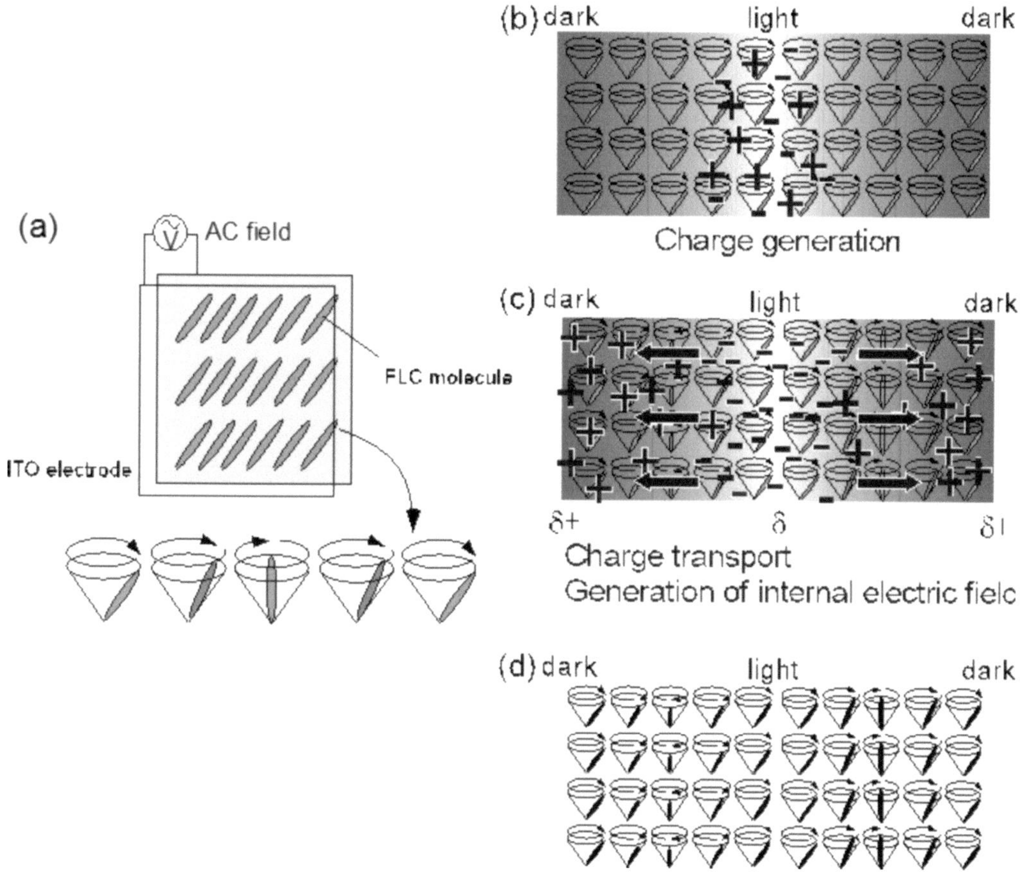

Figure 23. (a) Electro-optical switching of an FLC. (b) The rotational motion of FLC molecules under the application of an alternating electric field. (c) Positive and negative charges appear at the light positions of the interference fringe. (d) An internal electric field develops in the area between the light and dark positions of the interference fringes. The rotational motion of the FLC molecules in the corresponding area is biased by the internal electric field.

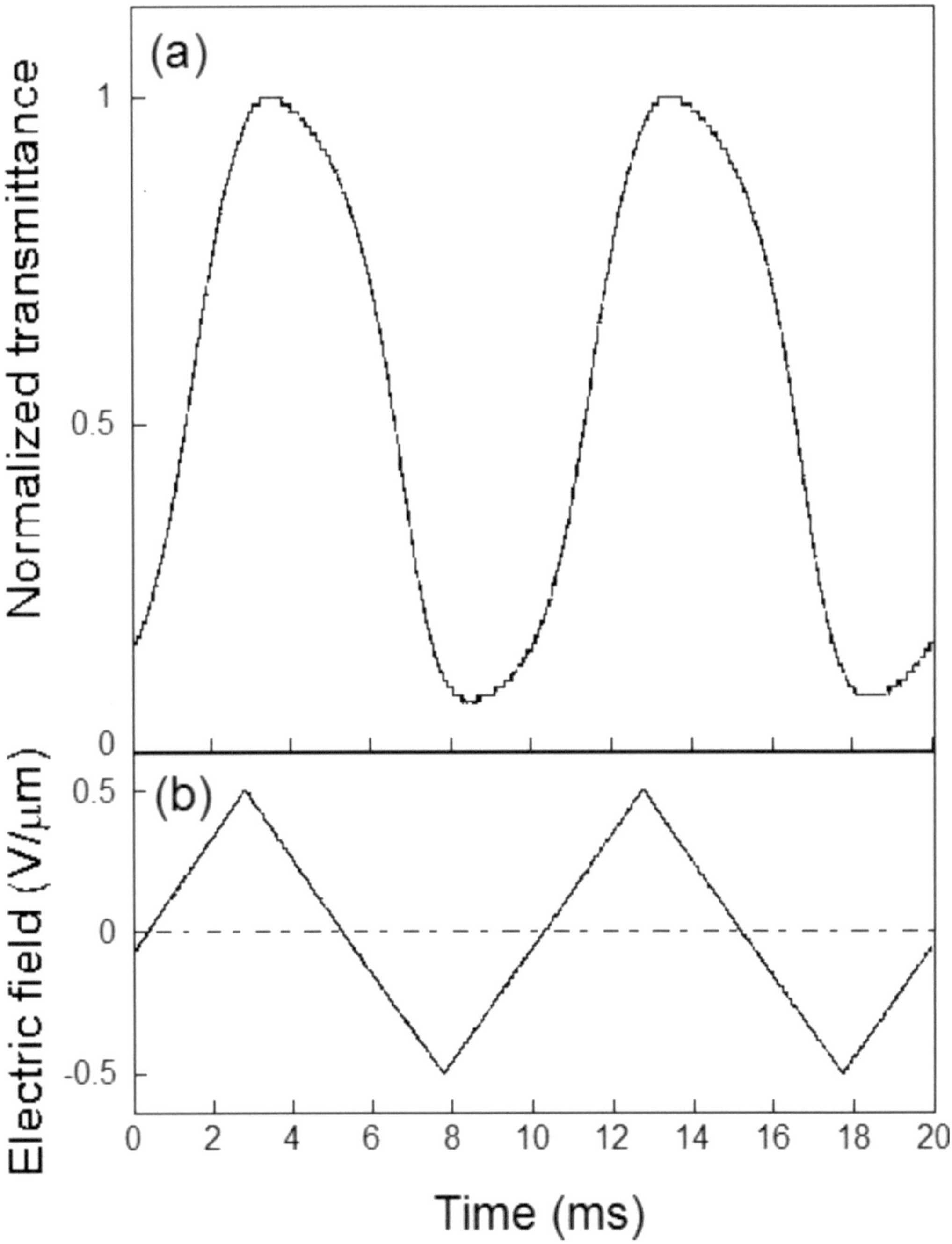

Figure 24. Switching behaviour of SCE8 mixed with 2 wt% CDH and 0.1 wt% TNF under a 100 Hz, ± 0.5 Vm/μm triangular wave electric field. (a) Transmittance of light through the FLC under a polarizing microscope. (b) Magnitude of the applied electric field.

http://dx.doi.org/10.5772/52921

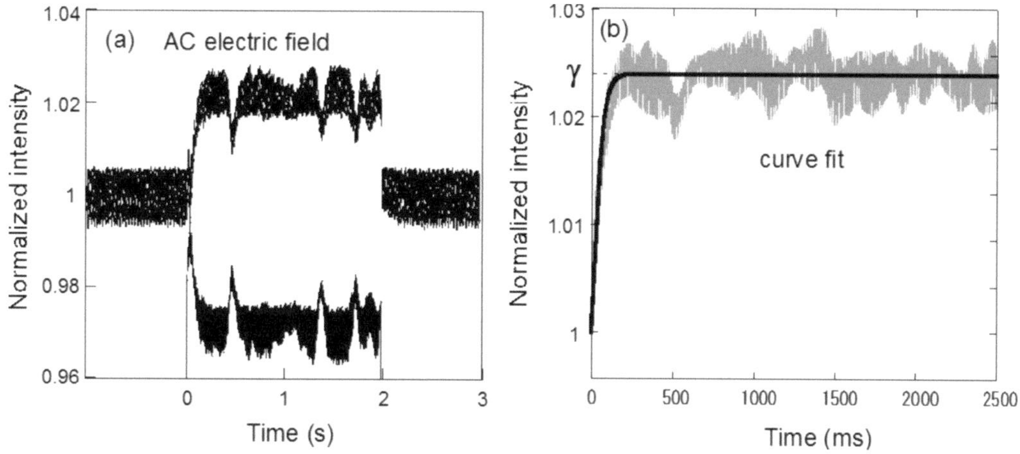

Figure 25. a) Typical example of the asymmetric energy exchange observed in two-beam coupling experiments. A triangular waveform AC electric field of ± 0.5 Vm/μm, 100 Hz was applied in this case. The sample angle α was 50° and the intersection angle ϕ was 20°. The shutter was opened at t = 0 s and closed at t = 2 s. (b) An example of the curve fit.

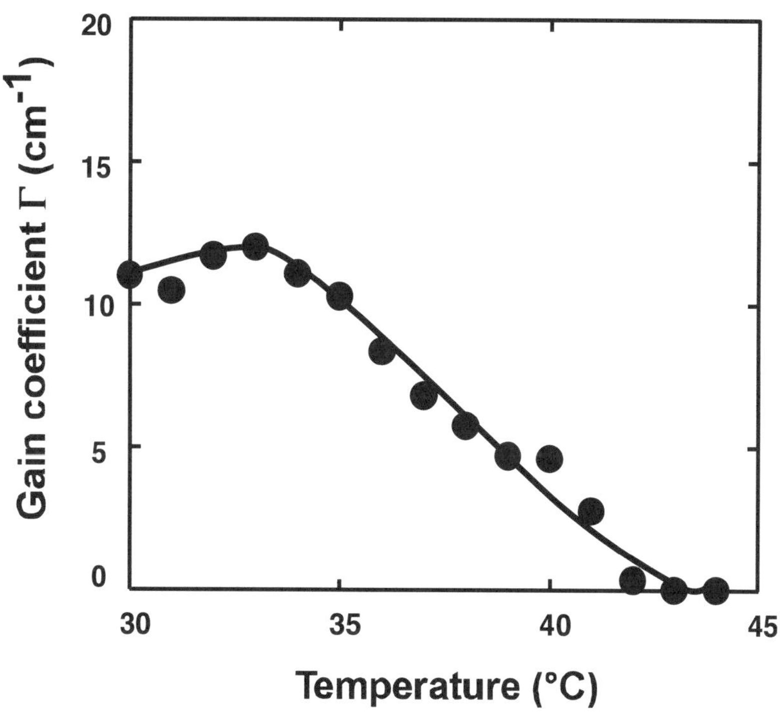

Figure 26. Gain coefficient of the two-beam coupling under an AC electric field as a function of temperature. An AC electric field of ± 0.5 V/μm, 100 Hz was applied.

Figure 27 shows the gain coefficient plotted as a function of the electric field strength. Asymmetric energy exchange was observed at electric field strengths higher than ±0.1 V/μm. Furthermore, no asymmetric energy exchange was observed without an external electric field. This eliminates the possibility that beam coupling resulted from either a thermal grating or from a photochemically-formed grating. The gain coefficient was independent of the external electric field strength for fields higher than ±0.1 V/μm. This behaviour differs from that reported previously for the photorefractive effect of FLCs under an applied DC electric field, wherein the gain coefficient decreased with the increasing strength of the external DC electric field. The refractive index grating formation time was measured based on the simplest single-carrier model for photorefractivity, whereby the gain transient is exponential. The grating formation time τ was obtained from the fitted curve.

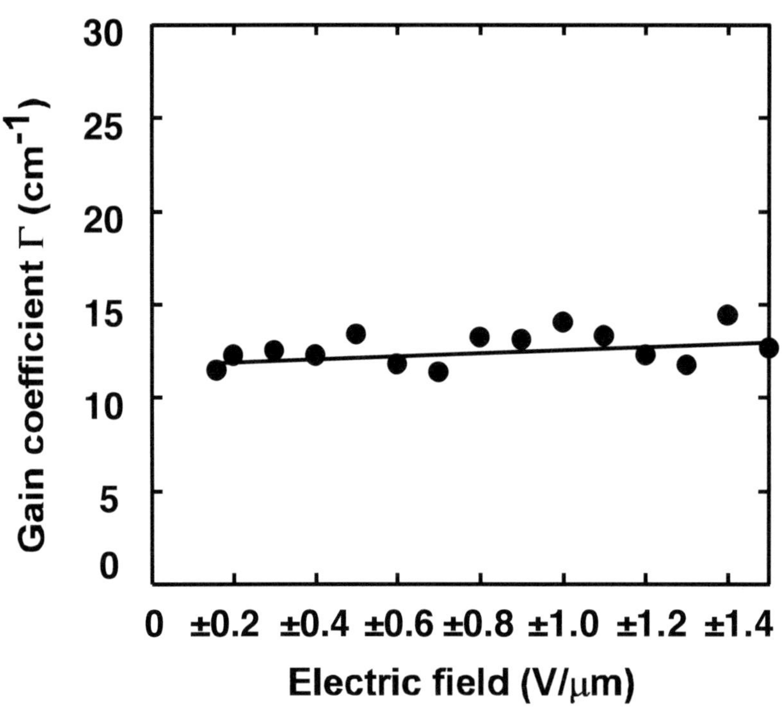

Figure 27. Gain coefficient of the two-beam coupling as a function of the applied AC electric field strength. The frequency of the field was 100 Hz.

http://dx.doi.org/10.5772/52921

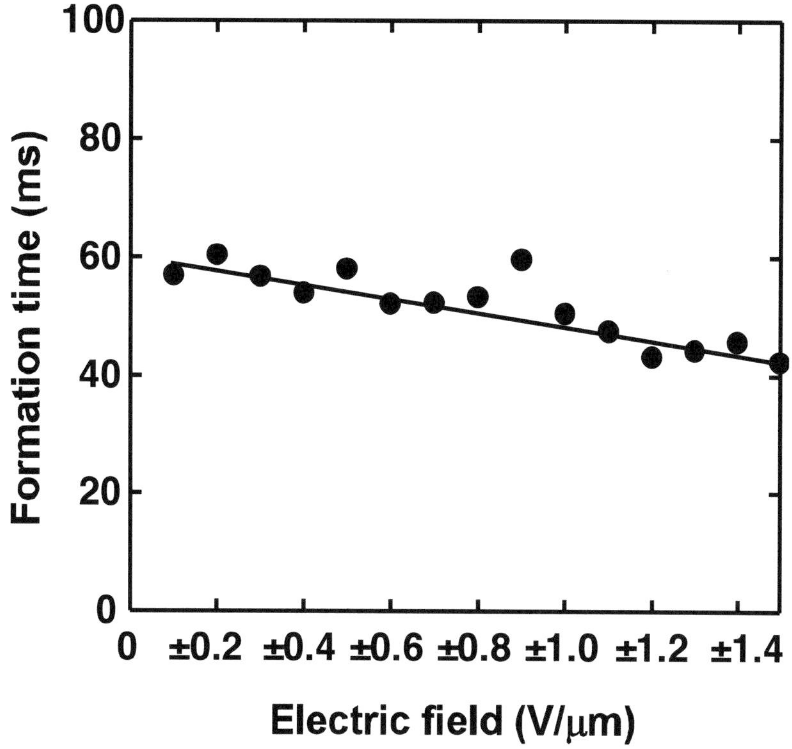

Figure 28. Grating formation time as a function of the AC electric field strength. The frequency of the AC field was 100 Hz.

As seen in Figure 28, the grating formation time decreased with an increased applied electric field strength as a result of the increased efficiency of charge generation. The grating formation time was determined to be 30-40 ms in the present case.

5.2. Frequency dependence of the gain coefficient and the grating formation time

The gain coefficients are plotted as a function of the frequency of the applied external electric field in Figure 29.

The gain coefficient was observed to increase with an increase in frequency within the range of 1 Hz to 100 Hz. The FLC exhibited switching based on polarization reversal at a frequency lower than 500 Hz. The gain coefficient reached a maximum at frequencies within the range of 60 Hz to 500 Hz. However, the magnitude of the gain coefficient decreased as the frequency was increased to higher than 500 Hz. This is thought to be because the FLC cannot perform a complete switching motion at frequencies higher than 500 Hz; therefore, the FLC molecules showed vibrational motion. Moreover, at frequencies from 500 Hz to 2000 Hz, some of the FLC molecules go through a rotational switching motion, whereas others undergo vibrational motion. When the frequency exceeded 2000 Hz, the magnitude of the gain coefficient reaches a constant value, regardless of the frequency. Thus, the value of the gain coefficient at frequencies lower than 500 Hz can be attributed to the photorefractive effect based on the complete switching of FLC molecules and, at frequencies higher than 2000 Hz, this represents the photorefractive effect based on the vibrational motion of FLC molecules caused by an alternating electric field. Figure 30 shows the frequency dependence of the grating formation time.

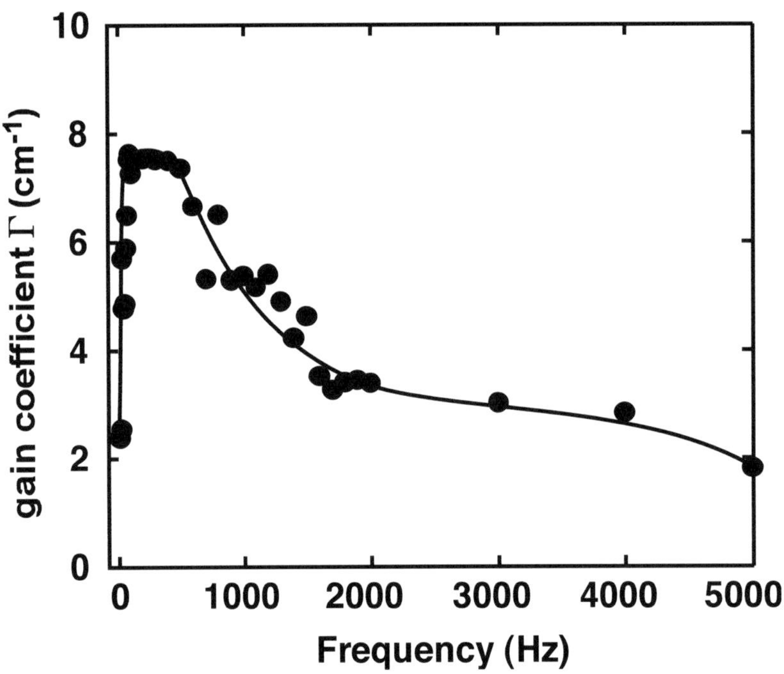

Figure 29. Gain coefficient as a function of the frequency of the applied electric field. The frequency was varied from 1 to 5000 Hz. The strength of the field was ± 0.5 V/μm.

http://dx.doi.org/10.5772/52921

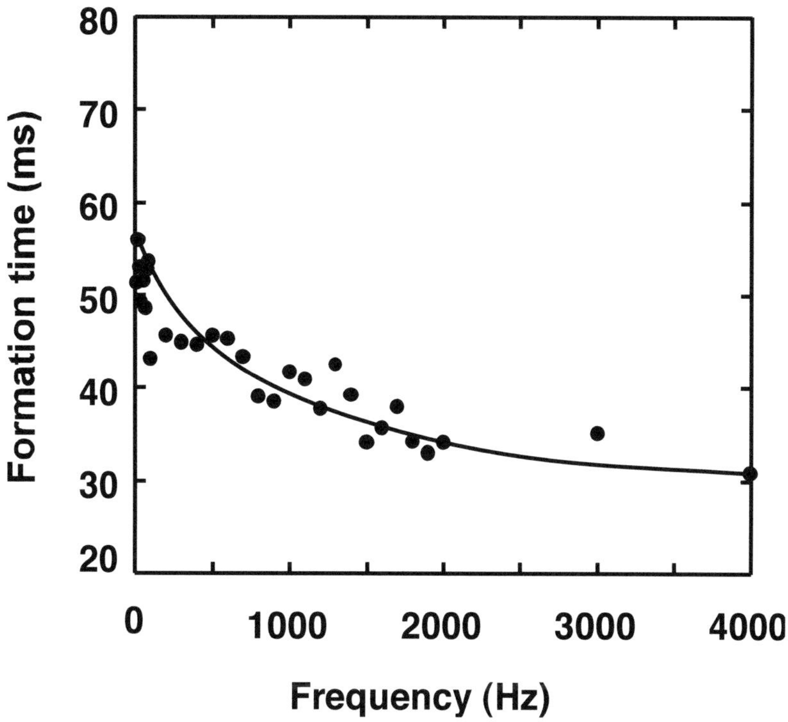

Figure 30. Grating formation time as a function of the frequency of the applied electric field. The frequency was varied from 1 Hz to 4000 Hz. The strength of the field was ± 0.5 V/μm.

The grating formation time was shortened as the frequency increased. In addition, the frequency dependence of the formation time was not correlated with the dependence of the gain coefficient. As the frequency of the applied electric field increased, the switching and vibrational motion of the FLC molecules is thought to be accelerated and the affect on the internal electric field becomes apparent. Asymmetric energy exchange in FLCs under a static DC field is dominated by the direction of the DC field. The increase/decrease in the beam intensity switches when the direction of the field is reversed. This originates from the direction of charge separation. Thus, an internal electric field is thought to be difficult to form under an AC electric field. However, asymmetric energy exchange was observed with good reproducibility, suggesting that an internal electric field is formed even under an applied AC electric field. In addition, no energy exchange was observed in any FLC lacking a photoconductive dopant under an AC electric field (Figure 31). This observation indicates that photoconductivity is necessary for two-beam coupling under an AC field.

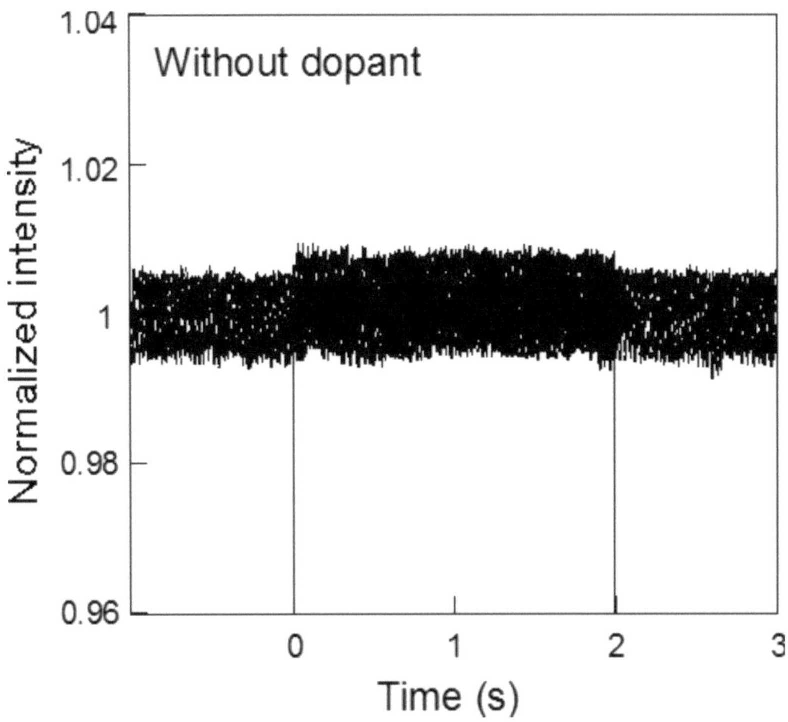

Figure 31. An example of the results of the two-beam coupling experiment under an AC electric field for SCE8 without a photoconductive dopant. An AC electric field of ± 0.5 V/μm, 100 Hz was applied. The shutter was opened at t = 0 s and closed at t = 2 s.

It is necessary to consider both hole transport by a hopping mechanism as well as ionic conduction in order to explain the formation of the space-charge field under an AC field. In addition, a number of experimental results support the large contribution of ionic conduction to the FLC medium. In addition, the photorefractive effect in SCE8 doped with 2 wt% ECz under an AC field was very small. If ionic conduction is the major contributor to the formation of the space-charge field, the anisotropic mobility of ionic species in the LC medium may affect the formation of the field. As shown in Figure 3, interference fringes are formed across the smectic layer so that migration of ionic species occurs in the inter-layer direction. The asymmetric structure of the surface stabilized state of the FLC may lead to the asymmetric mobility of cations and anions, which is thought to be one possible model explaining the charge separation under an AC electric field. However, the mechanism of the formation of the space-charge field under an AC field requires further investigation.

5.3. Two beam coupling experiment with the application of a biased AC field

An asymmetric energy exchange was observed upon the application of an AC field. This grating was interpreted as being based on the spatial difference in the molecular motion of the FLC molecules. The response time was on the order of a few tens of milliseconds and was dominated by the formation of the internal electric field. The photorefractivity in FLCs

http://dx.doi.org/10.5772/52921

is accomplished through charge generation and diffusion. The application of an AC field to an FLC results in a very stable photorefractive response. However, it is obvious that an AC field is not advantageous for charge separation. Thus, the effect of a biased AC electric field on the photorefractivity of SS-FLCs was investigated. Figure 32 shows the concept of applying a biased AC field.

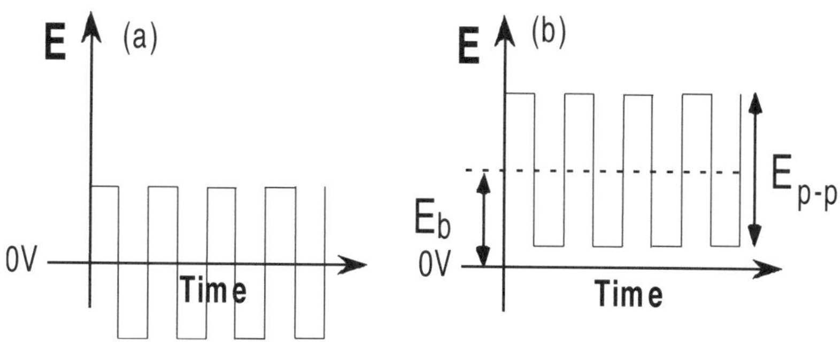

Figure 32. Concept of a biased AC field application. (a) Alternating electric field. (b) Biased alternating electric field.

A typical example of the two-beam coupling signal is observed with the application of a 2.0 $V_{p-p}\mu m^{-1}$, 100 Hz square waveform electric field without a bias field (E_b = 0), as shown in Figure 33(a). The transmitted intensities of the laser beams through the FLC/CDH/TNF mixture upon the application of an alternating electric field as a function of time is shown in the figure. The interference of the divided beams in the sample resulted in the increased transmittance of one of the beams and the decreased transmittance of the other. Although the transmitted intensity of the laser beam oscillates due to the switching motion of the FLC molecules, the average intensities of the beams were symmetrically changed, as shown in Figure 33(a). This indicates that a grating based on the spatial difference in the rotational switching motion of FLC molecules was formed and acted as a diffraction grating.

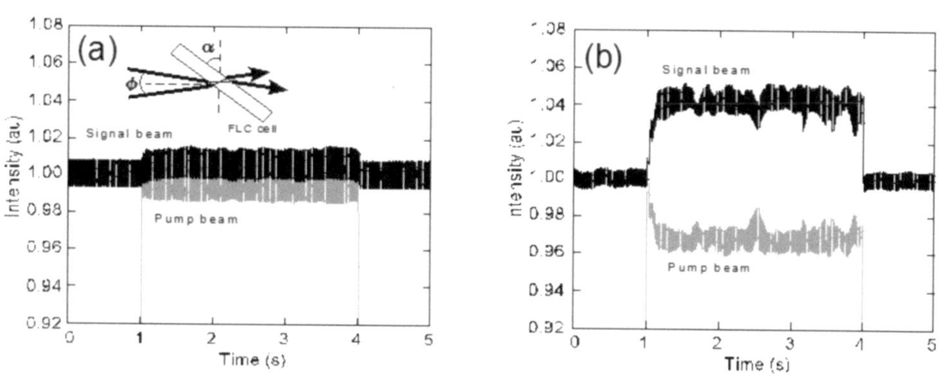

Figure 33. FigAsymmetric energy exchange with a square waveform of 2.0 $V_{p-p}\mu m^{-1}$, 100 Hz (a) without bias, and (b) with a bias of 0.5 V μm^{-1}. The intersection angle ϕ was 20° and the sample angle α was 50°.

The asymmetric energy exchange between the signal and the pump beams proves the formation of a diffraction grating phase-shifted from the interference fringe. The grating is considered to be based on the spatial difference in the rotational switching motion of FLC molecules. When the biased electric field was set to E_b = 0.5 V μm^{-1}, the two-beam coupling signal was dramatically enhanced (Figure 33(b)). The gain coefficient increased from 10 cm^{-1} (E_b = 0) to 70 cm^{-1} (E_b = 0.5 V μm^{-1}). Obviously, the biased field contributes to the formation of the diffraction grating. The internal electric field is thought to be difficult to form under an AC electric field. It has been reported that ionic conduction is the major contributor to the formation of the space-charge field in the photorefractive effect of an FLC. The anisotropic mobility of ionic species in an LC medium affects the formation of the field. In the experimental conditions reported, the interference fringe is formed across the smectic layer so that migration of ionic species occurs in the inter-layer direction. The asymmetric structure of the surface stabilized state of the FLC may lead to the asymmetric mobility of cations and anions. It was considered that the charge separated state is more effectively formed under a biased AC field.

6. Conclusion

The reorientational photorefractive effect based on the response of bulk polarization was observed in dye-doped FLC samples. Photorefractivity was observed only in the ferroelectric phase of these samples and the refractive index formation time was found to be shorter than that of nematic LCs. The response time was in the order of ms and was dominated by the formation of the internal electric field. These results indicate that the mechanism responsible for refractive index grating formation in FLCs is different from that for non-ferroelectric materials and that it is clearly related to the ferroelectric properties of the material. The photorefractivity of FLCs was strongly affected by the properties of the FLCs themselves. Besides, with properties such as spontaneous polarization, viscosity and phase transition temperature, the homogeneity of the SS-state was also found to be a major factor. The gain coefficient, refractive index grating formation time (response time) and stability of the two-beam coupling signal were all strongly affected by the homogeneity of the SS-state. Therefore, a highly homogeneous SS-state is necessary to create a photorefractive device. The techniques employed recently in the development of fine LC display panels will be utilized in the future in the fabrication of photorefractive devices.

Acknowledgment

The authors would like to thank the Japan Science and Technology Agency (JST) S-innovation and the Canon Foundation for support.

http://dx.doi.org/10.5772/52921

Author details

Takeo Sasaki

Tokyo University of Science, Japan

References

[1] Yeh, P. Introduction to Photorefractive Nonlinear Optics; John Wiley: New York, 1993.

[2] Solymar, L.; Webb, J. D.; Grunnet-Jepsen, A. The Physics and Applications of Photorefractive Materials; Oxford: New York, 1996.

[3] Moerner, W. E.; Silence, S. M. Chem. Rev. 1994, 94, 127-155.

[4] Meerholz, K.; Volodin, B. L.; Kippelen, B.; Peyghambarian, N. Nature 1994, 371, 497-500.

[5] Kippelen, B.; Marder, S. R.; Hendrickx, E.; Maldonado, J. L.; Guillemet, G.; Volodin, B. L.; Steele, D. D.; Enami, Y.; Sandalphon; Yao, Y. J.; Wang, J. F.; Röckel, H.; Erskine, L.; Peyghambarian, N. Science 1998, 279, 54-56.

[6] Kippelen, B.; Peyghambarian, N. Advances in Polymer Science, Polymers for Photonics Applications II, Springer, 2002, 87-156.

[7] Ostroverkhova, O.; Moerner, W. E. Chem. Rev., 2004, 104, 3267-3314.

[8] Sasaki, T. Polymer Journal, 2005, 37, 797-812.

[9] Tay, S.; Blanche, P. A.; Voorakaranam, R.; Tunc, A. V.; Lin, W.; Rokutanda, S.; Gu, T.; Flores, D.; Wang, P.; Li, G.; Hilarie, P.; Thomas, J.; Norwood, R. A.; Yamamoto, M.; Peyghambarian, N. Nature, 2008, 451, 694-698.

[10] Blanche, P. A.; Bablumian, A.; Voorakaranam, R.; Christenson, C.; Lin, W.; Gu, T.; Flores, D.; Wang, P.; Hsieh, W. Y.; Kathaperumal, M.; Rachwal, B.; Siddiqui, O.; Thomas, J.; Norwood, R. A.; Yamamoto, M.; Peyghambarian, N. Nature, 2010, 468, 80-83.

[11] Wiederrecht, G. P., Yoon, B. A., Wasielewski, M. R. Adv. Materials 2000, 12, 1533-1536.

[12] Sasaki, T.; Katsuragi, A.; Mochizuki, O.; Nakazawa, Y. J. Phys. Chem. B, 2003, 107, 7659 -7665.

[13] Talarico, M.; Goelemme, A. Nature Mater., 2006, 5, 185-188.

[14] Sasaki, T; Miyazaki, D.; Akaike, K.; Ikegami, M.; Naka, Y. J. Mater. Chem., 2011, 21, 8678-8686.

[15] Skarp, K.; Handschy, M. A. Mol. Cryst. Liq. Cryst. 1988, 165, 439-569.

[16] Sasaki, T.; Mochizuki, O.; Nakazawa, N.; Fukunaga, G.; Nakamura, T.; Noborio, K. Appl. Phys. Lett., 2004, 85, 1329-1331.

[17] Sasaki, T.; Mochizuki, O.; Nakazawa, Y.; Noborio, K. J. Phys. Chem. B, 2004, 108, 17083-17088.